AF763557

PROGRESS IN

Molecular Biology and Translational Science

Volume 100

PROGRESS IN

Molecular Biology and Translational Science

Animal Models of Human Disease

edited by

Karen T. Chang

Department of Cell & Neurobiology
Zilkha Neurogentic Institute, Keck School of Medicine
University of Southern California
Los Angeles, California

Kyung-Tai Min

Department of Biology
Indiana University, Bloomington, Indiana

Volume 100

ELSEVIER

AMSTERDAM • BOSTON • HEIDELBERG • LONDON
NEW YORK • OXFORD • PARIS • SAN DIEGO
SAN FRANCISCO • SINGAPORE • SYDNEY • TOKYO
Academic Press is an imprint of Elsevier

Academic Press is an imprint of Elsevier
32 Jamestown Road, London, NW1 7BY, UK
Radarweg 29, PO Box 211, 1000 AE Amsterdam, The Netherlands
30 Corporate Drive, Suite 400, Burlington, MA 01803, USA
525 B Street, Suite 1900, San Diego, CA 92101-4495, USA

This book is printed on acid-free paper. ∞

Copyright © 2011, Elsevier Inc. All Rights Reserved

No part of this publication may be reproduced, stored in a retrieval system or transmitted in any form or by any means electronic, mechanical, photocopying, recording or otherwise without the prior written permission of the Publisher

Permissions may be sought directly from Elsevier's Science & Technology Rights Department in Oxford, UK: phone (+44) (0) 1865 843830; fax (+44) (0) 1865 853333; email: permissions@elsevier.com. Alternatively you can submit your request online by visiting the Elsevier web site at http://elsevier.com/locate/permissions, and selecting *Obtaining permission to use Elsevier material*

Notice
No responsibility is assumed by the publisher for any injury and/or damage to persons or property as a matter of products liability, negligence or otherwise, or from any use or operation of any methods, products, instructions or ideas contained in the material herein. Because of rapid advances in the medical sciences, in particular, independent verification of diagnoses and drug dosages should be made

Library of Congress Cataloging-in-Publication Data
A catalog record for this book is available from the Library of Congress

British Library Cataloguing in Publication Data
A catalogue record for this book is available from the British Library

ISBN: 978-0-12-384878-9
ISSN: 1877-1173

For information on all Academic Press publications
visit our website at elsevierdirect.com

Working together to grow
libraries in developing countries

www.elsevier.com | www.bookaid.org | www.sabre.org

ELSEVIER BOOK AID International Sabre Foundation

Contents

Drosophila Models of Cardiac Disease 155

Nicole Piazza and R.J. Wessells

Animal Models of Retinal Disease 211

Erica L. Fletcher, Andrew I. Jobling, Kirstan A. Vessey, Chi Luu, Robyn H. Guymer, and Paul N. Baird

Contributors

Numbers in parentheses indicate the pages on which the authors' contributions begin.

H. Orhan Akman, Department of Molecular and Human Genetics, Baylor College of Medicine, Houston, Texas, USA (369)

Paul N. Baird, Centre For Eye Research Australia, University of Melbourne, Royal Victorian Eye and Ear Hospital, East Melbourne, Victoria, Australia (211)

Mathieu F. Bakhoum, Mitchell Center for Neurodegenerative Diseases, Department of Neurology, The University of Texas Medical Branch, Galveston, Texas, USA (483)

Jonathan Bibliowicz, University of Texas at Austin, Section of Molecular Cell and Developmental Biology; and Institute for Cellular & Molecular Biology, Austin, Texas, USA (287)

Joanne Chan, Vascular Biology Program; and Department of Surgery, Children's Hospital Boston, Harvard Medical School, Boston, Massachusetts, USA (111)

Simon J. Conway, Developmental Biology and Neonatal Medicine Program, HB Wells Center for Pediatric Research, Indiana University School of Medicine, Indianapolis, Indiana, USA (83)

T. Cook, Department of Pediatric Ophthalmology, Division of Developmental Biology Cincinnati Children's Hospital Medical Center, Cincinnati, Ohio, USA (331)

William J. Craigen, Department of Molecular and Human Genetics; and Department of Pediatrics, Baylor College of Medicine, Houston, Texas, USA (369)

Atsushi Ebata, Department of Pharmacology; and Department of Pathology, Boston University School of Medicine, Boston, Massachusetts, USA (499)

Andrew Ferree, Department of Pharmacology, Boston University School of Medicine, Boston, Massachusetts, USA (499)

Erica L. Fletcher, Department of Anatomy and Cell Biology, The University of Melbourne, Parkville, Victoria, Australia (211)

Christopher Gabel, Department of Pharmacology; and Department of Physiology and Biophysics, Boston University School of Medicine, Boston, Massachusetts, USA (499)

Vanessa Gobert, Université de Toulouse, UPS, CBD (Centre de Biologie du Développement), Bâtiment 4R3, 118 route de Narbonne; and CNRS, CBD UMR5547, F-31062 Toulouse, France (51)

Eric S. Goetzman, Department of Pediatrics, University of Pittsburgh School of Medicine, Pittsburgh, Pennsylvania, USA (389)

Jeffrey M. Gross, University of Texas at Austin, Section of Molecular Cell and Developmental Biology and Institute for Cellular & Molecular Biology; and Institute for Neuroscience, Austin, Texas, USA (287)

Maria Guillily, Department of Pharmacology, Boston University School of Medicine, Boston, Massachusetts, USA (499)

Robyn H. Guymer, Centre For Eye Research Australia, University of Melbourne, Royal Victorian Eye and Ear Hospital, East Melbourne, Victoria, Australia (211)

Marc Haenlin, Université de Toulouse, UPS, CBD (Centre de Biologie du Développement), Bâtiment 4R3, 118 route de Narbonne; and CNRS, CBD UMR5547, F-31062 Toulouse, France (51)

George R. Jackson, Mitchell Center for Neurodegenerative Diseases, Department of Neurology, The University of Texas Medical Branch, Galveston, Texas, USA (483)

Andrew I. Jobling, Department of Anatomy and Cell Biology, The University of Melbourne, Parkville, Victoria, Australia (211)

Chi Luu, Centre For Eye Research Australia, University of Melbourne, Royal Victorian Eye and Ear Hospital, East Melbourne, Victoria, Australia (211)

John D. Mably, Department of Cardiology, Children's Hospital Boston; and Department of Genetics, Harvard Medical School, Boston, Massachusetts, USA (111)

M. Mishra, Department of Biology, Indiana University, Bloomington, Indiana, USA (331)

Darren J. Moore, Laboratory of Molecular Neurodegenerative Research, Ecole Polytechnique Fédérale de Lausanne (EPFL), School of Life Sciences, Brain Mind Institute, 1015, Lausanne, Switzerland (419)

J. Nie, Department of Biology, Indiana University, Bloomington, Indiana, USA (331)

Nicole Piazza, University of Michigan Medical School, Ann Arbor, Michigan, USA (155)

Cédric Polesello, Université de Toulouse, UPS, CBD (Centre de Biologie du Développement), Bâtiment 4R3, 118 route de Narbonne; and CNRS, CBD UMR5547, F-31062 Toulouse, France (51)

Adithya Raghavan, Department of Molecular and Human Genetics, Baylor College of Medicine, Houston, Texas, USA (369)

David Ramonet, Laboratory of Molecular Neurodegenerative Research, Ecole Polytechnique Fédérale de Lausanne (EPFL), School of Life Sciences, Brain Mind Institute, 1015, Lausanne, Switzerland (419)

Fernando Roch, Université de Toulouse, UPS, CBD (Centre de Biologie du Développement), Bâtiment 4R3, 118 route de Narbonne; and CNRS, CBD UMR5547, F-31062 Toulouse, France (51)

Paige Snider, Developmental Biology and Neonatal Medicine Program, HB Wells Center for Pediatric Research, Indiana University School of Medicine, Indianapolis, Indiana, USA (83)

Rachel K. Tittle, University of Texas at Austin, Section of Molecular Cell and Developmental Biology, Austin, Texas, USA (287)

Alzbeta Trancikova, Laboratory of Molecular Neurodegenerative Research, Ecole Polytechnique Fédérale de Lausanne (EPFL), School of Life Sciences, Brain Mind Institute, 1015, Lausanne, Switzerland (419)

Kirstan A. Vessey, Department of Anatomy and Cell Biology, The University of Melbourne, Parkville, Victoria, Australia (211)

Lucas Waltzer, Université de Toulouse, UPS, CBD (Centre de Biologie du Développement), Bâtiment 4R3, 118 route de Narbonne; and CNRS, CBD UMR5547, F-31062 Toulouse, France (51)

Fen Wang, Center for Cancer and Stem Cell Biology, Institute of Biosciences and Technology, Texas A&M Health Science Center, Houston, Texas, USA (1)

R.J. Wessells, University of Michigan Medical School, Ann Arbor, Michigan, USA (155)

Benjamin Wolozin, Department of Pharmacology; and Department of Neurology, Boston University School of Medicine, Boston, Massachusetts, USA (499)

A. Zelhof, Department of Biology, Indiana University, Bloomington, Indiana, USA (331)

Preface

Various human diseases caused by genetics and environmental factors significantly impact the quality of life of the patients and their families, as well as impose a heavy economic burden on the society. We believe that the ultimate goal of understanding human biology is to prevent and cure diseases, and to alleviate pain and suffering, thus allowing individuals to maintain a healthy and active lifestyle. The sophisticated biological processes and complexity of the human body made it challenging for scientists to elucidate the underlying molecular and cellular mechanisms of various diseases. This is further complicated by ethical concerns and difficulties in obtaining proper tissue samples at various stages of disease progression.

The focus of this volume of Progress in Molecular Biology and Translational Science is on animal models of human diseases. Many biological processes and signaling pathways are evolutionarily conserved between animals, including humans. With the powerful genetics and ease of manipulation, animal models have greatly facilitated our understanding of the basic molecular and cellular mechanisms underlying various complex biological processes and human diseases. Here, we highlight the use of animal models to study various disease pathogenesis and their contribution to therapeutic development. We present some of what we believe to be the most common health issues faced by the general public, including cancer, cardiovascular, eye, metabolic, and neurological diseases. In addition to mouse models, this volume has included, when possible, chapters on vertebrate and invertebrate models such as zebra fish and Drosophila that are currently used by scientists to model the respective diseases.

Karen T. Chang and Kyung-Tai Min
Los Angeles, California
Bloomington, Indiana

Modeling Human Prostate Cancer in Genetically Engineered Mice

FEN WANG

Center for Cancer and Stem Cell Biology, Institute of Biosciences and Technology, Texas A&M Health Science Center, Houston, Texas, USA

Progress in Molecular Biology
and Translational Science, Vol. 100
DOI: 10.1016/B978-0-12-384878-9.00001-7

Copyright 2011, Elsevier Inc.
All rights reserved.
1877-1173/11 $35.00

The progression of prostate cancer is a slow and multiple-step process; clinically detectable prostate cancer normally manifest in aged men, although the lesions may have originated much earlier in life. Animal models that mimic the initiation, progression, and metastasis of human prostate cancer are needed to understand the etiology of prostate cancer and to develop new treatments. Recent progress in mouse genetic engineering technology has led to generation of a series of mouse models for prostate cancer research, which have been widely used for testing impacts of a single or combinations of several gene alterations on the onset, progression, and metastasis of prostate tumors, as well as for assessing the effects of environmental, clinical, and preclinical drugs for prostate cancer prevention and treatment. Although it is possible that no single "perfect" model can recapitulate every aspects of this highly heterogeneous disease, it is expected that the models mimicking certain aspects of prostate cancers will continue to provide preclinical guide to treat this prevalent disease.

I. Introduction

The prostate is an accessory gland of the mammalian male reproductive system, which produces prostatic fluid that contributes to 25–30% of the volume of the semen. Its morphology varies considerably among mammals. Human prostate is an acorn-shape gland that sits under the bladder and in front of the rectum. Adenocarcinoma of the prostate has become the most common cancer in American men; more than 217,000 new cases are diagnosed every year in the USA alone. Prostate cancer is responsible for more than 32,000 deaths in America per year; the mortality rate is second only to lung cancer.[1] The human prostate has three histologically distinct regions: the peripheral zone, the transition zone, and the central zone. About 85% of human prostate cancer arises in the peripheral zone, whereas benign prostatic hyperplasia (BPH), a nonmalignant overgrowth that is fairly common among aging men, occurs mainly in the transition zone.[1] The progression of prostate cancer is a slow and multiple-step process. Early prostate tumor is organ-confined, responsive to androgen deprivation, and is often surgically curative. At advanced stages, however, the tumors frequently metastasize primarily to bones, lymph nodes, and the lung and become castration resistant. Metastatic prostate cancer usually is lethal, and there is still no cure for men with this advanced disease. Although prostate cancer may originate as localized lesions early in life, most prostate cancer patients may have clinical symptoms only after they are over 60 years of age. Therefore, animal models that mimic the processes of onset, progression, metastasis, and escape from hormone therapies in human prostate cancer are needed to develop new therapeutic strategies for prostate cancer prevention and intervention.

Unlike human prostate that has an acorn-shaped morphology, rodent prostates have four pairs of lobes: the anterior prostate (AP, also known as the coagulating gland), dorsal prostate (DP), lateral prostate (LP), and ventral prostate (VP) lobes. The dorsal and lateral lobes are often collectively referred to as the dorsolateral prostate (DLP) lobes. These morphologically and histologically distinct lobes are arranged circumferentially surrounding the urethra. They display characteristic patterns of ductal network and produce lobe-specific sets of secretory proteins. Although the human and rodent prostates are morphologically different, they have an overall similar histology structure, which consists of epithelial and stromal compartments separated by basement membranes. Furthermore, histology and gene expression similarities between human peripheral zone and the rodent DLP suggest that they are structurally and functionally equivalent.[2,3]

Despite the differences in details of organ morphology and tissue histology, mouse and human prostates share extensive similarities in basic cellular and molecular biological features (Fig. 1). The rodent and human prostates have overall similar intimate two-way regulatory communications and symbiosis between epithelial and stromal compartments. Similar to human prostate, the mouse prostatic epithelium has three major cell types, luminal cells, basal cells, and neuroendocrine (NE) cells. These three major cell types can be distinguished by their morphological characteristics, molecular markers, secretory proteins, and relevance to progression of prostate tumors. Luminal cells are the predominant cell in the prostate epithelium, which are androgen-dependent and produce secretory proteins. The luminal epithelial cells are characterized by the expression of the androgen receptor (AR), cytokeratins 8 and 18, and the cell surface marker CD57, which are also exhibited in most human prostate cancer cells. The basal cells are located between the luminal cells and the basement membrane, which are characterized by the expression of cytokeratin 5, cytokeratin 14, CD44, and p63. Although some basal cells are AR positive, most basal cells do not express the AR. Although the function and the cell lineage of basal cells remain controversial, evidence reveals that epithelial stem cells and transient amplifying cells of the prostate reside in the basal cell compartment.[4,5] The NE cells are a minor population of uncertain embryological origin believed to provide paracrine signals to support the growth of luminal cells.[6,7] NE cells can be identified by expression of synaptophysin, chromogranin A, and synaptic vesicle protein 2. Regardless that most prostate cancer exhibits luminal epithelial markers and loses basal cell markers, both luminal and basal cells are reported to be the cells of origin of prostate cancer.[8,9]

The stromal to epithelial cell ratio is 5:1 in human and 1:1 in rodent prostates.[10] Despite this major histological difference between human and rodent prostates, both rodent and human prostate stroma are mainly composed

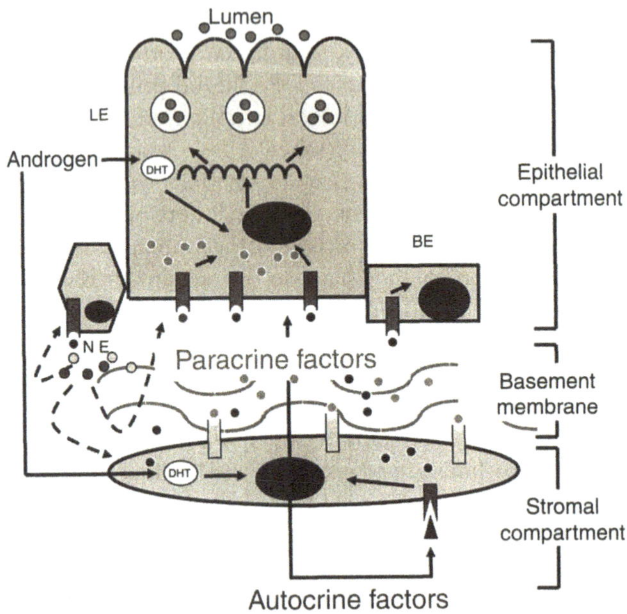

FIG. 1. Compartmentalization of the prostate. The prostate epithelium consists of secretory, basal, and neuroendocrine (N) epithelial (E) cells, separated from the stroma by the basement membrane. Testosterone, converted to dihydrotestosterone (DHT) within cells, controls direction-specific paracrine and autocrine factors in both the stroma and epithelium, which interact with their cognate receptors and mediate regulatory functions of the androgens. The growth factor signaling pathways also modulate the expression and transactivation activity of the androgen receptors. These regulatory communications between the stroma and epithelium control the growth, differentiation, and apoptosis of the prostate cells and are important for maintaining tissue homeostasis in the prostate and androgen-dependent nonmalignant tumors.

of specialized smooth muscle cell (SMC)-like and diverse other nonparenchymal cell types, including matrix-producing fibroblast-like cells, cells contributed by innervations, the immune system, and the circulatory system. The SMC-like cells, which express α-smooth muscle actin, intimately associated with and surround the epithelial cells. The fibroblast-like cells and other stromal cells are more dispersed within the stromal matrix. In rat prostate tumors, the reduction of SMC-like cells in respect to the number and degree of differentiation is found to be associated with the progression to malignancy.[11] Since loss of SMC-like cells and dominance of the undifferentiated fibroblast-like cells are found associating with tumor progression in rat prostate tumor models, it appears that the epithelial cells are essential for maintaining the properties of SMC-like cells as well as the overall cellular composition of the stroma.[11]

Although rodent prostates seldom develop spontaneous prostate tumors, Dr. W.F. Dunning of the University of Miami observed a spontaneous papillary adenocarcinoma of DP of a Copenhagen rat in 1961,[12] which was designated as the Dunning tumor. The tumor can be maintained in male syngeneic Copenhagen rats. After a series of passages in both intact and castrated hosts, several sublines of tumors had been developed, including both hormone-sensitive and -insensitive tumors.[13] The progression of the Dunning tumor model recapitulates what happens during human prostate tumor progression and has been widely used to understand how prostate tumor progress and become castration insensitive.[14] Yet, genetic manipulation in rats is not well established, which limits the development of new prostate tumor models in rats and hinders in-depth investigation of the onset, progression, metastasis, and relapse of prostate cancer at the molecular level.

Recent progress in mouse genetic engineering technology, including forced expression and gene ablation, has led to the generation of a series of genetically engineered mouse (GEM) models of human prostate cancer to mimic human prostate tumor initiation, progression, and metastasis. These models are generated based on potential etiological factors of human prostate cancer, such as inducing genetic instability or deregulating cell signaling, which cause dysregulation of proliferation, differentiation, apoptosis, homeostasis control, and other cellular activities. The major advantage of GEM models is that they provide a model system for evaluating the role of a single gene and its interaction with other genes or environmental factors in prostate cancer initiation, progression, and metastasis. These models can be used for testing new therapeutic or preventive agents in intervention and prevention of prostate tumor progression at various stages and for identifying molecular mechanisms by which these therapeutic and preventive agents exert their actions.

II. Technologies for Creating GEM Models

A. Overexpression

A wide spectrum of genes has been shown aberrantly expressed or mutated in prostate cancer cells, implying their roles in the onset, progression, metastasis, and relapse of prostate cancer. Therefore, forced expression of these genes in the prostate will provide *in vivo* assay systems to scrutinize their roles in prostate cancer. Generally, the mice that overexpress genes of interests in targeted tissues can be categorized into two types, transgenic mice and knockin mice. The conventional transgenic mice are the most common ones, which are referred to mice carrying an artificial gene or exogenous cDNA that is introduced into the mouse genome by microinjection into the pronucleus of fertilized eggs. A typical transgene includes a promoter that target the expression to specific cell types, a short intronic sequence that is required to ensure the integrity of mature mRNAs, a cDNA that encodes the protein of interest, and polyadenosine addition sites. Normally, multiple copies of the transgene are integrated into the genome randomly in tandem repeat manner. Although expression of the transgene is driven by the promoter, the chromatin structure of the insertion site strongly affects the expression level. The impact of the chromosome structure, however, at least partly, can be overcome by inclusion of an insulator element in the transgene. An insulator is a DNA fragment that associates with strong DNase 1 hypersensitive sites and tends to separate chromatin domains with different degrees of condensation, thus, minimizes negative effect of chromosome structures.[15] The transgenic technology provides an easy and time-saving way to create overexpression models. A highly active and tissue-specific promoter is critical for targeting the expression to a specific cell type. Yet, the expression levels vary with each strain, and insertion of the transgene may impact or disrupt normal function of the inserted allele. Thus, precautions have to be taken in phenotype analyses.

The "knockin" technology is another way to create forced expression models in which the coding sequence of a protein of interest is inserted precisely into a desired genomic location by genetically engineering technology in mouse embryonic stem (ES) cells. The expression of the knockin coding sequence is controlled by the normal transcriptional machinery of the knockin allele. The advantage of knockin models is that the expression pattern and the outcome of disruption of the host allele are predictable. One common allele for ubiquitous expression without obvious side effects is the ROSA26 allele.[16] Yet, this is a time-consuming and tedious way to generate knockin mice, and sometimes, the expression level may not be high enough. Together with either the Cre/loxP recombination or other gene expression regulatory mechanisms, such as the tetracycline controlled gene expression system, more sophisticated expression systems have been used to express genes of interest in mice. Instead of being directly controlled by the promoter only as in conventional transgenes, expression of these conditional expressing transgenes is further regulated both temporally and spatially either by Cre/loxP mediated recombination or by tetracycline regulatory transcription factors.

1. Prostate-Specific Promoters

The rat probasin (PB) promoter and its derivatives, which include the minimal PB promoter,[17] the long 12 kb PB promoter,[18] and the composite ARR2PB promoter,[19] are the most commonly used promoters for targeting expression of genes of interest to the prostate epithelium. PB is an abundant protein that belongs to the lipocalin superfamily. PB is located in the nucleus of prostate and seminal vesicle epithelial cells, as well as in prostatic secretion fluids. The LP has the highest expression level of PB, followed by the dorsal, anterior, VP, and seminal vesicles. The expression of PB reaches the maximum level when the mice become sexually mature; androgen ablation quickly decreases PB expression, indicating that the expression is regulated by androgens. Detailed characterization reveals that two distinct AR-binding sites, ARBS-1 and ARBS-2, are required for maximal androgen-induced gene expression.[20]

The minimal PB promoter includes 426 basepairs of the PB gene immediately upstream of the translational initiation sites and 28 basepairs of the 5′-uncoding region downstream of the transcription initiation site. It targets transgene expression specifically to epithelial cells in LP, DP, and VP at moderate levels and the AP and seminal vesicles at low levels.[17] The transcription activity of the minimal PB promoter can be detected as early as 2 weeks after birth. The expression reaches maximal level by 7 weeks, which is correspondent to the sexual maturation. The minimal PB promoter has been widely used for expressing transgene in the prostate epithelium. The modest level of transcription of the minimal PB promoter makes it suitable for expressing gene products with high impact even at low expression levels, such as viral oncogenes. However, it is unsuitable for expressing genes that need high expression levels to have impacts. A large PB (LPB) promoter fragment composed of 12 kb upstream sequence of the PB transcription initiation site is used to achieve high level expression in the prostate epithelium.[21] Similar to the minimal PB promoter, the expression activity of the LPB promoter is also androgen regulated. However, its bulky size makes it difficult to handle and reduces the efficiency of cloning and genomic integration.

A composite ARR2PB promoter has been made in which the inhibitory sequence between −426 and −287 in the PB promoter was removed and replaced with two copies of the AR response region of the PB promoter.[19] The composite promoter, although small in size, confers high androgen-dependent expression of transgenes in the prostate epithelium. Compared to previous PB promoter, ARR2PB driven expression is more consistent and at high levels. However, low level activity has been noted in some other tissues, suggesting that the specificity may have been compromised as a cost of high expression levels. *In vitro* experiments also show that the ARR2PB promoter gives basal expressions in PC-3, LNCaP, and DU145 prostate cancer cell lines and the expression is significantly induced by androgens, whereas in nonprostatic cell lines, the transcription activity is very low and not androgen responsive.

Prostate-specific antigen (PSA) is a kallikrein-like serine protease that is almost exclusively expressed in human prostate luminal epithelial cells, and its expression is androgen regulated. The 4 kb immediately upstream sequence of the PSA coding sequence, which includes the proximal promoter and a strong enhancer region, has been used to direct expression of transgenes in the prostate epithelium. The promoter delivers transgene expression to the LP by 8 weeks after birth at an androgen-dependent manner. The expression declines after androgen deprivation and can be restored by androgen replenishment.[22,23] In addition to the PB and PSA, several other promoters have been used to target expression of transgenes to the prostate epithelium with various degrees of specificities, which include the C3(1) promoter,[24] the fetal G-globin promoter,[25] the gp91-phox promoter,[26] the cryptdin 2 promoter,[27] MMTV,[28] and the 3.8-kb fragment of the PSP94 promoter.[29] Currently, no prostate basal cell- or stromal cell-specific promoters have been reported. Although not prostate specific, the p63 promoter is an obvious candidate for the basal cells since p63 is specifically expressed in basal cells, but not in luminal epithelial and stromal cells of the prostate. The generally stromal-specific fibroblast-specific protein-1 (fsp1) promoter has been used to express Cre recombinase in stroma cells, which has been shown to effectively ablate target genes in prostate stroma.[30,31]

B. Gene Targeting

Loss of function of tumor suppressors, such as Pten and p53, etc., through mutation, deletion, or epigenetic modification, is also common in prostate cancer. Analyzing mutant mice deficient in genes of interest is important in determining the function of these genes in the onset, progression, metastasis, and relapse of prostate cancer. Gene ablation based on homologous recombination in mouse ES cells enables the production of mutant ES cells that carry one mutant allele of gene of interest, which are subsequently used to generate chimeric mutant mice via microinjection to blastocysts.[32] If the ES cells have germline integration, the heterozygous mutants can be generated via breeding. In some cases heterozygous mutant mice are sufficient, but, in most cases, homozygous mutant mice are needed to produce phenotypic changes. The convention method is to directly delete a segment of DNA via homologous recombination, which has been widely used to generate genome-wide knockout, also known as germline knockout. This conventional method has been used to investigate genes that contribute to the onset and progression of prostate cancer. However, if their disruption leads to embryonic lethality or severe abnormalities in mice, this approach does not permit evaluation of function of those genes in prostate cancer since prostate diseases only occur in adult animals. Furthermore, dissecting the role of genes specifically in the prostate is often complicated by systemic defects since germline knockout mice have the deletions in all cells and tissues. Abnormalities observed in the prostate may arise indirectly from developmental defects or functional defects in other organs.

The Cre/loxP- and the Flp/FRT (Flp recognition target)-mediated DNA recombination systems provide a more sophisticated technology for spatially and temporally specific gene ablation. Together with prostate specifically expressed Cre or Flp recombinase, this technology greatly increases the power to functionally study the genes of interest in the prostate, especially those needed for embryonic development. The Cre/loxP system is derived from the bacteriophage P1 in which the Cre recombinase binds to two loxP sites and mediates recombination of the sequence flanked by two loxP sites.[33] If the two loxP sites are inserted in the genomic DNA in the same orientation, the recombination results in looping out the sequence between the two sites, leaving a single loxP site in the original DNA. If the two loxP sites are inserted in opposite directions, the recombination results in changing the direction of the intervening sequence. The loxP elements and a selection marker are normally inserted in the intronic sequences flanking the target coding sequence, which usually may not affect expression of the alleles. Therefore, together with transgenic mice that express the Cre recombinase in a tissue-specific manner, the Cre/loxP recombination is widely used for a temporally and/or spatially specific ablation of genes of interests. In addition to the Cre/loxP system, the Flp/FRT system has also been used for conditional gene ablations. The Flp recombinase is encoded by the yeast plasmid and catalyses a site-specific recombination reaction between two FRT sites.[34,35] Adding temporal regulation of Cre or Flp expression or activity further allows us to control gene ablation at a specific time point. The tetracycline-regulatable expression system has been increasingly used to control the Cre expression.[36] Fusion of the Cre recombinase and a mutated hormone-binding domain of the human estrogen receptor results in a chimeric protein, Cre-ER^{TM}. Since the Cre-ER^{TM} fusion protein can only be translocated to the nucleus and elicits its recombination activity in the presence of tamoxifen, the Cre-ER^{TM} mediated recombination is strictly under control of 4-hydroxytamoxifen.[37] The combination of tissue-specific expression and ligand-dependent Cre recombination will further provide precise timing and cell type-specific controls of gene disruptions.

1. Prostate-Specific Cre Lines

The first prostate epithelial-specific Cre line, PB-Cre, was generated by using the minimal PB promoter to target expression of the Cre in prostate epithelial cells.[38] Although the recombination efficiency is not high, the Cre driver is able to mediate the deletion of the floxed retinoblastoma (Rb) alleles in all lobes of the prostate, with the highest in VP and the lowest in AP. A second line of prostate-specific Cre transgenic line was generated subsequently, named PB-Cre4,[39] which carries the Cre cDNA under the control of the ARR2PB composite promoter. Expression of PB-Cre4 is postnatal and prostatic epithelium specific. Although the Cre recombination is detected in all lobes of the mouse prostate, the expression levels vary significantly in different lobes, being highest in the LP, followed by the VP, and then the DP and AP. Except for a few scattered areas in the gonads and the stroma of the seminal vesicle, no other organs in adult PB-Cre4 mice demonstrate significant Cre expression. A third prostate epithelium-specific Cre driver, ARR2PBi-Cre, was generated with the similar strategy,[40] in which the expression of modified Cre recombinase[41] is driven by the ARR2PB composite promoter. An insulator element from the chicken globulin locus is inserted at the 3′-end of the transgene to minimize negative effects on transcription of the transgene imposed by chromosome structure. The ARR2PBi-Cre transgenic mouse specifically and uniformly expressed Cre recombinase in all lobes of the prostate, seminal vesicles, and ductus deferens. Compared with the other two prostate-specific Cre strains, the ARR2PBi-Cre strain exhibits higher and more uniform expression of Cre recombinase in the prostate, although it is also expressed in seminal vesicles and ductus deferens.[40]

The human PSA promoter is also used to deliver expressions of Cre[42] and Cre-ERT2 fusion proteins to the prostate epithelium.[43] The line PSA-CreD4 shows high, prostate-specific Cre activity in all lobes. PCR analysis shows that no other tested tissues exhibit Cre expression, which has been used to disrupt Pten alleles in the prostate epithelium.[42,44] The PSA-ERT2 is also specifically expressed in the prostate, with the highest expression in the DLP and lowest expression in the AP.

In addition, two knockin Cre lines, *Nkx3.1*^{Cre} and *Nkx3.1*$^{Cre\mathrm{ERT2}}$, have been generated by knockin of the cDNA for Cre and Cre-ERT2 fusion proteins into the *Nkx3.1* allele, which are expressed in the precursors for prostate epithelial cells at embryonic day 17.0 when the prostate bud is formed. Although both Cre drivers are expressed in many organs, in the prostate, they are only expressed in the epithelium, which provide useful tools for studying genes of interests in prostate development[45,46] and in prostate stem cell research.[8]

III. Prostate Tumor Models Driven by SV40 T Antigens

The T/t antigens are the early genes of simian virus 40 (SV40), which include that large T and small t antigens. The large T effectively abrogates function of tumor suppressors, p53 and Rb, and causes genetic instability. In addition to suppressing p53 and Rb, the small t antigen also inhibits PP2A protein phosphatase activity, leading to overactivation of the MAP kinase pathway. Although the T/t antigens are not natural causes of human prostate cancer, they induce onset and progression of prostate tumors by disruption of genetic stability and unleashing cell signaling pathways that promote the tumor phenotype. In fact, several T/t antigen-driven mouse models resemble the development and progression of human prostate cancer in many aspects at an accelerated pace, which greatly facilitate mechanistic and preclinical studies of prostate cancer. These models have been widely used for screening potential dietary factors and drugs for prevention and intervention of the onset, progression, and metastasis of prostate cancer.

A. The Autochthonous TRAMP Model

The autochthonous TRAMP (transgenic adenocarcinoma of the mouse prostate) model was developed based on the minimal PB promoter to direct expression of the T/t antigens to the mouse prostatic epithelium.[47] The T/t antigens are expressed in epithelial cells of VP and DP at moderate levels,[47] which can be detected at the age of 4 weeks when the mice are reaching sexual maturity. The combination of repression of tumor suppressors p53 and Rb and overactivation of the MAP kinase pathway causes prostate adenocarcinoma in a high incident rate. The TRAMP mice develop early prostatic intraepithelial neoplasia (PIN) by 6 weeks, and mild- to high-grade PIN by 12 weeks. At 24 weeks, approximately 100% of male mice have poorly differentiated and invasive adenocarcinomas. Metastases mainly to the periaortic lymph nodes and lungs can be detected as early as 12 weeks of age. Although the bone is the most common metastasis site of human prostate cancer, the TRAMP tumors, as other GEM prostate tumor models, rarely metastasize to bone.[48] The molecular mechanism underlying this difference is unknown. Therefore, to study the differences between human and mouse prostate tumors and to develop mouse prostate tumor models that have bone metastasis will provide hints for screening new therapeutic reagents for bone metastasis of human prostate cancer.

At early stages, the TRAMP tumors are androgen sensitive; castration of TRAMP mice at 12 weeks causes a regression of prostate tumors along with the normal prostate.[49,50] Similar to the human disease, although the castration at 12 weeks is curative in about 20% of TRAMP mice, the majority of the TRAMP tumors progress to poorly differentiated and androgen-insensitive cancers with frequent metastases by 24 weeks. As observed in human prostate cancer, TRAMP cells also undergo the epithelial-mesenchymal transition (EMT) marked by the loss of E-cadherin at later stages, as the primary tumors become poorly differentiated and metastasize. Therefore, the TRAMP model provides a consistent source of primary and metastatic tumors for histopathobiological and molecular analysis to further define molecular events involved in the onset, progression, metastasis, and relapse of prostate cancer.

TRAMP-C1, TRAMP-C2, and TRAMP-C3 are the immortalized epithelial cell lines derived from primary TRAMP tumors.[51] All three cell lines express cytokeratin, E-cadherin, and AR. TRAMP-C1 and TRAMP-C2, but not TRAMP-C3, cell lines are tumorigenic when grafted into syngeneic C57BL/6 hosts. The three cell lines represent various stages of cellular transformation and progression to androgen-independent metastatic disease, which have been widely used as an *in vitro* system parallel to the original mouse model for investigating molecular and cellular mechanisms underlying prostate tumor progression, and for screening therapeutic agents for prostate tumor treatments.

It has been suggested that TRAMP tumors may represent prostatic NE tumors, since the NE differentiation markers may occur in early stages of PIN lesions, especially in mice with the FVB genetic background.[52] Yet, systematic studies of NE differentiation in the TRAMP mice reveal that foci of synaptophysin, a NE differentiation marker, are detected in 2.5% of a total of 162 PIN lesions. No synaptophysin is detected in any of the well-differentiated or phylloides-like lesions. Synaptophysin can be detected in 92% of poorly differentiated regions. However, when mice are castrated, synaptophysin is detected in 100% of poorly differentiated regions.[53] Of the seven metastatic lesions analyzed, four pelvic lymph node metastasis expressed synaptophysin, but the other metastatic lesions (liver, salivary gland, and mesenteric lymph node) did not. Thus, it is likely that the emergence of the NE differentiation is a stochastic event in the TRAMP model and is generally associated with end-stage and high-grade prostate cancer. Focal NE differentiation has been detected in nearly all clinical prostatic adenocarcinomas at late stages and has been associated with aggressive human prostate tumors and hormone refractory disease, which is thought to be of prognostic value. Approximately 10% of prostate cancers show more extensive NE features even at the early stages, which are classified as NE tumors. Similarly, all transgenic mouse prostate tumor models that have high metastasis rates exhibit NE differentiation at different extents. It is interesting that the most late-stage TRAMP tumors display the NE differentiation and that the genetic backgrounds of the mice affect NE differentiation in TRAMP tumors. Comparing the TRAMP tumors in different genetic backgrounds to identify genes that can modify the NE differentiation should shed new light on whether NE differentiation contributes to prostate tumor progression.

B. The LADY Model

The LADY prostate tumor model was created by using the 12-kb LPB promoter to deliver SV40 large T antigen expression to the prostate epithelium.[54,55] Using the large T antigen alone eliminates influence of the small t antigen that causes overactivation of the MAP kinase pathway. The LPB-Tag mice developed multifocal epithelial hyperplasia in the prostate by 10 weeks, which continued to progress to hyperplasia across the entire epithelium and then to low-grade dysplasia. With advancing age, the mice gradually developed the following lesions in sequence: low-grade PIN, high-grade PIN, microinvasion, invasive carcinoma, and poorly or undifferentiated carcinoma with NE differentiation, which is similar to that observed in human prostates. Initially, the LADY tumors are androgen dependent since the tumors regress quickly after the mice are castrated, and androgen replenishment restores both the original epithelial/stromal cell ratio and tumor growth. Yet, a small population of prostatic epithelial cells in castrated animals continues to proliferate after castration, indicating the potential for castration-resistant growth.

Although local invasions are commonly observed, the LADY mice generally develop prostate adenocarcinoma without metastasizing to distal organs. However, one LADY line, 12T-10, frequently has metastasized tumors. Similar to other LADY lines, the 12T-10 line develops PIN lesions first and invasive carcinoma at late stages. Metastatic frequency is 66% in mice at 6 months of age, and 88% at 9 months. The common targets are regional lymph nodes, liver, and lung. Interestingly, the high mitotic activity of NE cells in the 12T-10 line resembles those seen in human small cell carcinoma of the prostate. This contrasts to the relatively low mitotic rate of NE cells in prostate adenocarcinoma, raising the possibility that the 12T-10 line may be a useful model for elucidating the processes involved in initiation and progression of small cell carcinoma.

C. The T_{121} Model

The T_{121} model was created by using the ARR2PB promoter to drive expression of a truncated NH_2-terminal domain of SV40 large T antigen that includes 121 amino acids (T_{121}) in the prostate epithelium. The T_{121} mutant only specifically inactivates pRb and leaves p53 and other T antigen targets functionally intact. The $TgAPT_{121}$ transgenic mice exhibit highly abnormal epithelial proliferations accompanied by Pten-dependent apoptosis and develop widespread PIN lesions by 2 months of age. The epithelium is characterized by nuclear atypia and loss of single layer architecture. By 4 months, the PIN lesions slowly reach the microinvasive stage and then well-differentiated prostate adenocarcinoma in all animals. Pten hemizygosity reduces apoptosis by 50% and accelerates progression to adenocarcinomas in the $TgAPT_{121}$ transgenic mice.[56] It is suggested that inactivation of pRb is sufficient to induce aberrant proliferation and abundant apoptosis in prostatic luminal epithelial cells and that *$TgAPT_{121}$* mice likely mimic the initiation of prostate cancer caused by dysfunctional pRb pathways that are observed with high frequency in early stages of the human disease.

Interestingly, disruption of p53 results in increased mesenchymal cell proliferation that leads to an extensive hypercellular mesenchyme in the $TgAPT_{121}$ prostate during early onset and progression of the tumors, while epithelial apoptosis and proliferation is unaffected.[57] The results reveal that mesenchymal response in $TgAPT_{121}$ prostate is increased due to p53 deficiency, and that development of tumors in $TgAPT_{121}$ mice is facilitated by p53 deficiency. Furthermore, disruption of the epithelial cell cycle via inhibiting pRb induces p53 response in the stroma and suppresses stromal proliferation, suggesting that the interplay among multiple cell types imposes strong a selective pressure yielding a highly proliferative mesenchyme that loses p53.

D. The PSP94-TGMAP Model

The PSP-TGMAP is generated by using the 3.8-kb promoter fragment of PSP94 (prostate secretory protein of 94 amino acids) to target T antigens to the prostate epithelium.[58] At about 10 weeks of age, all male mice develop prostate hyperplasia that progress to well-differentiated adenocarcinoma by 24 weeks. The tumors further progress to castration insensitive and metastasize to periphery organs as the age advances. A T antigen knockin model has been created by inserting the T antigen-coding sequence to the PSP94 locus (PSP-KIMAP).[59] Compared with the original transgenic model, PSP-TGMAP, the PSP-KIMAP mice develop tumors at a younger age, and the tumor development is more predictive and has no founder line variations.

E. Nonprostate-Specific T Antigen-Driven Models

In addition to using prostate-specific promoters, several other promoters that are not prostate epithelium specific are also used to drive expression of T antigens in multiple organs. The Ggamma/T-15 transgenic mice were generated using the fetal globin promoter to direct T antigen expression in adult prostates.[60] The mice develop tumors within VP and DP at first, rapidly grow in the pelvic region, and then metastasis to lymph nodes and distant sites. About 75% of heterozygous and 100% of homozygous transgenic males develop prostate tumors. The tumorigenesis and progression is androgen independent, since castration shortly after prostate morphogenesis does not prevent tumor development. The advanced tumors exhibit both NE and epithelial phenotypes, but the epithelial differentiation markers are often lost early in tumorigenesis. Interestingly, expressions of the T antigen-binding tumor suppressers (p53 and Rb) are upregulated in Ggamma/T-15 tumors, and the expression of the B-cell lymphoma-2 (Bcl-2) proto-oncogene is also slightly upregulated. As the tumor arises from both NE and luminal epithelial cells, the Ggamma/T-15 transgenic mice provide a good model system for studying NE tumors and how NE differentiation contributes to prostate tumor progression.

C3(1)/SV40 T mice have been generated by using a 4.5-kb fragment of the rat [C3(1)] promoter to drive expression of T antigens.[24] The VP and DLP develop low-grade PINs at 2 months and high-grade PINs by 5 months. Prostate carcinomas, appearing to arise from PIN lesions, are normally found by 7 months in the VP and 11 months in the DLP. Most adenocarcinomas in the C3(1)/SV40 T mice are locally invasive. The main disadvantage of the C3(1)/SV40 T model is that the transgene is also expressed in other tissues including mammary, thyroid, and salivary glands. Nevertheless, the progression of PIN to invasive prostate carcinoma in the C3(1)/SV40 T mice appears to be similar to the process occurred in humans.

Cryptdin-2/SV40 T mice have been generated by using the 6-kb mouse cryptdin-2 promoter to direct expression of T antigen to a subset of NE cells in the prostate.[27] Transgene expression is initiated between 7 and 8 weeks of age in all prostatic lobes. The mice develop PIN lesions within a week afterward, which progress rapidly and the tumors invade lymph nodes, liver, lung, and bone by 6 months in an androgen-independent manner.

Although cytochrome b heavy chain (gp91-phox) is expressed exclusively in terminally differentiating cells of the myelomonocytic lineage, male gp91-phox T transgenic mice that carry T antigens driven by the gp91-phox promoter unexpectedly express the T antigens in the prostate and develop prostate tumors,[26] which are characterized as androgen-independent neuroblastomas.

IV. Protooncogene-Driven Models

A. c-Myc Overexpression Models

The Myc family includes c-Myc, N-Myc, and L-Myc, which are basic helix-loop-helix/leucine zipper (bHLH/LZ) transcription factors that are involved in regulating expression of up to 15% of all genes. c-Myc functions as a master regulator of genes that regulates diverse cellular processes including cell proliferation, differentiation, and apoptosis. Gain of function mutant of Myc has been shown to have strong oncogenic properties. Myc amplification and overexpression has been detected in many human cancers. Up to 30% of prostate tumors exhibit an increased c-Myc gene copy number or expression levels. In addition, elevated c-Myc expression is frequently observed in early stages of PIN lesions. Transgenic mice that express human c-Myc have been generated, in which the expression of the Myc transgene is controlled either by the PB promoter or by the composite ARR2PB promoter.[61] The PB-Myc transgenic mice have a low level of Myc expression and are designated as Lo-Myc. The ARR2PB-Myc mice express Myc at high levels and are designated as Hi-Myc.

The prostates expressing c-Myc exhibit a high rate of proliferation and apoptosis. However, c-Myc induced high rate proliferation overcomes the apoptosis and results in rapid development of PIN. In addition, expression of c-Myc also increases angiogenesis; increased vascular density has been detected in the prostate as early as in 2-month-old transgenic animals. The PIN lesions progress to invasive adenocarcinomas by 3–6 months in the Hi-Myc mice and by 10–12 months in the Lo-Myc mice, suggesting that progression of the disease is c-Myc expression level dependent. Frequently, the atypical epithelial cells are often found penetrating through the basement membrane and the fibromuscular layer in these mice, indicating local invasion of the lesion. Furthermore, lymphovascular invasion also has been detected in some cases. Castration of the Myc transgenic mice at 2 months of age causes complete regression of PIN within 1 month after the castration, which is correlated with the absence of detectable c-Myc expression. This indicates that the PIN phenotype requires constant expression of Myc and is reversible. Although castrating c-Myc transgenic mice at 8 months of age results in regression and fibrosis, all these mice have residual tumors present even after 5 months after castration. The AR signaling is inactive and the Myc transgene is not expressed in these castration-resisting tumors. Yet, the tumors fail to fully regress, suggesting that genetic events have occurred that maintain survival of tumor cells independent of c-Myc and that the lesion is irreversible at this stage. A similar strategy using the rat C(3)1 promoter to drive c-Myc expression only leads to PIN development without causing tumor development in the prostate.[62] The discrepancy in phenotypes likely is due to different expression levels.

B. Ras Overexpression Models

As a small GTPase, the activation of Ras depends on the binding to guanosine nucleotides, GTP (guanosine triphosphate). Like other members of the G-protein family, Ras functions as a binary molecular switch that controls a number of intracellular signaling networks regulating cellular processes, including proliferation, differentiation, apoptosis, adhesion, and migration. Three highly homologous genes encode human RAS, designated as H-Ras, K-Ras4A and K-Ras4B (the two K-Ras proteins arise from alternative gene splicing), and N-Ras, since they are first identified in Harvey, Kirsten sarcoma viruses, and neuroblastoma, respectively. Ras-regulated signal pathways are often deregulated in cancers, leading to increased invasion and metastasis accompanied by decreased apoptosis. Probasin-H-RAS^{val12} (PB-H-RAS^{val12}) transgenic mice have been generated in which the prostate epithelial expression of an activated RAS^{val12} mutant is driven by the minimal PB promoter. The prostate initially is normal, but develops atypical hyperplasia within 3 months. Consistent with activating RAS increases proliferation, a significant proportion of the prostates display PIN foci, which cover from 57% of the prostate in young adult mice to 19% of the prostate in old mice.[63] The PB-RAS transgenic prostates exhibit multifocal intestinal metaplasia wherein the epithelial cells are vacuolated cells similar to goblet cells of the intestinal epithelium. High levels of the goblet cell-specific peptide Itf/Tff3 in these transgenic prostates is in accordance with recent microarray studies showing that ITF/TFF3 is upregulated in human prostate cancer. In addition, the PB-RAS prostates have a thickened fibromuscular stroma. Thus, the PB-RAS mouse model can be useful for elucidating the early events in prostate carcinogenesis, as well as for studying the mechanisms and potential prostate cancer relevance of intestinal metaplasia.

Raf is a direct downstream target of the Ras family, which is a serine/threonine kinase and relays the signals to the MAP pathway. Once activated, Raf phosphorylates and activates the dual specificity protein kinases MEK1 and MEK2, which phosphorylate and activate ERK1 and ERK2. Activated ERKs then relay the signals to multiple downstream effectors and regulate gene expression involved in the cell division cycle, apoptosis, cell differentiation, and cell migration. The Raf family includes three members: A-Raf, B-Raf, and C-Raf. Unlike A-Raf and C-Raf that engages in multiple signaling pathways, B-Raf only activates MEK1 and MEK2. Transgenic mice expressing constitutively active B-Raf^{V600E} mutant under control of a tyrosinase promoter/enhancer have aberrant proliferation in p63+ basal epithelial cells and develop prostate hyperplasia, which progress to rapidly growing invasive adenocarcinoma.[64] The model demonstrates that activation of the MAP kinase signaling pathway alone is sufficient to drive prostate tumorigenesis. Interesting, although activation of B-Raf is sufficient to initiate development of invasive prostate adenocarcinoma, it is not required for the maintenance of tumors.

C. Constitutively Active Wnt/β-Catenin Models

Wnts are a family of secreted glycoproteins that regulate a broad spectrum of cellular processes. The canonical pathway of Wnt is mediated by binding to a cell surface Frizzled (Fzd) receptor together with LDL-receptor-related protein (LRP), and activating the β-catenin pathway. In the absence of Wnt signaling, cytoplasmic β-catenin is rapidly degraded by a complex that contains adenomatous polyposis coli (APC), Axin, and glycogen synthase kinase 3 (GSK3). Binding of Wnt to Fzd activates Dishevelled, blocks the association of β-catenin and GSK3, and, therefore, blocks the degradation of β-catenin. β-Catenin is then translocated into the nucleus, where it acts as a transcriptional coactivator and activates expression of target genes. Mutations that stabilize β-catenin have been shown associated with approximately 5% of primary human prostate tumors[65]; nuclear accumulation of β-catenin has been found strongly correlated with advanced prostate cancer and recurrence of the disease.[66,67] Prostate epithelium-specific deletion of the Apc alleles with the PB-Cre leads to prostate hyperplasia in mice within 4.5 weeks of age, and adenocarcinoma by 7 months.[68] Continued tumor growth usually necessitates sacrifice of the mice between 12 and 15 months of age. Despite the high proliferation rate, no metastasis of these tumors to the lymph nodes or other organs has been observed. Although castration of 6-week-old mice inhibits tumor formation, castration of mice with more advanced tumors does not eradicate the tumor, suggesting that tumors induced by loss of Apc function are capable of growing under conditions of androgen depletion. Similarly, conditional deletion of exon 3 of β-catenin in prostate epithelial cells, which leads to activation of the Wnt/β-catenin signaling, also results in epithelial hyperplasia followed by PIN.[69] Interestingly, once high-grade PIN develops, the prostate continues to grow even after androgen signaling blockade. The results suggest that constitutively active β-catenin may also play a role in compromising androgen dependency of prostate cancer.

D. Caveolin-1-Deficient Models

Caveolae are a special type of small invaginations of the plasma membrane in many cells, which is involved in molecular transport, endocytosis, cell adhesion, and signal transduction. Caveolin-1 (Cav-1), is a major protein of caveolae, often found overexpressed in human prostate cancer cells and in metastatic mouse prostate cancer. Tumor cell-secreted Cav-1 has proangiogenic activities in prostate cancer, and the expression is positively correlated with tumor angiogenesis. Interestingly, suppression of Cav-1 expression induces androgen sensitivity in high caveolin, androgen-insensitive mouse prostate cancer cells derived from metastases. On the other hand, overexpression of Cav-1 leads to androgen insensitivity in low caveolin, androgen-sensitive mouse prostate cancer cells, implying that Cav-1 may be a candidate related to tumor metastasis and hormone-resistant properties of prostate cancer.[70] Both mouse and human prostate cancer cells secrete Cav-1 that promotes cancer cell survival and clonal growth *in vitro*. It also promotes proangiogenic activities in prostate cancer through the PI3K–Akt–eNOS signaling module independent of caveolae function. Disruption of Cav-1 attenuates the tumor development and progression in TRAMP mice.[71] Knockdown of Cav-1 expression in TRAMP tumor cells significantly reduces their tumorigenic and metastatic potential, further demonstrating a positive correlation between Cav-1 expression and tumorigenicity.[71]

E. The PB-Bcl-2/TRAMP (BxT) Model

The antiapoptotic Bcl-2 family includes Bcl-2, Bcl-XL, Bcl-w, Mcl-1, Bfl1/A-1, and Bcl-B, which bind to and inactivate BH3-domain proapoptotic proteins. Overexpression of the Bcl-2 family members has been reported to be associated with therapeutic resistance in various human cancers. The PB-Bcl-2/TRAMP (BxT) bigenic mouse model has been generated to assess the impact of Bcl-2 on the pathogenesis and progression of prostate cancer.[72] The prostate of BxT bigenic mice exhibit an enlarged prostate that has increased in proliferation, attributable to the T/t antigens, and decreased apoptosis attributable to Bcl-2. Compared to the TRAMP mice, the BxT mice also have shortened tumor latency. Although the incidence of castration-resistant cancer is reduced in the BxT mice compared to the TRAMP mice, the incidence of metastases is identical in both the TRAMP and BxT mice. Interestingly, androgen deprivation significantly increases the expression of Bcl-2 and Grp78 and decreases the expression of Bax, Bcl-xl, and c-Myc in castration-responsive, but not in castration-resistant, TRAMP tumors.[73] Therefore, it appears that Bcl-2 has multiple roles in prostate tumor initiation and progression.

F. The Egr1-Deficient Model

Early growth response protein 1 (Egr1) is a transcription factor that is rapidly induced by growth factors, cytokines, and stress signals such as radiation, injury, or mechanical stress to transduce signals to the downstream cascades, which is involved in cell proliferation and apoptosis regulation. Egr1 overexpression occurs frequently in a majority of human prostate cancers and is implicated in the upregulation of several genes important for prostate tumor progression. In addition, loss of NAB2, an Egr1 corepressor, is frequently found in early human prostate carcinoma. Ablation of Egr1 in the T antigen-induced prostate tumor models in mice (CR2-T-Ag and TRAMP) significantly delays progression of the prostate lesion from PIN to invasive carcinoma, although tumor initiation and tumor growth rate are not affected by Egr1 ablation,[74] implying that aberrantly expressed Egr1 contributes to the progression, but not onset, of prostate tumors.

G. The Hepsin Overexpression Model

Hepsin is a type II transmembrane serine protease that represents a specialized group of cell surface proteolytic enzymes. Although the functional significance of this upregulation is unknown, overexpression of hepsin is frequently observed in human prostate cancer. Expression of hepsin in prostate epithelial cells under the control of the PB promoter promotes prostate tumor metastasis in LPB Tag mice.[75] Overexpression of hepsin in the prostate epithelium weakens the epithelial–stromal adhesion and disrupts the basement membrane. Unlike tumors in LPB-Tag mice that normally do not metastasize, up to 65% of the double transgenic LPB-Tag/PB-hepsin mice develop prostate cancer, which metastasizes to liver, lung, and bone by 21 weeks of age. These data indicate that hepsin promotes primary prostate cancer metastasis.

V. The Tumor Suppressor Models

A. Pten-Deficient Models

The tumor suppressor, PTEN (phosphatase and tensin homologue deleted from chromosome 10), is one of the most frequently mutated genes in human cancers, including prostate cancer. Loss-of-function mutation in Pten is found in about 30% of primary prostate cancer and 63% of metastatic tissues, and has been found correlating with increasing Gleason score and advanced histopathology in human prostate cancer. PTEN encodes a phosphatase that negatively regulates the PI3-kinase-AKT pathway; therefore, loss of function of PTEN results in overactivation of AKT and consequently elevates proliferation and reduces apoptosis in cancer cells. The loss of Pten is embryonic lethal in mice. The phenotypes of deleting one Pten allele are not consistent, ranging from developing hyperplastic–dysplastic changes in the prostate, skin, and colon, to developing tumors in multiple organs. Nevertheless, in Pten heterozygous mice, cancer progression and metastasis often happen with the loss of heterozygosity of Pten. Conditional deletions of Pten alleles selectively in the epithelium of mature prostate also have inconsistent phenotypes. One group has reported that ablation of Pten with PB-Cre4 leads to only noninvasive prostatic lesions[76]; the other has reported that the ablation with the same Cre driver leads to invasive prostate adenocarcinoma.[77] In addition, deletion of Pten alleles with PSA-Cre results in a nonlethal invasive prostate cancer after long latency[42]; deletion of Pten alleles with MMTV-LTR-Cre causes high-grade PIN lesions that frequently progress to focally invasive cancer, although the mice also develop neoplastic changes in skin and thymus.[78] Hyperplasia in the prostatic epithelium can be seen by 5 days of age and high-grade PIN in 2 weeks. Between 7 and 14 weeks, focal invasions appear in more than half of the mutant mice.

Deletions of the Pten alleles with PSA-Cre-ERT2 mice, which express the tamoxifen-dependent Cre-ERT2 fusion protein, cause prostate hyperplasia within 4 weeks after Pten ablation and PIN within 2–3 months.[43] Although PIN is found in all prostate lobes, the highest incidence of PIN is in the DLP lobes. Eight to ten months after Pten ablation, adenocarcinoma is observed in the DLP lobe that is thought to progress from PIN. No distant metastases have been found up to 20 months after Pten ablation, indicating that progression to metastasis requires additional alterations. Consistently, ablation of only one allele of Pten with the same Cre driver also leads to focal hyperplasia and PIN. The lesions are exclusively in the DLP, occur with a longer latency and much lower frequency, and do not progress to adenocarcinoma. Pten expression is undetectable in these epithelial cells in PIN lesions, suggesting that loss of Pten function appears to act as a permissive event for the uncontrolled cell proliferation.[43] The discrepancies in Pten conditional null phenotypes likely are attributed to differences in genetic background and the efficient of Cre-mediated recombination of each strain.

Conditionally expression of either luciferase or green fluorescent protein reporter has been combined with the Pten conditional null model to generate the cPten conditional null luciferase (cPten$^{-/-}$L) or the Pten conditional null-green fluorescent protein (cPten$^{-/-}$G) lines to facilitate monitoring tumor growth, regression after androgen deprivation, and relapse without invasive samplings.[79] Both cPten$^{-/-}$L and cPten$^{-/-}$G lines, in which the floxed Pten alleles are deleted with the PB-Cre4 driver, develop prostate tumors that occasionally metastasize to lymph nodes. Androgen deprivation induces tumors regress initially.[79] Yet, the tumors frequently relapse in 7–28 weeks after the castration. Like androgen-independent tumors in human, the relapsed tumors often have a significant increased population of cells with NE differentiation.[79] As in wild-type prostates, epithelial cells in Pten-null prostate tumors have an increased number of apoptotic cells after the androgen ablation, indicating that the tumor cells still respond to the androgen deprivation. However, the cancer cells continue to proliferate even after androgen withdrawal, which counteracts the effect of apoptosis induced by androgen deprivation. Thus, high proliferation due to Pten deficiency overcomes the apoptosis and contributes to hormone-resistant prostate cancer.[77]

Two pairs of isogenic, AR-positive epithelial cell lines have been isolated from the prostate of conditional Pten knockout mice.[80] The cells are either heterozygous or homozygous for Pten deletion, but, loss of both Pten alleles is associated with increased anchorage-independent growth *in vitro* and tumorigenesis *in vivo*. In addition, knockdown of Pten expression in the androgen-dependent Myc-CaP cells converts the cells become androgen independent, suggesting that Pten intrinsically controls androgen responsiveness, a critical step in the development of castration-resistant prostate cancer.[80] Knockdown of the AR in Pten null cells reverses androgen-independent growth *in vitro* and partially inhibits tumorigenesis *in vivo*, indicating that the AR is required for the Pten-controlled prostate tumorigenesis.[80] Furthermore, it has recently been shown in mouse models that the onset of hormone-deprivation resistance likely is related to progression of the prostate tumor,[81] and that inactivation of Pten is sufficient to cause androgen-deprivation resistance.[80] In the prostate, both epithelial and stromal homeostasis is controlled by androgens. It is likely that androgen deprivation causes stromal cells to produce proapoptotic signals, either as a direct or indirect consequence of reduced AR signaling, which induce epithelial apoptosis in prostate and androgen-dependent prostate tumors. In castration-resistant Pten-null tumors, however, strong survival/proliferation signals generated by overactivation of Akt and MAP kinase pathways due to Pten inactivation, at least in part, counterbalance the proapoptotic signals induced by androgen deprivation, and confer castration-resistant properties to the tumor cells.[81]

Consistently, the MPAKT transgenic mice that express membrane-anchored AKT in prostate epithelial cells driven by the minimal PB promoter develop hyperplasia and PIN lesions and often causes urinary obstructions. The MPAKT transgene is mainly expressed in the VP, and causes high p70S6K activation in VP epithelial cells. Interestingly, the VPs of MPAKT mice have similar mRNA expression profiles and angiogenic signature[82] as in human prostate cancer. The artificial fusion protein iAKT contains three structure domains, the membrane docking fragment (FRBL2), the PH domain-deleted AKT mutant, and a chemically induced dimerization domain. Thus, the activation of iAKT is controlled by the chemical dimerizer AP21967. PBCre-iAKT transgenic mice that express the iAKT in the prostate epithelial cells driven by the ARR2PB composite promoter have been generated.[83] Activation of iAKT stimulates epithelial cell proliferation of the prostate in gonad-intact mice and promotes survival of prostate epithelial cells in castrated mice. Yet, the PIN lesions in the AKT transgenic prostate do not progress to invasive prostate cancer.[84] Since the AKT transgenic prostate aberrantly express $p27^{Kip1}$, it has been suggested that a $p27^{Kip1}$-driven checkpoint limits progression of PIN to prostate cancer.[84] Thus, these results indicate that other cooperating pathways are involved in the tumorigenesis induced by inactivation of the Pten/Akt pathway.

Like other models, the Pten-null/conditional null models also have several weaknesses. First, unlike invasive human prostate cancer that primarily metastasizes to the bone, prostate tumors induced by Pten inactivation do not metastasize to bone. Second, although there are only limited cells in human prostate cancer that are CD44-positive, almost 100% of the Pten-deficient tumors cells are 100% CD44 positive.[85] Third, the onset of prostate cancer in the Pten-null model is associated with changes in p63+ cell number, morphology, and localization. Yet, in human prostate cancers, the presence of p63+ cells is negatively correlated with progression of adenocarcinoma.[86] In addition, not all human prostate tumors exhibit loss of function of Pten, and loss of Pten *per se* may not be sufficient to cause full blown tumors in mice, suggesting that multiple genetic "hits" are required for progression of prostate to invasive cancer and metastasis to other organs.

B. P53-Deficient Models

The tumor suppressor p53 regulates target gene expressions, including those involved in cell cycle arrest, apoptosis, senescence, DNA repair, or changes in metabolism in response to diverse cellular stresses, which constitutes of transcription activation, DNA binding, and oligomerization domains. Mutations in p53 that cause changes in the complex network of molecular events are often found to be associated with prostate tumor initiation, progression, metastasis, and androgen-deprivation resistance. Acute Pten inactivation induces growth arrest through the p53-dependent cellular senescence pathway. This Pten deficiency-induced senescence can be fully rescued by the combined loss of p53. The results demonstrate that the cellular senescence restricts tumorigenesis *in vivo* and that p53 is an essential fail-safe protein of Pten-deficient tumors.[76]

Transgenic mice (PB-53 mutant) that express a dominant-negative mutant of p53 (273Arg/His) in the prostate driven by the minimal PB promoter develop high-grade PIN by the age of 52 weeks without developing pathological change in the other organs.[87] Compared to the prostate of nontransgenic littermates, the transgenic prostate undergoes less apoptosis after castration, suggesting compromised apoptotic potential in the transgenic prostates. Although prostate epithelium-specific inactivation of p53 does not result in prostate tumorigenesis, it cooperates with loss of Rb to induce highly aggressive, poorly differentiated, and metastatic prostate cancer.[88] Interestingly, all malignant tumors arise from the proximal region of the prostatic ducts, where are enriched in prostate stem/progenitor cells.[89] At early neoplastic stages, the cells in the lesion foci express Sca1 (a mouse stem cell marker), luminal epithelium marker cytokeratin 8, NE markers synaptophysin and chromogranin A, and the AR, although the cells are not sensitive to androgen withdrawal. Noticeably, the PIN lesions developed from the distal regions of the prostatic ducts that consist of lineage-committed transit amplifying and/or differentiated cells that never progress to carcinoma, suggesting that the microenvironment may also play a role in progression of the lesions induced by loss of function of p53 and Rb.[89]

C. The Nkx3.1-Deficient Models

The Nkx3.1 is a homeobox transcription factor that is involved in cell fate determination and organogenesis regulation. Nkx3.1 is expressed in early stages of the prostate organogenesis. It is essential for proper branching morphogenesis of the prostate, and for prostatic epithelial cells to appropriately express secretory proteins. The human Nkx3.1 alleles are located in the chromosome 8p region. Deletion of this region occurs in as many as 80% of human prostate tumors and is one of the most common events in early prostate carcinogenesis. Indeed, emerging evidence indicates that Nkx3.1 may function as a prostate tumor suppressor.[90] Loss of NKX3.1 expression at the protein level in human prostate cancers due to loss of one allele in combination with hypermethylation of the 5′ UTR of the other allele, or point mutations in both coding and noncoding regions, has been shown to correlate with prostate tumor progression.

Homozygous Nkx3.1 null mice are viable, but exhibit defects in prostate organ structure and secretory function in young adult males. With increasing age, the Nkx3.1 mutant mice develop prostatic epithelial hyperplasia, dysplasia, and low- to high-grade PIN. The lesion foci exhibit significantly increased cell proliferation activity.[90,91] After 1 year of age, most homozygous Nkx3.1 null mice develop low- to high-grade PIN lesions, with the features of perturbed basal cell layer, disrupted stroma, and compromised expression of prostate epithelial differentiation markers. Although none of the aged Nkx3.1 null mice develop invasive carcinoma, tissue recombination experiments demonstrate that after serial transplantation in nude mice, prostatic tissue recombinants from Nkx3.1 null mice display neoplastic progression characterized by increasingly dysplastic histopathological alterations,[92] indicating that Nkx3.1 has tumor suppressor activity. Consistently, conditional ablation of one or both alleles of Nkx3.1 in the prostate epithelium with PSA-Cre leads to the development of PIN lesions.[93] The mechanisms underlying the tumor suppressor activity of Nkx3.1 are not fully understood. It has been proposed that Nkx3.1 binds to topoisomerase I via its homeodomain and enhances the enzymatic activity by accelerating binding of topoisomerase I to DNA.[94] In addition, Nkx3.1 upregulates expression of several antioxidant and prooxidant enzymes, including glutathione peroxidase 2 and 3 (GPx2, GPx3), peroxiredoxin 6 (Prdx6), and sulfhydryl oxidase Q6 (Qscn6), and, therefore, provides protection against oxidative damage and indirectly suppresses tumor initiation.[95]

Although loss of function of Nkx3.1 is not sufficient to cause prostate cancer, it cooperates with Pten loss to induce prostate carcinogenesis.[96,97] Nkx3.1/Pten double mutant mice display an increased incidence of high-grade PIN by 6 months of age. A majority of the Nkx3.1/Pten double mutant mice develop adenocarcinoma when they are over 1 year old, sometimes accompanied by metastases to lymph nodes and other distant organs.[96] The tumors are characterized as multifocal and poorly differentiated cells with prominent and multiple nucleoli, increased nuclear-to-cytoplasmic ratio, and active mitoses. Unlike wild-type prostates that have large-scale apoptosis within 2 days after the castration, the Nkx3.1/Pten mutant prostate continues to display a highly proliferative activity. At the age of 8 or 10 months, the castrated Nkx3.1/Pten double null mice develop castration-resistant high-grade PIN or carcinoma, which are similar to those found in noncastrated mice of the same ages and genotypes. Even mice castrated at the age of 3 weeks, which are not sexually mature, still develop PIN lesions at a high frequency. Interestingly, exposure of Pten/Nkx3.1 mutant mice to low androgen levels (1/10 of normal level) for a long period of time results in a marked acceleration of prostate tumorigenesis compared with those exposed to normal androgen levels.[98] Furthermore, the double mutant mice that have undergone both surgical castration and adrenalectomy still develop similar lesions, suggesting that the androgen independence is not simply just a consequence of adaptation to reduced androgen levels.

In addition to Pten, Nkx3.1 also cooperates with other tumor suppressors, including p53[99] and the cyclin-dependent kinase (CDK) inhibitor $p27^{kip1}$,[100,101] in suppressing prostate tumorigenesis. Nkx3.1 cooperates with $p27^{kip1}$ to suppress the proliferation of prostatic epithelial cells and the formation of PIN lesions. $P27^{kip1}$ mutant mice develop prostate hyperplasia and dysplasia at less extent than Nkx3.1 mutant mice. However, $P27^{kip1}$/Nkx3.1

double mutant mice develop extensive lesions with increasing nuclear crowding, nuclear elongation, and hyperchromasia.[101] The mice carrying triple heterozygous mutant alleles of Nkx3.1/p27^{kip1}/Pten exhibit even more severe prostate lesions.[100] Unexpectedly, homozygous inactivation of p27^{kip1} inhibits tumor progression, suggesting that p27^{kip1} possesses dosage-sensitive positive as well as negative roles in prostate cancer progression.[100] Interestingly, Pten loss causes reduction of Nkx3.1 expression in both murine and human prostate cancers. Restoration of Nkx3.1 expression in the epithelium of Pten null prostate leads to decreased cell proliferation and increased cell death and prevents tumor initiation. Restoration of Nkx3.1 expression also leads to increased p53 acetylation and half-life through MDM2-dependent mechanisms. The results suggest that Nkx3.1 stabilizes p53 and blocks prostate cancer initiation caused by Pten loss.[99]

D. p27^{Kip1} Downregulation Models

p27^{Kip1} controls the G1/S transition during the cell cycle via inhibiting cyclin/CDK; degradation of p27^{kip1} is required for the G1/S transition. In addition, p27^{Kip1} also regulates other cellular processes, including cell motility. Downregulation or loss of p27^{Kip1} expression is often found associated with prostate and other human cancers. The protein level of p27^{Kip1} in the cells is regulated primarily by the ubiquitin E3 ligase SCF^{SKP2}-mediated ubiquitin-dependent proteolysis, a process that is triggered by a small molecule called Cks1. Recently, we and others showed that elevated Cks1 expression is coincident with the reduction of p27^{kip1} proteins in tumor cells and is associated with the aggressive behavior of prostate cancer.[102,103] p27^{Kip1} null mice developed hyperplastic prostate.[104] SKP2 is a p27^{Kip1} downregulation regulator that mediates degradation of p27^{kip1}. Consistent with its critical role in triggering p27^{Kip1} proteolysis, overexpression of SKP2 in the prostate epithelium leads to significant downregulation of p27^{Kip1} and overproliferation in the epithelial cells, resulting in hyperplasia, dysplasia, and low-grade carcinoma in the prostate gland.[105] The results suggest that SKP2 can be an oncoprotein in the prostate via downregulating p27^{Kip1}.

E. pRb-Deficient Models

The Rb tumor suppressor pRb is involved in regulating cell cycle progression primarily through interactions with the E2F family of transcription factors and controlling the G1 to S-phase transition. Mutations in Rb have been found as early events in prostate cancer; loss of heterozygosity of the Rb gene has been reported to be associated with prostate cancer. Prostate epithelium-specific deletion of even one single Rb allele results in increased expression of E2F target genes concomitant with increases in prostate epithelial proliferation and causes focal prostate hyperplasia in mice,[106] demonstrating that loss of pRb disrupts cell cycle control and causes overproliferation in the prostate, but is insufficient to cause malignancy.

VI. Androgen Signaling Disruption Models

A. AR Mutant Models

The prostate is an androgen-dependent organ; androgen deprivation induces massive apoptosis in the epithelium within 2 days, and the prostate atrophies in 2 weeks. Androgen replenishment quickly induces prostate epithelial cells to undergo proliferation, and the atrophied prostate regenerates and recovers to its original volume and morphology within 2 weeks. Like a normal prostate, prostate cancers at early stages are also androgen responsive. Androgen deprivation makes prostate tumors shrink. Yet, the androgen therapy is only palliative; recurrence of the tumor is almost inevitable at some time after the androgen therapy. The recurrent tumors often become castration resistant and are highly malignant. To date, how androgens control prostate growth and development is still not well understood and, consequently, how prostate cancer cells escape from androgen regulation remains elusive. Tissue recombination experiments reveal that stromal ARs are required for the androgen to elicit mitogenic signals during development and for prostate branching morphogenesis.[107] The mice that have forced expression of the AR in luminal epithelial cells of the prostate driven by the minimal PB promoter are normal at young age, although they appear to have marked increase in epithelial proliferation in the VP and DLP.[108] At the age of 1 year or older, however, the transgenic prostate develops high-grade PIN. Interestingly, expression of the AR-E231G mutant, a mutation in the NH_2-terminal domain that influences interactions with coregulators, causes rapid development of PIN lesions, which progress to invasive and metastatic disease.[109]

On the other hand, ablation of the AR in epithelial cells of mature prostates reduces the glandular infolding, polarization of the cells, and expression of differentiation markers in the epithelium, including PB, PSP-94, and Nkx3.1.[110] Ablation of epithelial AR disrupts prostate differentiation and induces hyperproliferation and hyperapoptosis in the prostate. The results suggest that epithelial ARs control prostate homeostasis by suppressing both proliferation and apoptosis of the cells. Consistently, TRAMP mice lacking the epithelial AR in the prostate have increased luminal epithelial cell apoptosis and epithelial basal cell proliferation. The mice develop larger and more invasive metastatic tumors in lymph nodes and die earlier than littermates carrying wild-type AR.[85] Interestingly, ablation of the AR in both epithelial and stromal compartments of the TRAMP prostate results in smaller primary prostate tumors with lower proliferation rates,[85] which is in contrast to the epithelial AR ablation that results in larger primary prostate tumors with higher proliferation rates. The results suggest that stromal AR play a more dominant role than epithelial AR to promote tumor cell proliferation. Similar to AR knockout in the epithelium, AR knockout in both stroma and epithelium significantly increases intermediate cell populations in the tumors, suggesting AR signals in the epithelium promotes epithelial cell differentiation. Consistently, restoring AR in PC3 human prostate cancer cells suppresses prostate cancer metastasis in tissue recombination experiments; knockdown AR in CWR22rv1 human prostate tumor cells results in increased cell invasion *in vitro* and *in vivo*.[85] These data further suggest that AR signals in the epithelium suppress proliferation of the cells.

B. SRC-3 Mutation Models

SRC-3 is a coactivator for the AR, which is expressed in basal epithelial and stromal, but not luminal epithelial cells of normal prostates. Some epithelial cells in the PIN foci and well-differentiated carcinoma aberrantly express SRC-3, and almost all poorly differentiated prostate carcinoma cells express SRC-3. SRC-3 is not essential for prostate organogenesis and growth; it is dispensable for androgen-dependent prostate regeneration.[111] Yet, ablation of SRC-3 in TRAMP mice arrests prostate tumor progression at the well-differentiated carcinoma stage. In addition, SRC-3 null TRAMP prostates show much lower cell proliferation than TRAMP prostates with wild-type SRC-3 alleles, indicating that aberrant SRC-3 expression in partially transformed epithelial cells is essential for progression to poorly differentiated carcinoma. Thus, inhibition of SRC-3 expression or function is potentially of therapeutic value for suppressing prostate cancer progression. Although both SRC-1 and SRC-3 are overexpressed in multiple human endocrine cancers and knockdown of either one of them in prostate cancer cell lines impedes cellular proliferation, ablation of SRC-1 in TRAMP mice does not affect prostate cancer initiation, progression, and metastasis.[112]

VII. Models with Aberrant Cell Signaling

A. Aberrant FGF Signaling Models

The fibroblast growth factor family (FGF) encoded by 22 distinct genes are widely expressed regulatory polypeptides that control a broad spectrum of cellular processes. Among them, 18 FGFs bind to and signal through the FGF receptor (FGFR) tyrosine kinase.[113] The other four structurally homologous members (FGF11-14, also known as FHFs for FGF homologues factors) do not bind to the FGFR and are involved in intracellular processes unrelated to canonical FGF signaling. All FGFs are composed of a single polypeptide, which has a highly homologous core split and flanked by areas of less homology (Fig. 2A). The FGFR is a transmembrane tyrosine kinase that is composed of a single glycosylated polypeptide. The extracellular ligand-binding domain contains 2 or 3 immunoglobulin-like (Ig) loops (Fig. 2B). The intracellular tyrosine kinase domain is split by a nonconserved kinase insertion domain and followed by a C-terminal sequence. Many FGF members need to have heparan sulfates (HSs) as cofactors to bind and activate FGFR, which are highly heterogeneous polysaccharide site-chains of proteoglycans on the cell surface. The HS has both FGF- and FGFR-binding motifs that are involved in the determination of ligand-binding and -signaling specificity of FGFR complexes.

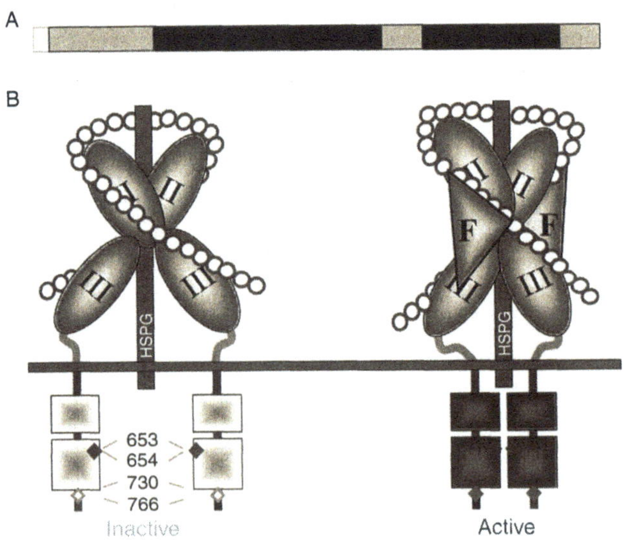

FIG. 2. The FGFR signaling axis. (A) Structure domains of the FGF. Boxes indicate the putative structure domains. Empty box, signal peptide; solid box, conserved domain; shaded box, nonconserved sequence. (B) Models of the FGFR complex. Docking of FGF releases FGFR from negative restrictions imposed by interaction with heparan sulfate chains of heparan sulfate proteoglycans (HSPG) resulting in an active conformation where adjacent kinases maintain an enzyme-substrate relationship. Positions of the four major tyrosine phosphorylation sites on the FGFR are indicated. II, Ig-loop II; III, Ig-loop III; F, FGF; K, kinase domain; circles, monosaccharides of heparan sulfate chains; solid bar, plasma membrane.

Four highly conserved genes (Fgfr1, Fgfr2, Fgfr3, and Fgfr4) encode the FGFR. Yet, the diversity of the FGFR is substantially expanded through alternative splicing, which produces a large repertoire of FGFR isotypes varying in both the extracellular and intracellular domains. These variants have distinct or fine-tuned biological functions, different tissue distribution, and ligand-binding specificity. Two major alternatively spliced isoforms in the second half of Ig-like domain III, namely IIIb and IIIc, in FGFR1, FGFR2, and FGFR3, bind a different subset of FGFs, resulting in different responses to different FGFs. FGFR2IIIb, a hallmark of epithelial cells in multicompartment parenchymal tissues, relays signals from stromal-derived FGFs, such as FGF7 and FGF10, promoting net intercompartmental homeostasis including restraints on malignant progression of tumor cells. This is in marked contrast to FGFR1IIIc and FGFR2IIIc isoforms that appear abnormally in epithelial cells due to ectopic gene activation, or splice switching. Ectopic expression of FGFR1IIIc appears to elicit a new set of abnormal signals in the epithelial cells whereas the switch to FGFR2IIIc cuts off epithelial cells from homeostasis-promoting stromal signals that activate FGFR2IIIb.

Although the kinase domains of the four FGFR isotypes are highly homologous (> 80%) in the amino acid sequences and share several similar phosphotyrosine sites, the four FGFRs elicit receptor- and cell-type-specific activities in cells.[113] Binding of FGF to the FGFR activates the FGFR kinases and leads to receptor autophosphorylation, creating binding sites for downstream effectors that contain a Src homology 2 (SH2) domain or a phosphotyrosine-binding (PTB) domain. Seven major tyrosine (Y) autophosphorylation sites have been identified in the FGFR1 kinase domain. Phosphorylated Y653 and Y654 are predominant in activation (derepression) of the receptor kinase activity. Phosphorylation of Y766 is required for recruiting PLCγ via its SH2 domains to the FGFR1 kinase. The function of other phosphorylation sites, including Y463, Y583, Y585, and Y730, has not been clearly established, despite some evidence that Y463, 583, and 585 contribute to the timed extent of mitogenic activity of FGFRs, Y430 is a CRK/CRKL-binding site, and phosphorylated Y730 is a P85-binding site. In some experimental systems, the four FGFR isotypes elicit similar and redundant effects on cell phenotypes and, in others, exert different effects. This explains why FGFR1 and FGFR2 have different biological endpoints in the prostate epithelial cell context. New approaches and rigidly selected and controlled experimental systems promise to dissect such differences.

The activation of the MAP kinase, PI3 kinase, and PLCγ pathways has been implicated in most FGFR1 responses (Fig. 3). Phosphorylation of the membrane-anchored FRS2α (also known as SNT1 for SUC1-associated *n*eurotrophic factor *t*arget proteins) by the FGFR1 kinase recruits and activates the GRB2/SOS1 complex that then interacts with *Ras* to activate the MAP kinase signaling pathway. In addition, Crk has also been proposed to be a functional

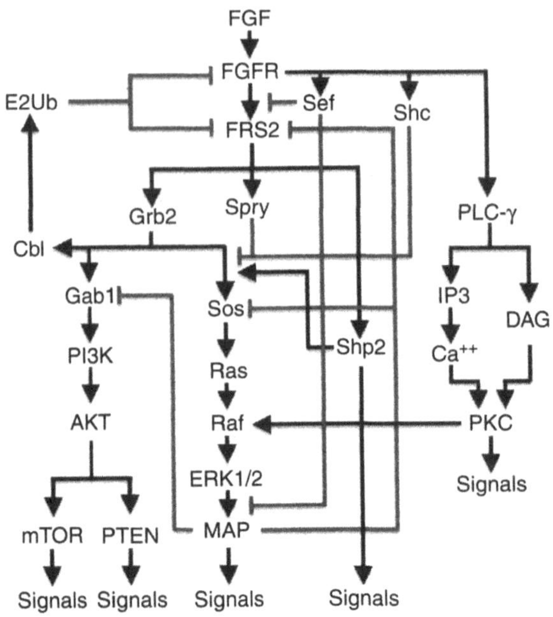

FIG. 3. The cascade of FGFR signaling. Blue lines represent positive regulation, and red lines represent negative effects. (See Color Insert.)

adaptor protein that binds to FGFR,[114] which has been reported to link the ERK pathway to FGFR1. The activation of the ERK pathway by the FGFR tyrosine kinase is tightly regulated by both positive and negative feedback loops at both transcription and posttranslation. Sprouty (Spry) is a feedback regulator of the FGF pathway at the posttranslational level, which consists of four conserved members, Spry1–4.[115] Tyrosine phosphorylation of Spry creates a decoy site that binds the docking molecule Grb2 and prevents SOS from activating Ras. Sef (similar expression of FGF) inhibits binding of FRS2α to the FGFR and prevents activation of ERK, and, therefore, negatively regulates the Ras-MAPK pathway. The expression of Sef is controlled by the FGF signaling axis.

Specific isoforms of FGF and FGFR are partitioned in the stroma and epithelium of the prostate, forming the basis of directional communications between the two compartments, which regulate development, function, and tissue homeostasis of the organ. Disruptions of these paracrine signals due to loss of gene expression, mutations, and changes in RNA splicing, or ectopic expression of other members of the FGF or FGFR are characteristic in diseases. FGF10 is essential for prostate development; ablation of *Fgf10* abrogates fetal prostate development and androgen responsiveness of prostatic rudiments when grafted to kidney capsules of wild-type mice.[116]

FGF2 is often overexpressed in human prostate cancer and BPH, which promotes cell motility and proliferation, increases tumor angiogenesis, and inhibits apoptosis. Forced expression of FGF2 under the control of the minimal PB promoter leads to male infertility and epithelial hyperplasia.[117] Compared with those of control littermate mice, acinar epithelial glands in the DP of FGF-2 transgenic mice are denser and show simple papillary hyperplasia. The VP of FGF-2 transgenic mice shows glandular and luminal enlargement without epithelial overgrowth. On the other hand, inactivation of the *Fgf2* in TRAMP mice significantly increases survival time, decreases metastasis, and inhibits progression to the poorly differentiated phenotype in primary prostatic tumors.[118] Furthermore, inactivation of one *Fgf2* allele also significantly suppresses TRAMP tumorigenesis and progression, suggesting that FGF2 promotes prostate tumorigenesis and progression in a gene dosage-dependent manner.

FGF3 has been overexpressed in the prostate and Wolffian duct derivatives by using the MMTV-transactivator.[119] The FGF3 transgenic mice develop extensive prostate hyperplasia; all prostatic lobes exhibit epithelial stratification, cribriform structures, and papillary tufts. The cell proliferation activity in the epithelial compartment is significantly increased. The luminal epithelial cells display increased nuclear-to-cytoplasmic ratios but retain relatively uniform nuclear AR and the tumor suppressor C-CAM1 staining. Although dysmorphogenic, no epithelial cell invasion to the basement membrane or stromal compartment has been found, indicating that the lesions stall at PIN stages. Together, the data suggest that the ectopically expressed FGF-3 severely perturbs prostate development and leads to prostatic hyperplasia. In addition, profound disorders of the Wolffian duct derivatives have been observed. The ampullary glands and vas deferens are extremely cystic, hypertrophic, and hyperplastic; the enlarged epididymi show a reduction of spermatozoa, and the seminal vesicles exhibit a dramatic reduction of seminal secretions.

The FGF7 transgenic mice (PKS) that express human FGF7 in the prostatic epithelium driven by the minimal PB promoter develop prostate hyperplasia[120]; many epithelial ducts are filled with secretory luminal epithelial cells that are tightly associated with highly enfolded basement membranes. Distortions of the ductal smooth muscle layer are also observed. Consistently, target expression of a truncated FGFR2IIIb receptor (KDNR), which functions as a dominant-negative inhibitor of the FGF7 receptor, to the prostatic epithelium with the similar strategy leads to a smaller prostate in mice.[120] The epithelium in many of the prostatic ducts is disorganized and contains numerous rounded cytokeratin-positive cells that are not tightly associated with the basement membrane. The stroma is also disorganized and does not form a tight layer of smooth muscle around the epithelial ducts, suggesting that the FGF7–FGFR2 signaling axe plays an important role in the communication between epithelial and stromal compartments, which is required for maintaining tissue structures of the prostate.

Expression of the b isoform of fibroblast growth factor 8 (FGF8b) is often upregulated premalignant and malignant lesions in men prostate. Targeted overexpression of FGF8b in prostate epithelial cells driven by the ARR2PB composite promoter causes prostatic hyperplasia in the LP and VP in as early as 2–3 months, and in all lobes between 6 and 16 months. The mice develop low-grade PIN by the age of 5–7 months. During the first 14 months, all transgenic animals exhibit multifocal epithelial hyperplasia; about 1/3 of the animals also have areas of low-grade PIN. Between 15 and 24 months, about two-thirds of the mice develop low-grade PIN and about half of the mice develop high-grade PIN lesions. Stromal cells have elevated proliferation activities. The affected stroma consists primarily of the SMC component. The epithelial cells exhibit papillary hyperplasia with atypia displayed in the FGF8b transgenic mice. The transgenic mice also have chronic inflammation, mostly involving T cells, in the prostate.[121] Mice harboring the FGF8b transgene and haploinsufficiency in Pten develop prostatic adenocarcinoma, which frequently metastasizes to the lymph node, whereas the lesions in mice carrying either the FGF8b transgene or Pten heterozygous null mutation generally only develop PIN in the prostate.[122] In addition to late age-related development of typical adenocarcinoma, the FGF8b/Pten$^{+/-}$ mice also displays a low incidence of mucinous adenocarcinoma, a rare variant type of human prostatic adenocarcinoma. Thus, the results demonstrate the cooperation between FGF8b overexpression and Pten deficiency in prostate tumorigenesis.

Among the four FGFRs, FGFR2IIIb is the resident FGFR isoform in the prostate epithelium, which is important for prostate tissue homeostasis.[123] Loss of the resident FGFR2 accompanied by gain of ectopic FGFR1 in prostate tumor cells is a hallmark change in prostate tumor progression.[124,125] Forced expression of FGFR1 in premalignant rat prostate tumor cells, DT3 from the Dunning 3327 tumors, accelerates progression of the cells to malignancy in a time-dependent manner. To test the effect of chronic activity of ectopic FGFR1 in the prostate epithelium, a constitutively active mutant of FGFR1 (caFGFR1) has been expressed in prostate epithelial cells using the minimal PB promoter.[126] The minimal PB promoter drives expression of caFGFR1 at low levels, which allows analysis impact of chronic FGFR1 activity of physiological levels in its native transmembrane context. Relative to wild-type littermates, young adult transgenic animals frequently develop prostate hyperplasia in less than a year. The disease is severe enough to cause death due to urinary obstruction. The hyperplasia is also accompanied by foci of intermediate-grade PIN (PIN 2–3) in the DP, LP, and AP lobes of young adult mice (Fig. 4). At the age of 20 months or older, the transgenic mice have a prostate that is

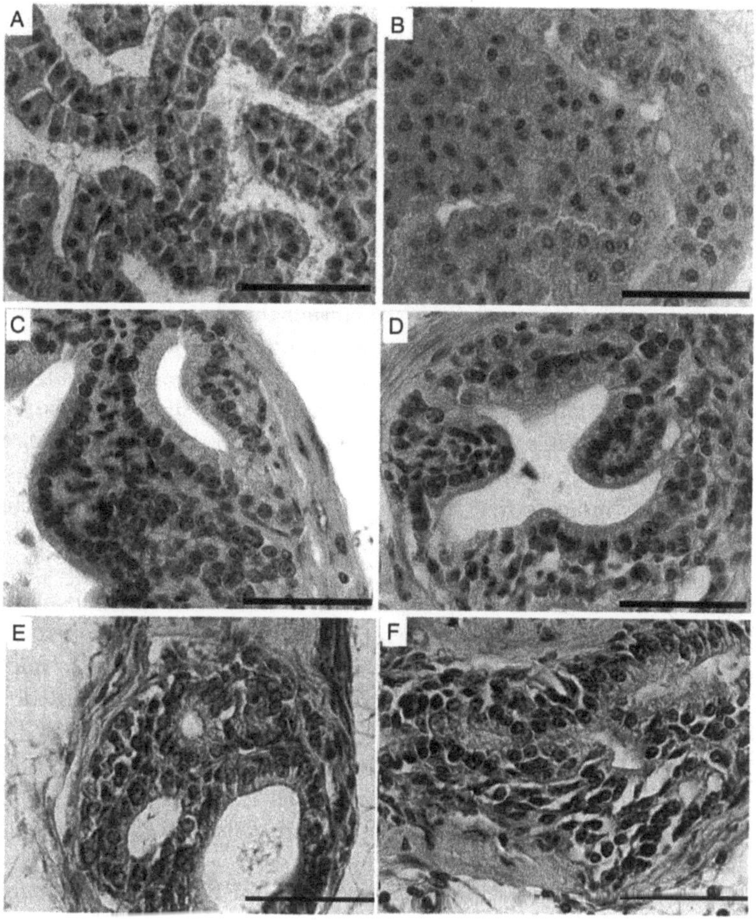

FIG. 4. Dosage-dependent activity of ectopic FGFR1 in induction of prostatic intraepithelial neoplasia. Paraffin sections prepared from the indicated wild-type or transgenic mice expressing constitutively active FGFR1 in the prostate epithelium were H&E stained to demonstrate tissue structures. (A) 30-week-old wild-type mouse; 30-week-old PB-caFGFR1 transgenic mouse; C&D, 25-month-old PB-caFGFR1 transgenic mice; E&F, 30-week-old ARR2PBi-caFGFR1 mice. Bar = 50 μm.

two- to six-times larger than that of normal littermates, which often exhibits high-grade PIN lesions throughout the prostate. It appears that the overall hyperplastic condition in the caFGFR1 prostate, in large part, is due to overpopulation of well-differentiated epithelial cells, although in areas of highest hyperplasia foci, the cells also exhibit a quantitative decrease in cytokeratin expression. No invasive carcinomas have been observed. Androgen-induced prostate regeneration after castration is significantly accelerated in caFGFR1 transgenic mice, suggesting that ectopic caFGFR1 promotes proliferation of prostate epithelial cells. Therefore, chronic exposure to the ectopic FGFR1 kinase activity alone appears to disrupt homeostasis between stroma and epithelium, but is insufficient to cause carcinoma.

Interestingly, ARR2PBi-caFGFR1 transgenic mice that express caFGFR1 at high levels with the ARR2PB promoter develop high-grade PIN within 8 months (Fig. 4). Foci with cribriform structures in acini lumen and epithelial cells pushing through stromal compartments are common features throughout the ARR2PBi-caFGFR1 prostate. Most epithelial cells in high-grade PIN foci fail to express cytokeratins that are differentiation markers of luminal epithelial cells of the prostate. In high-grade PIN foci, stromal cells surrounding the epithelial cells also fail to express α-actin, a characteristic marker associated with differentiated SMCs. The basement membranes surrounding the epithelial compartment are often disrupted, especially in the foci with high-grade PIN. The results indicate that the ectopic FGFR1 in prostate epithelial cells disrupts tissue homeostasis and induces preneoplastic lesions in a dosage-dependent manner.

Furthermore, expression of KDNR, a dominant-negative construct of FGFR2, with the PB promoter in prostate epithelial cells to depress FGFR2 signals promotes development of PIN in PB-caFGFR1 transgenic mice that express caFGFR1 at low levels. Interestingly, expression of KDNR does not accelerate PIN development induced by high levels of the caFGFR1 in ARR2PBi-caFGFR1 mice, although it increases the population of NE cells in the lesion foci. The results further confirm that loss of resident FGFR2 and gain of ectopic FGFR1 in prostate epithelial cells synergistically perturbs prostate homeostasis.

The membrane-anchored fusion protein of iFGFR1 is a chimeric molecule that contains a 12-Kd FK506-binding protein (FKBP12) fused in frame with the C-terminus of the FGFR1 intracellular kinase domain, which can be activated by a chemical induction of dimerization (CID) mechanism with FK506.[127] Transgenic mice JOCK1 that express iFGFR1 in luminal epithelial cells driven by the ARR2PB promoter have been generated to investigate the role of the FGFR1 kinase under conditions of maximal activation in the prostate.[128] Activation of iFGFR1 by the CID induces grade I–II of PIN in mice by 12 weeks, which is characterized by multiple layers of atypical epithelial cells. By 6 months, more extensive nuclear atypia, thickened "reactive" stroma, and basement membrane herniation occurs, which are the characteristics feature of high-grade PIN.[128] From 42 weeks onward, mice develop adenocarcinoma that penetrates into all lobes.[129] Moreover, up to 1 year activation of iFGFR1 causes the mice to develop a spectrum of prostate malignancies that have a low frequency of metastases to lymph nodes and liver. The induced hyperplasia is reversible by withdrawal of the CID activator at PIN stages. Yet, after extensive intraductal vascularization occurs, deactivation of iFGFR1 by CID withdrawal only reduces prostate adenocarcinoma progression. The results suggest that although continued progression requires prolonged FGFR1 signaling, the lesions are not reversible at late stages. Furthermore, activation of iFGFR1 in TRAMP mice increases vascular volume in 1 week,[130] evidence by increased vessel volume and branching. Interestingly, continuing activation of iFGFR1 is not required to maintain the new vasculature 6 weeks after the induction. Consistent upregulation of HIF-1α, vascular endothelial growth factor (VEGF), and angiopoietin 2 (Ang-2), and the loss of Ang-1 expression in the basal epithelium are associated with iFGFR1 activation. The results suggest that high level of aberrant FGFR1 signaling contributes to vascularization through the angiopoietins.

The results demonstrate that activation of truncated iFGFR1 kinase dimers alone is sufficient for causing prostate cancer in transgenic mice, which is quite different from ectopically expressed full-length transmembrane FGFR1 kinase physiological levels. The iFGFR1 fusion protein does not have an extracellular domain that binds to HS proteoglycans or other cofactors; it is artificially dimerized at its C-terminus. Normally the activity amplitude and substrate specificity of the FGFR kinase are subject to the restriction imposed by the interaction with HS proteoglycans and other cofactors. The truncated iFGFR1 is not subject to these restrictions due to lacking the ectodomain. Therefore, it may have a wider range of substrates coupled with more sustained and high extent activities. This may overlap sufficiently with pathways that promote the required genetic instability and tumorigenicity. The long latency of tumorigenesis and the independence of the more advanced tumors on iFGFR1 suggest other cooperating events are needed for prostate tumorigenesis induced by the tyrosine kinase.

Loss of the tissue homeostasis-promoting FGFR2-signaling axis has been associated with prostate tumor progression both in human prostate cancer and in animal models. Since *Fgfr2* ablation is early embryonically lethal, tissue-specific ablation of *Fgfr2* in mouse prostate epithelial progenitor cells during early prostate morphogenesis with the Cre/LoxP recombination system has been created to study the function of FGFR2 in the prostate epithelium.[45] Unlike a normal prostate that is composed of 4 pairs of AP, DP, LP, and VP, most young adult *Fgfr2* conditional null (*Fgfr2*^{cn}) mice develop a small prostate that often only has 2 pairs of DP and LP. In addition, branching morphogenesis of the *Fgfr2*^{cn} prostates is significantly impaired, which can be characterized by deficient intraluminal infolding, and the luminal epithelial cells are not well polarized. In contrast to wild-type prostates, maintenance of mature *Fgfr2*^{cn} prostates is not strictly androgen dependent. No significant prostatic atrophy is observed even at 2 weeks after castration in adult *Fgfr2*^{cn} mice. Similarly, androgen replenishment also fails to induce cell proliferation in the *Fgfr2*^{cn} prostates. The results showed that FGFR2 signals are essential for strict androgen dependency in adult prostates with respect to tissue homeostasis. In contrast, production of secretory proteins in *Fgfr2*^{cn} prostate remains to be strictly androgen dependent. Together, the results indicate that is androgens may regulate cellular homeostasis and functions of luminal epithelial cells by different mechanisms.

The FRS2α adaptor protein is extensively tyrosine phosphorylated by the FGFR kinase upon activation by the FGF ligand.[113] Phosphorylation of tyrosine on FRS2α creates binding sites for Grb2 (Y196, Y306, Y349, Y392) that mainly link to PI-3 kinase activation, and Shp2 (Y436 and Y471) that mainly links the FGFR kinase to the MAP kinase pathway. *Frs2α* is expressed in mouse embryos during early embryogenesis and almost ubiquitously in all fetal and adult tissues.[131] Disruption of *Frs2α* alleles abrogates the FGF-induced activation of MAP kinase and PI-3 kinase, chemotactic response, and cell proliferation, and severely impairs mouse development resulting in embryonic lethality at E7–E7.5.[132] The human Frs2α gene is located in 12q15; rearrangement in 12q15 is often found in human tumors, suggesting that aberrant expression of *Frs2α* may be involved in tumorigenesis.

The *Frs2α* alleles have been tissue-specifically disrupted in the progenitors for prostate epithelial cells in early prostate morphogenesis with the *Nkx3.1*^{Cre} driver (designated as *Frs2α*$^{cn/Nkx}$). Similar to wild-type prostates, the *Frs2α*$^{cn/Nkx}$ prostates have similar gross organ morphology, yet, have significantly less epithelial ducts than wild-type prostates. Interestingly, unlike the *Fgfr2*^{cn} prostate that has a compromised androgen dependency, *Frs2α*$^{cn/Nkx}$ prostates are strictly androgen dependent with respect to production of secretory proteins and tissue homeostasis, suggesting that the FGFR2 regulates androgen dependency of prostate epithelial cells independent of FRS2α. Although *Frs2α* is not expressed in luminal epithelial cells of mature prostates, it is highly expressed in prostate tumor epithelial cells in the TRAMP mouse, which is coincident with the ectopic appearance of FGFR1 (Fig. 5A), suggesting that the ectopic FRS2α-mediated FGF signaling axis is involved in prostate tumor progression.[133,134] The majority of TRAMP mice carrying the floxed *Frs2α* alleles develop low-to-high-grade PIN lesions within 10 weeks of age. PIN foci are apparent across the whole prostate with an estimated 80% of the prostate exhibiting various degrees and grades of PIN lesions (Fig. 5B). The PIN foci in TRAMP/*Frs2α*$^{cn/Nkx}$ prostates are fewer than in control TRAMP mice. The majority of the lesions are less severe, compared with the control TRAMP mice (Fig. 5B and C). By 24 weeks of age, most control TRAMP mice develop advanced, poorly differentiated tumors, whereas tumors in the TRAMP/*Frs2α*$^{cn/Nkx}$ mice are at well-differentiated stages. Furthermore, the TRAMP/*Frs2α*$^{cn/Nkx}$ mice have a significantly increased life span than those

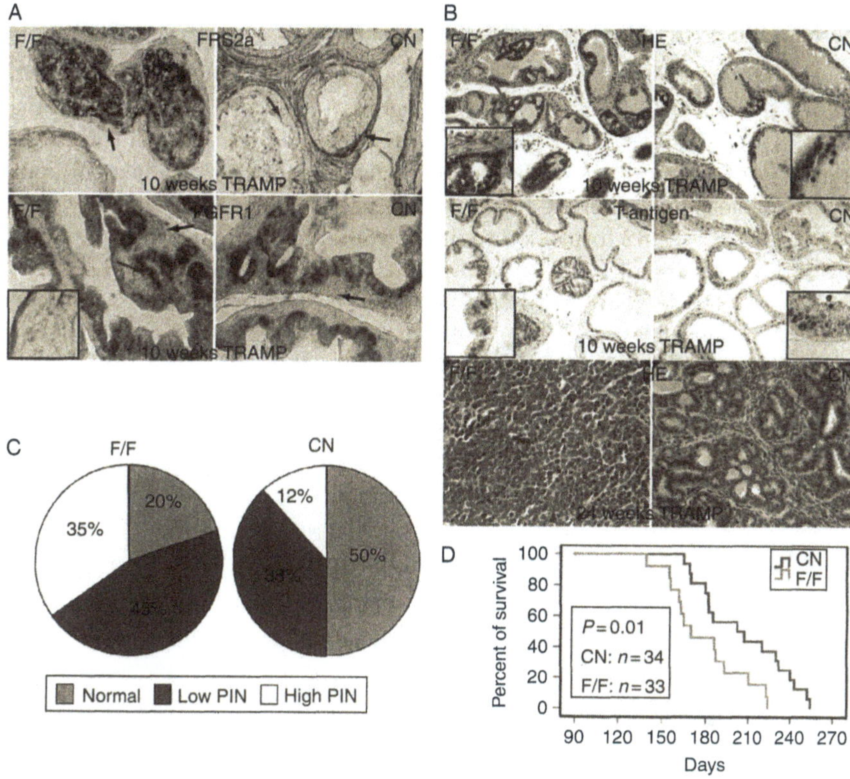

Fig. 5. Ablation of *Frs2α* in the prostatic epithelium inhibited tumorigenesis in the TRAMP prostatic tumor model. (A) Expression of *Frs2α* and *Fgfr1* at the mRNA level was assessed in TRAMP prostates by *in situ* hybridization. Red arrows indicate epithelial cells, and black arrows indicate stromal cells. *Inset*, wild-type control showing no expression of *Fgfr1* in the epithelial compartment. (B) Prostate tissue sections were prepared from TRAMP mice with or without *Frs2α* null alleles at the indicated ages and stained with HE or analyzed by immunohistochemical staining with anti-T-antigen antibody. Blue arrows indicate focal lesions, and green arrows indicate normal epithelial cells. (C) The percentage of areas occupied by the PIN foci in prostates of 10-week-old TRAMP mice. The data are mean and standard deviation from five prostates. (D) Mortality of TRAMP mice with the indicated *Frs2α* alleles was determined from daily observation over 250 days. The percentage of mice that survived to the respective age is shown. F/F, homozygous TRAMP-$Frs2\alpha^{flox}$ mice; CN, TRAMP-$Frs2\alpha^{cn}$ mice. (Adopted from Zhang *et al.*, 2008, Development. 135 (4): 782). (See Color Insert.)

TRAMP mice that have intact *Frs2α* alleles (Fig. 5D). These results strongly suggest that ablation of FRS2α-mediated signaling pathways in prostate epithelial cells inhibits prostate tumor initiation and progression in the TRAMP mice.

B. Models with HER2/neu Overexpression

HER2/neu (also known as ErbB-2) is a member of the epidermal growth factor receptor (EGFR) family, also known as the ErbB protein family that includes EGFR, HER2, HER3, and HER4. The HER2/neu is a transmembrane glycosylated tyrosine kinase that consists of an extracellular domain, which interacts with other EGFR family members and an intracellular tyrosine kinase domain. HER2 forms heterodimers with other EGFR family members, induced by binding of EGF to the EGFR. The interaction allows HER2/neu heterodimers to serve as a coreceptor of EGFR. The heterodimers are more stable and, therefore, elicit more potent signals than other EGFR dimers without HER-2/neu. HER2/neu and other members of the EGFR family have been implicated in human prostate cancer initiation and progression to castration-resistant stages.[135] Transgenic mice that overexpress HER2/neu in prostate epithelial cells with the PB promoter develop PIN by the age of 5 months, which progresses to invasive carcinoma after 1 year.[136]

C. Models with TGF-β Signaling Disruption

The transforming growth factor beta (TGF-β) superfamily is involved in regulation of many cellular processes both in adults and in embryos, including growth, differentiation, apoptosis, homeostasis, and other functions. TGF-β elicits signals by binding and activating to type II TGF-β receptor (TGF-βRII), which is a transmembrane serine/threonine kinase that phosphorylates the type I TGF-β receptor (TGF-βRI). The type I receptor then phosphorylates SMADs that are subsequently translocated to the nucleus where they function as transcription factors and participate in the regulation of gene expressions. The epithelium-to-stroma TGF-β signaling axis plays a critical role in the communication between the two compartments and maintenance of tissue homeostasis. Loss of the TGF-βRII receptor is often found associated with advanced human prostate cancer. Expression of dominant-negative TGF-βRII (DNIIR) mutant proteins, which attenuate TGF-β signaling, in the prostate epithelium of the LADY mice increases amounts of metastasis compared to age-matched control mice expressing only the large T antigens.[137] Yet, the sizes of the neoplastic prostates are not increased by the attenuation of TGF-β signaling. Knockout of the TGF-βRII in prostate stroma with FSP1-Cre (*Tgfβr2(fspKO)*) results in the development of PIN and progression to adenocarcinoma within 7 months.[30] Unlike prostates with functional TGF-βRII alleles, which quickly regress after androgen deprivation, prostates of *Tgfβr2 (fspKO)* mice regress to a lesser extent than the control mice and exhibit high levels of proliferation.[31] In the prostates with intact TGF-βRII, Wnt activity can be only detected in epithelial cells of proximal ducts after castration. In the prostate of *Tgfβr2(fspKO)* mice, however, Wnt activity can be detected throughout the glandular epithelia regardless of androgen status. *In vivo* tissue recombination of *Tgfβr2(fspKO)* prostatic stromal cells and wild-type or SV40 large T antigen expressing epithelial cells results in prostate-like tissues that are resistant to androgen deprivation. Furthermore, xenografts with human LNCaP prostate cancer cells show that prostatic stromal cells from the prostate *Tgfβr2(fspKO)* mice induce LNCaP cell proliferation and tumorigenesis as a result of elevated Wnt3a expression.[138] The results suggest that a TGF-beta, androgen, and Wnt paracrine signaling axis enables prostatic regression after androgen ablation and that the loss of stromal TGF-βRII expression compromises androgen dependency of the prostate.

D. Models with Disrupted IGF-I Signaling

The insulin-like growth factor (IGF) family has two highly homologous ligands, IGF-I and IGF-II, which elicit the regulatory activities by binding and activating their receptor, IGF-IR. The circulating IGF-I ligand is mainly synthesized and secreted by the liver under the control of growth hormone (GH). IGF-I is also produced as local a growth factor in many tissues and organs, including the prostate. Ligand-dependent activation of the IGF-IR kinase initiates multiple signaling events, including activation of the PI3K/Akt and MAP kinase pathways. Aberrant expression of the IGF-1 signaling axis has been found to be associated with prostate cancer risk, as well as associated with progression in the Dunning rat prostate tumor model.[139]

Although forced expression of IGF-1 in mouse prostate epithelial cells causes prostate hyperplasia,[140,141] forced expression of IGF-1 in the prostate epithelium of TRAMP mice unexpectedly delays progression of organ-confined tumors and emergence of metastatic lesions.[141] Consistently, ablation of IGF-IR in prostate epithelial cells activates ERK1/2 signaling, leads to cell autonomous proliferation and hyperplasia in the epithelium, and induces p53-regulated apoptosis and cellular senescence rescue programs.[142] Ablation of IGF- IR in the prostate epithelium on TRAMP mice, rather than inhibiting the tumorigenesis and progression, accelerates the emergence of aggressive prostate cancer. In line with this observation, reducing the serum IGF-1 level to 10% of control mice by tissue-specifically deletion of *Igf1* in the hepatocytes of TRAMP mice does not significantly affect the survival rates of TRAMP mice.[143] No difference in the pathologic stage of the prostate cancer can be detected between the *Igf1* conditional knockout and wild-type groups, suggesting that the reduction in systemic IGF-1 is not sufficient to inhibit prostate tumor progression in the TRAMP model. It has been proposed that the IGF signaling in prostate epithelial cells may suppress cellular senescence and apoptosis to allow epithelial cells to survive spontaneous transformation while enforcing a strong differentiation block. Thus, loss of IGF-I signaling, in combination with suppressed p53 function by the T antigens, provides selective advantage to transformed epithelial cells that overcome the IGF-IR-mediated differentiation block and progress toward more invasive and disseminated forms of carcinoma.[142] The results further suggest that the onset, progression, and invasion of prostate tumors in animals are composite outcomes resulting from multiple mutations and defects, which, sometimes, may not be consistent with data gathered from *in vitro* cell culture and biochemistry experiments. Recapitulation of tumor onset, progression, metastasis, and relapse in animal models is needed to validate *in vitro* findings, and will provide guidance for preclinical initiatives for prostate cancer therapies.

E. NF-κB Pathway Models

Nuclear factor-kappaB (NF-κB) is a family of transcription factors that regulate a broad spectrum of genes expression and control diverse functions. The activity of NF-κB is regulated by IκB (inhibitor of κB) that binds to and sequesters NF-κB in the cytoplasm. When the NF-κB pathway is activated, the IκB kinase phosphorylated IκB at serine residues 32 and 36. Phosphorylated IκB is then degraded via the ubiquitination-mediated proteasome degradation, resulting in the release of NF-κB, which is translocated to the nucleus where it functions as a transcription factor. Constitutive activation of NF-κB is often found in primary prostate cancer samples. The castration-resistant prostate tumor cell lines and androgen-independent prostate tumor xenografts often have higher NF-κB activities than androgen-dependent grafts. In addition, metastatic prostate tumors also have higher NF-κB activities than organ-confined ones.[144] In fact, elevated NF-κB activity correlates with recurrence after radical prostatectomy and has value as a prognostic risk factor. It has been shown that the prostatic NE cells activate the NF-κB pathway in LNCaP cells via secretory factors, resulting in increased levels of active AR. On the other hand, blocking NF-κB signaling inhibits *in vitro* activation of the AR. Ablation of IκBα, a major inhibitor of NF-κB, prevents regression of the mouse prostate after castration. Furthermore, ablation of one allele of IκBα in ARR2PB-c-Myc (Hi-Myc) mice significantly lowers the sensitivity of the tumor to androgen deprivation by castration, and the tumor cells continue to proliferate after the castration.[144] These results suggest that activation of NF-κB is sufficient to maintain androgen-independent growth of prostate and prostate cancer cells by upregulating AR activities.

F. The Vav3 Overexpression Model

Vav3 is a guanine-nucleotide-exchange-factor that relays signals to Rho-GTPases. Overexpression of Vav3 has been found to be associated with castration-resistant properties of prostate tumors both in human and animal models. Vav3-mediated Rho GTPase signaling has been shown to contribute to ligand-independent activation of AR and estrogen receptors in breast cancer.

Overexpression of constitutive active Vav3 in the prostate epithelium leads to development of PIN and prostate cancer at as early as 3 months of age.[145] The AR and PI3-kinase/Akt signaling pathways are both overactivated in Vav3 transgenic prostate. In addition, Vav3 transgenic mice developed nonbacterial chronic prostatitis in the prostate gland, which was associated with elevated incidence of prostate cancer. *In vitro* analysis showed that overexpression of Vav3 in prostate cancer cells enhanced NF-κB activity,[145] which may underlie the innate inflammatory response, as well as overactivation of the androgen signaling pathway induced by elevated Vav3 activity. Thus, the elevated Vav3 activity in the prostate epithelial cells enhances both AR signaling and the innate inflammatory response, and contributes to the onset and castration-resistant activity of prostate cancer.

VIII. Remarks

Although a definitive relationship remains to be proven, it is generally believed that epithelial hyperplasia and low-grade PIN are the early morphological lesions leading to prostate cancer. In addition, the proliferative inflammatory atrophy (PIA), a prostate epithelial response to microenvironmental stress, has also been shown to be precancerous lesions. These focal lesions gradually progress to higher and higher grades of PIN and then adenocarcinoma that will metastasize primarily to bone first and other organs at later stages. The process likely is a continuous change in gene expressions and mutations, which correlates with multiple changes in cellular properties, including proliferation, differentiation, migration, hormone dependency, and interaction with other cells in the microenvironment. Over the last decade, GEM models have been widely used for testing impacts of these changes on the onset, progression, and metastasis of prostate tumors, as well as for assessing the effects of environmental and preclinical drugs for prevention and treatment of prostate cancer. However, human prostate cancer is a complex disease and is highly heterogeneous. It is possible that no single mouse model will mimic all aspects of all human prostate cancers, although continuing efforts are taken to establish and characterize mouse models that best mimic human prostate cancers. Each model may only resemble certain properties of human prostate cancer. New models with selective alterations in diverse prostate stromal cell types are needed to better understand interactions among the stroma and epithelium and diverse cell types within each compartment. Nevertheless, specific mouse models displaying certain aspects of prostate cancers will continue to shed new light on understanding basic biological mechanisms as a preclinical guide to understanding and management of human prostate cancer.

References

1. Jemal A, Siegel R, Ward E, Hao Y, Xu J, Thun MJ. Cancer statistics, 2009. *CA Cancer J Clin* 2009;**59**:225–49.
2. Shappell SB, Thomas GV, Roberts RL, Herbert R, Ittmann MM, Rubin MA, et al. Prostate pathology of genetically engineered mice: definitions and classification. The consensus report from the Bar Harbor meeting of the Mouse Models of Human Cancer Consortium Prostate Pathology Committee. *Cancer Res* 2004;**64**:2270–305.
3. Berquin IM, Min Y, Wu R, Wu H, Chen YQ. Expression signature of the mouse prostate. *J Biol Chem* 2005;**280**:36442–51.
4. Leong KG, Wang BE, Johnson L, Gao WQ. Generation of a prostate from a single adult stem cell. *Nature* 2008;**456**:804–8.
5. Xin L, Lukacs RU, Lawson DA, Cheng D, Witte ON. Self-renewal and multilineage differentiation in vitro from murine prostate stem cells. *Stem Cells* 2007;**25**:2760–9.
6. di Sant' Agnese PA. Divergent neuroendocrine differentiation in prostatic carcinoma. *Semin Diagn Pathol* 2000;**17**:149–61.
7. Hansson J, Abrahamsson PA. Neuroendocrine differentiation in prostatic carcinoma. *Scand J Urol Nephrol Suppl* 2003;(212):28–36.
8. Wang X, Julio MK, Economides KD, Walker D, Yu H, Halili MV, et al. A luminal epithelial stem cell that is a cell of origin for prostate cancer. *Nature* 2009;**461**:495–500.
9. Goldstein AS, Huang J, Guo C, Garraway IP, Witte ON. Identification of a cell of origin for human prostate cancer. *Science* 2010;**329**:568–71.
10. Thomson AA, Marker PC. Branching morphogenesis in the prostate gland and seminal vesicles. *Differentiation* 2006;**74**:382–92.
11. Wu X, Jin C, Wang F, Yu C, McKeehan WL. Stromal cell heterogeneity in fibroblast growth factor-mediated stromal-epithelial cell cross-talk in premalignant prostate tumors. *Cancer Res* 2003;**63**:4936–44.
12. Dunning WF. Prostate cancer in the rat. *Natl Cancer Inst Monogr* 1963;**12**:351–69.
13. Isaacs JT, Heston WD, Weissman RM, Coffey DS. Animal models of the hormone-sensitive and -insensitive prostatic adenocarcinomas, Dunning R-3327-H, R-3327-HI, and R-3327-AT. *Cancer Res* 1978;**38**:4353–9.
14. Tennant TR, Kim H, Sokoloff M, Rinker-Schaeffer CW. The Dunning model. *Prostate* 2000;**43**:295–302.
15. Chung JH, Whiteley M, Felsenfeld G. A 5′ element of the chicken beta-globin domain serves as an insulator in human erythroid cells and protects against position effect in Drosophila. *Cell* 1993;**74**:505–14.
16. Soriano P. Generalized lacZ expression with the ROSA26 Cre reporter strain. *Nat Genet* 1999;**21**:70–1.
17. Greenberg NM, DeMayo FJ, Sheppard PC, Barrios R, Lebovitz R, Finegold M, et al. The rat probasin gene promoter directs hormonally and developmentally regulated expression of a heterologous gene specifically to the prostate in transgenic mice. *Mol Endocrinol* 1994;**8**:230–9.
18. Zhang J, Gao N, Kasper S, Reid K, Nelson C, Matusik RJ. An androgen-dependent upstream enhancer is essential for high levels of probasin gene expression. *Endocrinology* 2004;**145**:134–48.
19. Zhang J, Thomas TZ, Kasper S, Matusik RJ. A small composite probasin promoter confers high levels of prostate-specific gene expression through regulation by androgens and glucocorticoids in vitro and in vivo. *Endocrinology* 2000;**141**:4698–710.

20. Kasper S, Rennie PS, Bruchovsky N, Sheppard PC, Cheng H, Lin L, et al. Cooperative binding of androgen receptors to two DNA sequences is required for androgen induction of the probasin gene. *J Biol Chem* 1994;**269**:31763–9.
21. Yan Y, Sheppard PC, Kasper S, Lin L, Hoare S, Kapoor A, et al. Large fragment of the probasin promoter targets high levels of transgene expression to the prostate of transgenic mice. *Prostate* 1997;**32**:129–39.
22. Cleutjens KB, van der Korput HA, Ehren-van Eekelen CC, Sikes RA, Fasciana C, Chung LW, et al. A 6-kb promoter fragment mimics in transgenic mice the prostate-specific and androgen-regulated expression of the endogenous prostate-specific antigen gene in humans. *Mol Endocrinol* 1997;**11**:1256–65.
23. Cleutjens KB, van der Korput HA, van Eekelen CC, van Rooij HC, Faber PW, Trapman J. An androgen response element in a far upstream enhancer region is essential for high, androgen-regulated activity of the prostate-specific antigen promoter. *Mol Endocrinol* 1997;**11**:148–61.
24. Maroulakou IG, Anver M, Garrett L, Green JE. Prostate and mammary adenocarcinoma in transgenic mice carrying a rat C3(1) simian virus 40 large tumor antigen fusion gene. *Proc Natl Acad Sci USA* 1994;**91**:11236–40.
25. Perez-Stable CM, Schwartz GG, Farinas A, Finegold M, Binderup L, Howard GA, et al. The G gamma / T-15 transgenic mouse model of androgen-independent prostate cancer: target cells of carcinogenesis and the effect of the vitamin D analogue EB 1089. *Cancer Epidemiol Biomarkers Prev* 2002;**11**:555–63.
26. Skalnik DG, Dorfman DM, Williams DA, Orkin SH. Restriction of neuroblastoma to the prostate gland in transgenic mice. *Mol Cell Biol* 1991;**11**:4518–27.
27. Garabedian EM, Humphrey PA, Gordon JI. A transgenic mouse model of metastatic prostate cancer originating from neuroendocrine cells. *Proc Natl Acad Sci USA* 1998;**95**:15382–7.
28. Kitsberg DI, Leder P. Keratinocyte growth factor induces mammary and prostatic hyperplasia and mammary adenocarcinoma in transgenic mice. *Oncogene* 1996;**13**:2507–15.
29. Thota A, Karajgikar M, Duan W, Gabril MY, Chan FL, Wong YC, et al. Mouse PSP94 expression is prostate tissue-specific as demonstrated by a comparison of multiple antibodies against recombinant proteins. *J Cell Biochem* 2003;**88**:999–1011.
30. Bhowmick NA, Chytil A, Plieth D, Gorska AE, Dumont N, Shappell S, et al. TGF-beta signaling in fibroblasts modulates the oncogenic potential of adjacent epithelia. *Science* 2004;**303**:848–51.
31. Placencio VR, Sharif-Afshar AR, Li X, Huang H, Uwamariya C, Neilson EG, et al. Stromal transforming growth factor-beta signaling mediates prostatic response to androgen ablation by paracrine Wnt activity. *Cancer Res* 2008;**68**:4709–18.
32. Capecchi MR. Targeted gene replacement. *Sci Am* 1994;**270**:52–9.
33. Sternberg N, Hamilton D. Bacteriophage P1 site-specific recombination. I. Recombination between loxP sites. *J Mol Biol* 1981;**150**:467–86.
34. Golic KG, Lindquist S. The FLP recombinase of yeast catalyzes site-specific recombination in the Drosophila genome. *Cell* 1989;**59**:499–509.
35. Sadowski PD. The Flp recombinase of the 2-microns plasmid of *Saccharomyces cerevisiae*. *Prog Nucleic Acid Res Mol Biol* 1995;**51**:53–91.
36. St-Onge L, Furth PA, Gruss P. Temporal control of the Cre recombinase in transgenic mice by a tetracycline responsive promoter. *Nucleic Acids Res* 1996;**24**:3875–7.
37. Hayashi S, McMahon AP. Efficient recombination in diverse tissues by a tamoxifen-inducible form of Cre: a tool for temporally regulated gene activation/inactivation in the mouse. *Dev Biol* 2002;**244**:305–18.
38. Maddison LA, Nahm H, DeMayo F, Greenberg NM. Prostate specific expression of Cre recombinase in transgenic mice. *Genesis* 2000;**26**:154–6.

39. Wu X, Wu J, Huang J, Powell WC, Zhang J, Matusik RJ, et al. Generation of a prostate epithelial cell-specific Cre transgenic mouse model for tissue-specific gene ablation. *Mech Dev* 2001;**101**:61–9.
40. Jin C, McKeehan K, Wang F. Transgenic mouse with high Cre recombinase activity in all prostate lobes, seminal vesicle, and ductus deferens. *Prostate* 2003;**57**:160–4.
41. Lewandoski M, Sun X, Martin GR. Fgf8 signalling from the AER is essential for normal limb development. *Nat Genet* 2000;**26**:460–3.
42. Ma X, Ziel-van der Made AC, Autar B, van der Korput HA, Vermeij M, van Duijn P, et al. Targeted biallelic inactivation of Pten in the mouse prostate leads to prostate cancer accompanied by increased epithelial cell proliferation but not by reduced apoptosis. *Cancer Res* 2005;**65**:5730–9.
43. Ratnacaram CK, Teletin M, Jiang M, Meng X, Chambon P, Metzger D. Temporally controlled ablation of PTEN in adult mouse prostate epithelium generates a model of invasive prostatic adenocarcinoma. *Proc Natl Acad Sci USA* 2008;**105**:2521–6.
44. Korpershoek E, Loonen AJ, Corvers S, van Nederveen FH, Jonkers J, Ma X, et al. Conditional Pten knock-out mice: a model for metastatic phaeochromocytoma. *J Pathol* 2009;**217**:597–604.
45. Lin Y, Liu G, Zhang Y, Hu YP, Yu K, Lin C, et al. Fibroblast growth factor receptor 2 tyrosine kinase is required for prostatic morphogenesis and the acquisition of strict androgen dependency for adult tissue homeostasis. *Development* 2007;**134**:723–34.
46. Zhang Y, Zhang J, Lin Y, Lan Y, Lin C, Xuan JW, et al. Role of epithelial cell fibroblast growth factor receptor substrate 2{alpha} in prostate development, regeneration and tumorigenesis. *Development* 2008;**135**:775–84.
47. Greenberg NM, DeMayo F, Finegold MJ, Medina D, Tilley WD, Aspinall JO, et al. Prostate cancer in a transgenic mouse. *Proc Natl Acad Sci USA* 1995;**92**:3439–43.
48. Gingrich JR, Barrios RJ, Morton RA, Boyce BF, DeMayo FJ, Finegold MJ, et al. Metastatic prostate cancer in a transgenic mouse. *Cancer Res* 1996;**56**:4096–102.
49. Eng MH, Charles LG, Ross BD, Chrisp CE, Pienta KJ, Greenberg NM, et al. Early castration reduces prostatic carcinogenesis in transgenic mice. *Urology* 1999;**54**:1112–9.
50. Gingrich JR, Barrios RJ, Kattan MW, Nahm HS, Finegold MJ, Greenberg NM. Androgen-independent prostate cancer progression in the TRAMP model. *Cancer Res* 1997;**57**:4687–91.
51. Foster BA, Gingrich JR, Kwon ED, Madias C, Greenberg NM. Characterization of prostatic epithelial cell lines derived from transgenic adenocarcinoma of the mouse prostate (TRAMP) model. *Cancer Res* 1997;**57**:3325–30.
52. Chiaverotti T, Couto SS, Donjacour A, Mao JH, Nagase H, Cardiff RD, et al. Dissociation of epithelial and neuroendocrine carcinoma lineages in the transgenic adenocarcinoma of mouse prostate model of prostate cancer. *Am J Pathol* 2007;.
53. Kaplan-Lefko PJ, Chen TM, Ittmann MM, Barrios RJ, Ayala GE, Huss WJ, et al. Pathobiology of autochthonous prostate cancer in a pre-clinical transgenic mouse model. *Prostate* 2003;**55**:219–37.
54. Masumori N, Thomas TZ, Chaurand P, Case T, Paul M, Kasper S, et al. A probasin-large T antigen transgenic mouse line develops prostate adenocarcinoma and neuroendocrine carcinoma with metastatic potential. *Cancer Res* 2001;**61**:2239–49.
55. Kasper S, Sheppard PC, Yan Y, Pettigrew N, Borowsky AD, Prins GS, et al. Development, progression, and androgen-dependence of prostate tumors in probasin-large T antigen transgenic mice: a model for prostate cancer. *Lab Invest* 1998;**78**:i–vx.
56. Hill R, Song Y, Cardiff RD, Van Dyke T. Heterogeneous tumor evolution initiated by loss of pRb function in a preclinical prostate cancer model. *Cancer Res* 2005;**65**:10243–54.
57. Hill R, Song Y, Cardiff RD, Van Dyke T. Selective evolution of stromal mesenchyme with p53 loss in response to epithelial tumorigenesis. *Cell* 2005;**123**:1001–11.

58. Gabril MY, Onita T, Ji PG, Sakai H, Chan FL, Koropatnick J, et al. Prostate targeting: PSP94 gene promoter/enhancer region directed prostate tissue-specific expression in a transgenic mouse prostate cancer model. *Gene Ther* 2002;**9**:1589–99.
59. Duan W, Gabril MY, Moussa M, Chan FL, Sakai H, Fong G, et al. Knockin of SV40 Tag oncogene in a mouse adenocarcinoma of the prostate model demonstrates advantageous features over the transgenic model. *Oncogene* 2005;**24**:1510–24.
60. Perez-Stable C, Altman NH, Mehta PP, Deftos LJ, Roos BA. Prostate cancer progression, metastasis, and gene expression in transgenic mice. *Cancer Res* 1997;**57**:900–6.
61. Ellwood-Yen K, Graeber TG, Wongvipat J, Iruela-Arispe ML, Zhang J, Matusik R, et al. Myc-driven murine prostate cancer shares molecular features with human prostate tumors. *Cancer Cell* 2003;**4**:223–38.
62. Zhang X, Lee C, Ng PY, Rubin M, Shabsigh A, Buttyan R. Prostatic neoplasia in transgenic mice with prostate-directed overexpression of the c-myc oncoprotein. *Prostate* 2000;**43**:278–85.
63. Scherl A, Li JF, Cardiff RD, Schreiber-Agus N. Prostatic intraepithelial neoplasia and intestinal metaplasia in prostates of probasin-RAS transgenic mice. *Prostate* 2004;**59**:448–59.
64. Jeong JH, Wang Z, Guimaraes AS, Ouyang X, Figueiredo JL, Ding Z, et al. BRAF activation initiates but does not maintain invasive prostate adenocarcinoma. *PLoS ONE* 2008;**3**:e3949.
65. Voeller HJ, Truica CI, Gelmann EP. Beta-catenin mutations in human prostate cancer. *Cancer Res* 1998;**58**:2520–3.
66. Chen G, Shukeir N, Potti A, Sircar K, Aprikian A, Goltzman D, et al. Up-regulation of Wnt-1 and beta-catenin production in patients with advanced metastatic prostate carcinoma: potential pathogenetic and prognostic implications. *Cancer* 2004;**101**:1345–56.
67. de la Taille A, Rubin MA, Chen MW, Vacherot F, de Medina SG, Burchardt M, et al. Beta-catenin-related anomalies in apoptosis-resistant and hormone-refractory prostate cancer cells. *Clin Cancer Res* 2003;**9**:1801–7.
68. Bruxvoort KJ, Charbonneau HM, Giambernardi TA, Goolsby JC, Qian CN, Zylstra CR, et al. Inactivation of Apc in the mouse prostate causes prostate carcinoma. *Cancer Res* 2007;**67**:2490–6.
69. Yu X, Wang Y, Jiang M, Bierie B, Roy-Burman P, Shen MM, et al. Activation of beta-Catenin in mouse prostate causes HGPIN and continuous prostate growth after castration. *Prostate* 2009;**69**:249–62.
70. Thompson TC. Metastasis-related genes in prostate cancer: the role of caveolin-1. *Cancer Metastasis Rev* 1998;**17**:439–42.
71. Williams TM, Hassan GS, Li J, Cohen AW, Medina F, Frank PG, et al. Caveolin-1 promotes tumor progression in an autochthonous mouse model of prostate cancer: genetic ablation of Cav-1 delays advanced prostate tumor development in tramp mice. *J Biol Chem* 2005;**280**:25134–45.
72. Bruckheimer EM, Brisbay S, Johnson DJ, Gingrich JR, Greenberg N, McDonnell TJ. Bcl-2 accelerates multistep prostate carcinogenesis in vivo. *Oncogene* 2000;**19**:5251–8.
73. Tang Y, Wang L, Goloubeva O, Khan MA, Zhang B, Hussain A. Divergent effects of castration on prostate cancer in TRAMP mice: possible implications for therapy. *Clin Cancer Res* 2008;**14**:2936–43.
74. Abdulkadir SA, Qu Z, Garabedian E, Song SK, Peters TJ, Svaren J, et al. Impaired prostate tumorigenesis in Egr1-deficient mice. *Nat Med* 2001;**7**:101–7.
75. Klezovitch O, Chevillet J, Mirosevich J, Roberts RL, Matusik RJ, Vasioukhin V. Hepsin promotes prostate cancer progression and metastasis. *Cancer Cell* 2004;**6**:185–95.
76. Chen Z, Trotman LC, Shaffer D, Lin HK, Dotan ZA, Niki M, et al. Crucial role of p53-dependent cellular senescence in suppression of Pten-deficient tumorigenesis. *Nature* 2005;**436**:725–30.

77. Wang S, Gao J, Lei Q, Rozengurt N, Pritchard C, Jiao J, et al. Prostate-specific deletion of the murine Pten tumor suppressor gene leads to metastatic prostate cancer. *Cancer Cell* 2003;**4**:209–21.
78. Backman SA, Ghazarian D, So K, Sanchez O, Wagner KU, Hennighausen L, et al. Early onset of neoplasia in the prostate and skin of mice with tissue-specific deletion of Pten. *Proc Natl Acad Sci USA* 2004;**101**:1725–30.
79. Liao CP, Zhong C, Saribekyan G, Bading J, Park R, Conti PS, et al. Mouse models of prostate adenocarcinoma with the capacity to monitor spontaneous carcinogenesis by bioluminescence or fluorescence. *Cancer Res* 2007;**67**:7525–33.
80. Jiao J, Wang S, Qiao R, Vivanco I, Watson PA, Sawyers CL, et al. Murine cell lines derived from Pten null prostate cancer show the critical role of PTEN in hormone refractory prostate cancer development. *Cancer Res* 2007;**67**:6083–91.
81. Shen MM, Abate-Shen C. Pten inactivation and the emergence of androgen-independent prostate cancer. *Cancer Res* 2007;**67**:6535–8.
82. Majumder PK, Yeh JJ, George DJ, Febbo PG, Kum J, Xue Q, et al. Prostate intraepithelial neoplasia induced by prostate restricted Akt activation: the MPAKT model. *Proc Natl Acad Sci USA* 2003;**100**:7841–6.
83. Li B, Sun A, Youn H, Hong Y, Terranova PF, Thrasher JB, et al. Conditional Akt activation promotes androgen-independent progression of prostate cancer. *Carcinogenesis* 2007;**28**:572–83.
84. Majumder PK, Grisanzio C, O'Connell F, Barry M, Brito JM, Xu Q, et al. A prostatic intraepithelial neoplasia-dependent p27 Kip1 checkpoint induces senescence and inhibits cell proliferation and cancer progression. *Cancer Cell* 2008;**14**:146–55.
85. Niu Y, Altuwaijri S, Yeh S, Lai KP, Yu S, Chuang KH, et al. Targeting the stromal androgen receptor in primary prostate tumors at earlier stages. *Proc Natl Acad Sci USA* 2008;**105**:12188–93.
86. Wang S, Garcia AJ, Wu M, Lawson DA, Witte ON, Wu H. Pten deletion leads to the expansion of a prostatic stem/progenitor cell subpopulation and tumor initiation. *Proc Natl Acad Sci USA* 2006;**103**:1480–5.
87. Elgavish A, Wood PA, Pinkert CA, Eltoum IE, Cartee T, Wilbanks J, et al. Transgenic mouse with human mutant p53 expression in the prostate epithelium. *Prostate* 2004;**61**:26–34.
88. Zhou Z, Flesken-Nikitin A, Corney DC, Wang W, Goodrich DW, Roy-Burman P, et al. Synergy of p53 and Rb deficiency in a conditional mouse model for metastatic prostate cancer. *Cancer Res* 2006;**66**:7889–98.
89. Zhou Z, Flesken-Nikitin A, Nikitin AY. Prostate cancer associated with p53 and Rb deficiency arises from the stem/progenitor cell-enriched proximal region of prostatic ducts. *Cancer Res* 2007;**67**:5683–90.
90. Bhatia-Gaur R, Donjacour AA, Sciavolino PJ, Kim M, Desai N, Young P, et al. Roles for Nkx3.1 in prostate development and cancer. *Genes Dev* 1999;**13**:966–77.
91. Schneider A, Brand T, Zweigerdt R, Arnold H. Targeted disruption of the Nkx3.1 gene in mice results in morphogenetic defects of minor salivary glands: parallels to glandular duct morphogenesis in prostate. *Mech Dev* 2000;**95**:163–74.
92. Kim MJ, Bhatia-Gaur R, Banach-Petrosky WA, Desai N, Wang Y, Hayward SW, et al. Nkx3.1 mutant mice recapitulate early stages of prostate carcinogenesis. *Cancer Res* 2002;**62**:2999–3004.
93. Abdulkadir SA, Magee JA, Peters TJ, Kaleem Z, Naughton CK, Humphrey PA, et al. Conditional loss of Nkx3.1 in adult mice induces prostatic intraepithelial neoplasia. *Mol Cell Biol* 2002;**22**:1495–503.
94. Bowen C, Stuart A, Ju JH, Tuan J, Blonder J, Conrads TP, et al. NKX3.1 homeodomain protein binds to topoisomerase I and enhances its activity. *Cancer Res* 2007;**67**:455–64.

95. Ouyang X, DeWeese TL, Nelson WG, Abate-Shen C. Loss-of-function of Nkx3.1 promotes increased oxidative damage in prostate carcinogenesis. *Cancer Res* 2005;**65**:6773–9.
96. Abate-Shen C, Banach-Petrosky WA, Sun X, Economides KD, Desai N, Gregg JP, et al. Nkx3.1; Pten mutant mice develop invasive prostate adenocarcinoma and lymph node metastases. *Cancer Res* 2003;**63**:3886–90.
97. Kim MJ, Cardiff RD, Desai N, Banach-Petrosky WA, Parsons R, Shen MM, et al. Cooperativity of Nkx3.1 and Pten loss of function in a mouse model of prostate carcinogenesis. *Proc Natl Acad Sci USA* 2002;**99**:2884–9.
98. Banach-Petrosky W, Jessen WJ, Ouyang X, Gao H, Rao J, Quinn J, et al. Prolonged exposure to reduced levels of androgen accelerates prostate cancer progression in Nkx3.1; Pten mutant mice. *Cancer Res* 2007;**67**:9089–96.
99. Lei Q, Jiao J, Xin L, Chang CJ, Wang S, Gao J, et al. NKX3.1 stabilizes p53, inhibits AKT activation, and blocks prostate cancer initiation caused by PTEN loss. *Cancer Cell* 2006;**9**:367–78.
100. Gao H, Ouyang X, Banach-Petrosky W, Borowsky AD, Lin Y, Kim M, et al. A critical role for p27kip1 gene dosage in a mouse model of prostate carcinogenesis. *Proc Natl Acad Sci USA* 2004;**101**:17204–9.
101. Gary B, Azuero R, Mohanty GS, Bell WC, Eltoum IE, Abdulkadir SA. Interaction of Nkx3.1 and p27kip1 in prostate tumor initiation. *Am J Pathol* 2004;**164**:1607–14.
102. Lan Y, Zhang Y, Wang J, Lin C, Ittmann MM, Wang F. Aberrant expression of Cks1 and Cks2 contributes to prostate tumorigenesis by promoting proliferation and inhibiting programmed cell death. *Int J Cancer* 2008;**123**:543–51.
103. Shapira M, Ben-Izhak O, Slotky M, Goldin O, Lahav-Baratz S, Hershko DD. Expression of the ubiquitin ligase subunit cyclin kinase subunit 1 and its relationship to S-phase kinase protein 2 and p27Kip1 in prostate cancer. *J Urol* 2006;**176**:2285–9.
104. Cordon-Cardo C, Koff A, Drobnjak M, Capodieci P, Osman I, Millard SS, et al. Distinct altered patterns of p27KIP1 gene expression in benign prostatic hyperplasia and prostatic carcinoma. *J Natl Cancer Inst* 1998;**90**:1284–91.
105. Shim EH, Johnson L, Noh HL, Kim YJ, Sun H, Zeiss C, et al. Expression of the F-box protein SKP2 induces hyperplasia, dysplasia, and low-grade carcinoma in the mouse prostate. *Cancer Res* 2003;**63**:1583–8.
106. Maddison LA, Sutherland BW, Barrios RJ, Greenberg NM. Conditional deletion of Rb causes early stage prostate cancer. *Cancer Res* 2004;**64**:6018–25.
107. Cunha GR. Mesenchymal-epithelial interactions: past, present, and future. *Differentiation* 2008;**76**:578–86.
108. Stanbrough M, Leav I, Kwan PW, Bubley GJ, Balk SP. Prostatic intraepithelial neoplasia in mice expressing an androgen receptor transgene in prostate epithelium. *Proc Natl Acad Sci USA* 2001;**98**:10823–8.
109. Han G, Buchanan G, Ittmann M, Harris JM, Yu X, Demayo FJ, et al. Mutation of the androgen receptor causes oncogenic transformation of the prostate. *Proc Natl Acad Sci USA* 2005;**102**:1151–6.
110. Wu CT, Altuwaijri S, Ricke WA, Huang SP, Yeh S, Zhang C, et al. Increased prostate cell proliferation and loss of cell differentiation in mice lacking prostate epithelial androgen receptor. *Proc Natl Acad Sci USA* 2007;**104**:12679–84.
111. Chung AC, Zhou S, Liao L, Tien JC, Greenberg NM, Xu J. Genetic ablation of the amplified-in-breast cancer 1 inhibits spontaneous prostate cancer progression in mice. *Cancer Res* 2007;**67**:5965–75.
112. Tien JC, Zhou S, Xu J. The role of SRC-1 in murine prostate cancinogenesis is nonessential due to a possible compensation of SRC-3/AIB1 overexpression. *Int J Biol Sci* 2009;**5**:256–64.

113. McKeehan WL, Wang F, Luo Y. The fibroblast growth factor (FGF) signaling complex. In: Bradshaw R, Dennis E, editors. *Handbook of cell signaling*. 2nd ed. New York: Academic/Elsevier Press; 2009 [vol. I; chapter 38].
114. Larsson H, Klint P, Landgren E, Claesson-Welsh L. Fibroblast growth factor receptor-1-mediated endothelial cell proliferation is dependent on the Src homology (SH) 2/SH3 domain- containing adaptor protein Crk. *J Biol Chem* 1999;**274**:25726–34.
115. Guy GR, Wong ES, Yusoff P, Chandramouli S, Lo TL, Lim J, et al. Sprouty: how does the branch manager work? *J Cell Sci* 2003;**116**:3061–8.
116. Donjacour AA, Thomson AA, Cunha GR. FGF-10 plays an essential role in the growth of the fetal prostate. *Dev Biol* 2003;**261**:39–54.
117. Konno-Takahashi N, Takeuchi T, Nishimatsu H, Kamijo T, Tomita K, Schalken JA, et al. Engineered FGF-2 expression induces glandular epithelial hyperplasia in the murine prostatic dorsal lobe. *Eur Urol* 2004;**46**:126–32.
118. Polnaszek N, Kwabi-Addo B, Peterson LE, Ozen M, Greenberg NM, Ortega S, et al. Fibroblast growth factor 2 promotes tumor progression in an autochthonous mouse model of prostate cancer. *Cancer Res* 2003;**63**:5754–60.
119. Chua SS, Ma ZQ, Gong L, Lin SH, DeMayo FJ, Tsai SY. Ectopic expression of FGF-3 results in abnormal prostate and Wolffian duct development. *Oncogene* 2002;**21**:1899–908.
120. Foster BA, Evangelou A, Gingrich JR, Kaplan PJ, DeMayo F, Greenberg NM. Enforced expression of FGF-7 promotes epithelial hyperplasia whereas a dominant negative FGFR2iiib promotes the emergence of neuroendocrine phenotype in prostate glands of transgenic mice. *Differentiation* 2002;**70**:624–32.
121. Song Z, Wu X, Powell WC, Cardiff RD, Cohen MB, Tin RT, et al. Fibroblast Growth Factor 8 Isoform b Overexpression in Prostate Epithelium: A New Mouse Model for Prostatic Intraepithelial Neoplasia. *Cancer Res* 2002;**62**:5096–105.
122. Zhong C, Saribekyan G, Liao CP, Cohen MB, Roy-Burman P. Cooperation between FGF8b overexpression and PTEN deficiency in prostate tumorigenesis. *Cancer Res* 2006;**66**:2188–94.
123. McKeehan WL, Wang F, Kan M. The heparan sulfate-fibroblast growth factor family: diversity of structure and function. *Prog Nucleic Acid Res Mol Biol* 1998;**59**:135–76.
124. Feng S, Wang F, Matsubara A, Kan M, McKeehan WL. Fibroblast growth factor receptor 2 limits and receptor 1 accelerates tumorigenicity of prostate epithelial cells. *Cancer Res* 1997;**57**:5369–78.
125. Wang F, McKeehan K, Yu C, McKeehan WL. Fibroblast growth factor receptor 1 phosphotyrosine 766: molecular target for prevention of progression of prostate tumors to malignancy. *Cancer Res* 2002;**62**:1898–903.
126. Wang F, McKeehan K, Yu C, Ittmann M, McKeehan WL. Chronic activity of ectopic type 1 fibroblast growth factor receptor tyrosine kinase in prostate epithelium results in hyperplasia accompanied by intraepithelial neoplasia. *Prostate* 2004;**58**:1–12.
127. Freeman KW, Gangula RD, Welm BE, Ozen M, Foster BA, Rosen JM, et al. Conditional activation of fibroblast growth factor receptor (FGFR) 1, but not FGFR2, in prostate cancer cells leads to increased osteopontin induction, extracellular signal-regulated kinase activation, and in vivo proliferation. *Cancer Res* 2003;**63**:6237–43.
128. Freeman KW, Welm BE, Gangula RD, Rosen JM, Ittmann M, Greenberg NM, et al. Inducible prostate intraepithelial neoplasia with reversible hyperplasia in conditional FGFR1-expressing mice. *Cancer Res* 2003;**63**:8256–63.
129. Acevedo VD, Gangula RD, Freeman KW, Li R, Zhang Y, Wang F, et al. Inducible FGFR-1 activation leads to irreversible prostate adenocarcinoma and an epithelial-to-mesenchymal transition. *Cancer Cell* 2007;**12**:559–71.

130. Winter SF, Acevedo VD, Gangula RD, Freeman KW, Spencer DM, Greenberg NM. Conditional activation of FGFR1 in the prostate epithelium induces angiogenesis with concomitant differential regulation of Ang-1 and Ang-2. *Oncogene* 2007;**26**:4897–907.
131. McDougall K, Kubu C, Verdi JM, Meakin SO. Developmental expression patterns of the signaling adapters FRS-2 and FRS-3 during early embryogenesis. *Mech Dev* 2001;**103**:145–8.
132. Hadari YR, Gotoh N, Kouhara H, Lax I, Schlessinger J. Critical role for the docking-protein FRS2 alpha in FGF receptor-mediated signal transduction pathways. *Proc Natl Acad Sci USA* 2001;**98**:8578–83.
133. Jin C, McKeehan K, Guo W, Jauma S, Ittmann MM, Foster B, et al. Cooperation between ectopic FGFR1 and depression of FGFR2 in induction of prostatic intraepithelial neoplasia in the mouse prostate. *Cancer Res* 2003;**63**:8784–90.
134. Kwabi-Addo B, Ropiquet F, Giri D, Ittmann M. Alternative splicing of fibroblast growth factor receptors in human prostate cancer. *Prostate* 2001;**46**:163–72.
135. Montironi R, Mazzucchelli R. HER-2 expression and gene amplification in high-grade PIN and prostate cancer. *Arch Ital Urol Androl* 2006;**78**:135–9.
136. Li Z, Szabolcs M, Terwilliger JD, Efstratiadis A. Prostatic intraepithelial neoplasia and adenocarcinoma in mice expressing a probasin-Neu oncogenic transgene. *Carcinogenesis* 2006;**27**:1054–67.
137. Tu WH, Thomas TZ, Masumori N, Bhowmick NA, Gorska AE, Shyr Y, et al. The loss of TGF-beta signaling promotes prostate cancer metastasis. *Neoplasia* 2003;**5**:267–77.
138. Li X, Placencio V, Iturregui JM, Uwamariya C, Sharif-Afshar AR, Koyama T, et al. Prostate tumor progression is mediated by a paracrine TGF-beta/Wnt3a signaling axis. *Oncogene* 2008;**27**:7118–30.
139. Wang F, McKeehan WL. Insulin-like growth factor-1 messenger RNA isoforms in rat prostate and prostate tumors. *In Vitro Cell Dev Biol Anim* 1996;**32**:265–8.
140. Konno-Takahashi N, Takeuchi T, Shimizu T, Nishimatsu H, Fukuhara H, Kamijo T, et al. Engineered IGF-I expression induces glandular enlargement in the murine prostate. *J Endocrinol* 2003;**177**:389–98.
141. Kaplan-Lefko PJ, Sutherland BW, Evangelou AI, Hadsell DL, Barrios RJ, Foster BA, et al. Enforced epithelial expression of IGF-1 causes hyperplastic prostate growth while negative selection is requisite for spontaneous metastogenesis. *Oncogene* 2008;**27**:2868–76.
142. Sutherland BW, Knoblaugh SE, Kaplan-Lefko PJ, Wang F, Holzenberger M, Greenberg NM. Conditional deletion of insulin-like growth factor-I receptor in prostate epithelium. *Cancer Res* 2008;**68**:3495–504.
143. Anzo M, Cobb LJ, Hwang DL, Mehta H, Said JW, Yakar S, et al. Targeted deletion of hepatic Igf1 in TRAMP mice leads to dramatic alterations in the circulating insulin-like growth factor axis but does not reduce tumor progression. *Cancer Res* 2008;**68**:3342–9.
144. Jin RJ, Lho Y, Connelly L, Wang Y, Yu X, Saint Jean L, et al. The nuclear factor-kappaB pathway controls the progression of prostate cancer to androgen-independent growth. *Cancer Res* 2008;**68**:6762–9.
145. Lu W, Wang J, Wen T. Downregulation of Rho-GDI gamma promotes differentiation of neural stem cells. *Mol Cell Biochem* 2008;**311**:233–40.

Modeling Cancers in *Drosophila*

Cédric Polesello,[*,†] Fernando Roch,[*,†] Vanessa Gobert,[*,†] Marc Haenlin,[*,†] and Lucas Waltzer[*,†]

[*]*Université de Toulouse, UPS, CBD (Centre de Biologie du Développement), Bâtiment 4R3, 118 route de Narbonne, F-31062 Toulouse, France*

[†]*CNRS, CBD UMR5547, F-31062 Toulouse, France*

The basic cellular processes deregulated during carcinogenesis and the vast majority of the genes implicated in cancer appear conserved from humans to flies. This conservation, together with an ever-expanding fly genetic toolbox,

Progress in Molecular Biology
and Translational Science, Vol. 100
DOI: 10.1016/B978-0-12-384878-9.00002-9

Copyright 2011, Elsevier Inc.
All rights reserved.
1877-1173/11 $35.00

has made of *Drosophila melanogaster* a remarkably profitable model to study many fundamental aspects of carcinogenesis. In particular, *Drosophila* has played a major role in the identification of genes and pathways implicated in cancer and in disclosing novel functional relationships between cancer genes. It has also proved to be a genetically tractable system where to mimic cancer-like situations and characterize the mode of action of human oncogenes. Here, we outline some advances in the study of cancer, both at the basic and more translational levels, which have benefited from research carried out in flies.

I. Introduction

There are more than 200 different types of human cancers that, all together, are responsible for roughly one-fifth of all deaths in industrialized countries. Despite their diversity, cancer cells share, albeit to different extents, some fundamental properties: increased proliferation, decreased apoptosis, altered differentiation, altered metabolism, genomic instability, immortalization, alteration of the tissue structure, and invasive behavior.[1] At .the molecular level, cancers arise as a consequence of multiple genetic lesions that modify the normal function of oncogenes, tumor suppressor genes or miRNA genes, controlling the aforementioned processes.[2] Characterizing the functions of these genes as well as the key steps of cellular transformation is crucial to understand and fight tumorigenesis. Accordingly, much effort has been devoted to developing *ex vivo* and *in vivo* experimental models for cancers. Beside vertebrate models, the fruit fly *Drosophila melanogaster* has emerged as a powerful tool for cancer studies. Indeed, the key cellular processes (proliferation, apoptosis, migration, etc.) and the vast majority of the cancer-associated genes are conserved from *Drosophila* to humans.[3] Its relative simplicity, low level of gene redundancy, excellent genome annotation and one century of genetic characterization have also played a great deal in the success of *Drosophila*. Notably, the extensive genetic toolbox available in flies allows mimicking cancer-like situations and setting up sophisticated genetic screens.[4] Thereby, *Drosophila* has been instrumental for the identification of many signaling pathways deregulated in human cancers and the study of genetic interactions occurring between cancer genes.

However, not all the aspects of carcinogenesis can be studied in *Drosophila*. For instance, flies have an open circulatory system and no adaptive immunity; thus their tumors grow without angiogenesis and the immune surveillance of cancer cells does not operate in the same way. Also, flies have a very short lifespan whereas human cancers are fundamentally linked with aging. Finally, each human cancer cell develops in a genetic and microenvironmental context that is determinant for the outcome of the disease but whose complexity can hardly be recreated in any experimental model. Despite these limitations,

Drosophila has greatly contributed to our understanding of cancer. In this work, we summarize some *Drosophila* researches focusing either on the analysis of the general mechanisms underlying cell transformation or oriented to the design of cancer-specific model systems.

II. *Drosophila* and Cancer Cell Growth

The most prominent hallmark of cancer cells is their aberrant growth, as illustrated by the etymology of the word tumor (from the Latin "tumor," *i.e.*, swelling). This hyperplasia can be caused by increased proliferation and/or decreased cell death. *Drosophila* has helped characterizing many genes controlling these two processes.

A. Cell Proliferation

The core machinery controlling cell cycle progression is conserved from yeast to mammals, and *Drosophila* has been a successful model for studying cell cycle regulation *in vivo* in part because its E2F/CycE network is simpler than in mammals.[5] For instance, *Drosophila* has only two E2Fs (one activator of proliferation, E2F1, and one repressor, E2F2), one E2F heterodimeric partner (Dp), two RB family members (Rbf1 and Rbf2), one essential G1–S cyclin (Cyclin E), a single cyclin-dependent kinase inhibitor (Dacapo), and no INK homologue. Most work in flies has focused on the regulation of cell cycle during development.[6] Whereas S and M phases oscillate without gaps during early embryogenesis, imaginal disc cells, that will give rise to the adult structures, rely on canonical cell cycle during larval stages. In addition, a number of embryonic and larval structures undergo endoreplication and, in the adult gonads, the germ cells undergo meiosis. The comparative study of these different cell cycle types has brought important contributions to our understanding of cell cycle regulation.[7] For example, the G1–S transition, which primarily depends on Cyclin E (CycE) activity, has been extensively studied in *Drosophila*. Regulation of CycE turnover is critical for cell cycle progression and many cancers upregulate CycE in humans.[8] Works in flies showed that CycE is regulated both at the transcriptional and posttranscriptional levels *in vivo*,[6] notably by the E3 ubiquitin ligase Archipelago (Ago) that targets CycE to SCF-dependent degradation.[9] This finding prompted the discovery that Ago ortholog Fbw7 is a tumor suppressor.[9–11] Of note, beside *in vivo* genetic screens, RNA interference (RNAi) using *Drosophila* cell cultures provides new means to study cell proliferation. Accordingly, large-scale to genome-wide RNAi screens have been successfully conducted to identify regulators of cell cycle progression or of E2F activity.[12–14]

Cell proliferation is controlled by a variety of conserved signaling pathways, including RAS/MAPK, PI3K, and JAK/STAT pathways, whose deregulation can lead to cancer. Many components of these pathways were identified by genetic studies in *Drosophila* and they were shown to control proliferation also in this organism.[4] One salient example is the historical role played by *Drosophila* in the characterization of the RAS/MAPK cascade. In mammals, mitogenic growth factors activate receptor tyrosine kinases (RTKs) and the downstream RAS/MAPK pathway to induce cell division. Accordingly, activating mutations in *RAS* genes are among the most frequent genetic alteration in human cancers.[15] In *Drosophila*, changes in RTK/RAS signaling activity induce visible eye phenotypes that were used in elegant second-site modifier screens to unveil new components of this conserved pathway.[16] Together with analysis in *Caenorhabditis elegans*, these studies delineated the core RTK/RAS/MAPK signaling cascade from the cell surface to the nucleus. In parallel, works on *Drosophila* eye and wing development have shown that crosstalk between patterning signals such as Hedgehog (Hh), Decapentaplegic (Dpp), Wingless (Wg), or Notch (N) impinges directly on the activity of key cell cycle regulators. Moreover, their output on cell cycle seems to be tissue specific,[17–19] which implies that, depending on the developmental context, these signaling molecules can function either as tumor suppressors or oncogenes.

B. Cell Death

Cell death evasion is another central aspect of cancer and *Drosophila* models contributed to our understanding of cell death pathways, in particular apoptosis. The apoptotic program culminates with the sequential activation of initiator and effector caspases, a family of cysteine proteases essential for cell destruction.[20] During fly development, Dronc, the ortholog of Caspase-9, is the main initiator caspase. In response to apoptotic stimuli, it binds to the apoptosis promoting factor 1 (APAF1)-like protein Dark to activate effector caspases such as drICE.[21] Regulating caspase activity is critical for proper control of apoptosis and this is achieved both by members of the Bcl2 family and of the inhibitor of apoptosis (IAP) family. Actually IAPs were discovered in baculoviruses and their first cellular homologs were identified in a genetic screen in *Drosophila*, opening the way for the characterization of their function and mode of action.[22] Further studies showed that *Drosophila* IAPs (in particular DIAP1) rather than Bcl2 homologs play a predominant role in preventing caspase activation.[21] Actually, DIAP1 directly interacts with the initiator caspases and promotes their degradation by ubiquitinylating them. Conversely, DIAP1 is regulated by IAP antagonists, such as Reaper (Rpr), Head involution defective (Hid), and Grim, whose binding to DIAP1 releases the caspases and favors DIAP1 degradation. IAPs are overexpressed in many cancer cell lines and primary tumors and are now considered as potential targets for cancer treatment.[23]

In *Drosophila* as in cancer cells, apoptosis is controlled by developmental cues as well as by oncogenic signaling pathways and environmental stimuli such as irradiation. The two major upstream regulators activated in response to apoptotic stimuli in flies are the *p53* homolog *Dmp53* and the JNK pathway, which are critical regulators of apoptosis in humans.[24–26] Interestingly, Dmp53 also promotes apoptosis by activating Hippo signaling following irradiation, suggesting that this pathway may contribute to p53 tumor suppressor function.[27] Strikingly too, it has been known for a long time that irradiation can lead to the removal of more than 50% of cells in imaginal discs without affecting the size or the pattern of the final organ.[28] This intriguing feature relies on a process termed apoptosis-induced proliferation whereby compensatory divisions are induced in the neighboring nonapoptotic cells by the dying cells.[29] Depending on the tissue, compensatory proliferation actually requires the initiator caspase Dronc or the effector caspase DrICE and Dcp-1 to promote the nonapoptotic function of JNK or the accumulation of the mitogen Hh.[30–34] Of note, compensatory proliferation also occurs in mammals, for example, during intestinal and liver regeneration,[35,36] and might participate in tumorigenesis. Thus, delineation of this phenomenon in *Drosophila* could provide important information for our understanding of cancer progression.

C. Checkpoints Control

Most cancer cells are aneuploid, and T. Boveri proposed almost 100 years ago that abnormal chromosome constitution promotes tumor formation.[37] Aneuploidy in cancer cells is associated with increased chromosomal instability and defects in mitotic checkpoints. In addition, cells are exposed to exogenous and endogenous genotoxic agents and defects in DNA damage response lead to genomic instability and accumulation of potentially oncogenic mutations. *Drosophila* emerged as an appropriate model to study DNA damage and mitotic checkpoints due to the functional conservation of the molecular actors controlling these events.[38,39] For example, the Chk2/p53 cascade plays a conserved role in ionizing radiation-induced cell death in flies and mammals.[24,25,40–42] In addition to irradiation of embryos or imaginal discs, meiotic recombination and telomere maintenance have been extensively used to investigate DNA damage response *in vivo*.[38,43] Moreover, cell-based RNAi screens have been recently developed to identify genes important both for survival after DNA damage and for chromosome stability.[44,45] These different approaches helped defining conserved mechanisms controlling genome maintenance.

Besides, *Drosophila* can also serve to test the function of key regulators of chromosomal stability in cancer. For instance, centrosome amplification, which can lead to the formation of multiple spindle poles and unequal distribution of chromosomes in daughter cells, is a common phenotype in solid human tumors and has been linked to genetic instability and tumorigenesis.[37] But whereas evidence for a causal link between centrosome amplification and genomic instability is still missing, some recent findings in *Drosophila* challenge the whole notion. Indeed, flies presenting extra centrosomes are viable and can be maintained as stable diploid stocks for many generations.[46] Intriguingly, upon transplantation into adult host flies, larval brain cells with compromised centrosomal function generate metastatic tumors, supporting the idea that centrosome dysfunction is sufficient to promote tumorigenesis. However, these tumors seem to arise from a defect in asymmetric cell division rather than from increased genomic instability or aneuploidy, as usually assumed in mammals.[46,47] Furthermore, a genome-wide RNAi screen in *Drosophila* cells identified a set of genes that prevent the formation of multipolar spindles.[48] Importantly, reducing the levels of one of them, the nonessential kinesin motor HSET, selectively killed mammalian cancer cells with supernumerary centrosomes, suggesting that some tumors depend on centrosome clustering for survival, an observation that opens new avenues for cancer therapeutics.

III. *Drosophila*, Cancer and Tissue Growth

In most cases, mutations inducing ectopic proliferation or decreased apoptosis do not result in tissue overgrowth due to compensatory mechanisms that are likely to be also involved in the control of tumor growth. Some of the mechanisms and signaling pathways that coordinate tissue growth have been established in *Drosophila*.

A. The Hippo Pathway

In the past decade, a new conserved signaling pathway discovered in *Drosophila*, the Hippo pathway, has been shown to control organ size by coordinately regulating proliferation and apoptosis.[49] Importantly, deregulation of this pathway has been implicated in the development of many human cancers. The first components of the Hippo pathway, *warts* (*wts* also known as *lats*), *salvador* (*sav*), *hippo* (*hpo*), and *mob as tumor suppressor* (*mats*) have been identified in genetic screens looking for genes that normally restrict cell growth and proliferation.[50–57] Loss of function in any of these genes provokes a distinctive hyperplasia due to excessive proliferation and decreased apoptosis in response to the transcriptional activation of *cycE*, *DIAP1*, and *bantam* (Fig. 1). At the molecular level, these four core components form a molecular

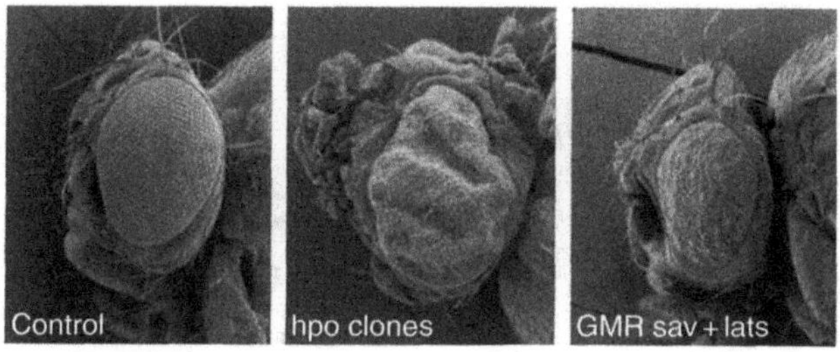

FIG. 1. The Hpo pathway controls both apoptosis and proliferation. Electron microscopy pictures of *Drosophila* eyes. A control wild-type eye is shown in the left panel. Downregulation of *hpo* activity leads to a massive increase in eye size due to both lack of apoptosis and increase in proliferation (middle panel). On the contrary, overexpression of *sav* and *wts* strongly reduces eye size (right panel).

complex that restrict tissue growth chiefly by phosphorylating the transcriptional coactivator Yorkie (Yki) (Fig. 2), thereby preventing its nuclear accumulation and the activation of the Hippo pathway transcriptional targets.[58,59]

Over the last few years, thanks to active researches by several fly labs, a number of molecules acting upstream or downstream of these four core components have been identified and their mode of action partly characterized (Fig. 2).[49] These include the planar cell polarity transmembrane proteins Fat and Dachsous, which seems to act as receptor and ligand, the Expanded–Merlin–Kibra complex, which participates in signal transduction from the cell cortex, the transcription factors Scalloped and Homothorax, which recruit Yki to DNA, and two negative regulators, dRASSF and Ajuba, which inhibit some of the core components. Most members of the Hippo pathway have functional homologs in mammals. Actually, human YAP (Yki homolog), Lats1 (Wts homolog), Mst2 (Hpo homolog), and Mob1 (Mats homolog) can rescue the corresponding *Drosophila* mutant *in vivo*[58,60–62] and interactions between core components of the Hpo pathway are largely conserved.[63]

The Hippo pathway was originally identified as a tumor suppressor pathway in *Drosophila* and its deregulation is emerging as an important step in mammalian carcinogenesis.[49] One striking example is the human tumor suppressor gene *NF2* (Merlin homolog) whose mutation has been causally linked to the development of neurofibromatosis type 2, a familial predisposition neoplasia syndrome that causes multiple tumors in the nervous system.[64] Also, mutation or downregulation of *lats* and *mst* genes has been found in several human cancer lines and genetic studies in mice demonstrated that *lats1* and *mst1/2* are tumor suppressor genes.[65,66] Conversely, *yap* was shown to control contact inhibition and has been implicated as a *bona fide* oncogene and a prognostic marker in human hepatocellular carcinoma.[67,68] Undoubtedly, ongoing investigations in flies will give a better understanding of Hippo pathway regulation during normal development and cancer.

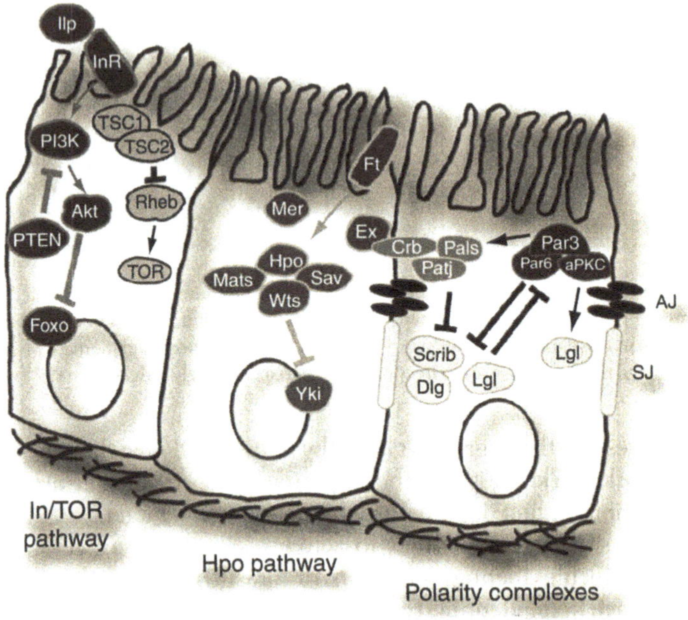

FIG. 2. Schematic representation of the Insulin/TOR pathway, Hippo pathway and epithelial polarity complex proteins in *Drosophila*. Binding of the insulin-like peptides (ILP) to the insulin receptor (InR) induces the activation of PI3K and Akt, which phosphorylates and inactivates the transcription factor FOXO, thereby reducing cell metabolism and proliferation. In response to nutrient sensing, the small GTPase Rheb is inhibited by the activity of the Tuberous sclerosis complex formed by TSC1 and TSC2. GTP-bound Rheb allows the activation of the TOR/Raptor complex, promoting translational initiation, ribosome biogenesis as well as nutrients storage. Activation of the Hpo pathway relies on a phosphorylation cascade involving the core components Hpo, Sav, Wts, and Mats, and leads to the cytoplasmic retention of the cofactor of transcription Yki. Ft and Crb somehow regulate Ex and Mer, which are required upstream of Hpo to control both proliferation and apoptosis. Apicobasal cell polarity is established and controlled by intricate cross-regulatory interactions between three major complexes. The Par3/Par6/αPKC and the Crb/Pals/Patj complexes are located in the subapical region of the cell and the Scrib/Dlg/Lgl complex is found in the lateral region. These interactions regulate the maintenance of subapical junctional complexes (adherens junctions, AJ, and septate junctions, SJ) that mediate cell–cell interactions.

B. Cell Competition

Cell competition is another important contributor to tissue growth and homeostasis. This process, which was first described in *Drosophila* using mutations in ribosomal protein genes called *Minutes*,[69,70] is thought to arise from differences in growth rates between neighboring cells.[71] As a consequence of this difference, apoptosis is triggered in the weaker cells, whereas the stronger cells are stimulated to proliferate. During cell competition, because the expansion of winners occurs at the expense of losers, total cell numbers does not change and the normal pattern of the organ is preserved. Growth regulators have been proposed to affect cell competition: for example, components of the Hippo tumor suppressor pathway[72] and most notably *dmyc*, the homolog of the proto-oncogene *c-myc*.[73,74] Mutations that reduce dMyc function lead to small flies but localized dMyc overexpression leads to the death of surrounding wild-type cells, even though they are perfectly healthy. Winner cells are thought to produce an unknown soluble killing signal that triggers apoptosis in the loser cells.[73–75] Reciprocally, losing cells induce an engulfment response in the winner cells.[76] How winners and losers are distinguished is an important but still unresolved question.

Cell competition also occurs in mammals, as illustrated by mutation in mouse *ribosomal protein L24* gene.[77] As this process takes place between different neighbor cells, it might be activated between normal and transformed cells and the fact that Myc and the Hippo pathway are both associated with tumorigenesis and cell competition suggests a possible link between these two processes. Accordingly, it has been proposed that cell competition may participate in tumor progression by allowing cancer cells to out-compete normal cells.[78]

C. The Insulin/TOR Pathway

One key conserved regulator of cell and organism growth is the Insulin/TOR pathway that couples dietary conditions to cell metabolism, growth, proliferation and survival. *Drosophila* has proven to be a powerful genetic model to decipher the Insulin/TOR pathway with direct implications for carcinogenesis.[79]

Briefly, in the first branch of the pathway, secreted *Drosophila* insulin-like peptides (Dilps) act in an endocrine manner by binding insulin receptor homolog InR to activate an intracellular signaling cascade implicating IRS1 homolog Chico, PI3K and Akt (Fig. 2).[80–82] Akt phosphorylates and inhibits the nuclear translocation of one major target of InR signaling: the transcription factor FOXO that controls the expression of several key actors of cell metabolism and cell cycle, including ribosomal proteins, dMyc, and the inhibitor of cap-dependant translation 4E-BP.[83–85] Diminishing InR signaling leads to reduced cell, organ, and body size by decreasing both proliferation and growth.[80] Conversely, reducing the activity of PTEN, which antagonizes PI3K activity, results in increased cell growth and proliferation.[86,87] The second branch, the target of rapamycin (TOR) pathway, directly senses the dietary status (amino acid and energy levels) at the individual cell level.[79] The complex formed by the proteins tuberous sclerosis complex 1 and 2 (TSC1 and TSC2) constitutes the upstream component of the pathway. Mutations in *Drosophila tsc1* or *tsc2* lead to larger cells and mild cell cycle acceleration.[88–91] The main known function of the TSC complex is to antagonize the small GTPase Rheb that activates the TOR–Raptor complex (TORC1) (Fig. 2).[92–94] TORC1 phosphorylates 4E-BP and S6K to activate protein biosynthesis and it also promotes ribosome biogenesis and Myc accumulation.[85,95,96] Importantly, multiple cross-regulations take place between the InR and TOR branches.[79]

In humans, cancer cells have an altered metabolism that is essential for their survival and growth. In addition, many components of the Insulin/TOR pathway are either tumor suppressors (e.g., *PTEN*) or oncogenes (e.g., *PI3KCA*, *AKT*).[97] Analysis of the InR/TOR pathway in flies has not only illuminated the intimate relationship between nutrient sensing and carcinogenesis but also helped define the function of these genes and their epistatic relationships. For example, tuberous sclerosis is an autosomal dominant disorder caused by mutation in *TSC1* or *TSC2* that leads to widespread development of benign tumor. Mutations in *gigas/tsc2* and *tsc1* in fly were recovered in genetic screens for genes that affect cell growth and their characterization was instrumental to link the tumor suppressor activity of *TSC1/TSC2* to the Insulin/TOR pathway.[88–91] Also, the TSC1/TSC2 target Rheb was first identified in *Drosophila*.[92–94] All together, these studies paved the way for the development of anticancer drugs targeting metabolic processes.

IV. *Drosophila*, Cancer, Differentiation, and Stemness

Another hallmark of cancer cells is their altered differentiation. Cell differentiation is usually associated with a low proliferating potential, and cancer can be envisaged as a "differentiation disease," as exemplified by the discovery of cancer stem cells.[98] *Drosophila* cancer models have brought valuable insights into the regulation of the balance between self-renewal and differentiation that goes awry in cancer cells.

For example, an analysis in flies demonstrated that loss of differentiation could act as a switch for tumor progression.[99] This work showed that the incidence of tumors and metastasis observed in a sensitized eye background could be, respectively, suppressed or enhanced by gain or loss of function in *atonal*, a gene controlling eye cell fate specification and differentiation.[99] Atonal suppresses tumor formation both by promoting differentiation and cell cycle exit and by inducing apoptosis through JNK signaling in cancer cells. Remarkably, an accompanying study demonstrated that the tumor suppressor function of *atonal* is conserved in mammals and also depends on the JNK pathway.[100] However, deregulation of differentiation master genes is not the only route to cancer progression. For instance, contrary to eye cells lacking either Rb or the Hippo pathway, the double mutant cells fail to exit the cell cycle and progressively dedifferentiate.[101] This dedifferentiation is not due to inappropriate proliferation,[101,102] but reflects an unexpected cooperation between Rb and Hippo pathways in maintaining differentiation, which depends neither on E2F or Yki.[101] Thus, beside their cell cycle defects, *Rb* mutant cells may be prone to dedifferentiate and generate cancer stem cells.

Whether cancer stem cells arise through the malignant transformation of normal stem cells or of committed downstream progenitors is a matter of controversy and probably depends on the cancer type.[103] However, strong evidence indicating that mutations affecting stem cells can induce tumors was obtained in flies. The best examples came from studies on the neural stem cells: the neuroblasts. Indeed, some of the first *Drosophila* tumor suppressors were identified by the massive growth of the brain in mutant larvae[104] and their subsequent analysis showed that this overgrowth is caused by defects in neural stem cell asymmetric division. Normally, a neuroblast divides into one neuroblast and one differentiating cell, the ganglion mother cell (GMC).[105] This asymmetric division relies on the precise orientation of the mitotic spindle and the distribution at opposed sides of the cell of different polarity protein complexes, which eventually restrict the localization of GMC fate determinants to the basal side of the dividing neuroblast. In a seminal study, Caussinus *et al.*[106] demonstrated that mutations altering several components of the neuroblast asymmetric division machinery can trigger brain tumors that can be indefinitely transplanted from host to host. It is tempting to speculate that a similar functional link between abnormal asymmetric division and cancer may be conserved in mammals.

Depending on the nature of the mutation, *Drosophila* brain tumors can arise from direct expansion of the neuroblasts, from misspecification of a GMC that reverts to a neuroblast fate, or from retention of the neuroblast in a transient amplifying state.[105] For example, in the absence of Prospero, which represses cell cycle and stem cell fate genes and activates neuronal differentiation in the GMC, differentiating daughter cells revert to a stem cell-like fate.[107,108] In contrast, in *brain tumor* (*brat*) mutants, neuroblasts are blocked in their maturation and are unable to produce differentiated progeny.[108] Brat seems to impinge both on the miRNA machinery and on Myc activity, to negatively regulate cell growth.[105] Thus, different aspects of stem cell biology can be altered to produce tumor stem cells.

Beside the central nervous system, *Drosophila* has other stem cell-like compartments, notably in the gonads, the gut and the hematopoietic system, which give good opportunities to study stem cell-derived tumors.[109] For instance, JAK/STAT signaling regulates *Drosophila* blood stem cell homeostasis[110] and activating mutations in the JAK kinase Hopscotch have been found to induce transplantable hematopoietic tumors more than 15 years ago.[111,112] Mirroring these results, it was recently found that activating mutations of

JAK2 in human blood stem cells are associated with various hematopoietic malignancies.[113] Finally, studies in *Drosophila* of the relationships between stem cells and their niche have helped defining many stem cell features relevant to carcinogenesis.[109]

V. *Drosophila*, Cancer and Cell Polarity

Malignant transformation and loss of cell polarity are tightly correlated in carcinoma, which represent around 80% of human cancers.[114] Studies in *Drosophila* pioneered the discovery of the role of cell polarity determinants in cancer. Indeed, *lethal giant larvae* (*lgl*) and *discs large* (*dlg*) were identified as tumor suppressor genes 50 years ago and it was subsequently shown that they act together with *scribble* (*scrib*) to control epithelial cell polarity.[115] *lgl*, *dlg*, or *scrib* mutant larvae exhibit hallmarks of human carcinoma: their imaginal discs, normally composed of a monolayer of epithelial cells, are grossly disorganized and overproliferate. These findings were then corroborated by expression studies showing that their orthologs are inactivated in human carcinoma and by functional assays demonstrating that they behave as tumor suppressors in vertebrates.[114]

Three major conserved complexes control epithelial cell polarity. The Crumbs/Pals1/Patj and the Par3/Par6/αPKC complexes are localized to the apical cortex, whereas the Scrib/Dlg/Lgl module is localized to the basolateral membrane (Fig. 2).[116] Intricate cross-regulatory interactions between these proteins control epithelial tissue integrity by regulating the maintenance of subapical junctional complexes that mediate cell–cell interactions. While members of the basolateral group consistently behave as tumor suppressors in many tissues, the function of apical components in human carcinoma seems instead context dependent.[114] Furthermore, different proteins of the endocytic machinery control epithelial cell polarity and behave as neoplastic tumor suppressors in *Drosophila*.[117] In humans, several regulators of endocytosis are abnormally expressed in tumors, although it is not clear whether they participate in tumorigenesis by controlling cell polarity or the trafficking of growth factor receptors.[118]

By loosening cell–cell contacts, disruption of epithelial polarity is directly linked to a critical step in cancer progression: the epithelial–mesenchymal transition (EMT). During this process, cells transit from a polarized, epithelial phenotype to a highly motile mesenchymal phenotype.[119] In *Drosophila* imaginal discs, cells lacking the so-called *scrib* group genes (*scrib*/*dlg*/*lgl*) lose their columnar shape and form multilayered aggregates of rounded cells.[120] In the ovary, their loss causes follicle cells to undergo EMT and to invade the germ line in a Par3-dependent manner.[121,122] Similarly, different EMT-promoting factors were found to target polarity complex components in mammals.[114]

However, the mechanisms leading to larval disc hyper-proliferation in *scrib* group mutants remain unclear. Loss of epithelial cell polarity, by affecting the localization of signaling proteins involved in growth control, may indirectly lead to increased proliferation. In addition, Lgl, αPKC, and Crumbs may regulate cell proliferation and survival by impinging on the Hippo pathway. Indeed, Crb seems to regulate Expanded localization, whereas αPKC and Lgl controls both Hpo and dRASSF localization.[123–125]

VI. *Drosophila* and Metastasis

Metastasis, the ultimate step of malignancy, is the main cause of death for cancer patients. It chiefly relies on the capacity of a tumor cell to migrate from the primary site and invade surrounding tissues.[126] In most cases, cancer cell metastasis implies EMT, disruption of the basement membrane, intravasation into the vasculature, extravasation into distant organs, survival and proliferation of the cancer cells into their new environment. Not all these aspects can be studied in *Drosophila*, but in compensation, the normal development of some fly tissues provides different models where to analyze *in vivo* the mechanism governing cell migration and thus gain insight into the behavior of metastatic cells.[127] Here, we shall focus on border cell migration, which represents an interesting paradigm to study cell invasion, and then discuss other more direct approaches aiming at modeling metastasis in *Drosophila*.

A. Border Cell Migration

Collective migration is involved in the propagation of several types of tumors, such as colorectal carcinoma, melanoma, and breast cancer.[128] Furthermore, some carcinoma metastatic tumors maintain a relatively well-differentiated epithelial morphology and a complete EMT is not mandatory for tumor cell dissemination into surrounding tissues.[119] Border cells are a group of 6–10 epithelial cells that delaminate from the anterior pole of the follicular epithelium surrounding the egg, migrate as a cell cluster between the nurse cells, and eventually reach the oocyte, located at the posterior of the egg chamber. Genetic screens and live imaging pinpointed the different cellular and molecular players governing this collective migration, an approach that has revealed unsuspected parallelisms with the mechanisms implicated in human metastasis.[129] For example, the transformation of the follicular epithelial cells into invasive cells is mediated by the JAK/STAT pathway that induces the

C/EBP transcription factor Slow border cell (Slbo).[130–132] Similarly, hyperactive JAT/STAT signaling and STAT3 target C/EBP∂ have been linked to metastatic progression in humans.[133–136]

Migrating border cells retain apicobasal polarity and several polarity components are required for proper detachment of border cells from neighboring epithelial cells and/or for their migration.[137–140] These results are consistent with the idea that polarity proteins, by regulating front-back polarity in migrating cells, are key regulators of metastasis.[141] Also, regulated DE-cadherin turnover stands as a critical feature to preserve cluster integrity and to allow migration.[139,142] Along this line, the discovery that DE-cadherin is stabilized by Myosin VI during border cell migration[143] prompted the finding that human Myosin VI is upregulated in some prostate and ovarian cancers and contributes to metastasis in a mouse xenograft model.[144,145]

The precise movement of the border cell cluster is guided by ligands for PDGF/VEGF and EGF receptors, which provide chemoattractant cues,[146–148] and by regulators of endocytosis, which spatially restrict receptor signaling.[149] Moreover, the fly homologs of two human metastasis tumor suppressor genes, *Nm23* and *missing in metastasis* (*mim*), were found to control border cell migration by impinging on endocytosis.[150,151] Thus, endocytosis may participate in metastasis by affecting both cell polarity and local response to guidance cues. Work on border cell migration also hinted that IAPs, beyond their well-described role in cell death, regulate metastasis. Indeed, DIAP1 mutant border cells fail to migrate but do not activate apoptosis.[152,153] Recently, human IAPs were also shown to promote tumor cell motility independently of their cytoprotective role.[154]

Finally, several lines of evidence indicate that hypoxia can trigger metastasis in mammals.[155] Remarkably, the hypoxic response controls border cell rate of invasion and migratory capacity through a conserved cascade involving HIF1α.[156] All together, the numerous similarities between border cell migration and metastasis strongly suggest that this model will continue to bring significant insights into this process.

B. Fly Models for Metastasis

As discussed above, mutations in several genes controlling *Drosophila* apicobasal cell polarity lead to overgrowth phenotypes but do not cause metastasis. However, seminal studies by Gateff[104] and then by Woodhouse *et al.*[157,158] showed that brain tumors coming from *lgl*, *dlg*, or *brat* mutant larvae and transplanted into the abdomen of adult host flies can give rise to cells that can cross basement membrane, invade distant organs, and form lethal secondary tumors. Additional parallels with metastasis in humans were brought up by the demonstration that matrix metalloproteinase (MMP) activity is critical for the invasive behavior of the transplanted cells.[159] This technique has been particularly useful to demonstrate the tumorigenic nature of several *Drosophila* mutations and, to a lesser extent, to identify genes potentially modulating metastasis.[106,112,160,161]

Another decisive contribution to the study of metastasis in flies was the design of refined genetic techniques allowing generation of cell clones carrying multiple genetic alterations in mosaic animals. These techniques, such as the MARCM system, made possible to study the behavior in an otherwise wild-type context of a clone of cells that are mutant for a particular gene and simultaneously overexpress any desired factor.[162] This has allowed screening for cooperative effects between oncogenes and tumor suppressor, closely mimicking the two-hit situation observed in cancers. In two parallel studies, it was shown that disruption of polarity (loss of *scrib*, *dlg*, or *lgl*) and the expression of an activated form of Ras or Notch are sufficient to induce full malignant transformation of epithelial cells as shown by the appearance of metastatic tumors in distant organs.[120,163] These tumors exhibited several features of metastatic mammalian tumors, such as loss of cell adhesion, decreased E-cadherin expression and the ability to degrade the basement membrane.[163] Importantly, oncogenic Ras and loss of Scrib were subsequently found to cooperate to promote cell invasion in mammals.[164]

This type of studies also showed that the JNK pathway is an important effector of the oncogenic Ras activity in this cooperative context. Normally, JNK induces apoptosis in *scrib* mutant cells[120] but, in the presence of RasV12, JNK activity assumes instead a growth-promoting activity and contributes to metastasis by activating MMP expression, basement membrane remodeling and JAK/STAT signaling.[165–168] How JNK signaling switches roles is still unclear but cooperation between Ras and JNK has also been observed in human carcinoma.[169] Yet, blocking apoptosis and promoting cell proliferation in *scrib* mutant cells is not sufficient to induce metastasis,[163] indicating that Ras/Scrib cooperation involves additional targets.

Another infamous proto-oncogene is *c-Src* whose overactivation has been associated with metastasis[170], a feature that seems conserved in flies. Indeed, whereas high levels of Src signaling normally induce cell death, they cooperate with RasV12 to induce invasive eye cell tumors.[171] Furthermore, mutations in *polyhomeotic*, a member of the *Polycomb Group* (*PcG*) of epigenetic silencers, also cooperate with RasV12 to induce metastatic tumors.[160] Similar connections may exist in human cancer cells where PcG proteins have pleiotropic functions. Finally, cooperation with RasV12 can be used in genetic screens to identify new genes controlling metastasis.[172] Thus these approaches stand as a strong paradigm to study conserved aspects of metastasis.

VII. *Drosophila* and Cancer Cell Environment

Mutations in cancer genes not only affect mutant cells themselves but also the behavior of the surrounding tissue. Conversely, wild-type cells also influence tumor course.[173] The analysis of tumor suppressor mutant clones in *Drosophila* mosaics showed that the behavior of cancer cells is clearly non cell-autonomous. For instance, mutations in *scrib* or *dCsk* lead to ectopic growth when mutant cells cover an entire epithelium, whereas mutant clones surrounded by normal cells are extruded from the epithelium, migrate and die by apoptosis.[120,174] As it has been shown for *lgl* clones, cell competition is likely to be involved in the elimination of these mutant cells.[175] Thus, the interface between normal and transformed cells may influence their invasive behavior.[176] In line with this hypothesis, Src or Ras-transformed mammalian cells extrude from cultured epithelial sheets only when juxtaposed to normal cells.[177,178] Moreover, in *Drosophila*, mutant clones for different endosomal components or for the E1 ubiquitin ligase induce overgrowth of the adjacent tissue by paracrine signals.[117,179] So, cell signals operating in both directions occur at the interface between normal and transformed cells.

An unexpected discovery made in flies is the existence of interclonal cooperation between two oncogenic events. Indeed, *scrib* loss and RasV12 also cooperate for invasive tumor induction when they occur in adjacent cells.[168] This interclonal cooperation relies on a two-tier process: first the propagation of JNK activation from *scrib* mutant cells to RasV12 cells and then JNK-induced upregulation of the JAK/STAT-activating cytokines. Thereby, JNK-dependent JAK/STAT signaling sustains RasV12 cell growth even when *scrib* mutant cells, which are the original source of JNK activity, have disappeared. It will be of particular interest to determine whether interclonal cooperation also contributes to human cancers.

Finally, components of the mammalian immune system play a dual role in cancer as they can either eradicate cancerous cell or promote tumor development.[180] *Drosophila* innate immune system also displays antitumor and protumor activities.[181,182] In the presence of *scrib* mutant clones, macrophages proliferate and are recruited to the tumor surface to restrain its growth. Macrophages are recruited by the tumoral disruption of the basement membrane, while their proliferation is activated by the initial expression of the JAK/STAT ligand Unpaired in the damaged tissue.[181] Efficient elimination of *scrib* mutant cells seems to rely on the production of the TNF ligand Eiger by the macrophages.[182] However, activation of TNF signaling by the blood cells promotes invasive growth of the tumor cells in the case of oncogenic cooperation between RasV12 and *scrib*.[182] Thus, as in mammals, malignant evolution may rely on the hijacking of the immune response. These results also suggest that *Drosophila*, which is devoid of acquired immunity, is an attractive model to study the relationship between cancer cells and the innate immune system.

VIII. *Drosophila* and Specific Cancer Models

In addition to providing essential clues about the fundamental cellular processes underlying tumorigenesis, *Drosophila* has been used to characterize certain oncogenes. Below, we highlight some of the main findings obtained with these specific cancer models.

A. BCR–ABL

The first human oncogene whose function was investigated in fly is the product of the notorious Philadelphia chromosome, BCR–ABL, which is responsible for almost all cases of chronic myeloid leukemia (CML) and some cases of acute lymphoid leukemia (ALL).[183] *BCR–ABL* is generated by a balanced translocation between *c-Abelson* (*Abl*) on chromosome 9 and the *breakpoint cluster region* (*bcr*) of chromosome 22. The location of the breakpoint within *bcr* is variable and gives rise to slightly different fusion proteins: p210 in most CML and p185 in most ALL. In both cases, the dimerization domain provided by BCR N-terminal moiety induces the constitutive activation of the tyrosine kinase ABL. To gain insights into BCR–ABL isoforms mode of action, Fogert *et al.*[184] generated transgenic flies expressing p210 or p185 human/fly chimeric proteins: BCR and the N-terminal ABL sequences were derived from human, whereas the more divergent C-terminal ABL tail derived from *Drosophila*. Although p185 and p210 rescued the lethality of *dAbl* mutant flies and activated the ABL signaling cascade, their overexpression generated distinct dominant phenotypes, notably in the eye, and also activated signaling pathways not normally employed by ABL.[184,185] Further work using this model may thus help identifying components of the BCR–ABL signaling cascades and the differences underlying the distinct clinical features of p210 and p185-associated leukemia.

B. Mixed Lineage Leukemia (MLL)-Associated Translocations

A similar approach was used to study oncogenic fusions involving *MLL*, the human homolog of *trithorax* (*trx*), which is translocated in 5–10% of patients with ALL or acute myeloid leukemia (AML).[186] *MLL* has been found in 73 different translocations and 54 partner genes have been cloned.[187] Invariably, the resulting fusion protein replaces the PHD fingers and SET domain of MLL with the C-terminal portion of the partner. MLL–AF9, which is associated with AML, and MLL–AF4, which is found in ALL, are among the most frequent translocation products. Although MLL and most MLL-fusion proteins are incorporated into macromolecular nuclear complexes carrying histone-modifying activities, how MLL-fusions promote leukemia is largely unknown. Expressing either MLL–AF4 or MLL–AF9 in flies induced pupal lethality.[186] Interestingly though, these two fusions had different effects on proliferation and chromosome condensation in larval brains and displayed largely nonoverlapping binding patterns on polytene chromosomes, suggesting that they interfere with distinct pathways. Hence, the C-terminal partners of the MLL fusion proteins may have a more important contribution to the chimera's aberrant activities than generally assumed.

C. ErbB-2 and Breast Cancer

Besides comparing the activity of related human oncogenes in an integrated system, *Drosophila* can be used to perform refined structure/function analysis of human oncoproteins *in vivo*. For instance, ErbB-2 is a member of the EGF receptor family frequently overactivated in breast cancer and in non-small cell lung cancer.[188] Cell culture assays showed that autophosphorylation of any four of the five tyrosine (pTyr) present in ErbB-2 C-terminal tail is sufficient for ErbB-2-induced transformation.[189] A study in flies using an activated ErbB-2 carrying a single mutation on each of the five pTyr showed that these proteins induced dominant phenotypes similar to those seen in D-EGFR gain of function alleles. Yet, these phenotypes can be distinctively suppressed by different second site mutations, allowing identification of adaptors and second messengers specifically functioning downstream of particular pTyrs.[190] In addition, a recent study revealed that ErbB-2 is structurally more related to D-EGFR than to other mammalian EGFRs,[191] emphasizing the relevance of *Drosophila* model to study ErbB-2 oncogenic activity.

D. CDH1 and Hereditary Diffuse Gastric Cancer

Hereditary diffuse gastric cancer (HDGC) accounts for 1–3% of all gastric cancer cases. It is an autosomal dominant inherited gastric cancer susceptibility syndrome that is causally linked to mutations in the human E-cadherin gene *CDH1*.[192] In HDGC patients, most mutations affecting *CDH1* lead to the production of nonfunctional E-cadherin, but some are missense mutations affecting different domains of the protein and their impact on E-cadherin function remains more elusive. Expression of two CDH1 missense mutants (A634V or V832M) in the *Drosophila* wing epithelium suggests that they retain some E-Cad function but differentially disturb epithelial cell.[193] Indeed, although both wild-type and mutant CDH1 localized properly and interacted with ß-catenin, the two mutants promoted cell extrusion towards the basal side of the epithelium. In addition, whereas A634V cells moved in connected groups, V832M cells escaped the epithelium as small clusters or as isolated cells. Additional studies in flies may thus reveal the common and specific functions affected by the different *CDH1* missense mutations.

E. SV40 Small T

Viruses are one of the main etiologic agents in human cancers and the study of oncoviruses led to critical discoveries in cancer biology. The simian virus SV40 is an archetypal DNA tumor virus that encodes two oncoproteins: large T, which inhibits Rb and p53, and small T (ST), which interferes with the protein phosphatase PP2A.[194] To gain insights into the cellular processes affected by ST, Kotadia *et al.*[195] expressed ST in *Drosophila* embryos. Their analysis revealed several new functions for ST that were subsequently verified in human cells. Notably, ST caused multiple defects in cytoskeleton organization, centrosome duplication and cell cleavage, all of which might reflect mechanisms by which ST induces transformation. Consistent with this hypothesis, these phenotypes were enhanced by decreasing PP2A activity and required the ST interaction domain with PP2A that is also necessary for ST oncogenic activity. These data extend the utility of *Drosophila* as a model for studying viral oncoproteins.

F. EGFR, PI3K, and Glioma

Malignant gliomas are the most frequent tumors of the central nervous system in adults.[196] These glial cell neoplasms are highly aggressive and respond poorly to therapeutic treatment. A growing body of evidence suggests that enhanced RTK signaling is an initial oncogenic event in gliomagenesis and that these signaling are mediated in large part by the PI3K/Akt/mTOR and Ras/MAPK pathways. The similarities between *Drosophila* and vertebrate glial cell development[197] enticed two groups to develop models for gliomagenesis in fly.

Witte *et al.*[198] analyzed transgenic lines overexpressing activated *Drosophila* EGFR, PI3K and also other tyrosine kinase receptors (PDGFR/VEGFR, InR, and FGFR) in the glial cells of the larval eye disc. These transgenes increased glial cell proliferation and, in addition, both EGFR and PI3K induced ectopic migration of the glial cells along the optic nerve. Hence, these models recapitulate key histological features of human gliomas, including the invasion along nerve tracts of brain structures. Interestingly, the invasive properties of EGFR/PI3K-expressing cells were partly reverted by feeding the larvae with the EGFR inhibitor gefitinib and completely rescued by the PI3K inhibitor wortmannin or the Akt inhibitor triciribine.

Read *et al.*[199] investigated the consequence of activating the D-EGFR and/or the PI3K signaling in the glial cells of the larval brain. By modifying the activity of different components of these two pathways, they showed that EGFR and PI3K pathways synergize to induce glial cell transformation. Upon transplantation, EGFR/PI3K-activated glial cells generated large invasive tumors that seemed to stimulate growth of new trachea in a process analogous to tumor angiogenesis. In addition, genetic analysis pinpointed key factors necessary for EGFR/PI3K-induced glial neoplasia in *Drosophila*, such as Myc, CycE, Cdk4, and Rictor, which may stimulate in concert cell cycle entry and progression, protein translation, cellular growth and migration in human gliogenesis.

Together, these two studies show that *Drosophila* models can be useful to decipher the signaling cascades implicated in gliomagenesis and search for therapeutic compounds.

G. PAX7–FKHR and Rhabdomyosarcoma

Alveolar rhabdomyosarcoma (ARMS) is an aggressive childhood muscle cancer causally linked to two different chromosomal translocations that produce chimeric proteins between the DNA binding domain of either PAX3 or PAX7 and the transcriptional activation domain of FKHR/FOXO1.[200] The PAX–FKHR fusions are believed to act as an oncogene by perturbing skeletal muscle differentiation, which is normally controlled by PAX3 and PAX7. Intriguingly, in a mouse model, PAX3–FKHR produced ARMS when expressed in differentiating myofibers but not in muscle stem cells,[201,202] suggesting that PAX3–FKHR malignant cells may arise from postmitotic, syncytial muscular tissue. Yet, which cell type is at the origin of ARMS remains a matter of controversy.[200] The parallels between fly and vertebrate myogenic programs[203] and the accessibility of *Drosophila* muscle to live imaging led Galindo *et al.*[204] to assess PAX–FKHR activity in *Drosophila* muscles. Strikingly PAX7–FKHR expression in differentiated muscles caused budding off individual cells from the syncytial myofibers and their dissemination to other tissues. In addition, increasing or decreasing Ras activity respectively enhanced or suppressed PAX7–FKHR-associated phenotypes. Thus, PAX–FKHR fusions may promote tumorigenesis by "reversing" or inhibiting muscle cell terminal differentiation by acting on Ras signaling. Interestingly too, PAX7–FKHR expression induced a gene-dosage sensitive larval lethality that could be used in a genetic screen to identify its functional partners.

H. c-Ret and Multiple Endocrine Neoplasia Type 2

Multiple endocrine neoplasia type 2 (MEN2) is a rare autosomal-dominant disorder characterized by the formation of primary tumors in multiple endocrine tissues. MEN2 is caused by activating mutations in *c-Ret*.[205] Different MEN2 clinical subtypes are linked with distinct *c-Ret* mutations: almost all MEN2A patients harbor mutations in the extracellular region, whereas MEN2B patients carry mutations within the tyrosine kinase domain. *Drosophila* and vertebrate Rets have similar expression patterns and Ret^{MEN2} mutant residues are conserved in flies.[206] Expression of $dRet^{MEN2A}$ or $dRet^{MEN2B}$ analogue mutants in the developing eye induces several defects that are related to human MEN2 tumors, such as increased proliferation, compensatory cell death, and abnormal cell differentiation.[207] These eye defects were sensitive to gene dosage and a large-scale genetic interaction screen pinpointed a number of Ret^{MEN2} modifier loci. Yet, all the modifiers interacted with both dRet^{MEN2A} and dRet^{MEN2B}, suggesting that the phenotypic differences between MEN2A and MEN2B diseases may be due to differential alteration of the Ret kinase activity rather than differences in signaling specificity. Interestingly, human orthologs of two modifiers loci, *TNIK* and *CHD3/Mi-2a* showed loss of heterozygosity in MEN2-associated pheochromocytomas and thus may participate in adrenal tissue tumorigenesis. Moreover, feeding *Drosophila* larvae with ZD6474, which blocks the tyrosine kinase activity of RET-derived oncoproteins in cell culture assays, also suppressed the dRet^{MEN2}-associated phenotypes.[208] Hence this *Drosophila* cancer model might also be used to screen for therapeutic compounds.

I. AML1–ETO and AML

Finally, we recently developed a fly model for AML1–ETO, the product of the t(8;21) translocation associated to ±12% of all cases of AML.[209] In AML1–ETO, the DNA binding domain of the RUNX transcription factor AML1 is fused to almost the entire ETO transcriptional corepressor.[210] It has been proposed that AML1–ETO promotes leukemia by interfering with the function of AML1 in myeloid differentiation, but its precise mechanism of action remains to be determined. In flies, the RUNX factor Lozenge (LZ) controls the emergence and the differentiation of one of the two main blood cell lineages.[211,212] Moreover, several features of blood cell development appear conserved between humans and flies, suggesting that *Drosophila* could be a model to investigate the mode of action of AML1–ETO. Expressing AML1–ETO specifically in the LZ blood cell lineage induced a preleukemic-like phenotype characterized by impaired differentiation and increased number of blood cell progenitors.[209] Furthermore, AML1–ETO expression induced lethality at the pupal stage. Using an *in vivo* tissue-specific RNAi screen strategy, we identified the Ca^{2+}-dependant protease CalpainB as required for AML1–ETO-induced phenotypes in *Drosophila*. Importantly, inhibition of human calpains caused AML1–ETO degradation and strongly decreased the clonogenic growth of a $t(8;21)^{+}$ leukemia cell line, indicating that calpain inhibitors might serve in leukemia treatment. This demonstrates that *Drosophila* stands as a genuine model to identify conserved regulators of AML1–ETO and potentially new avenues for AML therapy.

IX. Conclusion

As illustrated above, research on *Drosophila* has provided a wealth of data relevant to human cancers. These results favored the genetic decryption of signaling pathways implicated in cancer and the identification of the basic processes deregulated during cellular transformation, such as control of cell growth, self renewal, differentiation, and mobility. In addition, they have also facilitated the functional analysis of human oncogenes, the identification of new biomarkers and, ultimately, potential therapeutic drugs. There is no doubt that the development of new genetic techniques and additional dedicated cancer models in *Drosophila* will continue to fuel the field of cancer research with groundbreaking concepts. Finally, it is expected that multidisciplinary studies tightly combining *Drosophila* and mammalian models will facilitate translational research and hasten the fight against cancer.

Acknowledgments

We apologize to all researchers whose work could not be cited owing to space limitation and to our biased selection. We are grateful to N. Tapon and A. Weston at the London Research Institute for SEMs pictures. Work in our laboratory was supported by the CNRS and by grants from the Agence Nationale pour la Recherche, Association pour la Recherche sur le Cancer, and Association for International Cancer Research.

References

1. Hanahan D, Weinberg RA. The hallmarks of cancer. *Cell* 2000;**100**:57–70.
2. Croce CM. Oncogenes and cancer. *N Engl J Med* 2008;**358**:502–11.
3. Doronkin S, Reiter LT. *Drosophila* orthologues to human disease genes: an update on progress. *Prog Nucleic Acid Res Mol Biol* 2008;**82**:1–32.
4. Brumby AM, Richardson HE. Using *Drosophila melanogaster* to map human cancer pathways. *Nat Rev Cancer* 2005;**5**:626–39.
5. Budirahardja Y, Gonczy P. Coupling the cell cycle to development. *Development* 2009;**136**:2861–72.
6. Lee LA, Orr-Weaver TL. Regulation of cell cycles in *Drosophila* development: intrinsic and extrinsic cues. *Annu Rev Genet* 2003;**37**:545–78.
7. Garcia K, Duncan T, Su TT. Analysis of the cell division cycle in *Drosophila*. *Methods* 2007;**41**:198–205.
8. Hwang HC, Clurman BE. Cyclin E in normal and neoplastic cell cycles. *Oncogene* 2005;**24**:2776–86.
9. Moberg KH, Bell DW, Wahrer DC, Haber DA, Hariharan IK. Archipelago regulates Cyclin E levels in *Drosophila* and is mutated in human cancer cell lines. *Nature* 2001;**413**:311–6.
10. Strohmaier H, Spruck CH, Kaiser P, Won KA, Sangfelt O, Reed SI. Human F-box protein hCdc4 targets Cyclin E for proteolysis and is mutated in a breast cancer cell line. *Nature* 2001;**413**:316–22.

11. Welcker M, Clurman BE. FBW7 ubiquitin ligase: a tumour suppressor at the crossroads of cell division, growth and differentiation. *Nat Rev Cancer* 2008;**8**:83–93.
12. Bjorklund M, Taipale M, Varjosalo M, Saharinen J, Lahdenpera J, Taipale J. Identification of pathways regulating cell size and cell-cycle progression by RNAi. *Nature* 2006;**439**:1009–13.
13. Bettencourt-Dias M, Giet R, Sinka R, Mazumdar A, Lock WG, Balloux F, et al. Genome-wide survey of protein kinases required for cell cycle progression. *Nature* 2004;**432**:980–7.
14. Lu J, Ruhf ML, Perrimon N, Leder P. A genome-wide RNA interference screen identifies putative chromatin regulators essential for E2F repression. *Proc Natl Acad Sci USA* 2007;**104**:9381–6.
15. Karreth FA, Tuveson DA. Modelling oncogenic Ras/Raf signalling in the mouse. *Curr Opin Genet Dev* 2009;**19**:4–11.
16. Nagaraj R, Banerjee U. The little R cell that could. *Int J Dev Biol* 2004;**48**:755–60.
17. Firth LC, Baker NE. Extracellular signals responsible for spatially regulated proliferation in the differentiating *Drosophila* eye. *Dev Cell* 2005;**8**:541–51.
18. Baonza A, Freeman M. Control of cell proliferation in the *Drosophila* eye by Notch signaling. *Dev Cell* 2005;**8**:529–39.
19. Herranz H, Perez L, Martin FA, Milan M. A Wingless and Notch double-repression mechanism regulates G1–S transition in the *Drosophila* wing. *EMBO J* 2008;**27**:1633–45.
20. Conradt B. Genetic control of programmed cell death during animal development. *Annu Rev Genet* 2009;**43**:493–523.
21. Orme M, Meier P. Inhibitor of apoptosis proteins in *Drosophila*: gatekeepers of death. *Apoptosis* 2009;**14**:950–60.
22. Hay BA, Wassarman DA, Rubin GM. *Drosophila* homologs of baculovirus inhibitor of apoptosis proteins function to block cell death. *Cell* 1995;**83**:1253–62.
23. Hunter AM, LaCasse EC, Korneluk RG. The inhibitors of apoptosis (IAPs) as cancer targets. *Apoptosis* 2007;**12**:1543–68.
24. Ollmann M, Young LM, Di Como CJ, Karim F, Belvin M, Robertson S, et al. *Drosophila* p53 is a structural and functional homolog of the tumor suppressor p53. *Cell* 2000;**101**:91–101.
25. Brodsky MH, Nordstrom W, Tsang G, Kwan E, Rubin GM, Abrams JM. *Drosophila* p53 binds a damage response element at the *reaper* locus. *Cell* 2000;**101**:103–13.
26. Adachi-Yamada T, Fujimura-Kamada K, Nishida Y, Matsumoto K. Distortion of proximodistal information causes JNK-dependent apoptosis in *Drosophila* wing. *Nature* 1999;**400**:166–9.
27. Colombani J, Polesello C, Josue F, Tapon N. Dmp53 activates the Hippo pathway to promote cell death in response to DNA damage. *Curr Biol* 2006;**16**:1453–8.
28. James AA, Bryant PJ. A quantitative study of cell death and mitotic inhibition in gamma-irradiated imaginal wing discs of *Drosophila melanogaster. Radiat Res* 1981;**87**:552–64.
29. Fan Y, Bergmann A. Apoptosis-induced compensatory proliferation. The Cell is dead. Long live the Cell!. *Trends Cell Biol* 2008;**18**:467–73.
30. Huh JR, Guo M, Hay BA. Compensatory proliferation induced by cell death in the *Drosophila* wing disc requires activity of the apical cell death caspase Dronc in a nonapoptotic role. *Curr Biol* 2004;**14**:1262–6.
31. Fan Y, Bergmann A. Distinct mechanisms of apoptosis-induced compensatory proliferation in proliferating and differentiating tissues in the *Drosophila* eye. *Dev Cell* 2008;**14**:399–410.
32. Perez-Garijo A, Martin FA, Morata G. Caspase inhibition during apoptosis causes abnormal signalling and developmental aberrations in *Drosophila*. *Development* 2004;**131**:5591–8.
33. Ryoo HD, Gorenc T, Steller H. Apoptotic cells can induce compensatory cell proliferation through the JNK and the Wingless signaling pathways. *Dev Cell* 2004;**7**:491–501.
34. Perez-Garijo A, Shlevkov E, Morata G. The role of Dpp and Wg in compensatory proliferation and in the formation of hyperplastic overgrowths caused by apoptotic cells in the *Drosophila* wing disc. *Development* 2009;**136**:1169–77.

35. Valentin-Vega YA, Okano H, Lozano G. The intestinal epithelium compensates for p53-mediated cell death and guarantees organismal survival. *Cell Death Differ* 2008;**15**:1772–81.
36. Maeda S, Kamata H, Luo JL, Leffert H, Karin M. IKKbeta couples hepatocyte death to cytokine-driven compensatory proliferation that promotes chemical hepatocarcinogenesis. *Cell* 2005;**121**:977–90.
37. Holland AJ, Cleveland DW. Boveri revisited: chromosomal instability, aneuploidy and tumorigenesis. *Nat Rev Mol Cell Biol* 2009;**10**:478–87.
38. Song YH. *Drosophila melanogaster*: a model for the study of DNA damage checkpoint response. *Mol Cells* 2005;**19**:167–79.
39. Lu Y, Wang Z, Ge L, Chen N, Liu H. The RZZ complex and the spindle assembly checkpoint. *Cell Struct Funct* 2009;**34**:31–45.
40. Brodsky MH, Weinert BT, Tsang G, Rong YS, McGinnis NM, Golic KG, et al. *Drosophila melanogaster* MNK/Chk2 and p53 regulate multiple DNA repair and apoptotic pathways following DNA damage. *Mol Cell Biol* 2004;**24**:1219–31.
41. Sogame N, Kim M, Abrams JM. *Drosophila* p53 preserves genomic stability by regulating cell death. *Proc Natl Acad Sci USA* 2003;**100**:4696–701.
42. Wichmann A, Jaklevic B, Su TT. Ionizing radiation induces caspase-dependent but Chk2- and p53-independent cell death in *Drosophila melanogaster. Proc Natl Acad Sci USA* 2006;**103**:9952–7.
43. Ciapponi L, Cenci G. Telomere capping and cellular checkpoints: clues from fruit flies. *Cytogenet Genome Res* 2008;**122**:365–73.
44. Somma MP, Ceprani F, Bucciarelli E, Naim V, De Arcangelis V, Piergentili R, et al. Identification of *Drosophila* mitotic genes by combining co-expression analysis and RNA interference. *PLoS Genet* 2008;**4**:e1000126.
45. Ravi D, Wiles AM, Bhavani S, Ruan J, Leder P, Bishop AJ. A network of conserved damage survival pathways revealed by a genomic RNAi screen. *PLoS Genet* 2009;**5**:e1000527.
46. Basto R, Brunk K, Vinadogrova T, Peel N, Franz A, Khodjakov A, et al. Centrosome amplification can initiate tumorigenesis in flies. *Cell* 2008;**133**:1032–42.
47. Castellanos E, Dominguez P, Gonzalez C. Centrosome dysfunction in *Drosophila* neural stem cells causes tumors that are not due to genome instability. *Curr Biol* 2008;**18**:1209–14.
48. Kwon M, Godinho SA, Chandhok NS, Ganem NJ, Azioune A, Thery M, et al. Mechanisms to suppress multipolar divisions in cancer cells with extra centrosomes. *Genes Dev* 2008;**22**:2189–203.
49. Zhao B, Li L, Lei Q, Guan KL. The Hippo-YAP pathway in organ size control and tumorigenesis: an updated version. *Genes Dev* 2010;**24**:862–74.
50. Xu T, Wang W, Zhang S, Stewart RA, Yu W. Identifying tumor suppressors in genetic mosaics: the *Drosophila lats* gene encodes a putative protein kinase. *Development* 1995;**121**:1053–63.
51. Justice RW, Zilian O, Woods DF, Noll M, Bryant PJ. The *Drosophila* tumor suppressor gene *warts* encodes a homolog of human *myotonic dystrophy kinase* and is required for the control of cell shape and proliferation. *Genes Dev* 1995;**9**:534–46.
52. Tapon N, Harvey KF, Bell DW, Wahrer DC, Schiripo TA, Haber DA, et al. *salvador* promotes both cell cycle exit and apoptosis in *Drosophila* and is mutated in human cancer cell lines. *Cell* 2002;**110**:467–78.
53. Kango-Singh M, Nolo R, Tao C, Verstreken P, Hiesinger PR, Bellen HJ, et al. Shar-pei mediates cell proliferation arrest during imaginal disc growth in *Drosophila*. *Development* 2002;**129**:5719–30.
54. Harvey KF, Pfleger CM, Hariharan IK. The *Drosophila* Mst ortholog, hippo, restricts growth and cell proliferation and promotes apoptosis. *Cell* 2003;**114**:457–67.

55. Jia J, Zhang W, Wang B, Trinko R, Jiang J. The *Drosophila* Ste20 family kinase dMST functions as a tumor suppressor by restricting cell proliferation and promoting apoptosis. *Genes Dev* 2003;**17**:2514–9.
56. Pantalacci S, Tapon N, Leopold P. The Salvador partner Hippo promotes apoptosis and cell-cycle exit in *Drosophila*. *Nat Cell Biol* 2003;**5**:921–7.
57. Udan RS, Kango-Singh M, Nolo R, Tao C, Halder G. Hippo promotes proliferation arrest and apoptosis in the Salvador/Warts pathway. *Nat Cell Biol* 2003;**5**:914–20.
58. Huang J, Wu S, Barrera J, Matthews K, Pan D. The Hippo signaling pathway coordinately regulates cell proliferation and apoptosis by inactivating Yorkie, the *Drosophila* homolog of YAP. *Cell* 2005;**122**:421–34.
59. Dong J, Feldmann G, Huang J, Wu S, Zhang N, Comerford SA, et al. Elucidation of a universal size-control mechanism in *Drosophila* and mammals. *Cell* 2007;**130**:1120–33.
60. Tao W, Zhang S, Turenchalk GS, Stewart RA, St John MA, Chen W, et al. Human homologue of the *Drosophila melanogaster lats* tumour suppressor modulates CDC2 activity. *Nat Genet* 1999;**21**:177–81.
61. Lai ZC, Wei X, Shimizu T, Ramos E, Rohrbaugh M, Nikolaidis N, et al. Control of cell proliferation and apoptosis by *mob as tumor suppressor, mats*. *Cell* 2005;**120**:675–85.
62. Wu S, Huang J, Dong J, Pan D. *hippo* encodes a Ste-20 family protein kinase that restricts cell proliferation and promotes apoptosis in conjunction with *salvador* and *warts*. *Cell* 2003;**114**:445–56.
63. Guo C, Tommasi S, Liu L, Yee JK, Dammann R, Pfeifer GP. RASSF1A is part of a complex similar to the *Drosophila* Hippo/Salvador/Lats tumor-suppressor network. *Curr Biol* 2007;**17**:700–5.
64. Asthagiri AR, Parry DM, Butman JA, Kim HJ, Tsilou ET, Zhuang Z, et al. Neurofibromatosis type 2. *Lancet* 2009;**373**:1974–86.
65. St John MA, Tao W, Fei X, Fukumoto R, Carcangiu ML, Brownstein DG, et al. Mice deficient of Lats1 develop soft-tissue sarcomas, ovarian tumours and pituitary dysfunction. *Nat Genet* 1999;**21**:182–6.
66. Zhou D, Conrad C, Xia F, Park JS, Payer B, Yin Y, et al. Mst1 and Mst2 maintain hepatocyte quiescence and suppress hepatocellular carcinoma development through inactivation of the Yap1 oncogene. *Cancer Cell* 2009;**16**:425–38.
67. Xu MZ, Yao TJ, Lee NP, Ng IO, Chan YT, Zender L, et al. Yes-associated protein is an independent prognostic marker in hepatocellular carcinoma. *Cancer* 2000;**115**:4576–85.
68. Zhao B, Wei X, Li W, Udan RS, Yang Q, Kim J, et al. Inactivation of YAP oncoprotein by the Hippo pathway is involved in cell contact inhibition and tissue growth control. *Genes Dev* 2007;**21**:2747–61.
69. Morata G, Ripoll P. Minutes: mutants of *Drosophila* autonomously affecting cell division rate. *Dev Biol* 1975;**42**:211–21.
70. Simpson P, Morata G. Differential mitotic rates and patterns of growth in compartments in the *Drosophila* wing. *Dev Biol* 1981;**85**:299–308.
71. Johnston LA. Competitive interactions between cells: death, growth, and geography. *Science* 2009;**324**:1679–82.
72. Tyler DM, Li W, Zhuo N, Pellock B, Baker NE. Genes affecting cell competition in *Drosophila*. *Genetics* 2007;**175**:643–57.
73. de la Cova C, Abril M, Bellosta P, Gallant P, Johnston LA. *Drosophila myc* regulates organ size by inducing cell competition. *Cell* 2004;**117**:107–16.
74. Moreno E, Basler K. dMyc transforms cells into super-competitors. *Cell* 2004;**117**:117–29.
75. Senoo-Matsuda N, Johnston LA. Soluble factors mediate competitive and cooperative interactions between cells expressing different levels of *Drosophila* Myc. *Proc Natl Acad Sci USA* 2007;**104**:18543–8.

76. Li W, Baker NE. Engulfment is required for cell competition. *Cell* 2007;**129**:1215–25.
77. Oliver ER, Saunders TL, Tarle SA, Glaser T. Ribosomal protein L24 defect in *belly spot and tail* (Bst), a mouse Minute. *Development* 2004;**131**:3907–20.
78. Rhiner C, Moreno E. Super competition as a possible mechanism to pioneer precancerous fields. *Carcinogenesis* 2009;**30**:723–8.
79. Teleman AA. Molecular mechanisms of metabolic regulation by insulin in *Drosophila*. *Biochem J* 2010;**425**:13–26.
80. Bohni R, Riesgo-Escovar J, Oldham S, Brogiolo W, Stocker H, Andruss BF, et al. Autonomous control of cell and organ size by CHICO, a *Drosophila* homolog of vertebrate IRS1-4. *Cell* 1999;**97**:865–75.
81. Leevers SJ, Weinkove D, MacDougall LK, Hafen E, Waterfield MD. The *Drosophila* phosphoinositide 3-kinase Dp110 promotes cell growth. *EMBO J* 1996;**15**:6584–94.
82. Brogiolo W, Stocker H, Ikeya T, Rintelen F, Fernandez R, Hafen E. An evolutionarily conserved function of the *Drosophila* insulin receptor and insulin-like peptides in growth control. *Curr Biol* 2001;**11**:213–21.
83. Junger MA, Rintelen F, Stocker H, Wasserman JD, Vegh M, Radimerski T, et al. The *Drosophila* forkhead transcription factor FOXO mediates the reduction in cell number associated with reduced insulin signaling. *J Biol* 2003;**2**:20.
84. Puig O, Marr MT, Ruhf ML, Tjian R. Control of cell number by *Drosophila* FOXO: downstream and feedback regulation of the insulin receptor pathway. *Genes Dev* 2003;**17**:2006–20.
85. Teleman AA, Hietakangas V, Sayadian AC, Cohen SM. Nutritional control of protein biosynthetic capacity by insulin via Myc in *Drosophila*. *Cell Metab* 2008;**7**:21–32.
86. Goberdhan DC, Paricio N, Goodman EC, Mlodzik M, Wilson C. *Drosophila* tumor suppressor PTEN controls cell size and number by antagonizing the Chico/PI3-kinase signaling pathway. *Genes Dev* 1999;**13**:3244–58.
87. Huang H, Potter CJ, Tao W, Li DM, Brogiolo W, Hafen E, et al. PTEN affects cell size, cell proliferation and apoptosis during *Drosophila* eye development. *Development* 1999;**126**:5365–72.
88. Ito N, Rubin GM. *gigas*, a *Drosophila* homolog of *tuberous sclerosis gene product-2*, regulates the cell cycle. *Cell* 1999;**96**:529–39.
89. Gao X, Pan D. TSC1 and TSC2 tumor suppressors antagonize insulin signaling in cell growth. *Genes Dev* 2001;**15**:1383–92.
90. Potter CJ, Huang H, Xu T. *Drosophila* Tsc1 functions with Tsc2 to antagonize insulin signaling in regulating cell growth, cell proliferation, and organ size. *Cell* 2001;**105**:357–68.
91. Tapon N, Ito N, Dickson BJ, Treisman JE, Hariharan IK. The *Drosophila* tuberous sclerosis complex gene homologs restrict cell growth and cell proliferation. *Cell* 2001;**105**:345–55.
92. Saucedo LJ, Gao X, Chiarelli DA, Li L, Pan D, Edgar BA. Rheb promotes cell growth as a component of the insulin/TOR signalling network. *Nat Cell Biol* 2003;**5**:566–71.
93. Zhang Y, Gao X, Saucedo LJ, Ru B, Edgar BA, Pan D. Rheb is a direct target of the tuberous sclerosis tumour suppressor proteins. *Nat Cell Biol* 2003;**5**:578–81.
94. Stocker H, Radimerski T, Schindelholz B, Wittwer F, Belawat P, Daram P, et al. Rheb is an essential regulator of S6K in controlling cell growth in *Drosophila*. *Nat Cell Biol* 2003;**5**:559–65.
95. Miron M, Lasko P, Sonenberg N. Signaling from Akt to FRAP/TOR targets both 4E-BP and S6K in *Drosophila melanogaster*. *Mol Cell Biol* 2003;**23**:9117–26.
96. Grewal SS, Evans JR, Edgar BA. *Drosophila* TIF-IA is required for ribosome synthesis and cell growth and is regulated by the TOR pathway. *J Cell Biol* 2007;**179**:1105–13.
97. Yuan TL, Cantley LC. PI3K pathway alterations in cancer: variations on a theme. *Oncogene* 2008;**27**:5497–510.

98. Trosko JE. Review paper: cancer stem cells and cancer nonstem cells: from adult stem cells or from reprogramming of differentiated somatic cells. *Vet Pathol* 2009;**46**:176–93.
99. Bossuyt W, De Geest N, Aerts S, Leenaerts I, Marynen P, Hassan BA. The atonal proneural transcription factor links differentiation and tumor formation in *Drosophila*. *PLoS Biol* 2009;**7**:e40.
100. Bossuyt W, Kazanjian A, De Geest N, Van Kelst S, De Hertogh G, Geboes K, et al. *Atonal homolog 1* is a tumor suppressor gene. *PLoS Biol* 2009;**7**:e39.
101. Nicolay BN, Bayarmagnai B, Moon NS, Benevolenskaya EV, Frolov MV. Combined inactivation of pRB and hippo pathways induces dedifferentiation in the *Drosophila* retina. *PLoS Genet* 2010;**6**:e1000918.
102. Buttitta LA, Katzaroff AJ, Perez CL, de la Cruz A, Edgar BA. A double-assurance mechanism controls cell cycle exit upon terminal differentiation in *Drosophila*. *Dev Cell* 2007;**12**:631–43.
103. Wang JC. Good cells gone bad: the cellular origins of cancer. *Trends Mol Med* 2009;**16**:145–51.
104. Gateff E. Malignant neoplasms of genetic origin in *Drosophila melanogaster*. *Science* 1978;**200**:1448–59.
105. Neumuller RA, Knoblich JA. Dividing cellular asymmetry: asymmetric cell division and its implications for stem cells and cancer. *Genes Dev* 2009;**23**:2675–99.
106. Caussinus E, Gonzalez C. Induction of tumor growth by altered stem-cell asymmetric division in *Drosophila melanogaster*. *Nat Genet* 2005;**37**:1125–9.
107. Choksi SP, Southall TD, Bossing T, Edoff K, de Wit E, Fischer BE, et al. Prospero acts as a binary switch between self-renewal and differentiation in *Drosophila* neural stem cells. *Dev Cell* 2006;**11**:775–89.
108. Bowman SK, Rolland V, Betschinger J, Kinsey KA, Emery G, Knoblich JA. The tumor suppressors Brat and Numb regulate transit-amplifying neuroblast lineages in *Drosophila*. *Dev Cell* 2008;**14**:535–46.
109. Pearson J, Lopez-Onieva L, Rojas-Rios P, Gonzalez-Reyes A. Recent advances in *Drosophila* stem cell biology. *Int J Dev Biol* 2009;**53**:1329–39.
110. Krzemien J, Dubois L, Makki R, Meister M, Vincent A, Crozatier M. Control of blood cell homeostasis in *Drosophila* larvae by the posterior signalling centre. *Nature* 2007;**446**:325–8.
111. Harrison DA, Binari R, Nahreini TS, Gilman M, Perrimon N. Activation of a *Drosophila* Janus kinase (JAK) causes hematopoietic neoplasia and developmental defects. *EMBO J* 1995;**14**:2857–65.
112. Luo H, Hanratty WP, Dearolf CR. An amino acid substitution in the *Drosophila* hopTum-l Jak kinase causes leukemia-like hematopoietic defects. *EMBO J* 1995;**14**:1412–20.
113. Vainchenker W, Dusa A, Constantinescu SN. JAKs in pathology: role of Janus kinases in hematopoietic malignancies and immunodeficiencies. *Semin Cell Dev Biol* 2008;**19**:385–93.
114. Huang L, Muthuswamy SK. Polarity protein alterations in carcinoma: a focus on emerging roles for polarity regulators. *Curr Opin Genet Dev* 2010;**20**:41–50.
115. Bilder D. Epithelial polarity and proliferation control: links from the *Drosophila* neoplastic tumor suppressors. *Genes Dev* 2004;**18**:1909–25.
116. St Johnston D, Ahringer J. Cell polarity in eggs and epithelia: parallels and diversity. *Cell* 2010;**141**:757–74.
117. Vaccari T, Bilder D. At the crossroads of polarity, proliferation and apoptosis: the use of *Drosophila* to unravel the multifaceted role of endocytosis in tumor suppression. *Mol Oncol* 2009;**3**:354–65.
118. Mosesson Y, Mills GB, Yarden Y. Derailed endocytosis: an emerging feature of cancer. *Nat Rev Cancer* 2008;**8**:835–50.
119. Christiansen JJ, Rajasekaran AK. Reassessing epithelial to mesenchymal transition as a prerequisite for carcinoma invasion and metastasis. *Cancer Res* 2006;**66**:8319–26.

120. Brumby AM, Richardson HE. *scribble* mutants cooperate with oncogenic Ras or Notch to cause neoplastic overgrowth in *Drosophila*. *EMBO J* 2003;**22**:5769–79.
121. Bilder D, Li M, Perrimon N. Cooperative regulation of cell polarity and growth by *Drosophila* tumor suppressors. *Science* 2000;**289**:113–6.
122. Abdelilah-Seyfried S, Cox DN, Jan YN. Bazooka is a permissive factor for the invasive behavior of discs large tumor cells in *Drosophila* ovarian follicular epithelia. *Development* 2003;**130**:1927–35.
123. Ling C, Zheng Y, Yin F, Yu J, Huang J, Hong Y, et al. The apical transmembrane protein Crumbs functions as a tumor suppressor that regulates Hippo signaling by binding to Expanded. *Proc Natl Acad Sci USA* 2010;**107**:10532–7.
124. Grzeschik NA, Parsons LM, Allott ML, Harvey KF, Richardson HE. Lgl, aPKC, and Crumbs regulate the Salvador/Warts/Hippo pathway through two distinct mechanisms. *Curr Biol* 2010;**20**:573–81.
125. Robinson BS, Huang J, Hong Y, Moberg KH. Crumbs regulates Salvador/Warts/Hippo signaling in *Drosophila* via the FERM-domain protein expanded. *Curr Biol* 2010;**20**:582–90.
126. Molloy T, Van 't Veer LJ. Recent advances in metastasis research. *Curr Opin Genet Dev* 2008;**18**:35–41.
127. Starz-Gaiano M, Montell DJ. Genes that drive invasion and migration in *Drosophila*. *Curr Opin Genet Dev* 2004;**14**:86–91.
128. Friedl P, Gilmour D. Collective cell migration in morphogenesis, regeneration and cancer. *Nat Rev Mol Cell Biol* 2009;**10**:445–57.
129. Jang AC, Starz-Gaiano M, Montell DJ. Modeling migration and metastasis in *Drosophila*. *J Mammary Gland Biol Neoplasia* 2007;**12**:103–14.
130. Silver DL, Montell DJ. Paracrine signaling through the JAK/STAT pathway activates invasive behavior of ovarian epithelial cells in *Drosophila*. *Cell* 2001;**107**:831–41.
131. Borghese L, Fletcher G, Mathieu J, Atzberger A, Eades WC, Cagan RL, et al. Systematic analysis of the transcriptional switch inducing migration of border cells. *Dev Cell* 2006;**10**:497–508.
132. Wang X, Bo J, Bridges T, Dugan KD, Pan TC, Chodosh LA, et al. Analysis of cell migration using whole-genome expression profiling of migratory cells in the *Drosophila* ovary. *Dev Cell* 2006;**10**:483–95.
133. Wei D, Le X, Zheng L, Wang L, Frey JA, Gao AC, et al. Stat3 activation regulates the expression of vascular endothelial growth factor and human pancreatic cancer angiogenesis and metastasis. *Oncogene* 2003;**22**:319–29.
134. Silver DL, Naora H, Liu J, Cheng W, Montell DJ. Activated signal transducer and activator of transcription (STAT) 3: localization in focal adhesions and function in ovarian cancer cell motility. *Cancer Res* 2004;**64**:3550–8.
135. Abdulghani J, Gu L, Dagvadorj A, Lutz J, Leiby B, Bonuccelli G, et al. Stat3 promotes metastatic progression of prostate cancer. *Am J Pathol* 2008;**172**:1717–28.
136. Ranger JJ, Levy DE, Shahalizadeh S, Hallett M, Muller WJ. Identification of a Stat3-dependent transcription regulatory network involved in metastatic progression. *Cancer Res* 2009;**69**:6823–30.
137. McDonald JA, Khodyakova A, Aranjuez G, Dudley C, Montell DJ. PAR-1 kinase regulates epithelial detachment and directional protrusion of migrating border cells. *Curr Biol* 2008;**18**:1659–67.
138. Niewiadomska P, Godt D, Tepass U. DE-Cadherin is required for intercellular motility during *Drosophila* oogenesis. *J Cell Biol* 1999;**144**:533–47.
139. Schober M, Rebay I, Perrimon N. Function of the ETS transcription factor Yan in border cell migration. *Development* 2005;**132**:3493–504.

140. Pinheiro EM, Montell DJ. Requirement for Par-6 and Bazooka in *Drosophila* border cell migration. *Development* 2004;**131**:5243–51.
141. Etienne-Manneville S. Polarity proteins in migration and invasion. *Oncogene* 2008;**27**:6970–80.
142. Bai J, Uehara Y, Montell DJ. Regulation of invasive cell behavior by taiman, a *Drosophila* protein related to AIB1, a steroid receptor coactivator amplified in breast cancer. *Cell* 2000;**103**:1047–58.
143. Geisbrecht ER, Montell DJ. Myosin VI is required for E-cadherin-mediated border cell migration. *Nat Cell Biol* 2002;**4**:616–20.
144. Dunn TA, Chen S, Faith DA, Hicks JL, Platz EA, Chen Y, et al. A novel role of myosin VI in human prostate cancer. *Am J Pathol* 2006;**169**:1843–54.
145. Yoshida H, Cheng W, Hung J, Montell D, Geisbrecht E, Rosen D, et al. Lessons from border cell migration in the *Drosophila* ovary: A role for myosin VI in dissemination of human ovarian cancer. *Proc Natl Acad Sci USA* 2004;**101**:8144–9.
146. Bianco A, Poukkula M, Cliffe A, Mathieu J, Luque CM, Fulga TA, et al. Two distinct modes of guidance signalling during collective migration of border cells. *Nature* 2007;**448**:362–5.
147. Duchek P, Somogyi K, Jekely G, Beccari S, Rorth P. Guidance of cell migration by the *Drosophila* PDGF/VEGF receptor. *Cell* 2001;**107**:17–26.
148. Duchek P, Rorth P. Guidance of cell migration by EGF receptor signaling during *Drosophila* oogenesis. *Science* 2001;**291**:131–3.
149. Jekely G, Sung HH, Luque CM, Rorth P. Regulators of endocytosis maintain localized receptor tyrosine kinase signaling in guided migration. *Dev Cell* 2005;**9**:197–207.
150. Nallamothu G, Woolworth JA, Dammai V, Hsu T. *Awd*, the homolog of metastasis suppressor gene *Nm23*, regulates *Drosophila* epithelial cell invasion. *Mol Cell Biol* 2008;**28**:1964–73.
151. Quinones GA, Jin J, Oro AE. I-BAR protein antagonism of endocytosis mediates directional sensing during guided cell migration. *J Cell Biol* 2010;**189**:353–67.
152. Geisbrecht ER, Montell DJ. A role for *Drosophila* IAP1-mediated caspase inhibition in Rac-dependent cell migration. *Cell* 2004;**118**:111–25.
153. Oshima K, Takeda M, Kuranaga E, Ueda R, Aigaki T, Miura M, et al. IKK epsilon regulates F actin assembly and interacts with *Drosophila* IAP1 in cellular morphogenesis. *Curr Biol* 2006;**16**:1531–7.
154. Mehrotra S, Languino LR, Raskett CM, Mercurio AM, Dohi T, Altieri DC. IAP regulation of metastasis. *Cancer Cell* 2010;**17**:53–64.
155. Ruan K, Song G, Ouyang G. Role of hypoxia in the hallmarks of human cancer. *J Cell Biochem* 2009;**107**:1053–62.
156. Doronkin S, Djagaeva I, Nagle ME, Reiter LT, Seagroves TN. Dose-dependent modulation of HIF-1alpha/sima controls the rate of cell migration and invasion in *Drosophila* ovary border cells. *Oncogene* 2010;**29**:1123–34.
157. Woodhouse E, Hersperger E, Stetler-Stevenson WG, Liotta LA, Shearn A. Increased type IV collagenase in lgl-induced invasive tumors of *Drosophila*. *Cell Growth Differ* 1994;**5**:151–9.
158. Woodhouse E, Hersperger E, Shearn A. Growth, metastasis, and invasiveness of *Drosophila* tumors caused by mutations in specific tumor suppressor genes. *Dev Genes Evol* 1998;**207**:542–50.
159. Beaucher M, Hersperger E, Page-McCaw A, Shearn A. Metastatic ability of *Drosophila* tumors depends on MMP activity. *Dev Biol* 2007;**303**:625–34.
160. Martinez AM, Schuettengruber B, Sakr S, Janic A, Gonzalez C, Cavalli G. Polyhomeotic has a tumor suppressor activity mediated by repression of Notch signaling. *Nat Genet* 2009;**41**:1076–82.

161. Woodhouse EC, Fisher A, Bandle RW, Bryant-Greenwood B, Charboneau L, Petricoin 3rd EF, et al. *Drosophila* screening model for metastasis: Semaphorin 5c is required for l(2)gl cancer phenotype. *Proc Natl Acad Sci USA* 2003;**100**:11463–8.
162. Lee T, Luo L. Mosaic analysis with a repressible cell marker (MARCM) for *Drosophila* neural development. *Trends Neurosci* 2001;**24**:251–4.
163. Pagliarini RA, Xu T. A genetic screen in *Drosophila* for metastatic behavior. *Science* 2003;**302**:1227–31.
164. Dow LE, Elsum IA, King CL, Kinross KM, Richardson HE, Humbert PO. Loss of human Scribble cooperates with H-Ras to promote cell invasion through deregulation of MAPK signalling. *Oncogene* 2008;**27**:5988–6001.
165. Uhlirova M, Jasper H, Bohmann D. Non-cell-autonomous induction of tissue overgrowth by JNK/Ras cooperation in a *Drosophila* tumor model. *Proc Natl Acad Sci USA* 2005;**102**:13123–8.
166. Igaki T, Pagliarini RA, Xu T. Loss of cell polarity drives tumor growth and invasion through JNK activation in *Drosophila*. *Curr Biol* 2006;**16**:1139–46.
167. Srivastava A, Pastor-Pareja JC, Igaki T, Pagliarini R, Xu T. Basement membrane remodeling is essential for *Drosophila* disc eversion and tumor invasion. *Proc Natl Acad Sci USA* 2007;**104**:2721–6.
168. Wu M, Pastor-Pareja JC, Xu T. Interaction between Ras(V12) and scribbled clones induces tumour growth and invasion. *Nature* 2010;**463**:545–8.
169. Ke H, Harris R, Coloff JL, Jin JY, Leshin B, Miliani de Marval P, et al. The c-Jun NH2-terminal kinase 2 plays a dominant role in human epidermal neoplasia. *Cancer Res* 2010;**70**:3080–8.
170. Kim LC, Song L, Haura EB. Src kinases as therapeutic targets for cancer. *Nat Rev Clin Oncol* 2009;**6**:587–95.
171. Vidal M, Warner S, Read R, Cagan RL. Differing Src signaling levels have distinct outcomes in *Drosophila*. *Cancer Res* 2007;**67**:10278–85.
172. Chi C, Zhu H, Han M, Zhuang Y, Wu X, Xu T. Disruption of lysosome function promotes tumor growth and metastasis in Drosophila. *J Biol Chem* 2010; **285**:21817–23.
173. Witz IP. Tumor-microenvironment interactions: dangerous liaisons. *Adv Cancer Res* 2008;**100**:203–29.
174. Vidal M, Larson DE, Cagan RL. Csk-deficient boundary cells are eliminated from normal *Drosophila* epithelia by exclusion, migration, and apoptosis. *Dev Cell* 2006;**10**:33–44.
175. Froldi F, Ziosi M, Garoia F, Pession A, Grzeschik NA, Bellosta P, et al. The lethal giant larvae tumour suppressor mutation requires dMyc oncoprotein to promote clonal malignancy. *BMC Biol* 2010;**8**:33.
176. Vidal M, Salavaggione L, Ylagan L, Wilkins M, Watson M, Weilbaecher K, Cagan R. A role for the epithelial microenvironment at tumor boundaries. Evidence from drosophila and human squamous cell carcinomas. *Am J Pathol* 2010;**176**:3007–14.
177. Hogan C, Dupre-Crochet S, Norman M, Kajita M, Zimmermann C, Pelling AE, et al. Characterization of the interface between normal and transformed epithelial cells. *Nat Cell Biol* 2009;**11**:460–7.
178. Kajita M, Hogan C, Harris AR, Dupre-Crochet S, Itasaki N, Kawakami K, et al. Interaction with surrounding normal epithelial cells influences signalling pathways and behaviour of Src-transformed cells. *J Cell Sci* 2010;**123**:171–80.
179. Yan H, Chin ML, Horvath EA, Kane EA, Pfleger CM. Impairment of ubiquitylation by mutation in *Drosophila* E1 promotes both cell-autonomous and non-cell-autonomous Ras-ERK activation in vivo. *J Cell Sci* 2009;**122**:1461–70.
180. de Visser KE, Eichten A, Coussens LM. Paradoxical roles of the immune system during cancer development. *Nat Rev Cancer* 2006;**6**:24–37.

181. Pastor-Pareja JC, Wu M, Xu T. An innate immune response of blood cells to tumors and tissue damage in Drosophila. *Dis Model Mech* 2008;**1**:144–54 [Discussion: 153].
182. Cordero JB, Macagno JP, Stefanatos RK, Strathdee KE, Cagan RL, Vidal M. Oncogenic Ras diverts a host TNF tumor suppressor activity into tumor promoter. *Dev Cell* 2010;**18**:999–1011.
183. Quintas-Cardama A, Cortes J. Molecular biology of bcr–abl1-positive chronic myeloid leukemia. *Blood* 2009;**113**:1619–30.
184. Fogerty FJ, Juang JL, Petersen J, Clark MJ, Hoffmann FM, Mosher DF. Dominant effects of the *bcr–abl* oncogene on *Drosophila* morphogenesis. *Oncogene* 1999;**18**:219–32.
185. Stevens TL, Rogers EM, Koontz LM, Fox DT, Homem CC, Nowotarski SH, et al. Using Bcr–Abl to examine mechanisms by which abl kinase regulates morphogenesis in *Drosophila*. *Mol Biol Cell* 2008;**19**:378–93.
186. Muyrers-Chen I, Rozovskaia T, Lee N, Kersey JH, Nakamura T, Canaani E, et al. Expression of leukemic MLL fusion proteins in *Drosophila* affects cell cycle control and chromosome morphology. *Oncogene* 2004;**23**:8639–48.
187. Slany RK. The molecular biology of mixed lineage leukemia. *Haematologica* 2009;**94**:984–93.
188. Hynes NE, MacDonald G. ErbB receptors and signaling pathways in cancer. *Curr Opin Cell Biol* 2009;**21**:177–84.
189. Dankort DL, Wang Z, Blackmore V, Moran MF, Muller WJ. Distinct tyrosine autophosphorylation sites negatively and positively modulate neu-mediated transformation. *Mol Cell Biol* 1997;**17**:5410–25.
190. Settle M, Gordon MD, Nadella M, Dankort D, Muller W, Jacobs JR. Genetic identification of effectors downstream of Neu (ErbB-2) autophosphorylation sites in a *Drosophila* model. *Oncogene* 2003;**22**:1916–26.
191. Alvarado D, Klein DE, Lemmon MA. ErbB2 resembles an autoinhibited invertebrate epidermal growth factor receptor. *Nature* 2009;**461**:287–91.
192. Humar B, Guilford P. Hereditary diffuse gastric cancer: a manifestation of lost cell polarity. *Cancer Sci* 2009;**100**:1151–7.
193. Pereira PS, Teixeira A, Pinho S, Ferreira P, Fernandes J, Oliveira C, et al. E-cadherin missense mutations, associated with hereditary diffuse gastric cancer (HDGC) syndrome, display distinct invasive behaviors and genetic interactions with the Wnt and Notch pathways in *Drosophila* epithelia. *Hum Mol Genet* 2006;**15**:1704–12.
194. Cheng J, DeCaprio JA, Fluck MM, Schaffhausen BS. Cellular transformation by Simian Virus 40 and Murine Polyoma Virus T antigens. *Semin Cancer Biol* 2009;**19**:218–28.
195. Kotadia S, Kao LR, Comerford SA, Jones RT, Hammer RE, Megraw TL. PP2A-dependent disruption of centrosome replication and cytoskeleton organization in *Drosophila* by SV40 small tumor antigen. *Oncogene* 2008;**27**:6334–46.
196. Huse JT, Holland EC. Targeting brain cancer: advances in the molecular pathology of malignant glioma and medulloblastoma. *Nat Rev Cancer* 2010;**10**:319–31.
197. Freeman MR, Doherty J. Glial cell biology in *Drosophila* and vertebrates. *Trends Neurosci* 2006;**29**:82–90.
198. Witte HT, Jeibmann A, Klambt C, Paulus W. Modeling glioma growth and invasion in *Drosophila melanogaster*. *Neoplasia* 2009;**11**:882–8.
199. Read RD, Cavenee WK, Furnari FB, Thomas JB. A *Drosophila* model for EGFR-Ras and PI3K-dependent human glioma. *PLoS Genet* 2009;**5**:e1000374.
200. Charytonowicz E, Cordon-Cardo C, Matushansky I, Ziman M. Alveolar rhabdomyosarcoma: is the cell of origin a mesenchymal stem cell? *Cancer Lett* 2009;**279**:126–36.
201. Keller C, Hansen MS, Coffin CM, Capecchi MR. Pax3:Fkhr interferes with embryonic Pax3 and Pax7 function: implications for alveolar rhabdomyosarcoma cell of origin. *Genes Dev* 2004;**18**:2608–13.

202. Keller C, Arenkiel BR, Coffin CM, El-Bardeesy N, DePinho RA, Capecchi MR. Alveolar rhabdomyosarcomas in conditional Pax3:Fkhr mice: cooperativity of Ink4a/ARF and Trp53 loss of function. *Genes Dev* 2004;**18**:2614–26.
203. Maqbool T, Jagla K. Genetic control of muscle development: learning from *Drosophila*. *J Muscle Res Cell Motil* 2007;**28**:397–407.
204. Galindo RL, Allport JA, Olson EN. A *Drosophila* model of the rhabdomyosarcoma initiator PAX7–FKHR. *Proc Natl Acad Sci USA* 2006;**103**:13439–44.
205. Falchetti A, Marini F, Luzi E, Tonelli F, Brandi ML. Multiple endocrine neoplasms. *Best Pract Res Clin Rheumatol* 2008;**22**:149–63.
206. Hahn M, Bishop J. Expression pattern of *Drosophila ret* suggests a common ancestral origin between the metamorphosis precursors in insect endoderm and the vertebrate enteric neurons. *Proc Natl Acad Sci USA* 2001;**98**:1053–8.
207. Read RD, Goodfellow PJ, Mardis ER, Novak N, Armstrong JR, Cagan RL. A *Drosophila* model of multiple endocrine neoplasia type 2. *Genetics* 2005;**171**:1057–81.
208. Vidal M, Wells S, Ryan A, Cagan R. ZD6474 suppresses oncogenic RET isoforms in a *Drosophila* model for type 2 multiple endocrine neoplasia syndromes and papillary thyroid carcinoma. *Cancer Res* 2005;**65**:3538–41.
209. Osman D, Gobert V, Ponthan F, Heidenreich O, Haenlin M, Waltzer L. A *Drosophila* model identifies calpains as modulators of the human leukemogenic fusion protein AML1–ETO. *Proc Natl Acad Sci USA* 2009;**106**:12043–8.
210. Niebuhr B, Fischer M, Tager M, Cammenga J, Stocking C. Gatekeeper function of the RUNX1 transcription factor in acute leukemia. *Blood Cells Mol Dis* 2008;**40**:211–8.
211. Ferjoux G, Auge B, Boyer K, Haenlin M, Waltzer L. A GATA/RUNX cis-regulatory module couples *Drosophila* blood cell commitment and differentiation into crystal cells. *Dev Biol* 2007;**305**:726–34.
212. Waltzer L, Ferjoux G, Bataille L, Haenlin M. Cooperation between the GATA and RUNX factors Serpent and Lozenge during *Drosophila* hematopoiesis. *EMBO J* 2003;**22**:6516–25.

Probing Human Cardiovascular Congenital Disease Using Transgenic Mouse Models

Paige Snider and
Simon J. Conway

Developmental Biology and Neonatal Medicine Program, HB Wells Center for Pediatric Research, Indiana University School of Medicine, Indianapolis, Indiana, USA

Congenital heart defects (CHDs) impact *in utero* embryonic viability, children, and surviving adults. Since the first transfer of genes into mice, transgenic mouse models have enabled researchers to experimentally study and genetically test the roles of genes in development, physiology, and disease progression. Transgenic mice have become a bona fide human CHD pathology model and their use has dramatically increased within the past two decades. Now that the entire mouse and human genomes are known, it is possible to knock out, mutate, misexpress, and/or replace every gene. Not only have transgenic mouse models changed our understanding of normal development, CHD processes, and the complex interactions of genes and pathways required during heart development, but they are also being used to identify new avenues for medical therapy.

Progress in Molecular Biology
and Translational Science, Vol. 100
DOI: 10.1016/B978-0-12-384878-9.00003-0

Copyright 2011, Elsevier Inc.
All rights reserved.
1877-1173/11 $35.00

I. Introduction

Congenital heart defects (CHDs) are the most common of all human birth defects and are the leading cause of neonatal and infant morbidity and mortality,[1,2] yet their underlying molecular, biochemical, cellular, and physiological causes as well as etiologies remain poorly understood. It has been estimated that at least 60% of human embryos die *in utero*, the majority due to early postfertilization and/or implantation anomalies but approximately 15% undergo gastrulation and die during pregnancy or prenatally.[3] Examination of the myriad of mouse knockouts, gene traps, transgenic alleles, and chemically induced and spontaneous mouse mutants reveals that defects within the heart and vascular system are largely responsible for the observed *in utero* lethality of the embryo and early fetus.[4] If a transgenically altered mouse embryo survives implantation but fails to be born, it usually indicates that there is some form of lethal cardiovascular defect present. Significantly, the heart is the first organ to develop in the mammalian embryo and remarkably it functions even before it is even fully formed. Most CHDs are the result of anatomical and/or functional malformations that result from errors during embryogenesis, which implies that they are due to alterations in genes involved in cardiac development that underlie or predispose the embryo toward congenital disease. Although an endeavor in itself, merely identifying the genetic mutations resulting in CHDs does not help the patients; however, it may help in risk stratification.[5] Designing successful therapies often requires a thorough understanding of the underlying pathological mechanisms and potential target pathways within *in vivo* animal models.[6] This review summarizes the types and use of mouse model systems currently being used to understand the embryology, pathogenesis, classification, and molecular mechanisms involved in cardiac development in order to clarify the genetic basis of CHDs. The hope is that if these complex CHDs can be prevented and/or nullified in genetically defined mouse mutant models, then the knowledge gained will help engineer potential treatments for pediatric patients.

II. Congenital Heart Defects

The term CHDs can describe a number of different problems affecting the heart, but usually refers to problems with the heart's structure which are present at birth and can often lead to inadequate cardiac performance (reviewed by Hoffman and Kaplan[1] and Hoffman[7]). Although just as problematic but less appreciated, congenital functional defects[8–11] can also result in deleterious changes to the developing heart. CHDs are the most common type of birth defect and are responsible for more deaths in the first year of life than any other birth defects.[1,12,13] Despite advances in detection and treatment, congenital heart disease accounts for 3% of all infant deaths and 46% of those deaths result from congenital malformations.[14] All infants are at risk for having a congenital defect, regardless of age, race, or socioeconomic status.[15] The Centers for Disease Control reports cyanotic (blue discoloration caused by a relative lack of oxygen) heart defects occurred in 56.9 per 100,000 live births in the United States in 2005, with higher rates noted when maternal age exceeded 40 years.[14] The incidence of CHD in premature infants is 1.25%,[1] excluding isolated patent ductus arteriosus and atrial septal defect (ASD). It has been estimated that 10% of the children born with heart defects will require surgery in infancy or childhood.[13] CHDs can be severe and complex with life-threatening symptoms and require immediate medical care after birth, or they can be simple defects that have few if any symptoms and can be fixed easily or require no treatment. The occurrence of less severe CHDs, such as small ventricular septal defects (VSDs), within the newborn population should not be trivialized. The impact of stress on the family and the cumulative impact on healthcare expenditures of cardiovascular defects that generally have a favorable outcome can be considerable. Studies on the incidence of congenital heart disease have generally been from collection and reporting of recognized cases rather than careful population screening with currently available diagnostic technologies. Significantly, the incidence of CHDs has remained relatively constant over time and between populations, suggesting a greater contribution from genetic factors and a smaller etiological environmental contribution.[12] A genetic basis for etiology of CHD was founded by a combination of twin studies, demonstration of familial aggregation, and animal homologies.[16]

CHDs may occur alone or together as part of a wide ranging spectrum of other congenital defects. The majority of CHDs occur as isolated (or nonsyndromic) CHDs.[12] However, they can also be found in association with various genetic and chromosomal syndromes such as Down syndrome, Trisomy 13, Turner syndrome, Marfan syndrome, Noonan syndrome, and 22q11 DiGeorge syndrome.[12,17] Microdeletions of chromosomal regions are a significant cause of syndromal CHDs,[12] and extracardiac manifestations occur in approximately 30% of children with CHDs.[17] This is not surprising considering multiple organs and individual cell lineages have common growth and differentiation signals.[17,18] This often requires multiple surgeries to correct the various defects, resulting in both detrimental postoperative sequelae and compromised quality of life. Moreover, infants with either isolated and/or syndromic CHDs can develop developmental and psychological problems, even without having undergone surgery, indicating a significant secondary effect of CHDs *in utero*.[19] Although much still needs to be learned of genotype–phenotype correlations, it is already clear that CHDs represent a problem that ultimately affects a great number of families. Gratifyingly, the survival of patients with CHDs has significantly improved during the past several decades, particularly in neonates but unfortunately less so in children.[20] This achievement is jointly attributable to a growth in basic knowledge, individual therapeutic successes, advances in clinical management, and the commitment of patients and families. However, adult CHDs represent an increasing proportion of heart transplant recipients, and compared with adult recipients, patients with adult CHD experience higher post-heart transplantation mortality and retransplantation.[21] Moreover, immunosuppression differs among patients with both grown-up patients with congenital heart disease (GUCH) and adult recipients.[20,22] Further research directed at the cause and ultimately the prevention of structural birth defects of the heart is an essential component of any agenda to reduce premature death and morbidity due to cardiovascular disease.

III. Genetic Basis of Disease

The genetic basis of disease can be summarized into three major categories: single gene, chromosomal, and multifactorial causes.[12,16,23] First, single gene mutations conform to standard Mendelian inheritance patterns and can be assessed as autosomal or X-linked dominant mutations, as well as recessive mutations.[12,16,23] For example, single gene mutations have been demonstrated to result in both isolated CHDs (i.e., *NKX2.5*, *PROSIT240*, *TFAP2B*) and syndromic CHDs (i.e., *JAGGED1*, *NKX2.5*, *TBX5*, *ZIC3*) in patients.[24] One of the first single gene mutations demonstrated to give rise to an inherited CHD was in the T-box transcription factor gene family.[25,26] Holt–Oram syndrome was shown to be caused by mutations in *TBX5*, in both Holt–Oram syndrome families and sporadic Holt–Oram syndrome cases.[26,27] Like Holt–Oram patients, adult *Tbx5* heterozygous mutant mice were found to have ASD CHDs and conduction system abnormalities including atrioventricular conduction delay.[28] CHDs that fail to follow clear Mendelian inheritance are often assumed to result from sporadic mutations and/or multiple genetic insults, and identification of these CHD-causing genetic determinants is extremely difficult.[29]

The second category of genetic disorders is chromosomal aberrations.[12,16,23] When chromosomal regions are identified, they tend to be large with multiple genes localized to that region and this has made fine mapping and gene identification challenging.[29] Until the recent and dramatic growth in mouse genetics and increased understanding of molecular biology of cardiac development, chromosomal causes of CHDs were only investigated if syndromic anomalies or other organ systems were involved.[30] Since the development of sophisticated karyotyping techniques and advancements in localization, sequencing, and detection of loci critical for cardiac development, many associations between CHDs and chromosomal aberrations have started to be delineated.[13,30] For example, molecular advances in the 22q11 microdeletion syndrome (also known as DiGeorge syndrome) have led to greater understanding of the basic knowledge as well as advances in clinical management of young patients.[30,31] The 22q11 microdeletion syndrome is the most common human deletion syndrome, and the spectrum of cardiac defects was initially outlined in the mid-1970s.[32] Initially, the entire proximal region of mouse chromosome 16 (analogous to human chromosome 22) was deleted, and these heterozygous deletion mice mutants exhibited CHDs similar to interrupted aortic arch (IAA) CHDs.[33] Subsequently, transgenic complementation and gene targeting studies confirmed that the loss of *Tbx1* was the gene responsible for the structural abnormalities observed.[34–36] Significantly, the *Tbx1* mutant mice have gross anatomical defects that are thought to reflect a severe 22q11 microdeletion syndrome phenotype.[37] More recently, a hypomorphic *Tbx1* mouse mutant has been used to demonstrate that Tbx1 is required to activate expression of two key downstream target genes, namely the *Forkhead transcription factor (Foxa2)* and *Fgf8* growth factor.[38] In addition to the congenital aspects, the 22q11 microdeletion syndrome is also associated with late-onset features and an association with psychological problems, such as schizophrenia.[19,39] Initially, it was impossible to determine whether these developmental difficulties were due to genetic causes or postoperative complications, but further understanding of the molecular biology, mutation spectrum, and clinical subclassifications, in conjunction with availability of complementary animal models, is helping to distinguish primary and secondary genotype–phenotype correlations and underlying pathology.

The third category is multifactorial whereby genes interact with environmental influences to result in disease.[12,16,23] Estimations suggest that approximately 8% of CHDs are due to genetic factors, approximately 2% are due to environmental factors, and the rest are multifactorial.[40] Although epidemiological data have highlighted environmental influences,[41] these studies have mostly suggested risk rather than pinpointing underlying CHD mechanisms.[18] The joint and/or synergistic effects associated with the interplay between genes and the environment are termed "gene–environment interactions" and do not exhibit any recognizable Mendelian pattern of inheritance and are not associated with any identifiable chromosomal abnormalities.[15,23] The description "multifactorial" is used as it refers to the additive effects of several genes that can be modified to a greater or lesser extent by environmental factors, and that an individual who is genetically susceptible, may or may not develop a CHD depending on the interaction of risk factors, both genetic and environmental.[12,15] Examples of multifactorial causes of CHDs have largely come from mice mutant studies, specifically the findings that TBX5 and NKX2.5 proteins interact physically to activate target genes[28] and that defective interactions between GATA4/TBX5 and GATA4/NKX2.5 may underlie CHDs caused by *GATA4* mutations.[42] The recent discovery that many developmental events have to be additionally synchronized via posttranscriptional microRNA (miRNA) regulation, specifically *miR-1*, *miR-138*, and *miR-133*, and that each participates in multiple aspects of cardiac development and homeostasis,[43,44] suggests that a mutation within a single miRNAs may aberrantly affect multiple signaling pathways. These genetic studies have provided two important insights. First, that developmentally important transcription factors are involved in the etiology of CHD. Second, that the dosage of key cardiac developmental factors must be precisely regulated, as most disease causing mutations result in haploinsufficiency and some are the result of activating mutations.

Routine testing for the many genes that have been shown to be involved in causing CHDs (see Table I) is presently not a viable cost-effective patient option except on a research basis, although there is significant interest in transitioning this testing from the laboratory to clinic.[24] However, a recent pilot "personalized genome sequencing" report,[45] in which the complete genome of an individual was determined by massive parallel DNA sequencing, suggests that identifying the key genome sequences that may be associated with disease is predictive of the response to medication and would enable individual genotype–phenotype correlations may be possible in the near future.

IV. The Mouse as a Genetic Model

The development of animal models to study cardiac embryology has been driven by a need to understand both normal and abnormal heart formation during *in utero* development.[4,29,46] Transgenic technology is invaluable as it can be used to answer physiological questions that can only be answered within the intact organism.[47] Significantly, both mice and humans have evolutionary similarity,[48] share 90% identity in their amino acid sequence of homologous proteins, and their genomes are of similar size (~20,000–25,000 genes; The International Human Genome Sequencing Consortium, 2004; www.genome.gov). Moreover, mouse and human genes are similarly ordered along the chromosomes.[48] Importantly, mice have been extensively studied, and their genetics, embryology, and physiology are very well understood.[5,48,49] Additionally, mice are decidedly fertile, easy to feed and house, and are highly amenable to assisted reproductive techniques including cryopreservation which can preserve the sperm from genetically manipulated mouse lines.[50] Since the sequencing of the entire mouse genome, it is now feasible to remove and/or mutate every gene.[51,52] Furthermore, embryonic stem (ES) technology now allows for the generation of hypomorphic (partial loss of gene function), spatiotemporal

TABLE I

THREE SUBTYPES OF THE MOST COMMON CHDs ARE LISTED: SEPTATION, CYANOTIC, AND LEFT-SIDED OBSTRUCTION DEFECTS

Congenital heart defect	Category	Mouse mutants
Atrial septal defect (ASD)	Septation defect	*Activin type II6 receptor, Bmp6/7 dKO, Cited2, Eve, Gata4, MyHC, Nkx2.5, Nt3, Tbx5, Zic3*
Atrioventricular septal defect (AVSD)	Septation defect	*Alk2 cKO, Cryptic, Cx40/45, Fog2, Gata4, Has2, Lefty-1, Nfatc, Nkx2.5, Pitx2c, RxRa, Trisomy 16*
Aortic coarctation	Left-sided obstruction	*Foxc1/Foxc2 double hets, Vangl2*
Aortic stenosis (AS)	Left-sided obstruction	*Elastin, Hox1.5*
Bicuspid aortic valve (BV)	Other	*Foxc1/Foxc2 double hets*
Double outlet right ventricle (DORV)	Cyanotic heart defect	*Bmp2/4 compound KOs, Cited2, Cx40, Disheveled compound KOs, Gata3, Gdf1, Hand1-Hd2 replacement K1, Hey2, Jumonji, Lefty-1, Meis1, Nf1, Nfatc, Pbx2/3, Pitx2c, PlexinA2, Rb/p107 dcKO, Sema3c, SOX-4, Tolloid-like 1, Vangl2*
Ebstein's anomaly	Cyanotic heart defect	*Alk3 cKO*
Hypoplastic left heart (HLH)	Left-sided obstruction	*Ets1*
Interrupted aortic arch (IAA)	Left-sided obstruction	*Bmp receptor2, Chrd, Eta, Foxc2, Foxc1/Foxc2 double hets, Gata3, Ltbp, Pax3, Sema3c, TGFB receptor type II cKO, Tbx1, Vegf 120/188 Kls, Zic3*
Mitral stenosis	Left-sided obstruction	*Foxc1/Foxc2 double hets*
Patent ductus arteriosus (PDA)	Other	*Hand1–Hand2 replacement K1, Tfap28*
Pulmonary atersia	Cyanotic heart defect	*Jag1, Ptpn11*
Persistent truncus arteriosus (PTA)	Cyanotic heart defect	*Alk2 cKO, Alk5 cKO, Bmp receptor IA cKO, Chrd, Cited2, Fak cKO, Fox2c/Eta dKOs, Gata6 cKO, Gata 3, Hox1.5, Ltbp1, NCadherin cKO, Neuropilin1, Nt3, Pax3, Pbx1, Pdgf receptora, Pdgf receptor6, Ptpn11 cKO, Raldh2, Rac1 cKO, RXRa, Smad 4 cKO, Smad7 overexpressor, TGF6 receptor type II cKO, Tbx1, Tbx20 hypomorph, Tfap2, RARa1, RAR6, Trisomy 16, Wnt5a*
Single ventricle	Cyanotic heart defect	*ENU mutant (chr. 7), Tbx20*
Tetralogy of Fallot (TOF)	Cyanotic heart defect	*Cx40, Fog2, Gdf1, Hey2, Jag1, Pbx2/3, Meis1, Tbx1, Vegf 120/188 Kls*
Transposition of the great arteries (TGA)	Cyanotic heart defect	*Activin type II6 receptor, Cryptic, Gdf1, Lefty1, Neuropilin1, Perlecan, Pitx2c, Zic3*
Tricuspid atresia	Cyanotic heart defect	*Bmp6/7 dKO, Hey2, Lefty1, Pax3-Fkhr Kl*
Ventricular septal defect	Septation defect	*Alk2 cKO, Alk5 cKO, Bmp6/7 dKO, Chrd, Cited2, Cx40/45, eNos, Et1, Eta, Ets1, Fak cKO, Fgf receptor 2-IIIb, Foxc1/Foxc2 double hets, Foxc2, Fog2, Gata3, Gata4, Gdf1, Hey2, Jumonji, Ltbp1, Meis1, MyHc, NCadherin cKO, Neturopilin1, Nf1, Nfatc, Nkx2.5, Nt3, Pax3, Pax3-Fkhr K1, Pbx1, Pbx2/3, Pdgf receptora, Pdgf6, Pdgf, Receptor6, Raldh2, RARa1, RAR6, RxRa, Smad 4 cKO, Smad7 overexpressor, Sox4, Tbx5, Tbx20 hypomorph, Tfap2, TGF6 receptor type I lcKO, Tolloid-like 1, Vegf 120/188 kls, Vangl2, wnt5a, zic3*

The types of CHD that occur within each subtype and a partial listing of the mouse mutants that exhibit them are indicated. Note that the majority of mouse models are conventional systemic knockouts, unless otherwise indicated.

lineage-restricted gene replacement, and/or overexpression in normal or ectopic locations in mice. If a human sequence is in hand, the homologous mouse gene can be mutated by gene targeting and/or transgenically forced to be expressed, and the consequent physiological outcome determined.[48] Moreover, known human mutations can be introduced in mice to examine their *in utero* effects and assess genotype–phenotype correlations.[53,54] Additionally, precise regions of the *in utero* developing cardiovascular system can be genetically ablated.[55] In addition, mice have the advantage over flies, worms, and fish in that they have the mammalian four-chambered heart, making the data more applicable to human CHDs.[5] Knowledge of a mutation that predisposes or causes CHDs alone cannot translate directly into treatment, but an understanding of pathological processes will be crucial to the development of eventual treatments.[5] Animal models are one of the few approaches that we have to help with our understanding of the multifactorial basis for CHDs, including genetic predisposition to disease and environmentally acquired factors such as viral infection.[5]

Human diseases are mostly polygenic with the clinical phenotype being a result of interplay between multiple genes and environmental influences.[56] Animal models of cardiovascular physiology have mostly been large animals which have limitation due to genetic variance and difficulties of molecular–genetic manipulation.[56] However, mice are the most genetically manipulative vertebrate model system and are available on approximately 500 strains of inbred (http://www.informatics.jax.org/external/festing/search_form.cgi) and numerous mixed genetic backgrounds. In fact, strain-dependent phenotype variability can be used to examine the multifactorial basis of congenital defects. For example, *TGFβ1*-deficient mice placed on various inbred genetic backgrounds can exhibit phenotypes ranging from preimplantation to weaning-age lethality.[49,57] This strain-dependent variability facilitates the potential mapping of modifier genes.[58] Thus, researchers now have to tools to modify and control the protein content of the developing mammalian heart, resulting in mouse models that can potentially provide the critical links between mutated or absent protein and CHD pathogenesis at molecular, biochemical, and cellular levels.[59]

V. Transgenic Mice

A transgenic line is one in which artificially introduced genetic material is microinjected *in vitro* into the large male pronucleus fertilized mouse eggs.[60–62] Following random chromosomal integration of the transgene, injected eggs are subsequently implanted in the oviduct of pseudopregnant surrogate mothers and their progeny is screened for transgene integration. The microinjected DNA tends to integrate as multiple tandemly arranged copies at a random position in the genome, often after one or two cell divisions have occurred. Thus, the resulting mouse is only partially transgenic. If the transgenic cells contribute to the germ line, then some transgenic eggs or sperm will be produced and the next generation of mice will be fully transgenic. These are referred to as "stable transgenic" lines as the transgene has stably integrated in both somatic and germline cells so that the foreign construct is disseminated across future generations.[62] However, the use of transient generation of transgenic mice has been employed.[63,64] Here, no breeding is required because females with potential transgenic pups *in utero* are sacrificed during gestation and expression and/or effects of the transgene directly assessed. In order to express a recombinant protein in the transgenic animal, the DNA that is microinjected must contain the region that encodes a coding sequence of interest. This is placed downstream of a selected cell-specific promoter/enhancer (transcriptional regulatory region) which directs gene expression to a specific tissue, cell type, or developmental stage and upstream of a polyadenylation signal sequence.[47,64] Where possible, it is best to select promoter sequences with proven tissue-specific expression in transgenic mice, as this will increase likelihood of experimental success.

Transgenesis enables several types of experiments to be performed.[47]

> *Overexpression*: wherein a protein in a cell that normally expresses that protein is overexpressed, in order to experimentally address the consequences of producing an abnormally high level of that protein in a particular cell or tissue. Transgenic expression cassettes that use genomic DNA rather than cDNA are generally more efficiently expressed in mice. It is usually a good idea to include (where possible) generic introns when cDNA is used and levels of expression are important.
>
> *Ectopic expression*: wherein the ramifications of ectopic expression in which a protein is expressed in cell type that does not normally express it are addressed.
>
> *Gain-of-function*: wherein a mutation can be studied by designing a transgene that encodes a mutated protein that is constitutively active.
>
> *Loss-of-function*: wherein a dominant negative form of a specific protein is studied.
>
> *Reporter expression*: a reporter gene does not influence the normal biological activity of the cells in which it is expressed, but rather is used to explore the pattern of the promoter as well as to identify any *cis*-acting elements required for function of the transcriptional regulatory sequence attached to the reporter gene.[47,56]

The most commonly used reporter genes are *lacZ* and *green fluorescent protein* (*GFP*). *LacZ* is an enzyme which catalyzes the hydrolysis of X-gal (5-bromo-4-chloro-indolyl-β-D-galactosidase) and forms a bright blue precipitate within a single cell. In order to stain for *lacZ*, a tissue must be fixed and a chromogenic substrate is added. Although not preset in the heart, mammals have endogenous β-galactosidase activity which can result in background staining. By cloning a nuclear localization signal into the *β-galactosidase* gene, the nuclear transgene staining can be easily differentiated from the endogenous cytosolic signal.[56] Another popular reporter is GFP which can be used to analyze living (rather than fixed) cells. GFP is a small protein cloned from jelly fish that functions as a fusion protein and upon long-wave UV stimulation, emits green light.[56]

Like all molecular biological techniques, in addition to their power and flexibility, there are also several caveats. Once the injected DNA has been integrated into the murine genome, it can theoretically exhibit its function. However, it must be pointed out that since the insertion occurs randomly, positional/insertional effects must be considered as both the function of the inserted gene and endogenous genes may be altered.[64] Variegation effects must also be considered, since the expression of the transgene may be altered by surrounding elements. It has been estimated that 5–10% of transgene integration events cause disruptions, deletions, or translocations that cause a mutant phenotype.[65] In order to bypass the variegation position effects and study transcriptional regulatory elements, transient transgenic approaches can be employed,[63,64] as a comparison of multiple founders will ensure that any phenotypes due to positional effects of the transgene are readily apparent.[64] An additional disadvantage of this technology is the inability to control the number of transgene copies inserted into the host genome.[59] Therefore, many of the mouse models express the transgene at very high levels and raises questions of nonphysiological consequences due to atypical levels of gene expression.[5] Most times, the transgene is transmitted in a Mendelian fashion. If you have a mosaic mouse, transmission may be much sparser than Mendelian, and you may have to breed several litters to find your transgene. Once you have a positive offspring, breeding from that one should be Mendelian. At other times (rarely), you may have more than one integration site, in which case your transgene will be transmitted in greater than Mendelian numbers. Very rarely, the transgene is not in the germ line and is not transmitted. Additionally, the endogenous gene and protein product are present and may make interpretation of the data perplexing and inconclusive.[59]

Since the first transgenic animals were mice created by Rudolf Jaenisch[66] and Gordon *et al.*,[60] who demonstrated that a microinjected a recombinant plasmid in pronuclei of fertilized mouse oocytes could be passed onto their offspring to generate large numbers of transformed animals, there has been an explosion in the use of transgenic mice models in biomedical research and industry. Recombinant DNA technology of transgenesis has become vital in studying mammalian embryonic development, patterns of gene expression, physiological gene function, as well as the problems of gene regulation and cell differentiation in the complex mammalian system.[47,60,62] Other uses include the production of human hormones (e.g., insulin) and for testing in biomedical research. Furthermore, transgenic mice are routinely employed to mimic human diseases by cloning human genes that are known to underlie various human disorders, thus making them an invaluable research tool.[62] Additional transgenic species are also beginning to be generated, including sheep, goats, pigs, some chickens, and most recently a stable line of breeding transgenic marmosets. Marmosets are primates and thus our closest relatives (so far) to be genetically engineered.[67]

VI. Gene Knockout and Gene Knockin Mice

A knockout mouse is a genetically engineered mouse in which one or more genes have been turned off through a targeted mutation. Targeting is achieved via either "homologous recombination" or "gene trapping" within ES cells. ES cells are pluripotent and have the capacity to differentiate into nearly any type of adult cell and generate a mouse.[68,69] The targeting vector is electroporated into the ES cells *in vitro* and various drug resistance cassettes are used to select for the targeted lines that have undergone targeting.[59] ES cell colonies are then screened for the presence of the targeting gene, then once confirmed, the genetically modified ES cells are injected into a blastocyst harvested from a black mouse and implanted into a pseudopregnant agouti mouse.[59,69] The "chimeric" offspring will have patches of both the black and agouti coat color. Male mice with the highest percentages of chimerism are then selected based on coat color, and subsequently out crossed with wild-type black females to confirm germline transmission.[59,69] If the animals are viable, the mice can be bred to homozygosity and analysis of the heterozygotes and homozygotes can assess the effects of gene dosage. Knockout mice usually follow Mendelian inheritance strictly, and thus it is relatively straightforward to generate double, triple and even quadruple compound heterozygous and/or homozygous mutants to examine genetic interactions and redundancy amongst family members and coexpressed genes.

For homologous recombination (also called gene targeting or knockouts), a targeting vector is generated that shares identical, or homologous, sequence to the gene being targeted.[48,64,70] This homologous sequence flanks the existing gene's DNA sequence both upstream and downstream of the gene's location on the chromosome, and then the cells' own nuclear machinery automatically recognizes the identical stretches of sequence and swaps out the existing gene or portion of a gene with the artificial piece of DNA. Gene targeting constructs also contain a positive selection cassette conferring antibiotic resistance (typically neomycin) and a core region with the appurtenant genetic modification flanked by homology arms.[48,64] As homologous recombination is rare compared to random DNA integrations, strategies are used to enrich for recombinant ES cell clones.[64] These include use of negative selection markers such as *Diphtheria toxin α-chain* coding region or *herpes simplex virus type-I Thymidine* gene located outside of the targeting construct.[71,72] Thus, ES cells that have correctly undergone homologous recombination will not contain a negative selection marker, as nonhomologous flanking sequences will be removed prior to integration; however, randomly integrated constructs will retain the negative selection marker and be eliminated.[64] Given the usefulness of knockout mice lines, the National Institutes of Health has established The Knockout Mouse Project (KOMP) that aims to generate a comprehensive and public resource comprising mice containing a null mutation in every gene in the mouse genome (http://www.nih.gov/science/models/mouse/knockout/index.html). For gene trapping, instead of directly targeting a gene of interest via homologous recombination, random insertion in ES cells is used. Gene trap vectors principal element is a gene trapping cassette consisting of a promoterless reporter gene, and/or selectable marker flanked by an upstream 3′ splice site (splice acceptor, SA) and a downstream transcriptional termination/polyadenylation sequence.[73,74] The inserted piece of artificial DNA prevents the cell's RNA "splicing" machinery from working properly, thus preventing the existing gene from producing its designated protein and knocking out its function. As the vector simultaneously inactivates and reports the expression of the trapped gene at the insertion site and provides a gene-trap sequence tag (GTST), it is possible to quickly identify the location of the disrupted gene. Thus, using this approach, it is possible to track the activity of the artificial reporter gene to figure out the existing gene's normal pattern of activity in mouse tissues.[74] An international public consortium provides a central registry of the all lines available (http://www.genetrap.org/).

One advantage gene targeting has over transgenic techniques is that ES cell homologous recombination clearly defines the integration site allowing for precise genetic changes to be introduced.[64,75,76] "Knockout" mouse models are also able to provide useful information on gene function as well as being useful to study human diseases caused by either total or partial protein loss-of-function mutations.[69] This is particularly evident, when either reporters (such as *lacZ* or GFP) are inserted into specific sites and used to monitor gene expression under the control of the gene's own *cis* and *trans* transcriptional machinery. Moreover, "knockin" mice, wherein desired sequences are targeted to specific loci, not only allow the study of gain-of-function mutations through targeted mutagenesis but also allow the replacement of one gene by another functional and/or mutated gene, enabling genetic testing of the subtle variations between genes by replacing one gene with another.[69] Knockin mice can be used to model human CHDs that are a result of gain-of-function mutations. Several of the caveats with knockout models include early embryonic lethality, which prevents studying the role of a gene at later stages, and if the target gene has many functions in multiple cell types obscuring its cell autonomous role.[48] However, both of these problems can be readily overcome using conditional lineage-specific gene targeting (see section VII below). Additional problems can be encountered if there is duplication of genes during evolution, which can lead to functional redundancy. In this case, a deficient animal may overexpress an alternative gene product in order to compensate for the deficiency, thus severely limiting phenotypic characterizations.[48] However, this too can be circumvented via generation of double-knockout mice and removal of the compensating, parallel pathway and/or synergistic gene targets. Finally, the presence of a positive selection marker in the targeted gene can affect the gene of interest and result in "hypomorphic phenotype."[77] However, hypomorphic alleles can be very useful in the study of somatic-based diseases that generate a phenotype as a consequence of gradual gene function deterioration.[48] Another advantage of generating a hypomorphic allele is the ability to make compound heterozygotes that carry one hypomorphic allele and one null allele following the proper mouse mating and often have more severe phenotypes than hypomorphic alleles alone.[48] Despite these drawbacks, knockout mice offer one of the most powerful means now available for studying gene function in a living animal. Such studies will accelerate efforts to translate newfound knowledge of the human and mouse genomes into better strategies for diagnosing, treating, and preventing human CHDs.

VII. Conditional Knockout Mice

Conventional gene targeting generates a mutant allele in all cells of the mouse following fertilization. However, difficulties can be encountered, such as embryonic lethality, and analysis can be complicated due to indirect effects from ablating the gene in all tissues. This can be overcome by using a two-component mouse system employing recombinase to inactivate the gene of interest.[78] Cre and Flp are members of the α-integrase family of site-specific recombinases.[78,79] Application of this technology results in deletion of genetic material at a specific time and in specific cells, depending upon the promoter used to drive Cre or Flp recombinase.[78] Cre and Flp catalyse DNA strand exchange between recombination sites and depending on the number and orientation of the recombination sites, deletions, duplications, inversions, or integrations can be generated.[78] Cre and Flp remove segments of intervening DNA between two unique 34 base pair recognition sites termed loxP and FRT sites, respectively. They each comprise two 13 base pair palindromes separated by an asymmetric 8 base pair core.[47,78,80] In order to study CHDs, a mouse is prepared in which the targeted allele is flanked by loxP/FRT sites, and then this mouse is bred to a mouse expressing Cre or Flp which are expressed under the control of heart- or lineage-specific tissue-restricted promoter/enhancers.[78] The recombinase then excises the DNA segment between the recombination loxP/FRT sites, and the gene is rendered inactive.[47] Conditional knockouts are advantageous because both the timing at which the cells are affected and its spatial restriction are under the control of the investigator, thereby circumventing the embryonic lethality and/or multiple sites of expression concerns sometimes encountered with conventional knockouts.[47] In conditional technology, it is vital to make sure the recombinase recognition sites are placed around an essential part of the gene of interest, without compromising its function.[64] Several methods have been developed to help with this. Often the entire coding region will be flanked by loxP or FRT sites, or the loxP/FRT sites can be positioned within an intronic sequence located around an essential exon. Additionally, the promoter region of the gene can be targeted.[64] Further development of the targeted Cre-*loxP* technology now provides researchers with the ability to genetically ablate specific cardiovascular cell lineages at various different stages of heart development. Biological functions can now be analyzed *in vivo* and *in utero* via targeting the toxic gene *Diphtheria toxin A subunit* (*DTA*) to specific cell lineages and forcing the intended cells to undergo apoptotic cell death, without any bystander effects.[55,81,82] Given the enormous strides made using cell ablation within the chick embryo model,[83,84] it is more than likely that the combined power of mouse transgenesis and toxigenics will also provide a boon to the modeling of mouse mutants for studying CHDs. To date, the Cre-*loxP* system is the best characterized means of achieving conditional gene inactivation in mice and has contributed extensively as a tool for altering the mammalian genome.

Studies on CHD have been immeasurably enhanced with the development of cardiac-specific promoters that can now be used to direct expression of an engineered protein within only the heart at different developmental stages.[5] Cardiac-specific promoters in transgenic animals have given researchers a powerful tool to eliminate complicated effects of other organs or muscle systems.[5] Transgenes influencing structural aspects of cardiac function, physiology, metabolic processes, calcium cycling, ion channels, transport, mechanical stimulation, and electrical properties have been expressed with the use of these cardiac-specific promoters paving the way for murine models to mimic human CHD.[5] Ultimately, the development of the Cre-*loxP* recombinase system has revolutionized mouse genetics by proving to be an essential tool for conditional genetic alteration in mice.

VIII. Inducible Mice

In a clinical setting, physicians are often interested in the reversibility of cardiac phenotypes, including whether the cardiac pathology is controllable or reversible.[5] A more precise control of transgene induction is needed to address these questions, and pharmacological methods have proven the answer.[5] Tetracycline (tet) or its analog doxycycline (dox) has been most widely used to express a transgene at a precise developmental stage, allowing a phenotype to develop and then shut off expression to see if the heart returns to normalcy.[59] This tet-inducible system requires two different constructs. One encodes a tet-repressor herpes virus Vp16 termed rTA which is expressed either from a generic promoter that is nontissue-specific or from a developmentally regulated tissue-specific promoter.[85] The second construct (tetO) consists of a minimum promoter with no basal activity linked to the gene of interest.[85] The tet-V16 protein has a very high binding coefficient for the tetO.[85,86] There are two varieties of tet-regulatory sequences, one in which a gene is repressed by tet binding (rTA or "Tet-Off") and the other where the gene is activated by tet binding (rtTA or "Tet-On"). Repression of a gene with this system occurs because in the presence of tet or its analog, tet binds rTA and the confirmation changes so that it can no longer bind to the tetO sequence leading to downregulation of expression. A mutation in the tet activator (rtTA) promoter causes the mutant protein not to bind the tetO sequence in the target gene unless tet is introduced into the system.[85] Gene induction can be detected within a few hours and maximal expression occurs within a day.[85] The constructs can be microinjected separately, and animal with each construct can be mated, or the constructs can be coinjected. The tet-Off (rTA) and tet-On (rtTA) systems have been successfully used in mice to control gene expression, both in adults and *in utero* within the developing heart.[86,87]

IX. Forward Genetic Approaches

A "reverse genetics" strategy occurs when a gene is altered first and then the phenotype resulting from this alteration is subsequently studied, is the principle behind knockin and knockouts. However, it has been recognized that this strategy is somewhat limited by the fact that it examines genes already thought to be involved in the physiological process under study.[50] Thus, a "forward genetics" strategy which employs an unbiased approach, that more closely mimics the work of clinicians who tend to identify a disease first, and then try to isolate the genes responsible has been developed.[50,88] This approach uses *N*-ethyl-*N*-nitrosurea (ENU) as a potent means of inducing random point mutations across the genome.[50,89] ENU is an alkylating agent that causes base substitutions in DNA, creating allelic series of mutations that has a mutation rate of 1 in 1000 gametes in mice.[90] Mutations generated in mouse ES cells which retain germline competency can be used to study partial loss, complete loss, and/or gain-of-function alleles.[90,91] Although ENU is an effective random chemical mutagen capable of generating novel phenotype, the ENU-induced single-base-pair changes responsible for a given mutant phenotype are often difficult to detect and require laborious techniques such as positional cloning.[92] Upon intronic insertion, gene-trap transposons are capable of reducing wild-type transcript levels by mutating disrupted endogenous genes and producing novel phenotypes, making them useful as a random germline insertional mutagen in mice.[92] Transposons catalyze the insertion of foreign DNA into host chromosomes, which is the most important step in gene transfer applications.[93] This forward genetics approach has the added benefit of generating countless new phenotypes, several which exhibit unique CHDs including the single ventricle phenotype that has been underrepresented using reverse genetic approaches (Table I; Ref. 94).

X. Common CHDs

CHDs can be divided into three major subtypes (Table I): cyanotic (blue discoloration caused by a relative lack of oxygen), left-sided obstruction defects, and septation defects.[1,18,95] Tetralogy of Fallot (TOF), transposition of the great vessels, tricuspid atresia, total anomalous pulmonary venous return, truncus arteriosus, pulmonary atresia, Ebstein's anomaly, and some forms of total anomalous pulmonary venous return are the most common cyanotic CHDs. Left-sided obstructive lesions include hypoplastic left heart (HLH), mitral stenosis, aortic stenosis, aortic coarctation, and IAA. Septation defects, which constitute the most prevalent CHD subtype, include ASDs, VSDs, and atrioventricular septal defects. Other relatively common CHDs that lie outside these subtypes include patent ductus arteriosus (PDA) where a neonate's ductus arteriosus fails to close after birth and bicuspid aortic valve where only two valve leaflets instead of the normal three are formed in the aortic valve. Each of these CHDs may occur alone or together and the severity (and morbidity) of these CHDs varies greatly. The following sections highlight one of each of the subtypes' most common defects and some of the insights provided via mouse transgenics.

A. Ventricular Septal Defect

VSDs are the most common congenital cardiac defects in human infants, and there are both muscular and perimembranous VSDs. Clinically, isolated VSDs can lead to the failure of survival, but the effects of isolated VSD disease are quite variable and are dependent upon anatomical location and size. The larger the communication between the left and right ventricles, the greater the hemodynamic alteration and increased load on the heart, the worse the effects. The genetics of ventricular septation are poorly understood, and in human populations, no single gene disorder primarily causing isolated VSD exists.[87,96] Isolated VSD can occur in patients heterozygous for mutations in NKX2.5[97] or TBX5[25]; however, this phenotype is less common. More frequently, VSDs are seen in combination with other subtypes of CHDs, particularly outflow tract alignment and septation defects.

The morphogenesis of atrioventricular septation is complex,[98] as multiple primordia contribute to the formation of the interatrial and interventricular septa, including endocardial cushion tissue (derived from both colonizing extracardiac neural crest and intrinsic epithelial-to-mesenchymally derived lineages), endothelium, and myocardium. The muscular portion of the interventricular septum arises from ingrowth and folding of the myocardial wall of the developing ventricle, whereas the membranous portion of the interventricular septum is derived primarily from endocardial cushion tissue. Despite being the foremost CHD found in patients, isolated VSD in mice mutants is usually found in association with altered myocardial growth, OFT/AAA abnormalities, and/or valvular defects.[4,96] VSD pathogenesis is complex and multifactorial and may include underlying defects within the cardiomyocyte, cardiac neural crest, endothelial, endocardial cushion, epicardial, and/or cardiac fibroblast lineages that can each alter left–right ventricular morphogenesis, chamber formation, and septal positioning. Although a large number of transgenic models exhibit VSDs, very few have been shown to affect the ventricular septum exclusively, but many, such as *Bmp6/7* double knockouts, *Cited2*, *Fog2*, *Nkx2.5*, *Pax3*, *Sema3c*, *Tbx5*, and *Tll1*, have been shown to affect ventricular septation as part of a larger constellation of congenital cardiac anomalies (Table I). Initially, the *Hey2* knockout phenotype was reported as an isolated VSD,[96] but subsequent analysis has revealed that *Hey2* mutants exhibit a variable penetrance and spectrum of CHDs, including TOF, deformed valves, and cardiomyopathy.[99,100] These variable phenotypes probably reflect the fact that Hey2 is expressed in multiple cardiovascular lineages throughout development, including robust expression within the myocardium.[100] Despite the suggestion in several reviews and textbooks that increased apoptosis within the developing interventricular septum may underlie VSD pathogenesis, there is little experimental evident to support this implication. Using *Hey2* knockout mice as the model, both apoptosis and cell proliferation can readily be examined in knockout and age-matched littermate controls on the identical genetic background. While *Hey2* mutants exhibit both DORV and VSDs that are structurally similar to those observed in patients with CHDs (Fig. 1), apoptosis was unchanged within the developing heart (examined from

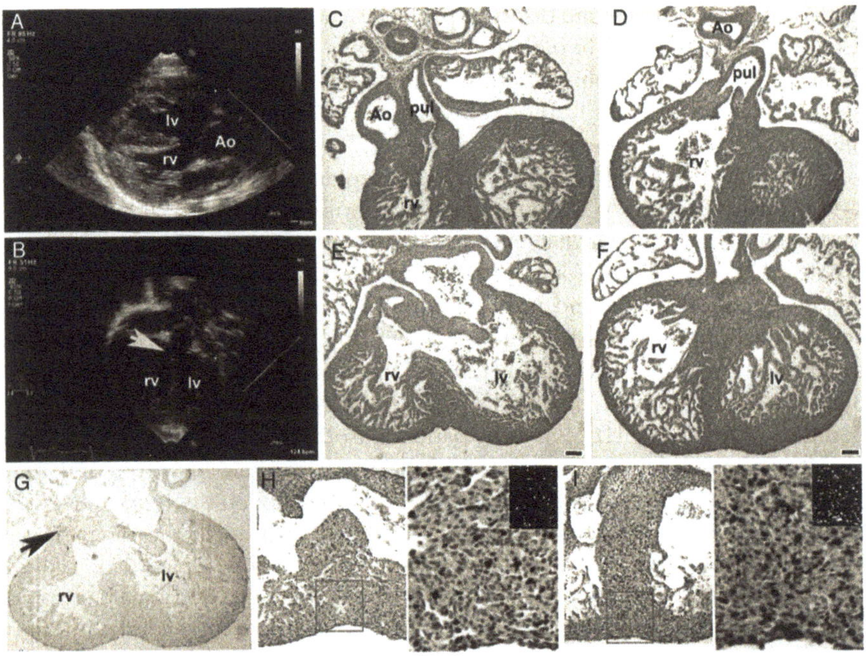

FIG. 1. (A) Echocardiography of 2-day-old newborn patient with DORV. Note aorta (Ao) is exiting from the right ventricle (rv). (B) Echocardiography of 5-month-old infant with perimembranous VSD. Note large hole connecting right and left ventricles (indicated via yellow arrow). (C–F) Histological hematoxylin/eosin stained sections of E14 *Hey2* knockout mutant and control littermate hearts. Note large VSD in (E) (indicated via yellow arrow) and DORV in (C) (as both the aorta and pulmonary (pul)) vessels exit the right ventricle. (G) Analysis of apoptosis via TUNEL immunohistochemistry in E14 Hety2 knockout heart. Note that there is no elevation of cell death associated with VSD, but there are a few apoptotic (brown) cells localized at the base of the DORV. (H and I) Characterization of cell proliferation in E14 *Hey2* cardiovascular system via BrdU immunohistochemistry. *Hey2* knockouts (H) exhibit VSDs, while the septum is intact in wild-type littermates (I). Heat map inserts (in H and I) indicated the relative differences between the mutant (~13% proliferating cells) when compared to normal control hearts (~27% proliferating cells). Note that BrdU staining is reduced by approximately 50% in the stunted mutant interventricular septum. Abbreviation: lv, left ventricle. Bar in (E) and (F) is 100 μM. (See Color Insert.)

embryonic days (E) 9 to 14 (only E14 shown in Fig. 1G), during ontogeny of the interventricular septum and septation of the mouse embryo heart. However, examination of cell proliferation revealed a significant reduction in cardiomyocyte proliferation at the base of the interventricular septum (Fig. 1H). These types of analyses illustrate the usefulness of the mouse transgenic analysis approaches and the type of detailed informative data that can be obtained. While most of the mouse models exhibit perimembranous VSDs, several mutants also exhibit muscular VSDs. Analysis of *Hand1–Hand2* knockin chimeras[101] revealed that stillborn high-percentage chimeric neonatal *Hand1–Hand2* knockins exhibited DORV with concomitant VSDs and PDA. Intriguingly, their hearts had a thinned myocardium and the myocardial architecture was extensively disorganized, with both multiple muscular and obligatory large perimembranous VSDs evident.[101] As *Hand1* and *Hand2* transcription factors are both expressed in the early cardiomyocytes, this suggests that growth of septal myocytes, elevated cell death, or positioning of the interventricular septum could underlie the *Hand1–Hand2* knockin chimeras CHDs. Evidence that *Hand1* is indeed an important regulator of the interventricular boundary is observed in *Mlc2v–Hand1* ectopic expressing transgenic mice that die at midgestation lacking a septum.[102] The multiple causes underlying VSD causation are further illustrated via the recent finding that suppression of the TGFβ superfamily intracellular signaling pathway within post-migratory cardiac neural crest cells in the outflow tract endocardial cushions can result in isolated VSD formation.[87] Using a three-component triple transgenic system, expression of the inhibitory Smad7 was induced via doxycycline within the neural crest lineage at post-migratory stages, resulting in isolated VSD and neonatal lethality. Although there is a paucity of data regarding the function of postmigratory neural crest within the heart itself, it has been shown via Cre-loxP that neural crest-specific N-cadherin, RhoA, and Cx43 expression are each required for normal OFT morphogenesis, cell shape, alignment, and cell–cell communication.[87] Thus, further studies are required to understand the link between lineage commitment and the many changes in cushion cell shape, cell–matrix adhesion, and cell–cell adhesion that occur within the heart itself during interventricular septal morphogenesis. One important disparity between the mouse models of VSDs and those seen in human infants is the high newborn mortality of mice compared with patients. In most mouse models, the VSDs are localized to the perimembranous region, are generally large, and are predicted to have significant left-to-right shunting.[96] It is also thought that mice are particularly sensitive to left-to-right shunting, given their high neonatal heart rates (~ 600+), as mouse models of PDA are also neonatal lethal[103,104] despite PDAs being well tolerated in neonatal patients.

B. Tetralogy of Fallot

TOF is the most common cyanotic CHD and the most common cause of the "blue baby syndrome." Approximately, 9–14% of babies with CHDs will have TOF (American Heart Association, 2010; www.heart.org). In TOF, an anterior placement of the conotruncal septum results in unequal division of the conus, which produces four cardiovascular alterations.[40] The four CHDs associated with TOF are VSD, right ventricular hypertrophy, pulmonary stenosis, and overriding aorta.[40,105,106] In TOF, the VSD is present because of an anterocephalad malalignment of the developing ventricular septum and a failure of the OFT cushions to muscularize.[40,106] Right ventricular hypertrophy is a hemodynamic consequence of the deviated outlet septum.[106] Pulmonary stenosis results from abnormal morphology of the septoparietal trabeculations surrounding the subpulmonary outflow tract, causing a narrowing of the right ventricular outflow and obstructing blood flow.[40,106] Overriding aorta is a result of the displacement of the outlet septum into the right ventricle, which causes the aortic root to be positioned directly over the VSD.[106] Cyanosis is the major hallmark of TOF and can be present in neonates based on the degree of the blood flow obstruction to the lungs.[105,106] Surgical intervention can result in complete repair, although adults have chronic issues such as pulmonary regurgitation, recurrence of pulmonary stenosis, and ventricular arrhythmias.[106] TOF is thought to be a multifactorial CHD and genetic associations of TOF are chromosomal aberrations such as trisomies 21, 18, 13 as well as microdeletions in chromosome 22.[106] Specific genetic associations via both mouse mutant analysis and patient screening also include *FOG2*, *JAG1*, *NKX2.5*, and *VEGF*. Cooperative morphological data analysis from mice and humans shows that TOF may be caused because the lack of longitudinal growth of the OFT prohibits the normal counterclockwise rotation of the OFT region.[107]

Several mouse mutants have been shown to exhibit TOF. The isoform-specific *Vascular endothelial growth factor* (*Vegf*)*Vegf120/120* knockin mice only express the 120 isoform and are highly predisposed to develop TOF.[108,109] The proposed etiology is that a localized increase in VEGF signaling within the secondary heart field-derived myocardium results in alterations of Notch signaling, causing OFT cushion hypoplasia, and subpulmonary myocardial apoptosis.[109] Significantly, *VEGF* haplotype correlates with increase risk for TOF in patients.[109,110] Similarly, as knockouts of mouse *Growth differentiation factor-1* (*Gdf1*) left–right patterning gene exhibited CHDs, a population study was performed to determine if a spectrum of CHDs can be attributed to human *GDF1*.[111,112] Indeed, a heterozygous loss-of-function in the human *GDF1* gene contributed to a distinct class of CHD affecting the conotruncus.[112] Thus, these studies confirmed that cardiac development in humans can be affected by left–right patterning signals and mutations in the entire pathway of left–right determination may be considered as candidates for CHD manifestations in humans on the basis of identifying a similar mechanism.[112] Finally, as mutations in the *FOG-2* (*friend of GATA 2*) transcriptional cofactor that can physically interact with GATA-4/5/6, were shown to occur in patients with TOF,[113,114] a mouse knockout was generated. Significantly, *Fog2* null mice die midgestation and display several cardiac malformations, including thin ventricular myocardium and TOF.[115] These *Fog2* knockout mice are currently being used to examine Fog2's ability to act in a dose-dependent manner and decrease Fog2-related pathways.[114] Similarly, transgenic expression of truncated *Fog2* alleles is being used to examine its *in utero* function.[114] Thus, via combining transgenic mice models and clinical and genetic characterization of patients, it is possible to gain a better understanding of the signaling pathways affecting cardiac development and the underlying causes of CHDs.

C. Hypoplastic Left Heart Syndrome

HLH is a relatively rare CHD in which the left side of the heart is severely underdeveloped and the causes are presumed to be genetic, although the exact genetic mechanism is unknown. Approximately, 4–8% of CHDs are HLH and the left ventricle is small and nonfunctional, while the right ventricle maintains both the pulmonary and systemic circulation.[105,116] HLH is a heterogeneous group of abnormalities and in addition to the small or missing left ventricle, there is atresia of the mitral and aortic valves and hypoplasia of the ascending aorta.[105,116] In patients with HLH, extensive reconstructive surgery is required for survival.[116] Constriction of blood flow to or from the left ventricle can significantly affect left ventricular development and it is supposed that HLH may occur secondarily to embryonic hemodynamic disruptions, rather than a primary defect intrinsic to the left ventricle.[116,117] It has been observed that the left ventricular growth worsens as fetal growth proceeds which support the theory that obstructed flow is suppressing myocardial growth.[117]

Although generating a mouse model has been challenging, as HLH is likely to be dependent on both genetic and epigenetic factors,[116] a model may have been developed. Recently, screening of Jacobsen syndrome patients who often exhibit HLH as part of their syndrome identified a 1.2-Mb region of overlap that contains six candidate genes.[118] One of the genes in that cardiac critical region is the transcription factor *ETS-1*, shown to be important in many biological functions, including regulation of cellular growth and differentiation.[118] Critically, *Ets1* knockout mice exhibit various cardiac defects, but a subset has non-apex-forming left ventricle, which is a hallmark of HLH, although the other structures associated with HLHS were normal.[118] Via mouse transgenic analysis, we have established the beginnings of a road map to understand how ventricular cells become specified, differentiate, and expand into a functional cardiac chamber.[119] The key is to now take this molecular and genetic information and devise innovative methods to augment or modify the effects of gene mutations on ventricular development.

XI. Conclusion

Significant advances in mouse genetics have led to a vast improvement in our understanding of cardiac morphogenesis. Multiple pathways and genetic networks have been indentified, and clear evidence accumulated to demonstrate which genes disrupt cardiac morphogenesis and can cause CHDs. Now the goal is to develop new concepts and models aimed at improving the diagnosis, treatment, and intervention of CHDs *in utero* and in newborns.

Acknowledgments

We are grateful to Dr. Tiffanie Johnson and the Riley Hospital for Children Congenital/Pediatric Cardiac MRI Program for the echocardiographic images and members of the Conway laboratory for their comments. This work was supported, in part, by Indiana University Department of Pediatrics and NIH funding (S. J. C.).

References

1. Hoffman JI, Kaplan S. The incidence of congenital heart disease. *J Am Coll Cardiol* 2002;**39**:1890–900.
2. Clark KL, Yutzey KE, Benson DW. Transcription factors and congenital heart defects. *Annu Rev Physiol* 2006;**68**:97–121.
3. Copp AJ. Death before birth: clues from gene knockouts and mutations. *Trends Genet* 1995;**11**:87–93.
4. Conway SJ, Kruzynska-Frejtag A, Kneer PL, Machnicki M, Koushik SV. What cardiovascular defect does my prenatal mouse mutant have, and why? *Genesis* 2003;**35**:1–21.
5. Yutzey KE, Robbins J. Principles of genetic murine models for cardiac disease. *Circulation* 2007;**115**:792–9.
6. Olaopa M, Caldwell RL, Barnes RM. Riley Heart Center Symposium on cardiac development 2009: transcriptional unification of heart morphogenesis. *Pediatr Cardiol* 2010;**31**:315–7.
7. Hoffman JI. Incidence of congenital heart disease: II. Prenatal incidence. *Pediatr Cardiol* 1995;**16**:155–65.
8. Conway SJ, Godt RE, Hatcher CJ, Leatherbury L, Zolotouchnikov VV, Brotto MA, et al. Neural crest is involved in development of abnormal myocardial function. *J Mol Cell Cardiol* 1997;**29**:2675–85.
9. Sedmera D, Pexieder T, Rychterova V, Hu N, Clark EB. Remodeling of chick embryonic ventricular myoarchitecture under experimentally changed loading conditions. *Anat Rec* 1999;**254**:238–52.

10. Ishiwata T, Nakazawa M, Pu WT, Tevosian SG, Izumo S. Developmental changes in ventricular diastolic function correlate with changes in ventricular myoarchitecture in normal mouse embryos. *Circ Res* 2003;**93**:857–65.
11. Phoon CK. Imaging tools for the developmental biologist: ultrasound biomicroscopy of mouse embryonic development. *Pediatr Res* 2006;**60**:14–21.
12. Brennan P, Young ID. Congenital heart malformations: aetiology and associations. *Semin Neonatol* 2001;**6**:17–25.
13. Andelfinger G. Genetic factors in congenital heart malformation. *Clin Genet* 2008;**73**:516–27.
14. Sadowski SL. Congenital cardiac disease in the newborn infant: past, present, and future. *Crit Care Nurs Clin North Am* 2009;**21**:37–48.
15. Zhu H, Kartiko S, Finnell RH. Importance of gene-environment interactions in the etiology of selected birth defects. *Clin Genet* 2009;**75**:409–23.
16. Nora JJ. Multifactorial inheritance hypothesis for the etiology of congenital heart diseases. The genetic-environmental interaction. *Circulation* 1968;**38**:604–17.
17. Belmont JW. Recent progress in the molecular genetics of congenital heart defects. *Clin Genet* 1998;**54**:11–9.
18. Bruneau BG. The developmental genetics of congenital heart disease. *Nature* 2008; **451**:943–8.
19. Miller SP, McQuillen PS, Hamrick S, Xu D, Glidden DV, Charlton N, et al. Abnormal brain development in newborns with congenital heart disease. *N Engl J Med* 2007;**357**:1928–38.
20. Kaltman JR, Andropoulos DB, Checchia PA, Gaynor JW, Hoffman TM, Laussen PC, et al. Perioperative Working Group. Report of the pediatric heart network and national heart, lung, and blood institute working group on the perioperative management of congenital heart disease. *Circulation* 2010;**121**:2766–72.
21. Dearani JA, Connolly HM, Martinez R, Fontanet H, Webb GD. Caring for adults with congenital cardiac disease: successes and challenges for 2007 and beyond. *Cardiol Young Suppl* 2007;**2**:87–96.
22. Somerville J. Grown-up congenital heart disease–medical demands look back, look forward 2000. *Thorac Cardiovasc Surg* 2001;**49**:21–6.
23. Nevin NC. Prevention and avoidance of congenital malformations. *Philos Trans R Soc Lond B Biol Sci* 1988;**319**:309–14.
24. Pierpont ME, Basson CT, Benson Jr. DW, Gelb BD, Giglia TM, Goldmuntz E, et al. Genetic basis for congenital heart defects: current knowledge: a scientific statement from the American Heart Association Congenital Cardiac Defects Committee, Council on Cardiovascular Disease in the Young: endorsed by the American Academy of Pediatrics. *Circulation* 2007;**115**:3015–38.
25. Basson CT, Cowley GS, Solomon SD, Weissman B, Poznanski AK, Traill TA, et al. The clinical and genetic spectrum of the Holt-Oram syndrome (heart-hand syndrome). *N Engl J Med* 1994;**330**:885–91.
26. Li QY, Newbury-Ecob RA, Terrett JA, Wilson DI, Curtis AR, Yi CH, et al. Holt-Oram syndrome is caused by mutations in TBX5, a member of the Brachyury (T) gene family. *Nat Genet* 1997;**15**:21–9.
27. Basson CT, Bachinsky DR, Lin RC, Levi T, Elkins JA, Soults J, et al. Mutations in human TBX5 cause limb and cardiac malformation in Holt-Oram syndrome. *Nat Genet* 1997;**15**:30–5.
28. Bruneau BG, Nemer G, Schmitt JP, Charron F, Robitaille L, Caron S, et al. A murine model of Holt-Oram syndrome defines roles of the T-box transcription factor Tbx5 in cardiogenesis and disease. *Cell* 2001;**106**:709–21.
29. Pelech AN, Broeckel U. Toward the etiologies of congenital heart diseases. *Clin Perinatol* 2005;**32**:825–44, vii.

30. Johnson MC, Hing A, Wood MK, Watson MS. Chromosome abnormalities in congenital heart disease. *Am J Med Genet* 1997;**70**:292–8.
31. Ryan AK, Goodship JA, Wilson DI, Philip N, Levy A, Seidel H, et al. Spectrum of clinical features associated with interstitial chromosome 22q11 deletions: a European collaborative study. *J Med Genet* 1997;**34**:798–804.
32. Lischner HW. DiGeorge syndrome(s). *J Pediatr* 1972;**81**:1042–4.
33. Lindsay EA, Botta A, Jurecic V, Carattini-Rivera S, Cheah Y-C, Rosenblatt HM, et al. Congenital heart disease in mice deficient for the DiGeorge syndrome region. *Nature* 1999;**401**:379–83.
34. Jerome LA, Papaioannou VE. DiGeorge syndrome phenotype in mice mutant for the T-box gene, Tbx1. *Nat Genet* 2001;**27**:286–91.
35. Lindsay EA, Vitelli F, Su H, Morishima M, Huynh T, Pramparo T, et al. Tbx1 haploinsufficiency identified by functional scanning of the DiGeorge syndrome region is the cause of aortic arch defects in mice. *Nature* 2001;**401**:97–101.
36. Merscher S, Funke B, Epstein JA, Heyer J, Puech A, Lu MM, et al. TBX1 is responsible for the cardiovascular defects in velo-cardio-facial/DiGeorge syndrome. *Cell* 2001;**104**:619–29.
37. Scambler PJ. 22q11 deletion syndrome: a role for TBX1 in pharyngeal and cardiovascular development. *Pediatr Cardiol* 2010;**31**:378–90.
38. Hu T, Yamagishi H, Maeda J, McAnally J, Yamagishi C, Srivastava D. Tbx1 regulates fibroblast growth factors in the anterior heart field through a reinforcing autoregulatory loop involving forkhead transcription factors. *Development* 2004;**131**:5491–502.
39. Bassett AS, Marshall CR, Lionel AC, Chow EW, Scherer SW. Copy number variations and risk for schizophrenia in 22q11.2 deletion syndrome. *Hum Mol Genet* 2008;**17**:4045–53.
40. Sadler TW. *Langman's Medical Embryology: Seventh Edition.* Baltimore, MD: Williams and Wilkins; 1995.
41. Jenkins KJ, Correa A, Feinstein JA, Botto L, Britt AE, Daniels SR, et al. American Heart Association Council on Cardiovascular Disease in the Young. Noninherited risk factors and congenital cardiovascular defects: current knowledge: a scientific statement from the American Heart Association Council on Cardiovascular Disease in the Young: endorsed by the American Academy of Pediatrics. *Circulation* 2007;**115**:2995–3014.
42. Garg V, Kathiriya IS, Barnes R, Schluterman MK, King IN, Butler CA, et al. GATA4 mutations cause human congenital heart defects and reveal an interaction with TBX5. *Nature* 2003;**424**:443–7.
43. Morton SU, Scherz PJ, Cordes KR, Ivey KN, Stainier DY, Srivastava D. MicroRNA-138 modulates cardiac patterning during embryonic development. *Proc Natl Acad Sci USA* 2008;**105**:17830–5.
44. Cordes KR, Srivastava D, Ivey KN. MicroRNAs in cardiac development. *Pediatr Cardiol* 2010;**31**:349–56.
45. Wheeler DA, Srinivasan M, Egholm M, Shen Y, Chen L, McGuire A, et al. The complete genome of an individual by massively parallel DNA sequencing. *Nature* 2008;**452**:872–6.
46. Doetschman T. Interpretation of phenotype in genetically engineered mice. *Lab Anim Sci* 1999;**49**:137–43.
47. Williams RS, Wagner PD. Transgenic animals in integrative biology: approaches and interpretations of outcome. *J Appl Physiol* 2000;**88**:1119–26.
48. Volarević S, Pende M, Pullen N. Manipulating mammalian genome by gene targeting. *Croat Med J* 1999;**40**:368–74.
49. Doetschman T. Influence of genetic background on genetically engineered mouse phenotypes. *Methods Mol Biol* 2009;**530**:423–33.
50. Hacking DF. Knock, and it shall be opened: knocking out and knocking in to reveal mechanisms of disease and novel therapies. *Early Hum Dev* 2008;**84**:821–7.

51. Waterson RH, Lindblad-Toh K, Birney E, Rogers J, Abril JF. Mouse genome sequencing consortium. *Nature* 2002;**420**:520–62.
52. Glaser S, Anastassiadis K, Stewart AF. Current issues in mouse genome engineering. *Nat Genet* 2005;**37**:1187–93.
53. Teng GQ, Zhao X, Lees-Miller RJP, Quinn FR, Li P, Rancourt DE, et al. Homozygous missense N629D hERG (KCNH2) potassium channel mutation causes developmental defects in the right ventricle and its outflow tract and embryonic lethality. *Circ Res* 2008;**103**:1483–91.
54. Granados-Riveron JT, Ghosh TK, Pope M, Bu'Lock F, Thornborough C, Eason J, et al. Alpha-cardiac myosin heavy chain (MYH6) mutations affecting myofibril formation are associated with congenital heart defects. *Hum Mol Genet* 2010;**19**(20):4007–16.
55. Ivanova A, Signore M, Caro N, Greene ND, Copp AJ, Martinez-Barbera JP. In vivo genetic ablation by Cre-mediated expression of diphtheria toxin fragment A. *Genesis* 2005;**43**:129–35.
56. Becker KD, Gottshall KR, Chien KR. Strategies for studying cardiovascular phenotypes in genetically manipulated mice. *Hypertension* 1996;**27**:495–501.
57. Kallapur S, Ormsby I, Doetschman T. Strain dependency of TGFbeta1 function during embryogenesis. *Mol Reprod Dev* 1999;**52**:341–9.
58. Bonyadi M, Rusholme SA, Cousins FM, Su HC, Biron CA, Farrall M, et al. Mapping of a major genetic modifier of embryonic lethality in TGF beta 1 knockout mice. *Nat Genet* 1997;**15**:207–11.
59. Moga MA, Nakamura T, Robbins J. Genetic approaches for changing the heart and dissecting complex syndromes. *J Mol Cell Cardiol* 2008;**45**:148–55.
60. Gordon JW, Scangos GA, Plotkin DJ, Barbosa JA, Ruddle FH. Genetic transformation of mouse embryos by microinjection of purified DNA. *Proc Natl Acad Sci USA* 1980;**77**:7380–4.
61. Hogan B, Williams J. Integration of foreign genes into the mammalian germ line: genetic engineering enters a new era. *Nature* 1981;**294**:9–10.
62. Wehner JM, Bowers BJ. Use of transgenics, null mutants, and antisense approaches to study ethanol's actions. *Alcohol Clin Exp Res* 1995;**19**:811–20.
63. Sinclair AM, Göttgens B, Barton LM, Stanley ML, Pardanaud L, Klaine M, et al. Distinct 5' SCL enhancers direct transcription to developing brain, spinal cord, and endothelium: neural expression is mediated by GATA factor binding sites. *Dev Biol* 1999;**209**:128–42.
64. Bockamp E, Maringer M, Spangenberg C, Fees S, Fraser S, Eshkind L, et al. Of mice and models: improved animal models for biomedical research. *Physiol Genomics* 2002;**11**:115–32.
65. Bronson SK, Plaehn EG, Kluckman KD, Hagaman JR, Maeda N, Smithies O. Single-copy transgenic mice with chosen-site integration. *Proc Natl Acad Sci USA* 1996;**93**:9067–72.
66. Jaenisch R, Mintz B. Simian virus 40 DNA sequences in DNA of healthy adult mice derived from preimplantation blastocysts injected with viral DNA. *Proc Natl Acad Sci USA* 1974;**71**:1250–4.
67. Sasaki E, Suemizu H, Shimada A, Hanazawa K, Oiwa R, Kamioka M, et al. Generation of transgenic non-human primates with germline transmission. *Nature* 2009;**459**:523–7.
68. Evans MJ, Kaufman MH. Establishment in culture of pluripotential cells from mouse embryos. *Nature* 1981;**292**:154–6.
69. Nilles KM, London B. Knockin animal models of inherited arrhythmogenic diseases: what have we learned from them? *J Cardiovasc Electrophysiol* 2007;**18**:1117–25.
70. Galli-Taliadoros LA, Sedgwick JD, Wood SA, Körner H. Gene knock-out technology: a methodological overview for the interested novice. *J Immunol Methods* 1995;**181**:1–15.
71. Yanagawa Y, Kobayashi T, Ohnishi M, Kobayashi T, Tamura S, Tsuzuki T, et al. Enrichment and efficient screening of ES cells containing a targeted mutation: the use of DT-A gene with the polyadenylation signal as a negative selection maker. *Transgenic Res* 1999;**8**:215–21.

72. Chen YT, Bradley A. A new positive/negative selectable marker, puDeltatk, for use in embryonic stem cells. *Genesis* 2000;**28**:31–5.
73. Friedrich G, Soriano P. Promoter traps in embryonic stem cells: a genetic screen to identify and mutate developmental genes in mice. *Genes Dev* 1991;**5**:1413–523.
74. Zambrowicz BP, Friedrich GA, Buxton EC, Lilleberg SL, Person C, Sands AT. Disruption and sequence identification of 2,000 genes in mouse embryonic stem cells. *Nature* 1998;**392**:608–11.
75. Smithies O, Gregg RG, Boggs SS, Koralewski MA, Kucherlapati RS. Insertion of DNA sequences into the human chromosomal beta-globin locus by homologous recombination. *Nature* 1985;**317**:230–4.
76. Capecchi MR. Altering the genome by homologous recombination. *Science* 1989; **244**:1288–92.
77. Olson EN, Arnold HH, Rigby PW, Wold BJ. Know your neighbors: three phenotypes in null mutants of the myogenic bHLH gene MRF4. *Cell* 1996;**85**:1–4.
78. Ryding AD, Sharp MG, Mullins JJ. Conditional transgenic technologies. *J Endocrinol* 2001;**171**:1–14.
79. Sauer B, Henderson N. Site-specific DNA recombination in mammalian cells by the Cre recombinase of bacteriophage P1. *Proc Natl Acad Sci USA* 1988;**85**:5166–70.
80. Sternberg N, Hamilton D. Bacteriophage P1 site-specific recombination. Recombination between loxP sites. *J Mol Biol* 1981;**150**:467–86.
81. Evans GA. Dissecting mouse development with toxigenics. *Genes Dev* 1989;**3**:259–63.
82. Sung YH, Song J, Lee HW. Functional genomics approach using mice. *J Biochem Mol Biol* 2004;**37**:122–32.
83. Hutson MR, Kirby ML. Model systems for the study of heart development and disease. Cardiac neural crest and conotruncal malformations. *Semin Cell Dev Biol* 2007;**18**:101–10.
84. Abu-Issa R, Kirby ML. Heart field: from mesoderm to heart tube. *Annu Rev Cell Dev Biol* 2007;**23**:45–68.
85. Rosenberg MP. Gene knockout and transgenic technologies in risk assessment: the next generation. *Mol Carcinog* 1997;**20**:262–74.
86. DeMayo FJ, Tsai SY. Targeted gene regulation and gene ablation. *Trends Endocrinol Metab* 2001;**12**:348–53.
87. Tang S, Snider P, Firulli AB, Conway SJ. Trigenic neural crest-restricted Smad7 over-expression results in congenital craniofacial and cardiovascular defects. *Dev Biol* 2010;**344**:233–47.
88. Kile BT, Hilton DJ. The art and design of genetic screens: mouse. *Nat Rev Genet* 2005;**6**:557–67.
89. Russell WL, Kelly EM, Hunsicker PR, Bangham JW, Maddux SC, Phipps EL. Specific-locus test shows ethylnitrosourea to be the most potent mutagen in the mouse. *Proc Natl Acad Sci USA* 1979;**76**:5818–9.
90. Chen Y, Yee D, Dains K, Chatterjee A, Cavalcoli J, Schneider E, et al. Genotype-based screen for ENU-induced mutations in mouse embryonic stem cells. *Nat Genet* 2000;**24**:314–7.
91. Greber B, Lehrach H, Himmelbauer H. Mouse splice mutant generation from ENU-treated ES cells-a gene-driven approach. *Genomics* 2005;**85**:557–62.
92. Carlson CM, Dupuy AJ, Fritz S, Roberg-Perez KJ, Fletcher CF, Largaespada DA. Transposon mutagenesis of the mouse germline. *Genetics* 2003;**165**:243–56.
93. Largaespada DA. Transposon mutagenesis in mice. *Methods Mol Biol* 2009;**530**:379–90.
94. Aune CN, Chatterjee B, Zhao XQ, Francis R, Bracero L, Yu Q, et al. Mouse model of heterotaxy with single ventricle spectrum of cardiac anomalies. *Pediatr Res* 2008;**63**:9–14.
95. Hoffman JI, Kaplan S, Liberthson RR. Prevalence of congenital heart disease. *Am Heart J* 2004;**147**:425–39.

96. Sakata Y, Kamei CN, Nakagami H, Bronson R, Liao JK, Chin MT. Ventricular septal defect and cardiomyopathy in mice lacking the transcription factor CHF1/Hey2. *Proc Natl Acad Sci USA* 2002;**99**:16197–202.
97. Benson DW, Silberbach GM, Kavanaugh-McHugh A, Cottrill C, Zhang Y, Riggs S, et al. Mutations in the cardiac transcription factor NKX2.5 affect diverse cardiac developmental pathways. *J Clin Invest* 1999;**104**:1567–73.
98. Webb S, Brown NA, Anderson RH. Formation of the atrioventricular septal structures in the normal mouse. *Circ Res* 1998;**82**:645–56.
99. Donovan J, Kordylewska A, Jan YN, Utset MF. Tetralogy of fallot and other congenital heart defects in Hey2 mutant mice. *Curr Biol* 2002;**12**:1605–10.
100. Watanabe T, Koibuchi N, Chin MT. Transcription factor CHF1/Hey2 regulates coronary vascular maturation. *Mech Dev* 2010;**127**(9–12):418–27.
101. Firulli AB, Firulli BA, Wang J, Rogers RH, Conway SJ. Gene replacement strategies to test the functional redundancy of basic helix-loop-helix transcription factor. *Pediatr Cardiol* 2010;**31**:438–48.
102. Togi K, Kawamoto T, Yamauchi R, Yoshida Y, Kita T, Tanaka M. Role of Hand1/eHAND in the dorsoventral patterning and interventricular septum formation in the embryonic heart. *Mol Cell Biol* 2004;**24**:4627–35.
103. Nguyen M, Camenisch T, Snouwaert JN, Hicks E, Coffman TM, Anderson PA, et al. The prostaglandin receptor EP4 triggers remodelling of the cardiovascular system at birth. *Nature* 1997;**390**:78–81.
104. Loftin CD, Trivedi DB, Langenbach R. Cyclooxygenase-1-selective inhibition prolongs gestation in mice without adverse effects on the ductus arteriosus. *J Clin Invest* 2002;**110**:549–57.
105. Moore KL, Presaud TVN. *The Developing Human: Clinically Oriented Embryology.* Philadelphia, PA: Saunders; 2003.
106. Bailliard F, Anderson RH. Tetralogy of Fallot. *Orphanet J Rare Dis* 2009;**4**:2.
107. Restivo A, Piacentini G, Placidi S, Saffirio C, Marino B. Cardiac outflow tract: a review of some embryogenetic aspects of the conotruncal region of the heart. *Anatom Rec A Discover Mol Cell Evol Biol* 2006;**288**:936–43.
108. Stalmans I, Lambrechts D, De Smet F, Jansen S, Wang J, Maity S, et al. VEGF: a modifier of the del22q11 (DiGeorge) syndrome? *Nat Med* 2003;**9**:173–82.
109. van den Akker NM, Molin DG, Peters PP, Maas S, Wisse LJ, van Brempt R, et al. Tetralogy of fallot and alterations in vascular endothelial growth factor-A signaling and notch signaling in mouse embryos solely expressing the VEGF120 isoform. *Circ Res* 2007;**100**:842–9.
110. Lambrechts D, Devriendt K, Driscoll DA, Goldmuntz E, Gewillig M, Vlietinck R, et al. Low expression VEGF haplotype increases the risk for tetralogy of Fallot: a family based association study. *J Med Genet* 2005;**42**:519–22.
111. Rankin CT, Bunton T, Lawler AM, Lee SJ. Regulation of left-right patterning in mice by growth/differentiation factor-1. *Nat Genet* 2000;**24**:262–5.
112. Karkera JD, Lee JS, Roessler E, Banerjee-Basu S, Ouspenskaia MV, Mez J, et al. Loss-of-function mutations in growth differentiation factor-1 (GDF1) are associated with congenital heart defects in humans. *Am J Hum Genet* 2007;**81**:987–94.
113. Pizzuti A, Sarkozy A, Newton AL, Conti E, Flex E, Digilio MC, et al. Mutations of ZFPM2/FOG2 gene in sporadic cases of tetralogy of Fallot. *Hum Mutat* 2003;**22**:372–7.
114. Finelli P, Pincelli AI, Russo S, Bonati MT, Recalcati MP, Masciadri M, et al. Disruption of friend of GATA 2 gene (FOG-2) by a de novo t(8;10) chromosomal translocation is associated with heart defects and gonadal dysgenesis. *Clin Genet* 2007;**71**:195–204.

115. Tevosian SG, Deconinck AE, Tanaka M, Schinke M, Litovsky SH, Izumo S, et al. FOG-2, a cofactor for GATA transcription factors, is essential for heart morphogenesis and development of coronary vessels from epicardium. *Cell* 2000;**101**:729–39.
116. Gruber PJ, Epstein JA. Development gone awry: congenital heart disease. *Circ Res* 2004;**94**:273–83.
117. Towbin JA, Casey B, Belmont J. The molecular basis of vascular disorders. *Am J Hum Genet* 1999;**64**:678–84.
118. Ye M, Coldren C, Liang X, Mattina T, Goldmuntz E, Benson DW, et al. Deletion of ETS-1, a gene in the Jacobsen syndrome critical region, causes ventricular septal defects and abnormal ventricular morphology in mice. *Hum Mol Genet* 2010;**19**:648–56.
119. Srivastava D, Gottlieb PD, Olson EN. Molecular mechanisms of ventricular hypoplasia. *Cold Spring Harb Symp Quant Biol* 2002;**67**:121–5.

Dissection of Cardiovascular Development and Disease Pathways in Zebrafish

JOANNE CHAN[*,†] AND JOHN D. MABLY[‡,§]

[*]*Vascular Biology Program, Children's Hospital Boston, Harvard Medical School, Boston, Massachusetts, USA*

[†]*Department of Surgery, Children's Hospital Boston, Harvard Medical School, Boston, Massachusetts, USA*

[‡]*Department of Cardiology, Children's Hospital Boston, Boston, Massachusetts, USA*

[§]*Department of Genetics, Harvard Medical School, Boston, Massachusetts, USA*

The use of animal models in medicine has contributed significantly to the development of drug treatments and surgical procedures for the last century, in particular for cardiovascular disease. In order to model human disease in an

Progress in Molecular Biology
and Translational Science, Vol. 100
DOI: 10.1016/B978-0-12-384878-9.00004-2

Copyright 2011, Elsevier Inc.
All rights reserved.
1877-1173/11 $35.00

animal, an appreciation of the strengths and limitations of the system are required to interpret results and design the logical sequence of steps toward clinical translation. As the world's population ages, cardiovascular disease will become even more prominent and further progress will be essential to stave off what seems destined to become a massive public health issue. Future treatments will require the imaginative application of current models as well as the generation of new ones. In this review, we discuss the resources available for modeling cardiovascular disease in zebrafish and the varied attributes of this system. We then discuss current zebrafish disease models and their potential that has yet to be exploited.

I. The Impact of Cardiovascular Disease and Need for Disease Models

Cardiovascular disease (CVD) is responsible for more deaths each year than the next most common diseases combined (sources, American Heart Association and Center for Disease Control). In the United States, 81.1 million citizens suffer from some form of CVD, which range from inherited birth defects to conditions such as coronary artery disease, high blood pressure, congestive heart failure, and also stroke (based on data from the National Health and Nutrition Examination Survey (NHANES) 2003–2006). This translates into a substantial economic burden as well; in 2010, it is estimated that heart disease will cost the United States $316.4 billion.[1] In addition, the lifespan of individuals in Western nations continues to increase as a result of multiple factors, such as better nutrition. Improvements in healthcare over the last century have contributed to this phenomenon and continued advancements in medical care will be crucial to combat the diseases that will afflict this large aging population. This includes the prevention and treatment of CVD since heart failure still remains a common affliction among the elderly.

The etiology of the different types of CVD is complex due to the confounding influence of environmental factors, such as diet. However, some forms of disease demonstrate a clear heritable component.[2–4] Now that many of the simple Mendelian traits have been mapped and genes identified (such as those responsible for familial hypertrophic and dilated cardiomyopathy, and long QT syndrome[5,6]), the next step becomes the determination of the consequence of interactions between different genetic factors. Epigenetic modifications and environmental exposures as well as genetic factors predisposing individuals to coronary artery disease, aneurysms, and stroke are also now starting to be identified.[7–10] With such a high prevalence of CVD, the importance of establishing better models for these conditions becomes essential to help sort out the relevance of these different factors as well as to improve and increase speed of diagnosis and treatment.

II. Current Animal Models of Cardiovascular Disease

Animal models have long been employed for the study of CVD and the testing of new and improved surgical techniques. Large animal models, such as pig and sheep, have traditionally been used for the study of CVD. They also closely mimic the hemodynamic stresses and physiological parameters seen in humans. Unfortunately, they are difficult to house and their size and associated costs prevent their use in a true high-throughput capacity. This niche was initially filled by the use of the rat model but with the advent of homologous recombination and the ability to specifically ablate individual genes, the mouse became the favored system. The trade-off was that rodent physiology and cardiovascular function are not as similar to humans as were large mammals, despite the genetic similarities. Many of the difficulties in measuring heart function in such a small animal have been overcome but still require substantial expertise.

These animal models have also been employed to study the pathogenesis of disease. More recently, with the completion of the sequencing of several vertebrate genomes, new molecular studies have defined several key signaling pathways and genes that were previously understudied in cardiac disease. Ideally, these animal models would serve as surrogates for different forms of CVD and recapitulate the varied symptoms associated with these conditions. Such a model would facilitate studies of approaches for management of the disease with small molecules or other, nonsurgical, cellular therapies. The interpretation of findings from small rodent models must be accompanied by caveats, however, since some attributes of cardiac physiology are dramatically different from humans (e.g., the high heart rate of the mouse compared to humans (500–600 vs. 60–90 beats per minute, respectively) requires a different profile of sarcomeric and channel proteins). As a result, the utility of findings in such a system may ultimately not translate well to larger mammals. The determination of a loss of function phenotype in mouse will lead to essential insights into gene function; however, the implications for human cardiovascular function have to be assessed carefully and will inevitably require validation in large animal models.

III. Molecular Tools to Study Zebrafish

Although the zebrafish is a relatively new animal model for disease, a wealth of resources is already available to scientists wishing to employ this system. ZFIN (The Zebrafish Model Organism Database) serves as an entry point and attempts to bring together general information from a vast variety of sources, including mRNA expression data and gene mutation/knockdown phenotypes. Numerous zebrafish lines, as well as cDNAs/ESTs and antibodies, are maintained by the Zebrafish International Resource Center (ZIRC) for distribution to the research community. ZFIN, as well as both the Sanger and Ensembl sites, also serves as portals to the chromosomal maps (both meiotic[11–14] and radiation hybrid maps[15–17]), as well as the data from the Sanger zebrafish genome sequence initiative.[18] These resources also provide access to gene/protein sequence information in addition to the traditional NCBI sites.

IV. Approaches to Generate Zebrafish Models for Disease

Traditional forward genetic screens involved the identification of important developmental genes based on the detection of a mutant phenotype and are now complemented by the ability to knockdown gene expression with morpholinos.[19–21] However, the availability of near complete genome sequence data has enabled efforts to systematically generate mutations in protein coding genes using new approaches such as "TILLING" (for *T*argeting *I*nduced *L*ocal *L*esions *in* *G*enomes).[22–24] The same ENU mutagenesis strategy is used to chemically induce mutations within the genome[25,26]; however, instead of a phenotype-driven assay, high-throughput sequencing is employed to directly detect point mutations in genes of interest. Several institutes, including the Sanger Centre, have initiated efforts to identify mutations in genes previously not targeted by the initial screens as a service to the research community. Another technique to induce genetic lesions that also shows great promise for zebrafish investigators is the use of customized zinc-finger nucleases (ZFNs).[27,28] This approach has been shown to be effective in introducing targeted frame-shift mutations with high efficiency.[27,28] Although still technically daunting, detailed protocols for constructing expression constructs to generate and then inject RNAs of ZFNs to induce mutations in target genes are available[29] and the vectors and reagents are freely accessible to the scientific community.

The traditional mutagenesis screens have led to the identification and characterization of numerous mutants while the characterization of those derived from the newer, gene-targeted, approaches is still ongoing. As a result, many of the zebrafish models being used to study CVD were derived from the large-scale ENU mutagenesis screens performed in Tubingen[30] and Boston,[31] as well as smaller screens at other institutes.[32,33] This approach provided an unbiased and genome-wide, large-scale approach for identifying novel genes required for cardiovascular function. New disease models generated using this traditional approach will also become available from the screening of 6000 mutagenized genomes as part of the ZF-MODELS (Zebrafish Models for Human Development and Disease, EU) consortium. This effort was specifically devised to target mutations that disrupt organogenesis and tissue formation with the objective of identifying new models for disease. One novel aspect of this initiative was the inclusion of screens to detect defects affecting tissue formation in the adult, past the typically studied timeframe of embryogenesis.

V. Zebrafish as a Model System for the Study of Cardiovascular Disease

The advantages of the zebrafish system that were initially exploited for developmental studies have also established it as a first-line model system to study cardiac and vascular biology and disease. These include the optical transparency of the developing embryo which allows real-time observation of the heart and vessels as they develop and begin to function. More importantly, because zebrafish can survive in the absence of heart function and blood circulation for the first several days of life, defects in cardiac and vascular formation may be studied during periods of growth not feasible in other model systems due to embryonic lethality.[30,31,34–40] This attribute of the zebrafish embryo is a result of its aqueous surroundings, which facilitate the diffusion of oxygen and nutrients directly from the water[41,42] unlike mammalian models such as the mouse which suffers hypoxic deterioration under similar conditions. Another advantage of the aquatic environment of zebrafish is the ease of application of drugs to animals during early stages of growth. Water soluble compounds are readily taken up by the embryo and this property has been exploited for the identification of new small molecules affecting cardiovascular growth; in theory, this approach could also be used for the identification of compounds which reverse pathological conditions.[43–45]

Although there are concerns regarding the use of any small animal to study defects that affect large mammals such as humans, zebrafish possess other unique attributes that make it an attractive model system. For example, the normal embryonic heart rate is 120–180 beats per minute, much closer to humans than mouse.[46] From a broader perspective, since zebrafish are vertebrates, they possess a strong genetic relationship to mammalian model systems, including gene conservation and regions of chromosomal synteny. This has permitted the use of zebrafish to study the genetic factors regulating both cardiovascular development and function.

VI. Early Cardiovascular Development in Zebrafish is Similar to Humans

Over the past 15 years, much of the groundwork was established to define the molecular determinants of cardiomyocyte cell fate and vessel growth during zebrafish development.[47–50] In fact, since the fundamental mechanisms of cardiac growth and function are highly conserved in zebrafish as in other widely used vertebrate models, key developmental steps in vertebrate cardiovascular development have been determined through genetic and cell biology studies in this system (see Fig. 1; reviewed in Refs. 51–53). In addition, the highly conserved architecture of the vertebrate vascular tree has provided a strong basis for use of the zebrafish model to dissect molecular mechanisms underlying developmental angiogenesis.[50,54–56] As we learn more about the genes and signaling pathways controlling vertebrate blood and lymphatic vessel development, more parallels can be drawn between zebrafish and mammalian models.[57,58] This has significantly increased the use of the zebrafish model as an alternative discovery model for vascular biology in the last few years.

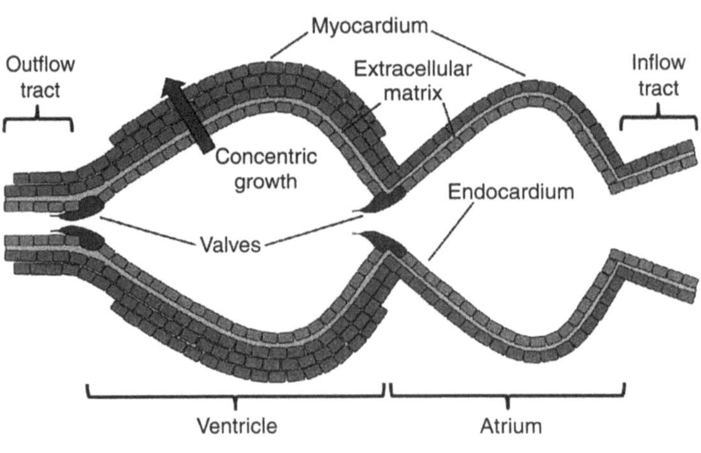

FIG. 1. Structure of the zebrafish embryonic heart. The zebrafish heart consists of two chambers and develops rapidly. Many of the distinguishing attributes of a vertebrate heart are detectable within 2–3 dpf, as indicated in the diagram. Mature valves do not form until later larval development but valve-like structures that allow a unidirectional flow of blood are present at this stage of development. The migration of the pericardium over the heart and the development of the zebrafish coronary system have not yet occurred by 3 dpf but does occur during later development.

VII. Conserved Gene Programs, Cell Behavior, and Molecular Mechanisms

The cardiomyocytes of zebrafish hearts originate in the ventral-marginal zone. These cells then migrate through the lateral plate mesoderm to generate the midline heart tube which resembles a cone with the preventricular cells pointing apically.[47,48,59] The ventricular and atrial precursors then undergo a medial–lateral sequestration with endodermal signals assisting in the proper midline alignment, as in other species.[60]

Cardiovascular development during early stages of zebrafish embryonic growth proceeds rapidly. Cardiomyocyte differentiation begins prior to heart-tube formation, during which time cells in the lateral plate mesoderm express key transcription factors (e.g., GATA4, dHAND, Nkx2.5) as well as genes for ion channels, calcium handling, and myofibrillar structure.[61] Myofibrillar arrays are not evident until the heart-tube stage although by 20 h postfertilization (hpf), the zebrafish heart tube has already formed.[47,60] Rhythmic, peristaltic contractions initiate by 24 hpf, with onset of contraction being essential for the development of proper sarcomeres.[47,62] These peristaltic waves soon transition to the sequential contractions of atrium and ventricle characteristic of other vertebrate hearts[46]; this occurs prior to the onset of innervation, organized conducting tissue, or definitive sinoatrial or atrioventricular nodes [reviewed in Ref. 53]. Mammals subsequently undergo septation between the chambers and within the outflow tract because of the requirement for a separate pulmonary circulation but otherwise, the zebrafish demonstrates a remarkable conservation of form and function with its vertebrate relatives. This is a key attribute for any disease model if results of studies are to be extended to other mammalian model systems, and ultimately humans.

The endothelium of the vessels and endocardium (the inner, endothelial lining of the heart) are derived from endothelial precursors in the lateral mesoderm of the embryo which arise during gastrulation.[47,48,63] Detailed analyses of blood and lymphatic vessel development have defined the important events in the development of the vertebrate vasculature as well, which include (i) the differentiation of ventral mesodermal cells into blood and endothelial lineages, (ii) the development of the major axial vessels (the aorta and cardinal vein (CV)), (iii) the mechanisms for angiogenic sprouting and determination of the guidance cues necessary to form secondary vessels, (iv) lymphangioblast formation and sprouting from venous endothelial cells to initiate the lymphatic vascular system, and (v) the development and recruitment of perivascular support cells or mural cells during the maturation of blood and lymphatic vessels (see Fig. 2).[64–68]

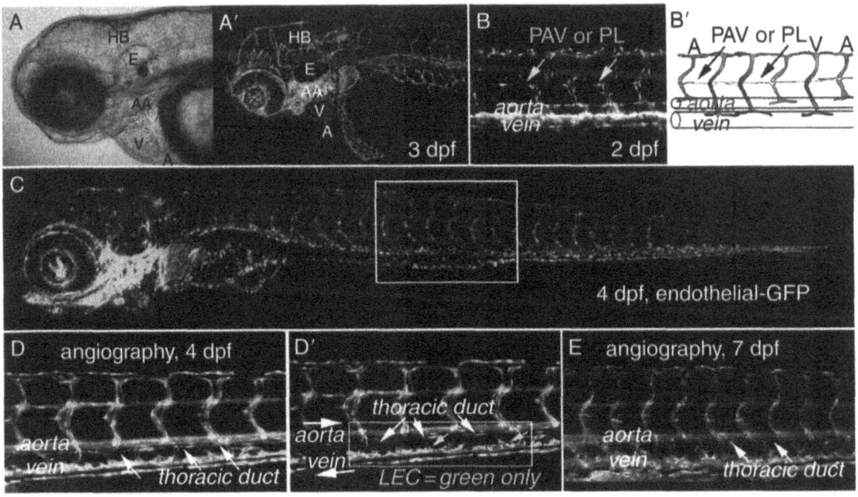

FIG. 2. Blood and lymphatic vessel development. Lateral views shown with stages in days postfertilization (dpf). Head region in phase (A), or by endothelial-GFP-labeling of cardiovascular system (A′). Anatomical locations for hindbrain (HB), ear (E), aortic arches (AA), ventricle (V), and atrium (A) are indicated. (B) Fluorescent and schematic views of intersegmental arteries (red), veins (blue), parachordal vessel (PAV), or lymphangioblasts (PL). (C) Endothelially labeled-GFP indicates vascular development at 4 dpf (box indicates trunk region). (D–E) Microangiography using red-fluorescent dextran to indicate blood flow in vessels, green indicates both blood and lymphatic endothelial cells. The thoracic duct forms just ventral to the aorta by 4 dpf. (E) At 7 dpf, the thoracic duct is patent. (See Color Insert.)

Early investigators postulated that blood and endothelial cells differentiate from the same progenitor cell type because erythrocytes are found next to blood vessels in embryonic development. This common progenitor cell type, termed the "hemangioblast," gives rise to both blood and vascular lineages. These precursor cells respond to spatiotemporal signals that allow them to differentiate into either hematopoietic cells or angioblasts. In the zebrafish model, lineage-tracing experiments demonstrated that a single cell type in the ventral mesoderm could differentiate into both lineages.[47,69] In addition, identification of the *cloche* mutant has provided further genetic evidence, as both blood and endothelial cells are significantly reduced in the autosomal recessive mutant embryos.[69] These precursor cells define two strips of lateral expression of hemangioblast markers beginning at somitic stages that coalesce at the midline to form endothelial and blood cells.[70,71]

In the last 10 years, intense effort has been placed on elucidation of the critical steps from hemangioblast formation to the divergence of the blood and endothelial cell types.[72] At early somite stages, both lineages share common molecular markers such as scl/tal1 and lmo.[70,73,74] Separation is presumed to coincide with changes in expression pattern. For example, angioblasts express flk1 and scl early on and then differentiate into flk1-positive and scl-negative cells. On the other hand, hematopoietic cells feature scl-positive and flk1-negative expression. For the blood cell lineage, progressive differentiation of cell types has been discussed in a recent review.[75] The head vasculature and heart are derived from the anterior lateral mesoderm of the zebrafish. These cells are able to acquire either a myocardial or an endothelial fate with the choice between these two fates guided by repression of the cardiac specification pathway by blood and/or vessel precursors (angioblasts).[63,76] This accounts for the absence of blood and angioblasts in the *cloche* mutant, although the size of the heart is increased.[76] Conversely, overexpression of a combination of the hemangioblast specifying transcription factors scl/tal and lmo2 increases the number of vessel precursors and subsequently reduces the number of cardiomyocyte precursors.[77]

Cells within the posterior mesoderm give rise to trunk vasculature and blood,[63] adjacent to intermediate mesoderm that forms the kidney.[78] Several genes regulate the number of cells acquiring vessel, blood, or kidney precursor fates. For example, knockdown of the transcription factor odd-skipped 1 leads to an expansion of the lateral angioblast population with consequent loss of kidney precursors.[79] BMP signaling is also a crucial component required for this balancing of cell types; restriction of expression of a dominant-negative BMP receptor to hemangioblasts leads to an increase in hematopoietic cells and fewer pronephric cells.[80] Conversely, overexpression of the transcription factor scl leads to an increased predisposition for lateral and intermediate mesoderm fating toward hematopoietic cells.[80] During subsequent vascular development, angioblasts migrate from the lateral plate mesoderm to the midline to form the major axial vessels, the dorsal aorta (DA), and the cardinal vein (CV), mostly through the process of vasculogenesis.[50,81] By about 24–26 hpf, circulation begins in most embryos with a slow movement of erythrocytes through a primary circuit from the heart's outflow tract through the aorta to the vein, then into the atrium and back to the single ventricle.[50] By the 2nd day of development, at about 48 hpf, intersegmental vessels (ISVs) are formed along the trunk of the embryo where a balanced number of arteries and veins ensure efficient blood flow in this region.[54] The beginning of lymphatic development also occurs by this time. As in other vertebrates, zebrafish lymphangioblasts also originate from the posterior cardinal vein (PCV) endothelial cells. Using *in vivo* imaging and transgenic lines, the migration route of early lymphangioblasts has been tracked in real time, to show that they contribute to the formation of thoracic duct, the major axial lymphatic vessel across all vertebrates.[55,56,68]

VIII. Zebrafish Cardiovascular Mutants as Disease Models

As a result of the unique survival qualities of the zebrafish in its aqueous environment, development of the cardiovascular system of the zebrafish embryo has been intensely studied and is well understood[61]; this has facilitated new approaches to exploit this organism to establish models of CVD and is now commonly used to complement studies in the mouse or other animal models. Another strength of the zebrafish as a genetic model system is that it enables the rapid and systematic dissection of interactions between genes in an intact vertebrate organism. Construction of a strong developmental framework has identified conserved cellular processes and molecular mechanisms that may be perturbed in human CVD. For example, in congenital and ischemic heart disease, cancer, and diabetes, reactivation of embryonic vascular genes and signaling pathways in an uncontrolled manner leads to the formation of a poorly functioning musculature and vasculature.[58,82–84]

The heart and vasculature are among the most prominent features of the developing zebrafish embryo because of the contraction of the heart which initiates and maintains circulation of blood through the embryo. Since the heart begins to contract by 24 hpf, defects that affect contractility are easily detectable early during embryonic growth, prior to the requirement for circulation of blood through the embryo. The unique ability of the zebrafish embryo to survive early growth without circulation of blood means that these studies can be continued through developmental stages past which mammalian embryos would not be viable. As a result, gene defects and small molecules that impair cardiac contractility, and thereby flow, can be easily monitored by visual inspection or by automated microscopic techniques.[85] In addition, certain zebrafish phenotypes could be correlated with defects within a single pathway (single phenotype, multiple genes, single/overlapping pathway[86–90]) providing a unique method to dissect and define genes with similar and overlapping functions.

The original large-scale zebrafish genetic screens were envisioned as a means to determine genes essential for development and organogenesis. Many mutations were identified that resulted in defects affecting the morphogenesis, and thereby, function, of the cardiovascular system.[30,31,35] It was soon apparent that these mutants would be useful for examining the importance of specific genes on cardiovascular function and subsequent growth of the embryo. However, some of these mutants cause severe developmental abnormalities that would limit their usefulness for this purpose (e.g., mutants with cardia bifida,[91,92] loss[93] or misorientation of the chambers,[94,95] or loss of the entire endocardium and most endothelium[31,69]). We will limit our discussion to mutants with phenotypes possessing attributes typical of CVD and to the gene-targeted analyses of pathways implicated in these human conditions.

IX. Models for Studying Hemodynamic Forces and Valve Disease

Before the vertebrate heart has completed development, contraction has already begun due to the requirement of blood flow for the growth of other tissues and organs. This onset of function generates forces; wall stress and stretch from contraction of the chambers is exerted on both endocardial and myocardial cells while shear stress from flow of blood through the heart is exerted at the fluid–endocardial interface (typically perpendicular to wall stress in the direction of flow). These forces are key regulators of the endocardial and myocardial transcriptomes.[96,97] As a result, much of cardiac morphogenesis must occur in the presence of active blood flow. Disruption of the fluid forces associated with flow has been previously linked to cardiac developmental abnormalities in humans and model vertebrates.[98–100] However, little is known about the mechanism through which cardiac flow is sensed and transduced into developmental signals. The zebrafish is particularly well suited to serve as an animal model to study the genetic pathways involved in cardiac mechanosensation because embryonic development proceeds normally in the absence of cardiac function and circulation.

The mechanisms by which these forces are detected are not known, although multiple genetic pathways have been implicated as downstream targets. One possible mediator of force could be primary cilia. Monocilia are present on the luminal surface of the endocardium in several species[101,102] and the role of cilia is well documented for the development of the early kidney pronephros and Kupffer's vesicle.[103] A relationship between flow and potential signaling through primary cilia in cardiovascular morphogenesis has not yet been investigated in the zebrafish but impaired ciliary function in the mouse heart has a dramatic effect on myocardial growth.[102] In addition to thinning of the myocardial wall, cells within the endocardial cushions (the valvular precursors) fail to undergo the proper epithelial–mesenchymal transition (EMT).

Several mutants with defects in early valve formation were identified in the original mutagenesis screens.[31] Many of these, however, appeared to reflect complex genetic influences from multiple loci, since subsequent generations did not produce affected embryos in the expected Mendelian ratios. Despite this setback and even though mature valves have not yet formed in the zebrafish heart by 16 dpf (days post fertilization), multiple genes that influence early steps in valve growth have been studied.[104] The steps in zebrafish valve development are also slightly different from other vertebrates; data suggest that early endocardial cushion formation may not occur in the traditional sense. Instead, invagination of a single sheet of endocardial cells occurs to directly form the leaflet between the two cardiac chambers (atrioventricular valve).[105]

Subsequent valve maturation occurs by elongation of the valvular structures through the first stage, with extensive extracellular matrix deposition continuing past 16 dpf.[104] Interestingly, the rate of valve maturation within the hearts of different zebrafish larvae varied with the size of the animal, supporting a role for biomechanical forces in this process.[104] The importance of the matrix deposition during this time is also illustrated by the loss of normal valve formation associated with mutation of the UDP-Glucose Dehydrogenase gene (UGDH).[106] Extending from this work, expression of chondroitin sulfate during embryonic growth was shown to be essential for formation of the atrioventricular region of the heart.[107]

The mutants *silent heart* (*sih*) and *weak atrium* (*wea*) (discussed in more detail later) have also been employed as models to study the role of hemodynamic forces on heart growth since they lack blood flow as a result of contractility defects. In particular, *sih* has been used extensively as a means to ablate flow in conjunction with loss of function analysis of a second gene.[86,108] The *sih* mutant itself provides evidence for the role of contractile force generation in the regulation of myocardial proliferation since one of the attributes of the mutant hearts is a thin-ventricular wall.[109] However, since both chambers are noncontractile, it is difficult to dissect the distinct roles of contractile from flow induced forces on the growth of the heart. Since the mutant *wea* lacks only atrial contractility and retains some flow, it has been more fully examined to determine the potential influence of these epigenetic factors on cardiac morphogenesis. Loss of atrial function has a severe detrimental effect on growth of the ventricular chamber; defective morphogenesis in the cardiac chamber not directly affected by the primary mutation is a common observation in chamber-specific contractile mutants.[96,110,111] Although the *wea* ventricle is able to continue to contract and maintain a weak circulation, the myocardial wall becomes thickened and changes in the transcriptional profile are detected.[96] These alterations in the shape of the organ proceed from alterations in cellular morphology. Both flow and contractility are feasible sources for the force generation that affects the shape of ventricular cells. However, by using a combination of crosses in and between wea and a mutant with impaired ventricular contractility (*half-hearted* (*haf*), mutation in ventricular myosin heavy chain, vmhc) it was possible to clarify that contractility was the key determinant of this cell size and shape transformation, and that it was independent of the influence of flow.[112]

In addition to the complexities of dissecting out the influences of the various biomechanical forces on heart growth, there are also important signals between cells in the adjacent tissue layers of the heart. This type of cell–cell communication is a common paradigm seen during tissue and organ morphogenesis, and important roles for endocardial proteins and epicardial tissue on myocardial development and maintenance have been described.[86,87,113–119] The role of hemodynamic forces on heart chamber morphogenesis is further complicated by its influence on valve growth. An initial defect in valve formation can alter flow through the cardiac chambers and thus create a feedback loop of impaired flow that would impact chamber and valve development further. Alternatively, a defect that affects myocardial growth could impair chamber contraction, thereby reducing flow and causing what might be perceived as a primary valve defect. Shear stress on the endocardial lining which is induced by flow through the heart has also been shown to be required for looping and formation of the bulbus arteriosus in zebrafish.[100] Hypoplastic Left Heart Syndrome (HLHS) is a pediatric disease characterized by an underdeveloped left ventricle that has been proposed to be linked, in part, to abnormal valve and/or flow dynamics. Although dramatic improvements have been made in the treatment of this condition,[120] the genetics remain unclear, even though it demonstrates strong heritability.[121] A good animal model for the study of this condition would be extremely valuable and some studies demonstrate that altered flow regulates ventricular development and can cause HLHS-like and valve defects.[99,100,122]

The presence of these mechanical forces on cells in the heart and vasculature requires the maintenance of strong cellular interactions. The desmosome is a specialized junctional complex that is present in cardiac and skeletal muscle, and mutations in genes encoding desmosomal proteins have been identified as a major cause of arrhythmogenic right ventricular cardiomyopathy (ARVC).[123,124] The knockdown of the plakoglobin gene in zebrafish results in cardiac defects that include reduced heart size and rate of contraction.[125] Although analysis of endothelial cell morphology and *in situ* expression within the atrioventricular boundary suggest a defect in valve formation, this may be a secondary phenotype associated with loss of normal cell junction formation. However, consistent with evidence of the alteration of canonical Wnt/β-catenin signaling,[126] inhibition of Wnt signaling is able to rescue the cardiac phenotype, demonstrating conservation of the molecular pathway in zebrafish as well.[125]

X. Contribution of Blood Flow to Vessel Development

The contribution of blood flow to blood vessel formation and remodeling has been a major topic of study. Mechanical forces generated by laminar or turbulent blood flow are known to induce cytoskeletal changes and alter endothelial gene expression patterns.[127,128] Indeed, blood vessels were presumed to acquire arterial or venous identities after the establishment of directional blood flow until genetic determinants were first identified through the mouse null models for ephrinB2 and EphB4.[129,130] Although AV specification and patterning is established before circulation begins, blood flow modifies early embryonic signals to refine and remodel the vasculature.[57,131] The optical transparency of the zebrafish embryo has provided another advantage here where the mechanical effects of blood or fluid flow may be examined within intact vessels.

Integration of molecular pathways regulating mechanical forces and vascular endothelial growth factor (VEGF) signaling on blood vessel development defined a novel role for miRNAs.[66] These authors focused on the remodeling of aortic arch blood vessels in response to blood flow using the zebrafish model and determined that flow induced expression of the mechanosensitive zinc-finger transcription factor, klf2a. This induces the expression of an endothelial-specific microRNA, mir-126, which in turn activates VEGF signaling. This study revealed a novel mechanical activation of a microRNA to integrate blood flow dynamics with angiogenic signaling during blood vessel development.[66] As the adult zebrafish tail fin is semitransparent, the contribution of blood flow and contractile forces to regenerative angiogenesis might also provide additional insights. The transition from an avascular regenerated tissue through an initial vascular plexus to the final reassembly of functional vessels occurs within 14 days in regenerated fin.[132] Future studies might be able to take advantage of genetic and transgenic lines to define mechanical and molecular regulators for adult vessel regeneration.

XI. Genes Associated with Cardiomyopathies Are Required for Contractile Function

Mutants with defects affecting either ventricular (*silent partner, sil*[111]; *pickwick, pik*[110]) or atrial contraction (*weak atrium, wea*[96]), or both (*silent heart, sih*[109]) were identified from the original large-scale ENU mutagenesis screens and the genetic lesions were identified by positional cloning. These studies focused on the loss of cardiomyocyte contractility in the autosomal recessive zebrafish mutants, proposed to be the manifestation of a loss of protein function. Interestingly, these four mutations were located in the zebrafish homologs of sarcomeric genes associated with the pathogenesis of cardiomyopathy in humans (cardiac troponin C, TTNC1 (*sil*; *tnnc1a*), titin (TTN) (*pik*; *ttna*), alpha myosin heavy chain, MYH6 (*wea*; *myh6*), and cardiac troponin T2 (*sih*; *tnnt2a*). The potential use of these mutants as models for cardiomyopathy remains to be fully exploited, in part because of the dearth of available assays for the study of adult phenotypes and measurement of physiological parameters in zebrafish. However, since these genes are well conserved across vertebrate species, the zebrafish mutant can be employed in assays to determine the affect of the human mutations on protein function. The cDNAs with mutations associated with the human cardiomyopathies could be cloned into expression vectors for injection into zebrafish embryos. If the human mutant proteins retain some function, they may rescue the autosomal recessive zebrafish mutant phenotype while other, more severe lesions, may not.

The analysis of the developmental role of the nexilin (nexn) gene in zebrafish is an example of this approach and demonstrates the power of this model system to rapidly validate mutations and model CVD.[133] The determination of mutations in the human gene (NEXN) as a source of human dilated cardiomyopathy was preceded by the analysis in zebrafish. Nexilin is expressed in both cardiac and skeletal muscle at the site of Z-disks, which are essential for the cross-linking of thin filaments and transmission of force generated by the myofilaments. By knocking down nexilin function in zebrafish, an essential role for this protein in stabilization of the Z-disk was shown. This decreased Z-disk function manifested as heart failure in zebrafish embryos and increased mechanical strain worsened the Z-disk damage.[133] When patients with dilated cardiomyopathy were examined and found with nexilin mutations, analysis of their tissue revealed the same cardiac Z-disk pathology as predicted from the nexilin-deficient zebrafish embryos.[133] The human mutations were confirmed as pathological by expressing the mutant forms of the nexilin proteins in zebrafish. These mutant proteins, when expressed in wild-type embryos, induced Z-disk damage and heart failure which also supported a dominant-negative mechanism, consistent with the human disease.[133]

The mutations in the genes themselves may also provide information regarding the importance of different protein motifs for function akin to deletion analyses performed with cDNAs. The original description of the positional cloning of the *pickwick* mutant described several alleles, including one with both skeletal and cardiac defects.[110] The distinction in the phenotypes between these different alleles suggested that regions of the titin protein exist that are more important for its function in cardiac muscle. This is consistent with the description of the cardiac-specific N2B element, a unique 572-residue sequence in the extensible region of cardiac titin that is involved in Ca^{2+} sensitivity.[134,135]

The characterization of the *silent partner* mutant reveals another aspect of the zebrafish model system that can be a useful advantage for the study of gene function in a specific tissue. While the gene encoding cardiac Troponin C is highly conserved in all vertebrate species, from human through to zebrafish, the genome of the latter contains two paralogs of this gene.[111] Due to an ancestral genome duplication event, many genes in the zebrafish genome have duplicates that demonstrate a partitioning of expression and function.[136,137] Although it is true that this duplication may sometimes make the genetic analysis of a specific gene difficult due to compensation of function by the other paralog, the zebrafish *tnnc1a* gene is expressed predominantly in the heart while the paralog (*tnnc1b*) is expressed at higher levels in skeletal muscle. This permitted an analysis of cTnC function in either tissue type by suppression of a single paralog.[111] Since few mutations associated with human familial cardiomyopathies have been demonstrated in the cardiac troponin C gene,[138,139] the *silent partner* mutant provides a model to assess the effects of the reported human mutations and their impact on protein function specifically in the heart.

Mutations in both the essential and regulatory light chain genes (MYL3 and MYL2, respectively) have been determined in patients with familial hypertrophic cardiomyopathy.[140,141] In addition, analysis of mutations in the myosin light chain kinase 2 (MYLK2) links defects in this gene with increased disease severity.[142] In zebrafish, two additional light chains have been determined to be important for cardiac function although there is no direct link between these specific genes and a human familial defect. In the mutant *tell tale heart* (*tel*), the regulatory cardiac myosin light chain gene 2 (myl7/cmlc2) is disrupted, causing the heart to contract weakly.[143] The mutant *lazy susan* (*laz*) also displays a similar phenotype as a result of a mutation in the atrial essential myosin light chain gene (cmlc1/similar to human MYL4).[144] These genes have also been studied in zebrafish by morpholino knockdown, revealing a similar defect to that observed in the mutant. However, analysis of *myl4* and *myl7* morphants also revealed a disruption of sarcomere structure. The changes in sarcomere length were distinct between the morphants; it was increased as a result of *myl4* loss while it was decreased in *myl7* morphants.[145] This change in sarcomere length translates into alterations in cardiomyocyte size, with larger cells in the myl4 morphants and smaller cells in the myl7 morphants.[145] These distinctions between the defects associated with loss of an essential versus regulatory myosin light chain protein reflect different cellular pathologies and may have implications for the progression of disease in human cardiomyopathy patients with defects in other myosin light chain genes.

In addition to disease models derived from the identification of zebrafish mutants from phenotype-based screens, it is also feasible to employ a gene-specific approach. Potential candidate genes can be examined using morpholino antisense technology to knockdown protein function by either inhibition of translation (by targeting the ATG start site)[21] or by blocking pre-mRNA splicing (by targeting splice donor and acceptor sites).[19] This approach has been employed to examine several genes with known linkage to human dilated cardiomyopathy.[146] Loss of function in these genes caused a consistent impairment of cardiac function with gene-specific variations in cellular architecture mimicking that detected in the human disease.[146]

XII. Gene Defects that Cause Arrhythmias

In zebrafish, the loss of function of the sodium–calcium exchanger 1 (slc8a1a/ncx1/ncx1h; *tremblor* (*tre*) mutant) causes a unique defect in contraction of the embryonic heart that resembles cardiac fibrillation.[147,148] Although the human homolog is not directly linked to simple Mendelian forms of cardiac disease, the exchanger is upregulated in cardiac hypertrophy, ischemia, and failure.[149] In the zebrafish mutant hearts, the calcium transients are clearly disrupted, in contrast to the normal currents seen in other mutants that exhibit impaired contractility, such as *sih*.[147] Similarly, altered Ca^{2+} transients in human hearts may contribute to an increased risk of arrhythmias and diastolic dysfunction in patients. The importance of the NCX1 protein in calcium homeostasis is evident from the sensitivity of gene dosage when rescuing the mutant phenotype by injection of mRNA.[147] This gene, like the cardiac troponin C gene discussed earlier, is also duplicated in zebrafish but the paralog slc8a1b is not highly expressed in the heart and is instead required for the extrusion of Ca^{2+} from fish gill cells.[150]

The mutant *island beat* (*isl*) exhibits a similar phenotype and was determined to have a genetic lesion within the L-type Ca^{2+} channel (cacna1d; cav1.3a).[151] Although the ventricle is noncontractile, individual cells within the atrium beat in an asynchronous fashion resembling atrial fibrillation in humans, consistent with the role of this gene in the coordination of cardiomyocyte contraction.[31,151] However, while the atrium appears relatively normal morphologically, the ventricle comprises fewer cardiomyocytes than wild-type.[151] This type of defect was not predicted from the identity of the mutant gene and implicated the cacna1d gene in a previously uncharacterized role in the regulation of cellular proliferation. Similarly, the ventricular contractility defect in the mutant *deadbeat* (*ded*)[152] was determined to result from mutation in the gene phospholipase C gamma 1 (plcg1), a gene previously implicated in arterial development.[153] The phenotype of the mutant can be rescued by plcg1 expression in cardiomyocytes while elimination of VEGF interferes with cardiac contractility[152]: these data implicate this pathway in a paracrine mechanism regulating cardiac contractility as well as vessel development.

Long QT syndrome is a cardiac defect detected as an increase in the QT interval on an electrocardiogram. This abnormality results in an abnormal heart beat that may lead to sudden death as a result of cardiac arrhythmia.[154] Although several genes have been identified in familial forms of this disease, 30–40% of cases display no mutations linked to these genetic loci.[155,156] One of the genes demonstrating a clear involvement in the disease pathology is the ether-à-go-go-related channel (KCNH2; ERG1).[155] In zebrafish, kcnh2 loss of function induced by genetic mutation or morpholino knockdown with a translational blocking morpholino results in a complete loss of ventricular contraction.[157]

However, when some residual protein activity remains, the hearts of zebrafish embryos exhibit a 2:1 heart block (two atrial contractions for a single ventricular contraction) and also bradycardia (slower heart rate), as revealed by a dose-dependent examination with high and low-morpholino doses, respectively.[158]

The 2:1 heart block induced with morpholino doses that do not cause a complete loss of function is similar to the phenotype of the mutant *breakdance* (*bre*).[158,159] This mutant does have a mutation in the kcnh2 gene but unlike the null alleles (s213 and s290),[157] it is a hypomorph, reducing but not eliminating activity.[158] The similarity in the biophysical properties of the zebrafish and human KCNH2 channels and the attributes of these two zebrafish mutant alleles provide support for their use as models for human long QT syndrome.[160] A third mutant that has been determined to be an activating missense mutation in kcnh2 (*reggae* (*reg*) mutant[31,161]) is characterized by an initial loss of contraction throughout the early heart tube (24 hpf) with only random waves of contraction. Small patches of beating cardiomyocytes are detectable near the inflow tract (sinus venosus) of the heart but are rarely propagated through the adjacent atrial chamber. When this occurs, the uncoordinated contraction resembles an atrial fibrillation, akin to that seen in *tre*. However, rare pulses of contraction that proceed in normal atrial–ventricular sequence suggest a brief relief in the atrioventricular block. These properties are commonly associated with sinus node exit block and suggest that the *reg* mutant may actually serve as a better model for human short QT syndrome.

In addition to zebrafish models with mutations in genes that are known to be associated with human cardiac disease, there are others with defects yet to be identified; *slow mo* (*smo*), a spontaneous mutant identified from the Hong Kong/Singapore background that is characterized by a low-heart rate is one such mutant. The resting heart rate is able to increase as the embryo develops but it remains slower than wild-type from the onset of contraction at 24 hpf. Dissociation of cells and analysis in culture reveal a similarly decreased contractile rate,[46] providing evidence that the bradycardia was an inherent property of the cardiomyocyte. The hyperpolarization-activated cation current (I_h, the fast kinetic component of this inward rectifying current) in *smo* hearts is abnormal. Although a defect in this pacemaker current would explain the phenotype, the channel associated with this current is not responsible for the mutation[162] and no alteration in the slow component was observed.[46] The observation that the *smo* mutants show no increase in heart rate in response to temperature elevation suggests that the mutation may not be a result of a defect in a channel protein.[46]

Although this mutant is autosomal recessive, a small number of mutants are able to survive to adulthood, unlike most of the other cardiovascular mutants. As a result, it is possible to generate clutches of embryos that are entirely homozygous for the mutation. This is a notable advantage for potential genetic studies and to identify small molecule modifiers since all embryos would exhibit the phenotype. Also, some of the adult homozygotes are viable and the hearts are functional. However, the effects of this mutation on cardiac function in the adult remain to be studied. The identification of the genetic lesion will eventually provide an obvious candidate gene to examine in humans with arrhythmias and associated contractile dysfunction.

XIII. Shared Pathways in Zebrafish Development and Human Vascular Diseases

The crucial role of the flk1 receptor in VEGF signaling is conserved in the zebrafish model and mutations affecting the zebrafish flk1/VEGFR2 gene (kinase insert domain receptor like; kdrl)[163] have been described.[164,165] The mutations reported by Habeck *et al.*[37] are nonsense mutations which allow development up to about 4 dpf with only mild pericardial edema.[37] Alkaline phosphatase staining of blood vessels revealed abnormal architecture of the arterial ISVs and subintestinal veins due to reduced VEGFR2 signaling. These defects are partially rescued by overexpression of the VEGF-A ligand.[37] A kinase inactive allele with a stop codon in the receptor's tyrosine kinase domain has also been identified and characterized. The recessive mutant appears phenotypically normal up to about 24 hpf, with minor reduction in the midcerebral vein[165] but develops severe edema by 2–3 dpf. Formation of the heart, DA, and CV all appear normal and functional. However, disruptions in arterial ISV sprouts from the DA lead to malformed primary segments[54,165] thereby interrupting ISV development and intersegmental blood flow. A remarkably similar phenotype is seen when the VEGFR effector gene, plcg1, is disrupted[153] and mirrors results from mouse flk1/VEGFR2 knockout and knockin studies.[166–168] In the zebrafish model, the plcg1 gene is preferentially expressed in arterial endothelial cells; thus, reduced function affects vascular development.[153]

The pathway signaling downstream of VEGF activation is also conserved in zebrafish, an important prerequisite for a system being proposed as a model for human disease. Ligand binding to VEGFR2 activates both the MAPK/Erk and the PI3K-AKT signaling pathways,[166,169,170] and functional conservation of both has also been demonstrated in the zebrafish model.[132,153,171,172] Using a VEGFR inhibitor for a 6-hour period, ISV formation can be reduced to about 30%.[171] However, in the presence of overexpressed activated-AKT but not its kinase inactive version, about 80% of these vessels can form despite functional blockade at the receptor level.[171] Similarly, a study using phospho-specific antibody staining identified a crucial period between the 18-somite and 24-somite stage at which the pERK (phosphorylated, active form) signal can be transiently detected only in the dorsal cells of the aorta.[172]

XIV. Molecules Required for Vessel Formation and Repair

It is interesting to note that the delineation of the genetic programs governing arteriovenous endothelial cell specification has been defined through the use of animal models.[57] This began with the discovery in mice that ephrinB2 and its receptor, EphB4, are distinctly expressed in arterial or venous endothelial cells, respectively.[129,130,173] Using mouse knockout and LacZ knockin models, loss of ephrinB2 or EphB4 function led to embryonic lethality due to a failure in vascular remodeling. These studies provided the first evidence of a genetic component in the determination of arterial or venous endothelial cell fate before the onset of blood flow.

An important contribution from zebrafish vascular studies has been the delineation of genes governing the arterial endothelial specification program.[57] Studies using genetic lines, chemical inhibitors, and morpholino approaches helped define a genetic hierarchy leading to the expression of ephrinB2 in arterial endothelial cells.[174–176] Major genes in this program include shh (sonic hedgehog), which induces VEGF-A expression, leading to precise level of Notch signaling in arterial endothelial cells, which regulates ephrinB2 expression. Parallel studies in the mouse and fish models implicated several genes in the Notch pathway as well as the downstream target gene gridlock/hey2 in this specification pathway.[57] From these studies, a consensus was beginning to form in which the venous endothelial cell fate was presumed to be a default pathway during embryonic development. However, this idea was challenged by the determination of the orphan transcription factor, COUP-TFII, as a major determinant of venous cell fate.[177] The current view is that alterations in the expression level of upstream regulators such as shh, VEGF-A, COUP-TFII, and Notch signaling can cause inappropriate expression of arterial or venous endothelial markers at vessel locations for veins or arteries, respectively.[174–177]

The expression of ephrinB2 or EphB4 has been widely accepted as definitive markers for arterial or venous endothelial cells.[57] The proposed cellular mechanism for their participation in arteriovenous distinction has been a traditional role in adhesive/repulsive interactions previously defined for this unique family of cell-attached ligands and receptors, known to mediate cell sorting events during development.[178] Repulsive cell movements are largely responsible for the segregation of endothelial cells into aorta or CV.[179] Additional endothelial cell sorting and migration between the aorta and CV have been suggested as another mechanism for refinement.[180] However, the role of ephrinB2–EphB4 interactions in the activation of distinct endothelial pathways to mediate these cellular events is currently unknown. Interestingly, the expression of arteriovenous markers is also seen in adult vessels during normal regeneration and in those that form under pathological conditions. It was noted in early studies that arterial expression of ephrinB2 was found in adult human arteries and also those in tumors.[181,182] With the identification of this important arterial gene program, researchers began to search for the activation of this genetic pathway in ischemic vascular models such as coronary vascular reperfusion.[183]

XV. From Vascular Biology Research to AntiAngiogenic Therapy

The normal growth of new blood vessels is essential for embryonic development and in adults for neovascularization during injury and repair. Interestingly, key molecules regulating developmental angiogenesis are deregulated in the formation of pathological vessels, including those found in ischemia, cancer, diabetes, and macular degeneration.[184] Whereas normal neovascularization proceeds by a precise interplay between proangiogenic and antiangiogenic regulators, pathological angiogenesis is characterized by an imbalance between these factors, leading to abnormal or nonfunctional blood vessels.[185] These observations provided the molecular basis for developing antiangiogenic therapies over that last 30 years. Studies on zebrafish vascular development in the last 15 years have shown that many cellular processes, genetic programs, and signaling pathways are highly conserved with mammalian vasculature. Thus, it is likely that studies in the fish model will contribute to and expedite the process from defining molecular mechanisms to their clinical application.

Historically, the development of antiangiogenic therapies has largely depended on mouse and cell-based models. Among angiogenesis-dependent diseases, cancer treatment has been the first to benefit. Although cancer is a complex disease process, with genetic predispositions and preferred sites of metastasis for each tumor type, its progression is dependent upon its blood supply.[186] Deregulation of growth control and reactivation of embryonic genes are known as hallmarks of cancer cells.[187] In a similar manner, tumor blood vessels also reactivate embryonic genes and signaling pathways. Among angiogenic factors, VEGF (referring to VEGF-A) has been most extensively studied due to its robust effects on blood vessel formation.[184,185,188] The ability to use human endothelial cells and mouse models led to the therapeutic use of an anti-VEGF antibody drug (Bevacizumab/Avastin) in combination with chemotherapy for the treatment of cancer[189]. Understanding the important role of VEGF in an age-dependent, vascularized form of macular degeneration (aberrant choroidal neovascularization[190]) provided the molecular basis for the use of a similar therapy (Ranibizumab/Lucentis).[191,192] In this case, the use of a single antiangiogenic agent effectively reduced the occurrence of abnormal vessels in a substantial proportion of cases.[193]

One consequence of increased VEGF signaling is the activation of its receptor tyrosine kinase (RTK) signaling pathway downstream of the VEGFRs. Thus, after tremendous effort in developing small molecule kinase inhibitors targeting the VEGFRs, two of them have also been approved for cancer treatment,[194] while additional inhibitors are at various stages of clinical trials with updated information through the National Cancer Institute website. It is also important to note that targeting VEGF can also lead to undesirable effects including hypertension, proteinuria, thrombosis, and hemorrhage, all related to the endogenous functions of this factor and its signaling pathway.[194] Another possible concern that could arise from anti-VEGF therapy is that it could alter the precise spatiotemporal expression of ephrinB2 and EphB4, needed for blood vessel regeneration and repair.

Eventually, the therapeutic regulation of abnormal angiogenesis may be extended to the treatment of additional human diseases where poor tissue perfusion and edema become progressively debilitating. For example, vascular dysfunction is an underlying factor in obesity, diabetes, and metabolic syndrome, contributing to the development of CVD.[195–199] Thus, as we learn more about the molecular and cellular mechanisms governing blood vessel development, additional therapeutic targets might be identified allowing combinations that could fine tune their use for a specific pathology. The developmental and molecular conservation of angiogenesis in zebrafish positions it as a versatile model system for such studies.

XVI. Vessel Guidance in Development and Disease

Angiogenic sprouting is largely responsible for the formation of smaller blood vessels in development and in disease processes. The identification of related guidance molecules in the navigation of neuronal axons and blood vessels provided an interesting biological parallel to advance our knowledge in both neural and vascular fields.[200,201] Neuronal guidance molecules such as ephrins, plexins, netrins, semaphorins, and Robo play similar roles in the proper wiring of blood vessels. In the zebrafish model, the formation of ISVs provided a system to dissect the molecular players involved, as sprouts are limited to one per somite.[54,202] This was illustrated by the genetic mutant, out-of-bounds, due to a mutation in the gene plexinD1.[203] Multiple sprouts simultaneously extend from the DA into the intersegmental region due to an inability to sense restriction by somitic expression of the semaphorin3a ligands. However, the formation of sufficient functional connections can provide adequate blood flow to allow some mutants to survive to adulthood.[202] A common theme began to appear where a ligand–receptor pair from each guidance family seems to participate in the vascular navigation process.[200,201]

Although the molecular players used for the guidance of axons and vessels are related, a structural difference distinguishes neuronal and vascular navigation. Neurons are specialized cells with cell bodies located at some distance from axons. Growth cones at the tips of axons sense attractive or restrictive cues and then relays information back to extend the axon in a particular direction. In vessel guidance, the process is carried out by different cell types. Endothelial cells at the tip of a blood vessel are known as tip cells, which extend a number of filopodia to detect local cues. This information is transmitted to the adjacent endothelial cell, called a "stalk cell," which extends the blood vessel through proliferation. Dissection of these cellular processes has advanced our understanding of sprouting angiogenesis under developmental conditions and in tumor growth. High-resolution images and marker analysis have defined unique features of the guiding endothelial tip cell. The number of filopodia and invasive migratory behavior of tip cells depend upon Dll4 and Notch signaling. Reexpression of these molecules also occurs in tumor blood vessels.[65,204–206] Based on findings in the zebrafish and mouse models, the importance of a precise level of endothelial Notch signaling has provided a novel target pathway for antiangiogenic therapy.

A precise amount of Notch activation determines the correct proportion of arterial or venous endothelial cells,[153,179] while an abnormal amount, either too much or too little, can induce excessive angiogenesis, generating vessels that do not carry blood flow; this has been termed "nonproductive" angiogenesis.[205] A proper ratio of tip cells and stalk cells are fundamental for proper navigation of a new blood vessel, as the tip cell does not divide but functions to extend filopodia to interact with other cell types, guidance cues, and cell surface receptors in order to determine the directional growth of a blood vessel. The role of the stalk cell is to divide, allowing the vessel to grow in length, in order to reach tissues that require blood supply. This occurs during embryonic development and in retinal vessels.[65,206,207] However, these molecules and cellular interactions have also been observed in tumor angiogenesis,[204,205] suggesting that the targeting of Dll4 might be able to prune the excessive angiogenesis. Although early experimental trials suggest that this might be a possible target for antiangiogenic therapy,[208] a later study by the same group demonstrated that chronic alteration of Dll4 signaling leads to the formation of neoplasms.[209] However, an alternative strategy to target the Notch receptors appears quite promising. In this approach, the authors targeted the Notch receptor's negative regulatory region (NRR). Upon ligand biding, the NRR enables ADAM protease cleavage at a juxtamembrane site that otherwise lies buried within the quiescent NRR. Subsequent proteolysis by the gamma-secretase complex releases the intracellular domain (ICD) to initiate Notch-dependent changes in transcription. By generating specific antibodies to the NRR of Notch 1 or Notch 2 these receptors are essentially locked in an inactive state. Using mouse tumor models, this strategy reduced tumor size and its vascular supply. Thus, cellular and molecular mechanisms defined in animal models of angiogenesis have provided novel therapeutic targets.

XVII. Lymphangiogenesis and Lymphatic Vessel Dysfunction

Primary and secondary lymphangiogenesis and lymphedema are major causes of debilitation[210] and defects in lymphatic regulation also promote tumor metastasis, a major cause of cancer death. The existence of a lymphatic system in the zebrafish was recently revealed through the use of morpholinos, transgenic, and genetic lines.[55,56] The accessibility of the fish model to high-resolution imaging has already provided additional, previously unknown, cellular details at each step in primary and secondary lymphatic vessel formation.[211–213] Lymphatic vessel formation begins with venous sprouting of endothelial cells to define the ISVs by 2–3 dpf. During this time, venous sprouts either connect with a primary segment to form the venous intersomitic vessel (ISV) or contribute to the parachordal vessel if an arterial ISV is formed. This provides a migratory route for the lymphatic endothelial cell, which begins its journey from the PCV up to the horizontal myoseptum in a path defined by the location of the parachordal vessel. Lymphangioblasts migrate along this path to reside in the space between the two major axial blood vessels. These interspersed spots occupied by lymphatic endothelial cells eventually connect to form the thoracic duct.[55] The formation of the thoracic duct is complete by the first week of development.

In a more recent study, venous-derived lymphatic endothelial cells that travel to the horizontal myoseptum (called parachordal lymphangioblasts) were seen to migrate further and contribute to additional lymphatic vessels.[211–213] Owing to this elaborate route, a spatiotemporal connection between ISV and thoracic duct formation appears to hold true in a number of zebrafish vascular studies where missteps in ISV formation often block the later development of thoracic duct. Indeed, the absence of the parachordal vessel or parachordal lymphangioblasts has been strongly correlated with malformation of the thoracic duct.[55,56,68]

Gene programs driving lymphangiogenesis are similarly conserved between zebrafish and mammalian models, with perturbation of lymphatic gene expression affecting this process. For example, reduced function of Prox1 or VEGF-C induced by morpholino antisense targeting leads to the complete lack of lymphatic development and extensive edema.[55,56] In contrast, partial function of the Notch genes appears sufficient for blood but not lymphatic vessel development.[213] In a morpholino-based study, an AUG morpholino against Dll4 or Notch genes blocked ISV formation while a later acting splice-blocking morpholino allowed ISVs to form but not the lymphatic ISVs. Interestingly, only the arterial ISV provided guidance for navigation of the lymphatic ISVs. Since these secondary lymphatic vessels journey alongside the arterial but not the venous ISVs, they will occur at only half the number of sites and support a function for arterial gene pathways in lymphatic vessel guidance.[211–213] This is consistent with data from other model systems, such as mouse, where the ephrinB2 gene is expressed during later development in lymphatic endothelial cells despite its characterization as a definitive marker for arterial endothelial cells. Mouse knockin studies targeting the cytoplasmic domain of ephrinB2 demonstrated an important role in the formation of lymphatic vessels through functional interactions at its C-terminal PDZ binding site.[214]

Genes regulating lymphatic endothelial identity such as VEGF-C, VEGFR3, prox1, sox18, and foxc2 have conserved functions in zebrafish. However, a more comprehensive and detailed description of the genetic pathways regulating lymphangiogenesis will be essential to understanding lymphatic vessel dysfunction. In a forward genetic screen for defects in lymphatic duct development, a mutation called *full of fluid* (*fof*) was identified and positionally cloned.[68] The genetic lesion is due to a mutation in the secreted collagen and calcium-binding EGF-domain-1 protein (CCBE1[68]) gene. Although no function has been defined for this gene as yet, a recent study of familial lymphedema and lymphatic malformations identified rare mutations in the human homolog among large families where recessive mutations in CCBE1 correlated with severe lymphedemas.[67] In this case, a novel gene first identified in the zebrafish model led to a clinically relevant discovery of a human genetic defect in lymphangiogenesis. Thus, expanded use of the zebrafish model to examine genes and signaling pathways affecting lymphangiogenesis could be valuable.[211–213]

XVIII. Models for Vascular Anomalies

The *heart of glass* (*heg*), *santa* (*san*), and *valentine* (*vtn*) mutants share a similar phenotype which is characterized by a hugely dilated heart.[30,31] This defect is caused by a failure of the myocardium to thicken beyond a single cell layer.[86,87] Positional cloning revealed that the genetic lesions in two of these mutants (*san* and *vtn*)[86,87] are within the zebrafish homologs of two genes associated with a genetic cerebral vascular disease in humans (Cerebral Cavernous Malformations, CCM; in humans, CCM1 and CCM2, respectively).[215–217] The human dominant disease is associated with abnormally dilated vascular channels within the CNS which are believed to arise as a result of localized defects in blood vessel development. The phenotypic similarities of the two zebrafish autosomal recessive mutations considered together with the presence of dominant mutations in the human homologs were strong evidence of a conserved common pathway. Interactions between these proteins, as well as with the transmembrane protein Heg (Heg1 in mammals), have now been demonstrated in multiple species.[87,88,218,219]

The Heg-CCM pathway is crucial to early cardiogenesis, so humans with a complete loss of signaling through this pathway may not be viable.[86–88] The human disease is dominant and, therefore, at least partial function is retained during early development even if the proposed two-hit mechanism for CCM is correct; this second genetic "hit" would cause a localized loss of function within specific cells or a cell lineage.[220–222] Even though the phenotype of the zebrafish mutant is a primary patterning defect that affects morphogenesis of the heart, the underlying cellular pathology of the zebrafish and human malformations may be similar. The zebrafish mutants reflect a defect in signaling from the endocardial, inner layer of the heart to the adjacent, myocardial cell layer.[86,87] Similarly, the defective cells in human patients that give rise to the vascular anomalies could also reflect a defect in signaling between an endothelial cell layer and an adjacent layer of tissue.

Both *san* and *vtn* mutant embryos show a reduced arterial and increased venous lumen size, as well as abnormalities in subintestinal vein (SIV) formation[223] but similar defects have been observed in mutants lacking flow, as previously described for the cardiac contractility mutant *sih*.[81,109] Since the gene disrupted in *sih*, tnnt2a, is restricted to myocardial muscle, the vascular defects in these mutants are most likely a result of loss of circulation in the embryo. The observed vascular defects in the *san* and *vtn* mutants also occur relatively late during embryonic development, well after the cardiac defect which prevents the onset of circulation, suggesting that these are secondary aspects of the phenotype and may not reflect defects in CCM signaling *per se*.[87]

Hereditary hemorrhagic telangiectasia type 2 (HHT) is a related vascular anomaly in humans that is associated with mutations in the activin receptor-like kinase 1 (Acvrl1; Alk1), a TGFβ type I receptor that is expressed in the vascular endothelium.[224] Disruption of the zebrafish homolog of Acvrl1 in the zebrafish mutant *violet beauregarde* (*vbg*) results in an abnormal circulation pattern detectable by 2 dpf; while some flow through the cranial vessels is present, no caudal circulation is seen through the trunk of the embryo.[225] The cranial vessels are dilated, a result of increased endothelial cell number within these structures.[225] This aspect of the zebrafish mutant is similar to the cranial vessel malformations seen in HHT, suggesting that *vbg* would be a useful model for this human autosomal dominant disorder. The *gridlock* (*grl*) mutant, a read-through mutation in the hey2 gene, is also characterized by a lack of circulation to the dorsal region of the embryo. However, unlike *vbg*, this loss of flow is a result of a localized blockage at the base of the DA at the site in the embryo where the two lateral dorsal vessels merge to form a single aorta. Because of this defect, the mutant has been proposed as a model for Coarctation of the Aorta, a common congenital condition seen in newborns.

Other vascular anomalies have reported the involvement of the PTEN and RASA1 genes.[226,227] In addition, the VEGFR3 signaling pathway is involved in about 90% of patients with Milroy's disease,[228] while the other genes required for lymphatic endothelial cell differentiation have all been associated with hereditary forms of lymphedema or lymphatic malformations (e.g., sox18, foxc2[210]). Since communication between cell types is an element of the pathogenesis of these diseases, the zebrafish can facilitate the definition of the molecular and cellular interactions between endothelial and mural cells in an intact organism. In addition, VEGF-A to VEGFR2 signaling requires the intracellular activation of both the MAPK/ERK and PI3K-AKT pathways. Deregulation of these downstream effectors for endothelial signaling could result in similar angiogenic consequences. Thus, in addition to developing antiangiogenic therapies targeting the extracellular molecules VEGF and Notch NRR, it may be possible to use small molecule inhibitors against intracellular effectors (e.g., MEK1/2 or PI3K); in fact, such approaches are being developed for cancer treatment. Since adverse effects have already been reported for prolonged use of the anti-VEGF drugs in cancer, the ability to target an endothelial pathway through a receptor and its effectors could reduce the dose required for both drugs, while minimizing the unwarranted effects.

Mutations in the von Hippel–Lindau (VHL) gene predispose patients to the development of highly vascularized neoplasms in multiple organ systems. Similarly, a zebrafish vhl mutant mirrors aspects of the human disease and provides a model for a number of human vascular pathologies associated with hypoxia.[229,230] Like the VHL patients, the zebrafish model is also susceptible to highly vascularized tumors in many organs, including the kidney. In addition, a conserved molecular mechanism affecting VEGF signaling is also responsible in the fish model, providing a clinically relevant model for hypoxia-induced pathological angiogenesis.

XIX. New Directions for Cardiovascular Biology

There remain major questions to be addressed in the field of cardiovascular biology that zebrafish is well suited to help answer. These include (i) the determination of the precise role of blood flow and mechanical forces in refining blood vessel and cardiac chamber development, (ii) the definition of the interactions between endothelial and mural cells in the vasculature and endocardium and myocardium in the heart, (iii) the determination of the molecules responsible for blood vessel heterogeneity between types and locations, and (iv) description of the genes and pathways involved in vessel and cardiac regeneration. However, there are also novel approaches that can be taken to study other types of diseases in this model system. For example, a number of endothelial receptors are usurped by pathogens to mediate entry into host cells. These include ephrinB2, the arterial endothelial marker used as entry receptor for the Nipah/Hendra viruses, as well as ANTXR1/TEM8 (tumor endothelial factor 8) and ANTXR2/CMG2 (capillary morphogenesis gene 2), the receptors for anthrax toxin entry. These and other infectious agents induce vascular damage in humans that leads to vascular leakage, edema, and hemorrhage. Using the zebrafish model and vascular delivery of toxin proteins, the host cell components mediating the effects of a particular toxin have been directly correlated with the toxin's enzymatic activity.[231–233] Thus, by investigating pathogen action during zebrafish vascular development, insights into tissue damage may be revealed. This may be useful for the development of broad therapies, such as agents that improve vascular integrity and resilience, in addition to antibiotics and antiviral approaches.

The regeneration of vascular or cardiac tissue would be a tremendous achievement and substantially impact the course of human CVD. The zebrafish system has already become an important model for these studies since, unlike adult human hearts, zebrafish retain much of this regenerative capacity. Like humans, other commonly used mammalian animal model systems are also refractory to spontaneous myocardial repair in the adult although recent work suggests that some proliferative capacity remains [Refs. 234,235 reviewed in Ref. 236]. The zebrafish are also able to regenerate other tissues including their vascularized caudal fin. Within weeks after amputation, regeneration of blood vessels, bone, and fin tissue is completed. Mutation of the early growth receptor 1 gene (egr1) leads to a complete lack of fin regeneration[237] which can be overcome with application of SKF91488, an inhibitor of histidine methyltransferase.[238] Fin regeneration has a strong dependence on angiogenesis and chemical inhibition of VEGF signaling blocks the growth of fin tissue at less than 1 mm. The inhibition is temporary, however, and fin growth resumes after the inhibitor is removed (Table I).[132]

XX. Summary

The zebrafish is a highly versatile model for exploring the development of the cardiovascular system with great potential for modeling human disease. As molecular disease pathways are better defined, the zebrafish provides a system in which their role in organ and tissue growth, and the consequences of their disruption, can be rapidly defined. Furthermore, the characterization of these

TABLE I

MODELS FOR HUMAN CARDIOVASCULAR DISEASE USING ZEBRAFISH

Gene (mutant)	Expression	Loss of function phenotype	Potential disease association
tnnt2 (*sih*)	Myocardium	Loss of both atrial and ventricular contraction[109]	Developmental influence of hemodynamics; not yet assessed as a model for cardiomyopathy in the adult
myh6 (*wea*)	Myocardium	Loss of atrial contraction[96]	Developmental influence of hemodynamics; not yet assessed as a model for cardiomyopathy in the adult
vmhc (*haf*)	Myocardium	Loss of ventricular contraction[112]	Developmental influence of hemodynamics; not yet assessed as a model for cardiomyopathy in the adult
tnnc1a (*sil*)	Myocardium	Loss of ventricular contraction[111]	Developmental influence of hemodynamics; not yet assessed as a model for cardiomyopathy in the adult
jup/plakoglobin	Endocardium, myocardium	Reduced heart size, impaired contractility[125]	Arrhythmogenic right ventricular cardiomyopathy (ARVC)
nexn/nexilin	Myocardium	Z-disc destabilization, heart failure[133]	Dilated cardiomyopathy
ttna (*pik*)	Myocardium, skeletal muscle	Weak ventricular contraction[110]	Cardiomyopathy
myl7/cmlc2 (*tel*)	Myocardium	Weak contraction of both chambers[143]	Cardiomyopathy
cmlc1/myl4 (*laz*)	Myocardium	Weak contraction of both chambers[144]	Cardiomyopathy
ncx1/ncx1h (*tre*)	Myocardium	Inability to synchronize contraction of the heart[147,148]	Atrial fibrillation, arrhythmias
cacna1d/cav1.3a (*isl*)	Myocardium	Noncontractile ventricle (some cells with uncoordinated contraction).[151]	Arrhythmias
kcnh2 (*bre*)	Myocardium	Silent ventricle in null alleles, 2:1 heart block in hypomorph[157,158]	Long QT syndrome
kcnh2 (*reg^a*)	Myocardium	Random waves of cardiac contraction[161]	Short QT syndrome
unknown (*smo*)		Decrease heart rate[46,162]	Arrhythmias (bradycardia)
VEGFR2/flk1/KDR-like (*t20257, t21588, y17*)	Endothelium, endocardium	Abnormal arterial development; reduced midcerebral vein[37,165]	Vascular disease

(*Continues*)

TABLE I (*Continued*)

Gene (mutant)	Expression	Loss of function phenotype	Potential disease association
plcg1 (*y10*)	Endothelium	Artery-side defect, vein formation normal[153]	Vascular disease
plcg1 (*ded*[b])	Endothelium (myocardium[b])	Progressive loss of ventricle-specific cardiac contractility	Arrhythmias
ccbe1 (*fof*)	Nonendothelial° (but required for lymphatic vessel development)	Defective lymphangioblast budding from venous endothelium[68]	Generalized lymph vessel dysplasia
heg (*heg*)	Endocardium, endothelium	Dilation of atrial and ventricular chambers due to myocardial growth defect[86]	Cerebral cavernous malformations
ccm1 (*san*)	Endocardium, endothelium	Dilation of atrial and ventricular chambers due to myocardial growth defect[87]	Cerebral cavernous malformations
ccm2 (*vtn*)	Endocardium, endothelium	Dilation of atrial and ventricular chambers due to myocardial growth defect[87]	Cerebral cavernous malformations
Acvrl1/Alk1 (*vbg*)	Endothelium	Dilated cranial vessels, lack of caudal circulation[225]	Hereditary hemorrhagic telangiectasia, type 2 (HHT)
hey2 (*grl*)	Endothelium	Lack of caudal circulation (hypomorphic phenotype)[239]	Coarctation of the aorta
vhl (*hu2081*, *hu2117*)	Endothelium, blood	Angiogenic defects; reduction of hemangioblast markers	Von Hippel–Lindau disease (VHL)

Genes with cardiovascular phenotypes generated by protein knockdown (morpholino) or isolation of a genetic mutant are listed along with tissue of expression (where known) and phenotype. Some loss of function phenotypes are associated with defects that may be useful for studying a disease pathology not inferred from the gene. In other cases, the phenotype associated with loss of function may serve as a useful surrogate to study the cellular and molecular pathology, even though the overt phenotype may not closely resemble the disease.

[a]The kcnh2 gene defect in the reggae mutant allele is an activating mutation.

[b]The *ded* allele of plcg1 is a nonsense mutation that results in loss of function. Expression of plcg1 is detected throughout the embryonic heart and may not be specific to the myocardium.

phenotypes may provide assays for the identification of small molecules that can help guide the development of therapies. An improved understanding of the genetic mechanism of a disease pathway may also reveal molecules that have been effectively targeted with treatment strategies already available. The increasing repertoire of tools available to zebrafish biologists will help to establish it as a fundamental model system for the study of other disease types as well.

Acknowledgements/Grant Information

American Heart Association, Scientist Development Grant #0635363N and Innovative Research Grant #0970267N, and the Manton Foundation for Orphan Disease Research (JDM); Department of Defense, Grant #TS093079, and the Manton Foundation for Orphan Disease Research (JC).

References

1. Lloyd-Jones D, Adams RJ, Brown TM, Carnethon M, Dai S, De Simone G, et al. Heart disease and stroke statistics—2010 update: a report from the American Heart Association. *Circulation* 2010;**121**:e46–e215.
2. Chico TJ, Milo M, Crossman DC. The genetics of cardiovascular disease: new insights from emerging approaches. *J Pathol* 2010;**220**:186–97.
3. Manace LC, Godiwala TN, Babyatsky MW. Genomics of cardiovascular disease. *Mt Sinai J Med* 2009;**76**:613–23.
4. Epstein JA, Rader DJ, Parmacek MS. Perspective: cardiovascular disease in the postgenomic era—lessons learned and challenges ahead. *Endocrinology* 2002;**143**:2045–50.
5. Paul M, Zumhagen S, Stallmeyer B, Koopmann M, Spieker T, Schulze-Bahr E. Genes causing inherited forms of cardiomyopathies. A current compendium. *Herz* 2009;**34**:98–109.
6. Hedley PL, Jorgensen P, Schlamowitz S, Wangari R, Moolman-Smook J, Brink PA, et al. The genetic basis of long QT and short QT syndromes: a mutation update. *Hum Mutat* 2009;**30**:1486–511.
7. El-Hamamsy I, Yacoub MH. Cellular and molecular mechanisms of thoracic aortic aneurysms. *Nat Rev Cardiol* 2009;**6**:771–86.
8. Roberts R. A customized genetic approach to the number one killer: coronary artery disease. *Curr Opin Cardiol* 2008;**23**:629–33.
9. Lanktree MB, Dichgans M, Hegele RA. Advances in genomic analysis of stroke: what have we learned and where are we headed? *Stroke* 2010;**41**:825–32.
10. Gluckman PD, Hanson MA, Buklijas T, Low FM, Beedle AS. Epigenetic mechanisms that underpin metabolic and cardiovascular diseases. *Nat Rev Endocrinol* 2009;**5**:401–8.
11. Knapik EW, Goodman A, Atkinson OS, Roberts CT, Shiozawa M, Sim CU, et al. A reference cross DNA panel for zebrafish (*Danio rerio*) anchored with simple sequence length polymorphisms. *Development* 1996;**123**:451–60.
12. Knapik EW, Goodman A, Ekker M, Chevrette M, Delgado J, Neuhauss S, et al. A microsatellite genetic linkage map for zebrafish (*Danio rerio*). *Nat Genet* 1998;**18**:338–43.

13. Fornzler D, Her H, Knapik EW, Clark M, Lehrach H, Postlethwait JH, et al. Gene mapping in zebrafish using single-strand conformation polymorphism analysis. *Genomics* 1998;**51**:216–22.
14. Gates MA, Kim L, Egan ES, Cardozo T, Sirotkin HI, Dougan ST, et al. A genetic linkage map for zebrafish: comparative analysis and localization of genes and expressed sequences. *Genome Res* 1999;**9**:334–47.
15. Geisler R, Rauch GJ, Baier H, van Bebber F, Brobeta L, Dekens MP, et al. A radiation hybrid map of the zebrafish genome. *Nat Genet* 1999;**23**:86–9.
16. Hukriede N, Fisher D, Epstein J, Joly L, Tellis P, Zhou Y, et al. The LN54 radiation hybrid map of zebrafish expressed sequences. *Genome Res* 2001;**11**:2127–32.
17. Hukriede NA, Joly L, Tsang M, Miles J, Tellis P, Epstein JA, et al. Radiation hybrid mapping of the zebrafish genome. *Proc Natl Acad Sci USA* 1999;**96**:9745–50.
18. Vogel G. Genomics. Sanger will sequence zebrafish genome. *Science* 2000;**290**:1671.
19. Draper BW, Morcos PA, Kimmel CB. Inhibition of zebrafish fgf8 pre-mRNA splicing with morpholino oligos: a quantifiable method for gene knockdown. *Genesis* 2001;**30**:154–6.
20. Ekker SC. Morphants: a new systematic vertebrate functional genomics approach. *Yeast* 2000;**17**:302–6.
21. Nasevicius A, Ekker SC. Effective targeted gene 'knockdown' in zebrafish. *Nat Genet* 2000;**26**:216–20.
22. Wienholds E, van Eeden F, Kosters M, Mudde J, Plasterk RH, Cuppen E. Efficient target-selected mutagenesis in zebrafish. *Genome Res* 2003;**13**:2700–7.
23. Moens CB, Donn TM, Wolf-Saxon ER, Ma TP. Reverse genetics in zebrafish by TILLING. *Brief Funct Genomic Proteomic* 2008;**7**:454–9.
24. Henikoff S, Till BJ, Comai L. Tilling. Traditional mutagenesis meets functional genomics. *Plant Physiol* 2004;**135**:630–6.
25. Solnica-Krezel L, Schier AF, Driever W. Efficient recovery of ENU-induced mutations from the zebrafish germline. *Genetics* 1994;**136**:1401–20.
26. Haffter P, Granato M, Brand M, Mullins MC, Hammerschmidt M, Kane DA, et al. The identification of genes with unique and essential functions in the development of the zebrafish, *Danio rerio*. *Development* 1996;**123**:1–36.
27. Meng X, Noyes MB, Zhu LJ, Lawson ND, Wolfe SA. Targeted gene inactivation in zebrafish using engineered zinc-finger nucleases. *Nat Biotechnol* 2008;**26**:695–701.
28. Doyon Y, McCammon JM, Miller JC, Faraji F, Ngo C, Katibah GE, et al. Heritable targeted gene disruption in zebrafish using designed zinc-finger nucleases. *Nat Biotechnol* 2008;**26**:702–8.
29. Foley JE, Maeder ML, Pearlberg J, Joung JK, Peterson RT, Yeh JR. Targeted mutagenesis in zebrafish using customized zinc-finger nucleases. *Nat Protoc* 2009;**4**:1855–67.
30. Chen JN, Haffter P, Odenthal J, Vogelsang E, Brand M, van Eeden FJ, et al. Mutations affecting the cardiovascular system and other internal organs in zebrafish. *Development* 1996;**123**:293–302.
31. Stainier DY, Fouquet B, Chen JN, Warren KS, Weinstein BM, Meiler SE, et al. Mutations affecting the formation and function of the cardiovascular system in the zebrafish embryo. *Development* 1996;**123**:285–92.
32. Alexander J, Stainier DY, Yelon D. Screening mosaic F1 females for mutations affecting zebrafish heart induction and patterning. *Dev Genet* 1998;**22**:288–99.
33. Patton EE, Zon LI. The art and design of genetic screens: zebrafish. *Nat Rev Genet* 2001;**2**:956–66.
34. Jin SW, Herzog W, Santoro MM, Mitchell TS, Frantsve J, Jungblut B, et al. A transgene-assisted genetic screen identifies essential regulators of vascular development in vertebrate embryos. *Dev Biol* 2007;**12**:5208–14.

35. Chen JN, van Bebber F, Goldstein AM, Serluca FC, Jackson D, Childs S, et al. Genetic steps to organ laterality in zebrafish. *Comp Funct Genomics* 2001;**2**:60–8.
36. Covassin LD, Siekmann AF, Kacergis MC, Laver E, Moore JC, Villefranc JA, et al. A genetic screen for vascular mutants in zebrafish reveals dynamic roles for Vegf/Plcg1 signaling during artery development. *Dev Biol* 2009;**329**:212–26.
37. Habeck H, Odenthal J, Walderich B, Maischein H, Schulte-Merker S. Analysis of a zebrafish VEGF receptor mutant reveals specific disruption of angiogenesis. *Curr Biol* 2002;**12**:1405–12.
38. Pendeville H, Winandy M, Manfroid I, Nivelles O, Motte P, Pasque V, et al. Zebrafish Sox7 and Sox18 function together to control arterial–venous identity. *Dev Biol* 2008;**317**:405–16.
39. Cermenati S, Moleri S, Cimbro S, Corti P, Del Giacco L, Amodeo R, et al. Sox18 and Sox7 play redundant roles in vascular development. *Blood* 2008;**111**:2657–66.
40. Herpers R, van de Kamp E, Duckers HJ, Schulte-Merker S. Redundant roles for sox7 and sox18 in arteriovenous specification in zebrafish. *Circ Res* 2008;**102**:12–5.
41. Pelster B, Burggren WW. Disruption of hemoglobin oxygen transport does not impact oxygen-dependent physiological processes in developing embryos of zebra fish (*Danio rerio*). *Circ Res* 1996;**79**:358–62.
42. Barrionuevo WR, Burggren WW. O_2 consumption and heart rate in developing zebrafish (*Danio rerio*): influence of temperature and ambient O_2. *Am J Physiol* 1999;**276**:R505–13.
43. Kalen M, Wallgard E, Asker N, Nasevicius A, Athley E, Billgren E, et al. Combination of reverse and chemical genetic screens reveals angiogenesis inhibitors and targets. *Chem Biol* 2009;**16**:432–41.
44. Mukhopadhyay A, Peterson RT. Deciphering arterial identity through gene expression, genetics, and chemical biology. *Curr Opin Hematol* 2008;**15**:221–7.
45. Pichler FB, Laurenson S, Williams LC, Dodd A, Copp BR, Love DR. Chemical discovery and global gene expression analysis in zebrafish. *Nat Biotechnol* 2003;**21**:879–83.
46. Baker K, Warren KS, Yellen G, Fishman MC. Defective "pacemaker" current (Ih) in a zebrafish mutant with a slow heart rate. *Proc Natl Acad Sci USA* 1997;**94**:4554–9.
47. Stainier DY, Lee RK, Fishman MC. Cardiovascular development in the zebrafish. I. Myocardial fate map and heart tube formation. *Development* 1993;**119**:31–40.
48. Lee RK, Stainier DY, Weinstein BM, Fishman MC. Cardiovascular development in the zebrafish. II. Endocardial progenitors are sequestered within the heart field. *Development* 1994;**120**:3361–6.
49. Vogel AM, Weinstein BM. Studying vascular development in the zebrafish. *Trends Cardiovasc Med* 2000;**10**:352–60.
50. Isogai S, Horiguchi M, Weinstein BM. The vascular anatomy of the developing zebrafish: an atlas of embryonic and early larval development. *Dev Biol* 2001;**230**:278–301.
51. Chen JN, Cowan DB, Mably JD. Cardiogenesis and the regulation of cardiac-specific gene expression. *Heart Fail Clin* 2005;**1**:157–70.
52. Fishman MC, Chien KR. Fashioning the vertebrate heart: earliest embryonic decisions. *Development* 1997;**124**:2099–117.
53. Stainier DY. Zebrafish genetics and vertebrate heart formation. *Nat Rev Genet* 2001;**2**:39–48.
54. Isogai S, Lawson ND, Torrealday S, Horiguchi M, Weinstein BM. Angiogenic network formation in the developing vertebrate trunk. *Development* 2003;**130**:5281–90.
55. Yaniv K, Isogai S, Castranova D, Dye L, Hitomi J, Weinstein BM. Live imaging of lymphatic development in the zebrafish. *Nat Med* 2006;**12**:711–6.
56. Kuchler AM, Gjini E, Peterson-Maduro J, Cancilla B, Wolburg H, Schulte-Merker S. Development of the zebrafish lymphatic system requires VEGFC signaling. *Curr Biol* 2006;**16**:1244–8.

57. Swift MR, Weinstein BM. Arterial–venous specification during development. *Circ Res* 2009;**104**:576–88.
58. Chung AS, Lee J, Ferrara N. Targeting the tumour vasculature: insights from physiological angiogenesis. *Nat Rev Cancer* 2010;**10**:505–14.
59. Stainier DY, Fishman MC. Patterning the zebrafish heart tube: acquisition of anteroposterior polarity. *Dev Biol* 1992;**153**:91–101.
60. Yelon D, Horne SA, Stainier DY. Restricted expression of cardiac myosin genes reveals regulated aspects of heart tube assembly in zebrafish. *Dev Biol* 1999;**214**:23–37.
61. Chen JN, Fishman MC. Genetics of heart development. *Trends Genet* 2000;**16**:383–8.
62. Ehler E, Rothen BM, Hammerle SP, Komiyama M, Perriard JC. Myofibrillogenesis in the developing chicken heart: assembly of Z-disk, M-line and the thick filaments. *J Cell Sci* 1999;**112**(Pt 10):1529–39.
63. Fouquet B, Weinstein BM, Serluca FC, Fishman MC. Vessel patterning in the embryo of the zebrafish: guidance by notochord. *Dev Biol* 1997;**183**:37–48.
64. Lieschke GJ, Currie PD. Animal models of human disease: zebrafish swim into view. *Nat Rev Genet* 2007;**8**:353–67.
65. Siekmann AF, Lawson ND. Notch signalling limits angiogenic cell behaviour in developing zebrafish arteries. *Nature* 2007;**445**:781–4.
66. Nicoli S, Standley C, Walker P, Hurlstone A, Fogarty KE, Lawson ND. MicroRNA-mediated integration of haemodynamics and Vegf signalling during angiogenesis. *Nature* 2010;**464**:1196–200.
67. Alders M, Hogan BM, Gjini E, Salehi F, Al-Gazali L, Hennekam EA, et al. Mutations in CCBE1 cause generalized lymph vessel dysplasia in humans. *Nat Genet* 2009;**41**:1272–4.
68. Hogan BM, Bos FL, Bussmann J, Witte M, Chi NC, Duckers HJ, et al. Ccbe1 is required for embryonic lymphangiogenesis and venous sprouting. *Nat Genet* 2009;**41**:396–8.
69. Stainier DY, Weinstein BM, Detrich 3rd HW, Zon LI, Fishman MC. Cloche, an early acting zebrafish gene, is required by both the endothelial and hematopoietic lineages. *Development* 1995;**121**:3141–50.
70. Thompson MA, Ransom DG, Pratt SJ, MacLennan H, Kieran MW, Detrich 3rd HW, et al. The cloche and spadetail genes differentially affect hematopoiesis and vasculogenesis. *Dev Biol* 1998;**197**:248–69.
71. Pham VN, Lawson ND, Mugford JW, Dye L, Castranova D, Lo B, et al. Combinatorial function of ETS transcription factors in the developing vasculature. *Dev Biol* 2007;**303**:772–83.
72. Xiong JW. Molecular and developmental biology of the hemangioblast. *Dev Dyn* 2008;**237**:1218–31.
73. Liao EC, Paw BH, Oates AC, Pratt SJ, Postlethwait JH, Zon LI. SCL/Tal-1 transcription factor acts downstream of cloche to specify hematopoietic and vascular progenitors in zebrafish. *Genes Dev* 1998;**12**:621–6.
74. Dooley KA, Davidson AJ, Zon LI. Zebrafish scl functions independently in hematopoietic and endothelial development. *Dev Biol* 2005;**277**:522–36.
75. Paik EJ, Zon LI. Hematopoietic development in the zebrafish. *Int J Dev Biol* 2010;**54**:1127–37.
76. Schoenebeck JJ, Keegan BR, Yelon D. Vessel and blood specification override cardiac potential in anterior mesoderm. *Dev Cell* 2007;**13**:254–67.
77. Gering M, Yamada Y, Rabbitts TH, Patient RK. Lmo2 and Scl/Tal1 convert non-axial mesoderm into haemangioblasts which differentiate into endothelial cells in the absence of Gata1. *Development* 2003;**130**:6187–99.
78. Serluca FC, Fishman MC. Pre-pattern in the pronephric kidney field of zebrafish. *Development* 2001;**128**:2233–41.

79. Mudumana SP, Hentschel D, Liu Y, Vasilyev A, Drummond IA. Odd skipped related1 reveals a novel role for endoderm in regulating kidney versus vascular cell fate. *Development* 2008;**135**:3355–67.
80. Gupta S, Zhu H, Zon LI, Evans T. BMP signaling restricts hemato-vascular development from lateral mesoderm during somitogenesis. *Development* 2006;**133**:2177–87.
81. Parker LH, Schmidt M, Jin SW, Gray AM, Beis D, Pham T, et al. The endothelial-cell-derived secreted factor Egfl7 regulates vascular tube formation. *Nature* 2004;**428**:754–8.
82. Mansson-Broberg A, Siddiqui AJ, Genander M, Grinnemo KH, Hao X, Andersson AB, et al. Modulation of ephrinB2 leads to increased angiogenesis in ischemic myocardium and endothelial cell proliferation. *Biochem Biophys Res Commun* 2008;**373**:355–9.
83. Izumo S, Lompre AM, Matsuoka R, Koren G, Schwartz K, Nadal-Ginard B, et al. Myosin heavy chain messenger RNA and protein isoform transitions during cardiac hypertrophy. Interaction between hemodynamic and thyroid hormone-induced signals. *J Clin Invest* 1987;**79**:970–7.
84. Izumo S, Nadal-Ginard B, Mahdavi V. Protooncogene induction and reprogramming of cardiac gene expression produced by pressure overload. *Proc Natl Acad Sci USA* 1988;**85**:339–43.
85. Burns CG, Milan DJ, Grande EJ, Rottbauer W, MacRae CA, Fishman MC. High-throughput assay for small molecules that modulate zebrafish embryonic heart rate. *Nat Chem Biol* 2005;**1**:263–4.
86. Mably JD, Mohideen MA, Burns CG, Chen JN, Fishman MC. heart of glass regulates the concentric growth of the heart in zebrafish. *Curr Biol* 2003;**13**:2138–47.
87. Mably JD, Chuang LP, Serluca FC, Mohideen MA, Chen JN, Fishman MC. santa and valentine pattern concentric growth of cardiac myocardium in the zebrafish. *Development* 2006;**133**:3139–46.
88. Kleaveland B, Zheng X, Liu JJ, Blum Y, Tung JJ, Zou Z, et al. Regulation of cardiovascular development and integrity by the heart of glass-cerebral cavernous malformation protein pathway. *Nat Med* 2009;**15**:169–76.
89. Buchner DA, Su F, Yamaoka JS, Kamei M, Shavit JA, Barthel LK, et al. pak2a Mutations cause cerebral hemorrhage in redhead zebrafish. *Proc Natl Acad Sci USA* 2007;**104**:13996–4001.
90. Liu J, Fraser SD, Faloon PW, Rollins EL, Vom Berg J, Starovic-Subota O, et al. A βPix-Pak2a signaling pathway regulates cerebral vascular stability in zebrafish. *Proc Natl Acad Sci USA* 2007;**104**:13990–5.
91. Kikuchi Y, Trinh LA, Reiter JF, Alexander J, Yelon D, Stainier DY. The zebrafish bonnie and clyde gene encodes a Mix family homeodomain protein that regulates the generation of endodermal precursors. *Genes Dev* 2000;**14**:1279–89.
92. Kupperman E, An S, Osborne N, Waldron S, Stainier DY. A sphingosine-1-phosphate receptor regulates cell migration during vertebrate heart development. *Nature* 2000;**406**:192–5.
93. Keegan BR, Feldman JL, Lee DH, Koos DS, Ho RK, Stainier DY, et al. The elongation factors Pandora/Spt6 and Foggy/Spt5 promote transcription in the zebrafish embryo. *Development* 2002;**129**:1623–32.
94. Horne-Badovinac S, Lin D, Waldron S, Schwarz M, Mbamalu G, Pawson T, et al. Positional cloning of heart and soul reveals multiple roles for PKC lambda in zebrafish organogenesis. *Curr Biol* 2001;**11**:1492–502.
95. Peterson RT, Mably JD, Chen JN, Fishman MC. Convergence of distinct pathways to heart patterning revealed by the small molecule concentramide and the mutation heart-and-soul. *Curr Biol* 2001;**11**:1481–91.

96. Berdougo E, Coleman H, Lee DH, Stainier DY, Yelon D. Mutation of weak atrium/atrial myosin heavy chain disrupts atrial function and influences ventricular morphogenesis in zebrafish. *Development* 2003;**130**:6121–9.
97. Brooks AR, Lelkes PI, Rubanyi GM. Gene expression profiling of vascular endothelial cells exposed to fluid mechanical forces: relevance for focal susceptibility to atherosclerosis. *Endothelium* 2004;**11**:45–57.
98. Grossfeld P. Hypoplastic left heart syndrome: new insights. *Circ Res* 2007;**100**:1246–8.
99. Sedmera D, Hu N, Weiss KM, Keller BB, Denslow S, Thompson RP. Cellular changes in experimental left heart hypoplasia. *Anat Rec* 2002;**267**:137–45.
100. Hove JR, Koster RW, Forouhar AS, Acevedo-Bolton G, Fraser SE, Gharib M. Intracardiac fluid forces are an essential epigenetic factor for embryonic cardiogenesis. *Nature* 2003;**421**:172–7.
101. Van der Heiden K, Hierck BP, Krams R, de Crom R, Cheng C, Baiker M, et al. Endothelial primary cilia in areas of disturbed flow are at the base of atherosclerosis. *Atherosclerosis* 2008;**196**:542–50.
102. Slough J, Cooney L, Brueckner M. Monocilia in the embryonic mouse heart suggest a direct role for cilia in cardiac morphogenesis. *Dev Dyn* 2008;**237**:2304–14.
103. Kramer-Zucker AG, Olale F, Haycraft CJ, Yoder BK, Schier AF, Drummond IA. Cilia-driven fluid flow in the zebrafish pronephros, brain and Kupffer's vesicle is required for normal organogenesis. *Development* 2005;**132**:1907–21.
104. Martin RT, Bartman T. Analysis of heart valve development in larval zebrafish. *Dev Dyn* 2009;**238**:1796–802.
105. Scherz PJ, Huisken J, Sahai-Hernandez P, Stainier DY. High-speed imaging of developing heart valves reveals interplay of morphogenesis and function. *Development* 2008;**135**:1179–87.
106. Walsh EC, Stainier DY. UDP-glucose dehydrogenase required for cardiac valve formation in zebrafish. *Science* 2001;**293**:1670–3.
107. Peal DS, Burns CG, Macrae CA, Milan D. Chondroitin sulfate expression is required for cardiac atrioventricular canal formation. *Dev Dyn* 2009;**238**:3103–10.
108. North TE, Goessling W, Peeters M, Li P, Ceol C, Lord AM, et al. Hematopoietic stem cell development is dependent on blood flow. *Cell* 2009;**137**:736–48.
109. Sehnert AJ, Huq A, Weinstein BM, Walker C, Fishman M, Stainier DY. Cardiac troponin T is essential in sarcomere assembly and cardiac contractility. *Nat Genet* 2002;**31**:106–10.
110. Xu X, Meiler SE, Zhong TP, Mohideen M, Crossley DA, Burggren WW, et al. Cardiomyopathy in zebrafish due to mutation in an alternatively spliced exon of titin. *Nat Genet* 2002;**30**:205–9.
111. Sogah VM, Serluca FC, Fishman MC, Yelon DL, MacRae CA, Mably JD. Distinct troponin C isoform requirements in cardiac and skeletal muscle. *Dev Dyn* 2010;**239**:3115–23.
112. Auman HJ, Coleman H, Riley HE, Olale F, Tsai HJ, Yelon D. Functional modulation of cardiac form through regionally confined cell shape changes. *PLoS Biol* 2007;**5**:e53.
113. Smith TK, Bader DM. Signals from both sides: control of cardiac development by the endocardium and epicardium. *Semin Cell Dev Biol* 2007;**18**:84–9.
114. Brutsaert DL. Cardiac endothelial-myocardial signaling: its role in cardiac growth, contractile performance, and rhythmicity. *Physiol Rev* 2003;**83**:59–115.
115. Meyer D, Birchmeier C. Multiple essential functions of neuregulin in development. *Nature* 1995;**378**:386–90.
116. Gassmann M, Casagranda F, Orioli D, Simon H, Lai C, Klein R, et al. Aberrant neural and cardiac development in mice lacking the ErbB4 neuregulin receptor. *Nature* 1995;**378**:390–4.
117. Lee KF, Simon H, Chen H, Bates B, Hung MC, Hauser C. Requirement for neuregulin receptor erbB2 in neural and cardiac development. *Nature* 1995;**378**:394–8.

118. Milan DJ, Giokas AC, Serluca FC, Peterson RT, Macrae CA. Notch1b and neuregulin are required for specification of central cardiac conduction tissue. *Development* 2006;**133**:1125–32.
119. Zhou B, Ma Q, Rajagopal S, Wu SM, Domian I, Rivera-Feliciano J, et al. Epicardial progenitors contribute to the cardiomyocyte lineage in the developing heart. *Nature* 2008;**454**:109–13.
120. McElhinney DB, Tworetzky W, Lock JE. Current status of fetal cardiac intervention. *Circulation* 2010;**121**:1256–63.
121. Hinton Jr. RB, Martin LJ, Tabangin ME, Mazwi ML, Cripe LH, Benson DW. Hypoplastic left heart syndrome is heritable. *J Am Coll Cardiol* 2007;**50**:1590–5.
122. deAlmeida A, McQuinn T, Sedmera D. Increased ventricular preload is compensated by myocyte proliferation in normal and hypoplastic fetal chick left ventricle. *Circ Res* 2007;**100**:1363–70.
123. Gerull B, Heuser A, Wichter T, Paul M, Basson CT, McDermott DA, et al. Mutations in the desmosomal protein plakophilin-2 are common in arrhythmogenic right ventricular cardiomyopathy. *Nat Genet* 2004;**36**:1162–4.
124. MacRae CA, Birchmeier W, Thierfelder L. Arrhythmogenic right ventricular cardiomyopathy: moving toward mechanism. *J Clin Invest* 2006;**116**:1825–8.
125. Martin ED, Moriarty MA, Byrnes L, Grealy M. Plakoglobin has both structural and signalling roles in zebrafish development. *Dev Biol* 2009;**327**:83–96.
126. Garcia-Gras E, Lombardi R, Giocondo MJ, Willerson JT, Schneider MD, Khoury DS, et al. Suppression of canonical Wnt/beta-catenin signaling by nuclear plakoglobin recapitulates phenotype of arrhythmogenic right ventricular cardiomyopathy. *J Clin Invest* 2006;**116**:2012–21.
127. Gimbrone Jr. MA. Endothelial dysfunction, hemodynamic forces, and atherosclerosis. *Thromb Haemost* 1999;**82**:722–6.
128. Mammoto A, Mammoto T, Ingber DE. Rho signaling and mechanical control of vascular development. *Curr Opin Hematol* 2008;**15**:228–34.
129. Wang HU, Chen ZF, Anderson DJ. Molecular distinction and angiogenic interaction between embryonic arteries and veins revealed by ephrin-B2 and its receptor Eph-B4. *Cell* 1998;**93**:741–53.
130. Gerety SS, Wang HU, Chen ZF, Anderson DJ. Symmetrical mutant phenotypes of the receptor EphB4 and its specific transmembrane ligand ephrin-B2 in cardiovascular development. *Mol Cell* 1999;**4**:403–14.
131. le Noble F, Moyon D, Pardanaud L, Yuan L, Djonov V, Matthijsen R, et al. Flow regulates arterial–venous differentiation in the chick embryo yolk sac. *Development* 2004;**131**:361–75.
132. Bayliss PE, Bellavance KL, Whitehead GG, Abrams JM, Aegerter S, Robbins HS, et al. Chemical modulation of receptor signaling inhibits regenerative angiogenesis in adult zebrafish. *Nat Chem Biol* 2006;**2**:265–73.
133. Hassel D, Dahme T, Erdmann J, Meder B, Huge A, Stoll M, et al. Nexilin mutations destabilize cardiac Z-disks and lead to dilated cardiomyopathy. *Nat Med* 2009;**15**:1281–8.
134. Trombitas K, Freiburg A, Centner T, Labeit S, Granzier H. Molecular dissection of N2B cardiac titin's extensibility. *Biophys J* 1999;**77**:3189–96.
135. Lee EJ, Peng J, Radke M, Gotthardt M, Granzier HL. Calcium sensitivity and the Frank–Starling mechanism of the heart are increased in titin N2B region-deficient mice. *J Mol Cell Cardiol* 2010;**49**:449–58.
136. Serluca FC, Sidow A, Mably JD, Fishman MC. Partitioning of tissue expression accompanies multiple duplications of the Na+/K+ ATPase alpha subunit gene. *Genome Res* 2001;**11**:1625–31.

137. Amores A, Force A, Yan YL, Joly L, Amemiya C, Fritz A, et al. Zebrafish hox clusters and vertebrate genome evolution. *Science* 1998;**282**:1711–4.
138. Hoffmann B, Schmidt-Traub H, Perrot A, Osterziel KJ, Gessner R. First mutation in cardiac troponin C, L29Q, in a patient with hypertrophic cardiomyopathy. *Hum Mutat* 2001;**17**:524.
139. Mogensen J, Murphy RT, Shaw T, Bahl A, Redwood C, Watkins H, et al. Severe disease expression of cardiac troponin C and T mutations in patients with idiopathic dilated cardiomyopathy. *J Am Coll Cardiol* 2004;**44**:2033–40.
140. Kabaeva ZT, Perrot A, Wolter B, Dietz R, Cardim N, Correia JM, et al. Systematic analysis of the regulatory and essential myosin light chain genes: genetic variants and mutations in hypertrophic cardiomyopathy. *Eur J Hum Genet* 2002;**10**:741–8.
141. Poetter K, Jiang H, Hassanzadeh S, Master SR, Chang A, Dalakas MC, et al. Mutations in either the essential or regulatory light chains of myosin are associated with a rare myopathy in human heart and skeletal muscle. *Nat Genet* 1996;**13**:63–9.
142. Davis JS, Hassanzadeh S, Winitsky S, Lin H, Satorius C, Vemuri R, et al. The overall pattern of cardiac contraction depends on a spatial gradient of myosin regulatory light chain phosphorylation. *Cell* 2001;**107**:631–41.
143. Rottbauer W, Wessels G, Dahme T, Just S, Trano N, Hassel D, et al. Cardiac myosin light chain-2: a novel essential component of thick-myofilament assembly and contractility of the heart. *Circ Res* 2006;**99**:323–31.
144. Meder B, Laufer C, Hassel D, Just S, Marquart S, Vogel B, et al. A single serine in the carboxyl terminus of cardiac essential myosin light chain-1 controls cardiomyocyte contractility in vivo. *Circ Res* 2009;**104**:650–9.
145. Chen Z, Huang W, Dahme T, Rottbauer W, Ackerman MJ, Xu X. Depletion of zebrafish essential and regulatory myosin light chains reduces cardiac function through distinct mechanisms. *Cardiovasc Res* 2008;**79**:97–108.
146. Vogel B, Meder B, Just S, Laufer C, Berger I, Weber S, et al. In-vivo characterization of human dilated cardiomyopathy genes in zebrafish. *Biochem Biophys Res Commun* 2009;**390**:516–22.
147. Langenbacher AD, Dong Y, Shu X, Choi J, Nicoll DA, Goldhaber JI, et al. Mutation in sodium–calcium exchanger 1 (NCX1) causes cardiac fibrillation in zebrafish. *Proc Natl Acad Sci USA* 2005;**102**:17699–704.
148. Ebert AM, Hume GL, Warren KS, Cook NP, Burns CG, Mohideen MA, et al. Calcium extrusion is critical for cardiac morphogenesis and rhythm in embryonic zebrafish hearts. *Proc Natl Acad Sci USA* 2005;**102**:17705–10.
149. Menick DR, Renaud L, Buchholz A, Muller JG, Zhou H, Kappler CS, et al. Regulation of Ncx1 gene expression in the normal and hypertrophic heart. *Ann NY Acad Sci* 2007;**1099**:195–203.
150. Liao BK, Deng AN, Chen SC, Chou MY, Hwang PP. Expression and water calcium dependence of calcium transporter isoforms in zebrafish gill mitochondrion-rich cells. *BMC Genomics* 2007;**8**:354.
151. Rottbauer W, Baker K, Wo ZG, Mohideen MA, Cantiello HF, Fishman MC. Growth and function of the embryonic heart depend upon the cardiac-specific L-type calcium channel alpha1 subunit. *Dev Cell* 2001;**1**:265–75.
152. Rottbauer W, Just S, Wessels G, Trano N, Most P, Katus HA, et al. VEGF-PLCgamma1 pathway controls cardiac contractility in the embryonic heart. *Genes Dev* 2005;**19**:1624–34.
153. Lawson ND, Mugford JW, Diamond BA, Weinstein BM. phospholipase C gamma-1 is required downstream of vascular endothelial growth factor during arterial development. *Genes Dev* 2003;**17**:1346–51.
154. Zareba W, Cygankiewicz I. Long QT syndrome and short QT syndrome. *Prog Cardiovasc Dis* 2008;**51**:264–78.

155. Bokil NJ, Baisden JM, Radford DJ, Summers KM. Molecular genetics of long QT syndrome. *Mol Genet Metab* 2010;**101**:1–8.

156. Yang Y, Liang B, Liu J, Li J, Grunnet M, Olesen SP, et al. Identification of a Kir3.4 Mutation in Congenital Long QT Syndrome. *Am J Hum Genet* 2010;**6**:872–80.

157. Arnaout R, Ferrer T, Huisken J, Spitzer K, Stainier DY, Tristani-Firouzi M, et al. Zebrafish model for human long QT syndrome. *Proc Natl Acad Sci USA* 2007;**104**:11316–21.

158. Langheinrich U, Vacun G, Wagner T. Zebrafish embryos express an orthologue of HERG and are sensitive toward a range of QT-prolonging drugs inducing severe arrhythmia. *Toxicol Appl Pharmacol* 2003;**193**:370–82.

159. Kopp R, Schwerte T, Pelster B. Cardiac performance in the zebrafish breakdance mutant. *J Exp Biol* 2005;**208**:2123–34.

160. Scholz EP, Niemer N, Hassel D, Zitron E, Burgers HF, Bloehs R, et al. Biophysical properties of zebrafish ether-a-go-go related gene potassium channels. *Biochem Biophys Res Commun* 2009;**381**:159–64.

161. Hassel D, Scholz EP, Trano N, Friedrich O, Just S, Meder B, et al. Deficient zebrafish ether-a-go-go-related gene channel gating causes short-QT syndrome in zebrafish reggae mutants. *Circulation* 2008;**117**:866–75.

162. Warren KS, Baker K, Fishman MC. The slow mo mutation reduces pacemaker current and heart rate in adult zebrafish. *Am J Physiol Heart Circ Physiol* 2001;**281**:H1711–9.

163. Bussmann J, Lawson N, Zon L, Schulte-Merker S. Zebrafish VEGF receptors: a guideline to nomenclature. *PLoS Genet* 2008;**4**:e1000064.

164. Habeck H. Analysis of zebrafish VEGF receptor mutant reveals specific disruption of angiogenesis. *Curr Biol* 2002;**12**:1405–12.

165. Covassin LD, Villefranc JA, Kacergis MC, Weinstein BM, Lawson ND. Distinct genetic interactions between multiple Vegf receptors are required for development of different blood vessel types in zebrafish. *Proc Natl Acad Sci USA* 2006;**103**:6554–9.

166. Takahashi T, Yamaguchi S, Chida K, Shibuya M. A single autophosphorylation site on KDR/Flk-1 is essential for VEGF-A-dependent activation of PLC-gamma and DNA synthesis in vascular endothelial cells. *EMBO J* 2001;**20**:2768–78.

167. Sakurai Y, Ohgimoto K, Kataoka Y, Yoshida N, Shibuya M. Essential role of Flk-1 (VEGF receptor 2) tyrosine residue 1173 in vasculogenesis in mice. *Proc Natl Acad Sci USA* 2005;**102**:1076–81.

168. Shalaby F, Rossant J, Yamaguchi TP, Gertsenstein M, Wu XF, Breitman ML, et al. Failure of blood-island formation and vasculogenesis in Flk-1-deficient mice. *Nature* 1995;**376**:62–6.

169. Gerber HP, McMurtrey A, Kowalski J, Yan M, Keyt BA, Dixit V, et al. Vascular endothelial growth factor regulates endothelial cell survival through the phosphatidylinositol 3′-kinase/Akt signal transduction pathway. Requirement for Flk-1/KDR activation. *J Biol Chem* 1998;**273**:30336–43.

170. Jiang BH, Zheng JZ, Aoki M, Vogt PK. Phosphatidylinositol 3-kinase signaling mediates angiogenesis and expression of vascular endothelial growth factor in endothelial cells. *Proc Natl Acad Sci USA* 2000;**97**:1749–53.

171. Chan J, Bayliss PE, Wood JM, Roberts TM. Dissection of angiogenic signaling in zebrafish using a chemical genetic approach. *Cancer Cell* 2002;**1**:257–67.

172. Hong CC, Peterson QP, Hong JY, Peterson RT. Artery/vein specification is governed by opposing phosphatidylinositol-3 kinase and MAP kinase/ERK signaling. *Curr Biol* 2006;**16**:1366–72.

173. Adams RH, Wilkinson GA, Weiss C, Diella F, Gale NW, Deutsch U, et al. Roles of ephrinB ligands and EphB receptors in cardiovascular development: demarcation of arterial/venous domains, vascular morphogenesis, and sprouting angiogenesis. *Genes Dev* 1999;**13**:295–306.

174. Lawson ND, Scheer N, Pham VN, Kim CH, Chitnis AB, Campos-Ortega JA, et al. Notch signaling is required for arterial–venous differentiation during embryonic vascular development. *Development* 2001;**128**:3675–83.
175. Lawson ND, Vogel AM, Weinstein BM. Sonic hedgehog and vascular endothelial growth factor act upstream of the Notch pathway during arterial endothelial differentiation. *Dev Cell* 2002;**3**:127–36.
176. Zhong TP, Childs S, Leu JP, Fishman MC. Gridlock signalling pathway fashions the first embryonic artery. *Nature* 2001;**414**:216–20.
177. You LR, Lin FJ, Lee CT, DeMayo FJ, Tsai MJ, Tsai SY. Suppression of Notch signalling by the COUP-TFII transcription factor regulates vein identity. *Nature* 2005;**435**:98–104.
178. Arvanitis D, Davy A. Eph/ephrin signaling: networks. *Genes Dev* 2008;**22**:416–29.
179. Kim YH, Hu H, Guevara-Gallardo S, Lam MT, Fong SY, Wang RA. Artery and vein size is balanced by Notch and ephrin B2/EphB4 during angiogenesis. *Development* 2008;**135**:3755–64.
180. Herbert SP, Huisken J, Kim TN, Feldman ME, Houseman BT, Wang RA, et al. Arterial-venous segregation by selective cell sprouting: an alternative mode of blood vessel formation. *Science* 2009;**326**:294–8.
181. Gale NW, Baluk P, Pan L, Kwan M, Holash J, DeChiara TM, et al. Ephrin-B2 selectively marks arterial vessels and neovascularization sites in the adult, with expression in both endothelial and smooth-muscle cells. *Dev Biol* 2001;**230**:151–60.
182. Shin D, Garcia-Cardena G, Hayashi S, Gerety S, Asahara T, Stavrakis G, et al. Expression of ephrinB2 identifies a stable genetic difference between arterial and venous vascular smooth muscle as well as endothelial cells, and marks subsets of microvessels at sites of adult neovascularization. *Dev Biol* 2001;**230**:139–50.
183. van den Akker NM, Caolo V, Wisse LJ, Peters PP, Poelmann RE, Carmeliet P, et al. Developmental coronary maturation is disturbed by aberrant cardiac vascular endothelial growth factor expression and Notch signalling. *Cardiovasc Res* 2008;**78**:366–75.
184. Carmeliet P. Angiogenesis in life, disease and medicine. *Nature* 2005;**438**:932–6.
185. Duda DG, Batchelor TT, Willett CG, Jain RK. VEGF-targeted cancer therapy strategies: current progress, hurdles and future prospects. *Trends Mol Med* 2007;**13**:223–30.
186. Folkman J. Angiogenesis: an organizing principle for drug discovery? *Nat Rev Drug Discov* 2007;**6**:273–86.
187. Yang J, Weinberg RA. Epithelial–mesenchymal transition: at the crossroads of development and tumor metastasis. *Dev Cell* 2008;**14**:818–29.
188. Ferrara N. History of discovery: vascular endothelial growth factor. *Arterioscler Thromb Vasc Biol* 2009;**29**:789–91.
189. Crawford Y, Ferrara N. VEGF inhibition: insights from preclinical and clinical studies. *Cell Tissue Res* 2008;**335**:261–9.
190. Ambati J, Ambati BK, Yoo SH, Ianchulev S, Adamis AP. Age-related macular degeneration: etiology, pathogenesis, and therapeutic strategies. *Surv Ophthalmol* 2003;**48**:257–93.
191. Marneros AG, Fan J, Yokoyama Y, Gerber HP, Ferrara N, Crouch RK, et al. Vascular endothelial growth factor expression in the retinal pigment epithelium is essential for choriocapillaris development and visual function. *Am J Pathol* 2005;**167**:1451–9.
192. Abete P, Cacciatore F, Testa G, Della-Morte D, Galizia G, de Santis D, et al. Ischemic preconditioning in the aging heart: from bench to bedside. *Ageing Res Rev* 2010;**9**:153–62.
193. Hubschman JP, Reddy S, Schwartz SD. Age-related macular degeneration: current treatments. *Clin Ophthalmol (Auckland, N.Z)* 2009;**3**:155–66.
194. Kamba T, McDonald DM. Mechanisms of adverse effects of anti-VEGF therapy for cancer. *Br J Cancer* 2007;**96**:1788–95.

195. Trayhurn P, Wood IS. Signalling role of adipose tissue: adipokines and inflammation in obesity. *Biochem Soc Trans* 2005;**33**:1078–81.
196. Kern TS. Contributions of inflammatory processes to the development of the early stages of diabetic retinopathy. *Exp Diab Res* 2007;**2007**:95103.
197. Hsueh WA, Quinones MJ. Role of endothelial dysfunction in insulin resistance. *Am J Cardiol* 2003;**92**:10J–17J.
198. Singhal A. Endothelial dysfunction: role in obesity-related disorders and the early origins of CVD. *Proc Nutr Soc* 2005;**64**:15–22.
199. Woo KS, Chook P, Yu CW, Sung RY, Qiao M, Leung SS, et al. Effects of diet and exercise on obesity-related vascular dysfunction in children. *Circulation* 2004;**109**:1981–6.
200. Carmeliet P, Tessier-Lavigne M. Common mechanisms of nerve and blood vessel wiring. *Nature* 2005;**436**:193–200.
201. Zacchigna S, Ruiz de Almodovar C, Carmeliet P. Similarities between angiogenesis and neural development: what small animal models can tell us. *Curr Top Dev Biol* 2008;**80**:1–55.
202. Childs S, Chen JN, Garrity DM, Fishman MC. Patterning of angiogenesis in the zebrafish embryo. *Development* 2002;**129**:973–82.
203. Torres-Vazquez J, Gitler AD, Fraser SD, Berk JD, Van NP, Fishman MC, et al. Semaphorin–plexin signaling guides patterning of the developing vasculature. *Dev Cell* 2004;**7**:117–23.
204. Ridgway J, Zhang G, Wu Y, Stawicki S, Liang WC, Chanthery Y, et al. Inhibition of Dll4 signalling inhibits tumour growth by deregulating angiogenesis. *Nature* 2006;**444**:1083–7.
205. Noguera-Troise I, Daly C, Papadopoulos NJ, Coetzee S, Boland P, Gale NW, et al. Blockade of Dll4 inhibits tumour growth by promoting non-productive angiogenesis. *Nature* 2006;**444**:1032–7.
206. Hellstrom M, Phng LK, Hofmann JJ, Wallgard E, Coultas L, Lindblom P, et al. Dll4 signalling through Notch1 regulates formation of tip cells during angiogenesis. *Nature* 2007;**445**:776–80.
207. Leslie JD, Ariza-McNaughton L, Bermange AL, McAdow R, Johnson SL, Lewis J. Endothelial signalling by the Notch ligand delta-like 4 restricts angiogenesis. *Development* 2007;**134**:839–44.
208. Yan M, Plowman GD. Delta-like 4/Notch signaling and its therapeutic implications. *Clin Cancer Res* 2007;**13**:7243–6.
209. Yan M, Callahan CA, Beyer JC, Allamneni KP, Zhang G, Ridgway JB, et al. Chronic DLL4 blockade induces vascular neoplasms. *Nature* 2010;**463**:E6–7.
210. Tammela T, Alitalo K. Lymphangiogenesis: molecular mechanisms and future promise. *Cell* 2010;**140**:460–76.
211. Bussmann J, Bos FL, Urasaki A, Kawakami K, Duckers HJ, Schulte-Merker S. Arteries provide essential guidance cues for lymphatic endothelial cells in the zebrafish trunk. *Development* 2010;**137**:2653–7.
212. Flores MV, Hall CJ, Crosier KE, Crosier PS. Visualization of embryonic lymphangiogenesis advances the use of the zebrafish model for research in cancer and lymphatic pathologies. *Dev Dyn* 2010;**239**:2128–35.
213. Geudens I, Herpers R, Hermans K, Segura I, Ruiz de Almodovar C, Bussmann J, et al. Role of delta-like-4/Notch in the formation and wiring of the lymphatic network in zebrafish. *Arterioscler Thromb Vasc Biol* 2010;**30**:1695–702.
214. Makinen T, Adams RH, Bailey J, Lu Q, Ziemiecki A, Alitalo K, et al. PDZ interaction site in ephrinB2 is required for the remodeling of lymphatic vasculature. *Genes Dev* 2005;**19**:397–410.
215. Sahoo T, Johnson EW, Thomas JW, Kuehl PM, Jones TL, Dokken CG, et al. Mutations in the gene encoding KRIT1, a Krev-1/rap1a binding protein, cause cerebral cavernous malformations (CCM1). *Hum Mol Genet* 1999;**8**:2325–33.

216. Liquori CL, Berg MJ, Siegel AM, Huang E, Zawistowski JS, Stoffer T, et al. Mutations in a gene encoding a novel protein containing a phosphotyrosine-binding domain cause type 2 cerebral cavernous malformations. *Am J Hum Genet* 2003;**73**:1459–64.
217. Denier C, Goutagny S, Labauge P, Krivosic V, Arnoult M, Cousin A, et al. Mutations within the MGC4607 gene cause cerebral cavernous malformations. *Am J Hum Genet* 2004;**74**:326–37.
218. Zhang J, Rigamonti D, Dietz HC, Clatterbuck RE. Interaction between krit1 and malcavernin: implications for the pathogenesis of cerebral cavernous malformations. *Neurosurgery* 2007;**60**:353–9 discussion 359.
219. Zawistowski JS, Stalheim L, Uhlik MT, Abell AN, Ancrile BB, Johnson GL, et al. CCM1 and CCM2 protein interactions in cell signaling: implications for cerebral cavernous malformations pathogenesis. *Hum Mol Genet* 2005;**14**:2521–31.
220. Gault J, Shenkar R, Recksiek P, Awad IA. Biallelic somatic and germ line CCM1 truncating mutations in a cerebral cavernous malformation lesion. *Stroke* 2005;**36**:872–4.
221. Labauge P, Denier C, Bergametti F, Tournier-Lasserve E. Genetics of cavernous angiomas. *Lancet Neurol* 2007;**6**:237–44.
222. Akers AL, Johnson E, Steinberg GK, Zabramski JM, Marchuk DA. Biallelic somatic and germline mutations in cerebral cavernous malformations (CCMs): evidence for a two-hit mechanism of CCM pathogenesis. *Hum Mol Genet* 2009;**18**:919–30.
223. Hogan BM, Bussmann J, Wolburg H, Schulte-Merker S. ccm1 cell autonomously regulates endothelial cellular morphogenesis and vascular tubulogenesis in zebrafish. *Hum Mol Genet* 2008;**17**:2424–32.
224. Marchuk DA, Srinivasan S, Squire TL, Zawistowski JS. Vascular morphogenesis: tales of two syndromes. *Hum Mol Genet* 2003;**12**(Spec No 1):R97–R112.
225. Roman BL, Pham VN, Lawson ND, Kulik M, Childs S, Lekven AC, et al. Disruption of acvrl1 increases endothelial cell number in zebrafish cranial vessels. *Development* 2002;**129**: 3009–19.
226. Eerola I, Boon LM, Mulliken JB, Burrows PE, Dompmartin A, Watanabe S, et al. Capillary malformation–arteriovenous malformation, a new clinical and genetic disorder caused by RASA1 mutations. *Am J Hum Genet* 2003;**73**:1240–9.
227. Tan W, Palmby TR, Gavard J, Amornphimoltham P, Zheng Y, Gutkind JS. An essential role for Rac1 in endothelial cell function and vascular development. *Faseb J* 2008;**22**:1829–38.
228. Brice G, Child AH, Evans A, Bell R, Mansour S, Burnand K, et al. Milroy disease and the VEGFR-3 mutation phenotype. *J Med Genet* 2005;**42**:98–102.
229. van Rooijen E, Voest EE, Logister I, Korving J, Schwerte T, Schulte-Merker S, et al. Zebrafish mutants in the von Hippel–Lindau tumor suppressor display a hypoxic response and recapitulate key aspects of Chuvash polycythemia. *Blood* 2009;**113**:6449–60.
230. van Rooijen E, Voest EE, Logister I, Bussmann J, Korving J, van Eeden FJ, et al. von Hippel–Lindau tumor suppressor mutants faithfully model pathological hypoxia-driven angiogenesis and vascular retinopathies in zebrafish. *Dis Model Mech* 2010;**3**:343–53.
231. Juris SJ, Melnyk RA, Bolcome 3rd RE, Chan J, Collier RJ. Cross-linked forms of the isolated N-terminal domain of the lethal factor are potent inhibitors of anthrax toxin. *Infect Immun* 2007;**75**:5052–8.
232. Bolcome 3rd RE, Sullivan SE, Zeller R, Barker AP, Collier RJ, Chan J. Anthrax lethal toxin induces cell death-independent permeability in zebrafish vasculature. *Proc Natl Acad Sci USA* 2008;**105**:2439–44.
233. Vasil ML, Stonehouse MJ, Vasil AI, Wadsworth SJ, Goldfine H, Bolcome 3rd RE, et al. A complex extracellular sphingomyelinase of *Pseudomonas aeruginosa* inhibits angiogenesis by selective cytotoxicity to endothelial cells. *PLoS Pathog* 2009;**5**:e1000420.

234. Beltrami AP, Urbanek K, Kajstura J, Yan SM, Finato N, Bussani R, et al. Evidence that human cardiac myocytes divide after myocardial infarction. *N Engl J Med* 2001;**344**:1750–7.
235. Engel FB, Schebesta M, Duong MT, Lu G, Ren S, Madwed JB, et al. p38 MAP kinase inhibition enables proliferation of adult mammalian cardiomyocytes. *Genes Dev* 2005;**19**:1175–87.
236. Anversa P, Leri A, Kajstura J. Cardiac regeneration. *J Am Coll Cardiol* 2006;**47**:1769–76.
237. Huang CC, Lawson ND, Weinstein BM, Johnson SL. reg6 is required for branching morphogenesis during blood vessel regeneration in zebrafish caudal fins. *Dev Biol* 2003;**264**:263–74.
238. Huang CC, Huang CW, Cheng YS, Yu J. Histamine metabolism influences blood vessel branching in zebrafish reg6 mutants. *BMC Dev Biol* 2008;**8**:31.
239. Zhong TP, Rosenberg M, Mohideen MA, Weinstein B, Fishman MC. Gridlock, an HLH gene required for assembly of the aorta in zebrafish. *Science* 2000;**287**:1820–4.

Drosophila Models of Cardiac Disease

Nicole Piazza and
R.J. Wessells

University of Michigan Medical School, Ann Arbor, Michigan, USA

The fruit fly *Drosophila melanogaster* has emerged as a useful model for cardiac diseases, both developmental abnormalities and adult functional impairment. Using the tools of both classical and molecular genetics, the study of the developing fly heart has been instrumental in identifying the major signaling events of cardiac field formation, cardiomyocyte specification, and the formation of the functioning heart tube. The larval stage of fly cardiac development has become an important model system for testing isolated preparations of living hearts for the effects of biological and pharmacological compounds on cardiac activity. Meanwhile, the recent development of effective techniques to study adult cardiac performance in the fly has opened new uses for the *Drosophila* model system. The fly system is now being used to study long-term alterations in adult performance caused by factors such as diet, exercise, and normal aging. The fly is a unique and valuable system for the study of such complex, long-term interactions, as it is the only invertebrate genetic model system with a working heart developmentally homologous to the vertebrate heart. Thus, the fly model combines the advantages of invertebrate genetics (such as large populations, facile molecular genetic techniques, and short lifespan) with physiological measurement

Progress in Molecular Biology
and Translational Science, Vol. 100
DOI: 10.1016/B978-0-12-384878-9.00005-4

Copyright 2011, Elsevier Inc.
All rights reserved.
1877-1173/11 $35.00

techniques that allow meaningful comparisons with data from vertebrate model systems. As such, the fly model is well situated to make important contributions to the understanding of complicated interactions between environmental factors and genetics in the long-term regulation of cardiac performance.

I. Introduction

In the late twentieth century, the rapid expansion of the field of developmental genetics has been one of the greatest triumphs of the biological sciences. Invertebrate genetic model organisms have been used to make startling progress in understanding the steps of patterning, cell specification, and organogenesis during normal development. More recently, invertebrate model systems have been employed as effective models for adult-onset and age-related diseases.

Since the fruit fly *Drosophila melanogaster* is the only major invertebrate model system that contains a working organ with developmental and functional homologies to the vertebrate heart, flies have been the only model system where the advantages of invertebrate genetics have been utilized to model cardiac development and disease.

The *Drosophila* heart undergoes developmental steps reminiscent of the early stages of vertebrate heart development. Following subdivision of the cardiac mesoderm from the broader visceral mesoderm, the cardiac field forms from two bilaterally symmetric, linear strands of cells. These two rows of presumptive cardiac cells then migrate to the midline, where they adhere to each other to form a tube with a hollow lumen. Meanwhile, cardiomyocytes are gradually specified by multiple signaling inputs that activate cell-autonomous expression of conserved transcription factors.[1,2]

Using molecular genetics, lineage tracing, and confocal imaging, the genetic pathways involved in fly cardiac specification have been identified and, in many cases, orthologs have been shown to be involved similarly in early vertebrate heart development.[3] In parallel, disruptions in these vertebrate orthologs have been associated with developmental disorders.[4,5]

The accessibility and simplicity of the *Drosophila* larval heart have made it a useful model for testing effects of natural or synthetic compounds that regulate cardiac performance. By introducing such factors into partially dissected open preparations, the direct effects of various neurotransmitters on cardiac rate and rhythm have been assayed. The transparency of the larva has facilitated optical measurements of rate and rhythm in intact animals under both normal and pathological conditions, while electrical field measurements have proved effective in describing defects in electrical conductance.[6]

Excitingly, the use of *Drosophila* genetics has begun to make a significant impact in the field of adult cardiac performance. As a short-lived model system with well-described characteristics of aging, flies represent an excellent opportunity to model long-term changes in cardiac function that can be traced throughout life, at both the population level and the individual level. Several techniques have been developed to facilitate this process, including generation of M-mode traces from intact or semi-intact preparations, external electrical pacing, and external field recording. Already, flies have been used to model the effects of nutrition, genetic disease, exercise-training, age-related pathology, and normal aging.[7]

This review will discuss the uses of the *Drosophila* model system for the study of cardiac developmental and adult disease. We will address, in turn, what has already been learned from the study of embryonic cardiac development, larval cardiac function, and the study of aging adult flies.

II. Embryonic Cardiac Development

Using traditional forward genetics supplemented with candidate gene approaches, a detailed portrait of the specification and organogenesis of the heart has emerged in flies. In this section, we will describe the development of the fly heart in four stages: cardiac field formation, cardiomyocyte specification, division of cardiomyocytes into subtypes, and migration/adhesion of the heart tube. In conclusion, we will outline the relationship between the genetic factors involved in these events in the fly and its vertebrate counterparts.

III. Cardiac Field

The field of presumptive cardiac cells is distinguished from other mesodermal-derived tissue types ultimately by the action of two conserved extracellular signaling molecules, the Wnt family homolog *wingless* (*wg*) and the BMP family homolog *decepentaplegic* (*dpp*) (Table I). *wg* is expressed in the embryo in a series of stripes which mark, and, indeed, help to define, the segments of the embryonic mesoderm along the anterior–posterior axis.[8–11] Meanwhile, *dpp* is an important component that signals position to the mesoderm along the dorsal–ventral axis.[12,13] Thus, these two signaling molecules act along perpendicular axes during embryogenesis.

The points of intersection between these two expression patterns are instructive and required for specification of cardiac cell types. Cardiac cell fate requires the combined exposure to these two signaling molecules not once but twice, at two separate stages of embryogenesis.[60] Mesoderm-specific context is provided to interpret this signal by the Nkx2.5 homolog *tinman* (*tin*).[14,15] Combined

TABLE I
Drosophila Homologs of Genes Involved in Cardiac Development

Vertebrate/animal homolog	*Drosophila* Homolog	Reference
Wnt family	*wingless (wg)*	8–11
BMP family	*decepentaplegic (dpp)*	12,13
Nkx2.5	*tinman (tin)*	14–16
NK family	*bagpipe*	17
FGF signaling	*heartless (htl)*	18–22
FOG family	*u-shaped (ush)*	23,24
COUP-TF	*seven-up (svp)*	25–27
GATA4	*pannier (pnr)*	28
T-box (Tbx5)	*nmr*	16
Lbx1	*ladybird*	29,30
Sarco-endoplasmic reticulum calcium ATPase (SERCA)	*SERCA*	31
Dystrophin	*dystrophin (dys)*	32,33
Sarcoglycan	*d-sarcoglycan*	34–36
Myosin transducer complex	*Myosin heavy chain (Mhc)*	37–41
KCNQ1	*KCNQ*	42–45
Evx2	*even-skipped (eve)*	46
ErbB2	EGF receptor	47–49
Opa1	*dopa1*	50,51
Superoxide dismutase 2	*Sod2*	52,53
Sestrin	*sesn*	54
4eBP	*4eBP*	55
Fatty acid transporter	*Fatp*	56,57
PGC1-α	*spargel*	58,59

misexpression of *wg*, *dpp*, and *tin* is sufficient to induce cardiac-specific gene expression ectopically.[60] Interestingly, visceral mesoderm is distinguished from dorsal mesodermal derivatives, such as heart, by the activity of another NK family homeobox protein, Bagpipe, which acts to provide context to more ventral mesodermal cells exposed to the combined signals of Wg and Dpp.[17]

This strategy of integrating broad extracellular signals with mesoderm-specific homeobox genes to specify cardiogenic mesoderm is broadly conserved in vertebrates as well. BMP family members are essential for induction of cardiac mesoderm in the mouse,[61] chick,[62] and frog,[63] while SMAD-family transcription factors that act downstream of Bmp signaling are autonomously required for cardiogenesis in mice.[64]

Moreover, as in flies, BMP signaling cooperates with cell-autonomous activity of the *tin* homolog Nkx2.5 to specify cardiomyocyte lineage in mouse. Nkx2.5 mutant mice have morphological defects resulting from lack of proper specification of the cardiac tube,[65] and Nkx2.5 expression is dependent on the activity of BMP-dependent SMAD transcription factors.[66]

The relationship between Wnt signaling and cardiac induction is more complicated in vertebrates than flies. In frog and chick embryos, it is necessary to inhibit Wnt signaling induced by ligands secreted from the neural tube to allow cardiac field formation.[67] This inhibition is accomplished by the action of two proteins from the organizer region, Dickkopf-1 and Crescent.[68,69] Dickkopf-1 requires the activity of the homeodomain transcription factor Hex in this context.[70]

However, work in conditional mouse knockouts has made it clear that at other stages, Wnt signaling is essential as a positive regulator of cardiomyocyte differentiation,[71,72] and becomes essential again as a positive regulator of cardiomyocyte proliferation postnatally.[73,74] A biphasic role, depending on timing, also exists in zebrafish, where Wnt signaling prior to gastrulation promotes cardiac differentiation, while signaling after gastrulation inhibits cardiac differentiation.[75] Further complicating the picture, signaling through noncanonical Wnt pathways is also required for cardiac specification, and this noncanonical signaling may itself act to dampen canonical signaling.[76]

In addition, proliferation of cardiac progenitors in both flies and vertebrates requires secreted FGF factors. In chick and zebrafish embryos, inhibition of FGF signaling reduces the size of the cardiac field.[18–20] In flies, the FGF Receptor, *heartless* (*htl*), as its name would suggest, is required for the formation of a functional heart tube.[21] Cardiac defects in *heartless* mutants may, however, be attributable primarily to a failure in the migration step of the cardiac mesoderm to the dorsal midline.[22]

Despite the greater complexity found in the vertebrate system, it seems clear that the signals involved in the early stages of fly cardiac development are well conserved in vertebrate models. Since these genes are also involved in many congenital cardiac abnormalities, a combination of research in fly and vertebrate models will continue to be of interest in elucidating these relationships in increasing detail.

IV. Cardiomyocyte Differentiation

Presumptive heart cells in the fly are subdivided by the coordinate action of several conserved transcription factors.[77,78] The heart becomes subdivided into a posterior pumping heart proper and a more anterior portion termed the aorta as a result of positional information provided by homeobox transcription factors (Fig. 1).[54,79] Later, during pupal metamorphosis, homeobox genes act in response to hormonal signals to control remodeling of cardiomyocytes into their adult form.[80] Within the cardiomyocytes forming the heart tube proper, several specific pairs of cells are fated to alter their shape[81] and form the valve-like ostia that allow inflow into the tube prior to pumping.[82]

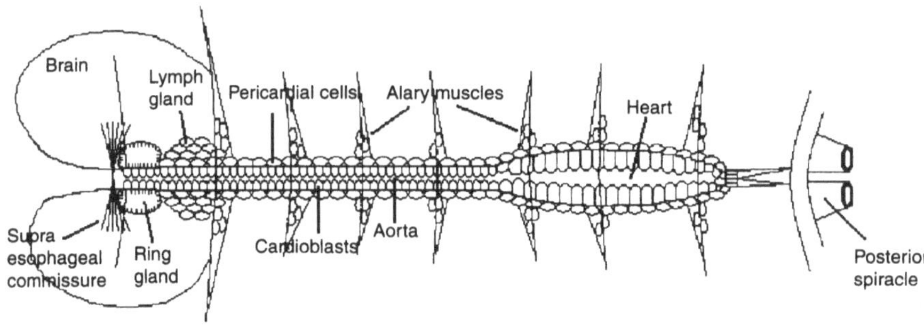

FIG. 1. Diagram of embryonic heart (adapted from Ref. 12).

After the early role of *tin* in specifying cardiac mesoderm, it is required again during the differentiation of myocardial cells, where it is necessary to define contractile cardiomyocytes and restrict formation of ostia to their proper location.[83] *tin* expression itself is maintained through the action of the T-box transcription factors *mid* and *H15* (aka *nmr1-2*),[84] and *mid*, *H15* and *tin* act coordinately to promote cardiomyocyte differentiation.[85,86]

Also acting coordinately as part of this regulatory network is the GATA factor encoding gene *pannier* (*pnr*), which acts in response to ectodermal *dpp* signaling to maintain *tin* expression and acts both coordinately and separately to activate target genes, including *tin* itself.[87–90] Interestingly, the *pnr* antagonist *u-shaped* (*ush*), a homolog of vertebrate FOG genes, antagonizes cardiogenesis early on,[23] but later acts to promote cardiogenesis by maintaining *tin* expression.[90]

Another important factor acting in conjunction with *tin* and *pnr* to promote cardiomyocyte differentiation is the T-box containing *Dorsocross* (*Doc*) family of transcription factors. While the three members of this family appear to have partially redundant function, a triple knockout results in failure of cardiomyocyte specification.[91] The combined activity of these genes activates genes directly involved in morphological specification, such as the *Drosophila* homolog of the vertebrate Hand genes.[92]

Although the combined activity of *tin*, *pnr*, and *Doc1-3* is required for cardiomyocyte specification, *Doc* genes then resolve themselves into a mutually exclusive expression pattern with *tin*, caused by mutual repression, in which *tin* is expressed in four out of six cells in each segment, and *Doc* genes are expressed in the remaining two. The two cells where *Doc* is expressed become ostia in the posterior heart tube, while the four cells where *tin* is expressed become contractile cardiomyocytes.[83]

Cardiomyocyte differentiation proceeds under the direction of differentiation genes, which, in some cases are direct *tin* targets, such as β-3 *tubulin*,[93] *dSur*,[94,95] or *dHand*[96] and in other cases are targets of transcription factors activated by *tin*, such as *dMef2*.[97] A key transcription factor in the cells fated to become ostia is the COUP-TF homolog *seven-up* (*svp*).[26] *svp* acts to repress *tin* expression in these cells and maintains expression of *wg*.[25,27]

Strikingly, many of the relevant transcription factors to *Drosophila* cardiac specification have vertebrate orthologs that are also involved in cardiac development and specification, and many of these have been associated with congenital heart diseases.[4] The vertebrate homolog of *tin*, *Nkx2.5*, has been linked to atrial and ventricular septal defects,[16] as well as tetralogy of Fallot.[98] Also linked to septal defects are the *pnr* homolog GATA4[28] and the *nmr* family homolog Tbx5.[16]

Even the combined coordinate and mutually repressive network between transcription factors elucidated in flies shows evidence of being reflected in vertebrate cardiogenesis as well. Genetic interactions have been observed between GATA4 and Tbx5 in mice,[99] and *Nkx2.5* along with *Nkx2.7* is necessary to downregulate Tbx factors later in cardiac development in zebrafish.[100] In general, there appears to be a greater degree of redundancy in vertebrate species, for example. GATA4 and GATA5 are both necessary to carry on the activities controlled by *pnr* in flies,[101] but the simplicity of the fly model will continue to be used as an important advantage to the identification of important regulatory families.

V. Migration/Adhesion

The formation of the final embryonic structure occurs in three steps. First, the bilaterally symmetric rows of cells undergo morphological changes to their membrane structure and align with each other. Then, the two rows of presumptive cardiomyocytes migrate to the dorsal midline. Lastly, the two rows adhere to each other and form a lumen between them (Fig. 2).

Prior to migration, the two rows of cardiac cells undergo a mesenchymal to epithelial transition, during which they establish contacts between themselves and form a continuous monolayer.[8,12] During this process, they begin to express various membrane markers and determinants of polarity.[102] This process is dependent initially on cell adhesion molecules, including the Ig-family protein Faint sausage (Fas)[103] and the transmembrane receptor protein Toll,[104] while maintenance of the ultrastructure becomes dependent on conserved extracellular matrix proteins, including laminins,[105,106] integrins,[107] and cadherins later.[103] In addition, the heterotrimeric G-protein encoding gene, *brokenheart* (*bkh*), is required in a cell-autonomous fashion to establish apical–basal polarity, and *bkh* embryos have polarity defects in tube formation resulting in dysfunctional hearts.[108]

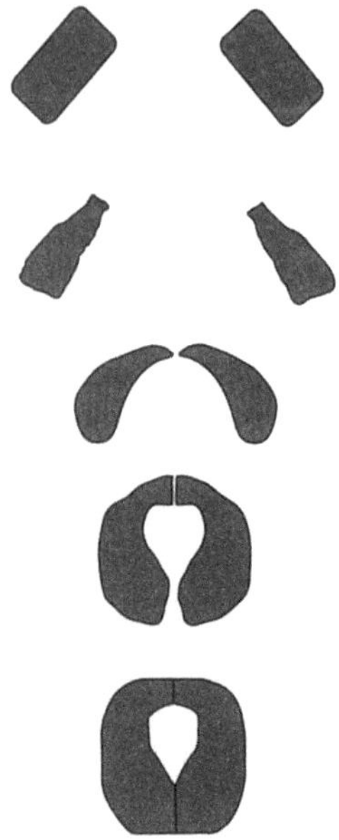

FIG. 2. Diagram of migration and alignment of embryonic cardioblasts (adapted from Ref. 2).

Once the two rows of cells are properly aligned, they must move together along their respective rows toward the dorsal midline where they will adhere to each other. As the cells migrate, they remain in contact with the "leading edge" of the overlying ectodermal sheet, which is simultaneously migrating toward the midline in the process of dorsal closure. A Type IV-collagen-like protein, Pericardin, mediates the coordinated movements of the two germ layers in this process and is essential for the migration of cardiac precursors to the midline.[109]

Signaling between the secreted protein Slit and its receptors Robo and Robo2 is essential for maintaining cell polarity during migration and for adhesion when the two bilateral rows of heart cells fuse at the midline.[110] In wild-type development, Slit accumulates at the points of connection between bilateral heart precursors and signals to Robo, which is localized at the apical surfaces of the cell pairs, although Robo2 can substitute for Robo in *robo* mutant flies.[110] While fusion of the bilateral heart precursors is occurring, Slit localization is absolutely required for proper localization of membrane proteins that govern the switch from basal–lateral to apical–lateral polarity necessary for initiation of proper cardiac function.[111] As a result, mutations in *slit* produce alignment defects reminiscent of those seen in flies carrying mutations in genes encoding the membrane proteins themselves, such as *discs large* and the E-cadherin-encoding *shotgun*.[2,112] Using live imaging, it has been demonstrated that the Slit/Robo interaction also plays an additional role to regulate the cell shape changes necessary for the heart cells to contact each other dorsally and ventrally to form a lumen.[2]

Recently, the morphogenesis of the anterior region of the heart near the border between the heart and aorta has been more fully described[29] and termed the outflow tract. Excitingly, this outflow tract has a distinctive derivation from the pharyngeal mesoderm, suggesting a direct developmental homology with the vertebrate outflow tract. In both vertebrates and flies, outflow tract is derived from cells in the pharyngeal mesoderm which express either the transcription factor *ladybird* in flies or the homologous transcription factor Lbx1 in vertebrates.[29,30] Interestingly, the migration and assembly of these cells from the pharyngeal mesoderm to form the outflow tract also require the interaction between Slit and Robo.[113]

Additional conserved factors have been identified in recent years that are responsible for maintenance of adhesion and/or polarity after the initial formation of the heart tube. In genetic screens for a phenotype in which myocardial cells lose their connections with the surrounding pericardial cells, several new players were identified. One such factor was the HMG-coA reductase, along with downstream effectors of the mevalonate pathway.[114] These factors act to regulate heterotrimeric G-proteins, which are also necessary for adhesion between myocardial and pericardial cells.[108,114] These proteins act, in part, by regulating protein components of the septate junction, including Neurexin-IV, Sinuous, and Coracle.[115]

In this context, studies of *Drosophila* cardiac development have led to productive use of the fly heart not only as a model for early stages of vertebrate heart development, but also as a model for tubulogenesis in general.[116]

Indeed, rather than thinking about conservation of genetic functions at the level of individual genes, it is becoming more common to think in terms of conservation of gene regulatory networks governing development of orthologous structures. As such, these networks involving interrelated genes involved in the formation and maintenance of similar structures, such as the heart, not only act as conserved regulators of the development of homologous structures but also provide a buffered system for the development and structure of conserved organ systems to be tweaked during speciation without losing the essential function of the organ. Bioinformatic evidence suggests that conservation of gene expression in homologous organ systems is relatively high across species,[117] and the gene network governing heart development continues to be elucidated in comparative experiments between flies and vertebrates.[77]

VI. Larval Heart Function

During the larval stages of *Drosophila* development, a tremendous degree of growth occurs, with a tiny embryo-sized first instar larva gradually reaching the size required for pupation and metamorphosis. The heart must increase its size proportionately during this process, but does so not by cell division, but rather by cell growth.[6] Thus, the final third instar larval heart is structurally similar to the embryonic heart, but much larger and more suitable for physiological assays. In addition, the larva is essentially transparent, making it an ideal stage for visualization of cardiac function in intact animals (Fig. 3).[82]

Another significant advantage of the larva as a model system is that it has a myogenic pacemaker that continues to promote regular contractile activity without neuronal input.[118] The larval fly heart is, however, innervated[119] and neuronal input regulates rate and rhythm of cardiac contractions.[120,121] Thus, partially dissected preps can be utilized to expose the heart to physiological medium and keep it beating regularly for extended periods of time. Using several different preparations and methods, the intact or semidissected larva has been used to test the effects of numerous biologically active compounds, both synthetic and naturally derived (e.g., hormones and neurotransmitters).

In this section, we will discuss the various methods developed to exploit the *Drosophila* larva as a model system, and summarize the findings with regard to cardiac susceptibility to various external bioactive inputs.

A. Methods

A breakthrough in automation and processing of measurements in intact larvae came with the adaption of photodiode-based measurements of late-stage, immobile larvae at the larva–pupa transition.[122] Measurements of transmitted light through the larva change rhythmically in consonance with the heart movements as the heart changes both its shape and the shape of the connected viscera, then returns to its previous shape in a cyclical fashion.

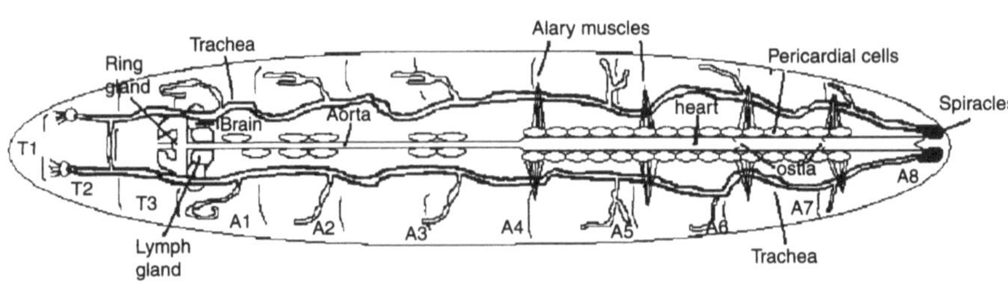

FIG. 3. Diagram of larval heart (adapted from Ref. 82).

The darkening and then lightening of the transparent larva thus produces a cycling trace of changing light exposure in a photodiode that receives input through the microscope field of view. This trace is driven by and mimics exactly the movements of the heart over the same period. Rate and rhythmicity can then be ascertained by mathematical application to the traces. Variations on this method have subsequently been employed in which pixel-tracing camera technology has replaced the photodiode, but acquisition and processing of the data are otherwise similar.[123]

The application of this technique to intact animals has made it suitable for genetic applications, and several candidate mutants have been analyzed in the fly model using this system.[122,124,125] In addition, a large-scale genetic screen has been conducted in which a collection of insertional mutants were screened for mutations that significantly accelerated or decelerated larval heartbeat, identifying several novel candidates for future study in the fly model.[123]

In addition to intact visualization, the larva is also the developmental stage most easily amenable to semi-intact dissection. Perhaps, the most common and fruitful usage of the *Drosophila* larva has been with preparations dissected in physiological media and then treated with various biological or pharmacological compounds. The accessibility of the larva to such treatments in combination with the many genetic tools available in the fly system has made this model a useful one for identifying and describing minimum components necessary for generation and propagation of cardioactive electrical signals, as well as for examining how neuronal inputs to cardiac activity are governed in a simple model system.

One method, which is suitable both for measurements of physiological function and as a preparation to perfuse the heart with pharmacological agents, is to perform a partial dissection, then pull a portion of the heart into a micropipette tip, which can then be employed to record spontaneous field potentials.[126]

Within the last 2 years, several novel ideas have emerged for preparation and measurement of larval heart activity, including two novel preparations for viewing and making recordings from intact larvae. One, known as the "ant farm" method, utilizes plastic spacers to imbed the larva in a small region where it is constrained from moving but unstressed and with both an air source and a thin layer of food available to the larva, thus avoiding potential changes to heart activity due to starvation or oxygen stress responses.[127] This method is well suited for examination of the response to genetic interventions, without potential complications from dissection, and can be employed for relatively lengthy recording times. Two different groups have introduced an additional method involving restraint by glue, which introduces some caveats with respect to the restraint method, but makes the animal easily available to the introduction of electrodes for field recording or pacing stimulus.[127,128] These preparations have been used in direct visual counting of heart rate[127] and also used in automated optical counting protocols.[129,130]

The introduction of the novel automated optical protocols represents a substantial advance building upon previous optical methods, expanding their possibilities. Importantly, these methods allow data to be quantitated not just at the level of patterns of cardiac activity over time, as in prior indirect measures, but allow direct data acquisition of several parameters of individual beats on a continuing basis.[129] Such parameters include systolic and diastolic volume, relaxation, and contraction velocity. Once these data have been acquired, they can be used in turn to derive other measurements such as fractional shortening. Interestingly, these methods have also been applied successfully to analyze cardiac movements in other systems, including vertebrate models such as the zebrafish.[117,129,130]

Another recent innovation with great potential, although requiring highly specialized equipment, is the combined use of optical coherence tomography (OCT) with laser scanning fluorescence microscopy. This technique requires the use of two dedicated imaging systems in combination, but does provide cross-sectional visualization in perpendicular planes of living animals.[131]

Section VII will describe how these various methods, past and present, have been employed to examine genetic, pharmacological, hormonal, and peptidergic regulation of cardiac rate and rhythm.

VII. Pacemaker Regulation

In the mid-1990s, early versions of these methods began to be employed to establish that the larval heart is moyogenic. Using partially dissected preps, treated with pharmacological compounds that block specific subsets of ion channels, it was established that calcium and potassium currents are essential for larval pacemaking, but that sodium channels are dispensable.[132] At the same time, these channel-blocking experiments lent molecular support to the earlier contention that the larval heartbeat is fully myogenic.[118,122,132]

Mutations in genes encoding components of ion channels closely support pharmacological results, as mutants deficient in potassium channel function, such as *slowpoke*, *shaker*, and *ether-a-go-go*, all have deficiencies in rhythmic control of heartbeat, while disruption of calcium or chloride channels has little effect.[125] The role of intracellular calcium storage and release in regulating contractile activity also appears to be conserved, as mutations in the ryanodine receptor ortholog in flies reduce contractile activity in cardiac muscle.[133]

Despite the myogenic nature of the insect pacemaker, it has been clearly established that regulation of the rate and rhythmicity of contractions is influenced by a variety of biological compounds. For example, injections of the insulin-signaling antagonist Adipokinetic Hormone into late larvae/prepupae had a cardioacceleratory effect.[134]

Using photodiodes to record heart movements, it was demonstrated that temperature-sensitive mutations in the gene *no action potential* (*nap*) caused arrhythmias which disappeared when animals were returned to the permissive temperature.[122] Lending further credence to the idea that neuronal regulation modulates the activity of the myogenic pacemaker, several neurotransmitters were demonstrated to accelerate heart rate without adverse effects on rhythmicity, including serotonin, octopamine, norepinephrine, dopamine, and acetylcholine.[124] Conversely, mutations affecting secretion or synthesis of these neurotransmitters tended to decelerate the heart.[124] Other reports have generated partially contradictory results[135] but some of these discrepancies may be explained by dose-dependent differences, as serotonin, for example, has more recently been shown to be cardioacceleratory or cardioinhibitory depending on dose of exposure.[136] Additionally, several FMRFamide-related peptides, including Dromyosuppressin, were demonstrated to decelerate the larval heart.[137]

The Crustacean Cardioacceleratory Peptide was also found to accelerate heart rate in *Drosophila* larvae.[138] Interestingly, release of Crustacean Cardioacceleratory Peptide is activated by Ecdysis Triggering Hormone just prior to eclosion and may be responsible for the temporary elevation in heartbeat that presages the emergence of the adult fly.[139,140] Other neuropeptides involved in the molting or adult emergence process are also cardioactive, including the pyrokinin-like peptide encoded by the *hugin* gene.[141]

Regulation by neuronally produced factors is necessary not only to adjust heart rate to changing conditions but also to maintain proper heart rate under normal resting conditions, as demonstrated by examining the heart rate in flies with a temperature-sensitive mutation in the dynamin-encoding gene *shibire*.[142] These mutants are deficient in the ability to recover endocytotic vesicles at the restrictive temperature[143] and are useful for temporary induction of phenotypes dependent on neurotransmitter communication. Both optical recording and electrocardiograms were employed on intact late larvae/prepupae to establish that blocking the functions of multiple neurotransmitters by inactivating *shibire* leads to defects in regulation of heart rate even under resting, unstressed conditions.[144]

An extensive series of tests of various exogenous factors in combination with genetics has led to significant progress in understanding the makeup of the various currents associated with the larval pacemaker. Multiple classes of G-proteins have been implicated as targets of pharmacological interventions, while cGMP and Protein Kinase G seem to be critical components of changing pacemaker responses to varying conditions.[145] Interestingly, some of the same genetic factors identified as key regulators in the involuntary response of heart rate to conditional change have also previously been shown to be important factors in behavioral regulation of activity levels and foraging behaviors, such as the *foraging* gene,[146,147] and the peptide encoded by the *flatline* gene. *flatline*, despite extensive sequence homology to *Manduca* allostatin, does not appear to act as a Juvenile Hormone antagonist in *Drosophila*, but instead has a potent myotropic activity in both cardiac and visceral muscle.[148]

More recently, a combination of genetic approaches and electrophysiology has identified a two-pore domain potassium channel Orkl that is essential for the regulation of heart rate and rhythm.[149] Reduction of *Orkl* expression throughout the animal or specifically in the heart tube leads to a proportional increase in heart rate while overexpression of *Orkl* blocks heart beat entirely and action potentials are unrecordable when *Orkl* is overexpressed.[149] Significantly, action potentials in wild-type hearts were recorded throughout the length of the heart, supporting the idea that *Drosophila* cardiomyocytes are more homogeneous than vertebrate cardiomyocytes. The highly conserved Sarco-Endoplasmic Reticulum Calcium ATPase (SERCA) protein has also recently been studied using the fly larva as a conserved readout. Flies carrying mutations in the *SERCA* gene have disrupted heart rate and extended periods of heart stoppage.[31]

The combination of evolving techniques for precise cardiac physiology in insect models with the sequenced genome in flies should result in continued importance of the fly model in addressing problems such as how currents are generated and regulated to precisely govern pacemaking. The *Drosophila* larva is well positioned to be an important model system for use in testing various combinations of genetic and pharmacological factors contributing to cardiac disease.

VIII. Adult Functional Models

During pupal morphogenesis, the heart is one of a few structures that persist without being completely degraded and remodeled.[119] However, a few modifications are made to the morphology of the heart before eclosion.[81,82] The formation of the familiar head/thorax/abdomen insect body plan is overlaid with the heart in such a way that the heart proper is located in the abdomen of the adult. A conical chamber is formed near the thorax/abdomen boundary that serves to collect haemolymph for expulsion through the aorta, which proceeds through the thorax and sends haemolymph forward toward the head (Fig. 4).

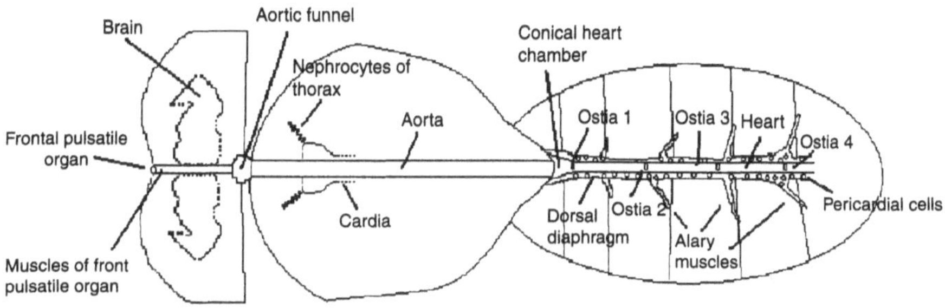

FIG. 4. Diagram of adult heart (adapted from Ref. 82).

Limited changes also occur to the structure of the heart itself. An additional layer of striated muscle is generated during pupation along the ventral surface of the heart, with striations that proceed in a longitudinal direction, rather than the transverse spiral "paper towel tube" shape of the fibers in the rest of the heart muscle. An additional pair of ostia are also generated during metamorphosis, such that the adult heart contains four pairs of ostia, as compared to the larval three.[81,82] In addition, the character of these ostia is morphologically different. In the adult, they are no longer simple openings, but take on a more valve-like appearance.

Perhaps, the most important difference from the point of view of disease modeling is the introduction of direct innervations to the heart at adult stages, adding a layer of complexity to the regulation of heart rate and rhythm.[121] Pairs of neurons in each segment innervate both the heart itself and the muscles attaching the heart to the body wall (alary muscles). Additional nerve projections known as Bipolar Neurons cluster toward the posterior of the heart where they serve as a point of release for Crustacean Cardioactive Peptide,[121] and application of CCAP at these sites instigates anterograde contractions, suggesting that these neurons are important regulators of "forward" heartbeat.[150] Meanwhile, bilateral, segmental innervations from glutaminergic neurons regulate the periodic reversal characteristic of insect heartbeat.[150]

Recently, another forward pair of ostia was discovered in two different species of Drosophilids using a recording technique from multiple sensor elements in sequence.[151] These forward ostia are essential for retrograde heartbeat, as they collect and separate haemolymph supply from specialized thoracic spaces, the implications of which for circulation are just beginning to be understood.[151] These separated flows of circulation may in the near future make it possible to use flies to model other circulatory disorders that had not previously been thought accessible using insect model systems.

Already, the fly system has been in use over the last few years as a model for various cardiac disorders, and offers several important advantages over vertebrate models. First, the rapidity and facility of the genetics available in the fly system are an obvious motivation to use the fly to examine complicated genetic interactions in multifactorial etiologies of disease. Second, the groundwork laid by the fly community and by the Drosophila genome project has made mutations, overexpression constructs, and inducible RNAi constructs available for the vast majority of the genome. Third, and perhaps most excitingly, the short lifespan and ability to monitor cardiac function longitudinally have made the fly system available as an important model for long-term studies, applicable either to progressive disease models, or even to changes to cardiac function brought about by the consequences of normal aging.[7]

In Section VIII.A, we will discuss the novel techniques developed in the last few years to make these studies possible, and then discuss the results and their application to vertebrate disease and aging.

A. Methods

An initial success in usage of adult flies to model cardiac functional disturbance was accomplished by examining maximal heart rate in cases of exogenous stress. Heart rate was increased either by exposing flies to increased temperature or to external currents, and the maximum tolerable increase in heart rate was measured at various ages.[152] It was determined that the ability of the heart to tolerate externally induced increases in heart rate is an age-dependent phenomenon, since genetically identical flies at older ages entered cardiac arrest when exposed to similar conditions to those tolerated easily by younger flies.[152]

Further progress was made when edge-detection techniques were adapted from usage in larvae (see above) to become feasible in adults as well.[123] This technique relies on pixel-tracing software to identify edges of the cardiac tube, identifiable by their differential contrast in high-resolution black and white video images. This technique allows long-term tracing of heart movements in unstressed and intact conditions. Importantly, this technique and others that do not rely on fluorescence offer the advantage of not requiring the introduction of additional stimuli that may affect endogenous cardiac regulation, such as ultraviolet light when detecting movement using Green Fluorescent Protein, for example.

Refinement of stress testing followed rapidly with a stress test protocol that relied on a defined condition to test the response flies of various genotypes, ages, and environmental exposures.[153] It offers the distinct advantage of being applicable to up to 10 flies at once, thus facilitating large study designs to take advantage of the large numbers common to fly studies, in comparison to vertebrates. Because this method does not depend on maximal tolerance of extreme stress, it is useful for picking up more subtle abnormalities and has been used effectively in screens to pick up mutations affecting short-term or long-term cardiac performance.[123] By testing the percentage of populations that respond to low-level stress by entering fibrillations or arrest events, as well as the percentage of flies experiencing a cardiac event that successfully recover normal heartbeat regulation, this method serves as a useful marker for physiological age and health that can pick up not just direct abnormalities, but, importantly, population tendencies toward abnormalities (Fig. 5). In combination with genetics, this method has served as an important first step toward identifying new genotypes that can be used as models of disease[154] or of aging.[42]

Further refinement to these techniques has involved partially dissected preps, bathed in physiological media that allows continued heart function for several hours.[155] Algorithms have been developed to analyze video traces taken from such preps[129] to extract several useful parameters, including fractional shortening, a two-dimensional measure that can be used proportionally to infer the volume expelled during contraction; arrhythmicity index, used to quantitate the frequency of arrhythmias; and period length, which is proportional to resting heart rate. The use of these measures in combination allows a significant degree of diagnostic precision, not just identifying cardiac abnormalities, but also classifying them into different subtypes of functional abnormalities.

Similar measures can also be achieved using video analysis of intact preps, although the use of intact preps involves some trade-offs in camera resolution as compared to dissected preps. In return, one gains the assurance that the

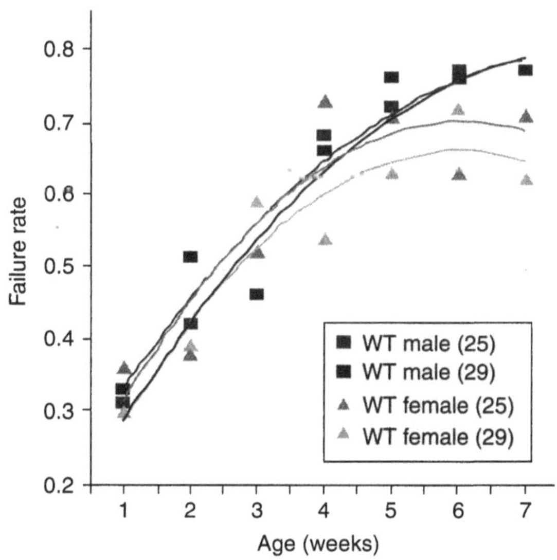

FIG. 5. Heart failure as a function of age after external electrical pacing from outbred wild-type offspring (WT; *yw* × Canton S). This parameter is highly age-dependent, but is not affected by temperature or gender during the first 5 weeks of adult life.[153]

animal is intact in its own physiological state rather than a mimicking media. Depending on experimental design, this trade-off may be worthwhile, in cases where longitudinal analysis of the same flies over time is critical[156] or where retention of neuronal input is desired.

Another option for acquiring parametric measurements of live cardiac performance in intact animals is the use of OCT technology. In this technique, the fly is enclosed in a small chamber to prevent large movements, and the internal space of the heart tube is detected in real time as it shortens and expands through an ultrasound-like technology.[157] This method combines the advantage of utilizing intact animals in their inherent physiological environment with some of the resolution advantages of the partial dissection/video method. The primary disadvantage of this powerful method is its requirement for specialized technology that is not readily available commercially.

Either OCT or video-based traces can be utilized to generate M-mode displays (Fig. 6), which take a thin slice of an ongoing display of movement and then display repeated short intervals of that same slice across time. In effect,

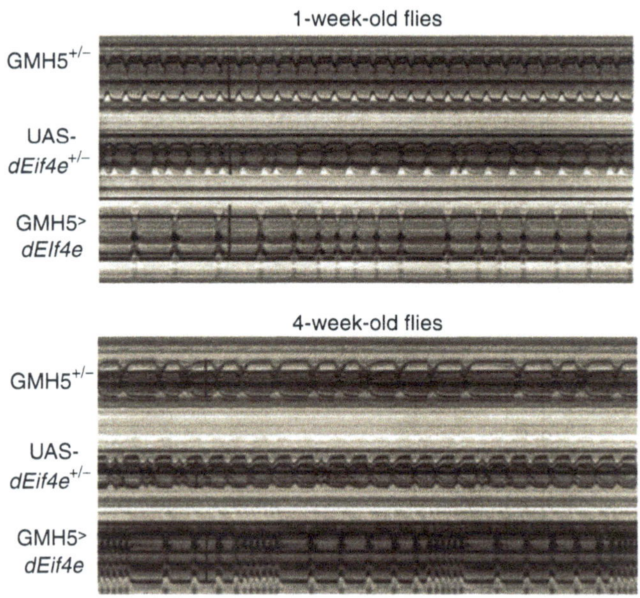

Fig. 6. Representative M-mode records showing the movement of the heart tube walls (y-axis) over time (x-axis). Blue bars indicate the diastolic diameter of each heart and the red bars indicate the systolic diameter. Records for 1-week-old flies (A) show predominately regular heart beat patterns as the GMH5 heterozygotes (as in Ref. 42), whereas most flies overexpressing UAS-*dEif4e* in the heart show arrhythmic heart beat patterns.[55] (See Color Insert.)

this creates a still-life visualization of the animated movement of that particular portion of the heart. This is valuable not just for visual purposes, but also can be used as input for analysis algorithms.[155]

While direct cardiac functional readouts have had a tremendous impact on the utility of the fly heart as a model, traditional genetic methods have also been effectively applied to generate tissue-specific data centered on the heart. For example, heart-specific microarrays have been generated,[158] screens using inducible RNAi constructs driven in the heart have been conducted[159] and genome-wide association studies have been conducted with cardiac performance as a quantitative output for QTL mapping.[155]

Excitingly, the fly heart has also been effective as a readout for testing human genes directly for effects on heart function. For example, genes from a human chromosomal region associated with congenital heart defects in Down's Syndrome patients have been tested in all permutations of single and combined overexpression in the fly heart. This study has been successful in narrowing down the potential effectors of this phenotype to two genes that, when expressed in combination, generate cardiac defects. The same genes are now being expressed in mice to confirm the phenotypic results directly in a vertebrate model.[160] Such studies are likely to become more common in the immediate future, as the power of fly genetics to test large numbers of factors in a short time will continue to make study designs possible that would be prohibitively expensive and time-consuming in vertebrate models. The information gained from bulk studies in the fly can then be used to intelligently design specific studies in vertebrate models limited to the best potential candidates.

In Section IX, we will describe in turn several recent studies that have effectively utilized the adult fly heart as a model for examining specific-genetic factors contributing to human disease.

IX. Single-Gene Disease Models

An effective entry point for modeling specific or general cardiac pathologies in flies is the candidate gene approach. A recent example of the effectiveness of such an approach is the recent work using the fly to model the cardiac dysfunction that occurs as a result of the progressive degeneration in muscular dystrophy patients.

The causal gene in multiple forms of the human disease muscular dystrophy has been identified as the structural protein Dystrophin.[161] Dystrophin plays a dual role, being necessary to provide the structural framework for contractile activity, and also being involved in signaling events inside muscle cells through its interactions with other proteins in the Dystrophin/Dystroglycan complex.[162,163]

In both flies and vertebrates, the *dystrophin(dys)* gene is expressed in multiple isoforms, with different isoforms being expressed in different tissue-specific patterns. Two different splice variants of the fly *dys* gene are expressed in the adult myocardium, suggesting that Dys is likely to play an endogenous role in cardiac function in insects as well.

In mouse models of muscular dystrophy, deficiency of *dys* gene expression results in structural degeneration of the myocardium in a progressive fashion that is aggravated by age-related structural derangement.[32] In flies mutant for the *Drosophila* homolog of *dys*, a similar structural degeneration of myocardial fiber arrays has been observed.[33]

Adult heart performance in *dys* mutant flies displays several abnormalities as a consequence of these structural defects, as measured by video tracking of the movements of dissected heart preparations.[33] Such flies have an increased resting heart rate, which comes about due to a decrease in fractional shortening, defined as a decrease in heart wall diameter change during contraction, a measure that is proportional to cardiac output. Since *dys* mutants exhibit both wider diastolic diameter and reduced fractional shortening, they are said to have a dilated cardiomyopathy phenotype.[33] This phenotype is a direct result of deficiency of Dys protein in cardiac tissue, since it can be rescued by cardiac-specific expression of the *dys* gene. Interestingly, the fly and mouse functional versions of these proteins are so similar that the fly phenotype can be rescued by cardiac expression of a truncated version of the homologous mouse gene.[33]

Another gene product that has been associated with muscular dystrophy in vertebrates is the Sarcoglycan protein, mutations in which are known to cause cardiomyopathy in vertebrates.[34,35] Deletions of the homologous fly gene have been examined, and shown to cause a progressive impairment of locomotor ability along with reduced cardiac function.[36] Cardiac phenotypes are qualitatively similar to those in *dys* mutants, with progressive structural derangement accompanied by increased diastolic diameter and reduced fractional shortening, characteristic of the dilated cardiomyopathy phenotype.[36]

These experiments, along with the others described below, serve well to illustrate several fundamental points. First, the fly model responds phenotypically to genetic interventions in ways that are recognizable and remarkably similar to vertebrates. Second, in some cases the genetic material involved is similar enough for cross-species experiments to produce useful data. Third, following upon the first two points, the essential similarity between vertebrate and insect models in isolated organ function makes the fly model a viable choice for structure–function analysis of vertebrate proteins. Indeed, the fly has already been effectively used to examine the role of various domains and splice variants of the Dys protein in skeletal muscle.[164]

The fly has also become a useful model for unraveling the effects of components of the contractile machinery in the myocardium. Due to genetic redundancy and multigene families in vertebrate lineages, the simpler fly model offers a desirable alternative for avoiding complications resulting from such redundancies. A recent example is found in the use of the fly system to study mutations in proteins from the myosin transducer complex.[37] The molecular motor formed by a complex of myosin light and heavy chains is a conserved molecular regulator of contractile activity in both skeletal and cardiac muscles. Both reduction or stimulation of transducer activity can cause muscle pathology in vertebrates.[38,39] In flies, there exists a single *Mhc* gene encoding all isoforms by means of differential splicing in different tissues.[40,41]

Using two existing point mutations in the *Mhc* gene, one that increases motor activity and one that inhibits motor activity, the effects of altering motor activity specifically in cardiac tissue were tested. As in vertebrates, either stimulation or inhibition of the myosin motor caused pathologies. Interestingly, the phenotypes generated were quite different, however.[37] Flies with reduced myosin transducer activity become dilated at diastole, have reduced systolic shortening, and have an increased incidence of arrhythmias that grows progressively worse. This constellation of phenotypes is characteristic of the dilated cardiomyopathy phenotype (Table II). Meanwhile, flies with increased transducer activity display morphological constrictions in isolated areas of the heart tube where full relaxation does not occur. As these flies undergo aging, these constrictions appear in more areas of the heart in a progressive fashion. This phenotype is one of several indices that are collectively described as restrictive cardiomyopathy (Table II).

TABLE II
PHENOTYPIC CHARACTERIZATION OF CARDIAC PATHOLOGIES ELICITED BY SINGLE-GENE MUTATION IN *DROSOPHILA*

Disease	Phenotypes
Dilated cardiomyopathy	Enlarged heart, abnormally functioning heart, reduced fractional shortening, enlarged tube diameters, impaired systolic function, increased diastolic diameter, increased systolic diameter, expanded heart period
Restrictive cardiomyopathy	Elevated arrhythmicity index, decreased resistance to external pacing stress, alterations to diastolic width and fractional shortening
Long QT syndrome	Increased risk of ventricular tachycardia and sudden death, sensitive to stress

A similar approach was employed to test the fly homolog of the vertebrate protein Muscle LIM Protein (MLP). This protein has been found associated with the Z-discs at the boundaries of sarcomeres in cardiac muscle,[165,166] and the collective assortment of proteins associated at the Z-disc has been demonstrated to be critical for cardiac contractile function.[167] Mutations in several of these components, including MLP, lead to dilated cardiomyopathy.[168]

A fly homolog, *mlp84B*, is also expressed in sarcomeric Z-discs of the cardiac muscle in flies.[169] RNAi knockdown of the fly gene did not cause overt structural defects, but did induce significant physiological deficiencies, including rhythmic defects and prolongation of the diastolic interval.[169]

Another grouping of conserved genetic factors associated with cardiac dysfunction in both vertebrates and insects is the group of genes encoding ion channels necessary for the generation or propagation of currents involved in electrical stimulation of cardiac contractions. Two recent studies have addressed specific single-gene mutations in flies affecting one such ion channel.

One such study focused on the fly homolog of the vertebrate *KCNQ1* gene. This gene encodes a subunit of a potassium channel necessary for cardiac repolarization.[43] Mutations in this gene have been associated in humans with a prolonged QT interval on electrocardiograms, a condition which, in turn, has been associated with Long QT Syndrome, a condition that causes increased risk of ventricular tachycardia and sudden death.[44,45]

Deletion mutants were generated for the fly homolog, also called *KCNQ*. Null mutants did not have overt structural abnormalities but, like animal models of Long QT syndrome,[170] were extremely sensitive to stress, when induced by external pacing.[42] These abnormalities could be rescued by cardiac-specific expression of a wild-type *KCNQ* transgene, indicating that the requirement for KCNQ protein is tissue autonomous in the myocardium.[155] Using M-Mode video traces from semi-intact preparations, it was determined that *KCNQ* mutant hearts begin to exhibit substantial irregularities in heart rhythm early in life, in contrast to wild-type hearts, which are quite regular until age-related deterioration starts to induce spontaneous arrhythmias.[155] Additionally, the period length of mutant heart beats rapidly degenerates to a much greater length, and hence, a much slower heart rate, than wild-type hearts of the same age. Again, as in vertebrates, electrical field potential recordings indicated a tendency toward failure to repolarize, suggesting that this is the underlying mechanism for the stress sensitivity and arrhythmias.[155] Similar effects were reported for other conserved channel genes, including the homolog of the vertebrate *HERG* gene.

Another such study examined mutations in the *dSUR* gene, vertebrate homologs of which form essential subunits of the K^{ATP} channel. These channels are known to be important in adult heart function, particularly in protection from ischemic stress.[171,172] Interestingly, cardiac expression of RNAi against *dSUR* rendered flies extremely susceptible to external pacing stress and to hypoxic stress.[95] Interestingly, this gene has also served as a pioneering example in flies of reiterated usage of developmental transcription factors essential for cardiac specification to also regulate cardiac differentiation and adult function. The *dSUR* gene is coordinately regulated directly by the GATA family homolog Pnr and by the Nkx family homolog Tin.[95]

Several developmentally important cardiogenic factors have been implicated in adult cardiac pathologies, particularly in the induction of pathological hypertrophy.[173] GATA4, for example, in addition to its association with several familial forms of congenital heart disease,[174,175] also has been strongly associated with adult-onset pathological hypertrophy[176] and is thought to contribute to hypertrophy by activating similar targets that are employed during initial cardiac growth during development.[177]

Using adult-specific expression drivers to drive RNAi constructs, the role of the *Drosophila* GATA family homolog *pnr* in maintenance of adult cardiac physiology was examined. Flies either expressing *pnr* RNAi, a dominant negative *pnr* construct, or carrying heterozygous genomic *pnr* mutations showed an enhanced sensitivity to external pacing stress, and these phenotypes could be rescued by cardiac expression of wild-type *pnr*.[178] Flies heterozygous for *pnr* mutations exhibited a high-cardiac arrhythmia index and a high-period length (equivalent to low heart rate). Both of these phenotypes were also susceptible to rescue by expression of a *pnr* construct with a cardiac-specific driver.[178] RT-PCR also revealed that expression of several structural components of muscle as well as essential ion channel subunits was downregulated in *pnr* heterozygotes.[178]

Heart-specific overexpression of the T-Box family homolog *nmr* was also able to rescue both pacing sensitivity and tendency to arrhythmia, suggesting that *nmr* is likely to be functionally downstream of *pnr* in this context.[178] The Nmr protein had previously been shown to act synergistically with the Nkx2.5 homolog Tin to regulate adult function.[179] Taken together, these results establish the fly as a model for conserved regulation of adult function by developmental genes. As such, the opportunity exists over the next few years to use the fly as a model to test alterations of such genes and their downstream targets, which are likely to involve similar targets to those in vertebrate pathological hypertrophy.

These studies are examples of how the fly heart can be used as a model system to unravel the etiology of diseases already known in humans, but difficult to test in vertebrate model systems. Such models, once developed, set the stage for testing of various interventions to rescue the disease phenotypes. Given the remarkable conservation of disease mechanisms, it seems likely that interventions identified in insect models will also find application in vertebrate models or human patients.

X. Disease Mechanisms

The usage of the fly model is not limited, of course, to forming models for the testing of interventions against known vertebrate diseases, however. The fly has also begun to emerge as a useful system for examining the role of various physiological interactions that are not yet well understood in vertebrates. A recent example is a study examining the role of the pericardial cells as a nonautonomous regulator of adult function. In vertebrates, the epicardial cells are known to be necessary for the proper maturation and adult function of the myocardium,[180,181] but the mechanism by which epicardial cells accomplish this is not yet well understood.

The pericardial cells surrounding the fly myocardium may act in a homologous fashion to the vertebrate epicardium in some respects. For example, the transcription factor Evx2 is expressed in the mouse epicardium, and its homolog *even-skipped* (*eve*) is expressed in a key subset of the fly pericardial cells.[46] If this expression is removed by genetic means, these cells become naïve and are deleted from the pericardial lineage, which results in several deleterious effects on both muscular mobility and myocardial performance, including reduced resting heart rate in both larvae and adults, as well as reduced resistance to pacing stress in adults.[46] Taken together, these results suggest the possibility that cells marked by expression of Eve-family transcription factors may be essential to signal to associated myocardial cells to maintain adult heart function in both vertebrates and insects.

Recently, a fusion construct, in which the pericardial enhancer for *eve* was fused to a protein that represses *eve* expression, was used to examine more closely the consequences of deletion of Eve-positive pericardial cells on adult myocardial function. It was found that such hearts from flies with Eve-positive pericardial cells deleted have an elevated arrhythmicity index, decreased resistance to external pacing stress, and alterations to diastolic width and fractional shortening characteristic of a restrictive cardiomyopathy phenotype.

These phenotypes are consistent with a conserved role for vertebrate epicardium with insect pericardial cells, and further study in the fly model will be instrumental in unlocking the mechanism by which these interactions take place.

More evidence for a conserved regulation of adult cardiac function has been recently provided by the identification of the EGF Receptor signaling pathway as an important postdevelopmental regulator of cardiac function. A mutation in the *rhomboid3* gene, which encodes a transmembrane protein necessary for the processing of a critical EGF-receptor ligand, was shown to cause a dilated cardiomyopathy phenotype.[182] This effect was specific to maintenance of adult function, as no developmental abnormalities were detected. Cardiac expression of an activated version of the EGF Receptor rescued this phenotype, as did expression of preprocessed ligand.[182] Furthermore, adult-specific expression of a dominant negative version of the EGF Receptor itself produced a progressive dilated cardiomyopathy phenotype as well. Taken together, these results argue strongly that active EGF Receptor is necessary to maintain postdevelopmental cardiac function in *Drosophila*.

This phenotypic role for EGF Receptor in the maintenance of postdevelopmental function appears to be conserved in vertebrates as well. This conservation has important consequences, particularly in the treatment of some forms of tumors. For example, inhibition of ErbB2, the closest human homolog of the *Drosophila* EGF Receptor, is a target of some chemotherapeutic treatments. Unfortunately, inhibition of ErbB2 has also been associated with dilated cardiomyopathy in humans.[47–49]

Another example of the usage of the fly model to shed light on mechanisms that are conceptually ambiguous in vertebrate comes from recent work on the role of oxidative stress in degenerative cardiac phenotypes resulting from disease or from normal aging. Oxidative stress has been proposed to be a causative agent in several cardiac disorders, including diabetic cardiomyopathy, where it may act indirectly through impairment of vascular function via induction of atherosclerosis or hypertension[183,184] or directly in the myocardium, where increased levels of reactive oxygen species (ROS) have been associated with heart failure and ischemia/reperfusion injury.[185,186] These effects are also exacerbated by a vicious cycle of coregulation with inflammatory cytokines.[187] The inflammation associated transcription factor NF-kappa B has also been associated with arrhythmias, including atrial fibrillation.[188] ROS levels in the myocardium itself have been shown to increase in response to inflammation and this process may be mediated by the proteins of the RAGE (receptor for advanced glycation end products) family, which is upregulated in response to hyperglycemia and triggers additional ROS producing activities.[189]

Evidence from a variety of vertebrate genetic models that increase ROS generation is in agreement that high levels of ROS are associated with pressure-induced cardiac hypertrophy[190–193] and sensitivity to ischemic stress.[194] Due to the cyclical nature of ROS as targets and regulators of numerous physiological states, it is difficult to resolve with certainty whether ROS are indeed causal factors in these pathologies or whether they are instead symptoms inevitably associated with dysfunctional cardiac tissue that exacerbate preexisting pathologies.

Two recent fly studies have used single-gene mutations to attempt to address these issues. The first of these used the fly homolog of the vertebrate Opa1 gene, whose product is expressed in the mitochondrial membrane in humans, and has been associated with several degenerative diseases of the eye.[50] Flies heterozygous for a null mutation in the *Drosophila* homolog, *dopa1*, exhibit vision defects, indicating a high degree of functional conservation with the human gene.[51] These flies also exhibit significant cardiac abnormalities, including a reduced resting heart rate, increased tendency toward arrhythmias, as well as decreases in fractional shortening and decreased resistance to external pacing stress.[51] Significantly, however, these heart phenotypes do not appear to be directly a consequence of increased oxidative damage, because treating the same flies with antioxidants fails to rescue the cardiac phenotypes. This failure is not due to ineffectiveness of the antioxidant treatment, because the same treatment effectively rescues the eye phenotypes observed in the same model.[195] Rather, the case is made that the eye phenotypes are a consequence of oxidative damage, but the cardiac phenotypes are related to deficiencies in mitochondrial respiratory function as a result of reduction in dOpa1 levels.

A common avenue for studying the effects of oxidative stress on specific functions is by examining the phenotypic effects of mutations or tissue-specific knockdowns of genes essential for scavenging ROS. A mouse knockout has been generated for superoxide dismutase 2 (SOD2),[52] a gene that encodes a mitochondrially localized protein that scavenges superoxide produced during mitochondrial reactions and converts it to hydrogen peroxide, from which other proteins can convert it to water. Mice homozygous for the SOD2 knockout display serious cardiac defects, consistent with a diagnosis of dilated cardiomyopathy,[52] in addition to other degenerative phenotypes and mitochondrial dysfunction.[196] Mice heterozygous for this knockout are overtly normal but develop cancers at a high rate due to sensitivity to endogenous and exogenous oxidative stress.[195,197] Conversely, treatment with an exogenous SOD2 mimetic can rescue some of these cardiac effects,[198] while expression of an *SOD2* transgene can alleviate effects of oxidative stress in cultured cardiomyocytes.[193]

Flies that carry a null mutation for *Sod2* are able to reach adulthood, but have an extremely short lifespan, averaging less than 2 days.[53] Since the lifespan is so short, flies were tracked using noninvasive video techniques throughout life to assess whether their progressive deterioration was qualitatively similar to the phenotypes seen in progressive human diseases or in aging.

Sod2 mutant hearts have a dramatic reduction in net heart rate, due to frequent pauses, followed by episodic resumption of rapid, shallow heartbeats.[156] During periods of beating, the heart rate is considerably faster than wild type, but with a reduced fractional shortening and presumably a lower contractile output as a result. Eventually, prior to the death of each individual animal, the heart ceases to beat at all, even in response to external stimuli. No evidence was seen for a progressive deterioration in high-oxidative stress conditions that resembled age-related or progressive disease-related pathology. Rather, there appeared to be a threshold effect, in which past a certain point of oxidative damage, the heart loses its ability to maintain effective function entirely, and up to that point, there is very little evidence of the progressive decline in resting function seen in aging wild-type flies (Fig. 7).[42,153,156]

Interestingly, when the same longitudinal tracking techniques were applied to wild-type flies at advanced ages, the same pattern was observed. Although there is a progressive decay in rhythmic control and ultrastructure over the lifespan of the animal (see aging section below), the chain of spontaneous functional failures that result in loss of effective cardiac performance do not actually begin until the last 36–48 h of life, and can, in fact, be used to predict which flies are in the process of dying 1–2 days prior to actual death[156] (Fig. 7).

A distinct advantage of flies as a cardiac model is the ability of flies, as evident in the *Sod2* mutant flies just described, to survive for relatively extended periods after extreme disruption of cardiac function. Since insects are not directly dependent on heart activity for oxygen, it is possible to observe the consequences of functional abnormalities that would result in sudden death in a vertebrate model. This advantage, in combination with the relative lack of redundancy already mentioned, has suggested the usage of flies to examine reiterative roles of gene products that are developmentally important to produce functional hearts and then are reused in temporally separable roles to maintain adult function.

XI. Unbiased Screens—Disease Phenotypes

The most historically important and obvious advantage of the *Drosophila* model system is the ability to do large-scale unbiased screens for a wide variety of phenotypes. Not only do the quick generation time and wide array of available techniques facilitate rapid identification of candidates, but the distance between primary screen candidates and specific retesting is also much shorter than in vertebrate model systems, allowing researchers to get from identification of loci to specific mechanisms within a reasonable time frame. The use of the fly model in this context has traditionally been associated with developmental biology, and indeed, most of the major developmental signaling pathways were first identified in this way. Only recently, as phenotypic characterization of physiological, rather than structural, abnormalities in flies has become more rapid and routine, has the concept of unbiased genetic screens become feasible for specific aspects of organ physiology, such as cardiac performance. Two recent examples demonstrate the inspiring possibilities now available for identifying novel regulators of adult cardiac function with the fly system.

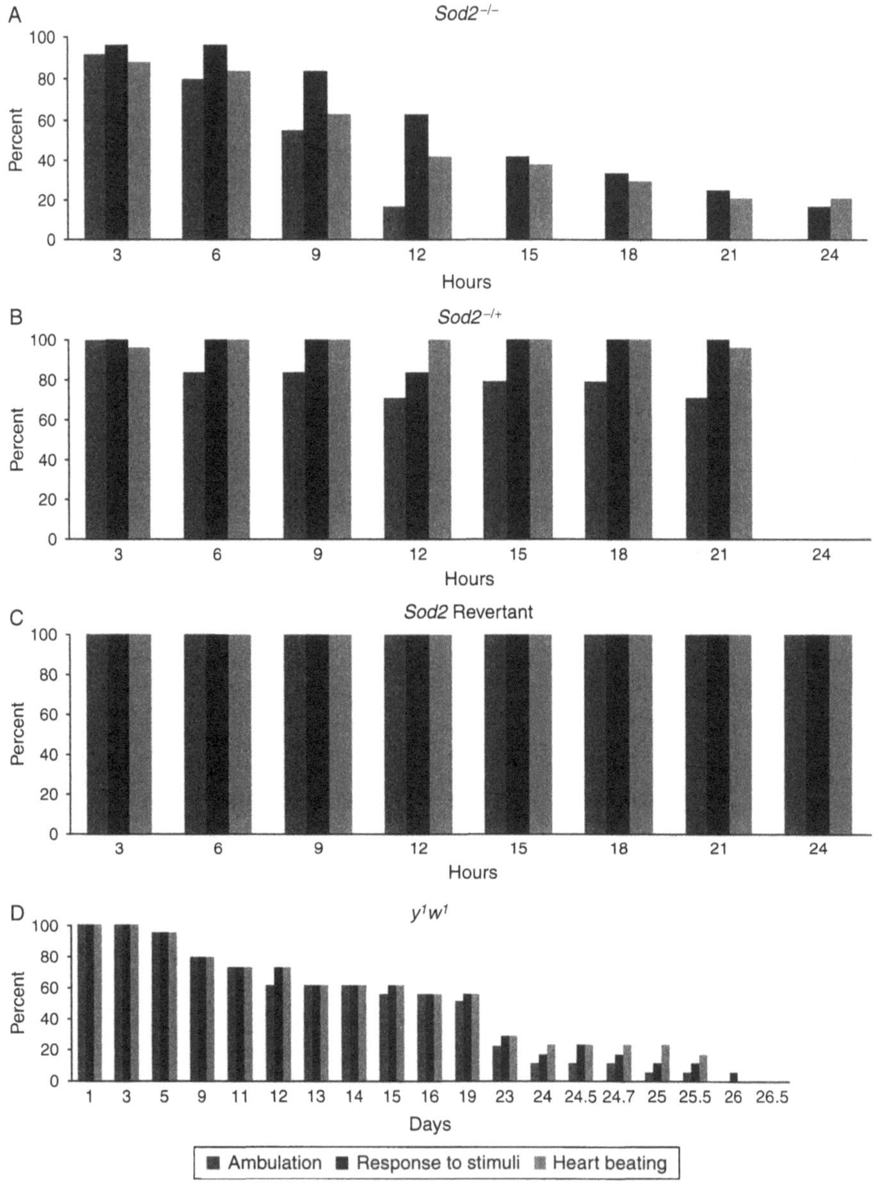

FIG. 7. Various measures of functionality decline at different rates in *Sod2*$^{-/-}$ flies. Percentage of flies observed to exhibit either spontaneous ambulation (red), response to gentle stimuli (blue), or spontaneous heartbeat (green) is charted for (A) *Sod2*$^{-/-}$, (B) *Sod2*$^{-/+}$, (C) *Sod2* revertants or (D) y^1w^1. (A), (B) and (C) are shown over a 24-hour time period, whereas (D) is shown throughout a lifetime. In (D), day 1 in the graph represents day 1 of the experiment, but day 64 of the age of flies.[156] (See Color Insert.)

One such screen has utilized the OCT technology recently developed to allow diagnosis of abnormalities in resting heart function in intact unanaesthetised animals (see Methods section above). Primary screening was done using a collection of deficiency stocks maintained at the public *Drosophila* stock center in Bloomington, Indiana. While these deficiencies are all homozygous lethal, they represent an excellent tool for identifying loci that are heteroinsufficient for physiological phenotypes. In this case, deficiencies on the second chromosome were screened for the dilated cardiomyopathy phenotype characteristic of adult flies with impaired adult functional regulation (Table II).

One gene identified in this screen, *weary* (*wry*), has already been further characterized and defined as a novel ligand for the conserved *Notch* receptor, based on cell aggregation assays.[199] After identifying the importance of *wry* for maintenance of adult function, the same study went further to demonstrate that other known ligands for Notch, as well as the downstream transcription factor associated with Notch signaling, Supressor of Hairless, are also required autonomously in the heart for adult function.[199] A role for Notch in the etiology of adult cardiac dysfunction has already been implicated in vertebrate studies.[200] Now that the conservation of this importance has been extended to flies, the fly model system can be used to further dissect pathway components and mechanistic interactions in the heart in upcoming years.

A second example took a quite different approach, using a genome-wide collection of inducible RNAi constructs, also publicly available from the Vienna *Drosophila* RNAi Center. Each RNAi was driven specifically in the adult heart, and primary screening was done not for heart function directly, but indirectly by assessing lethality at an increased temperature. The rationale for this is that since heart rate is temperature-dependent in insect species, passage of adults through a period of high temperature is equivalent to a cardiac stress test. This approach proved fruitful as ~6% of the genome displayed evidence of inability to tolerate this stress, and these showed a significant enrichment for genes previously classified as being part of conserved functional classes.[201]

One pathway that generated multiple hits in this analysis was the CCR4-Not complex, previously identified in yeast[202] as a chromatin regulator,[203,204] but never before associated with cardiac function in any organism. Secondary screens confirmed that knockdown of the *not3* gene in the myocardium by multiple methods generates flies with the classic dilated cardiomyopathy phenotype, including expanded heart period, reduced fractional shortening, and increased systolic diameter.[201] These phenotypes may be a consequence of structural alterations, since disruption of myofiber arrangement was also observed. However, RT-PCR from hearts with knockdown of *not3* revealed significant reduction in the gene expression from several genes previously identified to be critical for cardiac performance, including *Serc2a*, *Mhc*, and *KCNQ*,[201] suggesting that the phenotype of *not3* may be dependent on regulation of a cardiac gene program.

These observations were rapidly translated to vertebrate models by the construction of knockout mice for the mouse homolog of *not3*. Contractile abnormalities were observed in heterozygous knockout mice, and also in *ex vivo* explanted cardiomyocytes. Furthermore, hearts in heterozygous mice were extremely sensitive to cardiac stress, as assessed by transverse aortic constriction.[201] In support of the idea that Not3 acts through influencing the chromatin state of a cardiac genetic program, these phenotypes were able to be rescued by HDAC inhibitors. Expanding the relevance of these results to humans, a recent genome-wide association study for Long QT Syndrome in humans also identified a member of the Not complex, supporting a conserved role in the human population.[205]

In Section XII, we will discuss the developing use of the fly as a model of long-term changes in cardiac function brought about by consequences of aging, or by changes in "lifestyle," such as diet and exercise.

XII. Cardiac Aging

Invertebrate models have been instrumental in advancing the understanding of genetic regulation of lifespan. Both fly and worm models have been used extensively in screens and candidate approaches to identify single genes that are capable of extending lifespan when overexpressed, or when mutated.[206,207] Several major mechanisms of lifespan extension have been shown to be conserved in vertebrate and invertebrate models, including dietary restriction,[208] insulin/IGF signaling,[209,210] and TOR Kinase signaling.[211,212]

Until recently, however, invertebrate models have not been suitable for investigations of detailed organ physiology during aging, primarily because of a lack of suitable measurement techniques that can be applied in large numbers or in longitudinal study designs. In recent years, functional aging has begun to be an increasingly fertile topic of study in flies and worms, however. Worms have been used as a model for deterioration of mobility and integumental integrity during aging,[213] while flies have been used to analyze declines in functional mobility[214] and immune response.[215] Importantly, functional assays allow changes in lifespan associated with genetic modifications to be attributed causally to changes in actual function. Genetic alterations that improve lifespan may or may not also improve function,[216] while interventions that improve function during aging may not always extend lifespan.[217] Since extension of life without concomitant extension of function is of limited value, measurements of functional aging are likely to become an increasingly important function of invertebrate genetic model systems, such as worms and flies.

The development of increasingly powerful techniques for measurement of cardiac physiology in *Drosophila* in recent years has made cardiac aging an especially well-studied aspect of functional aging in flies. Given the extraordinary degree of conservation of functional genetics between fly and vertebrate hearts, as demonstrated by the multiple success stories in disease models discussed in the previous section, it seems likely that a similar degree of conservation will apply to mechanisms of cardiac age-related change as well.

Based on longitudinal human studies,[218] and on a variety of studies in vertebrate animal models,[218,219] conserved changes associated with aging cardiac function have been described. These fall into two categories, one being pathologies that are age-related simply because they are a result of accumulated damage that has had more time to accumulate in older individuals. These pathologies are often a result of prolonged exposure to various risk factors, such as high-fat diets, smoking, or family history, as well as environmental pathogens or irritants.[220–223] However, it is generally accepted that even in healthy individuals, a gradual deterioration in cardiac function occurs as a function of normal aging.

For example, older humans exhibit an increase in diastolic end-filling, resulting in an increase of contraction duration.[224,225] This increased contraction duration in turn leads to a reduction in resting heart rate, and an even more pronounced reduction in maximal heart rate during exercise. Reduction in maximal heart rate may also be a consequence of reduced number of functional atrial pacemaker cells, which contributes to loss of rhythmic homeostasis, as reflected in the age-related increase in atrial fibrillation events.[226] Aging humans and rodents also exhibit decreased resistance to a variety of cardiac stresses, especially ischemic reperfusion stress.[227]

Work in isolated cardiomyocytes has demonstrated that cardiomyocytes from aging individuals also have reduced contractile strength,[228] reflected in reduction of gene expression for components of the contractile machinery[229] and a reduction in mitochondrial gene expression and mitochondrial enzymatic activity.[230] Mitochondrial morphology is also altered in aged cardiomyocytes, as mitochondria from older cells appear swollen and fewer cristae.[231] As a consequence of reduced mitochondrial efficiency, higher amounts of ROS are also produced in these cells, causing further damage to mitochondria.[232,233] In addition, aging in humans has been associated with a progressive pathological hypertrophy, which may be driven in part by reiteration of a developmental program, since Nkx2.5 and GATA4 are reactivated during this process.[229]

Several aspects of cardiac function have been shown to decline progressively with age in a robust fashion detectable in widely divergent genetic backgrounds. Maximum heart rate declines steadily throughout life,[152] while resting heart rate undergoes a steady decline between 1 and 5 weeks of age[123,153] (Fig. 8). This reduction in beats per minute is largely a function of lengthening relaxation time, which gradually increases the period length of a contraction wave.[155] In addition, a gradual loss of rhythmic regularity leads to a gradual increase in incidence of spontaneous tachycardia-like events occurring in older flies.[155] Furthermore, a reduction in fractional shortening is evident during normal aging[33] which produces phenotypes similar to disease models in which contractile machinery is impaired.[36,37] This changing profile of performance is reminiscent of age-related dysfunctions in cardiac performance in the human population, who undergo increases in incidence of atrial fibrillation, reduction in maximal heart rate, and reduction in contractile output with age as well.[7,218]

Thus, flies are an excellent system to test the long-term effects of genetic or environmental antiaging interventions on cardiac function. The first such experiments sought to test the effects on cardiac performance of the best known signaling pathway that regulates lifespan, the insulin/IGF signaling pathway.

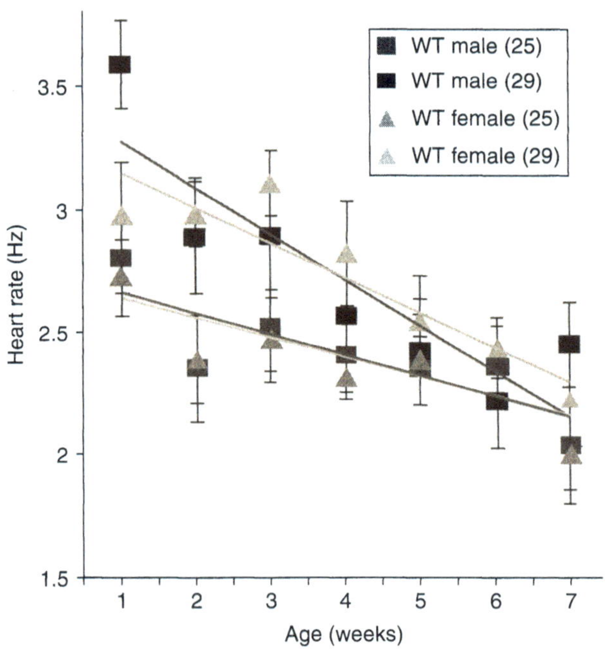

FIG. 8. Plot of heart rates from an outbred wild-type (WT; *yw* × Canton S) strain by age. Heart rate measurements were made at 25 and 29 °C for seven consecutive weeks. All flies were reared at the same temperature (25 °C). Heart rate measurements for both genders at each temperature are progressively lower with age (age main effect, F ratio = 54, $P < 0.001$).[153]

Reduction in insulin/IGF signaling, in some contexts, extends lifespan in flies, worms, and mice.[234–236] A series of alleles of the *Drosophila* Insulin Receptor (*InR*), some of which extend lifespan and some do not, were tested across ages for their effects on cardiac performance. Allelic combinations that did not extend lifespan also failed to extend cardiac performance, as measured by resistance to external pacing stress. By contrast, allelic combinations that did extend lifespan also delayed the age-related decrease in cardiac stress resistance.[153] Similar results were observed when insulin/IGF signaling was reduced by other means, including mutations in the insulin receptor substrate homolog *chico* and ablation of cells responsible for secretion of insulin-like peptide.[153] These results suggested that reduction of insulin/IGF signaling extends lifespan in part by extending functional characteristics of essential organ systems to more advanced ages.

Strikingly, it was also demonstrated that this function of insulin signaling acts autonomously in this context to regulate cardiac aging. Although blocking insulin signaling specifically in adipose tissue has been shown to extend lifespan in flies, mice, and worms,[235,237–239] cardiac-specific reduction of insulin signaling was capable of impeding cardiac aging even in animals where insulin signaling was unaltered in any other tissue[153] (Fig. 9). Taken together, these

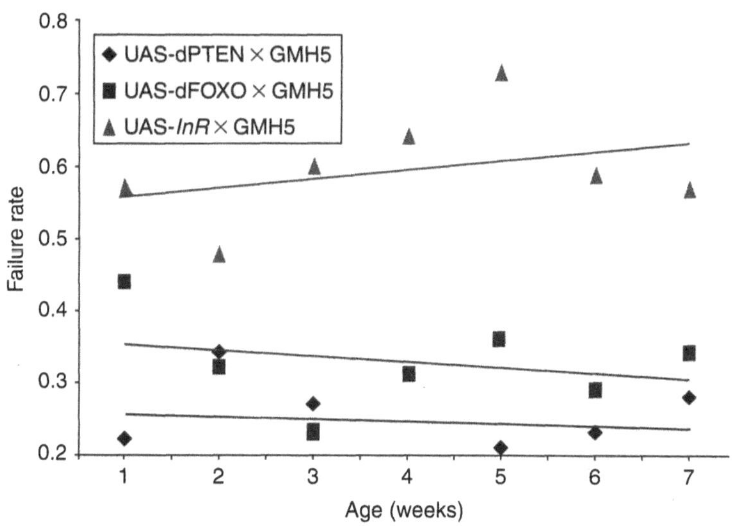

FIG. 9. Stress-induced cardiac failure rate over 7 weeks in flies with heart-specific overexpression of *InR*, dPTEN, or dFOXO. UAS-*InR* × *GMH5-Gal4* have a high failure rate at all ages (mean failure rate = 0.58). In contrast to heart-specific *InR* overexpression, both UAS-dPTEN × GMH5 and UAS-dFOXO × GMH5 progeny showed a low-failure rate at all ages (mean failure rate = 0.22 and 0.33, respectively), which does very with age ($\chi^2 = 0.17$, $P = 0.7$ and $\chi^2 = 2.90$, $P = 0.9$, respectively; $n > 160$ for each sample).[153]

results support a model in which insulin/IGF signaling activity plays a conserved role in adipose tissue, where it regulates the production of secondary signals that regulate organ-specific aging. Importantly, one of those secondary signals is insulin itself, and blockage of this signal alone in cardiac tissue is sufficient to severely impede cardiac age-related dysfunction.

Significantly, tissue-specific impedance of the aging process in the heart, despite its potent local effects, had no impact on overall lifespan,[153] presumably due to normal age-related decline of other essential organ systems. This observation highlights a substantial advantage of the *Drosophila* system as a model of cardiac aging, since it allows age-related change in cardiac function to be studied separately from lifespan, an approach that would be impossible in a vertebrate model system, where acute dependence on cardiac function for immediate oxygen supply renders cardiac function inextricably tied to lifespan.

Another signaling pathway known to influence lifespan in multiple organisms is the TOR Kinase signaling pathway.[211] TOR signaling interacts with, and experiences cross-talk with, insulin/IGF signaling at multiple points in the respective pathways, but nevertheless, produces phenotypes separable from those of the canonical insulin pathway.[240] A heteroallelic combination of *dTOR* mutations is capable of extending lifespan in *Drosophila*.[241] This same combination appears to alter the response to insulin-signaling downstream of the transcription factor *dFoxo*, as *dTOR* mutants block several effects of an activated *dFoxo* transgene.[241] Both loss of function mutations in *dTOR* and cardiac expression of TOR antagonists are capable of providing protection against age-related decline in both resting heart rate and cardiac stress resistance.[241]

Further support for this idea has come from recent study of a negative feedback regulator of TOR signaling, Sestrin. Flies carrying mutations in the *Drosophila* homolog of Sestrin, *dSesn*, exhibit many phenotypes characteristic of hyperactive TOR signaling, including impaired heart performance that is qualitatively similar to that seen in flies of advanced ages. *dSesn* mutants displayed increased period length, and increased incidence of arrhythmias, disorganization of structural myofibrils, and dilation of the heart diameter during both systolic and diastolic phases.[54] Slowing of heart rate and dilation of heart diameter could be rescued by feeding flies rapamycin, while rhythmic defects could be rescued by feeding of antioxidants.[54] This seems to support a model in which TOR activity leads directly to increased period length and structural defects, but rhythmic defects are a secondary consequence of increased oxidative stress in flies where TOR is hyperactive.

Thus, two major nutrient responsive signaling pathways, TOR and insulin/IGF, act to regulate cardiac aging and lifespan. The possibility that they do so through a common regulatory target was tested using a candidate gene approach. Since both of these signaling pathways have a conserved role in regulating translation through control of either expression or activity of the conserved

4eBP protein,[242–245] the role of 4eBP in controlling cardiac aging was tested directly. Cardiac-specific overexpression of *d4eBP* provided a similar degree of protection against age-related decline in rhythmicity and stress resistance to that provided by heart-specific impedance of TOR activity (Fig. 10).[55] By contrast, heart-specific overexpression of the regulatory target of 4eBP, the eukaryotic initiation factor, *deif4e*, caused hearts at young ages to display impaired stress resistance in a profile characteristic of aging wild-type hearts (Fig. 10).[55] Differential expression and/or activity of 4eBP seems to completely account for the tissue-autonomous effects of both insulin/IGF and TOR signaling, as a series of co-overexpressions consistently indicated that *d4eBP* and *deif4e* acted epistatically to both *dFoxo* and *dTOR*.[55] This role in regulating cardiac aging is specific to *d4eBP* and is not general to all regulators of translation levels, as other regulators of translation were not able to generate such phenotypes.

Another major signaling molecule that interacts with both insulin and TOR signaling to regulate translational levels, for example, is S6 Kinase, which regulates translation in part by phosphorylating ribosomal S6 to regulate its activity.[246] Hypomorphic mutations in the *Drosophila S6K* gene provided a potent extension of cardiac stress resistance to older ages, suggesting that this signaling molecule too plays an important role in the translation/cardiac aging axis. However, cardiac-specific expression of a dominant negative *dS6K*

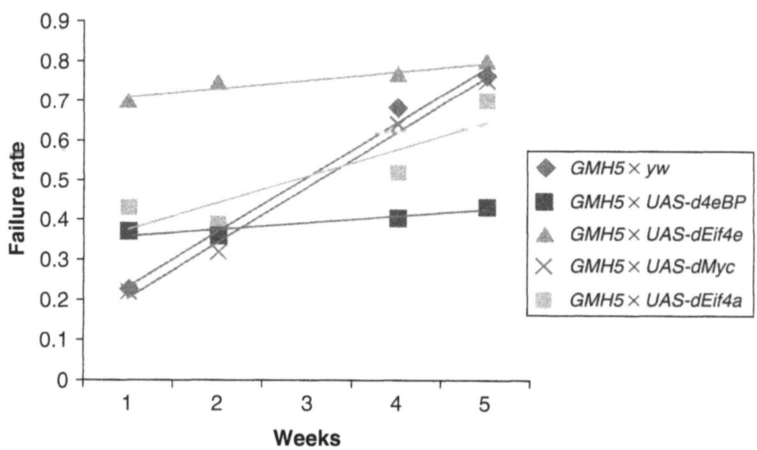

FIG. 10. Cardiac overexpression of d4eBP showed a significantly flattened slope of stress-induced failure. Indeed, the change in failure rate with age is no longer statistically significant (genotype-by-age, $\chi^2 = 1$, $P = 0.4$). Expression of dEif4e also flattened the slope throughout the time period, but showed a higher failure rate compared with controls at each time point (at week 1, $\chi^2 = 22$, $P < 0.0001$).[55]

transgene had no effect on cardiac aging.[55] Expression of dominant negative *dS6K* in the insulin producing cells (IPCs),[247] however, greatly extended cardiac stress resistance in aging flies. This effect is likely to be mediated by the observed reduction in secretion of insulin-like peptides in these flies,[55] since the cardiac aging profile of these flies is similar to flies in which the IPCs have been ablated.[153]

If TOR and insulin/IGF signaling regulate cardiac aging using a specific translational regulatory protein, 4eBP, as an effector, what are the likely targets for 4eBP that are of importance? A reasonable hypothesis is that genes regulating fatty-acid metabolism are likely to be crucial in this process. During vertebrate aging, both gene expression of genes encoding proteins involved in fatty-acid metabolism[248] and fatty-acid substrate utilization are altered.[249] Furthermore, both fatty-acid metabolism itself[241] and gene expression of genes involved in lipid metabolism[250] are altered as a consequence of dFoxo and dTOR activity. Thus, signaling pathways that modulate physiology in response to diet are likely to be of continuing importance for modeling not only effects of normal aging on cardiac physiology but also the effects of diets high in lipid on progressive cardiac disease states.

How does this compare with the effects of insulin/IGF axis on vertebrate cardiac aging? The role of insulin/IGF in vertebrates is more complicated than in flies, since the effects on aging related functional changes are superimposed on the potent effects of this pathway on growth and on repair mechanisms. The insulin receptor is required for postnatal cardiac growth in mice and also for hypertrophy in response to exercise.[251–254] Insulin/IGF also plays a very different role in damaged hearts than in healthy aging hearts. For example, although chronic administration of IGF-1 in murine hearts induces pathological hypertrophy, IGF-1 administration to hearts following myocardial infarction improves survival of cardiomyocytes.[255,256] Further complicating matters, IGF-1 exerts a protective effect on repair of damaged cardiac tissue, but these effects are dependent on which isoform is expressed in the heart, as some have tissue-specific and opposing effects.[257]

A significant advantage of the fly model in this context is that these effects can be removed from the equation, facilitating study of the direct effects of nutrient sensing pathways, such as insulin and TOR signaling, on cardiac functional aging, without complications from other effects of pathway manipulation. Since flies have an open circulatory system, indirect effects on the heart from alterations in vascular biology can be excluded. Since adult fly hearts are essentially postmitotic, effects derived from alterations in postnatal growth and repair can also be excluded.

In Section XIII, we will focus on recent work using the fly model to address genetic responses to "lifestyle choices" and their effects on cardiac physiology.

XIII. Diet and Exercise

Perhaps, the most important disease that can be modeled in flies is obesity. Obesity is arguably the most common contributing factor to cardiac disease worldwide,[258,259] and extensive evidence from human studies and from vertebrate models indicates that exposure to high-fat diets and/or obesity induces a variety of pathological states that affect cardiac function, either directly or indirectly, including atherosclerosis,[260,261] inflammation,[262] hypertension,[263,264] and dyslipidemia.[265]

Since the fly model system lacks a vasculature, it represents an excellent *in vivo* system for modeling direct effects of nutritional content and energy state on myocardial function, without complications from changes in vascular function. Fatty acids are the primary energy source for cardiomyocytes and during the aging process a decrease in fatty-acid usage has been observed,[266] suggesting that ATP production may be compromised in aging hearts. This idea is consistent with age-related changes in myocardial expression of genes required for beta-oxidation.[248,267]

Recent work has also demonstrated that altering fatty-acid metabolism has potent effects on cardiac physiology in the fly. A dominant hypomorphic mutation in a gene encoding a conserved fatty-acid transporter, *dFatp*, has been examined for effects on a variety of stresses. Homologs of *dFatp* are necessary for long-chain fatty-acid uptake in a variety of species, from yeast to human.[56,57] While null mutations for *dFatp* are lethal, hypomorphic mutations improve resistance to cold stress, heat stress, and starvation stress (Morley and Wessells, unpublished observation). Although *dFatp* hypomorphs display a dramatic increase in stored triglycerides and circulating fatty-acid levels, they nevertheless have an improved lifespan, not only on standard diets, but also on extreme high-fat diets, suggesting that they have an increased general resistance to lipotoxicity (Piazza and Wessells, unpublished observation).

The benefits of this mutation do not extend to cardiac function, however. Mutant flies have reduced cardiac contractility and reduced resistance to external pacing stress, consistent with the idea that they are compromised energetically. Defects in energetic function are tissue autonomous, since they can be reproduced by heart-specific expression of RNAi constructs against *dFatp* (Jennens and Wessells, unpublished observation) in otherwise wild-type flies. Overexpression of *dFatp* in the myocardium is also deleterious, producing phenotypes resembling lipotoxic cardiomyopathy. This is consistent with results in vertebrate models, where cardiac overexpression of the closest murine homolog of *dFatp* induces lipotoxicity and compromised cardiac function.[268] Interestingly, the cardiac defects induced by hypomorphic mutation of *dFatp* are rescued by knocking down one copy of a gene required for triglyceride storage, suggesting that these phenotypes are not governed solely by the amount of transport activity available to the heart, but by the balance of transport activity available in comparison to the amount of circulating free fatty acids and energy requirements in times of stress. Excitingly, the detrimental cardiac phenotypes of reduced *dFatp* expression can also be rescued by inducing an endurance exercise program, which tends to also reduce stored triglycerides. These results highlight the value of the fruit fly model as a way for examining the impact of environmental factors on genetic diseases.

Indeed, upcoming years are likely to see a substantial increase in the usage of the fly system as a model for the complex interactions between diet, function, and aging. It is already becoming clear, as advances in assays for organ physiology in flies continue to become available, that lifespan extending mutations or interventions do not necessarily improve function concomitantly. Lifespan is a highly combinatorial readout for a wide ranging number of changes to the physiology of various organ systems. Mathematically speaking, lifespan extension may depend most directly on whether a particular intervention happens to improve aspects of physiology that are relevant to caged, laboratory lifespan. The effects on overall physiology or on wild-dwelling animals may in fact be quite different. In the previous example, as an illustration, reductions in overall fatty-acid intake and usage powerfully extend lifespan and resistance to external stresses, but the same animals are unlikely to survive for long in the wild, since they have especially low tolerance for temporary cardiac stresses such as are commonly experienced by wild animals, or city-dwelling humans, every day. The fly system, since it combines invertebrate genetic tools with functional assays for specific organs, such as the heart, offers an exceptional model system for scrutinizing details and trade-offs involved in various interventions that alter lifespan or cardiac performance.

Another development that will further this process in upcoming years is the development of flies as a model for exercise training. High levels of habitual exercise serve multiple modeling purposes. First, the ability of flies to undergo a training regimen is a useful assay to measure overall functional fitness of animals undergoing a genetic or environmental intervention. Second, flies undergoing training can be a useful model system to test the effects of varying exercise and diet levels on cardiac function across ages. Third, flies undergoing training can be a model to use in screens to identify genes that are necessary for the heart to experience functional improvement as a result of training.

Endurance exercise has long been known to produce improvements both to skeletal muscle function and to cardiac function in vertebrates. In addition to its preventive role, exercise training can even benefit patients already experiencing heart failure[269,270] and protect against damage from ischemia and reperfusion.[271] Improvements to vertebrate cardiac function following endurance training have been documented in both young and old animals.[272,273]

The *Drosophila* model, due to its relatively short lifespan, offers an excellent opportunity to study the effects of age across ages in long-term, longitudinal study designs.

A protocol for long-term endurance training has been devised for *Drosophila*. It relies on the instinct for negative geotaxis that causes flies to involuntarily run upwards when knocked to the bottom of a container. A machine, called the Power Tower, named after an amusement park ride with a similar motion, was built to take advantage of this instinct by repeatedly knocking flies down at regular intervals, inducing them to continuously run upwards as long as the machine is running[217] (Fig. 11). A training paradigm was designed empirically that produces long-term improvements in mobility and cardiac stress resistance in multiple genotypes of wild-type flies.

Flies that undergo endurance training gradually begin to show greater resistance to external pacing stress during and following training. Although no difference in rhythmicity was detected between cohorts, exercised groups

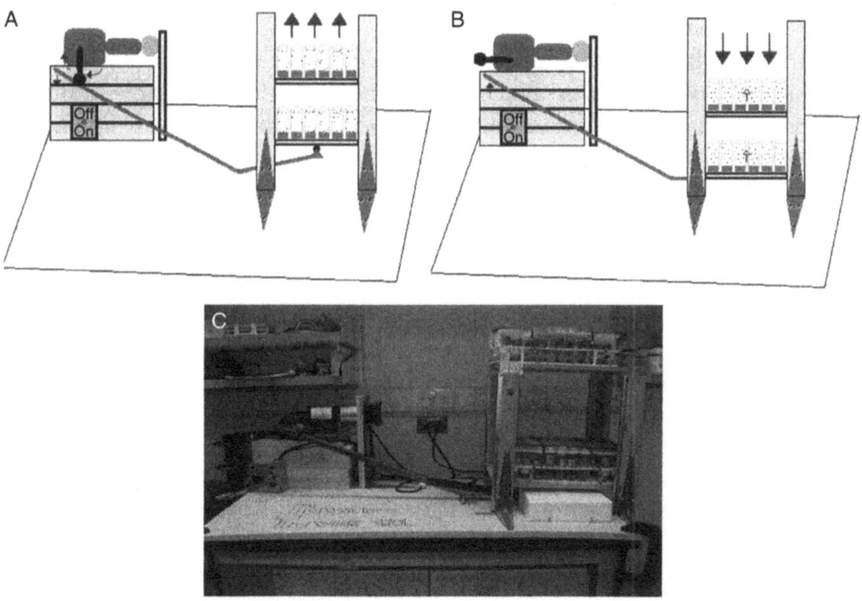

FIG. 11. Schematic representation of the Power Tower, a motor-operated enforced climbing apparatus that exercise-trains *Drosophila*. (A) As the rotation arm circles around in a clockwise direction, the lever pushes down and the two platforms of vial holders rise up. (B) As the rotating arm rolls-off, the lever lifts and the platform of vial holders drops down. The flies drop to the bottom of the vial, inducing their negative geotaxis instinct. The flies continue to repeat this process until the motor is shut-off. Exercise-training thus occurs through continuous climbing. (C) An image of the Power Tower.[217]

displayed improvements in contractility and experienced a lower incidence of cardiac arrest in response to externally induced stress.[217] In contrast to vertebrate studies, this improvement is only seen in flies that begin training at young ages. After a certain point of age-related decline is reached, flies are no longer able to execute the training program at the activity level necessary to induce benefits.

A hallmark of endurance training in humans and vertebrate models is a progressive alteration in mitochondrial number and efficiency.[274,275] Benefits to cardiac function acquired during exercise training in flies do not seem to be dependent on alterations in mitochondrial number or efficiency, however. A fly line with an extraordinary degree of mitochondrial activity without increased generation of ROS[276] displays dramatically improved mobility, with or without exercise, suggesting that improvements in mobility may be mediated in part by changes in mitochondrial activity in flies, as in vertebrates. However, these "mito-efficient" flies do not display a corresponding improvement in cardiac contractility or stress resistance, suggesting that the mechanisms by which endurance training improves cardiac function may be distinct from those by which it improves skeletal function.[217]

First steps have also been taken to begin to model the long-term effects of diet and exercise on cardiac function in combination. For example, different dietary constituents have been altered to form a matrix of varying dietary components. Flies were grown on each diet and resistance to external pacing stress tested throughout life as an indicator of cardiac functional health across ages on each diet (Morley and Wessells, unpublished). It was found that altering the level of carbohydrates in the diet had little, if any effect, on the normal age-related decline in cardiac stress resistance. However, reducing lipid content in the diet profoundly altered the aging profile of cardiac stress resistance. Reduction in yeast (source of both protein and lipid in the fly diet) has long been known to induce a lifespan extending dietary restriction process.[235] Interestingly, the level of yeast reduction that produces lifespan extension has no effect on cardiac aging. However, if yeast is reduced still further, to a level that actually reduces lifespan, presumably through nutritional insufficiency, cardiac function is dramatically improved. At this reduced yeast level, cardiac stress resistance decreases much more slowly with age than in wild type (Morley and Wessells, unpublished), suggesting that cardiac function, especially under stress conditions, has quite different energy requirements from those influencing overall lifespan.

This difference becomes more pronounced when intensive exercise is introduced into the experimental design. Lifespan extension in response to diet has been shown to be more a function of the ratio of carbohydrate calories to protein and lipid calories than a function of actual caloric reduction.[277] However, the ability to execute an endurance exercise program and derive benefits to cardiac function requires a much higher level of calories than the amount that is optimal for lifespan extension under inactive conditions. Furthermore, the amount of lipid contained in the diet is much more important to the ability to complete endurance exercise than the amount of carbohydrate.[217]

Significantly, there is also a qualitative difference between the effects of dietary restriction on cardiac performance and the effects of endurance exercise on cardiac performance. Dietary restriction benefits primarily late life control of rhythmicity, while endurance exercise benefits primarily late life retention of contractility and fractional shortening. This difference is reflected as well in the difference in resistance to external pacing between the two interventions, as dietary restricted flies have a reduced incidence of fibrillation events, while exercise trained flies have a reduced incidence of arrest events when paced (Morley and Wessells, unpublished observation[217]). Another important difference is that, at least in flies, dietary restriction does not improve measures of age-related functional decline such as mobility assays[278] or cold tolerance,[279] although it does extend lifespan.[208] By contrast, endurance exercise does not increase maximal lifespan, but does improve several indices of cardiac and skeletal muscle function.[217,280]

Although endurance training in flies produces several physiological responses that resemble those observed in humans and vertebrate animal models, whether or not these effects occur through mechanisms that are genetically conserved is still an open question. A first step toward addressing this question has been to use a candidate gene approach to test whether genes known to be important in the vertebrate response to endurance exercise may also serve the same function in flies.

Perhaps, the best studied gene in the vertebrate exercise response is *PGC1-α*, a transcriptional cofactor that is strongly upregulated by exercise in both humans[58] and in rats.[59] Upregulation of *PGC1-α* causes an increase in mitochondrial biogenesis[280] and an upregulation of fatty-acid oxidation.[266]

Overexpression of the fly homolog of *PGC1-α*, known as *spargel*, appears to provide some of the same benefits as an endurance training program. Flies overexpressing *spargel* in muscle experience an upregulation in global mitochondrial activity, an improvement in mobility, as measured by climbing assays, and an increase in oxidative stress resistance (Jennens and Wessells, unpublished observation).

Flies overexpressing *spargel* in the myocardium display a dramatic improvement in resistance to external pacing stress across ages. Interestingly, although *spargel* overexpression and endurance exercise provide a similar magnitude of pacing stress resistance, if *spargel* overexpressing flies are also subjected to endurance training, they receive a further increase in the magnitude of stress resistance (Jennens and Wessells, unpublished observation). This is consistent with a model in which *spargel* is involved in a conserved genetic response to endurance exercise that provides benefits to cardiac stress resistance, but that other, as yet unidentified genes are also necessary to provide the full potential benefits of this intervention. Future work using this model will detail further which aspects of cardiac function are altered by exercise or by *spargel* expression. The way is now open, as well, to utilize the traditional strength of the fly model system, which is to do unbiased screens to identify novel loci that are necessary or sufficient to induce cardioprotective effects of endurance exercise.

The use of adult flies as a model for functional aging, diet, and exercise is only in its infancy, as is the newly feasible use of flies as a model for adult cardiac physiology. It is to be expected, however, that many more researchers will begin to utilize this model system in these contexts. Eventually, the fly system may prove to be just as important a model for adult physiology and aging as it has been historically in the study of development.

XIV. Summary

The *Drosophila* model system has evolved into a powerful model for multiple aspects of cardiac function and disease. Using classical genetics, lineage tracing, and confocal microscopy, flies have been used to define conserved signaling pathways regulating the formation of the cardiac field, controlling migration of cardiac cells to the midline, governing repolarization of the cellular arrangement and fusion of opposing cells to form the heart tube, and instituting a cardiogenic network of genes that specify specific functional subsets of cardioblasts.

The larval stage of *Drosophila* development has been a fruitful model for isolating the effects of various genetic alterations on current formation and propagation of contractile waves, as well as the maintenance of rhythmic control. The development of isolated preps with beating hearts under physiological conditions has made possible the use of the larva as a model to test the effects of various biological and pharmacological compounds on cardiac rate, rhythm, and contractile force.

Novel methods for visualizing adult cardiac function and for measuring cardiac stress resistance in adults have led to further use for the fly model. Several fly lines have been created that genetically model specific human diseases that impair cardiac function. In some cases, these are genomic mutants, while in others, cardiac-specific gene alteration produces cardiac phenotypes in isolation from the rest of the body.

The ability to track cardiac function across ages in adult flies has led to the use of the fly as a model for cardiac aging. Several conserved genetic pathways have been identified that alter the rate of age-related change in cardiac performance in flies. The ease of maintenance and short lifespan of the fly have made it possible to begin using the fly to model the effects of alterations and diet and exercise level on cardiac function across ages. The intersection between diet, exercise, and aging is of critical importance to addressing the continued increase in heart disease linked to obesity and insulin resistance in the world's population. As a model system, flies are well suited for dealing with multiple interventions simultaneously in large numbers. They are also well suited for tracking lifelong effects of such interventions. As such, we expect that flies will be used even more frequently in upcoming years as a valuable tool for understanding these interrelated problems.

References

1. Tao Y, Schulz RA. Heart development in *Drosophila*. *Semin Cell Dev Biol* 2007;**18**:3–15.
2. Medioni C, Astier M, Zmojdzian M, Jagla K, Sémériva M. Genetic control of cell morphogenesis during *Drosophila melanogaster* cardiac tube formation. *J Cell Biol* 2008;**182**:249–61.
3. Bodmer R. Heart development in *Drosophila* and its relationship to vertebrate systems. *Trends Cardiovasc Med* 1995;**5**:21–8.
4. Garg V. Insights into the genetic basis of congenital heart disease. *Cell Mol Life Sci* 2006;**63**:1141–8.
5. Bryantsev AL, Cripps RM. Cardiac gene regulatory networks in *Drosophila*. *Biochim Biophys Acta* 2009;**1789**:343–53.
6. Bodmer R, Wessells RJ, Johnson EC, Dowse H. Heart development and function. In: Gilbert LI, Iatrou K, Gills SS, editors. *Comprehensive molecular insect science*, Vol. 2. Amsterdam: Elsevier; 2005. p. 199–250, Volumes 107.
7. Wessells RJ, Bodmer R. Cardiac aging. *Semin Cell Dev Biol* 2007;**18**:111–6.
8. Dunin-Borkowski OM, Brown NH, Bate M. Anterior–posterior subdivision and the diversification of the mesoderm in *Drosophila*. *Development* 1995;**121**:4183–93.
9. Lawrence PA, Bodmer R, Vincent JP. Segmental patterning of heart precursors in *Drosophila*. *Development* 1995;**121**:4303–8.
10. Wu X, Golden K, Bodmer R. Heart development in *Drosophila* requires the segment polarity gene wingless. *Dev Biol* 1995;**169**:619–28.
11. Park M, Wu X, Golden K, Axelrod JD, Bodmer R. The wingless signaling pathway is directly involved in *Drosophila* heart development. *Dev Biol* 1996;**177**:104–16.
12. Rugendorff A, Younossi-Hartenstein A, Hartenstein V. Embryonic origin and differentiation of the *Drosophila* heart. *Rouxs Arch Dev Biol* 1994;**203**:266–80.
13. Frasch M. Induction of visceral and cardiac mesoderm by ectodermal Dpp in the early *Drosophila* embryo. *Nature* 1995;**374**:464–7.
14. Azpiazu N, Frasch M. Tinman and bagpipe: two homeo box genes that determine cell fates in the dorsal mesoderm of *Drosophila*. *Genes Dev* 1993;**7**:1325–40.
15. Bodmer R. The gene tinman is required for specification of the heart and visceral muscles in *Drosophila*. *Development* 1993;**118**:719–29.
16. Schott JJ, Benson DW, Basson CT, Pease W, Silberbach GM, Moak JP, et al. Congenital heart disease caused by mutations in the transcription factor NKX2-5. *Science* 1998;**281**:108–11.

17. Lee HH, Frasch M. Nuclear integration of positive Dpp signals, antagonistic Wg inputs and mesodermal competence factors during *Drosophila* visceral mesoderm induction. *Development* 2005;**132**:1429–42.
18. Reifers F, Walsh EC, Léger S, Stainier DY, Brand M. Induction and differentiation of the zebrafish heart require fibroblast growth factor 8 (fgf8/acerebellar). *Development* 2000;**127**:225–35.
19. Alsan BH, Schultheiss TM. Regulation of avian cardiogenesis by Fgf8 signaling. *Development* 2002;**129**:1935–43.
20. Marques SR, Lee Y, Poss KD, Yelon D. Reiterative roles for FGF signaling in the establishment of size and proportion of the zebrafish heart. *Dev Biol* 2008;**321**:397–406.
21. Shishido E, Ono N, Kojima T, Saigo K. Requirements of DFR1/heartless, a mesoderm-specific *Drosophila* FGF-receptor, for the formation of heart, visceral and somatic muscles, and ensheathing of longitudinal axon tracts in CNS. *Development* 1997;**124**:2119–28.
22. Gisselbrecht S, Skeath JB, Doe CQ, Michelson AM. Heartless encodes a fibroblast growth factor receptor (DFR1/DFGF-R2) involved in the directional migration of early mesodermal cells in the *Drosophila* embryo. *Genes Dev* 1996;**10**:3003–17.
23. Fossett N, Tevosian SG, Gajewski K, Zhang Q, Orkin SH, Schulz RA. The friend of GATA proteins U-shaped, FOG-1, and FOG-2 function as negative regulators of blood, heart, and eye development in *Drosophila*. *Proc Natl Acad Sci USA* 2001;**19**:7342–7.
24. Klinedinst SL, Bodmer R. Gata factor Pannier is required to establish competence for heart progenitor formation. *Development* 2003;**130**:3027–38.
25. Lo PC, Skeath JB, Gajewski K, Schulz RA, Frasch M. Homeotic genes autonomously specify the anteroposterior subdivision of the *Drosophila* dorsal vessel into aorta and heart. *Dev Biol* 2002;**251**:307–19.
26. Lo PC, Frasch M. A role for the COUP-TF-related gene seven-up in the diversification of cardioblast identities in the dorsal vessel of *Drosophila*. *Mech Dev* 2001;**104**:49–60.
27. Perrin L, Monier B, Ponzielli R, Astier M, Semeriva M. *Drosophila* cardiac tube organogenesis requires multiple phases of Hox activity. *Dev Biol* 2004;**272**:419–31.
28. Garg V, Kathiriya IS, Barnes R, Schluterman MK, King IN, Butler CA, et al. GATA4 mutations cause human congenital heart defects and reveal an interaction with TBX5. *Nature* 2003;**424**:443–7.
29. Zikova M, Da Ponte JP, Dastugue B, Jagla K. Patterning of the cardiac outflow region in *Drosophila*. *Proc Natl Acad Sci USA* 2003;**100**:12189–94.
30. Schäfer K, Neuhaus P, Kruse J, Braun T. The homeobox gene Lbx1 specifies a subpopulation of cardiac neural crest necessary for normal heart development. *Circ Res* 2003;**92**:73–80.
31. Sanyal S, Jennings T, Dowse H, Ramaswami M. Conditional mutations in SERCA, the sarco-endoplasmic reticulum Ca^{2+}-ATPase, alter heart rate and rhythmicity in *Drosophila*. *J Comp Physiol* 2006;**176**:252–63.
32. Chamberlain JS, Metzger J, Reyes M, Townsend D, Faulkner JA. Dystrophin-deficient mdx mice display a reduced life span and are susceptible to spontaneous rhabdomyosarcoma. *FASEB J* 2007;**21**:2195–204.
33. Taghli-Lamallem O, Akasaka T, Hogg G, Nudel U, Yaffe D, Chamberlain JS, et al. Dystrophin deficiency in *Drosophila* reduces lifespan and causes a dilated cardiomyopathy phenotype. *Aging Cell* 2008;**7**:237–49.
34. Zhu X, Wheeler MT, Hadhazy M, Lam MY, McNally EM. Cardiomyopathy is independent of skeletal muscle disease in muscular dystrophy. *FASEB J* 2002;**16**:1096–8.
35. Wheeler MT, Allikian MJ, Heydemann A, Hadhazy M, Zarnegar S, McNally EM. Smooth muscle cell-extrinsic vascular spasm arises from cardiomyocyte degeneration in sarcoglycan-deficient cardiomyopathy. *J Clin Invest* 2004;**113**:668–75.

36. Allikian MJ, Bhabha G, Dospoy P, Heydemann A, Ryder P, Earley JU, et al. Reduced life span with heart and muscle dysfunction in *Drosophila* sarcoglycan mutants. *Hum Mol Genet* 2007;**16**:2933–43.
37. Cammarato A, Dambacher CM, Knowles AF, Kronert WA, Bodmer R, Ocorr K, et al. Myosin transducer mutations differentially affect motor function, myofibril structure, and the performance of skeletal and cardiac muscles. *Mol Biol Cell* 2008;**19**:553–62.
38. Geeves MA, Holmes KC. The molecular mechanism of muscle contraction. *Adv Protein Chem* 2005;**71**:161–93.
39. Oldfors A. Hereditary myosin myopathies. *Neuromuscul Disord* 2007;**17**:355–67.
40. Bernstein SI, Mogami K, Donady JJ, Emerson Jr. CP. *Drosophila* muscle myosin heavy chain encoded by a single gene in a cluster of muscle mutations. *Nature* 1983;**302**:393–7.
41. Bernstein SI, Hansen CJ, Becker KD, Wassenberg 2nd DR, Roche ES, Donady JJ, et al. Alternative RNA splicing generates transcripts encoding a thorax-specific isoform of *Drosophila melanogaster* myosin heavy chain. *Mol Cell Biol* 1986;**1986**:2511–9.
42. Ocorr K, Akasaka T, Bodmer R. Age-related cardiac disease model of *Drosophila*. *Mech Ageing Dev* 2007;**1**:112–6.
43. Sanguinetti MC, Curran ME, Zou A, Shen J, Spector PS, Atkinson DL, et al. Coassembly of K (V)LQT1 and minK (IsK) proteins to form cardiac I(Ks) potassium channel. *Nature* 1996;**384**:80–3.
44. Priori SG, Napolitano C, Schwartz PJ, Grillo M, Bloise R, Ronchetti E, et al. Association of long QT syndrome loci and cardiac events among patients treated with beta-blockers. *JAMA* 2004;**292**:1341–4.
45. Roberts R. Genomics and cardiac arrhythmias. *J Am Coll Cardiol* 2006;**47**:9–21.
46. Fujioka M, Wessells RJ, Han Z, Liu J, Fitzgerald K, Yusibova GL, et al. Embryonic even skipped-dependent muscle and heart cell fates are required for normal adult activity, heart function, and lifespan. *Circ Res* 2005;**97**:1108–14.
47. Crone SA, Zhao YY, Fan L, Gu Y, Minamisawa S, Liu Y, et al. ErbB2 is essential in the prevention of dilated cardiomyopathy. *Nat Med* 2002;**8**:459–65.
48. Lemmens K, Doggen K, De Keulenaer GW. Role of neuregulin-1/ErbB signaling in cardiovascular physiology and disease: implications for therapy of heart failure. *Circulation* 2007;**116**:954–60.
49. Ozcelik C, Erdmann B, Pilz B, Wettschureck N, Britsch S, Hübner N, et al. Conditional mutation of the ErbB2 (HER2) receptor in cardiomyocytes leads to dilated cardiomyopathy. *Proc Natl Acad Sci USA* 2002;**99**:8880–5.
50. Carelli V, Ross-Cisneros FN, Sadun AA. Mitochondrial dysfunction as a cause of optic neuropathies. *Prog Retin Eye Res* 2004;**23**:53–89.
51. Shahrestani P, Leung HT, Le PK, Pak WL, Tse S, Ocorr K, et al. Heterozygous mutation of *Drosophila* Opa1 causes the development of multiple organ abnormalities in an age-dependent and organ-specific manner. *PLoS One* 2009;**4**:e6867.
52. Li Y, Huang TT, Carlson EJ, Meloy S, Ursell PC, Olson JL, et al. Dilated cardiomyopathy and neonatal lethality in mutant mice lacking manganese superoxide dismutase. *Nat Genet* 1995;**11**:376–81.
53. Duttaroy A, Paul A, Kundu M, Belton A. A Sod2 null mutation confers severely reduced adult life span in *Drosophila*. *Genetics* 2003;**165**:2295–9.
54. Lee JH, Budanov AV, Park EJ, Birse R, Kim TE, Perkins GA, et al. Sestrin as a feedback inhibitor of TOR that prevents age-related pathologies. *Science* 2010;**327**:1223–8.
55. Wessells R, Fitzgerald E, Piazza N, Ocorr K, Morley S, Davies C, et al. d4eBP acts downstream of both dTOR and dFoxo to modulate cardiac functional aging in *Drosophila*. *Aging Cell* 2009;**8**:542–52.

56. Faergeman NJ, DiRusso CC, Elberger A, Knudsen J, Black PN. Disruption of the *Saccharomyces cerevisiae* homologue to the murine fatty acid transport protein impairs uptake and growth on long-chain fatty acids. *J Biol Chem* 1997;**272**:8531–8.
57. Herrmann T, Buchkremer F, Gosch I, Hall AM, Berniohr DA, Stremmel W. Mouse fatty acid transport protein 4 (FAT4): characterization of the gene and functional assessment as a very long chain acyl-CoA synthetase. *Gene* 2001;**270**:31–40.
58. Pilegaard H, Saltin B, Neufer PD. Exercise induces transient transcriptional activation of the PGC-1alpha gene in human skeletal muscle. *J Physiol* 2003;**546**:851–8.
59. Goto M, Terada S, Kato M, Katoh M, Yokozeki T, Tabata I, et al. cDNA cloning and mRNA analysis of PGC1 in epitrochlearis muscle in swimming-exercised rats. *Biochem Biophys Res Commun* 2000;**274**:350–4.
60. Lockwood WK, Bodmer R. The patterns of wingless, decapentaplegic, and tinman position the *Drosophila* heart. *Mech Dev* 2002;**114**:13–26.
61. Winnier G, Blessing M, Labosky PA, Hogan BL. Bone morphogenetic protein-4 is required for mesoderm formation and patterning in the mouse. *Genes Dev* 1995;**9**:2105–16.
62. Schultheiss TM, Burch JB, Lassar AB. A role for bone morphogenetic proteins in the induction of cardiac myogenesis. *Genes Dev* 1997;**15**:451–62.
63. Shi Y, Katsev S, Cai C, Evans S. BMP signaling is required for heart formation in vertebrates. *Dev Biol* 2000;**224**:226–37.
64. Song L, Fässler R, Mishina Y, Jiao K, Baldwin HS. Essential functions of Alk3 during AV cushion morphogenesis in mouse embryonic hearts. *Dev Biol* 2007;**301**:276–86.
65. Lyons I, Parsons LM, Hartley L, Li R, Andrews JE, Robb L, et al. Myogenic and morphogenetic defects in the heart tubes of murine embryos lacking the homeo box gene Nkx2-5. *Genes Dev* 1995;**9**:1654–66.
66. Lien CL, McAnally J, Richardson JA, Olson EN. Cardiac-specific activity of an Nkx2-5 enhancer requires an evolutionarily conserved Smad binding site. *Dev Biol* 2002;**244**:257–66.
67. Tzahor E, Lassar AB. Wnt signals from the neural tube block ectopic cardiogenesis. *Genes Dev* 2001;**15**:255–60.
68. Marvin MJ, Di Rocco G, Gardiner A, Bush SM, Lassar AB. Inhibition of Wnt activity induces heart formation from posterior mesoderm. *Genes Dev* 2001;**15**:316–27.
69. Schneider VA, Mercola M. Wnt antagonism initiates cardiogenesis in *Xenopus laevis*. *Genes Dev* 2001;**15**:304–15.
70. Foley AC, Mercola M. Heart induction by Wnt antagonists depends on the homeodomain transcription factor Hex. *Genes Dev* 2005;**19**:387–96.
71. Tzahor E. Wnt/beta-catenin signaling and cardiogenesis: timing does matter. *Dev Cell* 2007;**13**:10–3.
72. Kwon C, Cordes KR, Srivastava D. Wnt/beta-catenin signaling acts at multiple developmental stages to promote mammalian cardiogenesis. *Cell Cycle* 2008;**7**:3815–8.
73. Jaspard B, Couffinhal T, Dufourcq P, Moreau C, Duplàa C. Expression pattern of mouse sFRP-1 and mWnt-8 gene during heart morphogenesis. *Mech Dev* 2000;**90**:263–7.
74. Garriock RJ, D'Agostino SL, Pilcher KC, Krieg PA. Wnt11-R, a protein closely related to mammalian Wnt11, is required for heart morphogenesis in *Xenopus*. *Dev Biol* 2005;**279**:179–92.
75. Ueno S, Weidinger G, Osugi T, Kohn AD, Golob JL, Pabon L, et al. Biphasic role for Wnt/beta-catenin signaling in cardiac specification in zebrafish and embryonic stem cells. *Proc Natl Acad Sci USA* 2007;**104**:9685–90.
76. Pandur P, Läsche M, Eisenberg LM, Kühl M. Wnt-11 activation of a non-canonical Wnt signalling pathway is required for cardiogenesis. *Nature* 2002;**418**:636–41.
77. Olson EN. Gene regulatory networks in the evolution and development of the heart. *Science* 2006;**313**:1922–7.

78. Reim I, Frasch M. Genetic and genomic dissection of cardiogenesis in the *Drosophila* model. *Pediatr Cardiol* 2010;**31**:325–34.
79. Ponzielli R, Astier M, Chartier A, Gallet A, Thérond P, Sémériva M. Heart tube patterning in *Drosophila* requires integration of axial and segmental information provided by the Bithorax Complex genes and hedgehog signaling. *Development* 2002;**129**:4509–21.
80. Monier B, Astier M, Sémériva M, Perrin L. Steroid-dependent modification of Hox function drives myocyte reprogramming in the *Drosophila* heart. *Development* 2005;**132**:5283–93.
81. Molina MR, Cripps RM. Ostia, the inflow tracts of the *Drosophila* heart, develop from a genetically distinct subset of cardial cells. *Mech Dev* 2001;**109**:51–9.
82. Curtis NJ, Ringo JM, Dowse HB. Morphology of the pupal heart, adult heart, and associated tissues in the fruit fly, *Drosophila melanogaster. J Morphol* 1999;**240**:225–35.
83. Zaffran S, Reim I, Qian L, Lo PC, Bodmer R, Frasch M. Cardioblast-intrinsic tinman activity controls proper diversification and differentiation of myocardial cells in *Drosophila*. *Development* 2006;**133**:4073–83.
84. Reim I, Mohler JP, Frasch M. Tbx20-related genes, mid and H15, are required for tinman expression, proper patterning, and normal differentiation of cardioblasts in *Drosophila*. *Mech Dev* 2005;**122**:1056–69.
85. Miskolczi-McCallum CM, Scavetta RJ, Svendsen PC, Soanes KH, Brook WJ. The *Drosophila melanogaster* T-box genes midline and H15 are conserved regulators of heart development. *Dev Biol* 2005;**278**:459–72.
86. Qian L, Liu J, Bodmer R. Neuromancer Tbx20-related genes (H15/midline) promote cell fate specification and morphogenesis of the *Drosophila* heart. *Dev Biol* 2005;**279**:509–24.
87. Gajewski K, Fossett N, Molkentin JD, Schulz RA. The zinc finger proteins Pannier and GATA4 function as cardiogenic factors in *Drosophila*. *Development* 1999;**126**:5679–88.
88. Gajewski K, Zhang Q, Choi CY, Fossett N, Dang A, Kim YH, et al. Pannier is a transcriptional target and partner of tinman during *Drosophila* cardiogenesis. *Dev Biol* 2001;**233**:425–36.
89. Alvarez AD, Shi W, Wilson BA, Skeath JB. Pannier and pointedP2 act sequentially to regulate *Drosophila* heart development. *Development* 2003;**130**:3015–26.
90. Klinedinst SL, Bodmer R. Gata factor Pannier is required to establish competence for heart progenitor formation. *Development* 2003;**130**:3027–38.
91. Reim I, Frasch M. The Dorsocross T-box genes are key components of the regulatory network controlling early cardiogenesis in *Drosophila*. *Development* 2005;**132**:4911–25.
92. Han Z, Yi P, Li X, Olson EN. Hand, an evolutionarily conserved bHLH transcription factor required for *Drosophila* cardiogenesis and hematopoiesis. *Development* 2006;**133**:1175–82.
93. Kremser T, Hasenpusch-Theil K, Wagner E, Buttgereit D, Renkawitz-Pohl R. Expression of the beta3 tubulin gene (beta Tub60D) in the visceral mesoderm of *Drosophila* is dependent on a complex enhancer that binds tinman and UBX. *Mol Gen Genet* 1999;**262**:643–58.
94. Nasonkin I, Alikasifoglu A, Ambrose C, Cahill P, Cheng M, Sarniak A, et al. A novel sulfonylurea receptor family member expressed in the embryonic *Drosophila* dorsal vessel and tracheal system. *J Biol Chem* 1999;**274**:29420–5.
95. Akasaka T, Klinedinst S, Ocorr K, Bustamante EL, Kim SK, Bodmer R. The ATP-sensitive potassium (KATP) channel-encoded dSUR gene is required for *Drosophila* heart function and is regulated by tinman. *Proc Natl Acad Sci USA* 2006;**103**:11999–2004.
96. Han Z, Olson EN. Hand is a direct target of tinman and GATA factors during *Drosophila* cardiogenesis and hematopoiesis. *Development* 2005;**132**:3525–36.
97. Gajewski K, Kim Y, Lee YW, Olson EN, Schulz RA. D-mef2 is a target for tinman activation during *Drosophila* heart development. *EMBO J* 1997;**16**:515–22.
98. Goldmuntz E, Geiger E, Benson DW. NKX2.5 mutations in patients with tetralogy of fallot. *Circulation* 2001;**104**:2565–8.

99. Maitra M, Schluterman MK, Nichols HA, Richardson JA, Lo CW, Srivastava D, et al. Interaction of Gata4 and Gata6 with Tbx5 is critical for normal cardiac development. *Dev Biol* 2009;**326**:368–77.
100. Tu CT, Yang TC, Tsai HJ. Nkx2.7 and Nkx2.5 function redundantly and are required for cardiac morphogenesis of zebrafish embryos. *PLoS One* 2009;**4**:e4249.
101. Singh MK, Li Y, Cobb RM, Zhou D, Lu MM, Epstein JA, et al. Gata4 and Gata5 cooperatively regulate cardiac myocyte proliferation in mice. *J Biol Chem* 2010;**285**:1765–72.
102. Drubin DG, Nelson WJ. Origins of cell polarity. *Cell* 1996;**84**:335–44.
103. Haag TA, Haag NP, Lekven AC, Hartenstein V. The role of cell adhesion molecules in *Drosophila* heart morphogenesis: faint sausage, shotgun/DE-cadherin, and laminin A are required for discrete stages in heart development. *Dev Biol* 1999;**208**:56–9.
104. Wang J, Tao Y, Reim I, Gajewski K, Frasch M, Schulz RA. Expression, regulation, and requirement of the toll transmembrane protein during dorsal vessel formation in *Drosophila melanogaster. Mol Cell Biol* 2005;**25**:4200–10.
105. Yarnitsky T, Volk T. Laminin is required for heart, somatic muscles, and gut development in the *Drosophila* embryo. *Dev Biol* 1995;**169**:609–18.
106. Tepass U. Epithelial differentiation in *Drosophila*. *Bioessays* 1997;**19**:673–82.
107. Stark KA, Yee GH, Roote CE, Williams EL, Zusman S, Hynes RO. A novel alpha integrin subunit associates with betaPS and functions in tissue morphogenesis and movement during *Drosophila* development. *Development* 1997;**124**:4583–94.
108. Frémion F, Astier M, Zaffran S, Guiellèn A, Homburger V, Sémériva M. The heterotrimeric protein Go is required for the formation of heart epithelium in *Drosophila*. *J Cell Biol* 1999;**145**:1063–76.
109. Chartier A, Zaffran S, Astier M, Sémériva M, Gratecos D. Pericardin, a *Drosophila* type IV collagen-like protein is involved in the morphogenesis and maintenance of the heart epithelium during dorsal ectoderm closure. *Development* 2002;**129**:3241–53.
110. Qian L, Liu J, Bodmer R. Slit and Robo control cardiac cell polarity and morphogenesis. *Curr Biol* 2005;**15**:2271–8.
111. Santiago-Martínez E, Soplop NH, Kramer SG. Lateral positioning at the dorsal midline: slit and roundabout receptors guide *Drosophila* heart cell migration. *Proc Natl Acad Sci USA* 2006;**103**:12441–6.
112. Santiago-Martínez E, Soplop NH, Patel R, Kramer SG. Repulsion by Slit and Roundabout prevents Shotgun/E-cadherin-mediated cell adhesion during *Drosophila* heart tube lumen formation. *J Cell Biol* 2008;**182**:241–8.
113. Zmojdzian M, Da Ponte JP, Jagla K. Cellular components and signals required for the cardiac outflow tract assembly in *Drosophila*. *Proc Natl Acad Sci USA* 2008;**105**:2475–80.
114. Yi P, Han Z, Li X, Olson EN. The mevalonate pathway controls heart formation in *Drosophila* by isoprenylation of Ggamma1. *Science* 2006;**313**:1301–3.
115. Yi P, Johnson AN, Han Z, Wu J, Olson EN. Heterotrimeric G proteins regulate a noncanonical function of septate junction proteins to maintain cardiac integrity in *Drosophila*. *Dev Cell* 2008;**15**:704–13.
116. Horowitz A, Simons M. Branching morphogenesis. *Circ Res* 2008;**103**:784–95.
117. Chan PK, Lin CC, Cheng SH. Noninvasive technique for measurement of heartbeat regularity in zebrafish (*Danio rerio*) embryos. *BMC Biotechnol* 2009;**19**:11.
118. Miller TA. Structure and physiology of the circulatory system. In: *Comprehensive insect physiology, biochemistry and pharmacology. Vol 3, Integument, respiration and circulation*. Oxford: Pergamon Press; 1985. p. 289–353.
119. Rizki TM. The circulatory system and associated cells and tissues. In: Ashbumer M, Wright TRF, editors. *The genetics and biology of Drosophila*. New York: Academic Press; 1978. p. 397–452.

120. Siegmund T, Korge G. Innervation of the ring gland of *Drosophila melanogaster*. *J Comp Neurol* 2001;**431**:481–91.

121. Dulcis D, Levine RB. Innervation of the heart of the adult fruit fly, *Drosophila melanogaster*. *J Comp Neurol* 2003;**465**:560–78.

122. Dowse H, Ringo J, Power J, Johnson E, Kinney K, White L. A congenital heart defect in *Drosophila* caused by an action-potential mutation. *J Neurogenet* 1995;**10**:153–68.

123. Wessells RJ, Bodmer R. Screening assays for heart function mutants in *Drosophila*. *Biotechniques* 2004;**37**:58–60, 62, 64.

124. Johnson E, Ringo J, Dowse H. Modulation of *Drosophila* heartbeat by neurotransmitters. *J Comp Physiol* 1997;**167**:89–97.

125. Johnson E, Ringo J, Bray N, Dowse H. Genetic and pharmacological identification of ion channels central to the *Drosophila* cardiac pacemake. *J Neurogenet* 1998;**12**:1–24.

126. Papaefthmiou C, Theophilidis G. An *in vitro* method for recording the electrical activity of the isolated heart of the adult *Drosophila melanogaster*. *In Vitro Cell Dev Biol Anim* 2001;**37**:445–9.

127. Cooper AS, Rymond KE, Ward MA, Bocook EL, Cooper RL. Monitoring heart function in larval *Drosophila melanogaster* for physiological studies. *J Vis Exp* 2009;**16**:1596.

128. Vogler G, Ocorr K. Visualizing the beating heart in *Drosophila*. *J Vis Exp* 2009;**28**:1425.

129. Fink M, Callol-Massot C, Chu A, Ruiz-Lozano P, Izpisua Belmonte JC, Giles W, et al. A new method for detection and quantification of heartbeat parameters in *Drosophila*, zebrafish and embryonic mouse hearts. *Biotechniques* 2009;**46**:101–13.

130. Ocorr K, Fink M, Cammarato A, Bernstein S, Bodmer R. Semi-automated optical heartbeat analysis of small hearts. *J Vis Exp* 2009;**16**:1435.

131. Bradu A, Ma L, Bloor JW, Podoleanu A. Dual optical coherence tomography/fluorescence microscopy for monitoring *Drosophila melanogaster* larval heart. *J Biophotonics* 2009;**2**:380–8.

132. Gu GG, Singh S. Pharmacological analysis of heartbeat in *Drosophila*. *J Neurobiol* 1995;**28**:269–80.

133. Sullivan KM, Scott K, Zuker CS, Rubin GM. The ryanodine receptor is essential for larval development in *Drosophila melanogaster*. *Proc Natl Acad Sci USA* 2000;**97**:5942–7.

134. Noyes BE, Katz FN, Schaffer MH. Identification and expression of the *Drosophila* adipokinetic hormone gene. *Mol Cell Endocrinol* 1995;**109**:133–41.

135. Zornik E, Paisley K, Nichols R. Neural transmitters and a peptide *Drosophila* heart rate. *Peptides* 1999;**20**:45–51.

136. Dasari S, Cooper RL. Direct influence of serotonin on the larval heart of *Drosophila melanogaster*. *J Comp Physiol* 2006;**176**:349–57.

137. Johnson E, Ringo J, Dowse H. Native and heterologous neuropeptides are cardioactive in *Drosophila melanogaster*. *J Insect Physiol* 2000;**46**:1229–36.

138. Nichols R, Kaminski S, Wallin E, Zornik E. Regulating the activity of a cardioacceleratory peptide. *Peptides* 1999;**20**:1153–8.

139. Baker JD, McNabb SL, Truman JW. The hormonal coordination of behavior and physiology at adult ecdysis in *Drosophila melanogaster*. *J Exp Biol* 1999;**202**:3037–48.

140. Gammie SC, Truman JW. Eclosion hormone provides a link between ecdysis-triggering hormone and crustacean cardioactive peptide in the neuroendocrine cascade that controls ecdysis behavior. *J Exp Biol* 1999;**202**:343–52.

141. Meng X, Wahlström G, Immonen T, Kolmer M, Tirronen M, Predel R, et al. The *Drosophila* hugin gene codes for myostimulatory and ecdysis-modifying neuropeptides. *Mech Dev* 2002;**117**:5–13.

142. Chen MS, Obar RA, Schroeder CC, Austin TW, Poodry CA, Wadsworth SC, et al. Multiple forms of dynamin are encoded by shibire, a *Drosophila* gene involved in endocytosis. *Nature* 1991;**351**:583–6.

143. Masur SK, Kim YT, Wu CF. Reversible inhibition of endocytosis in cultured neurons from the *Drosophila* temperature-sensitive mutant shibire. *J Neurogenet* 1990;**6**:191–206.

144. Johnson E, Ringo J, Dowse H. Dynamin, encoded by shibire, is central to cardiac function. *J Exp Zool* 2001;**289**:81–9.

145. Johnson E, Sherry T, Ringo J, Dowse H. Modulation of the cardiac pacemaker of *Drosophila*: cellular mechanisms. *J Comp Physiol* 2002;**172**:227–36.

146. Sokolowski MB. Foraging strategies of *Drosophila melanogaster*: a chromosomal anaylsis. *Behav Genet* 1980;**10**:291–302.

147. de Belle JS, Hilliker AJ, Sokolowski MB. Genetic localization of foraging (for): a major gene for larval behavior in *Drosophila melanogaster. Genetics* 1989;**123**:157–63.

148. Price MD, Merte J, Nichols R, Koladich PM, Tobe SS, Bendena WG. *Drosophila melanogaster* flatline encodes a myotropin orthologue to *Manduca sexta* allatostatin. *Peptides* 2002;**23**:787–94.

149. Lalevée N, Monier B, Sénatore S, Perrin L, Sémériva M. Control of cardiac rhythm by ORK1, a *Drosophila* two-pore domain potassium channel. *Curr Biol* 2006;**16**:1502–8.

150. Dulcis D, Levine RB. Glutamatergic innervation of the heart initiates retrograde contractions in adult *Drosophila melanogaster. J Neurosci* 2005;**25**:271–80.

151. Wasserthal LT. *Drosophila* flies combine periodic heartbeat reversal with a circulation in the anterior body mediated by a newly discovered anterior pair of ostial valves and 'venous' channels. *J Exp Biol* 2007;**210**:3707–19.

152. Paternostro G, Vignola C, Bartsch DU, Omens JH, McCulloch AD, Reed JC. Age-associated cardiac dysfunction in *Drosophila melanogaster. Circ Res* 2001;**88**:1053–8.

153. Wessells RJ, Fitzgerald E, Cypser JR, Tatar M, Bodmer R. Insulin regulation of heart function in aging. *Nat Genet* 2004;**36**:1275–81.

154. Bier E, Bodmer R. *Drosophila*, an emerging model for cardiac disease. *Gene* 2004;**342**:1–11.

155. Ocorr K, Reeves NL, Wessells RJ, Fink M, Chen HS, Akasaka T, et al. KCNQ potassium channel mutations cause cardiac arrhythmias in *Drosophila* that mimic the effects of aging. *Proc Natl Acad Sci USA* 2007;**104**:3943–8.

156. Piazza N, Hayes M, Martin I, Duttaroy A, Grotewiel M, Wessells R. Multiple measures of functionality exhibit progressive decline in a parallel, stochastic fashion in *Drosophila* Sod2 null mutants. *Biogerontology* 2009;**10**:637–48.

157. Wolf MJ, Amrein H, Izatt JA, Choma MA, Reedy MC, Rockman HA. *Drosophila* as a model for the identification of genes causing adult human heart disease. *Proc Natl Acad Sci USA* 2006;**31**:1394–9.

158. Zeitouni B, Sénatore S, Séverac D, Aknin C, Sémvériva M, Perrin L. Signalling pathways involved in adult heart formation revealed by gene expression profiling in *Drosophila*. *PLoS Genet* 2007;**3**:1907–21.

159. Kim YO, Park SJ, Balaban RS, Nirenberg M, Kim Y. A functional genomic screen for cardiogenic genes using RNA interference in developing *Drosophila* embryos. *Proc Natl Acad Sci USA* 2003;**10**:159–64.

160. Grossman TR, Gamliel A, Wessells RJ, Taghli-Lamallem O, Jepsen K, Ocorr K, et al. DSCAM and COL6A2 cooperate to produce down syndrome congenital heart defects (personal communication).

161. Koenig M, Monaco AP, Kunkel LM. The complete sequence of dystrophin predicts a rod-shaped cytoskeletal protein. *Cell* 1988;**53**:219–28.

162. Petrof BJ, Shrager JB, Stedman HH, Kelly AM, Sweeney HL. Dystrophin protects the sarcolemma from stresses developed during muscle contraction. *Proc Natl Acad Sci USA* 1993;**90**:3710–4.
163. Straub V, Campbell KP. Muscular dystrophies and the dystrophin-glycoprotein complex. *Curr Opin Neurol* 1997;**10**:168–75.
164. Shcherbata HR, Yatsenko AS, Patterson L, Sood VD, Nudel U, Yaffe D, et al. Dissecting muscle and neuronal disorders in a *Drosophila* model of muscular dystrophy. *EMBO J* 2007;**26**:481–93.
165. Geier C, Perrot A, Ozcelik C, Binner P, Counsell D, Hoffmann K, et al. Mutations in the human muscle LIM protein gene in families with hypertrophic cardiomyopathy. *Circulation* 2003;**107**:1390–5.
166. Mohapatra B, Jimenez S, Lin JH, Bowles KR, Coveler KJ, Marx JG, et al. Mutations in the muscle LIM protein and alpha-actinin-2 genes in dilated cardiomyopathy and endocardial fibroelastosis. *Mol Genet Metab* 2003;**80**:207–15.
167. Gutstein DE, Morley GE, Tamaddon H, Vaidya D, Schneider MD, Chen J, et al. Conduction slowing and sudden arrhythmic death in mice with cardiac-restricted inactivation of connexin43. *Circ Res* 2001;**88**:333–9.
168. Knöll R, Hoshijima M, Hoffman HM, Person V, Lorenzen-Schmidt I, Bang ML, et al. The cardiac mechanical stretch sensor machinery involves a Z disc complex that is defective in a subset of human dilated cardiomyopathy. *Cell* 2002;**111**:943–55.
169. Mery A, Taghli-Lamallem O, Clark KA, Beckerle MC, Wu X, Ocorr K, et al. The *Drosophila* muscle LIM protein, Mlp84B, is essential for cardiac function. *J Exp Biol* 2008;**211**:15–23.
170. Robbins J. KCNQ potassium channels: physiology, pathophysiology, and pharmacology. *Pharmacol Ther* 2001;**90**:1–19.
171. Suzuki M, Sasaki N, Miki T, Sakamoto N, Ohmoto-Sekine Y, Tamagawa M, et al. Role of sarcolemmal K(ATP) channels in cardioprotection against ischemia/reperfusion injury in mice. *J Clin Invest* 2002;**109**:509–16.
172. Hanley PJ, Gopalan KV, Lareau RA, Srivastava DK, von Meitzer M, Daut J. Beta-oxidation of 5-hydroxydecanoate, a putative blocker of mitochondrial ATP-sensitive potassium channels. *J Physiol* 2003;**547**:387–93.
173. Oka T, Xu J, Molkentin JD. Re-employment of developmental transcription factors in adult heart disease. *Semin Cell Dev Biol* 2007;**18**:117–31.
174. Pehlivan T, Pober BR, Brueckner M, Garrett S, Slaugh R, Van Rheeden R, et al. GATA4 haploinsufficiency in patients with interstitial deletion of chromosomes region 8p23.1 and congenital heart disease. *Am J Med Genet* 1999;**83**:201–6.
175. Rajagopal SK, Ma Q, Obler D, Shen J, Manichaikul A, Tomita-Mitchell A, et al. Spectrum of heart disease associated with murine and human GATA4 mutation. *J Mol Cell Cardiol* 2007;**43**:677–85.
176. Liang Q, Wiese RJ, Bueno OF, Dai YS, Markham BE, Molkentin JD. The transcription factor GATA4 is activated by extracellular signal-regulated kinase 1- and 2-mediated phosphorylation of serine 105 in cardiomyocytes. *Mol Cell Biol* 2001;**21**:7460–9.
177. Bisping E, Ikeda S, Kong SW, Tarnavski O, Bodyak N, McMullen JR, et al. Gata4 is required for maintenance of postnatal cardiac function and protection from pressure overload-induced heart failure. *Proc Natl Acad Sci USA* 2006;**103**:14471–6.
178. Qian L, Bodmer R. Partial loss of GATA factor Pannier impairs adult heart function in *Drosophila*. *Hum Mol Genet* 2009;**18**:3153–63.
179. Qian L, Mohapatra B, Akasaka T, Liu J, Ocorr K, Towbin JA, et al. Transcription factor neuromancer/TBX20 is required for cardiac function in *Drosophila* with implications for human heart disease. *Proc Natl Acad Sci USA* 2008;**105**:19833–8.

180. Cai CL, Martin JC, Sun Y, Cui L, Wang L, Ouyang K, et al. A myocardial lineage derives from Tbx18 epicardial cells. *Nature* 2008;**454**:104–8.
181. Zhou B, Ma Q, Rajagopal S, Wu SM, Domian I, Rivera-Feliciano J, et al. Epicardial progenitors contribute to the cardiomyocyte lineage in the developing heart. *Nature* 2008;**454**:109–13.
182. Yu L, Lee T, Lin N, Wolf MJ. Affecting Rhomboid-3 function causes a dilated heart in adult *Drosophila*. *PLoS Genet* 2010;**6**:e1000969.
183. Khullar M, Al-Shudiefat AA, Ludke A, Binepal G, Singal PK. Oxidative stress: a key contributor to diabetic cardiomyopathy. *Can J Physiol Pharmacol* 2010;**88**:233–40.
184. Subramanian S, Kalyanaraman B, Migrino RQ. Mitochondrially targeted antioxidants for the treatment of cardiovascular diseases. *Recent Pat Cardiovasc Drug Discov* 2010;**5**:54–65.
185. Lakshmi SV, Padmaja G, Kuppusamy P, Kutala VK. Oxidative stress in cardiovascular disease. *Indian J Biochem Biophys* 2009;**46**:421–40.
186. Misra MK, Sarwat M, Bhakuni P, Tuteja R, Tuteja N. Oxidative stress and ischemic myocardial syndromes. *Med Sci Monit* 2009;**15**:RA209–19.
187. Khaper N, Bryan S, Dhingra S, Singal R, Bajaj A, Pathak CM, et al. Targeting the vicious inflammation-oxidative stress cycle for the management of heart function. *Antioxid Redox Signal* 2010;**13**:1033–49.
188. Gao G, Dudley Jr. SC. Redox regulation, NF-kappaB, and atrial fibrillation. *Antioxid Redox Signal* 2009;**11**:2265–77.
189. Yan SF, Ramasamy R, Schmidt AM. The RAGE axis: a fundamental mechanism signaling danger to the vulnerable vasculature. *Circ Res* 2010;**106**:842–53.
190. Aikawa R, Nagai T, Tanaka M, Zou Y, Ishihara T, Takano H, et al. Reactive oxygen species in mechanical stress-induced cardiac hypertrophy. *Biochem Biophys Res Commun* 2001;**289**:901–7.
191. Date T, Belanger AJ, Mochizuki S, Sullivan JA, Liu LX, Scaria A, et al. Adenovirus-mediated expression of p35 prevents hypoxia/reoxygenation injury by reducing reactive oxygen species and caspase activity. *Cardiovasc Res* 2002;**55**:309–19.
192. Pimentel DR, Adachi T, Ido Y, Helbeck T, Jiang B, Lee Y, et al. Strain-stimulated hypertrophy in cardiac myocytes is mediated by reactive oxygen species-dependent Ras S-glutathiolation. *J Mol Cell Cardiol* 2006;**41**:613–22.
193. Adiga IK, Nair RR. Multiple signaling pathways coordinately mediate reactive oxygen species dependent cardiomyocyte hypertrophy. *Cell Biochem Funct* 2008;**26**:346–51.
194. Asimakis GK, Lick S, Patterson C. Postischemic recovery of contractile function is impaired in SOD2(+/−) but not SOD1(+/−) mouse hearts. *Circulation* 2002;**26**:981–6.
195. Van Remmen H, Ikeno Y, Hamilton M, Pahlavani M, Wolf N, Thorpe SR, et al. Life-long reduction in MnSOD activity results in increased DNA damage and higher incidence of cancer but does not accelerate aging. *Physiol Genomics* 2003;**16**:29–37.
196. Lebovitz RM, Zhang H, Vogel H, Cartwright Jr. J, Dionne L, Lu N, et al. Neurodegeneration, myocardial injury, and perinatal death in mitochondrial superoxide dismutase-deficient mice. *Proc Natl Acad Sci USA* 1996;**93**:9782–7.
197. Van Remmen H, Williams MD, Guo Z, Estlack L, Yang H, Carlson EJ, et al. Knockout mice heterozygous for Sod2 show alterations in cardiac mitochondrial function and apoptosis. *Am J Physiol Heart Circ Physiol* 2001;**281**:1422–32.
198. Ding G, Fu M, Qin Q, Lewis W, Kim HW, Fukai T, et al. Cardiac peroxisome proliferator-activated receptor gamma is essential in protecting cardiomyocytes from oxidative damage. *Cardiovasc Res* 2007;**76**:269–79.
199. Kim IM, Wolf MJ, Rockman HA. Gene deletion screen for cardiomyopathy in adult *Drosophila* identifies a new notch ligand. *Circ Res* 2010;**106**:1233–43.

200. High FA, Epstein JA. The multifaceted role of Notch in cardiac development and disease. *Nat Rev Genet* 2008;**9**:49–61.
201. Neely GG, Kuba K, Cammarato A, Isobe K, Amann S, Zhang L, et al. A global *in vivo Drosophila* RNAi screen identifies NOT3 as a conserved regulator of heart function. *Cell* 2010;**141**:142–53.
202. Denis CL. Identification of new genes involved in the regulation of yeast alcohol dehydrogenase II. *Genetics* 1984;**108**:833–44.
203. Laribee RN, Shibata Y, Mersman DP, Collins SR, Kemmeren P, Roguev A, et al. CCR4/NOT complex associates with the proteasome and regulates histone methylation. *Proc Natl Acad Sci USA* 2007;**104**:5836–41.
204. Peng W, Togawa C, Zhang K, Kurdistani SK. Regulators of cellular levels of histone acetylation in *Saccharomyces cerevisiae*. *Genetics* 2008;**179**:277–89.
205. Newton-Cheh C, Eligelsheim M, Rice KM, de Bakker PI, Yin X, Estrada K, et al. Common variants at ten loci influence QT interval duration in the QTGEN study. *Nat Genet* 2009;**41**:399–406.
206. Lepperdinger G, Berger P, Breitenbach M, Frohlich KU, Grillari J, Grubeck-Loebenstein B, et al. The use of genetically engineered model systems for research on human aging. *Front Biosci* 2008;**1**:7022–31.
207. Partridge L. Some highlights of research on aging with invertebrates. *Aging Cell* 2009;**8**:509–13.
208. Katewa SD, Kapahi P. Dietary restriction and aging, 2009. *Aging Cell* 2010;**9**:105–12.
209. Broughton S, Partridge L. Insulin/IGF-like signalling, the central nervous system and aging. *Biochem J* 2009;**418**:1–12.
210. Narasimhan SD, Yen K, Tissenbaum HA. Converging pathways in lifespan regulation. *Curr Biol* 2009;**19**:657–66.
211. Stanfel MN, Shamieh LS, Kaeberlein M, Kennedy BK. The TOR pathway comes of age. *Biochim Biophys Acta* 2009;**1790**:1067–74.
212. Sudarsanam S, Johnson DE. Functional consequences of mTOR inhibition. *Curr Opin Drug Discov Dev* 2010;**13**:31–40.
213. Maricq AV, Peckol E, Driscoll M, Bargmann CI. Mechanosensory signaling in *C. elegans* mediated by the GLR-1 glutamate receptor. *Nature* 1995;**378**:78–81.
214. Grotewiel MS, Martin I, Bhandari P, Cook-Wiens E. Functional senescence in *Drosophila melanogaster*. *Ageing Res Rev* 2005;**4**:372–97.
215. Libert S, Chao Y, Zwiener J, Pletcher SD. Realized immune response in enhanced in long-lived *puc* and *chico* mutants but is unaffected by dietary restriction. *Mol Immunol* 2008;**45**:810–7.
216. Rhodenizer D, Martin I, Bhandari P, Pletcher SD, Grotewiel M. Genetic and environmental factors impact age-related impairment of negative geotaxis in *Drosophila* by altering age-dependent climbing speed. *Exp Gerontol* 2008;**43**:739–48.
217. Piazza N, Gosangi B, Devilla S, Arking R, Wessells R. Exercise-training in young *Drosophila melanogaster* reduces age-related decline in mobility and cardiac performance. *PLoS One* 2009;**4**:e5886.
218. Lakatta EG, Levy D. Arterial and cardiac aging: major shareholders in cardiovascular disease enterprises: Part II: the aging heart in health: links to heart disease. *Circulation* 2003;**107**:346–54.
219. Dai DF, Wessells RJ, Bodmer R, Rabinovitch PS. Cardiac aging. In: Wolf N, editor. *The comparative biology of aging*. New York: Springer; 2009. p. 259–86.
220. Yarnell J, Yu S, Patterson C, Cambien F, Arveiler D, Amouyel P, et al. Family history, longevity, and risk of coronary heart disease: the PRIME study. *Int J Epidemiol* 2003;**32**:71–7.

221. Kratz M. Dietary cholesterol, atherosclerosis and coronary heart disease. *Handb Exp Pharmacol* 2005;**170**:195–213.
222. Kuller LH. Nutrition, lipids, and cardiovascular disease. *Nutr Rev* 2006;**64**:15–26.
223. Menotti A, Lanti M, Maiani G, Kromhout D. Determinants of longevity and all-cause mortality among middle-aged men. Role of 48 personal characteristics in a 40-year follow-up of Italian Rural Areas in the Seven Countries Study. *Aging Clin Exp Res* 2006;**18**:394–406.
224. Rodeheffer RJ, Gerstenblith G, Becker LC, Fleg JL, Weisfeldt ML, Lakatta EG. Exercise cardiac output is maintained with advancing age in healthy human subjects: cardiac dilation and increased stroke volume compensate for a diminished heart rate. *Circulation* 1984;**69**:203–13.
225. Lakatta EG, Gerstenblith G, Angell CS, Shock NW, Weisfeldt ML. Prolonged contraction duration in aged myocardium. *J Clin Invest* 1975;**55**:61–8.
226. Cheitlin MD. Cardiovascular physiology-changes with aging. *Am J Geriatr Cardiol* 2003;**12**:9–13.
227. Sadek HA, Nulton-Persson AC, Szweda PA, Szweda LI. Cardiac ischemia/reperfusion, aging, and redox-dependent alterations in mitochondrial function. *Arch Biochem Biophys* 2003;**420**:201–8.
228. Chiamvimonvat N. Diastolic dysfunction and the aging heart. *J Mol Cell Cardiol* 2002;**34**:607–10.
229. Bodyak N, Kang PM, Hiromura M, Sulijoadikusumo I, Horikoshi N, Khrapko K, et al. Gene expression profiling of the aging mouse cardiac myocytes. *Nucleic Acids Res* 2002;**30**:3788–94.
230. Hudson EK, Tsuchiya N, Hansford RG. Age-associated changes in mitochondrial mRNA expression and translation in the Wistar rat heart. *Mech Ageing Dev* 1998;**103**:179–93.
231. Coleman R, Silbermann M, Gershon D, Reznick AZ. Giant mitochondria in the myocardium of aging and endurance-trained mice. *Gerontology* 1987;**33**:34–9.
232. Muscari C, Guarnieri C, Biagetti L, Finelli C, Caldarera CM. Influence of age on oxidative damage in mitochondria of ischemic and reperfused rat hearts. *Cardioscience* 1990;**1**:275–8.
233. Sohal RS, Sohal BH. Hydrogen peroxide release by mitochondria increases during aging. *Mech Ageing Dev* 1991;**57**:187–202.
234. Richardson A, Liu F, Adamo ML, Van Remmen H, Nelson JF. The role of insulin and insulin-like growth factor-I in mammalian ageing. *Best Pract Res Clin Endocrinol Metab* 2004;**18**:393–406.
235. Partridge L, Piper MD, Mair W. Dietary restriction in *Drosophila*. *Mech Ageing Dev* 2005;**126**:938–50.
236. Walker G, Houthoofd K, Vanfleteren JR, Gems D. Dietary restriction in *C. elegans*: from rate-of-living effects to nutrient sensing pathways. *Mech Ageing Dev* 2005;**126**:929–37.
237. Tatar M. The neuroendocrine regulation of *Drosophila* aging. *Exp Gerontol* 2004;**39**:1745–50.
238. Katic M, Kennedy AR, Leykin I, Norris A, McGettrick A, Gesta S, et al. Mitochondrial gene expression and increased oxidative metabolism: role in increased lifespan of fat-specific insulin receptor knock-out mice. *Aging Cell* 2007;**8**:827–39.
239. Kenyon CJ. The genetics of ageing. *Nature* 2010;**464**:504–12.
240. Bjedov I, Toivonen JM, Kerr F, Slack C, Jacobson J, Foley A, et al. Mechanisms of life span extension by rapamycin in the fruit fly *Drosophila melanogaster*. *Cell Metab* 2010;**11**:35–46.
241. Luong N, Davies CR, Wessells RJ, Graham SM, King MT, Veech R, et al. Activated FOXO-mediated insulin resistance is blocked by reduction of TOR activity. *Cell Metab* 2006;**4**:133–42.
242. Jünger MA, Rintelen F, Stocker H, Wasserman JD, Végh M, Radimerski T, et al. The *Drosophila* forkhead transcription factor FOXO mediates the reduction in cell number associated with reduced insulin signaling. *J Biol* 2003;**2**:20.

243. Puig O, Marr MT, Ruhf ML, Tjlan R. Control of cell number by *Drosophila* FOXO: downstream and feedback regulation of the insulin receptor pathway. *Genes Dev* 2003;**17**:2006–20.
244. Hay N, Sonenberg N. Upstream and downstream of mTOR. *Genes Dev* 2004;**18**:1926–45.
245. Teleman AA, Chen YW, Cohen SM. 4E-BP functions as a metabolic brake used under stress conditions but not during normal growth. *Genes Dev* 2005;**19**:1844–8.
246. McNeill H, Craig GM, Bateman JM. Regulation of neurogenesis and epidermal growth factor receptor signaling by the insulin receptor/target of rapamycin pathway in *Drosophila*. *Genetics* 2008;**170**:843–53.
247. Rulifson EJ, Kim SK, Nusse R. Ablation of insulin-producing neurons in flies: growth and diabetic phenotypes. *Science* 2002;**296**:1118–20.
248. Linford NJ, Beyer RP, Gollahon K, Krajcik RA, Malloy VL, Demas V, et al. Transcriptional response to aging and caloric restriction in heart and adipose tissue. *Aging Cell* 2007;**6**:673–88.
249. Sample J, Cleland JG, Seymour AM. Metabolic remodeling in the aging heart. *J Mol Cell Cardiol* 2006;**40**:56–63.
250. Vihervaara T, Puig O. dFOXO regulates transcription of a *Drosophila* acid lipase. *J Mol Biol* 2008;**376**:1215–23.
251. Belke DD, Betuing S, Tuttle MJ, Graveleau C, Young ME, Pham M, et al. Insulin signaling coordinately regulates cardiac size, metabolism, and contractile protein isoform expression. *J Clin Invest* 2002;**109**:629–39.
252. Shioi T, McMullen JR, Kang PM, Douglas PS, Obata T, Franke TF, et al. Akt/protein kinase B promotes organ growth in transgenic mice. *Mol Cell Biol* 2002;**22**:2799–809.
253. McMullen JR, Shioi T, Zhang L, Tarnavski O, Sherwood MC, Dorfman AL, et al. Deletion of ribosomal S6 kinases does not attenuate pathological, physiological, or insulin-like growth factor 1 receptor-phosphoinositide 3-kinase-induced cardiac hypertrophy. *Mol Cell Biol* 2004;**24**:6231–40.
254. DeBosch B, Treskov I, Lupu TS, Weinheimer C, Kovacs A, Courtois M, et al. Akt1 is required for physiological cardiac growth. *Circulation* 2006;**113**:2097–104.
255. Li Q, Li B, Wang X, Leri A, Jana KP, Liu Y, et al. Overexpression of insulin-like growth factor-1 in mice protects from myocyte death after infarction, attenuating ventricular dilation, wall stress, and cardiac hypertrophy. *J Clin Invest* 1997;**15**:1991–9.
256. Delaughter MC, Taffet GE, Fiorotto ML, Entman ML, Schwartz RJ. Local insulin-like growth factor 1 expression induces physiologic, then pathologic, cardiac hypertrophy in transgenic mice. *FASEB J* 1999;**13**:1923–9.
257. Winn N, Paul A, Musaró A, Rosenthal N. Insulin-like growth factor isoforms in skeletal muscle aging, regeneration, and disease. *Cold Spring Harb Symp Quant Biol* 2002;**67**:507–18.
258. Higgins M, Kannel W, Garrison R, Pinsky J, Stokes 3rd J. Hazards of obesity—the Framingham experience. *Acta Med Scand Suppl* 1988;**723**:23–36.
259. Kannel WB, Cupples LA, Ramaswami R, Stokes 3rd J, Kreger BE, Higgins M. Regional obesity and risk of cardiovascular disease: the Framingham study. *J Clin Epidemiol* 1991;**44**:183–90.
260. McGill Jr. HC, McMahan CA, Herderick EE, Zieske AW, Malcom GT, Tracy RE, et al. Pathobiological Determinants of Atherosclerosis in Youth (PDAY) Research Group. Obesity accelerates the progression of coronary atherosclerosis in young men. *Circulation* 2002;**105**:2712–8.
261. King VL, Hatch NW, Chan HW, de Beer MC, de Beer FC, Tannock LR. A murine model of obesity with accelerated atherosclerosis. *Obesity* 2010;**18**:35–41.
262. Weisberg SP, McCann D, Desai M, Rosenbaum M, Leibel RL, Ferrante Jr. AW. Obesity is associated with macrophage accumulation in adipose tissue. *J Clin Invest* 2003;**112**:1796–808.

263. Krauss RM, Winston M, Fletcher BJ, Grundy SM. Obesity: impact on cardiovascular disease. *Circulation* 1998;**98**:1472–6.
264. Narkiewicz K. Obesity and hypertension—the issue is more complex than we thought. *Nephrol Dial Transplant* 2006;**21**:264–7.
265. Howard BV, Ruotolo G, Robbins DC. Obesity and dyslipidemia. *Endocrinol Metab Clin North Am* 2003;**32**:855–67.
266. Koves TR, Li P, An J, Akimoto T, Slentz D, Ilkayeva O, et al. Peroxisome proliferator-activated receptor-gamma co-activator 1alpha-mediated metabolic remodeling of skeletal myocytes mimics exercise training and reverses lipid-induced mitochondrial inefficiency. *J Biol Chem* 2005;**280**:33588–98.
267. Dhahbi JM, Tsuchiya T, Kim HJ, Mote PL, Spindler SR. Gene expression and physiologic responses of the heart to the initiation and withdrawal of caloric restriction. *J Gerontol A Biol Sci Med Sci* 2006;**61**:218–31.
268. Hsiu-Chiang C, Kovacs A, Blanton RM, Han X, Courtois M, Weinheimer CJ, et al. Transgenic expression of fatty acid transport protein 1 in the heart causes lipotoxic cardiomyopathy. *Circ Res* 2005;**96**:225–33.
269. Kavanagh T, Myers MG, Baigrie RS, Mertens DJ, Sawyer P, Shephard RJ. Quality of life and cardiorespiratory function in chronic heart failure: effects of 12 months' aerobic training. *Heart* 1996;**76**:42–9.
270. Erbs S, Höllriegel R, Linke A, Beck EB, Adams V, Gielen S, et al. Exercise training in patients with advanced chronic heart failure (NYHA IIIb) promotes restoration of peripheral vasomotor function, induction of endogenous regeneration, and improvement of left-ventricular function. *Circ Heart Fail* 2010;**3**:486–94.
271. Delbin MA, Antunes E, Zanesco A. Role of exercise training on pulmonary ischemia/reperfusion and inflammatory response. *Rev Bras Cir Cardiovasc* 2009;**24**:552–61.
272. Suvorava T, Lauer N, Kojda G. Physical inactivity causes endothelial dysfunction in healthy young mice. *J Am Coll Cardiol* 2004;**44**:1320–7.
273. Derumeaux G, Ichinose F, Raher MJ, Morgan JG, Coman T, Lee C, et al. Myocardial alterations in senescent mice and effect of exercise training: a strain rate imaging study. *Circ Cardiovasc Imaging* 2008;**1**:227–34.
274. Gollnick PD, Saltin B. Significance of skeletal muscle oxidative enzyme enhancement with endurance training. *Clin Physiol* 1982;**2**:1–12.
275. Davidson SR, Burnett M, Hoffman-Goetz L. Training effects in mice after long-term voluntary exercise. *Med Sci Sports Exerc* 2006;**38**:250–5.
276. Soh JW, Hotic S, Arking R. Dietary restriction in *Drosophila* is dependent on mitochondrial efficiency and constrained by pre-existing extended longevity. *Mech Ageing Dev* 2007;**128**:581–93.
277. Skorupa DA, Dervisefendic A, Zwiener J, Pletcher SD. Dietary composition specifies consumption, obesity, and lifespan in *Drosophila melanogaster*. *Aging Cell* 2008;**7**:478–90.
278. Bhandari P, Jones MA, Martin I, Grotewiel MS. Dietary restriction alters demographic but not behavioral aging in *Drosophila*. *Aging Cell* 2007;**6**:631–7.
279. Burger JM, Promislow DE. Are functional and demographic senescence genetically independent? *Exp Gerontol* 2006;**41**:1108–16.
280. Holloszy JO, Booth FW. Biochemical adaptations to endurance exercise in muscle. *Annu Rev Physiol* 1976;**38**:273–91.

Animal Models of Retinal Disease

Erica L. Fletcher,[*] Andrew I. Jobling,[*] Kirstan A. Vessey,[*] Chi Luu,[†] Robyn H. Guymer,[†] and Paul N. Baird[†]

[*]*Department of Anatomy and Cell Biology, The University of Melbourne, Parkville, Victoria, Australia*

[†]*Centre For Eye Research Australia, University of Melbourne, Royal Victorian Eye and Ear Hospital, East Melbourne, Victoria, Australia*

Progress in Molecular Biology and Translational Science, Vol. 100
DOI: 10.1016/B978-0-12-384878-9.00006-6

Copyright 2011, Elsevier Inc.
All rights reserved.
1877-1173/11 $35.00

Diseases of the retina are the leading causes of blindness in the industrialized world. The recognition that animals develop retinal diseases with similar traits to humans has led to not only a dramatic improvement in our understanding of the pathogenesis of retinal disease but also provided a means for testing possible treatment regimes and successful gene therapy trials. With the advent of genetic and molecular biological tools, the association between specific gene mutations and retinal signs has been made. Animals carrying natural mutations usually in one gene now provide well-established models for a host of inherited retinal diseases, including retinitis pigmentosa, Leber congenital amaurosis, inherited macular degeneration, and optic nerve diseases. In addition, the development of transgenic technologies has provided a means by which to study the effects of these and novel induced mutations on retinal structure and function. Despite these advances, there is a paucity of suitable animal models for complex diseases, including age-related macular degeneration (AMD) and diabetic retinopathy, largely because these diseases are not caused by single gene defects, but involve complex genetics and/or exacerbation through environmental factors, epigenetic, or other modes of genetic influence. In this review, we outline in detail the available animal models for inherited retinal diseases and how this information has furthered our understanding of retinal diseases. We also examine how transgenic technologies have helped to develop our understanding of the role of isolated genes or pathways in complex diseases like AMD, diabetes, and glaucoma.

I. Introduction

Over the last 100 years, major advances in understanding the pathogenesis of retinal diseases has occurred largely from initial studies on animals that show signs similar to human disease. A detailed understanding of retinal disease in animals has not only furthered our understanding of disease mechanisms, but has also provided a means with which to test novel therapeutic agents.

Like the variety of presentations of retinal diseases in humans, animals suffer from a similarly broad range of conditions. Since the initial observations over 40 years ago that prolonged or bright light exposure can induce retinal degeneration in rodents, many advances have been made in our understanding of the pathogenesis of these specific diseases. As will be documented below, studying the detailed cellular changes that occur in animals that exhibit similar diseases to humans has been critical to this understanding. Over the last few years, a range of more sophisticated genetic tools have also allowed a number of gene mutations causing retinal disease to be identified in both humans and animals. This has expanded our knowledge of the role that particular molecules or pathways play in cell dysfunction and loss.

Although the development of transgenic animals has led to a tremendous improvement in our knowledge of the cause of retinal disease, there remain many diseases, where animals do not adequately reflect the condition in humans. These typically involve common diseases in our society which present with complex etiologies manifesting through either complex genetic and environmental interactions. Moreover, it is now recognized that even in diseases caused by a single mutation, genetic variation at other sites within the same gene can lead to changed severity, penetrance, or disease type.

The aim of this review is to provide an overview of the characteristics of retinal diseases that affect animals and how this has helped us understand the underlying pathogenesis of disease. In particular, we examine the information gained from studying animals that carry naturally occurring mutations, or where there has been genetic modification so that specific molecules or pathways become dysfunctional. In addition, we consider animal models in complex diseases that involve multiple susceptibility genes and/or interactions with the environment. We evaluate how disease in these animals does not always replicate human disease, and what might be necessary to improve the development of the "ideal" animal model for some complex conditions.

II. Overview of Retinal Structure and Function

The retina is a multilayered outpouching of the central nervous system that consists of alternating layers of neurons and synapses (Fig. 1A). Light passes through all layers of the retina prior to being absorbed by photopigments located within the outer segments of photoreceptors.[1] Photoreceptors convert light into a neurochemical signal that is passed to second order neurons, called bipolar cells. Bipolar cells communicate in turn with the temporally acting amacrine cells and ganglion cells, with the latter being the main output neurons of the retina. Thus, the retinal "through" pathway consisting of photoreceptors, bipolar cells, and ganglion cells plays a central part in visual processing within the retina. Diseases of the retina are associated with the loss of function or death of one or more classes of neurons in the retinal through pathway. Visual impairment can arise from loss of photoreceptors, defects in synaptic transmission between photoreceptors and bipolar cells, or loss of ganglion cells.

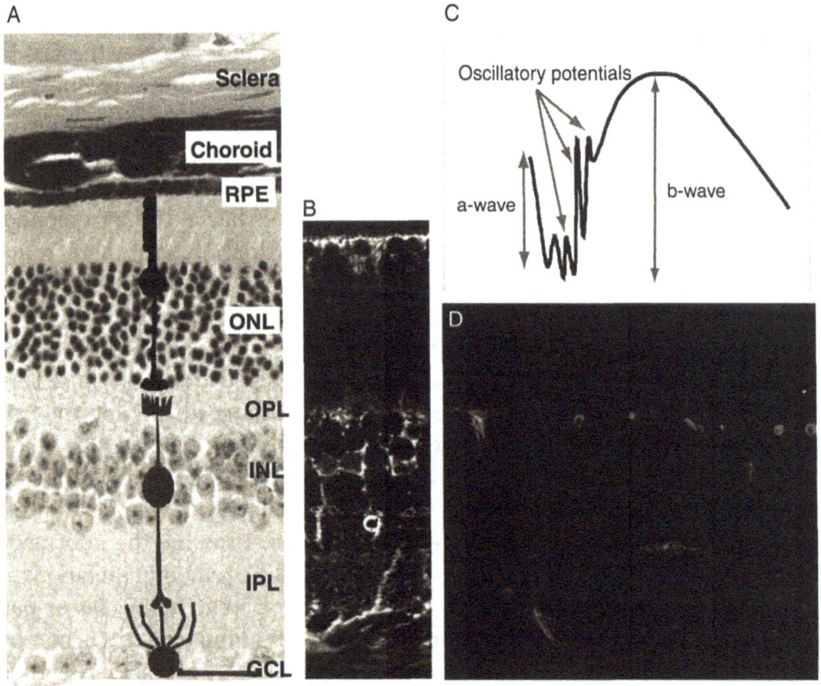

FIG. 1. Summary of the cellular structure and function of the retina. (A) A vertical section of the adult rat retina showing the retinal layers and major neuronal subclasses. The neuronal layers of the retina include the outer nuclear layer (ONL) corresponding to the location of photoreceptor nuclei (black cell), the inner nuclear layer (INL), where horizontal, bipolar and amacrine cells are located, and the ganglion cell layer (GCL), where nuclei of ganglion and displaced amacrine cells are located. The two synaptic layers include the outer plexiform layer (OPL) and the inner plexiform layer (IPL). Beneath the retina, are located the retinal pigment epithelium (RPE), choroid, and sclera. The major neuronal circuit that carries visual information through the retina is the retinal through pathway which comprises photoreceptors (indicated as a black cell), bipolar cells (indicated as a red cell), and ganglion cells (indicated as a blue cell). (B) Vertical section of rat retina immunolabeled for the glial cell marker, glutamine synthetase. Müller cells span the entire thickness of the retina and have their somata located in the middle of the inner nuclear layer. (C) Schematic diagram of the major wavelet components of the flash ERG, including the a-wave (derived from the activity of photoreceptors), b-wave (attributed to postreceptoral neurons especially ON Bipolar cells) and oscillatory potentials (attributed to the function of amacrine cells). (D) Vertical section of the rat retina immunolabeled for the microglial marker, IBA-1 (red) and a marker of blood vessels (green). Microglia, the main resident immune cell of the retina, are stellate cells located within the two plexiform layers. (See Color Insert.)

Maintenance of retinal integrity depends on support cells, including macroglia such as Müller cells and astrocytes,[2,3] or the retinal pigment epithelium (RPE), a monolayer of epithelial cells that sits below the retina and is integral to photoreceptor function[4] (Fig. 1). Retinal glia, including Müller cells take up and recycle the neurotransmitters, glutamate and GABA, siphon potassium from the extracellular space, shuttle energy metabolites between the vasculature and neurons, and also play a role in maintaining the blood retinal barrier.[2,3] If any of these functions is disrupted, deleterious effects on retinal neurons ensues.[2]

The RPE is an epithelial monolayer of cells that forms a barrier between the underlying choroidal vasculature and the neural retina[4] (Fig. 1). Photoreceptor outer segments sit within the microvillus projections of the apical surface of RPE cells. The RPE has an enormous range of functions but four of the most crucial are absorption of stray light by virtue of the melanin granules found with the cells, transport of retinoids between the choroidal vasculature and photoreceptor outer segments, phagocytoses photoreceptor outer segments and detoxification of photoreceptor outer segment debris.[4] Secondary loss of photoreceptors can occur if any of these functions of the RPE are disrupted.

Microglia are generally considered the resident immune cells of the retina.[5] However, owing to their rapid activation following retinal injury and their contribution to the development of inflammation they have been implicated in a variety of retinal diseases. They lie dormant within the inner retina, and in their quiescent state have a distinct stellate morphology (Fig. 1). Several studies suggest that microglia in this resident state extend and retract their fine processes so as to constantly survey their local environment.[6–8] Following injury they change morphology by retracting their processes and take on a more amoeboid morphology that is also associated with the release of a range of cytokines.[6,7]

A common method used to test the functional status of the retina is with electroretinogram (ERG) recording.[9] The flash ERG is an extracellular recording that derives from the activity of different cohorts of retinal neurons, that is often used in animal studies to gain an understanding of the onset and severity of disease.[9] Following a brief flash of light, recording electrodes placed on the cornea signal a serial waveform[9] (Fig. 1). As shown in Fig. 1C, the first negative deflection of the waveform is called the a-wave and reflects photoreceptor activity. The second positive going waveform is the b-wave and is attributed to the activity of ON bipolar cells. On the leading edge of the b-wave are small oscillations, called oscillatory potentials that reflect inner retinal processing, especially from amacrine cells. By examining the amplitude and timing of each of these waveforms it is possible to determine the retinal locus of disease.

In summary, visual impairment is caused by the loss or dysfunction of neurons that form the retinal through pathway, including photoreceptors, bipolar cells, and ganglion cells. Deficits can be caused by direct loss of these cells, or by secondary affects via loss of function affecting the range of support cells that retinal neurons rely on.

III. Genetics of Retinal Disease

A. Mendelian Inherited Conditions

The sequestration of light energy and its conversion into a neuronal response through the photoreceptors is based on both an intricate protein structure and a high energy demand, both of which require a large number of genes for normal function. This dependency may explain why in excess of 200 retinal genes have been mapped through linkage analysis and over 150 of these identified as disease causative in humans as of January 2010 (http://www.sph.uth.tmc.edu/RetNet/). Hereditary retinal disease genes show segregation with disease in a Mendelian pattern of inheritance (dominant, recessive, or X-linked) and mutations in many of the known genes have now been identified. However, it is clear that there exists extensive heterogeneity with over 50 genes implicated in retinitis pigmentosa (RP) alone (Fig. 2). These genes impact on a variety of events within the eye including the visual transduction cascade, structural proteins, retinoid cycle as well as splicing.[10,11]

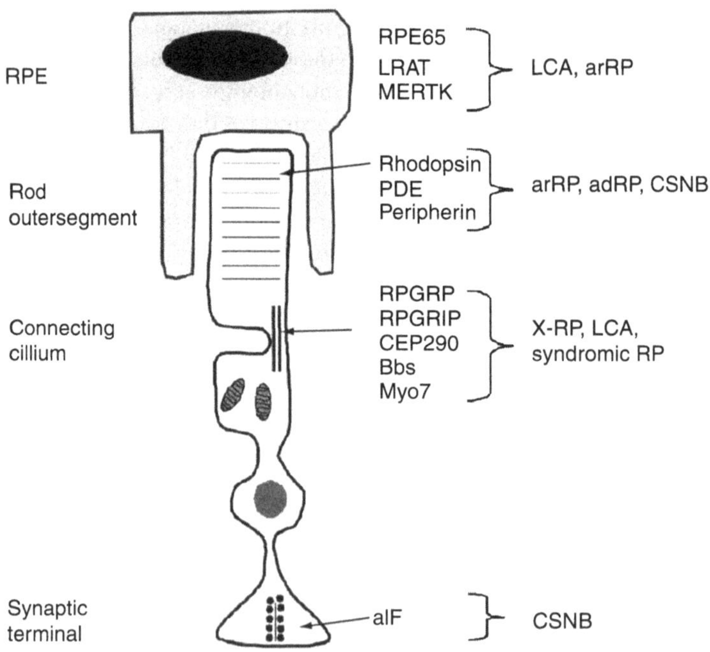

FIG. 2. Schematic diagram of the RPE/Photoreceptor complex showing the variety of proteins that are associated with outer retinal diseases including Leber congenital amaurosis, autosomal recessive Retinitis Pigmentosa (arRP), autosomal dominant Retinitis Pigmentosa (adRP), X-linked Retinitis Pigmentosa (X-RP), and congenital stationary night blindness (CSNB).

Genotype–phenotype correlations in identified retinal genes are not straightforward even once a specific retinal disease gene mutation is identified. For example in the peripherin/retinal degeneration slow (RDS) gene, involved in maintenance of proper disc alignment and disc shedding, over 34 different mutations have been implicated in autosomal dominant RP. While animal studies have allowed detailed examination of mutations within this gene, it is clear that they can affect rods and cones differently and that both haploinsufficiency and negative dominant mutations play a major role in pathogenesis. In addition, mutations in peripherin/RDS can also lead to a range of macular dystrophies and most intriguing is that the same mutation can give rise to completely different phenotypes within a family.[12] These may vary from an apparently normal fundus, due to incomplete penetrance, to various other macular dystrophies including cone-rod dystrophy as well as RP reflecting variable penetrance. Thus haplotypic background, other modifier genes such as the ROM1 gene, where digenic inheritance may occur, or environmental influences are likely to influence final presenting phenotype.[13]

The issue of heterogeneity and incomplete penetrance was thought to be a major stumbling block in the potential treatment of individuals with retinal disease. However, the safety and efficacy of gene therapy for the inherited retinal disease of Leber congenital amaurosis (LCA) type 2 has already been demonstrated in three separate studies.[14–16] The reported improvement in visual acuity or function reported in several patients in these studies lends hope that other retinal diseases may also be treated in the future using a similar strategy. None of this would have been possible without the use of animal studies. Successful gene transfer using the adeno-associated viral (AAV) vector was first demonstrated in experiments performed over 10 years ago in the mouse.[17] Preclinical follow up studies in the naturally occurring RPE65 gene Briard dog,[18] as well as mouse[19] and monkey[20] models of inherited RPE65 deficiency then followed. The choice of RPE65 for gene replacement reflected a general consensus that this represented an ideal target for retinal gene therapy based on its key functional role in the retina.

While great success has been shown with these clinical trials, a number of questions remain as to when the most appropriate time to undertake gene replacement therapy might be or how effective it might be at later stages of disease when many retinal cells have already been lost. It is also unclear if other mutations might manifest following viral vector replacement with the wild-type gene. It should also be noted that while animals provide attractive models for human retinal disease, they do not always fully mimic the human disease situation and thus interpretation of findings needs to be considered in this regard. The future expansion and success of retinal gene therapy in clinical practice and the use of appropriate animal models will therefore depend on making sense of the above issues.

B. Complex Inheritance

Animal models of retinal diseases are typically devised to model one or two gene mutations of interest to elucidate their role in disease. While this is particularly useful in understanding the etiology of monogenic or digenic disease, the vast majority of eye disease in developed societies arises from complex diseases such as age-related macular degeneration (AMD), diabetic retinopathy (DR) or glaucoma that affect the retina or optic nerve head. These diseases therefore present with additional challenges when considering the use of animal models. Complex diseases are polygenic with not only multiple genetic variants of varying degrees of risk (increased, neutral, or protective) but also epigenetic and other genetic changes that when coupled with appropriate exposure to a range of environmental risk factors, give rise to disease. These environmental factors could include such things as age, gender, smoking history, infectious agents, weight, insulin control, and diet. In terms of genes associated with DR, the aldose reductase (*AKR1B1*) gene has been shown to be significantly associated with disease as well as several other genes, in some but not all studies, other genes including *NOS3*, *VEGF*, *ITGA2*, and *ICAM1*, have been implicated through meta-analysis.[21] When considering AMD, considerable success has been made in identifying its genetic determinants with several genes (*CFH*, *CFHR1-5*, *C2/BF*, *C3*, *APOE*, *LOC387715/HTRA1*) being consistently associated with disease in multiple studies.[22] More recently other genes including *LIPC* and *TIMP3* have also been reported as associated with AMD following genome-wide association studies (GWAS).[23,24] A large number of other genes have also been reported as associated with AMD but typically in single studies without replication.[22] In the case of glaucoma, four genes have so far been identified as involved including *myocilin*, *cytochrome P4501B1*, *optineurin*, and *WDR36*. Of these genes, myocilin is the best characterized, being largely associated with juvenile or early adult onset primary open-angle glaucoma (POAG) but these genes so far only account for less than 10% of the disease.[25]

As well as having a genetic and environmental component, complex diseases also present with a range of clinical features that allows diagnosis into different clinical entities. In the case of DR this can be proliferative or non-proliferative entities, in the case of AMD this can be geographic atrophy or choroidal neovascular disease, and in the case of glaucoma this can be elevated or normal pressure, open or acute angle closure. While these subtypes represent broad definitions of disease, it is clear that other disease subtypes are also present . These include retinal angiomatous proliferation (RAP) and polypoidal choroidal neovascular (PCV), in the case of AMD with these clinical disease manifestations often being associated with individuals of distinct ethnicities such as PCV being first identified and possibly being more common in individuals of Asian ethnicity.

Whether the same combination of genes and environment in complex diseases, leads to either the initiation of the disease process, progression from early disease to more advanced disease states or finally to end stage disease is currently under debate. Thus, it is unclear if all three stages of disease are due to the same genetic and environmental risk factors, whether different combinations are responsible for each of these stages or whether cumulative risk exposures are necessary to drive this process. Taken together, the above issues provide a challenge in determining, what is an appropriate model for complex retinal disease, and thus how results from animal models can be interpreted back to the human condition.

IV. RP: A Family of Inherited Photoreceptor Degenerations

RP denotes a family of hereditary disorders that primarily affects rod photoreceptors and retinal pigment epithelial function and that lead to progressive vision loss and blindness. It has a prevalence of between 1 in 3000 and 1 in 5000 and in developed countries is ranked second behind DR as a cause of visual impairment in those between 16 and 50 years of age. The characteristic cellular features include gradual loss of rod photoreceptors, followed by cones at a later stage.

It is generally agreed that RP has a strong hereditary component. In typical RP, three major modes of inheritance have been identified, including sex-linked, autosomal recessive and autosomal dominant. As will be described below, a great deal has been learned about RP from studies on animal models of retinal degeneration. The criteria for considering the value of a particular animal model of retinal degeneration are at least twofold. First, the animal should express the same genetic mutation as in human disease. Secondly, they should show gradual loss of rods followed by cones. Animal models of retinal degeneration are all associated with genetic inactivation or dysfunction of a single gene, although environmental factors can accelerate the disease progression.

A. Animal Models with Genetic Mutations Affecting Rhodopsin

The diversity of presentation of RP is mirrored by the diversity of retinal degenerations observed in animals, with a number of reports of retinal degeneration occurring naturally in mice, rats, cats, and dogs. In general, they can be separated on the basis of which functional class of protein is affected in the disease. Mutations generally affect rhodopsin, proteins important for maintaining rod structure, proteins involved in phototransduction, or proteins important for maintaining RPE or glial function (Fig. 2). It is notable that mutations that affect photoreceptor associated proteins can also cause LCA, a severe retinal degeneration or congenital stationary night blindness (CSNB), a condition characterized by visual dysfunction in the absence of photoreceptor loss (Fig. 2).

Rhodopsin is the visual pigment found within rod photoreceptors and belongs to the large superfamily of seven transmembrane spanning G protein receptors.[26] It is a protein of 348 amino acids that span the discs within photoreceptor outer segments and binds to the vitamin A derivative, 11-*cis*-retinal. There are a number of amino acid residues within rhodopsin that are vital for maintaining its molecular structure and function. For example, Lysine 296 is the attachment site with which the chromophore, 11-*cis*-retinal, forms a Schiff-base linkage. Cysteines 110 and 187 are thought to be important in creating disulfide bonds to maintain proper structure. In addition, there are a number of amino acids (especially serines and threonines in the carboxyl terminal region) that provide phosphorylation sites or sites for interaction with cytoplasmic proteins. Some amino acid residues play a role in the trafficking of rhodopsin from its site of synthesis in the inner segment to the outer segment.

Investigations on the amino acid sequence of rhodopsin in a variety of species has revealed that some amino acid residues can be varied without functionally altering the protein, whereas, other changes in amino acids cause loss of function often with characteristics of hereditary retinal degeneration.[26] Thus, phenotype and severity of disease is to an extent linked to the particular amino acid altered by the gene mutation. Some cause CSNB (loss of rod function in the absence of cell death), some RP (gradual loss of rods then cones), and others more severe forms of retinal degeneration including LCA (blindness within the first year of life).

The mutations characteristic of autosomal dominant RP, have been classified into at least two classes, broadly depending on the affect of the mutation on rhodopsin function.[27] Class I mutations are predominantly associated with mutations affecting the C-terminus of the protein. In these cases, mutant rhodopsin resembles wild-type rhodopsin in their translocation to the plasma membrane and binding to chromophore.[27] However, the mutations affect trafficking through the Golgi apparatus and impaired targeting of disc outer segments. In contrast, Class II anomalies are characterized by mutations in the intradiscal, transmembrane, or cytoplasmic domains of rhodopsin and result in misfolding of the protein and an inability to bind to chromophore.

With the exception of some types of dog, there are few naturally occurring animal models of retinal degenerations associated with rhodopsin mutation. However, since the first reported mutation in human rhodopsin, described in 1990,[28] several transgenic rodent models have been developed, that carry either Class I or Class II mutations.[29–33] The generation of these transgenic mice has led to a great deal of information relating to the mechanisms by which aberrant rhodopsin affect photoreceptor viability.

The most common mutation affecting those with autosomal dominant RP in the USA is a substitution of proline for histidine in codon 23 near the N-terminus of rhodopsin.[28] Confirmation that this Class II rhodopsin mutation leads to progressive loss of photoreceptors has been accomplished with the development of several transgenic mice that express mutant human Pro23His or mouse Pro23His rhodopsin.[31,32] The first transgenic mouse lines developed, expressed mutant human Pro23His rhodopsin, in addition to normal mouse rhodopsin. These mice developed gradual photoreceptor death and severe loss of visual function,[32,34] with evidence for abnormal trafficking of rhodopsin to outer plexiform layer.[34] One of the unusual findings from the generation of this transgenic mouse line, was that photoreceptor degeneration developed even in control mice that over expressed normal human rhodopsin.[32] This suggests that the level of rhodopsin expression, whether normal or aberrant, may be an important contributing factor in the development of photoreceptor death.

More recently, a transgenic mouse line, called the VPP mouse, has been developed that expresses the Pro23His mutant form of *mouse* rhodopsin in conjunction with a valine 20 for glycine substitution and proline 27 for leucine. VPP mice develop gradual loss of retinal function and photoreceptor loss from 3 weeks of age with many similarities to human autosomal dominant RP.[29,31] Like human autosomal dominant RP, expression of both normal mouse rhodopsin and mutant rhodopsin occurs in VPP mouse. Although, the rate of opsin synthesis and disc shedding is unchanged in VPP compared with wild-type mice, there is evidence of disorganization of basal discs and a gradual shortening of the outer segment at the earliest stages of disease.[35] Evidence for gradual degeneration of rods then cones has been described in transgenic rats that express the Pro23His form of rhodopsin[36] (Fig. 3) Photoreceptor loss in the Pro23His rat is slow and is associated with increased metabolic demands.[37] As shown in Fig. 3, after 3 months of age, the outer nuclear layer is around half its normal thickness, and a significant number of rods remain even after 1 year.

In order to characterize Class I mutants, a transgenic mouse that expressed a naturally occurring mutation in the stop codon was developed. In this mouse, glutamine is placed at position 344, causing a truncation of the last five amino acids of rhodopsin.[33] Sung *et al.*[33] showed that Q344ter mice develop progressive photoreceptor loss from postnatal day 10, and show impaired trafficking of mutant rhodopsin from the cell body to outer segment, resulting in an abnormal build up of mutant protein within the photoreceptor cell body. Similarly, aberrant rhodopsin transport has been implicated in photoreceptor degeneration in transgenic mice carrying the proline 347 to serine mutation,[30]

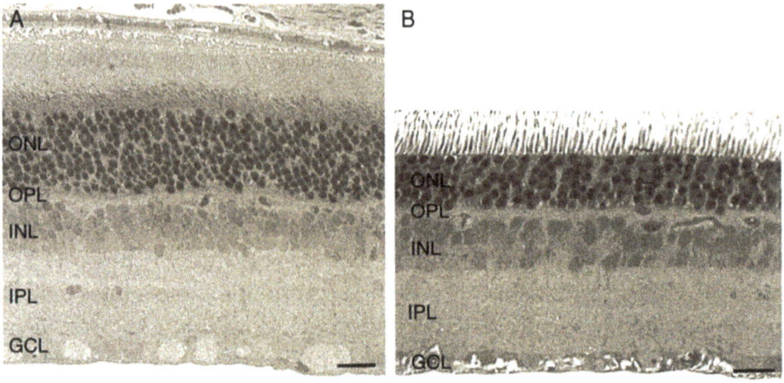

FIG. 3. Nissl stained sections of retinae from a 12-week-old (A) control Sprague Dawley rat and (B) Pro23His rat. Loss of photoreceptors occurs at a slow rate in the Pro23His rat, so that by 12 weeks of age, approximately half remain. Abbreviations. ONL, outer nuclear layer; OPL, outer plexiform layer; INL, inner nuclear layer; IPL, inner plexiform layer; GCL, ganglion cell layer. Scale bar = 20 μm.

a mutation affecting the C-terminus of rhodopsin that shows no functional abnormality in *in vitro* studies, but is associated with more severe disease in humans than other autosomal dominant rhodopsin mutations. In these animals, progressive photoreceptor loss occurs in combination with an accumulation of vesicles between the inner and outer segments that contain rhodopsin.[30]

Some rhodopsin mutations, especially Class II mutations, are predicted to cause photoreceptor death by being constitutively active. In these cases it is predicted that rhodopsin functions as if in continuous lighting, because rhodopsin has an inability to bind normally to its chromophore, 11-*cis*-retinal. The main rhodopsin mutation predicted to cause this type of effect is one where the lysine-296 residue to which chromophore normally binds is altered. Transgenic K296E mice carry a lysine-296 for glutamate substitution, and develop gradual photoreceptor loss. In order to examine whether photoreceptor loss in these mice was caused by continual activation of transducin, or because of inadequate rhodopsin deactivation K296E mice have been crossed with transducin null or transducin/arrestin double knockout mice. Chen *et al.*[38] showed that K296Emice display the same rate of photoreceptor loss compared to K296E/transducin null mice suggesting that continuous activation of transducin did not affect photoreceptor viability in this animal model. However, when K296E mice were crossed with arrestin/transducin double knockout mice, a substantial delay in photoreceptor loss was observed, suggesting that persistent rhodopsin/arrestin binding is a stimulus for photoreceptor loss in these mice.

The effect of transducin activation on rate of retinal degeneration in mouse models carrying other rhodopsin mutations has been considered with mixed results. On the one hand, Smardzija *et al.*[39] showed that photoreceptor loss was reduced in VPP mice that lacked transducin expression, whereas Brill *et al.*[40] showed that the absence of transducin expression had no affect on the rate of photoreceptor loss. The differences in these two studies highlight the role that other genetic modifiers have on disease progression. Light history in particular can have a profound role on severity of photoreceptor loss. The amino acid sequence of RPE65, an enzyme important in retinoid transport within the RPE is known to confer sensitivity or resistance to even moderate levels of lighting in rodents.

Finally, it is notable that the severity of photoreceptor loss depends on the relative proportions of mutant to wild-type rhodopsin. Indeed, VPP mice that were crossed with rhodopsin null mice display more severe loss of photoreceptors compared with VPP mice crossed with mice that are heterozygous or homozygous for wild-type rhodopsin.[29] This observation highlights the cytotoxicity of mutant rhodopsin, and suggest that ways of increasing expression of normal wild-type rhodopsin may slow photoreceptor loss.

To date the only animal identified as expressing a natural rhodopsin mutation is the dog. Inherited retinal degenerations in dogs, termed progressive retinal atrophies are widespread amongst different types of dogs; several canine models of autosomal recessive and X-linked RP have been described. Recently, Kijas[41] described a progressive retinal atrophy in English Mastiff dogs that mimics many of the features of human autosomal dominant RP that is caused by a point mutation in the intradiscal regions of rhodopsin (Thr4Arg). Like some forms of autosomal dominant RP, affected English Mastiff dogs develop a slowly progressive retinal degeneration that shares many features with human autosomal dominant RP. Importantly, mutant dogs are clinically normal for the first few months of life prior to disease onset, providing an ideal setting with which to develop gene therapy strategies in a large animal.[41]

One of the problems with using rodents to study cellular mechanisms in inherited retinal degeneration is the lack of a significant number of cones. Inherited retinal degenerations in humans is known to affect rods early in disease, but is followed by loss of cones. Indeed, ways to preserve cones could have a major impact on the quality of life for those with RP because complete loss of vision does not occur until a large percentage of cones are lost. In order to better understand how retinal degenerations affect cones, a transgenic pig has been developed, that carries the Pro347Leu rhodopsin mutation in the C-terminus.[42] Like rodents and humans carrying this mutation, progressive rod degeneration occurs, followed by gradual loss of cones. This porcine model may prove helpful for studies examining possible ways of restoring vision.

B. Mutations in Genes Encoding Components of the Phototransduction Cascade

In contrast to autosomal dominant RP, which are caused by mutations in rhodopsin, to date 21 autosomal recessive RP genes have been identified, most of which encode proteins important in the rod phototransduction cascade. Moreover, many animals with retinal degeneration carry these mutations.

Phototransduction refers to the cascade of reactions that occurs within photoreceptor outer segments following light exposure that ultimately leads to the hyperpolarization of the photoreceptor membrane potential. Several proteins contribute to the cascade including the G-protein, transduction, and phosphodiesterase (PDE), a major enzyme that catalyzes the breakdown of cGMP to GMP. Mutations in PDE in particular, are associated with retinal degeneration, whereas mutations affecting transducin or arrestin are less likely to cause photoreceptor degeneration. PDE is a heterotetrameric complex consisting of α, β, and two γ subunits. Each subunit is vital for normal photoreceptor function. Mutations in PDE6A and PDE6B have been reported to cause autosomal recessive RP in humans.[43,44] In addition, mutations in PDE6A and PDE6B are associated with progressive retinal atrophy in several types of dog and retinal degeneration in mice.[45–48] Mutations in PDE are also a cause of photoreceptor loss in mice.[49]

Retinal degenerations that have been observed in natural populations of mice are referred to as rd1–rd16 based on the order in which they were described.[50] The rd1 mouse was the first mouse model of retinal degeneration identified.[48,51] Mice that are homozygous for the rd1 mutation express a nonsense mutation in exon 7 of the *PDE6B* gene that encodes the β subunit of PDE.[52] It is a commonly expressed mutation amongst mouse strains and has also been identified amongst natural populations of mice in Japan, California, and the United Kingdom.[48] The rd1 mouse develops rapid rod photoreceptor loss from approximately P10 followed by loss of cones from approximately P35[48] (Fig. 4). The mutation in PDE is associated with an inability to breakdown cGMP within photoreceptor outer segments.[53] Consequently, sustained depolarization of the photoreceptors occurs, that may contribute to apoptotic cell death because of metabolic stress or unregulated entry of cations into the photoreceptor outer segement.[54]

A criticism of the use of the rd1 mouse as a model of human autosomal recessive RP, is the fact that disease in this model overlays with the normal developmental processes. Blanks *et al.*[55] showed that in developing rd1 mice, the dendrites of ON bipolar cells do not form normal invaginations within photoreceptor terminals. Moreover, photoreceptors fail to develop normal

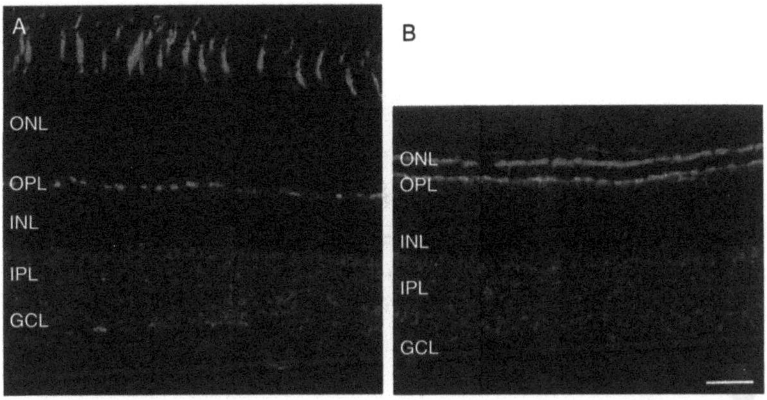

FIG. 4. Vertical sections of a (A) control C57Bl6 and (B) Postnatal day 25 rd1 mouse retina that have been labeled for the cone photoreceptor marker, Peanut Agglutin (green). Shown in red are rod bipolar cells immunolabeled for Protein Kinase C. By postnatal day 25, only a single layer of cones remain in the outer nuclear layer. All the inner layers of the retina remain intact. Abbreviations as per Fig. 3. Scale bar = 50 μm. (See Color Insert.)

outer segments. In contrast, the rd10 mouse model, first reported in 2002, carries a missense mutation in exon 13 of *PDE6B*, develops normal vision, prior to the gradual loss of rods from postnatal day 30.[56,57]

At least two dog species manifest retinal degenerations caused by mutations in *PDE6B*, including the Irish Setters and Sloughi dogs.[47,58] Progressive retinal atrophy is the canine equivalent of RP and is an autosomal recessive trait in Irish setters. It is caused by a nonsense mutation in the C-terminus of the β subunit of PDE leading to truncation and destabilization of the gene product and a nonfunctioning PDE.[47,49] Retinal development is normal in affected dogs until 13 days of age, when photoreceptor development is arrested, and degeneration ensues within 1 month. Rod photoreceptor loss occurs gradually over 5 months, followed by cone loss within 1 year. Like the rd1 mouse, this degeneration is associated with accumulation of cGMP, and sustained depolarization of rod photoreceptors.

In addition to mutations in the β subunit of *PDE*, mutations in both α and γ subunits are also associated with retinal degeneration in mice and Cardigan Welsh Corgi dogs.[45,46,59] Progressive retinal atrophy in Cardigan Welsh Corgis have a 1-bp deletion in *PDE6A* that leads to a string of 28 altered amino acids followed by a stop codon.[59] Notably, the altered protein is missing the catalytic domain and its membrane binding site, resulting in PDE activity that is near absent. Like dogs expressing a *PDE6B* mutation, photoreceptor development is arrested in affected Corgis at an early age. Rhodopsin is present in rod outer segments until 3 weeks of age. However, degeneration of rods occurs following 3 weeks of age.[59]

C. Mutations in Genes Encoding Photoreceptor Structural Proteins

Photoreceptor outer segments contain many flattened membranous structures called discs that contain rhodopsin and the proteins important for phototransduction. Peripherin is an important protein that regulates the formation and renewal of photoreceptor disc outer segments. It is an integral membrane protein consisting of 346 amino acids that is localized to the rim region of rod and cone outer segments.[60] It also complexes with a homologous protein called ROM1. Mutations in peripherin are associated with a variety of retinal degenerations, including autosomal dominant,[61,62] central areolar choroidal dystrophy, retinitis punctata albescens, cone-rod dystrophy and inherited macular dystrophies such as adult onset foveomacular vitelliform dystrophy and butterfly dystrophy.[13]With respect to RP, approximately 34 different peripherin/RDS mutations have been implicated in autosomal dominant RP, accounting for 3–9% of all mutations associated with autosomal dominant RP.[63]

The RDS mouse (rds or rd2) is one of the oldest and well-established animal models of retinal degeneration.[64,65] These mice carry a mutation in chromosome 17 encoding the *peripherin/rds* gene. In contrast to wild-type mice, development of photoreceptor outer segments is impaired in homozygous rds mice. In addition, rds mice show signs of degeneration starting form postnatal day 14. Loss of both rods and cones is slow with complete loss of photoreceptors taking 9–12 months, even in the central retina.[66,67]

More recently, a transgenic mouse has been created that expresses a point mutation, proline 216 to leucine (P216L) amino acid substitution in peripherin.[68] Stabilization of photoreceptor discs is thought to involve interactions across the intradiscal space with proteins such as ROM-1. The extracellular loop region of peripherin is thought to be an important region mediating these interactions, and the transgenic P216L mouse has a point mutation in this region. P216L mice show gradual photoreceptor loss that is proportional to the level of transgene expression.[68]

D. Mutations in Proteins Expressed in the Photoreceptor Cilium

Photoreceptors contain a number of cellular compartments. The inner segment region is known to contain a high density of mitochondria and Golgi apparatus, and is the site of synthesis of important proteins such as opsin, and the components that form the outer segment discs. Proteins formed within the inner segment or indeed anywhere else within the photoreceptor cell body are transported to the outer segment via a connecting cilium that is located between the inner and outer segments. Mutations in proteins that form the connecting cilium are associated with RP, especially X-linked RP, and also LCA.

X-linked RP is the most severe form of RP in terms of age of onset and progression, and predominantly affects males. Mutations in the *retinitis pigmentosa GTPase regulator* (*RPGR*) gene or the *retinitis pigmentosa GTPase regulator interacting protein* (*RPGRIP*) are the major causes accounting for at least 70% of all X-linked forms of RP.[69] Both RPGR and RPGRIP are expressed in the photoreceptor cilium and are thought to be important for regulating protein trafficking.[70]

Mouse models of X-linked RP (sometimes referred to as RP3) include *RPGRIP* null mice and *RPGR* transgenic mice.[70,71] *RPGR* null mice show gradual photoreceptor loss and retina dysfunction over a 6-month period, that is associated with abnormal accumulation of disc outer segment membranes at the base of photoreceptor outer segments. It is interesting to note that although RPGR is an important protein localized to the cilium, the morphology of the cilium in *RPGR* null mice appears normal. In contrast, *RPGRIP* null mice develop a more severe retinal degeneration, with gradual loss of photoreceptors and retinal dysfunction that is complete by 3 months of age. Although *RPGRIP* null mice develop the full complement of photoreceptors, their photoreceptor disc outer segments appear disorganized,[71] and there is evidence of abnormal trafficking of proteins to the outer segment. In particular, *RPGRIP* null mice demonstrate mislocation of RPGR, suggesting that RPGRIP is important for tethering RPGR within the connecting cilium. Consistent with the notion that RPGRIP is vital for regulating RPGR localization and maintaining normal disc morphogenesis, mice that lack both RPGRIP and RPGR show a similar phenotype to *RPGR* null mice.[71] Moreover, gene replacement of RPGRIP in *RPGRIP* null mice rescues photoreceptors, partially restoring retinal function, and trafficking of RPGR.[72,73]

Canine X-linked progressive retinal atrophy has been reported in Siberian Huskies and maps to the same region of the X chromosome as in humans.[74] Affected dogs are phenotypically variable depending on the type of mutation, with some displaying severe retinal degeneration that arises during development, or degeneration that develops at a later age.

E. Mutations in RPE-Associated Proteins that Affect Photoreceptor Viability

Maintenance of normal photoreceptor function depends on the integrity of a normal functioning RPE.[3,4] Anatomically, photoreceptors sit within the microvillus projections of the RPE.[4] The RPE is a single layer of epithelial cells that serves as part of the blood–retinal barrier that separates the choroid from the retina and regulates the flow of nutrients and metabolites into and out of the retina.[4] It contains transporters for a variety of neuroactive substances, as well as retinoids. It also plays a role in absorbing scattered light (due to melanin granules), creating adhesive forces that maintain retinal attachment to the RPE; and it phagocytoses photoreceptor outer segments important for photoreceptor disc renewal. The importance of the RPE for maintaining retinal integrity is seen in studies that have ablated these cells.[75] Injection of sodium iodoacetate leads to rapid photoreceptor loss.[76] Similarly, genetic ablation of approximately 60–80% RPE cells by crossing transgenic RPE^{CreER} mice with floxed mice carrying a diphtheria toxin A-chain leads to a substantial reduction in photoreceptor function as well as the formation of outer retinal rosettes.[75] Not surprisingly, mutations in genes encoding proteins in the RPE are a cause of photoreceptor degeneration and have been associated with human RP.

The Royal College of Surgeon's rat, and *mer* knockout mouse all display an accumulation of outer segment disc debris between photoreceptors and the RPE prior to a gradual loss of photoreceptors[77–79] (Fig. 5). In the RCS rat, accumulation of debris is evident from postnatal day 18 leading to gradual loss of photoreceptors.[80,81] As shown in Fig. 5, a large region of debris accumulates in the outer retina by postnatal day 60 in RCS rats, at a stage when few photoreceptor nuclei remain. Retinal degeneration in these animals is associated with an inability of the RPE to adequately phagocytose photoreceptor outer segments. A genetic defect in *Mertk*, has been identified as the causative gene defect in RCS rats.[82] *MERTK* encodes a transmembrane protein with an extracellular domain important for binding photoreceptor outer segments and an intracellular tyrosine kinase domain. This genetic defect leads to a failure of RPE phagocytosis of photoreceptor outer segments, a process crucial for renewal of photoreceptor outer segments. Evidence that RPE dysfunction plays a role in photoreceptor loss in the RCS rat comes from the observation that chimeras formed from the combination of normal and RCS rat embryos display localized retinal degeneration only over areas of RCS RPE.[83] Moreover, photoreceptor rescue is observed in animals that have received RPE transplants, or that have received a viral gene transfer of *MERTK*.[84]

It has been assumed that the inner retina remains largely intact following the loss of photoreceptors across a large number of models of retinal degeneration. More recently, Marc *et al.* have shown that there are significant changes throughout the retina that can be described by events in three major phases.[85,86] Phase I and II includes the period during which rods then cones die, whereas Phase III describes the changes that occur in the inner retina after the loss of cones. As shown in Fig. 6, there are plastic changes in the inner retina of RCS rats well after loss of photoreceptors that are characteristic of Phase III remodeling. Phase III remodeling is characterized by migration of

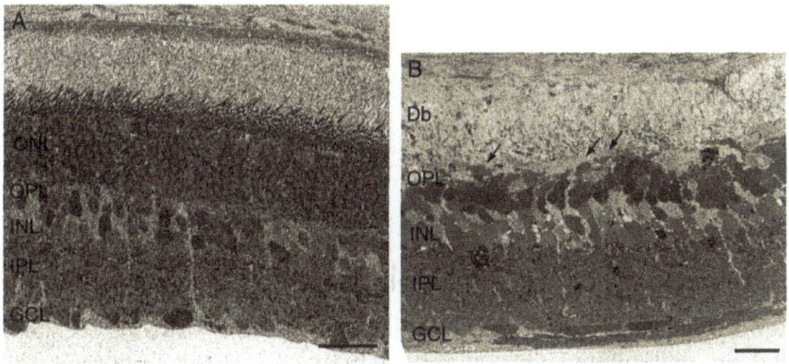

FIG. 5. Vertical sections of postnatal day 60 (A) normal and (B) RCS rat retina that have been immunolabeled for the neurotransmitter glutamate. Glutamate labels all neurons in the rat retina. In contrast to the control retina, there are only a few remaining photoreceptor nuclei by P60 in the RCS rat retina. There is a large amount of debris (Db) in the outer retina that has accumulated over time because of the inability of the RPE to phagocytose the photoreceptor outer segments. Abbreviations as in Fig. 3, Scale bar = 20 μm.

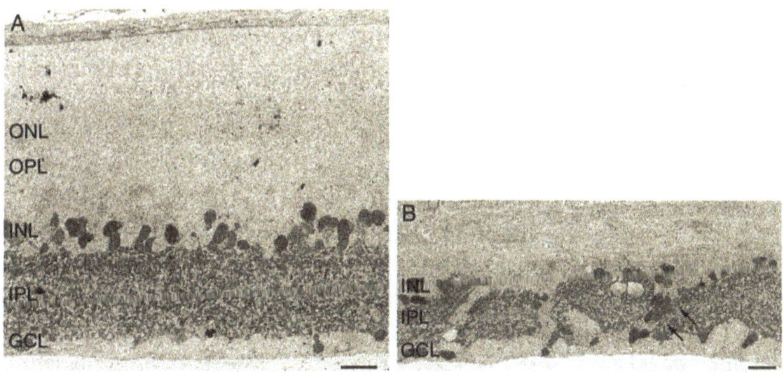

FIG. 6. Vertical sections of 1 year old control and RCS rat retinae that have been immunolabeled for the inhibitory neurotransmitter GABA. In the normal retina, GABA labels amacrine and displaced amacrine cells. In contrast to the normal rat retina, by 1 year of age, the inner retina of the RCS rat shows signs of remodeling. In particular, there are regions where amacrine cells appear to migrate through the inner plexiform layer (arrows). Abbreviations as per Fig. 3. Scale bar = 20 μm.

neurons to new locations, formation of corrupted retinal circuits, the gradual formation of a glial scar that entombs the remaining inner retina and cell death within the inner retina. In order to optimize treatments for those with RP, it will be necessary to better understand how the inner retina changes following loss of photoreceptors.

Mutations in *MERTK* have been identified in some patients with autosomal recessive forms of RP,[87] lending further strength to the importance of this gene in maintaining normal photoreceptor function. In addition, *mer* knockout mice show an almost identical phenotype to the RCS rat, with debris accumulation and loss of photoreceptors because of an inability to adequately phagocytose photoreceptor outer segments.[78]

F. Mutations Affecting Glial Cell Function

Retinal glia, which include both Müller cells and astrocytes are recognized to be vital for maintaining normal retinal function, having important roles in neurotransmitter uptake and recycling, potassium siphoning, shuttling of energy metabolites and maintenance of the blood retinal barrier.[3] There is considerable evidence that glia are altered from an early stage of retinal degeneration.[80,88] In addition, some studies have suggested that photoreceptor death may be exacerbated by glial dysfunction such as by abnormal glutamate uptake and recycling.[80,89]

Müller cells are thought to play an important role in maintaining the structural integrity of the retina. Within the outer retina, Müller cell processes form adherens junctions with neighboring photoreceptor inner segments, to form the outer limiting membrane. Approximately 4% of autosomal recessive RP patients carry a mutation in the *Crumbs homolog 1* (*CRB1*) gene.[90] CRB1 localizes to the outer limiting membrane, a region where glial cells, photoreceptors, and RPE cells interdigitate to form the outer limiting membrane.[91] *CRB1* null mice, lack expression of CRB1 protein and display marked retinal disorganization especially affecting photoreceptors.[92] Localized regions within the retina show rosettes of photoreceptors, attributable to the loss of maintenance of adhesion between photoreceptors and Müller cells.[92] In contrast to *CRB1* null mice, rd8 mice express a truncated CRB1 protein. Disorganization of photoreceptors and the development of rosettes occur in rd8 mice; however, the rate of photoreceptor loss is less severe than in the *CRB1* null mouse. In view of the identification of *CRB1* mutations as a cause of both the severe retinal degeneration, LCA, as well as RP, it has been suggested that *CRB1* null mice represents an animal model of LCA, whereas rd8 mice a model of RP.

G. Animal Models of Syndromic Forms of RP

In rare cases, RP is associated with neurological and/or other systemic anomalies. These forms of RP are referred to as RP syndromes. Below, we describe two of the common families of syndromic RP including Usher's syndrome, and Bardet–Biedl syndrome for which there are animal models.

These two families of diseases highlight that there are common proteins important for maintaining photoreceptor integrity and cell functions in other regions of the body.

Usher's syndrome is the most common cause of deaf-blindness, accounting for 17% of all cases of RP and some 20% of all cases of deaf-blindness.[93] Like other forms of RP, it is a heterogeneous family of diseases, broadly classified into three forms that are distinguished by the extent and age of onset of deafness, extent of retinal degeneration and extent of vestibular dysfunction.[93] Usher Type I patients are profoundly deaf from birth, and have vestibular dysfunction which results in retarded motor development. In addition, Usher Type I is characterized by gradual retinal degeneration from adolescence. Usher Type II is characterized by less severe deafness, and normal vestibular function. Finally, those with Usher Type III have less severe hearing and vestibular dysfunction, but their hearing loss is progressive. Genetically, mutations in any of six genes can cause Usher I (called USH1B-H), whereas a restricted number of mutations are associated with Usher 2 (USH2A or B) and Usher 3 (USH 3A).[93]

Several animal models of Usher's syndrome have been characterized.[93] Spontaneous mutations in the mouse homolog of Ush1 have been identified and include genetic mutations encoding *Myo7a*, *Ush1c*, *Cdh23*, and *Pch15*. These mutant mice, including the shaker 1, deaf circler, waltzer, Ames waltzer, or Jackson shaker mice all develop deafness and vestibular dysfunction. However, none develops retinal degeneration, although some show signs of mild photoreceptor dysfunction, as detected by the ERG.[94,95] Similarly, *Ush1C* null and *Ush1d* mutant mice display profound hearing loss and vestibular dysfunction in the absence of retinal degeneration.[96,97] Recently, a knockin mouse has been created that expresses the human *USH1C* c216 G > A mutation.[98] These mice show signs of profound hearing loss and vestibular dysfunction from birth, and also retinal degeneration. In particular, rod dysfunction was detected from 1 month of age, with a significant reduction in the thickness of the ONL by 12 months of age.

There are some similarities in structure and function of inner hair cells within the cochlear and retinal photoreceptors. Both inner hair cells and photoreceptors are ciliated cells that express common proteins.[99] Photoreceptor outer segments communicate with the remaining parts of the photoreceptor by way of a connecting cilium that is located between the inner and outer segments. The connecting cilium is recognized as an important region for transport of proteins from the inner to outer segment. Stereocilia are located on the apical surface of hair cells within the cochlear and are deflected in response to a sound wave.[99] Gene mutations associated with Usher's syndrome encode proteins, such as Myosin VIIa, harmonin, Cdh23, that are expressed by both stereocilia of inner hair cells and cilia within photoreceptors.[100] For example, myosin VIIa is located within the connecting cilium of photoreceptors and has been shown to be important for transport of opsin from the photoreceptor inner segment to outer segment.[95] Similarly, myosin VIIa is thought to be important for transport of proteins to their intended destination within inner hair cells.[100]

Bardet–Biedl syndrome is another severe RP syndrome that is associated with mutations encoding proteins important for the function of cilia. Bardet–Biedl syndrome is a systemic condition characterized by dysfunction in a variety of body systems, characterized by obesity, retinopathy, polydactyly, renal disorders, cardiac malformations, cognitive disabilities, and hypogenitalism. It is caused by mutations in one of 14 genes, that all encode proteins expressed by ciliated cells, such as photoreceptors, inner hair cells, lung epithelia, olfactory neurons, kidney cells to name a few. The gene mutations causing Bardet–Biedl syndrome all encode proteins (called Bbs proteins) localized in the basal body and ciliary axoneme.[101] Although Bardet–Biedl syndrome shows Mendelian inheritance patterns, it also shows complex traits. There is considerable phenotypic variation, and at least one Single Nucleotide Polymorphism is associated with worsening vision in some patients.

There are a variety of mice in which Bbs genes have been modified. These mice share many of the phenotypic features of human disease. Notably, *Bbs1*-, *Bbs2*-, *Bbs4*-, and *Bbs6*-null mice and well as a knockin *Bbs1* model all show gradual photoreceptor loss and retinal dysfunction within 6–8 weeks of age.[102–104] In addition, *Bbs*-null mice show aberrant transport of proteins important for phototransduction to photoreceptor outer segments.[103] They also display a variety of systemic signs related to anomalies in cilia function in other body systems.[102]

H. Variation in Disease Severity Across Different Models of Retinal Degeneration

A common feature of RP, is the variability in disease presentation that can occur even in members of the same family.[105] Individuals within a family who have inherited a particular gene mutation can display a number of disease phenotypes from the nonprogressive CSNB to autosomal dominant RP. This heterogeneity is likely to arise from gene products that interact with the expression of opsin and/or alterations in environmental conditions.

There is considerable evidence that prolonged exposure to moderate levels of light causes photoreceptor loss in wild-type animals.[106–108] There is considerable heterogeneity in response to light across different strains of mice; C57Bl6 mice are resistant to light induced retinal degeneration, whereas Balb/c mice display rapid loss of photoreceptors when exposed to the same level of light.[109] Wenzel *et al.* showed that the susceptibility to light damage correlated with the kinetics of rhodopsin regeneration and the availability of rhodopsin during light exposure. An essential protein in the rhodopsin regeneration cycle is RPE65. This protein isomerases all-*trans*-retinal to 11-*cis*-retinal and is a crucial step in the recycling of retinoids important for visual function. Wenzel *et al.*[109] showed that sequence variations in RPE65 confer resistance or susceptibility to light, and was the basis for the difference in light sensitivity across three strains of mice.

Light exposure is known to accelerate photoreceptor loss in rodents expressing rhodopsin mutations.[108,110] Indeed, there are substantial differences in the rate of photoreceptor loss in animals reared in the dark compared to those exposed to normal lighting.[110] Moreover, variations in RPE65 modify the severity of photoreceptor loss in the VPP mouse model of retinal degeneration.[39] Notably, exposure to light even of short duration has been linked with accelerated photoreceptor loss in English Mastiff dogs, carrying a rhodopsin mutation.[111] In these animals, photoreceptor loss and retinal thinning within 4 weeks was observed in regions that had been exposed to 60-s duration light flash from a conventional fundus camera (total energy exposure of retina was 60 mJ cm^{-2}). Moreover, when the light intensity was reduced 10-fold, photoreceptor loss and retinal thinning proceeded more slowly. A similar sensitivity to light has been shown in Pro23His rats.[108]

The role that light plays in accelerating photoreceptor loss in animal models of inherited retinal degenerations has been attributed to increased phototransduction, and the burden this places on photoreceptors. Hao *et al.*[112] examined the role that different steps in the phototransduction cascade play in modulating light induced retinal degeneration. They examined the extent of photoreceptor loss in transgenic mice that either (i) lacked expression of rhodopsin kinase and arrestin; these mice display sustained levels of activated rhodopsin or (ii) lacked expression of the α-subunit of transducin (Gnat1), the key G-protein that mediates the signaling between activated rhodopsin and activation of PDE in rods. They showed that light induced photoreceptor loss was dependent on activation of rhodopsin, and independent of transducin activation.

Level of pigmentation is another important contributor to severity of retinal degeneration. In particular, VPP mice on an albino background display greater more rapid photoreceptor loss than VPP mice on a pigmented background.[113] Moreover, this difference in severity developed irrespective of light exposure. Albinism is associated with recessive inheritance of a mutation in tyrosinase, an enzyme important for melanin trafficking. These studies highlight that the presence of a single gene mutation alone does not entirely explain the severity of photoreceptor loss, and that other gene variations, especially affecting RPE function and retinoid recycling are important determinants in the severity of disease.

V. LCA: A Severe Retinal Degeneration Caused by Anomalies in RPE, Glial or Photoreceptor Dysfunction

LCA refers to a family of inherited retinal degenerations that cause visual impairment before the age of 1 year.[114] It is characterized by severe and early vision loss, sensory nystagmus, amaurotic pupils, and absent retinal function as tested by electroretinography. It is recognized as the most severe form of retinal degeneration. To date, it is associated with mutations in 14 genes, many of which also cause various types of RP. The genes that cause LCA encode a variety of proteins important for maintenance of normal function of photoreceptors. Like RP, mutations in genes such as Crb1 affect Müller cell function and normal developmental of retinal structure; other mutations affect RPE function (e.g., *RPE65*, *LRAT*, *MERTK*) and finally, LCA is associated with mutations in genes that directly affect photoreceptor function (e.g., *CRX*, *RPGRIP*, *CEP290*).[50] Mutations in some of these genes (e.g., *CRB1*, *MERTK*, *RPGRIP*) also cause less severe forms of retinal degeneration including RP. The reason for the difference in phenotypic severity is likely to reflect the importance of the mutant region for protein function or the level of expression. Studies on animal models of LCA have been instrumental in the development of the first gene therapy trials in humans to treat retinal disease.[115]

The most widely studied animal models of LCA are those associated with RPE65 mutations.[115] RPE65 is an important enzyme in the retinoid cycle and is highly expressed in the RPE.[116] For continuing operation of photoreceptors, a source of 11-*cis*-retinal is necessary. Two pathways are known to regulate the supply of 11-*cis*-retinal to rods and cones, and both depend on retinal isomerase to enzymatically convert all-*trans*-retinal to 11-*cis*-retinal. The first pathway involves transport of all-*trans*-retinal from photoreceptors to the RPE where conversion takes place via a series of steps including isomerization via RPE65. The second pathway involves transport and conversion to 11-*cis* within Müller cells via an as yet unidentified retinal isomerase.[117] Mutations in RPE65 cause a severe reduction in the availability of all-*trans*-retinal, and consequent photoreceptor dysfunction and loss.[115,118]

Three mouse models have been identified that carry mutations in RPE65, including the natural mutant rd12 mouse model,[119] *RPE65* null mouse[120] and the *RPE65R91W* transgenic mouse.[121] *RPE65* null and rd12 mice demonstrate a slow loss of photoreceptors, with the photoreceptor layer thinning by approximately 50% by 12 months of age. It is interesting to note, however, that cones are preferentially affected in *RPE65* null mice, with almost complete loss of short wavelength cones from the inferior retina within 1 month. *RPE65R91W* knockin mice were generated to express one of the common missense mutations observed in patients with LCA. *RPE65R91W* mice show partial expression of mutant

RPE65 and 10% generation of 11-*cis*-retinal. Despite differences in the level of RPE65 expression compared with rd12 and *RPE65* null mice, the phenotype is remarkably similar, showing gradual loss of photoreceptors over 1 year.

The Briard sheep dog is known to carry an autosomal recessive form of LCA caused by a *RPE65* mutation.[122,123] These animals are homozygous for a mutation in *RPE65*, and retain normal vision for 1.5 years and show complete degeneration of photoreceptors in the peripheral retina by 5–7 years of age, depending on the strain. Like mouse strains with RPE65 mutations, retinal function is severely affected in Briard dogs from an early age.

It is interesting to note that in comparison to other mouse models (e.g., rd1) of retinal degeneration, *RPE65* mutant mice show slow and very late photoreceptor degeneration. Indeed, there is a disproportionate loss of retinal function compared with the degree of photoreceptor death.[119,120] This is mostly likely explained by the cell targeted by the gene mutation-namely the RPE. As mentioned earlier, RPE65 is highly expressed by the RPE and is crucial for retinoid recycling, in particular replenishing photoreceptor levels of 11-*cis*-retinal. One might therefore predict that photoreceptor dysfunction would occur from an early age well before photoreceptor death. Retinal function, as measured by the ERG, shows severe reductions in all three strains of *RPE65* mutant mice[119,120]; rd12 display severely reduced a- and b-waves from 3 weeks of age, well before the loss of photoreceptors.[120] The disproportionate loss of retinal function in comparison to photoreceptor loss has important implications for treatments.

Gene therapy for LCA associated with RPE65 mutations has been developed over the last decade.[115] Gene therapy using AAV adenovirus and lentiviral vectors have been performed in *RPE65* null and rd12 mice. Subretinal injection of human RPE65 cDNA in *RPE65* null mice showed restoration of retinal function,[124] even when trialed in 2-year-old mice.[125] Moreover, retinal isomerase activity and 11-*cis*-retinal was found to be comparable to wild-type mice following subretinal injection of adenoviral vector in 2-week-old RPE65 null mice.[126] Similarly, subretinal delivery of human RPE65 to rd12 mice at 2 weeks of age resulted in photoreceptor rescue, restored retinal function and improvements in visually guided behavior in 6.5-month-old mice.[19]

A crucial step in the development of gene therapy for LCA was the successful treatment of Briard sheep dogs in 2001.[18] Recombinant canine RPE65 was injected either intravitreally or subretinally at approximately 4 months of age and visual function tested 3 months later.[18] In a number of studies that followed it was shown that gene therapy in dogs restored RPE65 expression to RPE cells, restored retinal function, and prevented photoreceptor loss.[115,127,128]

Since 2007, three independent clinical trials of *RPE65* gene therapy in 18 patients with LCA have been conducted and have confirmed that subretinal injection of RPE65 cDNA is generally safe and without toxic side effects.[14,15,129,130] Moreover, encouraging improvements in perception of light, visual acuity, and reduced nystagmus have been reported in these studies, despite all the patients entering the study with very severe retinal degeneration.[14,15,129]

More recently, attention has turned to large animals that develop retinal degeneration because of the need to test treatment regimes on eyes of a more similar size and structure to humans. In this regard, a severe autosomal recessive retinal dysplasia has been described in cats. Cats, especially Abyssinian cats, can inherit a recessive mutation in *CEP290*, a protein that localizes to the cilium of both rods and cones, and that complexes with RPGRIP. Mutations in *CEP290* are known to mice (rd16), causing rapid photoreceptor loss within 6 weeks.[131] Similarly, affected cats display gradual photoreceptor loss and blindness within 1 year.[132,133] The recognition and availability of retinal degeneration in cats is attractive for evaluation of surgical treatments such as the implantation of electronic devices because the cat eye is of a similar size to a human eye, has an area centralis, similar retinal circuitry, and a great deal is known about visual processing in cats.

VI. CSNB: An Opportunity to Better Understand Outer Retinal Signaling

In contrast to the inherited retinal degenerations described above, some mutations affecting photoreceptor associated proteins can cause loss of rod function, in the absence of cell death. CSNB is a genetically and clinically heterogenous nonprogressive, inherited disorder of the retina that affects the visual function of patients. As the name suggests, inability to see at night (nyctalopia) is the most common symptom and is often picked up in children as a fear of the dark. Other symptoms that may be associated with the disorder include reduced visual acuity, myopia, nystagmus, and strabismus. By studying animal models of CSNB, a more detailed understanding of outer retinal processing has been achieved.

The clinical diagnosis of CSNB has generally been based on a finding of night blindness and an abnormal rod ERG that does not progressively worsen with age. CSNB is further classified into complete CSNB (cCSNB) or incomplete CSNB (icCSNB) based on whether photoreceptoral dysfunction or postreceptoral dysfunction is identified.

A. Complete CSNB

cCSNB is characterized by a scotopic ERG with larger than normal a-wave but complete absence of response from second order neurons, suggesting that photoreceptors hyperpolarize normally in response to light but that the postphotoreceptor response driven by second order neurons is absent. Genetic transmission of cCSNB has been found to be X-linked recessive and autosomal recessive. Many of the X-linked recessive cCSNB patients have been found to have a defect in the gene *NYX*, which encodes the protein nyctalopin.[134,135] Autosomal recessive cCSNB has been found to occur as a result of defects in the genes encoding the metabotropic glutamate receptor mGluR6 (*GRM6 gene*)[136,137] and the recently defined cation channel that the mGluR6 receptor gates, the Transient Receptor Potential Cation Channel, Subfamily M, Member 1 (TRPM1) channel.[138–140]

The first and only mouse model of X-linked recessive cCSNB was described in a naturally occurring mutant, the *nob* mouse.[141] The scotopic ERG of these mice is characteristic of the human form cCSNB, with a normal a-wave but no b-wave and so these mice were called *nob* (no b-wave) mutants. In addition, the gross and ultrastructural morphology of the retina of these mice was characteristic of CSNB with no loss of photoreceptors over time.[141,142] In 2003, the gene underlying the defect in these mutant mice was determined to be the *nxy* gene (*nyx*^{nob}), which encodes nyctalopin and is the same genetic fault that occurs in the human form.[143]

Three mouse models of autosomal recessive cCSNB due to genetic alteration in the *grm6* gene have been described.[144–146] The first of these was a true knockout (*grm6*^{Tm1Nak}) and demonstrated that *grm6* was the main glutamate receptor on ON bipolar cells, allowing them to depolarize in response to light.[145] The other two mouse models (*grm6*^{nob3} and *grm6*^{nob4}) were generated at Jackson laboratories from an N-ethyl-N-nitrosourea (ENU) mutagenesis study in C57Bl6 mice and were discovered from an ERG screen. These mice produce *grm6* mRNA but do not express the grm6 protein.[144,146] All of the *grm6* mice have a scotopic ERG that is characteristic of the human form of cCSNB, with a normal a-wave but no b-wave. In addition, the gross morphology of the retina of these mice is normal.

The role of mutations of the TRPM1 channel as a cause of autosomal recessive cCSNB has only recently been demonstrated.[147] Prior to this the cation channel gated by the mGluR6 on ON bipolar cells was unknown. In 2008, the TRPM1 channel was found to be the underlying defect in CSNB in Appoloosa horses and interestingly was also demonstrated to be the reason for their unusual coat spotting pattern.[147] In 2009, the first reports of TRPM1 knockout mice were published.[148,149] The ERG phenotype of *Trpm1* null mice is similar to that observed in humans with cCSNB, with a normal rod a-wave but no b-wave.[138–140] Like human cCSNB, the *Trpm1* null mice have normal retinal morphology.[148] Furthermore, the visual acuity and contrast sensitivity of the *Trpm1* null mice is reduced slightly, similar to that observed in human cCSNB. The role of the TRPM1 channel in the ON pathway has been confirmed by patch-clamp recordings from bipolar cells in the *Trpm1* null mice.

OFF bipolar cells are responsive to application of glutamate analogs, while the ON bipolar cell responses are severely hampered in the *Trpm1* null mice,[148] consistent with the vital role that TRPM1 plays in mediating ON bipolar cell activity.

B. Incomplete CSNB

icCSNB patients characteristically have a scotopic ERG with relatively normal a-wave and reduced, but not absent, responses from second order neurons.[150,151] Like cCSNB, genetic transmission of icCSNB has been found to be X-linked recessive and autosomal recessive. By far the most common genetic defect that leads to icCSNB is an X-linked recessive defect in the calcium channel gene, *CACNA1F*.[134,152,153] This gene encodes the voltage-gated L-type calcium channel (VGCC) alpha1F subunit (Cav1.4α), which is required to trigger calcium-dependent exocytosis of glutamatergic vesicles from photoreceptor terminals.[154] The α_{1F} subunit is critical to producing a VGCC on photoreceptor terminals that does not inactivate during sustained depolarization.[155] Another defect in calcium signaling in the photoreceptor terminal has also been implicated in icCSNB. Mutations in the gene *CABP4*, which encodes the protein, calcium-binding protein CABP4, have been found to result in autosomal recessive icCSNB.[156,157] CAPB4 is directly associated with the C-terminal domain of the Cav1.4α and acts to alter the voltage at which the calcium channel is open.[158] These proteins are crucial in mediating synaptic transmission of glutamate from photoreceptors.

Two mouse models of X-linked icCSNB due to mutations in *Cacna1f* have been described.[159,160] One is a knockout generated by an early stop codon in Exon 7, *Cacna1f*^{G305X}[160] and the other is a spontaneous mutant with an out-of-frame insertion, which produces an early stop codon in Exon 2, discovered in the colonies of Jackson labs, *Cacna1f*^{nob2}.[159] Both of these mouse models have icCSNB, and replicate the human phenotype however there are slight differences in the ERG phenotype between the models that may represent some of the phenotype heterogeneity between different people with the disorder. The *Cacna1f*^{G305X} mice have a scotopic ERG with a marginally reduced a-wave and absent postreceptoral responses. The photopic ERG and visual acuity response is also absent.[160,161] In contrast, the *Cacna1f*^{nob2} mice have a scotopic ERG with relatively normal a-wave and reduced, not absent, responses from second order neurons.[159] In addition the photopic ERG driven by the cone pathway is also present but greatly reduced[159] and they also have normal photopic visual acuity.[161] The reason for this difference appears to be that due to alternative splicing of the *Cacna1f* gene, such that in the *Cacna1f*^{nob2} mice there is still around 10% of functional Cav1.4α protein expressed.[161] One common feature of both the *Cacna1f*^{G305X} and the *Cacna1f*^{nob2} mouse models which has not been investigated in the human is the stark anatomical changes that occur in the retina of these mutants.[159,160] In both mice, the rod bipolar cells and horizontal cells extend their dendrites through the OPL and into the ONL to form ectopic synapses with the photoreceptors. However, the OFF bipolar cell dendrites are not altered.[159] Studies using these mice have shown that the Cav1.4α subunit is the primary VGCC α_1 subunit involved in mediating tonic glutamate release from rods and cones. In addition, these results indicate that the Cav1.4α and the resulting activity of the VGCC, is required for normal synapse formation and maintenance between photoreceptors and ON bipolar cells in the OPL.

Mutations in the gene *CABP4*, which encodes the protein, calcium-binding protein CABP4, have been found to result in autosomal recessive icCSNB.[156,157] A CaBP4 knockout mouse (*Cabp4* null) was generated and found to be a reasonable model of autosomal recessive CSNB prior to the finding that this mutation could lead to CSNB in humans.[158] Indeed, the role of the *CABP4*, protein in visual signaling was determined from these mice.[158] The ERG phenotype of these mice mirrors that observed in the human disorder. *Cabp4* null mice have a relatively normal scotopic a-wave, although it is slightly reduced, and a scotopic b-wave that is greatly reduced. The photopic ERG is also significantly reduced in these mice.[158] Like the *Cacna1f* knockout mice, the *Cabp4* null mice have abnormal retinal morphology, in that the dendrites of the rod bipolar cells and horizontal cells extend into the ONL, forming ectopic synapses with photoreceptors.[158] Overall, these studies suggest that CAPB4 may play a role in modulating the voltage at which the calcium channel, Cav1.4α, is open and that like Cav1.4α, CAPB4 is required for normal synapse formation in the OPL.

C. CSNB Caused by Alterations in Proteins Required for Phototransduction

In some patients with CSNB, the disorder is caused by a mutation in one of the proteins required for phototransduction in rod photoreceptors including rhodopsin, the transducin-α subunit GNAT1, rhodopsin kinase, and arrestin. As mentioned above, inheritance of mutations in genes encoding these proteins can also cause RP, or LCA.

Mutations in rhodopsin are associated with autosomal dominant CSNB and occur in the region important for activation of the G-protein, transducin.[162–168] Computer and *in vitro* modeling of these mutant rhodopsin proteins show that the region where transducin binds remains constitutively active, even when the rhodopsin molecule is not activated by light, but only at a low, pseudo "physiological" level. This level of activity is just enough for the patients to never fully dark adapt but is not enough to induce retinal degeneration.

A transgenic mouse model of one of these human rhodopsin mutations that lead to autosomal dominant CSNB due to constitutive activation of transducin has been generated.[164,167,169] These mice have a substitution of aspartic acid for glycine at position 90 in the rhodopsin protein (Rho^{G90D}) and were created such that the ratio of transgenic opsin to endogenous opsin could be varied.[164,169] In mice heterozygous for both Rho^{G90D} (+/−) and Rho^{WT} (+/−), retinal morphology is maintained up until at least 1 year, with no apparent loss of photoreceptors.[169]

Mutations in transducin, especially the α-subunit, GNAT1 can lead to CSNB.[170–173] Photoreceptor function in these patients is associated with inappropriate activation of the phototransduction cascade. One of these families, the Nougaret family, with a point mutation in the GTP-binding region, is one of the largest, with at least 12 generations recorded, and earliest recorded pedigrees of a dominantly inherited trait.[173] Indeed, CSNB as an autosomal dominant inherited trait has been known as the Nougaret form of CSNB. A mouse model of the Nougaret family mutation has been generated with a glycine substitution at amino acid 38 of the rod transducin-α subunit ($T\alpha^{G38D}$).[171] Mice heterozygous for the G38D mutation have an ERG phenotype similar to that of the Nougaret family, with reduced rod sensitivity and impaired phototransduction recovery. In addition the heterozygous $T\alpha^{G38D}$ mice display no apparent signs of retinal degeneration at 5 months of age.[171] It is through studies in these mice that it has been determined that the Nougaret rod transducin-α subunit mutation results in impaired GTP hydrolysis, as might be expected considering the point mutation occurs in the GTP-binding region.

In summary, CSNB is associated with a variety of gene mutations encoding proteins affecting photoreceptor to bipolar cell signaling. Studying animal models of has CSNB provided a unique opportunity to study retinal signaling, as well as to better understand how these mutations affect the retina.

VII. Inherited Macular Degenerations

The macula is a highly specialized region of the retina that provides high visual acuity central vision. It is defined by the region that contains the pigment, xanthophyll, beneath the retina. The fovea which is found at the center of the macula is highly specialized for maximizing visual acuity, consisting of a foveal pit, where the inner retinal layers are shifted to the side of a small depression. This region contains the highest density of cones, in particular red and green cones, and lacks rod photoreceptors. This highly specialized region is also avascular. In view of the high density of photoreceptors, this region of the retina has the highest metabolic demand. There is an increase in RPE cell density in the macula compared with more peripheral regions, and the underlying choriocapillaris is structurally different to more peripheral regions. In view of its importance for central vision, diseases that affect this specialized region can cause profound loss of central vision, while leaving the peripheral retina relatively unaffected. As will be described below, diseases of the macula can be inherited in a Mendelian fashion, or develop as a consequence of aging. Although rodents do not possess a macula, information about the effect that specific gene mutations have on the retina can be studied with the use of transgenic mice carrying the same mutations as in human inherited macular degenerations.

A. Stargardt's Disease

Stargardt's disease is an inherited form of macular degeneration affecting approximately 1 in 10,000 people, that manifests as loss of central vision and progressive degeneration of the macula.[174] It is most commonly an autosomal recessively inherited condition, although autosomal dominant Stargardt's disease has been reported and is associated with a later onset condition.[175] The RPE is typically affected in Stargardt's disease, showing signs of accumulation of fluorescent lipofuscin-like deposits.[174] Thus, the pathogenesis of Stargardt's disease has been thought to involve a failure of phagocytosis of photoreceptor outer segments leading to accumulation of lipofuscin. More recently, genetic mutational analysis has revealed that both autosomal recessive and autosomal dominant forms of Stargardt's disease are associated with mutations affecting the expression of proteins important in lipid transport.[176,177]

Up to 50 mutations in the *ABCA4* gene, a member of the ABC transporter family, have been attributed to autosomal recessive Stargardt's disease.[176] Like other members of the ABC family, ABCA transporters utilize ATP to transport a variety of substrates across cell membranes. The *ABCA4* gene encodes the rim protein RmP, a transporter exclusively localized within the rim region of rod outer segment discs, and is thought to be important for transport of vitamin A derivates such as N-retinylidene-phosphatidylethanolamine across the rod outer segment membrane.[176,178]

ABCR null mice are a commonly used mouse model of autosomal recessive Stargardt's disease.[176,178] These mice show delayed dark adaptation, abnormal accumulation of all-*trans*-retinol, abnormal accumulation of the phospholipid, phosphatidylethanolamine, within outer segments and evidence of increased lipofuscin in RPE cells.[178,179] However, these mice show very little evidence of photoreceptor loss, even following 44 weeks of age.[178,180]

Autosomal dominant inherited Stargardt's like disease leads to gradual loss of central vision from the second decade of life.[174,175] It has been associated with mutations in genes encoding very long chain fatty acids (*ELOV-4*). Transgenic mice carrying a mutant form of human *ELOV4* develop RPE changes, show reduced photoreceptor outer segment phagocytosis and debris accumulation within the subretinal space from 2 months of age.[181] In addition *ELOV4* mice show an increased number of lipofuscin granule within the RPE.[181]

Although a great deal has been learned about the pathogenesis of Stargardt's disease from the use of transgenic mouse models, the type of information gained is limited by the lack of a macula and the low cone to rod ratio. However, while the dog retina does not possess a specialized fovea, there is a central region that is associated with a high density of cones. Recently, retinal degeneration associated with an *ABCA4* mutation has been identified in pit bull terriers.[182] Affected animals develop rapid loss of cones followed by rods from between 3 and 6 months of age, making this a more severe disease than other canine progressive retinal atrophies. More work is needed to better characterize this large animal model of inherited cone-rod dystrophy.

B. Sorsby Fundus Dystrophy

Transport of retinoid from the blood stream to the retina is recognized as of vital importance for maintaining normal photoreceptor function. Indeed, vitamin deficiency is a well-known cause of photoreceptor dysfunction and loss. Sorsby Fundus Dystrophy is a rare autosomal dominant macular degeneration characterized by the accumulation of protein and lipid within Bruch's membrane, a thin connective tissue layer between the RPE and underlying choroid.[183] The condition is associated with rapid loss of central vision followed by loss of peripheral vision, and ultimately blindness. Despite its rarity, this disease has received some attention because its clinical features are similar to those of AMD. Notably, in addition to deposits within Bruch's membrane, patients with Sorsby Fundus Dystrophy also develop subretinal neovascularization and atrophy of the RPE and choroid. Affected individuals carry mutations in *TIMP3*, a gene encoding the tissue inhibitor of metalloproteinase 3, an important protein for regulating extracellular matrix turnover.[184]

The pathogenesis of Sorsby Fundus Dystrophy has been greatly enhanced with the development of a transgenic mouse that expressed a Ser156Cys mutation in murine *TIMP3*.[185] These mice display abnormalities within the RPE from 8 months of age, and an increase in thickness of Bruch's membrane from 30 months. In contrast to patients with Sorsby Fundus Dystrophy, mice carrying the Ser156Cys mutation in *TIMP3* did not display aberrant blood vessel growth into the retina. More recently, *TIMP3* null mice have been generated.[186] The thickness of Bruch's membrane and RPE structure generally was unaffected in these mice over an 18 month study period. However, TIMP3null mice showed significantly enlarged blood vessels within the choroid.

C. X-linked Juvenile Retinoschisis

X-linked juvenile retinoschisis is a common form of inherited macular degeneration in men.[187] It is characterized by a splitting of the retina at the level of the nerve fiber layer and ganglion cell layers, resulting in the formation of large vacuoles within the central retina and photoreceptor degeneration. Vision loss varies from complete blindness to moderate visual impairment. The gene causing retinoschisis is called *retinoschisin*, and produces a 24-kDa polypeptide that is secreted by photoreceptors and bipolar cells.[188,189] In view of the importance of Müller cells in maintaining normal retinal layering, it was originally thought that retinoschisis was caused by a defect in retinoschisin expression by Müller cells.[190] Although retinoschisin is found throughout the retina, mRNA for *retinoschisin* is located only within photoreceptor inner segments.[191]

The role that retinoschisin plays in modulating retinal structure and function has been determined by the generation of three different mouse models that lack expression of retinoschisin (*Rs1h*).[192] These mice display marked splitting of the inner nuclear layer, disorganization of the retinal layers, and subsequently degeneration of photoreceptors, especially cones.[192–194] Detailed analysis of changes in synapse associated proteins in *Rs1h* null mice revealed aberrant transport of both MAGUK and PSD95 to the outer and inner plexiform layers.[192] In addition, morphology of ribbon synapses was severely affected.[192] These results highlight that the synapses of the outer and inner plexiform layer are crucial for maintaining normal retinal structure, and that when synaptic connections are impaired, photoreceptor death develops.

VIII. Age-Related Macular Degeneration

AMD represents the leading cause of adult-onset blindness, affecting 1.5% of all people of Caucasian background over the age of 50, and 10% over 75 years of age.[195–197] In order to understand the mechanism of AMD, the contributions that the choroid, Bruch's membrane, RPE, and photoreceptors make to the disease needs to be understood. As shown in Fig. 1, the RPE forms a single layer of cells between the choroid and the photoreceptors. The basement membrane on which the RPE sits is called Bruch's membrane and in humans consists of 5 layers with the inner most layer being the basement membrane of the RPE, then a collagen layer, an elastic fiber layer, another collagen layer then the outer most layer is the vascular endothelium of the choriocapillaris. These four cell types are all known to change with age. Bruch's membrane increases in thickness with age, there is loss of elastic fibers within Bruch's membrane, the collagen cross-links and there is a reduction in RPE cell density.

In its early stages, the pathology associated with AMD is mostly found in the RPE and Bruch's membrane. The RPE accumulates lipofuscin, and shows evidence of partially digested photoreceptor outer segments. In addition, the number of RPE cells in the macula decreases. At the ultrastructural level, deposits are observed between the RPE and Bruch's membrane, called basal laminar deposits and basal linear deposits. It has been suggested that the formation of basal linear deposits is suggestive of progressing AMD, whereas the formation of basal laminar deposits is an indicator of RPE and photoreceptor degeneration.

The second major change that occurs in the early stages of AMD is an increased thickness of Bruch's membrane and drusen formation. Drusen are subretinal deposits composed of a large number of different components including acute phase proteins, lipid, b-amyloid, and complement proteins. In view of the importance of oxygen and nutrient transport across Bruch's membrane, a change in structure can have deleterious effects on photoreceptor integrity.

Clinically, patients can progress to one or both forms of advanced AMD.[198] Dry AMD is called geographic atrophy and is characterized by RPE atrophy and subsequent photoreceptor degeneration. It is notable that rod photoreceptors are initially affected, followed by cones. Approximately 10% of patients develop the Wet form of AMD which is associated with pathological growth of blood vessels from the choroid into the macula. Although the underlying mechanism for angiogenesis in AMD is complex, it is now well accepted that the angiogenic growth factor, vascular endothelial growth factor (VEGF), plays a key role.

A great deal has been learned about the factors that lead to many of the signs of AMD. For example, a great deal has been learned about the mechanisms of photoreceptor death from studies of retinal degeneration using the animal models described earlier. In addition, animal models of inherited macular degenerations have been important for understanding the contribution that identified gene mutations have in AMD. However, to date there is a paucity of animal models that show all the hallmark features of AMD progression.

Development of animal models of AMD has been challenging for a number of reasons. First, commonly used laboratory animals, such as mice, do not possess a macula, nor live for many years. Moreover, there are no reported cases of the development of an AMD-like condition in large animals like cats or dogs. There are also some significant differences in the structure and function of the mouse retina compared with humans. Notably, mice have fewer cones (3% in mice compared with 20% in humans). In addition, the transport of lipids across the RPE appears to be quite distinct. In humans, the transport of lysosomes and phagosomes proceeds across the RPE toward Bruch's membrane. Indeed, drusen may develop as a consequence of secretion of debris from the basal surface of the RPE. In contrast, lysosomes accumulate in the apical region of mouse RPE cells, with very little transport from the basal surface.[199] This has implications for the development of drusen, which in mice are rarely seen even in aged genetically modified mice.[200] Studying AMD in mice is also difficult because of the complex genetics that has been implicated in human AMD, and the strong environmental component. Nevertheless, much can be learned about the impact individual gene mutations have on retinal integrity with the use of transgenic mice carrying the same genetic mutations as those in humans. Below we summarize the retinal changes in transgenic mice that express similar genetic mutations to those with AMD.

A. Animal Models with Single Gene Mutations

The contribution of both genetics and the environment to the development of AMD is well known.[201] Twin studies show that genetic factors contribute to 46–71% of the overall variation in severity of AMD.[202] Concordance is 55% in monozygotic twins and only 25% in dizygotic twins. More recently, single nucleotide polymorphisms in several genes, especially complement factor H (CFH), have been linked with increased risk of developing severe AMD. Using animal models, a great deal has been learned about individual steps in the pathogenesis of AMD. For example, using the variety of models of retinal degeneration, a tremendous amount is now known about the mechanisms involved in photoreceptor death. Similarly, models of the inherited macular degenerations have aided our understanding of the role that RPE dysfunction plays in retinal dysfunction. More recently, animal models have been developed that carry mutations in similar genes to humans affected by AMD. The development of these newer animal models has aided our understanding of the role of isolated gene products in the disease process.

1. Animals with Mutations in Genes Affecting RPE Function

RPE dysfunction may play a role in the development of AMD, especially in its early stages.[203,204] With aging, RPE cells show signs of abnormal phagocytosis of photoreceptor outer segments, including build up of outer segment breakdown products. In addition, lipofuscin, a sign of RPE stress increases with age and AMD, and drusen and/or lipid deposition at the level of Bruch's membrane are all hallmarks of the disease. Recently, a homozygous mcd/mcd mouse has been described that shows signs of early AMD.[205] These mice carry a mutation in cathepsin D, an important protease for lysosomal digestion by the RPE. These mice show drusenoid lesions from 9 months of age when examined by ophthalmoscopy. In addition the RPE shows accumulation of lipofuscin, and outer segment debris. Signs of basal linear and basal laminar deposits and retinal dysfunction as measured by the ERG are observed by 12 months of age.

A second mouse has been generated that has additional deletions in the cathepsin D cleavage site.[206] Mcd2 mice cannot form active cathepsin D and develop earlier retinal degeneration that mcd/mcd mice.[206] Indeed, mcd2 mice show loss of photoreceptors from 3 months of age in concert with changes in Bruch's membrane and the formation of basal laminar deposits. These mouse models highlight the importance of normal RPE function for maintaining photoreceptor viability. But, neither animal model is reported to develop choroidal neovascularization, suggesting that other pathways may be important contributors to the development of more severe disease.

2. Animals with Mutations Affecting the Immune System

Over recent years there has been recognition that polymorphisms in genes encoding proteins important in regulating the immune response can substantially increase the risk of developing AMD. In 2005, a major risk allele, the Y402H polymorphism in CFH was reported, and is associated with approximately 50% risk.[207–209] Notably, the identification of this CFH polymorphism initiated studies investigating dysregulation of the complement pathway in the pathogenesis of AMD, and led to a view that inflammation plays an important part in the overall disease mechanism.

The complement system is an important component of the innate immune system. CFH is thought to negatively regulate the alternative complement pathway by inhibiting factor I-mediated inactivation of C3b. In particular, impaired CFH leads to an over activity of the alternative pathway, leading to excessive deposition of complement within tissues and subsequent injury.

Several transgenic mouse models of AMD that carry mutations in genes encoding proteins important for several components of the immune system have been developed. CFH null mice show deposition of C3a glomerular basement membrane of the kidney limiting their life expectancy. In animals that survive to 2 years, there is evidence of structural and functional anomalies in the retina with age.[210] In particular, CFH null mice show reduced visual acuity, reduced photoreceptoral function in 2-year-old mice. In addition, small autofluorescent spots are visible in the temporal retina, there was increased C3 deposition in the RPE and Bruch's membrane, and evidence of photoreceptor degeneration and mild inner retinal modeling is present.[210] Notably, there was a reduction in Bruch's membrane thickness and a reduction in the accumulation of lipid material at the level of the RPE/Bruch's membrane. These data suggest that although CFH is important for maintaining the long term integrity of the retina in mice, there was no evidence of progression of AMD even over a 2-year period. Moreover, lack of CFH in mice was associated with a phenotype quite distinct to those with a high risk CFH polymorphism.[210]

A second group of animal models have been developed that carry mutations affecting chemokines or chemokine receptors.[211,212] monocyte chemoattractant protein-1 (MCP-1) or Chemokine 2 (CCl2) is a chemokine that is important for monocyte signaling and recruitment.[213] It is known to be important in mediating adhesion of inflammatory cells to blood vessels, controlling extravasation.[214] It is upregulated early in response to light induced photoreceptor degeneration. However, polymorphisms in the genes controlling CCl2 expression are not associated with increased risk of developing AMD. Transgenic mice that lack expression of MCP-1 or its receptor, CCR2 have been reported to develop some of the cardinal features of AMD.[211] By fundoscopy, Ambati *et al.*[211] reported an increasing number of drusenoid like lesions in MCP-1 null and CCR2 null mice with age. In addition, MCP-null mice showed lipofuscin accumulation—thickening of Bruch's membrane, photoreceptor degeneration, and increased C3 deposition at the level of the RPE and Bruch's membrane. Importantly, approximately 25% of MCP-1-null mice progressed and developed choroidal neovascularization.[211]

More recently, the use of the MCP-1-null mouse as a model of AMD has been re-evaluated.[200] Using confocal scanning laser ophthalmoscopy, autofluorescent subretinal lesions were noted in both C57bl6 and CCl2-null mice. However, despite an increase with advancing age, there was no difference in the number of these subretinal deposits with age. In a careful comparison of the CCl2-null mice with age matched controls, few differences could be attributed to the lack of chemokine.[200] Notably, there were age related losses in photoreceptors, increased subretinal microglia and changes in the ultrastructure of Bruch's membrane. But no difference between the two strains was established, suggesting that the CCl2-null mice have limited value as a model of AMD.

Another family of chemokines receptor CX3CR1 has been recently linked with AMD.[212] CX3CR1 is the receptor for the chemokine, fractalkine, and is expressed by microglia.[7] CX3CR1 immunoreactive cells have been identified in the outer retina of those with AMD, close to drusen, and also within the drusen themselves.[212] Moreover, a T280M polymorphism in *C3XCR1* reduces monocyte chemotaxis and is associated with increased risk of developing AMD.[212,215] Mice that lack expression of CX3CR1 are known to develop neurodegeneration more rapidly, suggesting that fractalkine may be an important chemokine for limiting neuronal loss.[216] Retinae of aged *CX3CR1* null mice show progressive accumulation of microglia in the subretinal space that was accompanied by drusenoid lesions.[212] In addition, the outer nuclear layer of *CX3CR1* null is reduced, and they show a greater sensitivity to oxidative stress induced by exposure to light. These results suggest that subretinal accumulation of microglia may be an important event in the development of AMD.

It is interesting to note that when *CCL2* null and *CX3CR1* null mice are crossed to produce a *CCl2/CX3CR1* double knockout mouse, accelerated changes in the retina have been observed.[217] Notably, these double knockout mice show AMD like lesions from 6 weeks of age including drusen, lipofuscin accumulation within the RPE, an increase in Bruch's membrane thickness and photoreceptor loss. In addition, approximately 15% of mice develop choroidal neovascularization.[174,217]

3. Animals with Mutations Affecting Oxidative Stress

The retina has the highest oxygen demand of the body. It also has a high concentration of polyunsaturated fatty acids rendering it vulnerable to damage from oxidative stress. Oxidative stress refers to the cellular and molecular damage caused by reactive oxygen species, which are often byproducts of oxygen metabolism. One of the main anti-oxidant systems in the retina is made up of three superoxide dismutase isoenzymes (SOD1-3). SOD1 exists in the cytosol, mn-SOD2 in the mitochondrial matrix and extracellular SOD3 exists as a secretory form. SOD1 has the highest activity of any SOD in the retina, and has therefore been investigated for its possible role in AMD.[218] *SOD1* null mice show some key signs of human AMD from 9 months of age, including drusen, thickened Bruch's membrane and approximately 8% developed choroidal neovascularization.[219] Moreover, exposure to moderate levels of light increased the number of drusen like deposits in young *SOD1* null mice. These results implicate oxidative stress in the development of AMD like signs in mice.[219]

4. Animals with Mutations Affecting Lipid Synthesis

Anomalies in lipid transport across the RPE, and cholesterol deposition in Bruch's membrane are hallmarks of early AMD. Indeed, cholesterol, its transporter apoE, and other lipid components are major constituents of drusen[220] and also basal laminar deposits.[221] The APOE gene is known to encode three common alleles including APOEε2, APOEε3, and APOEε4, of which APOEε3 is the most common in Caucasians. Association studies have suggested that APOEε4 carriers are at lower risk of developing AMD, whereas those carrying APOEε2 are at greater risk.[221] Transgenic APOEε3-Leiden mice expressing the human APOEε3-Leiden gene which produces a dysfunctional form of human APOEE3.[222] When these mice are fed a diet rich in fat and cholesterol for 9 months they develop basal laminar deposits, but no drusen. Development of retinal changes in *ApoE* null mice on the other hand is controversial, with one study showing no change even when fed a high cholesterol diet for 9 months,[222] while another showed thickening of Bruch's membrane and an increase in membrane bound material to Bruch's membrane in *ApoE* null mice.[223] More recently, transgenic *ApoE* knockin mice have been created that express human APOEε2, APOEε3, or APOEε4.[224] Retinal pathology was only observed in older transgenic mice that had been raised on a high fat diet.[224] In particular, mice expressing human APOEε4 and fed a diet high in fat developed the most severe pathology, including Bruch's membrane thickening, and soft drusenoid deposits. Notably, approximately 18% of aged APOEε4 containing mice developed choroidal neovascularization.[224] These studies highlight that a combination of diet and genetic influences may be important in the development of AMD.

B. Animal Models of Choroidal Neovascularization

Choroidal neovascularization (CNV) is a devastating sequelae in those with AMD. As was described earlier, only a small number of transgenic mice with signs of AMD progress to the point of developing choroidal neovascularization. Thus, few studies have been performed specifically examining CNV in animal models of AMD such as *CCl2* null, *CX3CR1*null, *CX3CR1/CCl2* double knockout mice. In contrast, a great deal has been learned about the mechanism of CNV with the use of the laser induced model of choroidal neovascularization.[225] In this model, pigmented mice (e.g., C57Bl6) mice receive 4–6 laser burns from a photocoagulation laser. These laser burns have enough energy to induce thermal damage within the outer retina and RPE. Notably they also cause defects within Bruch's membrane. Within 7 days, vascular leakage and subsequent choroidal neovascularization develops in response to upregulation of VEGF expression. This model is a well-established model of choroidal neovascularization and has been used extensively to characterize the factors important in the development of CNV and has been instrumental in the development of anti-VEGF treatments.

IX. Animal Models of Retinal Vascular Diseases

Diseases of the retinal vasculature, which include conditions like retinopathy of prematurity (ROP) and DR are leading causes of blindness in both infants and those of working age. Vascular pathology, in particular pathological angiogenesis, is recognized as the main visually detectable sign of change in these two conditions. In addition to retinal vascular pathology, neuronal, and glial cell anomalies also develop, often prior to the development of overt vascular disease.[226]

A. Diabetic Retinopathy

DR is one of the most feared complications of diabetes mellitus, and is a major contributor to vision loss in the Western world. Recent studies indicate that virtually all patients with Type I (insulin dependent) and over 60% of those with Type II (noninsulin) diabetes will develop some form of DR.[227–230] Clinically, the earliest signs of DR include breakdown of the blood retinal barrier, loss of pericytes, and the formation of microaneuryms.[228] As the disease progresses, an increasing number of hemorrhages become apparent, as well as other signs of retinal ischemia including the presence of cotton wool spots. Proliferative DR is associated with growth of new blood vessels on the surface of the retina. Vision loss can occur from vitreous hemorrhaging or from tractional retinal detachment[228] that develops in late stages of the disease or from macular oedema.[228] In addition to the progressive vascular pathology that occurs in those with diabetes, there is considerable evidence that retinal neurons and glia are altered, often prior to the development of even the earliest vascular changes.[226,231,232]

Currently no rodent model of diabetes demonstrates the progressive vascular pathology in the same way as human disease. Hyperglycemia is commonly induced in animals by treatment with streptozotocin, a chemotherapy agent that selectively kills β-cells of the pancreas. In the rat STZ model, many early vascular changes have been reported as well as alterations in retinal function and in glial cells.[226] An increase in the expression of the angiogenic growth factor VEGF occurs within 1 week of onset of diabetes[233] and evidence for vascular leukostasis and breakdown of the blood retinal barrier occurs.[234,235] Accompanying these subtle vascular changes, evidence for retinal dysfunction and early glial changes has been reported from as early as 2 days following STZ injection.[236] However, even following 2 years of diabetes, only minor changes in capillaries are observed.[237] Therefore, it is generally accepted that the STZ rat model is a useful model for examining only the very earliest stages of DR.

A major secondary risk factor for progression of DR is systemic hypertension. Indeed, control of systemic blood pressure is known to reduce the progression of DR.[238] In view of the importance of hypertension in exacerbating DR, careful examination of the vascular, glial, and neuronal changes has been made in several models of hypertensive diabetes. Spontaneously hypertensive rats (SHR) have been rendered diabetic by a single injection of STZ and display greater levels of vascular pathology compared with normotensive controls, including thickening of the vascular basement membrane, loss of pericytes and increased permeability to albumin after 20 weeks of diabetes.[239] Notably, vascular compromise is prevented or reduced by anti-hypertensive treatments such as captopril.[239] Moreover, elevated expression of VEGF and its receptor VEGF-R2 have been reported in diabetic SHR retinae compared with normotensive diabetic control rats,[240] and oxidative stress has been shown to be worse in diabetic SHR rats compared with controls. However, like other rodent models of DR, no evidence for pathological growth of blood vessels on the surface of the retina has been reported, even after considerable duration of hyperglycemia.

The renin-angiotensin system (RAS) is well known for its role in controlling systemic blood pressure.[241] In addition, there is evidence that the posterior eye has its own independent RAS.[241,242] There is evidence that angiotensin II can act as an angiogenic growth factor, and is elevated in vascular diseases such as DR and ROP. The transgenic m(Ren-2)27 rat has been used as a model to study background DR and other complications related to over expression of components of the RAS.[241–243]

The transgenic m(Ren-2)27 rat have elevated renin and Ang II levels in extra-renal sites including the eye because of the insertion of the murine *Ren-2* gene into the genome.[244] When made diabetic, these rats display severe diabetic kidney disease after only 3 months of diabetes that has many features of human diabetic nephropathy.[245–247] In the retina, there is an increase in the number of acellular capillaries compared with wild-type rats, and also evidence of endothelial replication, and increased VEGF and VEGFR-2 expression.[243] Significant alterations in retinal function have also been noted.[248] As shown in Fig. 7, nondiabetic Ren2 rats show a reduction in amplitude of the a- and b-wave, as well as the Oscillatory Potentials and cone b-wave. However, there is an even greater loss of retinal function following 4 weeks of diabetes in Ren2 rats, a reduction that is considerably larger than in nonhypertensive diabetic animals.

Glial cells have been implicated in the development of retinal dysfunction during diabetes. One of the principal functions of Müller cells is the uptake and degradation of the inhibitory neurotransmitter GABA. Indeed, following

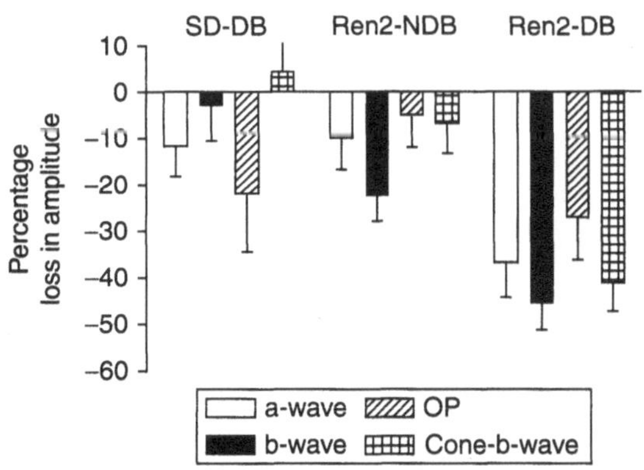

FIG. 7. Graph showing the percentage loss of amplitude in various ERG waveforms in diabetic Sprague Dawley rats, nondiabetic Ren-2 and diabetic Ren2 rats that have been diabetic for 4 weeks. Loss of photoreceptor function (a-wave) and oscillatory potentials is known to occur from an early stage of diabetes. Although systemic hypertension causes a reduction in retinal function, the combination of diabetes and hypertension significantly comprises retinal function within 4 weeks in diabetic Ren2 rats.

release of GABA from neurons, it is taken up via high affinity GABA transporters into Müller cells where it forms succinate, a key component of the citric acid cycle. In view of the link between GABA uptake and metabolism in retinal Müller cells, the level of GABA accumulation within glia can be used as a marker of glial cell dysfunction. As shown in Fig. 8, following 20 weeks of diabetes, there is a significant increase in the level of GABA immunoreactivity in Müller cell somata compared to nondiabetic Ren2 controls or nonhypertensive, nondiabetic control animals.

The use of genetically altered mice to aid our understanding of the pathogenesis of DR has been limited, largely because of the resistance of mice to the development of vascular pathology. Recently, the $Ins2^{Akita}$ mouse model of diabetes has been developed. These mice carry a point mutation in the *insulin 2* gene on chromosome 7, which leads to gradual β-cell death within the

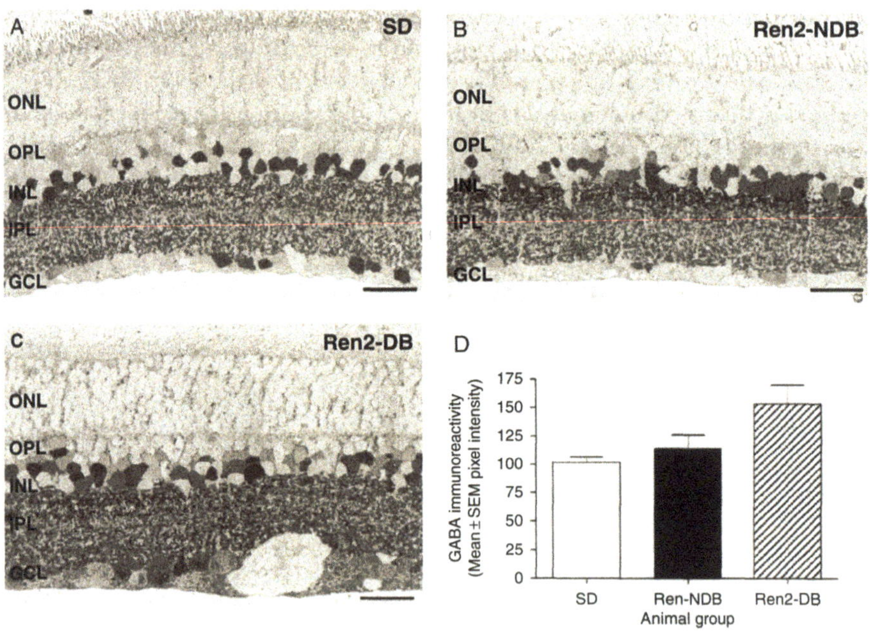

FIG. 8. Vertical sections of (A) control Sprague Dawley, (B) nondiabetic Ren2, and (C) diabetic Ren2 rats than have been diabetic for 20 weeks that have been immunolabeled for GABA. Müller cells are recognized as the major cell type important for removal and degradation of GABA following neural activity. A sign of Müller cell dysfunction is the accumulation of GABA within Müller cells, which is evident in diabetic Ren2 rats. (D) A graph of the mean + SEM GABA immunolabeling of Müller cell somata in control, nondiabetic Ren2 and diabetic Ren2 rats following 20 weeks. There is a statistically significant increase in GABA immunolabeling of Müller cell somata suggestive of Müller cell dysfunction. Abbreviations as per Fig. 3. Scale bar = 20 μm.

pancreas and spontaneous diabetes from 4 weeks of age. Careful evaluation of the retina has revealed early changes in the vasculature, including leukostasis and breakdown of the blood retinal barrier from 12 weeks of age.[249] In addition, loss of specific classes of amacrine cells has been reported.

The recent observations that *eNOS* null mice develop many of the hallmark signs of vascular pathology of DR make this transgenic mouse also a promising animal for future studies.[250] Diabetic *eNOS* null mice show severe increased vascular permeability within 3 weeks of the onset of diabetes, and also an increase in the development of acellular capillaries.[250]

Large animals are known to develop diabetes spontaneously, and this has enabled a limited number of studies to be performed examining the vascular pathology in diabetes (e.g., monkey).[11,251] The development of a more systematic large animal model of diabetes has only been achieved by inducing diabetes with alloxan or feeding beagles a diet rich in galactose. By inducing galactosemia with a diet rich in galactose, a series of biochemical changes are induced that induce vascular pathology similar to that seen in diabetes.[252,253] Galactose fed beagles develop vascular compromise within 3 years, that have similar features to background retinopathy, and evidence of capillary drop out and retinal ischemia following longer periods.[252] In addition, a small number of galactose fed beagles develop signs of proliferative retinopathy following extended periods of time on the galactose rich diet.[254]

B. Animal Models of Retinal Angiogenesis

Pathological growth of blood vessels is a significant problem in a number of retinal diseases including DR, ROP and retinal occlusive disease. Studying the mechanisms involved in proliferative DR has been hampered by the limited progression that rodent models of diabetes show. Thus, alternative models of retinal angiogenesis have been developed including inducing retinal capillary closure, studying models of oxygen induced retinopathy (OIR), or examining blood vessel changes in transgenic mice that overexpress the main angiogenic growth factor, VEGF.

1. Porcine Model of Selective Retinal Capillary Closure

Retinal hypoxia is recognized as a major stimulus for the aberrant growth of inner retinal blood vessels.[11,255] In experimental animals, techniques used to induce retinal ischemia/hypoxia have commonly affected the whole thickness of the retina by reducing retinal oxygenation from both the retinal and choroidal circulations. These methods include elevation of intraocular pressure above systemic blood pressure,[256–258] occlusion of carotid and/or vertebral arteries,[259,260] ligature of the ophthalmic artery or central retinal artery,[261–263] endovascular balloon tamponade of the ophthalmic and/or ciliary arteries[264] or systemic hypoxia induced by breathing an oxygen/nitrogen mixture with a low oxygen content.[265,266]

Most of these methods cause total retinal ischemia/hypoxia and not surprisingly, total retinal ischemia results in damage to all retinal layers including ganglion cell damage (or death) in the inner retina, mid-retinal cell damage to Müller cells and structural abnormalities in the RPE and photoreceptors.[267] These widespread abnormalities in all retinal layers do not resemble the pathological changes associated with retinal capillary closure commonly seen in the early stages of DR in which capillary closure affects the retinal capillaries of the inner and mid-retina. A new porcine model of selective retinal capillary embolisation, has been developed by Foulds and co-workers.[268]

The pig is a suitable species for the development of a survival model of inner retinal hypoxia/ischemia, because it is similar in size to the human eye, has both rods and cones and a pseudofovea (visual streak).[269] The pig retina like the human retina has a dual blood supply, the inner to mid-retina being supplied by superficial and deep capillary vascular beds and the outer retina from the choriocapillaris. However, unlike the human, the pig eye receives its blood supply from the external carotid artery and not the internal carotid.[269]

In the porcine model selective retinal capillary closure, microspheres of 10 microns diameter are injected into the common carotid artery via a cannula. Microspheres can be identified immediately after injection as small bright well-defined dots indicating fluorescence of microspheres in the retina and larger less well-defined spots indicating the presence of numerous microspheres in the choroid (Fig. 9). Fluorescein angiogram shows that in some areas microspheres appear to be located within individual capillaries with no significant downstream closure of the capillary bed while in other areas there appear to be closure of terminal precapillary arterioles with downstream capillary closure (Fig. 9B).

The combined rod-cone bright flash ERG responses show a significant reduction in the amplitude of the b-wave but not of the a-wave amplitude and thus a reduction of the b/a-wave amplitude ratio in the embolized eye (Fig. 9C). The oscillatory potentials are also decreased in amplitudes, in the embolized eye. These findings suggest that hypoxia induced by embolization is restricted to the inner and mid-retina and provides a useful model of retinal capillary closure that mimics the retinal conditions of many eye diseases including DR.

2. Animal Models of ROP

ROP is a major cause of visual impairment in children born prematurely, affecting approximately 68% of children born weighing less than 1250 g.[270] ROP is considered a vascular disease, caused by excessive growth of blood

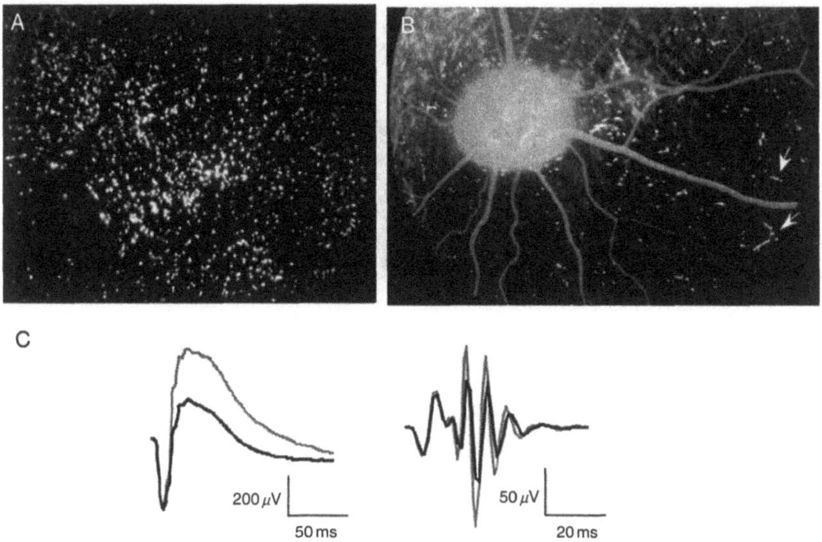

FIG. 9. (A) Fundus photograph immediately after delivery of microspheres shows numerous small bright well-defined dots indicating fluorescence of microspheres in the retina. (B) Fluorescein angiogram showing closure of terminal capillaries (arrows). (C) Combined rod-cone bright flash ERG responses and the oscillatory potentials of the embolized (black traces) and control eyes (gray traces). The b/a-wave amplitude ratio and the amplitude of the oscillatory potentials are significantly reduced in the embolized eye.

vessels on the surface of the retina in response to the high levels of oxygen used for critical care of neonates (especially those with very low birth weights and an immature retinae). Current treatments target the abnormal retinal angiogenesis. However despite treatment, many children suffer ongoing vision impairment. Rodent and feline models of ROP, which are more correctly referred to as OIR, have been developed and provide information relating to the pathogenesis of disease and also the underlying mechanisms of retinal angiogenesis.[271]

There are a number of experimental paradigms used to develop rodent models of OIR, differing in the mode of high oxygen delivery to newborn or infant rodents.[271] A commonly used model of rat OIR involves placing a newborn rat into a high oxygen environment (~80% oxygen) from birth until postnatal day 11.[271–273] This is the period of time when the superficial vasculature is developing in the rodent. Exposure to high oxygen during this developmental period attenuates the normal growth of vessels, leaving regions within the peripheral retina avascular. As shown in Fig. 10, there are significantly fewer blood vessels in the outer and inner retina in rats that have received OIR

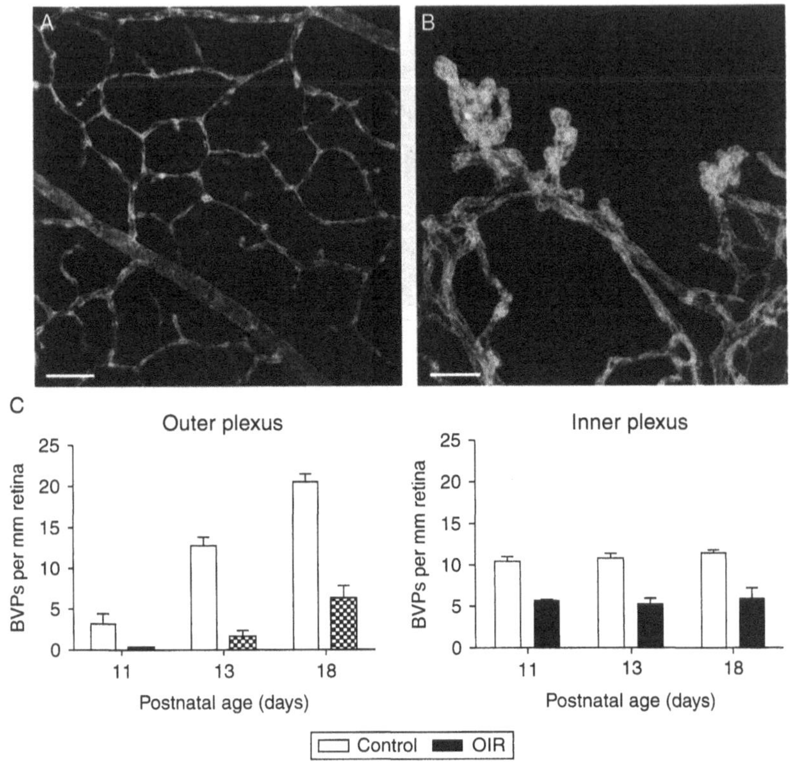

FIG. 10. Flatmounts of the superficial vascular plexus of a (A) P18 control rat retina and (B) a P18 rat retina that received oxygen induced retinopathy. OIR in the rat leads to the development of an avascular peripheral retina, and aberrant vessel growth extending from fronds in the mid peripheral retina (arrows). (C) Graph showing the number of blood vessel profiles (BVPs) in the peripheral deep and inner plexus at P11, P13, and P18, when animals are exposed to normal air. At all three ages, there is a significant reduction in the number of blood vessels in the peripheral retina following OIR. Aberrant vessel growth on the surface of the retina most likely develops in response to an increase in angiogenic growth factor expression in this region of retina.

compared to controls. From postnatal day 11, rodents are placed in a normal air environment until the conclusion of the study, usually around postnatal day 18, at the peak of angiogenic growth on the surface of the retina. It is well known that during the second normal oxygen phase, aberrant blood vessel growth occurs in response to the upregulation of angiogenic growth factors such as VEGF and IGF-1. At postnatal day 18, aberrant fronds form from the edge of the vasculature in the midperipheral/peripheral retina.

This rat model of OIR shares many similarities to human ROP, including the development of a peripheral avascular zone, and subsequent pathological angiogenesis into the periphery.[271] It has been argued that oxygen saturation of a premature infant is not always constantly at a high level because of respiratory distress and immaturity. Consequently, some authors have suggested that a model involving cyclic delivery of high and low oxygen is more representative of the neonatal environment. In this model, newborn pups are exposed to alternating levels of 50% or 10% oxygen.

A commonly used mouse models of OIR involves placing a postnatal day 5 mouse pup in 75% oxygen until postnatal day 11. Pathological angiogenesis occurs within 5 days. Using this model, a great deal is now known about the pathways and mechanisms of angiogenesis, because transgenic and knockout mice have been treated using this experimental paradigm. However, it should be noted that the location of the vascular changes that occur in the mouse model of OIR do not mirror the human condition. In particular, unlike the human and rat models, the vascular changes in the mouse model of OIR do not represent a developmental arrest, but rather regression of the already developed vasculature. Consequently, pathological angiogenesis occurs within the central and midperipheral retina (Fig. 10).

Recent studies suggest that in addition to the aberrant growth of blood vessels on the surface of the retina, there are significant alterations in retinal neurons and glia.[274–276] Astrocytes, which are normally located in the nerve fiber layer, and form a template over which blood vessel develop, are lost in the peripheral retina rat and cat OIR.[276,277] It has been suggested that in the absence of the astrocyte template, blood vessel growth occurs aberrantly on the surface of the retina at the expense normal growth into the deeper layers of the retina. Moreover, agents that prevent the loss of astrocytes are not only associated with reductions in angiogenesis, but promote re-vascularization of the retina[276,277] (Fig. 11).

3. Transgenic Mice Overexpressing Angiogenic Growth Factors

It is well accepted that the angiogenic growth factor, VEGF, is a key mediator of angiogenesis in the retina, and is a key stimulus for choroidal neovascularization in AMD as well as inner retinal neovascularization observed in DR.[225]

Several transgenic mice have been developed that involve the overexpression of VEGF. Subretinal injection of a recombinant adeno-associated virus VEGF construct into normal C57Bl6 mice has been shown to lead to the development of microaneurysms and increased vascular leakage in a small number of eyes.[278] In addition, transgenic trVEGF029 mice express human $VEGF_{165}$ driven by the rhodopsin promoter and show mild vascular changes with many similarities to those seen in early DR.[279] trVEGF029 mice show an early mild elevation of VEGF that normalizes by 6 weeks. This early spike in VEGF expression, however, is sufficient to induce increased vascular leakage, some microaneurysms, pericyte loss and the formation of acellular capillaries.

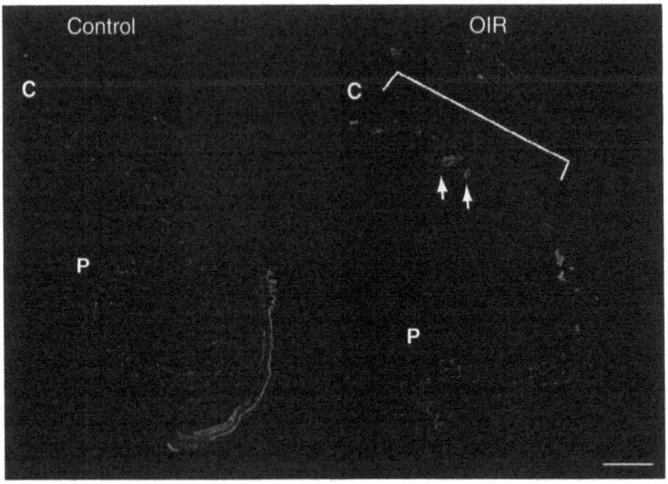

FIG. 11. Low power vertical sections of a control P18 mouse retina and a P18 mouse retina that has received oxygen induced retinopathy that have been immunolabeled to identify changes in blood vessels (green). In the normal mouse retina there are two main plexuses of blood vessels-a superficial layer located in the ganglion cell layer and a deep plexus in the Outer plexiform layer. In OIR mice, the central retina is rendered avascular (White line) and there is aberrant vessel growth in to the vitreous (white arrows). Scale bar = 200 μm. (See Color Insert.)

However, because VEGF expression is driven by the rhodopsin promoter, aberrant vascular development occurs from postnatal day 7 when the first rods are forming.[280] In particular, vascular pathology in both the deep and superficial vascular plexus occurs, which is in contrast to the vascular pathology that develops in diabetes. Overall, this mouse may prove a useful model of retinal angiogenesis for testing novel treatment approaches.

X. Animal Models of Optic Nerve Disease

A. Inheritable Optic Neuropathies

The inherited optic neuropathies are a group of diseases that are characterized by optic nerve dysfunction, primarily due to alterations in ganglion cell function and ultimately cell death. These disease pathologies exhibit hereditability based on familial expression or genetic analysis, however, their clinical phenotype is somewhat heterogeneous both within and between the various disease states. Despite this, the end point of the respective diseases is similar, exhibiting optic disc atrophy due to ganglion cell loss. The primary hereditary optic neuropathies consist of Leber's hereditary optic neuropathy (LHON) and the autosomal dominant, autosomal recessive and X-linked recessive forms of optic atrophy. The two most well detailed optic neuropathies, namely LHON and autosomal dominant optic atrophy (DOA) will be discussed.

1. Leber's Hereditary Optic Neuropathy

LHON is characterized by a rapid, painless loss of vision in one eye, with the second eye usually following weeks to months later.[281] Vision loss typically occurs between 15 and 35 years, and involves central vision. Fundus examination during the acute/subacute period often reveals circumpapillary changes in blood vessels, swelling of the nerve fiber layer, and an absence of fluorescein leakage.[282] Axonal loss, demyelination and optic nerve atrophy, in addition to a pale optic disc is often observed later in the disease process. The presence and severity of these clinical features may vary significantly between affected individuals and there is evidence of an age-dependent reversal of vision loss.

LHON is transmitted via maternal, mitochondrial inheritance.[283] While there have been several mutations described, three point mutations are believed to cause the majority of the cases of LHON and are thus termed the "primary" mutations (Fig. 12). They are located at positions 11778 (G11778A,

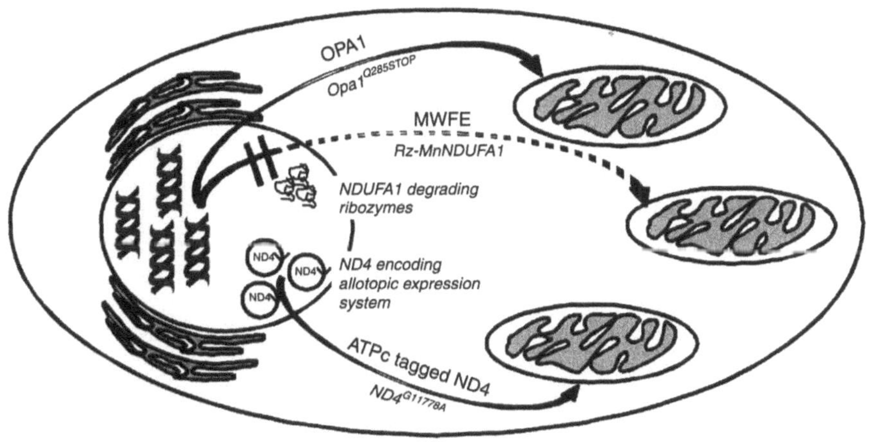

Fig. 12. Diagrammatic representation of the current strategies used to model diseases involving mitochondrial proteins. The inheritable optic neuropathies such as LHON and DOA involve mutations in the mitochondrial proteins OPA1 and ND4, respectively. In the case of LHON, the ND4 subunit of complex 1 is expressed within the mitochondria making it difficult to introduce this mutation into an animal model. To overcome this, researchers have either targeted another subunit of complex 1, one which is encoded by the cell nucleus (*NDUFA1*) or allotopically expressed the mutated ND4 protein within the nucleus and then imported it into the mitochondria (ATPc tagged ND4). Producing the model for DOA is more straight forward since OPA1 is normally expressed within the nucleus before being transported into the mitochondria.

69% of cases), 3460 (G3460A, 13% of cases), and 14484 (T14484C, 14% of cases) of the mitochondrial DNA[281,284] and target specific subunits[285–287] within complex I of the oxidative phosphorylation system, which provides most of the energy to eukaryotic cells. While these mutations have been shown to affect the activity of complex I to varying degrees,[288,289] factors such as additional mitochondrial mutations, mitochondrial haplotype, nuclear modifying agent(s), and environmental factors have been suggested to modulate overall phenotype. The only feature that appears to be mutation-dependent is the recovery of vision that is more frequent in those individuals with the 14484 mutation.[290,291]

The mitochondrial nature of LHON makes the generation of a true animal model somewhat problematic. While it is easier to target nuclear genes that encode for mitochondrial proteins, the specific LHON mutations are found within the mitochondrial genome. To date, there is no murine model which incorporates the specific mutation within the mitochondrial genome. To overcome these difficulties specialist neuronal cybrid (cytoplasmic hybrid) cell lines carrying specific LHON mutations have been used to model the disease *in vitro*, while injections of specific compounds (chloramphenicol[292,293]) and dietary manipulations (folate deficiency and methanol toxicity[292,294]) have been used to produce LHON-like effects in retinal structure. More recently, the delivery of adeno-associated vectors expressing allotopic mutations or ribozymes targeting mitochondrial proteins have been designed in an attempt to produce a true model of LHON (Fig. 12).

a. Rz-MnNDUFA1 Model. As mentioned mitochondrial proteins arise both from nuclear and mitochondrial genomes. Complex I, which is the mitochondrial protein complex affected in LHON, is made up of only seven subunits derived from mitochondrial DNA, while the remaining 35 subunits are nuclear-encoded proteins. Qi and colleagues took advantage of this fact to produce a model in which the gene responsible for the complex I nuclear-encoded protein, MWFE, was knocked down[295] (Fig. 12). Using a ribozyme (Rz) specifically designed to degrade the mRNA encoding the *NDUFA1* gene (which encodes for MWFE), the authors were able to specifically target complex I activity. *In vitro* delivery of the Rz-*NDUFA1* construct results in reduced *NDUFA1* expression (−64%), and an 81% decrease in complex I activity.[295] Intravitreal delivery of an AAV Rz-*NDUFA1* construct results in pathological retinal changes similar to that seen in LHON patients. Four months after delivery, the optic nerve head shows mild edema, due to swelling and mitochondrial accumulation within the axons, while the postretinal nerve exhibits axonal loss and demyelination. This axonal loss is reflected in a progressive loss of ganglion cells (18% and > 50% at 4 and 6 months postinjection, respectively), while the progressive nature of the pathology is also reflected in the loss of myelin (33% and 50% for 4 and 6 months, respectively). Ultrastructurally, mitochondria display a swollen and distended structure with dissolution of cristae.[295] While these changes are similar to that described for LHON, it must be emphasized that the targeting of the MWFE subunit is unlike the molecular alterations observed in LHON patients, where the ND4 subunit of complex I is typically affected.

b. Rz-MnSOD Model. Due to the alterations in complex 1 observed in LHON patients, alterations in respiratory chain function and reactive oxygen species are thought to play a critical role in the pathogenesis of this disease. Using a similar Rz strategy as above, Qi and colleagues targeted the nuclear expression of mitochondrial superoxide dismutase (SOD), resulting in the knocking down of mitochondrial *SOD2* gene expression.[296] *In vitro* assessment showed this construct to reduce SOD protein expression and decrease mitochondrial membrane potential, while increasing superoxide production and eventually resulting in apoptotic cell death. *In vivo* delivery of the virus associated Rz-Mn*SOD* construct intravitreally produces almost identical optic nerve/ganglion cell pathology to that described in the Rz-Mn*NDUFA1* model above. Postretinal optic nerve shows progressive axonal loss and demyelination between 4 and 6 months postinjection. The extent of demyelination is identical to that observed in the Rz-Mn*NDUFA1* model (33% after 4 months, 50% after 6 months), while there is a progressive loss of ganglion cells (27% and 54% after 4 and 6 months, respectively). Similarly there is an aberrant ultrastructure, with mitochondrial exhibiting swollen, distended structures in addition to loss of cristae.[296] While the Rz-Mn*SOD* model has a similar phenotype to the Rz-Mn*NDUFA1* model, there are some model-specific traits, with the optic nerve head exhibiting transient swelling at 6 months, and some mild inflammation after 6 months. Although this model does not harbor any of the mutations found in LHON patients, the similarities in retinal/optic nerve pathology suggest that the production of reactive oxygen species may be a major mechanism in LHON.

c. Allotopic ND4 Model. As highlighted, targeting mutations in mitochondrially expressed subunits of complex I to produce useful animal models of LHON is problematic. The previous two model systems used one approach, that being to mutate other mitochondrial proteins which are encoded by the nuclear DNA. Another method is to use the nuclear expression machinery to encode a gene normally expressed by the mitochondrial DNA and then import that protein into the mitochondria. In 2007, Qi *et al.* used such an allotopic expression system to produce an animal model for LHON (Fig. 12). A mitochondrial targeting sequence (ATPc) was added to the N-terminal of a nuclear ND4 sequence which had been mutated to introduce the major LHON mutation (G11778A, 69% of cases). Despite previous reports highlighting problems with ND4 mitochondrial targeting, *in vitro* studies showed the mutated ND4 protein was expressed within the mitochondria. Delivery of this construct results in reduced cell survival and mitochondrial membrane potential, while there is an increase in the production of reactive oxygen species, leading to apoptosis. *In vivo* delivery of the allotopically expressed ND4, results in a transient swelling of the optic nerve head within 1 month of injection. There is a progressive loss of postretinal nerve axons (17% by 6 months) in addition to a decrease in myelin area (35% at 6 months postinjection). Ganglion cell numbers are found to decrease at 6 months (~−33%), while mitochondria within the nerve show proliferation and are distended, with a smaller number of cristae present. The delivery of the mutant ND4 also results in a threefold increase in reactive oxygen species within the optic nerve, supporting a role for oxidation in the development of LHON.

d. Rotenone Model. Rotenone is a naturally occurring plant-derived pesticide that is a potent and specific inhibitor of mitochondrial complex I, binding to the ND1 subunit. In retinal cell lines, this compound has been shown to activate the integrated stress response in oligodendroglia, increase oxidative damage and interrupt glutamate transport in retinal cell cultures. In 2002, Zhang *et al.*[297] suggested that intravitreal injection of rotenone could provide a model system for optic neuropathies such as LHON. In this study, injection of rotenone produces a rapid reduction (≤ 1 h, approx. − 29%) in central ganglion cell layer thickness, that remains 48 hr postinjection (approx. − 19%). In addition, a transient reduction in inner plexiform layer thickness occurs 24 h after the injection. While the ganglion cell nature of this affect was likened to LHON, there have been no other studies investigating other similarities between the *in vivo* rotenone-induced effects and LHON.

2. Autosomal DOA

Autosomal DOA (also known as Kjer's optic atrophy) is the most common form of the hereditary optic neuropathies. Onset usually occurs within the first decade of life; however, the patient is generally unaware of the visual defect, since it is generally mild in nature.[281] DOA shows a slow progressive reduction in central vision; however, this is variable with 19–50% of patients exhibiting deterioration after follow up.[298] Patients also experience a generalized color defect, with optic disc pallor prevalent on the temporal side. Ultimately, the optic nerve shows axonal swelling, demyelination, and ganglion cell loss, similar to that observed in LHON.

The majority of DOA patients show linkage to the *OPA1* locus (3q28–q29), while a small number of DOA lines map to other chromosomal loci (OPA3, 4, 5, 7).[298] The *OPA1* gene encodes a dynamin-related GTPase that is involved in maintenance of the mitochondrial network, acting as a profusion protein. Over 200 mutations have been described within the *OPA1* gene, usually grouped within the GTPase and C-terminal regions.[298,299] The majority of these mutations give rise to truncated proteins and thus haploinsufficiency is thought to be a major mechanism in DOA. *OPA1* mutations typically result in mitochondrial network disruption, aberrant cristae and release of cytochrome c inducing apoptosis.[300,301] Unlike the mutation in LHON which is present in the mitochondrial genome, OPA1, despite being targeted to the mitochondria, is transcribed from the nuclear genome. Because of this the production of a true DOA model is more straight forward than for LHON (Fig. 12).

a. Early Models of DOA. The *Bst* (*belly spot and tail*) mouse was initially proposed as a model for DOA due to it showing a loss of retinal ganglion cells and atrophy of the optic nerve.[302,303] In addition, similarities in the mode of inheritance (i.e., dominant), an intrinsic variability in phenotype and the suggestion that Bst was a murine ortholog of OPA1, supported the use of the *Bst* mouse to model DOA. However, work undertaken by Delettre *et al.*[304] characterized the genomic structure and position of the mouse *OPA1* gene, thus discounting the Bst mouse as a true model of DOA.[304] Another proposed model of DOA utilized OPA1-directed small interference (si) RNA to silence gene expression within the retina. Intravitreal injection of the siRNA in C57BL6J mice results in an alteration in visually evoked potential (VEP), yet no change in the ERG. Animals taken out to 3 months show no axonal degeneration.[305] While this model has had limited use, OPA1-directed siRNA has been used to characterize the role of OPA1 in ganglion cell Ca^{2+} homeostasis and cell survival.[306]

Currently there are two established animal models for DOA which incorporate mutations within exon 8 (*Opa1*^{Q285STOP}) and intron 10 (*Opa1*$^{enu/+}$) of the murine *OPA1* gene.

b. Opa1^{Q285STOP} *Model.* Using an ENU mutagenized DNA archive, Davies and coworkers produced a heterozygous (*Opa1*^{Q285STOP}) mouse model of DOA that incorporated a C to T mutation at position 1051 (exon 8) in the murine *Opa1* gene.[307] This mutation truncates the protein at the beginning of exon 8, which is at the beginning of the dynamin GTPase domain. This is similar to the human condition where the dynamin GTPase domain is one of the mutational hotspots within the human *OPA1* gene and aberrantly truncated OPA1 protein occurs in approximately 50% of the mutations currently found in humans. The homozygous mutation was found to be lethal, while Western blot analysis of the heterozygote confirmed ~50% reduction in Opa1 protein within the retina.[307] This is an important consideration since haploinsufficiency due to reduced Opa1 protein expression is thought to be the major reason for DOA pathology. The model shows a mild phenotype with some variability and no gross change in retinal anatomy, similar to that observed in humans. Despite this, visual deficits are evident in the *Opa1*^{Q285STOP} mice after 12 months, with the extent of the deficit not worsening after 18 months.[307,308] Ultrastructurally, the optic nerve shows signs of degeneration after 9 months, while changes in myelin such as clumping and areas of demyelination can be observed after 18 and 24 months.[307,309] Again similar nerve changes have been described in individuals suffering from DOA. However, unlike in the human, the nerve changes in the *Opa1*^{Q285STOP} mice do not result in a significant decrease in retinal ganglion cells or their axons.[307,309] The mitochondria from the retinal ganglion cells show changes that possibly reflect increased density of cristae as early as 6 months,[309] while fibroblasts taken from adult *Opa1*^{Q285STOP} mice show increased mitochondrial fission and fragmentation of their network.[307] These changes in mitochondrial structure/function are thought to lead to an increase in the numbers of autophagosomes within the ganglion cell layer. Similar to most cases of DOA, the *Opa1*^{Q285STOP} mouse shows no other sign of neuromuscular deficit.[308]

c. Opa1$^{enu/+}$ Model. In the same year that the *Opa1*^{Q285STOP} model was produced, Alavi *et al.* characterized a heterozygous splice mutation in murine *Opa1* that produced features similar to DOA. This mutation in intron 10, induces a skipping of exon 10, producing an in frame deletion of 27 amino acids within the GTPase domain. As previously highlighted this domain contains a number of mutations in human suffers of DOA. In fact, the *Opa1*$^{enu/+}$ mouse represents a molecular model of DOA with several families showing a splice defect in exon 10 of the *OPA1*.[310] As with the previous mouse model, the *Opa1*$^{enu/+}$ mouse shows a ~50% reduction in Opa1 protein, again supporting haploinsufficiency as a major mechanism in the disease. Furthermore, the homozygote offspring die *in utero*, reflecting the importance of this gene in development. These mice show a mild progressive atrophy, with no gross change in retinal structure. There is no alteration in the scotopic or photopic ERG after 2–24 months,[311,312] while VEPs do show reduced scotopic amplitudes.[312] As with the human disease and the *Opa1*^{Q285STOP} model there is considerable variation in phenotype which has been suggested to be due to the fact that the Opa1$^{enu/+}$ model is not a congenic inbred line.[313] By 8 months, the optic nerve shows a reduced number of axons with the remaining exhibiting abnormal morphology and swelling.[311] As with the human disease, there is a progressive loss of ganglion cells by approximately 13 months, which becomes significant by 17 months. This ganglion cell loss was found throughout the retina, both peripherally (−40%) and centrally (−35%).[312] In common with the other mouse model of DOA, the ganglion cell mitochondria show disorganized ultrastructure. Finally, unlike the *Opa1*^{Q285SSTOP} mouse, there are additional neuromuscular and metabolic alterations within the *Opa1*$^{enu/+}$ model, with mice showing tremor, abnormal clutching reflex and reduced body weight.[313] While most cases of DOA do not have additional neurologic deficits, some individuals exhibit other pathology ("DOA plus").[314]

B. Glaucoma

Glaucoma is a heterogeneous group of disorders that are characterized by a progressive loss of ganglion cells, degeneration of the optic nerve and visual field defects. The disease may exhibit a primary or secondary etiology and show congenital, juvenile, or adult onset. POAG is the major form of adult onset glaucoma and is often associated with increased intraocular pressure (IOP). However, while still exhibiting ganglion cell and optic nerve pathology, a subset of individuals with POAG have pressures within the normal range. These individuals are classified as having normal tension glaucoma (NTG).

Because of the heterogeneous nature of glaucoma, the disease is likely to arise due to a complex interaction between genetic and environmental factors. On the genetic front, there are at least 24 different loci that have been linked to various forms of glaucoma, with four genes identified (*myocilin*, *cytochrome P4501B1*, *optineurin*, *WDR36*). Of these genes, myocilin is the best characterized, being largely associated with juvenile or early adult onset POAG and being the most common form of inherited glaucoma currently known. In addition to these genetic factors, other factors such as intraocular pressure, age, reduced ocular blood flow, systemic blood pressure alterations, and excessive stimulation of the glutamatergic system have all be linked to the development of glaucoma.[315]

Interestingly, the clinical phenotype of NTG resembles that of the inheritable optic neuropathies, LHON, and DOA. While mutations associated with LHON do not appear to predispose individuals to NTG, several studies have reported an association between the presence of selective *OPA1* mutations (IVS8 + 4C/T and IVS8 + 32T/C) and the development of NTG. Such an association suggests that mitochondrial dysfunction may also play a role in glaucoma. Despite this association with NTG, there appears to be no increased risk of developing POAG when individuals harbor *OPA1* mutations.

The complex etiology of glaucoma makes the development of suitable animal models difficult. Numerous acute and chronic models of increased IOP-related ganglion cell death have been developed utilizing techniques such hyperosmotic saline injection, episcleral vein cauterization, and induction of ocular hypertension (for recent review see Ref. 316). Such models investigate the role of increased IOP in the progression of ganglion cell and axonal degeneration observed in glaucoma patients. In addition to these, several genetic models have been developed which that result in elevated IOP and subsequent glaucoma-like pathology. However, as reducing pressure in NTG patients doesn't necessarily prevent progression of the disease, the role of IOP-independent mechanisms also needs to be investigated. It is only relatively recently that some promising animal model systems have been developed that result in NTG-like changes without elevating IOP.

1. The DBA/2J Mouse Model of Glaucoma

The DBA/2J mouse strain is a well-characterized model of secondary glaucoma arising due to iris pigment dispersion and atrophy.[317] This anterior chamber pathology is readily identified at 5 months, with significant iris atrophy by 18 months of age.[318] Iris cell death occurs due to mutations in the *Tyrp1* (*Tyrp1b*) and *Gpnmb* ($Gpnmb^{R150X}$) genes which encode melanosomal proteins. Generally, these animals show an increase in IOP by 9 months, with levels declining after 12 months, probably due to ciliary body atrophy. The increase in IOP is accompanied by an age-dependant decrease in axonal number (34% axonal loss at 10 months of age, 80% loss after 17–19 months), with optic nerve head excavation, and retinal ganglion cell loss also observed.[318] Studies have shown that a decrease in visual performance closely follows the period of elevated IOP and optic nerve degeneration (3–12 months).[319] The neuronal cell loss in this model is limited to the ganglion cell layer, with changes in other retinal neurons only rarely observed in older animals.[318,319] As these changes reflect those alterations observed in glaucoma patients, this model has been extensively used to investigate possible disease mechanisms. For example, changes in ganglion cell metabolism,[320] mitochondrial response[321] and inflammatory mediators[322] have all been investigated using the DBA/2J model.

2. $Myoc^{Y423H}$ and $MYOC^{Y437H}$ Models

As mentioned the *myocilin* (*MYOC*) gene is one of the four genes that have been linked with glaucoma. In 2006, Senatorov *et al.*[323] produced a model of POAG when they introduced a mutation in the mouse myocilin gene ($myoc^{Y423H}$) that corresponded to the human polymorphism found in glaucomatous individuals ($MYOC^{Y437H}$). Later in 2008 another group produced a transgenic animal that expressed the mutated human myocilin protein ($MYOC^{Y437H}$).[324] Both animal models produce a two- to threefold increase in MYOC protein expression within the trabecular meshwork, with the aberrant protein forming cytoplasmic accumulations and inhibiting normal secretion of the protein. Despite these changes there was no alteration in the structures at the iridocorneal angle. Both models exhibit only a mild increase in IOP (~2 mmHg), however, as would be expected in the mouse, the effect is more pronounced at night (~3.4 mmHg).[324] Despite this relatively small increase in IOP, there is a peripheral loss of retinal ganglion cells (~20% decrease), while there is no alteration in central densities. Optic nerves show a distorted axonal bundle structure indicative of degeneration, while the myelinated portion of the nerve shows a peripheral loss of axons and degradation of the myelin sheaths. This ganglion cell and optic nerve pathology is comparable to that observed in patients with glaucoma.

3. The *Col1a1*$^{R/R}$ Model

This mouse model was originally developed to investigate wound healing[325] and carried mutations (col1a1^{Q774P}, col1a1^{A777P,} and col1a1^{I776M}) around the consensus cleavage site for matrix metalloproteinase (MMP)-1.[326] These mutations render the mutant collagen protein resistant to MMP degradation and so these mice develop dermal thickening, aberrant uteri, and collagen I accumulation in the sclera. Because of these fibrotic changes in collagen structures it was suggested that changes in ocular collagenous structures may result in glaucomatous changes. The ocular phenotype of this model shows a gradual increase in IOP usually occurring after 16 weeks (21% increase in IOP) and reaching a plateau around 24 weeks (42%, 36 weeks 44%, 54 weeks 33%).[327] While some studies suggest the increase in IOP is sustained,[327,328] a relatively recent report highlighted that IOP returned to normal after 36 weeks.[329] The authors suggested that an alteration in the genetic background (B6;129S4 to C57BL6) to improve breeding ability may have contributed to the shorter duration of IOP elevation. While there is no evidence of gross anterior angle changes,[327,328] a ~25% reduction in outflow facility was observed during periods of increase IOP.[329] Optic nerves from the *Col1a1*$^{r/r}$ mice show lower axonal densities compared to control animals, with a 29% loss of axons observed after 54 weeks.[328] The period between 24 and 54 weeks exhibits the most rapid loss of axonal density (33%, decrease between these time points). During this period of axonal loss, there was an increase in the extent of inclusion bodies, as well as glial cells within the optic nerve. This model shows a more gradual increase in IOP than other animal models of glaucoma and in addition to the effect of the col1a1 mutation on the outflow facility of these mice, it has been suggested that an increase in the stiffness of the sclera canal may also directly affect the optic nerve leading to the glaucoma-like pathology.

4. The *Optn*^{E50K} Model

Optineurin (*OPTN*) is one of the four genes linked to glaucoma, with glaucoma patients showing a higher incidence of sequence alterations than that found in control groups. There are a number of *OPTN* mutations that have been linked to disease, with *OPTN*^{E50K} predisposing individuals to a more severe NTG phenotype. The phenotypic effect of alterations in OPTN has recently been assessed using transgenic technology. While the *OPTN*^{H486R} mutation (*optn*^{H489R} in the mouse) failed to produce a glaucoma-like phenotype, the *OPTN*^{E50K} transgenic (*optn*^{E50K}) developed significant retinal pathology. The retina exhibits a progressive thinning starting at 12 months of age, with a loss of retinal ganglion cells observed in the periphery and a thinning of the nerve fiber layer. These mice also showed general retinal degeneration, with amacrine and bipolar cell loss observed in the peripheral retina. The outer retina also showed alterations in outer plexiform and nuclear layers and shortening of the outer segments of the photoreceptors. Despite these retinal alterations, the anterior segments of the transgenic animals were normal and there was no alteration in IOP. Additional work has shown the E50K mutation to inhibit the movement of OPTN to the nucleus, inhibit interaction with downstream signaling molecules (Rab8), as well as cause an increase in oxidative stress.

XI. Conclusions

A great deal has been learned from the careful evaluation of animal models of retinal disease. Three major changes have led to significant advances in our understanding of human retinal disease. First, careful analysis of the retinal changes that occur naturally or can be induced in animals helped elucidate the animal models that share features with human disease. Secondly, with the advent of molecular tools and the unraveling of the human and mouse genomes, it was possible to correlate mutations in genes with changes in retinal structure and function. Lastly, our ability to develop transgenic mice with mutations known to be important in human disease has led to tremendous advances in our knowledge of the pathogenesis of disease. However, with the advances in molecular biology and animal husbandry, there is now a recognition that retinal disease is not only caused by mutations inherited in a simple Mendelian fashion. Rather, there are environmental and complex genetic factors that exacerbate or prevent the onset of disease. Certainly the use of animals will aid in allowing us to identify key mutations or genetic changes within the gene that may lead to altered gene function. These can then be incorporated into future diagnostic tests as well as providing for the evaluation of safe and efficient treatments. The challenge for the future will be to develop suitable animal models that allow modeling of these complex interactions. In addition, the use of mid-large chronic blind animal models can be used for studying safety and efficacy of ongoing efforts in development of retinal prosthesis (bionic eye).

Acknowledgments

This work was supported by the National Health and Medical Research Council (NHMRC) of Australia (No. 566814 and No. 566815 to ELF), NHMRC Practitioner Award (RHG), NHMRC Centre for Clinical Research Excellence No. 529923—Translational Clinical Research in Major Eye Diseases (PNB, RG), the Australia–India Strategic Research Fund (AISRF) jointly through the

Department of Biotechnology, Government of India and the Department of Innovation, Industry, Science and Research, Government of Australia (PNB and RHG), Diabetes Australia Research Trust (ELF), Retina Australia (ELF), Macular Degeneration Foundation (ELF), JACOM Foundation (PNB). The Centre for Eye Research Australia receives operational infrastructure support from the Victorian Government.

References

1. Wassle H. Parallel processing in the mammalian retina. *Nat Rev Neurosci* 2004;**5**(10):747–57.
2. Bringmann A, Iandiev I, Pannicke T, Wurm A, Hollborn M, Wiedemann P, et al. Cellular signaling and factors involved in Muller cell gliosis: neuroprotective and detrimental effects. *Prog Retin Eye Res* 2009;**28**(6):423–51.
3. Bringmann A, Pannicke T, Grosche J, Francke M, Wiedemann P, Skatchkov SN, et al. Muller cells in the healthy and diseased retina. *Prog Retin Eye Res* 2006;**25**(4):397–424.
4. Strauss O. The retinal pigment epithelium in visual function. *Physiol Rev* 2005;**85**(3):845–81.
5. Hanisch UK, Kettenmann H. Microglia: active sensor and versatile effector cells in the normal and pathologic brain. *Nat Neurosci* 2007;**10**(11):1387–94.
6. Davalos D, Grutzendler J, Yang G, Kim JV, Zuo Y, Jung S, et al. ATP mediates rapid microglial response to local brain injury in vivo. *Nat Neurosci* 2005;**8**(6):752–8.
7. Eter N, Engel DR, Meyer L, Helb HM, Roth F, Maurer J, et al. In vivo visualization of dendritic cells, macrophages, and microglial cells responding to laser-induced damage in the fundus of the eye. *Invest Ophthalmol Vis Sci* 2008;**49**(8):3649–58.
8. Nimmerjahn A, Kirchhoff F, Helmchen F. Resting microglial cells are highly dynamic surveillants of brain parenchyma in vivo. *Science* 2005;**308**(5726):1314–8.
9. Weymouth AE, Vingrys AJ. Rodent electroretinography: methods for extraction and interpretation of rod and cone responses. *Prog Retin Eye Res* 2008;**27**(1):1–44.
10. Jutte A, Deufrains A, Dietze U. Microcirculation disturbances at the onset of diabetic retinopathy. New therapeutic procedures? *Klin Monbl Augenheilkd* 1980;**177**(1):1–6, [author's transl].
11. Linsenmeier RA, Braun RD, McRipley MA, Padnick LB, Ahmed J, Hatchell DL, et al. Retinal hypoxia in long-term diabetic cats. *Invest Ophthalmol Vis Sci* 1998;**39**(9):1647–57.
12. Wong KK, DeLeeuw RJ, Dosanjh NS, Kimm LR, Cheng Z, Horsman DE, et al. A comprehensive analysis of common copy-number variations in the human genome. *Am J Hum Genet* 2007;**80**(1):91–104.
13. Boon CJ, den Hollander AI, Hoyng CB, Cremers FP, Klevering BJ, Keunen JE. The spectrum of retinal dystrophies caused by mutations in the peripherin/RDS gene. *Prog Retin Eye Res* 2008;**27**(2):213–35.
14. Bainbridge JW, Smith AJ, Barker SS, Robbie S, Henderson R, Balaggan K, et al. Effect of gene therapy on visual function in Leber's congenital amaurosis. *N Engl J Med* 2008;**358**(21):2231–9.
15. Hauswirth WW, Aleman TS, Kaushal S, Cideciyan AV, Schwartz SB, Wang L, et al. Treatment of leber congenital amaurosis due to RPE65 mutations by ocular subretinal injection of adeno-associated virus gene vector: short-term results of a phase I trial. *Hum Gene Ther* 2008;**19**(10):979–90.
16. Maguire AM, Simonelli F, Pierce EA, Pugh Jr. EN, Mingozzi F, Bennicelli J, et al. Safety and efficacy of gene transfer for Leber's congenital amaurosis. *N Engl J Med* 2008;**358**(21):2240–8.

17. Ali RR, Reichel MB, Thrasher AJ, Levinsky RJ, Kinnon C, Kanuga N, et al. Gene transfer into the mouse retina mediated by an adeno-associated viral vector. *Hum Mol Genet* 1996;**5**(5):591–4.
18. Acland GM, Aguirre GD, Ray J, Zhang Q, Aleman TS, Cideciyan AV, et al. Gene therapy restores vision in a canine model of childhood blindness. *Nat Genet* 2001;**28**(1):92–5.
19. Pang JJ, Chang B, Kumar A, Nusinowitz S, Noorwez SM, Li J, et al. Gene therapy restores vision-dependent behavior as well as retinal structure and function in a mouse model of RPE65 Leber congenital amaurosis. *Mol Ther* 2006;**13**(3):565–72.
20. Jacobson SG, Boye SL, Aleman TS, Conlon TJ, Zeiss CJ, Roman AJ, et al. Safety in nonhuman primates of ocular AAV2-RPE65, a candidate treatment for blindness in Leber congenital amaurosis. *Hum Gene Ther* 2006;**17**(8):845–58.
21. Abhary S, Hewitt AW, Burdon KP, Craig JE. A systematic meta-analysis of genetic association studies for diabetic retinopathy. *Diabetes* 2009;**58**(9):2137–47.
22. Baird PN, Hageman GS, Guymer RH. New era for personalized medicine: the diagnosis and management of age-related macular degeneration. *Clin Experiment Ophthalmol* 2009;**37**(8):814–21.
23. Chen W, Stambolian D, Edwards AO, Branham KE, Othman M, Jakobsdottir J, et al. Genetic variants near TIMP3 and high-density lipoprotein-associated loci influence susceptibility to age-related macular degeneration. *Proc Natl Acad Sci USA* 2010;**107**(16):7401–6.
24. Neale BM, Fagerness J, Reynolds R, Sobrin L, Parker M, Raychaudhuri S, et al. Genome-wide association study of advanced age-related macular degeneration identifies a role of the hepatic lipase gene (LIPC). *Proc Natl Acad Sci USA* 2010;**107**(16):7395–400.
25. Hewitt AW, Craig JE, Mackey DA. Complex genetics of complex traits: the case of primary open-angle glaucoma. *Clin Experiment Ophthalmol* 2006;**34**(5):472–84.
26. Gal A, Apfelstedt-Sylla E, Janecke AR, Zrenner E. Rhodopsin mutations in inherited retinal dystrophies and dysfunctions. *Prog Retin Eye Res* 1997;**16**:51–79.
27. Sung CH, Davenport CM, Nathans J. Rhodopsin mutations responsible for autosomal dominant retinitis pigmentosa. Clustering of functional classes along the polypeptide chain. *J Biol Chem* 1993;**268**(35):26645–9.
28. Dryja TP, McGee TL, Reichel E, Hahn LB, Cowley GS, Yandell DW, et al. A point mutation of the rhodopsin gene in one form of retinitis pigmentosa. *Nature* 1990;**343**(6256):364–6.
29. Frederick JM, Krasnoperova NV, Hoffmann K, Church-Kopish J, Ruther K, Howes K, et al. Mutant rhodopsin transgene expression on a null background. *Invest Ophthalmol Vis Sci* 2001;**42**(3):826–33.
30. Li T, Snyder WK, Olsson JE, Dryja TP. Transgenic mice carrying the dominant rhodopsin mutation P347S: evidence for defective vectorial transport of rhodopsin to the outer segments. *Proc Natl Acad Sci USA* 1996;**93**(24):14176–81.
31. Naash MI, Hollyfield JG, al-Ubaidi MR, Baehr W. Simulation of human autosomal dominant retinitis pigmentosa in transgenic mice expressing a mutated murine opsin gene. *Proc Natl Acad Sci USA* 1993;**90**(12):5499–503.
32. Olsson JE, Gordon JW, Pawlyk BS, Roof D, Hayes A, Molday RS, et al. Transgenic mice with a rhodopsin mutation (Pro23His): a mouse model of autosomal dominant retinitis pigmentosa. *Neuron* 1992;**9**(5):815–30.
33. Sung CH, Makino C, Baylor D, Nathans J. A rhodopsin gene mutation responsible for autosomal dominant retinitis pigmentosa results in a protein that is defective in localization to the photoreceptor outer segment. *J Neurosci* 1994;**14**(10):5818–33.
34. Roof DJ, Adamian M, Hayes A. Rhodopsin accumulation at abnormal sites in retinas of mice with a human P23H rhodopsin transgene. *Invest Ophthalmol Vis Sci* 1994;**35**(12):4049–62.

35. Liu X, Wu TH, Stowe S, Matsushita A, Arikawa K, Naash MI, et al. Defective phototransductive disk membrane morphogenesis in transgenic mice expressing opsin with a mutated N-terminal domain. *J Cell Sci* 1997;**110**(Pt 20):2589–97.
36. Machida S, Kondo M, Jamison JA, Khan NW, Kononen LT, Sugawara T, et al. P23H rhodopsin transgenic rat: correlation of retinal function with histopathology. *Invest Ophthalmol Vis Sci* 2000;**41**(10):3200–9.
37. Acosta ML, Shin YS, Ready S, Fletcher EL, Christie DL, Kalloniatis M. Retinal metabolic state of the proline-23-histidine rat model of retinitis pigmentosa. *Am J Physiol Cell Physiol* 2010;**298**(3):C764–74.
38. Chen J, Shi G, Concepcion FA, Xie G, Oprian D. Stable rhodopsin/arrestin complex leads to retinal degeneration in a transgenic mouse model of autosomal dominant retinitis pigmentosa. *J Neurosci* 2006;**26**(46):11929–37.
39. Samardzija M, Wenzel A, Naash M, Reme CE, Grimm C. Rpe65 as a modifier gene for inherited retinal degeneration. *Eur J Neurosci* 2006;**23**(4):1028–34.
40. Brill E, Malanson KM, Radu RA, Boukharov NV, Wang Z, Chung HY, et al. A novel form of transducin-dependent retinal degeneration: accelerated retinal degeneration in the absence of rod transducin. *Invest Ophthalmol Vis Sci* 2007;**48**(12):5445–53.
41. Kijas JW, Cideciyan AV, Aleman TS, Pianta MJ, Pearce-Kelling SE, Miller BJ, et al. Naturally occurring rhodopsin mutation in the dog causes retinal dysfunction and degeneration mimicking human dominant retinitis pigmentosa. *Proc Natl Acad Sci USA* 2002;**99**(9):6328–33.
42. Petters RM, Alexander CA, Wells KD, Collins EB, Sommer JR, Blanton MR, et al. Genetically engineered large animal model for studying cone photoreceptor survival and degeneration in retinitis pigmentosa. *Nat Biotechnol* 1997;**15**(10):965–70.
43. McLaughlin ME, Ehrhart TL, Berson EL, Dryja TP. Mutation spectrum of the gene encoding the beta subunit of rod phosphodiesterase among patients with autosomal recessive retinitis pigmentosa. *Proc Natl Acad Sci USA* 1995;**92**(8):3249–53.
44. Dryja TP, Rucinski DE, Chen SH, Berson EL. Frequency of mutations in the gene encoding the alpha subunit of rod cGMP-phosphodiesterase in autosomal recessive retinitis pigmentosa. *Invest Ophthalmol Vis Sci* 1999;**40**(8):1859–65.
45. Sakamoto K, McCluskey M, Wensel TG, Naggert JK, Nishina PM. New mouse models for recessive retinitis pigmentosa caused by mutations in the Pde6a gene. *Hum Mol Genet* 2009;**18**(1):178–92.
46. Petersen-Jones SM, Entz DD, Sargan DR. cGMP phosphodiesterase-alpha mutation causes progressive retinal atrophy in the Cardigan Welsh corgi dog. *Invest Ophthalmol Vis Sci* 1999;**40**(8):1637–44.
47. Suber ML, Pittler SJ, Qin N, Wright GC, Holcombe V, Lee RH, et al. Irish setter dogs affected with rod/cone dysplasia contain a nonsense mutation in the rod cGMP phosphodiesterase beta-subunit gene. *Proc Natl Acad Sci USA* 1993;**90**(9):3968–72.
48. Farber DB, Flannery JG, Bowes-Rickman C. The rd mouse story: seventy years of research on an animal model of inherited retinal degeneration. *Prog Retin Eye Res* 1994;**13**:31–64.
49. Farber DB, Danciger JS, Aguirre G. The beta subunit of cyclic GMP phosphodiesterase mRNA is deficient in canine rod-cone dysplasia 1. *Neuron* 1992;**9**(2):349–56.
50. Baehr W, Frederick JM. Naturally occurring animal models with outer retina phenotypes. *Vision Res* 2009;**49**(22):2636–52.
51. Keeler C. Retinal degeneration in the mouse is rodless retina. *J Hered* 1966;**57**(2):47–50.
52. Bowes C, Li T, Danciger M, Baxter LC, Applebury ML, Farber DB. Retinal degeneration in the rd mouse is caused by a defect in the beta subunit of rod cGMP-phosphodiesterase. *Nature* 1990;**347**(6294):677–80.

53. Farber DB, Lolley RN. Cyclic guanosine monophosphate: elevation in degenerating photoreceptor cells of the C3H mouse retina. *Science* 1974;**186**(4162):449–51.
54. Acosta ML, Fletcher EL, Azizoglu S, Foster LE, Farber DB, Kalloniatis M. Early markers of retinal degeneration in rd/rd mice. *Mol Vis* 2005;**11**:717–28.
55. Blanks JC, Adinolfi AM, Lolley RN. Photoreceptor degeneration and synaptogenesis in retinal-degenerative (rd) mice. *J Comp Neurol* 1974;**156**(1):95–106.
56. Gargini C, Terzibasi E, Mazzoni F, Strettoi E. Retinal organization in the retinal degeneration 10 (rd10) mutant mouse: a morphological and ERG study. *J Comp Neurol* 2007;**500**(2):222–38.
57. Chang B, Hawes NL, Hurd RE, Davisson MT, Nusinowitz S, Heckenlively JR. Retinal degeneration mutants in the mouse. *Vision Res* 2002;**42**(4):517–25.
58. Dekomien G, Runte M, Godde R, Epplen JT. Generalized progressive retinal atrophy of Sloughi dogs is due to an 8-bp insertion in exon 21 of the PDE6B gene. *Cytogenet Cell Genet* 2000;**90**(3–4):261–7.
59. Tuntivanich N, Pittler SJ, Fischer AJ, Omar G, Kiupel M, Weber A, et al. Characterization of a canine model of autosomal recessive retinitis pigmentosa due to a PDE6A mutation. *Invest Ophthalmol Vis Sci* 2009;**50**(2):801–13.
60. Molday RS, Hicks D, Molday L. Peripherin. A rim-specific membrane protein of rod outer segment discs. *Invest Ophthalmol Vis Sci* 1987;**28**(1):50–61.
61. Kajiwara K, Hahn LB, Mukai S, Travis GH, Berson EL, Dryja TP. Mutations in the human retinal degeneration slow gene in autosomal dominant retinitis pigmentosa. *Nature* 1991;**354**(6353):480–3.
62. Farrar GJ, Kenna P, Jordan SA, Kumar-Singh R, Humphries MM, Sharp EM, et al. A three-base-pair deletion in the peripherin-RDS gene in one form of retinitis pigmentosa. *Nature* 1991;**354**(6353):478–80.
63. Dryja TP, Hahn LB, Kajiwara K, Berson EL. Dominant and digenic mutations in the peripherin/RDS and ROM1 genes in retinitis pigmentosa. *Invest Ophthalmol Vis Sci* 1997;**38**(10):1972–82.
64. Travis GH, Brennan MB, Danielson PE, Kozak CA, Sutcliffe JG. Identification of a photoreceptor-specific mRNA encoded by the gene responsible for retinal degeneration slow (rds). *Nature* 1989;**338**(6210):70–3.
65. van Nie R, Ivanyi D, Demant P. A new H-2-linked mutation, rds, causing retinal degeneration in the mouse. *Tissue Antigens* 1978;**12**(2):106–8.
66. Sanyal S, De Ruiter A, Hawkins RK. Development and degeneration of retina in rds mutant mice: light microscopy. *J Comp Neurol* 1980;**194**(1):193–207.
67. Reuter JH, Sanyal S. Development and degeneration of retina in rds mutant mice: the electroretinogram. *Neurosci Lett* 1984;**48**(2):231–7.
68. Kedzierski W, Lloyd M, Birch DG, Bok D, Travis GH. Generation and analysis of transgenic mice expressing P216L-substituted rds/peripherin in rod photoreceptors. *Invest Ophthalmol Vis Sci* 1997;**38**(2):498–509.
69. Vervoort R, Lennon A, Bird AC, Tulloch B, Axton R, Miano MG, et al. Mutational hot spot within a new RPGR exon in X-linked retinitis pigmentosa. *Nat Genet* 2000;**25**(4):462–6.
70. Hong DH, Pawlyk BS, Shang J, Sandberg MA, Berson EL, Li T. A retinitis pigmentosa GTPase regulator (RPGR)-deficient mouse model for X-linked retinitis pigmentosa (RP3). *Proc Natl Acad Sci USA* 2000;**97**(7):3649–54.
71. Zhao Y, Hong DH, Pawlyk B, Yue G, Adamian M, Grynberg M, et al. The retinitis pigmentosa GTPase regulator (RPGR)- interacting protein: subserving RPGR function and participating in disk morphogenesis. *Proc Natl Acad Sci USA* 2003;**100**(7):3965–70.

72. Pawlyk BS, Smith AJ, Buch PK, Adamian M, Hong DH, Sandberg MA, et al. Gene replacement therapy rescues photoreceptor degeneration in a murine model of Leber congenital amaurosis lacking RPGRIP. *Invest Ophthalmol Vis Sci* 2005;**46**(9):3039–45.
73. Pawlyk BS, Bulgakov OV, Liu X, Xu X, Adamian M, Sun X, et al. Replacement gene therapy with a human RPGRIP1 sequence slows photoreceptor degeneration in a murine model of Leber Congenital Amaurosis. *Hum Gene Ther* 2010;**21**(8):993–1004.
74. Zhang Q, Acland GM, Wu WX, Johnson JL, Pearce-Kelling S, Tulloch B, et al. Different RPGR exon ORF15 mutations in Canids provide insights into photoreceptor cell degeneration. *Hum Mol Genet* 2002;**11**(9):993–1003.
75. Longbottom R, Fruttiger M, Douglas RH, Martinez-Barbera JP, Greenwood J, Moss SE. Genetic ablation of retinal pigment epithelial cells reveals the adaptive response of the epithelium and impact on photoreceptors. *Proc Natl Acad Sci USA* 2009;**106**(44):18728–33.
76. Orzalesi N, Calabria GA, Grignolo A. Experimental degeneration of the rabbit retina induced by iodoacetic acid. A study of the ultrastructure, the rhodopsin cycle and the uptake of 14C-labeled iodoacetic acid. *Exp Eye Res* 1970;**9**(2):246–53.
77. Dowling JE, Sidman RL. Inherited retinal dystrophy in the rat. *J Cell Biol* 1962;**14**:73–109.
78. Duncan JL, LaVail MM, Yasumura D, Matthes MT, Yang H, Trautmann N, et al. An RCS-like retinal dystrophy phenotype in mer knockout mice. *Invest Ophthalmol Vis Sci* 2003;**44**(2):826–38.
79. Sidman RL, Kosaras B, Tang M. Pigment epithelial and retinal phenotypes in the vitiligo mivit, mutant mouse. *Invest Ophthalmol Vis Sci* 1996;**37**(6):1097–115.
80. Fletcher EL, Kalloniatis M. Neurochemical architecture of the normal and degenerating rat retina. *J Comp Neurol* 1996;**376**(3):343–60.
81. Fletcher EL, Kalloniatis M. Neurochemical development of the degenerating rat retina. *J Comp Neurol* 1997;**388**(1):1–22.
82. D'Cruz PM, Yasumura D, Weir J, Matthes MT, Abderrahim H, LaVail MM, et al. Mutation of the receptor tyrosine kinase gene Mertk in the retinal dystrophic RCS rat. *Hum Mol Genet* 2000;**9**(4):645–51.
83. Mullen RJ, LaVail MM. Inherited retinal dystrophy: primary defect in pigment epithelium determined with experimental rat chimeras. *Science* 1976;**192**(4241):799–801.
84. Vollrath D, Feng W, Duncan JL, Yasumura D, D'Cruz PM, Chappelow A, et al. Correction of the retinal dystrophy phenotype of the RCS rat by viral gene transfer of Mertk. *Proc Natl Acad Sci USA* 2001;**98**(22):12584–9.
85. Jones BW, Watt CB, Frederick JM, Baehr W, Chen CK, Levine EM, et al. Retinal remodeling triggered by photoreceptor degenerations. *J Comp Neurol* 2003;**464**(1):1–16.
86. Marc RE, Jones BW, Watt CB, Strettoi E. Neural remodeling in retinal degeneration. *Prog Retin Eye Res* 2003;**22**(5):607–55.
87. Gal A, Li Y, Thompson DA, Weir J, Orth U, Jacobson SG, et al. Mutations in MERTK, the human orthologue of the RCS rat retinal dystrophy gene, cause retinitis pigmentosa. *Nat Genet* 2000;**26**(3):270–1.
88. Eisenfeld AJ, Bunt-Milam AH, Sarthy PV. Muller cell expression of glial fibrillary acidic protein after genetic and experimental photoreceptor degeneration in the rat retina. *Invest Ophthalmol Vis Sci* 1984;**25**(11):1321–8.
89. Delyfer MN, Forster V, Neveux N, Picaud S, Leveillard T, Sahel JA. Evidence for glutamate-mediated excitotoxic mechanisms during photoreceptor degeneration in the rd1 mouse retina. *Mol Vis* 2005;**11**:688–96.
90. den Hollander AI, Davis J, van der Velde-Visser SD, Zonneveld MN, Pierrottet CO, Koenekoop RK, et al. CRB1 mutation spectrum in inherited retinal dystrophies. *Hum Mutat* 2004;**24**(5):355–69.

91. van de Pavert SA, Kantardzhieva A, Malysheva A, Meuleman J, Versteeg I, Levelt C, et al. Crumbs homologue 1 is required for maintenance of photoreceptor cell polarization and adhesion during light exposure. *J Cell Sci* 2004;**117**(Pt 18):4169–77.
92. van de Pavert SA, Sanz AS, Aartsen WM, Vos RM, Versteeg I, Beck SC, et al. Crb1 is a determinant of retinal apical Muller glia cell features. *Glia* 2007;**55**(14):1486–97.
93. Williams DS. Usher syndrome: animal models, retinal function of Usher proteins, and prospects for gene therapy. *Vision Res* 2008;**48**(3):433–41.
94. Libby RT, Kitamoto J, Holme RH, Williams DS, Steel KP. Cdh23 mutations in the mouse are associated with retinal dysfunction but not retinal degeneration. *Exp Eye Res* 2003;**77**(6):731–9.
95. Liu X, Udovichenko IP, Brown SD, Steel KP, Williams DS. Myosin VIIa participates in opsin transport through the photoreceptor cilium. *J Neurosci* 1999;**19**(15):6267–74.
96. Lefevre G, Michel V, Weil D, Lepelletier L, Bizard E, Wolfrum U, et al. A core cochlear phenotype in USH1 mouse mutants implicates fibrous links of the hair bundle in its cohesion, orientation and differential growth. *Development* 2008;**135**(8):1427–37.
97. Di Palma F, Holme RH, Bryda EC, Belyantseva IA, Pellegrino R, Kachar B, et al. Mutations in Cdh23, encoding a new type of cadherin, cause stereocilia disorganization in waltzer, the mouse model for Usher syndrome type 1D. *Nat Genet* 2001;**27**(1):103–7.
98. Lentz JJ, Gordon WC, Farris HE, MacDonald GH, Cunningham DE, Robbins CA, et al. Deafness and retinal degeneration in a novel USH1C knock-in mouse model. *Dev Neurobiol* 2010;**70**(4):253–67.
99. Hudspeth AJ. Making an effort to listen: mechanical amplification in the ear. *Neuron* 2008;**59**(4):530–45.
100. Saihan Z, Webster AR, Luxon L, Bitner-Glindzicz M. Update on Usher syndrome. *Curr Opin Neurol* 2009;**22**(1):19–27.
101. Ansley SJ, Badano JL, Blacque OE, Hill J, Hoskins BE, Leitch CC, et al. Basal body dysfunction is a likely cause of pleiotropic Bardet–Biedl syndrome. *Nature* 2003;**425**(6958):628–33.
102. Nishimura DY, Fath M, Mullins RF, Searby C, Andrews M, Davis R, et al. Bbs2-null mice have neurosensory deficits, a defect in social dominance, and retinopathy associated with mislocalization of rhodopsin. *Proc Natl Acad Sci USA* 2004;**101**(47):16588–93.
103. Abd-El-Barr MM, Sykoudis K, Andrabi S, Eichers ER, Pennesi ME, Tan PL, et al. Impaired photoreceptor protein transport and synaptic transmission in a mouse model of Bardet–Biedl syndrome. *Vision Res* 2007;**47**(27):3394–407.
104. Eichers ER, Abd-El-Barr MM, Paylor R, Lewis RA, Bi W, Lin X, et al. Phenotypic characterization of Bbs4 null mice reveals age-dependent penetrance and variable expressivity. *Hum Genet* 2006;**120**(2):211–26.
105. Applebury ML. Variations in retinal degenerations. *Curr Biol* 1992;**2**(3):113–5.
106. LaVail MM, Gorrin GM, Repaci MA, Yasumura D. Light-induced retinal degeneration in albino mice and rats: strain and species differences. *Prog Clin Biol Res* 1987;**247**:439–54.
107. Wenzel A, Grimm C, Samardzija M, Reme CE. Molecular mechanisms of light-induced photoreceptor apoptosis and neuroprotection for retinal degeneration. *Prog Retin Eye Res* 2005;**24**(2):275–306.
108. Yu TY, Acosta ML, Ready S, Cheong YL, Kalloniatis M. Light exposure causes functional changes in the retina: increased photoreceptor cation channel permeability, photoreceptor apoptosis, and altered retinal metabolic function. *J Neurochem* 2007;**103**(2):714–24.
109. Wenzel A, Reme CE, Williams TP, Hafezi F, Grimm C. The Rpe65 Leu450Met variation increases retinal resistance against light-induced degeneration by slowing rhodopsin regeneration. *J Neurosci* 2001;**21**(1):53–8.

110. Organisciak DT, Darrow RM, Barsalou L, Kutty RK, Wiggert B. Susceptibility to retinal light damage in transgenic rats with rhodopsin mutations. *Invest Ophthalmol Vis Sci* 2003;**44** (2):486–92.
111. Cideciyan AV, Jacobson SG, Aleman TS, Gu D, Pearce-Kelling SE, Sumaroka A, et al. In vivo dynamics of retinal injury and repair in the rhodopsin mutant dog model of human retinitis pigmentosa. *Proc Natl Acad Sci USA* 2005;**102**(14):5233–8.
112. Hao W, Wenzel A, Obin MS, Chen CK, Brill E, Krasnoperova NV, et al. Evidence for two apoptotic pathways in light-induced retinal degeneration. *Nat Genet* 2002;**32**(2):254–60.
113. Naash MI, Ripps H, Li S, Goto Y, Peachey NS. Polygenic disease and retinitis pigmentosa: albinism exacerbates photoreceptor degeneration induced by the expression of a mutant opsin in transgenic mice. *J Neurosci* 1996;**16**(24):7853–8.
114. den Hollander AI, Roepman R, Koenekoop RK, Cremers FP. Leber congenital amaurosis: genes, proteins and disease mechanisms. *Prog Retin Eye Res* 2008;**27**(4):391–419.
115. Cideciyan AV. Leber congenital amaurosis due to RPE65 mutations and its treatment with gene therapy. *Prog Retin Eye Res* 2010;**29**(5):398–427.
116. Lamb TD, Pugh Jr. EN. Dark adaptation and the retinoid cycle of vision. *Prog Retin Eye Res* 2004;**23**(3):307–80.
117. Wang JS, Kefalov VJ. An alternative pathway mediates the mouse and human cone visual cycle. *Curr Biol* 2009;**19**(19):1665–9.
118. Gu SM, Thompson DA, Srikumari CR, Lorenz B, Finckh U, Nicoletti A, et al. Mutations in RPE65 cause autosomal recessive childhood-onset severe retinal dystrophy. *Nat Genet* 1997;**17**(2):194–7.
119. Pang JJ, Chang B, Hawes NL, Hurd RE, Davisson MT, Li J, et al. Retinal degeneration 12 (rd12): a new, spontaneously arising mouse model for human Leber congenital amaurosis (LCA). *Mol Vis* 2005;**11**:152–62.
120. Redmond TM, Yu S, Lee E, Bok D, Hamasaki D, Chen N, et al. Rpe65 is necessary for production of 11-*cis*-vitamin A in the retinal visual cycle. *Nat Genet* 1998;**20**(4):344–51.
121. Samardzija M, von Lintig J, Tanimoto N, Oberhauser V, Thiersch M, Reme CE, et al. R91W mutation in Rpe65 leads to milder early-onset retinal dystrophy due to the generation of low levels of 11-*cis*-retinal. *Hum Mol Genet* 2008;**17**(2):281–92.
122. Aguirre GD, Baldwin V, Pearce-Kelling S, Narfstrom K, Ray K, Acland GM. Congenital stationary night blindness in the dog: common mutation in the RPE65 gene indicates founder effect. *Mol Vis* 1998;**4**:23.
123. Veske A, Nilsson SE, Narfstrom K, Gal A. Retinal dystrophy of Swedish briard/briard-beagle dogs is due to a 4-bp deletion in RPE65. *Genomics* 1999;**57**(1):57–61.
124. Dejneka NS, Surace EM, Aleman TS, Cideciyan AV, Lyubarsky A, Savchenko A, et al. In utero gene therapy rescues vision in a murine model of congenital blindness. *Mol Ther* 2004;**9**(2):182–8.
125. Jacobson SG, Aleman TS, Cideciyan AV, Sumaroka A, Schwartz SB, Windsor EA, et al. Identifying photoreceptors in blind eyes caused by RPE65 mutations: prerequisite for human gene therapy success. *Proc Natl Acad Sci USA* 2005;**102**(17):6177–82.
126. Chen Y, Moiseyev G, Takahashi Y, Ma JX. RPE65 gene delivery restores isomerohydrolase activity and prevents early cone loss in Rpe65$^{-/-}$ mice. *Invest Ophthalmol Vis Sci* 2006;**47**(3):1177–84.
127. Narfstrom K, Katz ML, Bragadottir R, Seeliger M, Boulanger A, Redmond TM, et al. Functional and structural recovery of the retina after gene therapy in the RPE65 null mutation dog. *Invest Ophthalmol Vis Sci* 2003;**44**(4):1663–72.
128. Acland GM, Aguirre GD, Bennett J, Aleman TS, Cideciyan AV, Bennicelli J, et al. Long-term restoration of rod and cone vision by single dose rAAV-mediated gene transfer to the retina in a canine model of childhood blindness. *Mol Ther* 2005;**12**(6):1072–82.

129. Cideciyan AV, Aleman TS, Boye SL, Schwartz SB, Kaushal S, Roman AJ, et al. Human gene therapy for RPE65 isomerase deficiency activates the retinoid cycle of vision but with slow rod kinetics. *Proc Natl Acad Sci USA* 2008;**105**(39):15112–7.

130. Cideciyan AV, Hauswirth WW, Aleman TS, Kaushal S, Schwartz SB, Boye SL, et al. Human RPE65 gene therapy for Leber congenital amaurosis: persistence of early visual improvements and safety at 1 year. *Hum Gene Ther* 2009;**20**(9):999–1004.

131. Chang B, Khanna H, Hawes N, Jimeno D, He S, Lillo C, et al. In-frame deletion in a novel centrosomal/ciliary protein CEP290/NPHP6 perturbs its interaction with RPGR and results in early-onset retinal degeneration in the rd16 mouse. *Hum Mol Genet* 2006;**15**(11):1847–57.

132. Narfstrom K, Nilsson SE. Hereditary rod-cone degeneration in a strain of Abyssinian cats. *Prog Clin Biol Res* 1987;**247**:349–68.

133. Narfstrom K. Hereditary progressive retinal atrophy in the Abyssinian cat. *J Hered* 1983;**74**(4):273–6.

134. Bech-Hansen NT, Naylor MJ, Maybaum TA, Sparkes RL, Koop B, Birch DG, et al. Mutations in NYX, encoding the leucine-rich proteoglycan nyctalopin, cause X-linked complete congenital stationary night blindness. *Nat Genet* 2000;**26**(3):319–23.

135. Leroy BP, Budde BS, Wittmer M, De Baere E, Berger W, Zeitz C. A common NYX mutation in Flemish patients with X linked CSNB. *Br J Ophthalmol* 2009;**93**(5):692–6.

136. O'Connor E, Allen LE, Bradshaw K, Boylan J, Moore AT, Trump D. Congenital stationary night blindness associated with mutations in GRM6 encoding glutamate receptor MGluR6. *Br J Ophthalmol* 2006;**90**(5):653–4.

137. Zeitz C, van Genderen M, Neidhardt J, Luhmann UF, Hoeben F, Forster U, et al. Mutations in GRM6 cause autosomal recessive congenital stationary night blindness with a distinctive scotopic 15-Hz flicker electroretinogram. *Invest Ophthalmol Vis Sci* 2005;**46**(11):4328–35.

138. Audo I, Kohl S, Leroy BP, Munier FL, Guillonneau X, Mohand-Said S, et al. TRPM1 is mutated in patients with autosomal-recessive complete congenital stationary night blindness. *Am J Hum Genet* 2009;**85**(5):720–9.

139. Nakamura M, Sanuki R, Yasuma TR, Onishi A, Nishiguchi KM, Koike C, et al. TRPM1 mutations are associated with the complete form of congenital stationary night blindness. *Mol Vis* 2010;**16**:425–37.

140. van Genderen MM, Bijveld MM, Claassen YB, Florijn RJ, Pearring JN, Meire FM, et al. Mutations in TRPM1 are a common cause of complete congenital stationary night blindness. *Am J Hum Genet* 2009;**85**(5):730–6.

141. Pardue MT, McCall MA, LaVail MM, Gregg RG, Peachey NS. A naturally occurring mouse model of X-linked congenital stationary night blindness. *Invest Ophthalmol Vis Sci* 1998;**39**(12):2443–9.

142. Ball SL, Pardue MT, McCall MA, Gregg RG, Peachey NS. Immunohistochemical analysis of the outer plexiform layer in the nob mouse shows no abnormalities. *Vis Neurosci* 2003;**20**(3):267–72.

143. Gregg RG, Mukhopadhyay S, Candille SI, Ball SL, Pardue MT, McCall MA, et al. Identification of the gene and the mutation responsible for the mouse nob phenotype. *Invest Ophthalmol Vis Sci* 2003;**44**(1):378–84.

144. Maddox DM, Vessey KA, Yarbrough GL, Invergo BM, Cantrell DR, Inayat S, et al. Allelic variance between GRM6 mutants, Grm6nob3 and Grm6nob4 results in differences in retinal ganglion cell visual responses. *J Physiol* 2008;**586**(Pt 18):4409–24.

145. Masu M, Iwakabe H, Tagawa Y, Miyoshi T, Yamashita M, Fukuda Y, et al. Specific deficit of the ON response in visual transmission by targeted disruption of the mGluR6 gene. *Cell* 1995;**80**(5):757–65.

146. Pinto LH, Vitaterna MH, Shimomura K, Siepka SM, Balannik V, McDearmon EL, et al. Generation, identification and functional characterization of the nob4 mutation of Grm6 in the mouse. *Vis Neurosci* 2007;**24**(1):111–23.

147. Bellone RR, Brooks SA, Sandmeyer L, Murphy BA, Forsyth G, Archer S, et al. Differential gene expression of TRPM1, the potential cause of congenital stationary night blindness and coat spotting patterns (LP) in the Appaloosa horse (Equus caballus). *Genetics* 2008;**179**(4):1861–70.

148. Morgans CW, Zhang J, Jeffrey BG, Nelson SM, Burke NS, Duvoisin RM, et al. TRPM1 is required for the depolarizing light response in retinal ON-bipolar cells. *Proc Natl Acad Sci USA* 2009;**106**(45):19174–8.

149. Shen Y, Heimel JA, Kamermans M, Peachey NS, Gregg RG, Nawy S. A transient receptor potential-like channel mediates synaptic transmission in rod bipolar cells. *J Neurosci* 2009;**29**(19):6088–93.

150. Miyake Y, Horiguchi M, Terasaki H, Kondo M. Scotopic threshold response in complete and incomplete types of congenital stationary night blindness. *Invest Ophthalmol Vis Sci* 1994;**35**(10):3770–5.

151. Miyake Y, Yagasaki K, Horiguchi M, Kawase Y, Kanda T. Congenital stationary night blindness with negative electroretinogram. A new classification. *Arch Ophthalmol* 1986;**104**(7):1013–20.

152. Boycott KM, Maybaum TA, Naylor MJ, Weleber RG, Robitaille J, Miyake Y, et al. A summary of 20 CACNA1F mutations identified in 36 families with incomplete X-linked congenital stationary night blindness, and characterization of splice variants. *Hum Genet* 2001;**108**(2):91–7.

153. Strom TM, Nyakatura G, Apfelstedt-Sylla E, Hellebrand H, Lorenz B, Weber BH, et al. An L-type calcium-channel gene mutated in incomplete X-linked congenital stationary night blindness. *Nat Genet* 1998;**19**(3):260–3.

154. Doering CJ, Peloquin JB, McRory JE. The Ca(v)1.4 calcium channel: more than meets the eye. *Channels (Austin)* 2007;**1**(1):3–10.

155. Taylor WR, Morgans C. Localization and properties of voltage-gated calcium channels in cone photoreceptors of *Tupaia belangeri*. *Vis Neurosci* 1998;**15**(3):541–52.

156. Littink KW, van Genderen MM, Collin RW, Roosing S, de Brouwer AP, Riemslag FC, et al. A novel homozygous nonsense mutation in CABP4 causes congenital cone-rod synaptic disorder. *Invest Ophthalmol Vis Sci* 2009;**50**(5):2344–50.

157. Zeitz C, Kloeckener-Gruissem B, Forster U, Kohl S, Magyar I, Wissinger B, et al. Mutations in CABP4, the gene encoding the Ca2+-binding protein 4, cause autosomal recessive night blindness. *Am J Hum Genet* 2006;**79**(4):657–67.

158. Haeseleer F, Imanishi Y, Maeda T, Possin DE, Maeda A, Lee A, et al. Essential role of Ca2+-binding protein 4, a Cav1.4 channel regulator, in photoreceptor synaptic function. *Nat Neurosci* 2004;**7**(10):1079–87.

159. Chang B, Heckenlively JR, Bayley PR, Brecha NC, Davisson MT, Hawes NL, et al. The nob2 mouse, a null mutation in Cacna1f: anatomical and functional abnormalities in the outer retina and their consequences on ganglion cell visual responses. *Vis Neurosci* 2006;**23**(1):11–24.

160. Mansergh F, Orton NC, Vessey JP, Lalonde MR, Stell WK, Tremblay F, et al. Mutation of the calcium channel gene Cacna1f disrupts calcium signaling, synaptic transmission and cellular organization in mouse retina. *Hum Mol Genet* 2005;**14**(20):3035–46.

161. Lodha N, Bonfield S, Orton NC, Doering CJ, McRory JE, Mema SC, et al. Congenital stationary night blindness in mice—a tale of two cacna1f mutants. *Adv Exp Med Biol* 2010;**664**:549–58.

162. al-Jandal N, Farrar GJ, Kiang AS, Humphries MM, Bannon N, Findlay JB, et al. A novel mutation within the rhodopsin gene (Thr-94-Ile) causing autosomal dominant congenital stationary night blindness. *Hum Mutat* 1999;**13**(1):75–81.
163. Dryja TP, Berson EL, Rao VR, Oprian DD. Heterozygous missense mutation in the rhodopsin gene as a cause of congenital stationary night blindness. *Nat Genet* 1993;**4**(3):280–3.
164. Naash MI, Wu TH, Chakraborty D, Fliesler SJ, Ding XQ, Nour M, et al. Retinal abnormalities associated with the G90D mutation in opsin. *J Comp Neurol* 2004;**478**(2):149–63.
165. Rao VR, Cohen GB, Oprian DD. Rhodopsin mutation G90D and a molecular mechanism for congenital night blindness. *Nature* 1994;**367**(6464):639–42.
166. Sieving PA, Murayama K, Naarendorp F. Push-pull model of the primate photopic electroretinogram: a role for hyperpolarizing neurons in shaping the b-wave. *Vis Neurosci* 1994;**11** (3):519–32.
167. Sieving PA, Richards JE, Naarendorp F, Bingham EL, Scott K, Alpern M. Dark-light: model for nightblindness from the human rhodopsin Gly-90–>Asp mutation. *Proc Natl Acad Sci USA* 1995;**92**(3):880–4.
168. Zeitz C, Gross AK, Leifert D, Kloeckener-Gruissem B, McAlear SD, Lemke J, et al. Identification and functional characterization of a novel rhodopsin mutation associated with autosomal dominant CSNB. *Invest Ophthalmol Vis Sci* 2008;**49**(9):4105–14.
169. Sieving PA, Fowler ML, Bush RA, Machida S, Calvert PD, Green DG, et al. Constitutive "light" adaptation in rods from G90D rhodopsin: a mechanism for human congenital nightblindness without rod cell loss. *J Neurosci* 2001;**21**(15):5449–60.
170. Dryja TP, Hahn LB, Reboul T, Arnaud B. Missense mutation in the gene encoding the alpha subunit of rod transducin in the Nougaret form of congenital stationary night blindness. *Nat Genet* 1996;**13**(3):358–60.
171. Moussaif M, Rubin WW, Kerov V, Reh R, Chen D, Lem J, et al. Phototransduction in a transgenic mouse model of Nougaret night blindness. *J Neurosci* 2006;**26**(25):6863–72.
172. Muradov KG, Artemyev NO. Loss of the effector function in a transducin-alpha mutant associated with Nougaret night blindness. *J Biol Chem* 2000;**275**(10):6969–74.
173. Szabo V, Kreienkamp HJ, Rosenberg T, Gal A. p.Gln200Glu, a putative constitutively active mutant of rod alpha-transducin (GNAT1) in autosomal dominant congenital stationary night blindness. *Hum Mutat* 2007;**28**(7):741–2.
174. Ramkumar HL, Zhang J, Chan CC. Retinal ultrastructure of murine models of dry age-related macular degeneration (AMD). *Prog Retin Eye Res* 2010;**29**(3):169–90.
175. Vasireddy V, Wong P, Ayyagari R. Genetics and molecular pathology of Stargardt-like macular degeneration. *Prog Retin Eye Res* 2010;**29**(3):191–207.
176. Molday RS, Zhong M, Quazi F. The role of the photoreceptor ABC transporter ABCA4 in lipid transport and Stargardt macular degeneration. *Biochim Biophys Acta* 2009;**1791** (7):573–83.
177. Allikmets R, Singh N, Sun H, Shroyer NF, Hutchinson A, Chidambaram A, et al. A photoreceptor cell-specific ATP-binding transporter gene (ABCR) is mutated in recessive Stargardt macular dystrophy. *Nat Genet* 1997;**15**(3):236–46.
178. Weng J, Mata NL, Azarian SM, Tzekov RT, Birch DG, Travis GH. Insights into the function of Rim protein in photoreceptors and etiology of Stargardt's disease from the phenotype in abcr knockout mice. *Cell* 1999;**98**(1):13–23.
179. Mata NL, Weng J, Travis GH. Biosynthesis of a major lipofuscin fluorophore in mice and humans with ABCR-mediated retinal and macular degeneration. *Proc Natl Acad Sci USA* 2000;**97**(13):7154–9.
180. Mata NL, Tzekov RT, Liu X, Weng J, Birch DG, Travis GH. Delayed dark-adaptation and lipofuscin accumulation in abcr$^{+/-}$ mice: implications for involvement of ABCR in age-related macular degeneration. *Invest Ophthalmol Vis Sci* 2001;**42**(8):1685–90.

181. Vasireddy V, Jablonski MM, Mandal MN, Raz-Prag D, Wang XF, Nizol L, et al. Elovl4 5-bp-deletion knock-in mice develop progressive photoreceptor degeneration. *Invest Ophthalmol Vis Sci* 2006;**47**(10):4558–68.

182. Kijas JW, Zangerl B, Miller B, Nelson J, Kirkness EF, Aguirre GD, et al. Cloning of the canine ABCA4 gene and evaluation in canine cone-rod dystrophies and progressive retinal atrophies. *Mol Vis* 2004;**10**:223–32.

183. Hamilton WK, Ewing CC, Ives EJ, Carruthers JD. Sorsby's fundus dystrophy. *Ophthalmology* 1989;**96**(12):1755–62.

184. Weber BH, Vogt G, Wolz W, Ives EJ, Ewing CC. Sorsby's fundus dystrophy is genetically linked to chromosome 22q13-qter. *Nat Genet* 1994;**7**(2):158–61.

185. Weber BH, Lin B, White K, Kohler K, Soboleva G, Herterich S, et al. A mouse model for Sorsby fundus dystrophy. *Invest Ophthalmol Vis Sci* 2002;**43**(8):2732–40.

186. Janssen A, Hoellenriegel J, Fogarasi M, Schrewe H, Seeliger M, Tamm E, et al. Abnormal vessel formation in the choroid of mice lacking tissue inhibitor of metalloprotease-3. *Invest Ophthalmol Vis Sci* 2008;**49**(7):2812–22.

187. Sikkink SK, Biswas S, Parry NR, Stanga PE, Trump D. X-linked retinoschisis: an update. *J Med Genet* 2007;**44**(4):225–32.

188. Sauer CG, Gehrig A, Warneke-Wittstock R, Marquardt A, Ewing CC, Gibson A, et al. Positional cloning of the gene associated with X-linked juvenile retinoschisis. *Nat Genet* 1997;**17**(2):164–70.

189. Grayson C, Reid SN, Ellis JA, Rutherford A, Sowden JC, Yates JR, et al. Retinoschisin, the X-linked retinoschisis protein, is a secreted photoreceptor protein, and is expressed and released by Weri-Rb1 cells. *Hum Mol Genet* 2000;**9**(12):1873–9.

190. Condon GP, Brownstein S, Wang NS, Kearns JA, Ewing CC. Congenital hereditary (juvenile X-linked) retinoschisis. Histopathologic and ultrastructural findings in three eyes. *Arch Ophthalmol* 1986;**104**(4):576–83.

191. Reid SN, Akhmedov NB, Piriev NI, Kozak CA, Danciger M, Farber DB. The mouse X-linked juvenile retinoschisis cDNA: expression in photoreceptors. *Gene* 1999;**227**(2):257–66.

192. Weber BH, Schrewe H, Molday LL, Gehrig A, White KL, Seeliger MW, et al. Inactivation of the murine X-linked juvenile retinoschisis gene, Rs1h, suggests a role of retinoschisin in retinal cell layer organization and synaptic structure. *Proc Natl Acad Sci USA* 2002;**99**(9):6222–7.

193. Jablonski MM, Dalke C, Wang X, Lu L, Manly KF, Pretsch W, et al. An ENU-induced mutation in Rs1h causes disruption of retinal structure and function. *Mol Vis* 2005;**11**:569–81.

194. Zeng Y, Takada Y, Kjellstrom S, Hiriyanna K, Tanikawa A, Wawrousek E, et al. RS-1 gene delivery to an adult Rs1h knockout mouse model restores ERG b-wave with reversal of the electronegative waveform of X-linked retinoschisis. *Invest Ophthalmol Vis Sci* 2004;**45**(9):3279–85.

195. Attebo K, Mitchell P, Smith W. Visual acuity and the causes of visual loss in Australia. The Blue Mountains Eye Study. *Ophthalmology* 1996;**103**(3):357–64.

196. Klaver CC, Wolfs RC, Vingerling JR, Hofman A, de Jong PT. Age-specific prevalence and causes of blindness and visual impairment in an older population: the Rotterdam Study. *Arch Ophthalmol* 1998;**116**(5):653–8.

197. Rattner A, Nathans J. Macular degeneration: recent advances and therapeutic opportunities. *Nat Rev Neurosci* 2006;**7**(11):860–72.

198. Coleman HR, Chan CC, Ferris 3rd. FL, Chew EY. Age-related macular degeneration. *Lancet* 2008;**372**(9652):1835–45.

199. Mishima H, Kondo K. Extrusion of lysosomal bodies from apical mouse retinal pigment epithelium. *Albrecht Von Graefes Arch Klin Exp Ophthalmol* 1981;**216**(3):209–17.

200. Luhmann UF, Robbie S, Munro PM, Barker SE, Duran Y, Luong V, et al. The drusenlike phenotype in aging Ccl2-knockout mice is caused by an accelerated accumulation of swollen autofluorescent subretinal macrophages. *Invest Ophthalmol Vis Sci* 2009;**50**(12):5934–43.
201. Swaroop A, Branham KE, Chen W, Abecasis G. Genetic susceptibility to age-related macular degeneration: a paradigm for dissecting complex disease traits. *Hum Mol Genet* 2007;**16**(Spec No. 2):R174–82.
202. Seddon JM, Cote J, Page WF, Aggen SH, Neale MC. The US twin study of age-related macular degeneration: relative roles of genetic and environmental influences. *Arch Ophthalmol* 2005;**123**(3):321–7.
203. Chen H, Liu B, Lukas TJ, Neufeld AH. The aged retinal pigment epithelium/choroid: a potential substratum for the pathogenesis of age-related macular degeneration. *PLoS ONE* 2008;**3**(6):e2339.
204. Hogan MJ. Role of the retinal pigment epithelium in macular disease. *Trans Am Acad Ophthalmol Otolaryngol* 1972;**76**(1):64–80.
205. Rakoczy PE, Zhang D, Robertson T, Barnett NL, Papadimitriou J, Constable IJ, et al. Progressive age-related changes similar to age-related macular degeneration in a transgenic mouse model. *Am J Pathol* 2002;**161**(4):1515–24.
206. Zhang D, Brankov M, Makhija MT, Robertson T, Helmerhorst E, Papadimitriou JM, et al. Correlation between inactive cathepsin D expression and retinal changes in mcd2/mcd2 transgenic mice. *Invest Ophthalmol Vis Sci* 2005;**46**(9):3031–8.
207. Haines JL, Hauser MA, Schmidt S, Scott WK, Olson LM, Gallins P, et al. Complement factor H variant increases the risk of age-related macular degeneration. *Science* 2005;**308** (5720):419–21.
208. Klein RJ, Zeiss C, Chew EY, Tsai JY, Sackler RS, Haynes C, et al. Complement factor H polymorphism in age-related macular degeneration. *Science* 2005;**308**(5720):385–9.
209. Hageman GS, Anderson DH, Johnson LV, Hancox LS, Taiber AJ, Hardisty LI, et al. A common haplotype in the complement regulatory gene factor H (HF1/CFH) predisposes individuals to age-related macular degeneration. *Proc Natl Acad Sci USA* 2005;**102** (20):7227–32.
210. Coffey PJ, Gias C, McDermott CJ, Lundh P, Pickering MC, Sethi C, et al. Complement factor H deficiency in aged mice causes retinal abnormalities and visual dysfunction. *Proc Natl Acad Sci USA* 2007;**104**(42):16651–6.
211. Ambati J, Anand A, Fernandez S, Sakurai E, Lynn BC, Kuziel WA, et al. An animal model of age-related macular degeneration in senescent Ccl-2- or Ccr-2-deficient mice. *Nat Med* 2003;**9**(11):1390–7.
212. Combadiere C, Feumi C, Raoul W, Keller N, Rodero M, Pezard A, et al. CX3CR1-dependent subretinal microglia cell accumulation is associated with cardinal features of age-related macular degeneration. *J Clin Invest* 2007;**117**(10):2920–8.
213. Lu B, Rutledge BJ, Gu L, Fiorillo J, Lukacs NW, Kunkel SL, et al. Abnormalities in monocyte recruitment and cytokine expression in monocyte chemoattractant protein 1-deficient mice. *J Exp Med* 1998;**187**(4):601–8.
214. Kuziel WA, Morgan SJ, Dawson TC, Griffin S, Smithies O, Ley K, et al. Severe reduction in leukocyte adhesion and monocyte extravasation in mice deficient in CC chemokine receptor 2. *Proc Natl Acad Sci USA* 1997;**94**(22):12053–8.
215. Tuo J, Smith BC, Bojanowski CM, Meleth AD, Gery I, Csaky KG, et al. The involvement of sequence variation and expression of CX3CR1 in the pathogenesis of age-related macular degeneration. *FASEB J* 2004;**18**(11):1297–9.
216. Cardona AE, Pioro EP, Sasse ME, Kostenko V, Cardona SM, Dijkstra IM, et al. Control of microglial neurotoxicity by the fractalkine receptor. *Nat Neurosci* 2006;**9**(7):917–24.

217. Tuo J, Bojanowski CM, Zhou M, Shen D, Ross RJ, Rosenberg KI, et al. Murine ccl2/cx3cr1 deficiency results in retinal lesions mimicking human age-related macular degeneration. *Invest Ophthalmol Vis Sci* 2007;**48**(8):3827–36.
218. Behndig A, Svensson B, Marklund SL, Karlsson K. Superoxide dismutase isoenzymes in the human eye. *Invest Ophthalmol Vis Sci* 1998;**39**(3):471–5.
219. Imamura Y, Noda S, Hashizume K, Shinoda K, Yamaguchi M, Uchiyama S, et al. Drusen, choroidal neovascularization, and retinal pigment epithelium dysfunction in SOD1-deficient mice: a model of age-related macular degeneration. *Proc Natl Acad Sci USA* 2006;**103**(30):11282–7.
220. Wang L, Clark ME, Crossman DK, Kojima K, Messinger JD, Mobley JA, et al. Abundant lipid and protein components of drusen. *PLoS ONE* 2010;**5**(4):e10329.
221. Klaver CC, Kliffen M, van Duijn CM, Hofman A, Cruts M, Grobbee DE, et al. Genetic association of apolipoprotein E with age-related macular degeneration. *Am J Hum Genet* 1998;**63**(1):200–6.
222. Kliffen M, Lutgens E, Daemen MJ, de Muinck ED, Mooy CM, de Jong PT. The APO(*)E3-Leiden mouse as an animal model for basal laminar deposit. *Br J Ophthalmol* 2000;**84**(12):1415–9.
223. Dithmar S, Curcio CA, Le NA, Brown S, Grossniklaus HE. Ultrastructural changes in Bruch's membrane of apolipoprotein E-deficient mice. *Invest Ophthalmol Vis Sci* 2000;**41**(8):2035–42.
224. Malek G, Johnson LV, Mace BE, Saloupis P, Schmechel DE, Rickman DW, et al. Apolipoprotein E allele-dependent pathogenesis: a model for age-related retinal degeneration. *Proc Natl Acad Sci USA* 2005;**102**(33):11900–5.
225. Grossniklaus HE, Kang SJ, Berglin L. Animal models of choroidal and retinal neovascularization. *Prog Retin Eye Res* 2010;**29**(6):500–19.
226. Fletcher EL, Phipps JA, Ward MM, Puthussery T, Wilkinson-Berka JL. Neuronal and glial cell abnormality as predictors of progression of diabetic retinopathy. *Curr Pharm Des* 2007;**13**(26):2699–712.
227. Aiello LP, Gardner TW, King GL, Blankenship G, Cavallerano JD, Ferris 3rd. FL, et al. Diabetic retinopathy. *Diab Care* 1998;**21**(1):143–56.
228. Frank RN. Diabetic retinopathy. *N Engl J Med* 2004;**350**(1):48–58.
229. Klein R, Klein BE, Moss SE, Davis MD, DeMets DL. The Wisconsin epidemiologic study of diabetic retinopathy. III. Prevalence and risk of diabetic retinopathy when age at diagnosis is 30 or more years. *Arch Ophthalmol* 1984;**102**(4):527–32.
230. Klein R, Klein BE, Moss SE, Davis MD, DeMets DL. The Wisconsin epidemiologic study of diabetic retinopathy. II. Prevalence and risk of diabetic retinopathy when age at diagnosis is less than 30 years. *Arch Ophthalmol* 1984;**102**(4):520–6.
231. Bearse Jr. MA, Han Y, Schneck ME, Barez S, Jacobsen C, Adams AJ. Local multifocal oscillatory potential abnormalities in diabetes and early diabetic retinopathy. *Invest Ophthalmol Vis Sci* 2004;**45**(9):3259–65.
232. Fortune B, Schneck ME, Adams AJ. Multifocal electroretinogram delays reveal local retinal dysfunction in early diabetic retinopathy. *Invest Ophthalmol Vis Sci* 1999;**40**(11):2638–51.
233. Do carmo A, Ramos P, Reis A, Proenca R, Cunha-vaz JG. Breakdown of the inner and outer blood retinal barrier in streptozotocin-induced diabetes. *Exp Eye Res* 1998;**67**(5):569–75.
234. Qaum T, Xu Q, Joussen A, Clemens M, Qin W, Miyamoto K, et al. VEGF-initiated blood–retinal barrier breakdown in early diabetes. *Investig Ophthalmol Vis Sci* 2001;**42**(10):2408–13.
235. Miyamoto K, Khosrof S, Bursell SE, Rohan R, Murata T, Clermont AC, et al. Prevention of leukostasis and vascular leakage in streptozotocin-induced diabetic retinopathy via intercellular adhesion molecule-1 inhibition. *Proc Natl Acad Sci USA* 1999;**96**(19):10836–41.

236. Phipps JA, Fletcher EL, Vingrys AJ. Paired-flash identification of rod and cone dysfunction in the diabetic rat. *Invest Ophthalmol Vis Sci* 2004;**45**(12):4592–600.
237. Su E-N, Alder VA, Yu D-Y, Yu PK, Cringle SJ, Yogesan K. Continued progression of retinopathy despite spontaneous recovery to normoglycemia in a long-term study of stretozotocin-induced diabetes in rats. *Graefes Arch Clin Exp Ophthalmol* 2000;**238**:163–73.
238. UK Prospective Diabetes Study Group. Tight blood pressure control and risk of macrovascular and microvascular complications in type 2 diabetes: UKPDS 38. UK Prospective Diabetes Study Group. *BMJ* 1998;**317**(7160):703–13.
239. Dosso AA, Leuenberger PM, Rungger-Brandle E. Remodeling of retinal capillaries in the diabetic hypertensive rat. *Invest Ophthalmol Vis Sci* 1999;**40**(10):2405–10.
240. Suzuma I, Hata Y, Clermont A, Pokras F, Rook SL, Suzuma K, et al. Cyclic stretch and hypertension induce retinal expression of vascular endothelial growth factor and vascular endothelial growth factor receptor-2: potential mechanisms for exacerbation of diabetic retinopathy by hypertension. *Diabetes* 2001;**50**(2):444–54.
241. Wilkinson-Berka JL, Fletcher EL. Angiotensin and bradykinin: targets for the treatment of vascular and neuro-glial pathology in diabetic retinopathy. *Curr Pharm Des* 2004;**10**(27):3313–30.
242. Fletcher EL, Phipps JA, Ward MM, Vessey KA, Wilkinson-Berka JL. The renin-angiotensin system in retinal health and disease: its influence on neurons, glia and the vasculature. *Prog Retin Eye Res* 2010;**29**(4):284–311.
243. Moravski CJ, Skinner SL, Stubbs AJ, Sarlos S, Kelly DJ, Cooper ME, et al. The renin-angiotensin system influences ocular endothelial cell proliferation in diabetes: transgenic and interventional studies. *Am J Pathol* 2003;**162**(1):151–60.
244. Mullins JJ, Peters J, Ganten D. Fulminant hypertension in transgenic rats harbouring the mouse Ren-2 gene. *Nature* 1990;**344**(6266):541–4.
245. Kelly DJ, Skinner SL, Gilbert RE, Cox AJ, Cooper ME, Wilkinson-Berka JL. Effects of endothelin or angiotensin II receptor blockade on diabetes in the transgenic (mRen-2)27 rat. *Kidney Int* 2000;**57**(5):1882–94.
246. Kelly DJ, Stein-Oakley A, Zhang Y, Wassef L, Maguire J, Koji T, et al. Fas-induced apoptosis is a feature of progressive diabetic nephropathy in transgenic (mRen-2)27 rats: attenuation with renin-angiotensin blockade. *Nephrology (Carlton)* 2004;**9**(1):7–13.
247. Kelly DJ, Wilkinson-Berka JL, Allen TA, Cooper ME, Skinner SL. A new model of diabetic nephropathy with progressive renal impairment in the transgenic (mRen-2)27 rat. *Kidney Int* 1998;**54**:343–52.
248. Phipps JA, Wilkinson-Berka JL, Fletcher EL. Retinal dysfunction in diabetic ren-2 rats is ameliorated by treatment with valsartan but not atenolol. *Invest Ophthalmol Vis Sci* 2007;**48**(2):927–34.
249. Barber AJ, Antonetti DA, Kern TS, Reiter CE, Soans RS, Krady JK, et al. The Ins2Akita mouse as a model of early retinal complications in diabetes. *Invest Ophthalmol Vis Sci* 2005;**46**(6):2210–8.
250. Li Q, Verma A, Han PY, Nakagawa T, Johnson RJ, Grant MB, et al. Diabetic eNOS knockout mice develop accelerated retinopathy. *Invest Ophthalmol Vis Sci* 2010;**51**(10):5240–6.
251. Kim SY, Johnson MA, McLeod DS, Alexander T, Otsuji T, Steidl SM, et al. Retinopathy in monkeys with spontaneous type 2 diabetes. *Invest Ophthalmol Vis Sci* 2004;**45**(12):4543–53.
252. Takahashi Y, Wyman M, Ferris 3rd. F, Kador PF. Diabeteslike preproliferative retinal changes in galactose-fed dogs. *Arch Ophthalmol* 1992;**110**(9):1295–302.
253. Engerman RL, Kern TS. Retinopathy in galactosemic dogs continues to progress after cessation of galactosemia. *Arch Ophthalmol* 1995;**113**(3):355–8.
254. Kador PF, Takahashi Y, Wyman M, Ferris 3rd. F. Diabeteslike proliferative retinal changes in galactose-fed dogs. *Arch Ophthalmol* 1995;**113**(3):352–4.

255. Ditzel J, Standl E. The problem of tissue oxygenation in diabetes mellitus. *Acta Med Scand Suppl* 1975;**578**:59–68.
256. Smith GG, Baird CD. Survival time of retinal cells when deprived of their blood supply by increased intraocular pressure. *Am J Ophthalmol* 1952;**35**(5:2):133–6.
257. Foulds WS, Johnson NF. Rabbit electroretinogram during recovery from induced ischaemia. *Trans Ophthalmol Soc UK* 1974;**94**(2):383–93.
258. Johnson NF, Foulds WS, Strang R. Proceedings: the effect of acute ischaemia on the function and structure of the rabbit retina. *Exp Eye Res* 1973;**17**(4):390.
259. Osborne NN, Block F, Sontag KH. Reduction of ocular blood flow results in glial fibrillary acidic protein (GFAP) expression in rat retinal Muller cells. *Vis Neurosci* 1991;**7**(6):637–9.
260. Block F, Schwarz M, Sontag KH. Retinal ischemia induced by occlusion of both common carotid arteries in rats as demonstrated by electroretinography. *Neurosci Lett* 1992;**144** (1–2):124–6.
261. Stefansson E, Wilson CA, Schoen T, Kuwabara T. Experimental ischemia induces cell mitosis in the adult rat retina. *Invest Ophthalmol Vis Sci* 1988;**29**(7):1050–5.
262. Otori Y, Shimada S, Morimura H, Ishimoto I, Tohyama M, Tano Y. Expression of c-fos and c-jun mRNA following transient retinal ischemia: an approach using ligation of the retinal central artery in the rat. *Surv Ophthalmol* 1997;**42**(Suppl. 1):S96–S104.
263. Vidal-Sanz M, Lafuente MP, Mayor S, de Imperial JM, Villegas-Perez MP. Retinal ganglion cell death induced by retinal ischemia. Neuroprotective effects of two alpha-2 agonists. *Surv Ophthalmol* 2001;**45**(Suppl. 3):S261–7, [Discussion: S273–6].
264. Moren H, Undren P, Gesslein B, Olivecrona GK, Andreasson S, Malmsjo M. The porcine retinal vasculature accessed using an endovascular approach: a new experimental model for retinal ischemia. *Invest Ophthalmol Vis Sci* 2009;**50**(11):5504–10.
265. Kaur C, Foulds WS, Ling EA. Blood–retinal barrier in hypoxic ischaemic conditions: basic concepts, clinical features and management. *Prog Retin Eye Res* 2008;**27**(6):622–47.
266. Kaur C, Sivakumar V, Foulds WS. Early response of neurons and glial cells to hypoxia in the retina. *Invest Ophthalmol Vis Sci* 2006;**47**(3):1126–41.
267. Johnson NF, Foulds WS. The effects of total acute ischaemia on the structure of the rabbit retina. *Exp Eye Res* 1978;**27**(1):45–59.
268. Foulds WS, Kek WK, Luu CD, Song IC, Kaur C. A porcine model of selective retinal capillary closure induced by embolization with fluorescent microspheres. *Invest Ophthalmol Vis Sci* 2010;**51**(12):6700–9; Epub 2010 Jul 29.
269. Prince JH, Diesen CD, Eglitis L, Ruskell GL. *Anatomy and Histology of the eye and orbit in domestic animals*. Springfield, IL: Charles C Thomas; 1960.
270. Good WV, Hardy RJ, Dobson V, Palmer EA, Phelps DL, Quintos M, et al. The incidence and course of retinopathy of prematurity: findings from the early treatment for retinopathy of prematurity study. *Pediatrics* 2005;**116**(1):15–23.
271. Fletcher EL, Downie LE, Hatzopoulos K, Vessey KA, Ward MM, Chow CL, et al. The significance of neuronal and glial cell changes in the rat retina during oxygen-induced retinopathy. *Doc Ophthalmol* 2010;**120**(1):67–86.
272. Penn JS, Thum LA. The rat as an animal model for retinopathy of prematurity. *Prog Clin Biol Res* 1989;**314**:623–42.
273. Penn JS, Tolman BL, Lowery LA, Koutz CA. Oxygen-induced retinopathy in the rat: hemorrhages and dysplasias may lead to retinal detachment. *Curr Eye Res* 1992;**11**(10):939–53.
274. Downie LE, Hatzopoulos KM, Pianta MJ, Vingrys AJ, Wilkinson-Berka JL, Kalloniatis M, et al. Angiotensin type-1 receptor inhibition is neuroprotective to amacrine cells in a rat model of retinopathy of prematurity. *J Comp Neurol* 2010;**518**(1):41–63.

275. Downie LE, Pianta MJ, Vingrys AJ, Wilkinson-Berka JL, Fletcher EL. Neuronal and glial cell changes are determined by retinal vascularization in retinopathy of prematurity. *J Comp Neurol* 2007;**504**(4):404–17.
276. Downie LE, Pianta MJ, Vingrys AJ, Wilkinson-Berka JL, Fletcher EL. AT1 receptor inhibition prevents astrocyte degeneration and restores vascular growth in oxygen-induced retinopathy. *Glia* 2008;**56**(10):1076–90.
277. Zhang Y, Stone J. Role of astrocytes in the control of developing retinal vessels. *Invest Ophthalmol Vis Sci* 1997;**38**(9):1653–66.
278. Rakoczy PE, Brankov M, Fonceca A, Zaknich T, Rae BC, Lai CM. Enhanced recombinant adeno-associated virus-mediated vascular endothelial growth factor expression in the adult mouse retina: a potential model for diabetic retinopathy. *Diabetes* 2003;**52**(3):857–63.
279. Shen WY, Lai YK, Lai CM, Binz N, Beazley LD, Dunlop SA, et al. Pathological heterogeneity of vasoproliferative retinopathy in transgenic mice overexpressing vascular endothelial growth factor in photoreceptors. *Adv Exp Med Biol* 2006;**572**:187–93.
280. van Eeden PE, Tee L, Shen WY, Lukehurst S, Lai CM, Rakoczy PE, et al. Characterisation of a model for retinal neovascularisation. VEGF model characterisation. *Adv Exp Med Biol* 2006;**572**:163–8.
281. Newman NJ, Biousse V. Hereditary optic neuropathies. *Eye (Lond)* 2004;**18**(11):1144–60.
282. Smith JL, Hoyt WF, Susac JO. Ocular fundus in acute Leber optic neuropathy. *Arch Ophthalmol* 1973;**90**(5):349–54.
283. Erickson RP. Leber's optic atrophy, a possible example of maternal inheritance. *Am J Hum Genet* 1972;**24**(3):348–9.
284. Votruba M. Molecular genetic basis of primary inherited optic neuropathies. *Eye (Lond)* 2004;**18**(11):1126–32.
285. Howell N, Bindoff LA, McCullough DA, Kubacka I, Poulton J, Mackey D, et al. Leber hereditary optic neuropathy: identification of the same mitochondrial ND1 mutation in six pedigrees. *Am J Hum Genet* 1991;**49**(5):939–50.
286. Johns DR, Neufeld MJ, Park RD. An ND-6 mitochondrial DNA mutation associated with Leber hereditary optic neuropathy. *Biochem Biophys Res Commun* 1992;**187**(3):1551–7.
287. Wallace DC, Singh G, Lott MT, Hodge JA, Schurr TG, Lezza AM, et al. Mitochondrial DNA mutation associated with Leber's hereditary optic neuropathy. *Science* 1988;**242** (4884):1427–30.
288. Brown MD. The enigmatic relationship between mitochondrial dysfunction and Leber's hereditary optic neuropathy. *J Neurol Sci* 1999;**165**(1):1–5.
289. Majander A, Huoponen K, Savontaus ML, Nikoskelainen E, Wikstrom M. Electron transfer properties of NADH:ubiquinone reductase in the ND1/3460 and the ND4/11778 mutations of the Leber hereditary optic neuroretinopathy (LHON). *FEBS Lett* 1991;**292**(1–2):289–92.
290. Johns DR, Heher KL, Miller NR, Smith KH. Leber's hereditary optic neuropathy. Clinical manifestations of the 14484 mutation. *Arch Ophthalmol* 1993;**111**(4):495–8.
291. Mackey DA. Three subgroups of patients from the United Kingdom with Leber hereditary optic neuropathy. *Eye (Lond)* 1994;**8**(Pt 4):431–6.
292. Carelli V, Ross-Cisneros FN, Sadun AA. Optic nerve degeneration and mitochondrial dysfunction: genetic and acquired optic neuropathies. *Neurochem Int* 2002;**40**(6):573–84.
293. Harley RD, Huang NN, Macri CH, Green WR. Optic neuritis and optic atrophy following chloramphenicol in cystic fibrosis patients. *Trans Am Acad Ophthalmol Otolaryngol* 1970;**74** (5):1011–31.
294. Sadun A. Acquired mitochondrial impairment as a cause of optic nerve disease. *Trans Am Ophthalmol Soc* 1998;**96**:881–923.
295. Qi X, Lewin AS, Hauswirth WW, Guy J. Suppression of complex I gene expression induces optic neuropathy. *Ann Neurol* 2003;**53**(2):198–205.

296. Qi X, Lewin AS, Hauswirth WW, Guy J. Optic neuropathy induced by reductions in mitochondrial superoxide dismutase. *Invest Ophthalmol Vis Sci* 2003;**44**(3):1088–96.
297. Zhang X, Jones D, Gonzalez-Lima F. Mouse model of optic neuropathy caused by mitochondrial complex I dysfunction. *Neurosci Lett* 2002;**326**(2):97–100.
298. Yu-Wai-Man P, Griffiths PG, Hudson G, Chinnery PF. Inherited mitochondrial optic neuropathies. *J Med Genet* 2009;**46**(3):145–58.
299. Ferre M, Amati-Bonneau P, Tourmen Y, Malthiery Y, Reynier P. eOPA1: an online database for OPA1 mutations. *Hum Mutat* 2005;**25**(5):423–8.
300. Olichon A, Baricault L, Gas N, Guillou E, Valette A, Belenguer P, et al. Loss of OPA1 perturbates the mitochondrial inner membrane structure and integrity, leading to cytochrome c release and apoptosis. *J Biol Chem* 2003;**278**(10):7743–6.
301. Olichon A, Landes T, Arnaune-Pelloquin L, Emorine LJ, Mils V, Guichet A, et al. Effects of OPA1 mutations on mitochondrial morphology and apoptosis: relevance to ADOA pathogenesis. *J Cell Physiol* 2007;**211**(2):423–30.
302. Rice DS, Williams RW, Davisson MT, Harris B, Goldowitz D. A new mutant phenotype of retinal ganglion cell disgenesis discovered in the mouse. *Abstr Soc Neurosci* 1993;**19**:51.
303. Rice DS, Williams RW, Ward-Bailey P, Johnson KR, Harris BS, Davisson MT, et al. Mapping the Bst mutation on mouse chromosome 16: a model for human optic atrophy. *Mamm Genome* 1995;**6**(8):546–8.
304. Delettre C, Lenaers G, Belenguer P, Hamel CP. Gene structure and chromosomal localization of mouse Opa1: its exclusion from the Bst locus. *BMC Genet* 2003;**4**:8.
305. Depeyre C, Chen-Kuo-Chang M, Payet O, Cornille K, Muller A, Lenaers G, et al. A murine model of transitory optic neuropathy based on small interference RNA-induced OPA1 silencing in vivo (gene mutation associated with Kjer's disease). *J Fr Ophtalmol* 2006;**29**(8):875–80.
306. Dayanithi G, Chen-Kuo-Chang M, Viero C, Hamel C, Muller A, Lenaers G. Characterization of Ca2+ signalling in postnatal mouse retinal ganglion cells: involvement of OPA1 in Ca2+ clearance. *Ophthalmic Genet* 2010;**31**(2):53–65.
307. Davies VJ, Hollins AJ, Piechota MJ, Yip W, Davies JR, White KE, et al. Opa1 deficiency in a mouse model of autosomal dominant optic atrophy impairs mitochondrial morphology, optic nerve structure and visual function. *Hum Mol Genet* 2007;**16**(11):1307–18.
308. Yu-Wai-Man P, Davies VJ, Piechota MJ, Cree LM, Votruba M, Chinnery PF. Secondary mtDNA defects do not cause optic nerve dysfunction in a mouse model of dominant optic atrophy. *Invest Ophthalmol Vis Sci* 2009;**50**(10):4561–6.
309. White KE, Davies VJ, Hogan VE, Piechota MJ, Nichols PP, Turnbull DM, et al. OPA1 deficiency associated with increased autophagy in retinal ganglion cells in a murine model of dominant optic atrophy. *Invest Ophthalmol Vis Sci* 2009;**50**(6):2567–71.
310. Pesch UE, Leo-Kottler B, Mayer S, Jurklies B, Kellner U, Apfelstedt-Sylla E, et al. OPA1 mutations in patients with autosomal dominant optic atrophy and evidence for semi-dominant inheritance. *Hum Mol Genet* 2001;**10**(13):1359–68.
311. Alavi MV, Bette S, Schimpf S, Schuettauf F, Schraermeyer U, Wehrl HF, et al. A splice site mutation in the murine Opa1 gene features pathology of autosomal dominant optic atrophy. *Brain* 2007;**130**(Pt 4):1029–42.
312. Heiduschka P, Schnichels S, Fuhrmann N, Hofmeister S, Schraermeyer U, Wissinger B, et al. Electrophysiological and histologic assessment of retinal ganglion cell fate in a mouse model for OPA1-associated autosomal dominant optic atrophy. *Invest Ophthalmol Vis Sci* 2010;**51**(3):1424–31.
313. Alavi MV, Fuhrmann N, Nguyen HP, Yu-Wai-Man P, Heiduschka P, Chinnery PF, et al. Subtle neurological and metabolic abnormalities in an Opa1 mouse model of autosomal dominant optic atrophy. *Exp Neurol* 2009;**220**(2):404–9.

314. Amati-Bonneau P, Valentino ML, Reynier P, Gallardo ME, Bornstein B, Boissiere A, et al. OPA1 mutations induce mitochondrial DNA instability and optic atrophy 'plus' phenotypes. *Brain* 2008;**131**(Pt 2):338–51.
315. Ray K, Mookherjee S. Molecular complexity of primary open angle glaucoma: current concepts. *J Genet* 2009;**88**(4):451–67.
316. Johnson TV, Tomarev SI. Rodent models of glaucoma. *Brain Res Bull* 2010;**81**(2–3):349–58.
317. Anderson MG, Smith RS, Hawes NL, Zabaleta A, Chang B, Wiggs JL, et al. Mutations in genes encoding melanosomal proteins cause pigmentary glaucoma in DBA/2J mice. *Nat Genet* 2002;**30**(1):81–5.
318. Libby RT, Anderson MG, Pang IH, Robinson ZH, Savinova OV, Cosma IM, et al. Inherited glaucoma in DBA/2J mice: pertinent disease features for studying the neurodegeneration. *Vis Neurosci* 2005;**22**(5):637–48.
319. Saleh M, Nagaraju M, Porciatti V. Longitudinal evaluation of retinal ganglion cell function and IOP in the DBA/2J mouse model of glaucoma. *Invest Ophthalmol Vis Sci* 2007;**48**(10):4564–72.
320. Baltan S, Inman DM, Danilov CA, Morrison RS, Calkins DJ, Horner PJ. Metabolic vulnerability disposes retinal ganglion cell axons to dysfunction in a model of glaucomatous degeneration. *J Neurosci* 2010;**30**(16):5644–52.
321. Ju WK, Kim KY, Angert M, Duong-Polk KX, Lindsey JD, Ellisman MH, et al. Memantine blocks mitochondrial OPA1 and cytochrome c release and subsequent apoptotic cell death in glaucomatous retina. *Invest Ophthalmol Vis Sci* 2009;**50**(2):707–16.
322. Zhou X, Li F, Kong L, Tomita H, Li C, Cao W. Involvement of inflammation, degradation, and apoptosis in a mouse model of glaucoma. *J Biol Chem* 2005;**280**(35):31240–8.
323. Senatorov V, Malyukova I, Fariss R, Wawrousek EF, Swaminathan S, Sharan SK, et al. Expression of mutated mouse myocilin induces open-angle glaucoma in transgenic mice. *J Neurosci* 2006;**26**(46):11903–14.
324. Zhou Y, Grinchuk O, Tomarev SI. Transgenic mice expressing the Tyr437His mutant of human myocilin protein develop glaucoma. *Invest Ophthalmol Vis Sci* 2008;**49**(5):1932–9.
325. Liu X, Wu H, Byrne M, Jeffrey J, Krane S, Jaenisch R. A targeted mutation at the known collagenase cleavage site in mouse type I collagen impairs tissue remodeling. *J Cell Biol* 1995;**130**(1):227–37.
326. Wu H, Byrne MH, Stacey A, Goldring MB, Birkhead JR, Jaenisch R, et al. Generation of collagenase-resistant collagen by site-directed mutagenesis of murine pro alpha 1(I) collagen gene. *Proc Natl Acad Sci USA* 1990;**87**(15):5888–92.
327. Aihara M, Lindsey JD, Weinreb RN. Ocular hypertension in mice with a targeted type I collagen mutation. *Invest Ophthalmol Vis Sci* 2003;**44**(4):1581–5.
328. Mabuchi F, Lindsey JD, Aihara M, Mackey MR, Weinreb RN. Optic nerve damage in mice with a targeted type I collagen mutation. *Invest Ophthalmol Vis Sci* 2004;**45**(6):1841–5.
329. Dai Y, Lindsey JD, Duong-Polk X, Nguyen D, Hofer A, Weinreb RN. Outflow facility in mice with a targeted type I collagen mutation. *Invest Ophthalmol Vis Sci* 2009;**50**(12):5749–53.

Toward a Better Understanding of Human Eye Disease: Insights From the Zebrafish, *Danio rerio*

JONATHAN BIBLIOWICZ,[*,†,1]
RACHEL K. TITTLE,[*,1] AND
JEFFREY M. GROSS[*,†,‡]

[*]*University of Texas at Austin, Section of Molecular Cell and Developmental Biology, Austin, Texas, USA*

[†]*University of Texas at Austin, Institute for Cellular & Molecular Biology, Austin, Texas, USA*

[‡]*University of Texas at Austin, Institute for Neuroscience, Austin, Texas, USA*

Visual impairment and blindness is widespread across the human population, and the development of therapies for ocular pathologies is of high priority. The zebrafish represents a valuable model organism for studying human ocular disease; it is utilized in eye research to understand underlying developmental processes, to identify potential causative genes for human disorders, and to develop therapies. Zebrafish eyes are similar in morphology, physiology, gene

[1]These authors contributed equally to the work.

Progress in Molecular Biology
and Translational Science, Vol. 100
DOI: 10.1016/B978-0-12-384878-9.00007-8

Copyright 2011, Elsevier Inc.
All rights reserved.
1877-1173/11 $35.00

expression, and function to human eyes. Furthermore, zebrafish are highly amenable to laboratory research. This review outlines the use of zebrafish as a model for human ocular diseases such as colobomas, glaucoma, cataracts, photoreceptor degeneration, as well as dystrophies of the cornea and retinal pigmented epithelium.

I. Introduction

A. Overview

Visual impairments affect over 160 million people worldwide, of whom 37 million are blind.[1] Major causes of blindness include cataracts, glaucoma, age-related macular degeneration (ARMD) and diabetic retinopathy, and the development of effective therapies for these disorders is of high priority. Model organisms with similar physiology to humans are vital to understand underlying developmental processes, identify potential causative genes for human disorders, and develop therapies.

This review focuses on the zebrafish as a model organism for studying human eye diseases. The zebrafish is a remarkably amenable model organism for scientific research, and it combines advantages characteristic of invertebrate models with those inherent to vertebrates for modeling human physiology. As will be explained in the following sections, the zebrafish eye is similar in morphology, physiology, gene expression, and function to the human eye. Zebrafish researchers utilize mutant zebrafish recovered from forward genetic screens along with a number of reverse genetic techniques to model many of the major eye diseases which plague humans (Table I).

B. The Zebrafish as a Model Organism

The zebrafish, *Danio rerio*, is a common aquarium fish that originated in the Ganges region of India.[88] The rapid increase in its popularity for the study of vertebrate development and genetics is mainly due to its transparent embryos that develop *ex utero*, making the visualization of developmental events possible. In addition, its large clutch sizes and ease of maintenance, features that are commonly associated with invertebrate model organisms such as *Drosophila*, make it an attractive alternative to model species such as the mouse. Zebrafish have a short generation time of 2–4 months and a single mating pair can produce around 200 offspring on a weekly basis.[89] These, combined with the low relative cost of raising zebrafish, make it an ideal model organism as evidenced by the rapid growth of the zebrafish research community.

TABLE I

Zebrafish Mutants and Morphants with Ocular Phenotypes Relevant to Human Disorders and Associated Pathologies

Gene	Mutant/morpholino	Ocular phenotype	References	Associated human ocular disease[a]	OMIM
Coloboma					
adenomatous polyposis coli (*apc*)	Mutant	Coloboma; defects in optic vesicle patterning and optic fissure closure	2	Familial Adenomatous Polyposis	175100
bcl6 co-repressor (*bcor*)	Morpholino	Coloboma; microphthalmia	3	Oculofaciocardiodental and Lenz microphthalmia	300485
cadherin 2 neuronal (*cdh2*, *glass onion*)	Mutant	Coloboma; optic fissure closure defect	4	–	–
laminin, β1 (*lamb1*, *lamb1^{hi1113bTg}*), *laminin, γ1* (*lamc1*, *lamc1^{hi3890Tg}*)	Mutant	Coloboma; basement membrane defects	5	–	–
paired box gene 2a (*pax2a*, *no isthmus*)	Mutant	Coloboma	6	Renal-coloboma syndrome	167409
patched1 (*ptc1*, *blowout*)	Mutant	Coloboma; defects in optic stalk morphogenesis	7,8	–	–
thioredoxin-related transmembrane protein 3 (*tmx3*)	Morpholino	Microphthalmia; coloboma	9	Microphthalmia and coloboma	–
transcription factor ap2 alpha (*tfap2a*)	Morpholino	Coloboma	10	Branchio-oculo-facial syndrome	107580
zinc family member 2a (*zic2a*)	Morpholino	Coloboma; defects in optic stalk morphogenesis and optic vesicle patterning	11	–	–
zinc finger proteins 703 and 503 (*znf703,503, also known as nlz1 and nlz2*)	Morpholino	Coloboma; defects in optic vesicle patterning and optic fissure closure	12	–	–
Photoreceptors					
cone transducin α (*tcα*, *no optokinetic response f*)	Mutant	Reduced sensitivity to light by cones	13	–	–
crumbs homolog 2 (*crb2*, *oko meduzy*)	Mutant	Photoreceptor defects	14	–	–

(*Continues*)

TABLE I (*Continued*)

Gene	Mutant/morpholino	Ocular phenotype	References	Associated human ocular disease[a]	OMIM
crumbs homolog 2, like (*crb2l*)	Morpholino	Photoreceptor defects	14	–	–
dihydrolipoamide S-acetyltransferase (*pdhe2, no optokinetic response a*)	Mutant	Blindness; photoreceptor synaptic transmission defects	15,16	Pyruvate dehydrogenase deficiency	–
dynactin 1a (*dctn1a, mikre oko*)	Mutant	Retinal degeneration	17,18	–	–
dynactin 2 (*p50*), (*dctn2, ale oko*)	Mutant	Retinal degeneration	19	–	–
erythrocyte membrane protein band 4.1-like 5 (*epb41l5, mosaic eyes*)	Mutant	Photoreceptor defects	20,21	–	
fleer (*flr*)	Mutant	Rod outer-segment defects	22,23	–	–
intraflagellar transport proteins, 57, 80, 88, and 172 (*ift57^{hi3417Tg}, oval*(*ift88*), *ift172^{hi2211Tg}*)	Mutants (*ift57, 88, 172*), Morpholino (*ift80*)	Outer segment defects; retinal degeneration	24–27	–	–
membrane protein, palmitoylated 5a (*mpp5a, nagie oko*)	Mutant	Disrupted RPE; retinal lamination defects; photoreceptor defects	28,29	–	–
Novel protein (*partial optokinetic response b*)	Mutant	Cone degeneration	30,31	–	–
phosphodiesterase 6 alpha (*pde6α, eclipse*)	Mutant	Cone degeneration	32	–	–
protein kinase C iota (*prkci, heart and soul*)	Mutant	Photoreceptor morphogenesis defects	33	–	–
protocadherin 15b (*pcdh15b*)	Morpholino	Photoreceptor defects; visual function defects	34	Usher syndrome	605514
TNF receptor-associated factor 3 interacting protein (*traf3ip, elipsa*)	Mutant	Photoreceptor defects; visual defects	22,35	–	–
Unknown (*brudas*)	Mutant	Photoreceptor defects	22	–	–
Unknown (*niezerka*)	Mutant	Photoreceptor defects	22,36	–	–

Unknown (*nightblindness a, b, e, f, g*)	Mutant	Visual function defects; retinal degeneration (*nba*, *nbe*, *nbf*)	37–39	–	–
Other retinal phenotypes					
patched2 (*ptc2*, *leprechaun*)	Mutant	Müller glial reactivity; vitreoretinal abnormalities	40	Basal cell naevus syndrome (BCNS)	601309
phosphatase and tensin homolog b (*ptenb*)	Mutant	Ocular tumors	41	–	–
RPE					
choroideremia (*chm*)	Mutant	Retinal degeneration; RPE defects	42,43	Choroideremia	300390
protein kinase C iota (*prkci*, *heart and soul*)	Mutant	RPE morphogenesis defects	33	–	–
silver homolog a (*silva*, *fading vision*)	Mutant	RPE defects; photoreceptor defects	44	–	–
vacuolar protein sorting 18p (*vps18*, $vps18p^{hi2499aTg}$)	Mutant	Melanosome maturation defects; reduced visual function	45	–	–
vacuolar protein sorting 39 homolog (*vps39*, *leberknodel*)	Mutant	RPE vesicle traffic defects; PR defects	46	–	–
v-ATPase complex (multiple genes)	Mutants	Melanosome defects; photoreceptor outer segment defects;	47	–	–
Unknown (*bleached*)	Mutant	Pigmentation defect; blindness; retinal degeneration	48	–	–
Unknown (*fade out*)	Mutant	RPE defects; photoreceptor defects	49	–	–
Unknown (*gantenbein*)	Mutant	Cone dystrophy; RPE degeneration	50	–	–
Hyaloid Vasculature					
forkhead box C1a (*foxc1a*) + *forkhead box C1b* (*foxc1b*)	Morpholino (coinjection)	Reduced hyaloid basement membrane integrity	51	Axenfield–Reiger syndrome, glaucoma	601090
heparan sulfate 6-O-sulfotransferase 2 (*hs6st2*)	Morpholino	Aberrant patterning of hyaloid vasculature	52	–	–
laminin, α1 (*lama1*)	Morpholino	Hyaloid vasculature dysmorphogenesis	53	–	–
laminin, α1 (*lama1*, *bashful*)	Mutant	No hyaloid vasculature	52	–	–
mab-21-like 2 (*mab21l2*)	Morpholino	Reduced hyaloid vasculature	52	–	–

(*Continues*)

TABLE I (*Continued*)

Gene	Mutant/morpholino	Ocular phenotype	References	Associated human ocular disease[a]	OMIM
microfibrillar-associated protein 2 (*mfap2*)	Morpholino	Reduced hyaloid vasculature branching	52	–	–
plexin D1 (*plxnd1*, *out of bounds*)	Mutant	Aberrant patterning of hyaloid vasculature	52	–	–
syndecan 2 (*sdc2*)	Morpholino	No vasculature on the lens	52	–	–
Unknown (*fused eyes*)	Mutant	No hyaloid vasculature	52	–	–
Unknown (*margin affected*)	Mutant	Reduced, then absent hyaloid vasculature	52	–	–
Unknown (*platinum*)	Mutant	Premature detachment of hyaloid vasculature from lens	52	–	–
Lens					
cadherin 4, retinal (*cdh4*)	Morpholino	Small opaque lens	54	–	–
CDP-diacylglycerol-inositol 3-phosphatidyltransferase (*cdipt*, *lens opaque*)	Mutant	Lens cell hyperproliferation; lens degeneration	55–57	–	–
choroideremia (*chm*)	Mutant	Small opaque lens	58	Choroideremia	300390
coatomer protein complex, subunit ζ1 (*copz1*, $copz1^{hi528Tg}$)	Mutant	Cortical lens defects	24	–	–
connexin 48.5 (*cx48.5*)	Morpholino	Cataracts and small lens	59	Cataracts	121015
decapentaplegic and Vg-related 1 (*dvr1*)	Morpholino	Retention of nuclei in lens fibers	60	–	–
fibroblast growth factor 19 (*fgf19*)	Morpholino	Defective lens cell survival and differentiation	61,62	–	–
forkhead box E3 (*foxe3*)	Morpholino	Lens dysmorphogenesis; epithelial cell hyperproliferation; defective fiber differentiation	63,64	Congenital primary aphakia, Peter's anomaly, cataracts	601094
growth differentiation factor 6a (*gdf6a*)	Morpholino	Cortical lens defects; defective lens gene expression	65,66	Microphthalmia	601147

heat shock cognate 70-kd protein (hsp70)	Morpholino	Immature lens	67	–	–
heat shock transcription factor 1 (hsf1)	Morpholino	Small lens	68	–	–
integrator complex subunit 7 (ints7, ints7^{hi1548Tg}, ints7^{hi3649Tg})	Mutant	Devere lens disorganization	24	–	–
laminin, α1 (lama1)	Morpholino	Lens degeneration	69	–	–
laminin, α1 (lama1, bashful)	Mutant	Lens degeneration, focal corneal dysplasia	53,55,70–72	–	–
laminin, β1 (lamb1, grumpy, lamb1^{hi1113bTg})	Mutant	Cortical lens defects; lens dysplasia	5,58,72	–	–
laminin, γ1 (lamc1, sleepy, lamc1^{hi3890Tg})	Mutant	Cortical lens defects; lens dysplasia and degeneration	5,24,72	–	–
lengsin, lens protein with glutamine synthetase domain (lgsn)	Morpholino	Lens dysmorphogenesis; lens fiber defects	73	–	–
mab-21-like 2 (mab21l2)	Morpholino	Lens cell death	74	–	–
paired box gene 2a (pax2a, no isthmus)	Mutant	Small opaque lens by 6 dpf	58	Coloboma	167409
paired box gene 6b (pax6b)	Morpholino	Variable phenotype: small lens to absent lens	75	Peter's anomaly, aniridia	607108
paired box gene 6b (pax6b, sunrise)	Mutant	Small lens	76,77	Peter's anomaly, aniridia	607108
paired-like homeodomain transcription factor 3 (pitx3)	Morpholino	Lens dysmorphogenesis: disordered epithelial cells, defective fiber differentiation, fiber cell death	63,78,79	Cataracts, anterior segment dysgenesis	602669
patched 1 (ptc1)	Morpholino	Small, dysplastic lens	7	–	–
retinal homeobox gene 3 (rx3, chokh)	Mutant	Small lens, no retina	80	Anophthalmia/microphthalmia	601881
RNA binding motif 42 (rbm42, rbm42^{hi2735ATg})	Mutant	Cortical lens defects	24	–	–

(Continues)

TABLE I (*Continued*)

Gene	Mutant/morpholino	Ocular phenotype	References	Associated human ocular disease[a]	OMIM
syndecan 2 (*sdc2*)	Morpholino	Small lens	52	–	–
ubiquitin-like, containing PHD and RING finger domains, 1 (*uhrf1*, $uhrf1^{hi272Tg}$, $uhrf1^{hi3020Tg}$)	Mutant	Lens disorganization and degeneration	24	–	–
WD repeat domain 36 (*wdr36*, $wdr36^{hi3630aTg}$)	Mutant	Thickening of lens epithelium; cortical lens defects; lens degeneration	81	Primary open-angle glaucoma	609669
Unknown (*bumper*)	Mutant	Lens epithelial cell hyperproliferation, lens fiber degeneration	77,82,83	–	–
Unknown (*chiorny*)	Mutant	Small lens	84	–	–
Unknown (*cloche*)	Mutant	Opaque lens with retained nuclei	85	–	–
Unknown (*disrupted lens*)	Mutant	Disorganized lens fibers	55,70	–	–
Unknown (*dou yan*)	Mutant	Small lens	86	–	–
Unknown (*korinthe*)	Mutant	Lens degeneration	77	–	–
Unknown (*margin affected*)	Mutant	Small lens	52	–	–
Unknown (*platinum*)	Mutant	Small lens	52	–	–
Unknown (*rosine*)	Mutant	Lens degeneration	77	–	–
Unknown (*yol*)	Mutant	Lens degeneration	87	–	–

Due to space limitations, not all relevant mutant and morphant zebrafish are included in this table.

[a]Online Mendelian Inheritance in Man (OMIM) entries are listed if the disrupted zebrafish gene is orthologous to a human gene associated with an ocular disease.

Since its introduction as a model organism three decades ago, advances in available technologies pertaining to zebrafish research have quickly closed the gap with those available for other model organisms (see Section I.C). Importantly, the Zebrafish Information Network (ZFIN) maintains a central comprehensive online database that provides researchers with access to information and services that include gene expression pattern databases, mutant and transgenic lines, and links to a fully sequenced genome (www.zfin.org).

C. Tools and Techniques

External development of zebrafish embryos provides the opportunity for various manipulations from the time of fertilization. These include the injection of tracer dyes or of mRNA (for transient expression of protein). Established fate maps[90,91] enable the transplantation of labeled blastula- or gastrula-stage donor cells into specific regions of host embryos. When the donor and host are of different genetic backgrounds, the resulting chimeric zebrafish can be used to determine whether a gene whose mutation is embryonic lethal early in development has a role later in the life of the organism, or to establish the cell or tissue autonomy of a mutant phenotype.[92] Of particular relevance to this review, entire lenses can be transplanted between embryos to determine lens versus retinal contribution to an ocular phenotype.[93]

1. Targeted Knockdown of Gene Expression

Homologous recombination strategies utilized in other model organisms to introduce specific mutations into the genome are not yet available in zebrafish. However, specific and heritable targeted mutagenesis has recently been effected with zinc-finger nucleases.[94,95] Additionally, transient gene knockdown is commonly performed in zebrafish by injection of morpholinos to block gene expression.[96,97] Morpholinos are effective and inexpensive oligomeric constructs designed to bind mRNA and prevent proper splicing or translation. Morpholinos are effective for several days, and their transient activity is sufficient for the embryonic timeframe of many zebrafish experiments. Morpholinos can also be introduced into specific tissues later in development by electroporation,[98] and photoactivatable morpholinos (which can be activated in specific cells of transparent embryos by a laser) allow precise spatiotemporal control of gene knockdown.[99] Thus, zinc finger nucleases and morpholinos complement the catalog of mutant fish lines generated from mutagenesis screens.

2. Transgenic Techniques

Transient or stable introduction of transgenes in zebrafish is effected in a number of ways. Injection of DNA into a fertilized egg results in mosaic, transient expression in the embryo and low efficiency integration into the genome.[100–102] Tol2 transposon-mediated strategies result in much higher efficiency transgene integration,[103] and transgenic lines of zebrafish are routinely created by utilizing this technique. Specific spatiotemporal control of transgene expression is achieved by placing the transgene under the control of a ubiquitous promoter,[104] heat-shock promoter,[105] or tissue-specific enhancer/promoter elements.[106] In addition to spatiotemporal control over expression of genes of interest, these techniques may be utilized to express dominant negatives[107] or genetically encoded reporter constructs such as GFP.[104] Genetically encoded photoconvertible proteins, such as Kaede, may be utilized for lineage-tracing experiments.[108] Transgenes can also be used to ablate specific cells by expression of either diphtheria toxin[109] or nitroreductase (in which case cell-autonomous ablation is not elicited until the application of a prodrug).[110]

Many of the transgenic tools originally implemented in mouse and Drosophila research are also utilized in zebrafish. The Gal4–UAS (upstream activation sequence) system for ectopic gene expression involves spatiotemporally controlled transcription of *gal4* under a promoter of choice. The Gal4 protein then binds to and activates expression of genes downstream of its UAS.[106,111,112] Cre–LoxP recombination,[113] which allows for excision or inversion of a segment of transgene DNA upon activation of Cre, has also been utilized in zebrafish.[114] Cre recombinase can be activated in zebrafish in many ways, including most recently the photo-uncaging of 4-OH-cyclofen for activation of a ligand-inducible Cre.[115] Therefore, the future of zebrafish transgenesis is extremely exciting.

D. Genetic Screens

Large-scale genetic screens in *Drosophila* and *Caenorhabditis elegans* have identified numerous genes required for embryonic development.[116,117] Similar approaches were thought to not be feasible in vertebrates due to long generation times and small number of progeny of traditional vertebrate models such as the mouse and chick.[89] However, the pioneering work of George Streisinger nearly three decades ago established the zebrafish as a powerful genetic model organism for the identification of genes important for vertebrate development.[118,119] Two large-scale genetic screens performed in Christiane Nüsslein-Volhard and Wolfgang Driever's labs followed 15 years later and were published in a special issue of the journal *Development*, describing mutations affecting various aspects of vertebrate development.[120,121] Since then, multiple large- and small-scale mutagenesis efforts have produced numerous mutant lines that have been studied to better understand vertebrate development, as well as disease.

Most genetic screens in zebrafish utilize the chemical mutagen ethylnitrosourea (ENU) due to its high mutagenic efficiency (Fig. 1A).[125,126] Like ethyl methanesulphonate (EMS), which is commonly used in genetic screens in

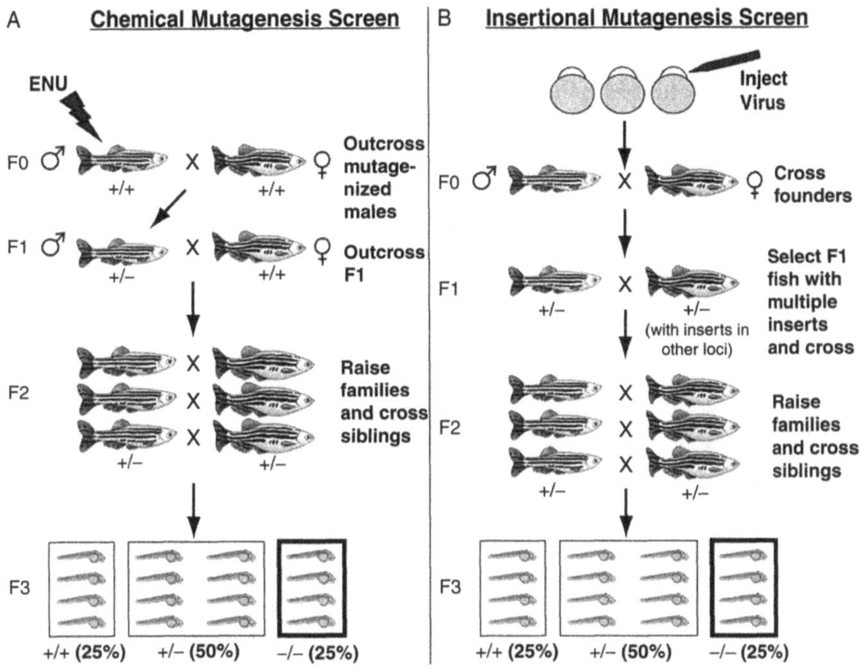

FIG. 1. Chemical versus insertional mutagenesis screens in zebrafish. In chemical screens (A), the mutagen ethlynitrosourea (ENU) is used to induce mutations in the adult male germline. The mutagenized males (F0 generation) are then outcrossed to generate a heterozygous F1 population. F1 progeny are crossed again to wild-type fish, giving rise to F2 families that carry a specific mutation. F2 siblings are bred to each other in order to generate an F3 generation out of which 25% will be homozygous for the mutation.[122,123] In an insertional screen (B), a retrovirus is injected into fertilized eggs to generate transgenic adults (F0) that will be subsequently bred to generate an F1 generation with multiple viral inserts. PCR and Southern blot analysis is then performed in order to identify F1 fish that carry multiple inserts. As in a chemical screen, F1 fish are bred to generate F2 families; those will be subsequently bred to siblings to give rise to F3 progeny that will be screened for physiological and/or behavioral phenotypes.[123,124]

Drosophila, ENU is an alkylating agent that produces point mutations in the adult germline. While ENU mutagenesis has proved to be a powerful method for the induction of point mutations in the zebrafish, the identification of the mutated gene can be laborious. Positional cloning of a single mutation requires the work of one researcher for approximately 6–12 months.[127] A separate screen performed in Nancy Hopkins' lab used retroviral insertional mutagenesis as an alternative (Fig. 1B). Since the genetic sequence of the insertion is known, the location of the retroviral insertion can be assayed using standard PCR techniques, quickly and reliably revealing the affected gene.[124,128] While the efficiency of mutagenesis using retroviral insertion is estimated to be about one-seventh that of ENU mutagenesis, the cloning of one mutation requires only 3–4 weeks to perform.[127] The main disadvantage of insertional mutagenesis is that, due to low efficiency, larger screens need to be performed as compared to ENU-based screens in order to identify equal numbers of mutants. As a result, the need for larger facilities make these screens impractical for most academic labs.

The identification of ENU-induced point mutations can be simplified using a technique known as TILLING (Targeted Induced Local Lesions in Genomes). Unlike traditional screens in which mutants are isolated in the F3 generation and then positionally cloned, with TILLING mutations can be identified in the F1 generation. Mutagenized males are bred to generate large populations of F1 offspring whose DNA is analyzed by PCR in order to identify mutations in a gene of interest. A mutation of interest is then isolated by outcrossing a single identified carrier.[129] This approach has generated over 150 loss-of-function alleles yet to be published.[130] In addition, a consortium has been recently established to consolidate TILLING efforts, greatly improving the likelihood of finding mutations of interest in the zebrafish genome. Researchers can now place requests for mutations in specific genes online (www.sanger.ac.uk/projects/D_rerio/mutres).

Mutants isolated from genetic screens can further be used to screen for small molecules that might suppress the mutant phenotype. These compounds can then be further studied as potential therapeutic agents for a particular defect or disease.[131] Small molecule screens, like genetic screens, utilize the already mentioned advantages of zebrafish: large clutch size, rapid development, and transparency of the embryos allowing for the rapid screening of compounds that affect certain aspects of development. The dosage, as well as the time during development of administration can be controlled and the effects on embryonic development assayed by various methods such as *in situ* hybridization, vital dyes, and transgenics to visualize effects on specific tissues, as well as behavioral assays.[132–135] Since the first large-scale small molecule screen was published 10 years ago,[136] multiple screening efforts have identified compounds that affect various biological processes including cell cycle and cancer, control of stem cell populations, and the formation of retinal vasculature.[133,134,137]

E. Eye Development and Anatomy

The zebrafish has long been recognized as a useful model for the study of human ocular development and disease.[138–141] Detailed characterization of the embryonic development of the posterior segment of the eye, which includes the neural retina[142] and the RPE,[143] and the anterior segment (which includes the lens, cornea, ciliary body, and the various tissues of the iridocorneal angle[144–156]), has not only shed light on the sequence of events in vertebrate eye development, but has also highlighted the similarities in the architecture of the zebrafish eye to that of the human eye.

In zebrafish, eye development is rapid. The optic vesicle, which will ultimately give rise to the neural retina and the retinal pigment epithelium, evaginates from the forebrain at around 12 hours postfertilization (hpf) and remains attached to and continuous with the forebrain through a transient structure called the optic stalk (Fig. 2). The optic vesicle then gives rise to the optic cup through a series of morphogenetic events that occur from about 16 to 20 hpf.[156] Morphogenesis of the optic cup continues as the optic fissure forms ventrally by 24 hpf and subsequently closes by 48 hpf. Neurogenesis begins at 28 hpf and by as early as 72 hpf zebrafish embryos exhibit visual function.[155]

The anterior segment of the embryonic eye develops concurrently with the events mentioned thus far. At 16 hpf, surface ectoderm cells overlying the optic cup thicken to form the lens placode (Fig. 3; Ref. 145), the lens mass delaminates from the surface ectoderm at approximately 24 hpf, and fully detaches by

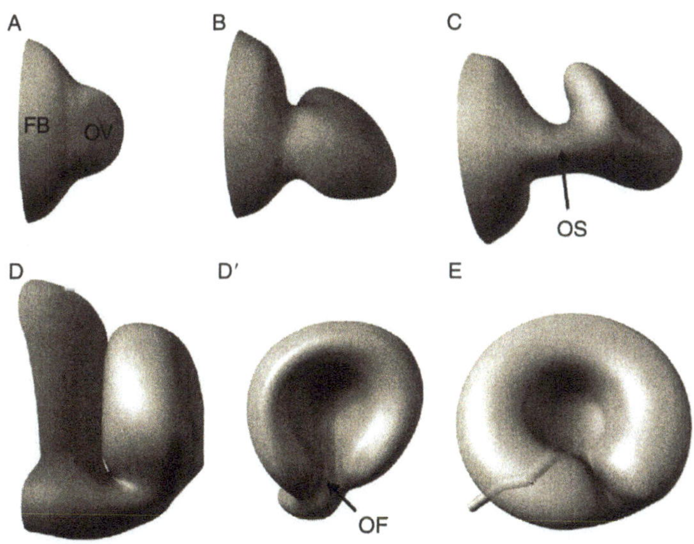

Fig. 2. Development and morphogenesis of the zebrafish eye. Eye development commences around 12 hpf as the optic vesicle (OV) evaginates from the forebrain (FB) (A). The optic vesicle then elongates into a flattened wing-like structure at around 16 hpf (B) that is attached to the forebrain through a transient structure called the optic stalk (OS in C). The eye subsequently rotates and invaginates (C) to form the "optic cup" at around 24 hpf as depicted in D (anterior view) and D′ (lateral view). Morphogenesis of the embryonic eye is mostly complete by 48 hpf as the optic fissure (OF in D′) is closed and neurogenesis of the retina is underway.[142,156,157]

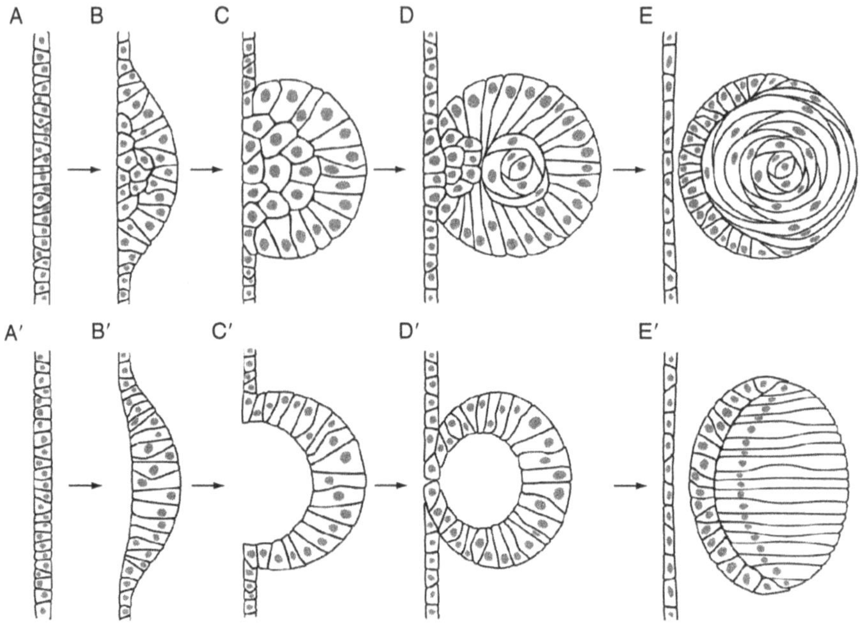

FIG. 3. Early lens development in zebrafish and mouse. After.[146] In zebrafish (A–E) and mouse (A′–E′), the surface ectoderm overlying the optic cup (A, A′) thickens to form the lens placode (B, B′). In zebrafish, the lens mass delaminates from the surface ectoderm (C, D), while in mouse the lens placode evaginates to form the lens vesicle (C′, D′). Primary lens fibers in the zebrafish elongate within the lens mass in a circular fashion (E), whereas the primary lens fibers in the mouse elongate to the anterior to fill the lens vesicle space (E′). The remaining surface ectoderm becomes the corneal epithelium in both zebrafish and mouse.[145,147,155,158] Anterior is to the left.

26 hpf.[145,147,148,155,156] The surface ectoderm overlying the lens becomes the corneal epithelium, which is two cell layers thick by 30 hpf.[150] Migratory periocular mesenchymal cells (which first enter the anterior chamber of the eye at 24 hpf) coalesce to form the corneal endothelium between 30 and 36 hpf.[148,150,155,156]

Humans are a diurnal species, and day-time vision is predominantly mediated by cone photoreceptors in the retina. In contrast to nocturnal mice and rats, whose retinas contain few cones, larval zebrafish vision is mediated almost entirely by cone photoreceptors.[159] As in humans, the mature zebrafish retina is composed of three nuclear layers separated by two plexiform layers (Fig. 4). Zebrafish possess four types of cones (blue, UV, and red/green double cones) and one rod cell type.[160] Rod and cone cell bodies reside in the outer nuclear layer (ONL), while the inner nuclear layer (INL) is occupied by amacrine, horizontal, bipolar cells, and Müller glia. Visual signals originating

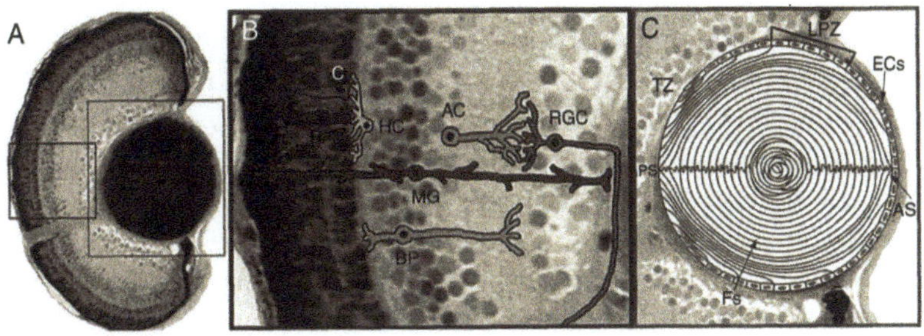

FIG. 4. Structure of the zebrafish eye. (A) Transverse histological section of a wild-type zebrafish eye at 5 dpf. (B) Illustration of the major neuronal and glial cell types in the zebrafish retina. Rods and cones relay sensory input to retinal interneurons (Horizontal and Bipolar Cells). Following synaptic interactions with Amacrine Cells, the information is passed to the output neurons, the Ganglion Cells. Müller glia perform multiple functions in the retina, including maintaining retinal health and structure (see Section II.D). C, Cones; R, Rods; HC, Horizontal Cells; BP, Bipolar; AC, Amacrine Cells; RGC, Ganglion Cells; MG, Müller Glia. (C) Diagram of the zebrafish lens at 5 dpf. Lens epithelial cells surround the anterior periphery; proliferation of these cells is mainly restricted to the lateral proliferative zone. Epithelial cells at the transition zone exit the cell cycle, migrate, elongate, and degrade their light-scattering organelles to become new secondary fibers. Tightly packed lens fibers which extend from the posterior suture to the anterior suture make up the bulk of the lens.[145–148,247–249] ECs, lens epithelial cells; LPZ, lateral proliferative zone; TZ, transition zone; Fs, lens fibers; AS, anterior suture; PS, posterior suture.

in the photoreceptors are transmitted through the retina to the ganglion cells, which make up the ganglion cell layer (GCL); their axons then relay the signal to the brain.[142,161]

II. Posterior Segment

A. Coloboma

Early during vertebrate eye development, the optic vesicle evaginates from the forebrain and ultimately forms the optic cup (see Section I). As a result of a series of morphogenetic processes that follow, a transient gap called the optic fissure forms ventrally and subsequently closes by the fusion of the surrounding tissue.[156,162] Failure of optic fissure closure is the underlying cause of colobomas, which ultimately affect one or more parts of the eye including retina, choroid, iris cornea, lens, ciliary body, and optic nerve.[163] In humans, the incidence of coloboma is estimated at around 1 per 10,000 births[164–166]; severe colobomas may cause as much as 10% of childhood blindness.[167] The clinical presentations of colobomas are varied, as numerous genetic loci have been associated with this diverse class of ocular pathologies.[168] While the causes underlying most human coloboma conditions remain unknown, studies in zebrafish have contributed to our understanding of the developmental bases of optic fissure closure.

A gene expression profiling study was conducted recently by Brown *et al.*[12] to identify genes whose transcript levels are enriched in the optic fissure of the mouse. Two related genes, *Nlz1* and *Nlz2*, were shown to be expressed in the optic fissure; subsequent morpholino knockdown of both in zebrafish resulted in coloboma. Brown *et al.* demonstrated that *nlz1*and *nlz2* may function in the optic fissure by directly regulating the expression of the transcription factor *pax2a*. In *nlz1/2*-deficient embryos, *pax2a* expression was drastically reduced, and both the nlz1 and nlz2 proteins bound the *pax2* promoter *in vitro*, suggesting direct transcriptional regulation. In humans, mutation in the *PAX2* gene result in coloboma as part of renal-coloboma syndrome (OMIM 120330),[169] and the human *PAX2* and zebrafish *pax2a* genes are expressed in the developing optic stalk and optic fissure.[6,170] Interestingly, the zebrafish *No-isthmus* (*noi*) mutant line, which carries a nonsense mutation in the *pax2a* gene, exhibit optic fissure closure defects, resulting in coloboma.[6]

Additional studies in zebrafish have shown that *pax2a* is transcriptionally regulated *by Sonic hedgehog* (*Shh*), a well-studied developmental morphogen. In the zebrafish mutant *cyclops*, in which a mutation affecting the nodal pathway results in a deficiency in *Shh* signaling, *pax2a* transcript levels are highly reduced in the optic stalk.[171] In *blowout* (*blw*) mutants, which possess a loss-of-function mutation in the *patched1* gene, the expression domain of *pax2a* is increased, resulting in an enlarged optic stalk and colobomas.[7,8] The Patched1 protein is the receptor and negative regulator of the Hedgehog pathway; mutations in patched genes are known to result in an overactive Hedgehog pathway.[172] Interestingly, colobomas have been described in a family carrying specific deletions in the human SHH gene,[173] suggesting that the genetic interaction between *Shh* and *pax2a* in zebrafish might be conserved in humans. This interaction was recently shown in zebrafish to be mediated by *zic2a*, a transcription factor whose activity is controlled by *Shh*. *zic2a* restricts *pax2a* expression to the optic stalk, and loss of *zic2a* activity results in the expansion of *pax2a* expression into the retina, resulting in coloboma.[11] While human *ZIC2* mutations have not yet been associated with the formation of coloboma, they have been linked to holoprosencephaly, a forebrain defect frequently associated with colobomas.[11,174] Studies in zebrafish have therefore identified *ZIC2* as an additional candidate gene in human colobomas.

The Hedgehog pathway has also been shown to control the expression of *vax* genes (*vax1* and *vax2*), whose function has also been shown to be important for the proper development and specification of the optic stalk and ventral retina, tissues in which they are normally expressed.[175,176] Morpholino knockdown of both *vax1* and *vax2* in zebrafish resulted in the loss of optic stalk and ventral retinal tissue identity, and the presence of colobomas. Overexpression of Shh results in the expansion of *vax* gene expression domains, while blocking the Hedgehog pathway reduces *vax* gene expression. While Shh appears to control the expression of both *pax2a* and *vax* genes, these likely act in parallel to specify optic stalk identity, as the experimental loss of *pax2*a does not lead to transcriptional loss of *vax* genes, and vice versa.[176] Since both *pax2a* and *vax* genes encode transcription factors, it remains to be seen what factors might act downstream of *pax2a* and to directly control the proper development of the optic stalk and ventral retina.

While the Hedgehog pathway seems to play a central role in dictating optic stalk identity and proper optic fissure closure, other developmental pathways have been implicated in proper morphogenesis of the eye. The requirement of retinoic acid (RA) during ocular development, for example, is well established. RA is derived from vitamin A, which has been linked to ocular pathologies in humans, including coloboma.[177] In zebrafish, exposure to exogenous RA early during eye development results in an enlarged optic stalk and, depending on the dose of RA, duplication of the ventral retina.[178,179] Conversely, embryonic exposure to citral, which blocks RA synthesis, results in optic stalk and ventral retinal loss and ocular coloboma.[180] In addition, morpholino knockdown of *bcox*, an enzyme critical for the synthesis of RA, results in numerous developmental phenotypes, including coloboma.[181]

A possible relationship between RA synthesis and ocular colobomas has emerged through the characterization of the zebrafish *adenomatous polyposis coli* (*apc*) mutant. Human mutations in *APC*, a well studied tumor suppressor, underlie familial adenomatous polyposis syndrome.[182] These patients suffer from failure in the development of the ventral retina and coloboma, phenotypes that are also found in zebrafish *apc* mutants. Through elegant gain- and loss-of-function experiments, Nadauld *et al.*[2] demonstrated that Apc regulates RA synthesis indirectly through the regulation of Rdh5, an enzyme responsible for RA synthesis in the ventral retina. These findings suggest that a deficiency of RA, a vitamin A derivative, might underlie ocular phenotypes in familial adenomatous polyposis syndrome.

The expression of RA synthesizing enzymes in the optic vesicle and optic cup in zebrafish, chick, and mouse suggests that RA is required autonomously in the eye for proper ocular development.[183–185] However, surprisingly, studies in mouse have demonstrated that RA activity in the neural crest-derived periocular mesenchyme (POM) is responsible for early eye morphogenesis.[186] Consistent with these findings, migration defects and subsequent apoptosis of POM in the zebrafish *lmx1b* morphant lead to morphogenetic defects in the optic fissure and colobomas. Loss of *lmx1b* results in a transient upregulation of Fgf signaling, which in turn at least partially affects proper patterning and morphogenesis of the embryonic eye.[187]

The developmental pathways discussed thus far most likely regulate optic fissure closure indirectly, through the regulation of other factors. Studies in zebrafish have also suggested that the molecular mechanisms underlying optic fissure closure are likely to involve interactions between the cells of the retina/optic stalk with the extracellular matrix (ECM). Mutations in *laminin β1* and *γ1* (*lamβ1* and *lamγ1*), for example, result in disruptions in basement membrane integrity and the presence of a number of ocular phenotypes including coloboma.[5,24] Laminins are major components of the ECM, which serves as a scaffold for tissue morphogenesis, as well as a site for the deposition of growth factors such as Fgf and Hedgehog ligands. In the future, it would be interesting to determine whether ocular phenotypes in laminin mutants might be a result of misregulation of these signaling pathways and/or a consequence of morphogenetic defects due to perturbed cell–ECM interactions.

B. Photoreceptors

Photoreceptor degenerations are the most common form of blindness in the Western world and involve the loss of vision due to dystrophy and/or death of retinal photoreceptors. These pathologies can be roughly divided into those conditions that initially affect rod photoreceptors, such as retinitis pigmentosa (RP), and those that initially affect cone photoreceptors, such as macular degeneration.[141] While photoreceptor degeneration can also be caused indirectly by primary defects in the RPE, which serves an important function in photoreceptor health and homeostasis, this topic will be discussed in a later section (see Section II.C).

Photoreceptors are highly polarized sensory cells that consist of an inner segment (IS) that is connected to an outer segment (OS) by a highly modified cilium (Fig. 5).[189] In mature photoreceptors, proteins that are required for growth and maintenance of photoreceptor OS's, as well as for phototransduction, are transported along polarized microtubules from the basal IS to the apical OS. Photoreceptors are similar to epithelial cells in that their surfaces are divided into apical and basolateral domains by cell junctions. Their centrosomes are located at the apical surface, while their nuclei reside basally.[190] Proper development and subsequent maintenance of vertebrate photoreceptors relies on the establishment of proper apical–basal polarity, as well as the function of the transport machinery. It is therefore not surprising that of the more than 100 genetic loci known to cause photoreceptor degenerations in vertebrates, most affect the structure and function of the OS.[191]

Genetic screens in zebrafish have shed light on the molecular bases of these cellular functions by the isolation of mutants in which photoreceptor development and/or maintenance are defective.[24,192,193] One such mutant, *mosaic eyes* (*moe*), was found to contain expanded apical features in photoreceptors.[20,21] The *moe* locus encodes a FERM-domain containing protein that forms a complex with Crumbs proteins, which in turn have been shown to be critical for apical–basal polarity in *Drosophila*, as well as vertebrates.[194–196] Interestingly, mutations in the human orthologue of Crumbs, *CRB1*, are associated

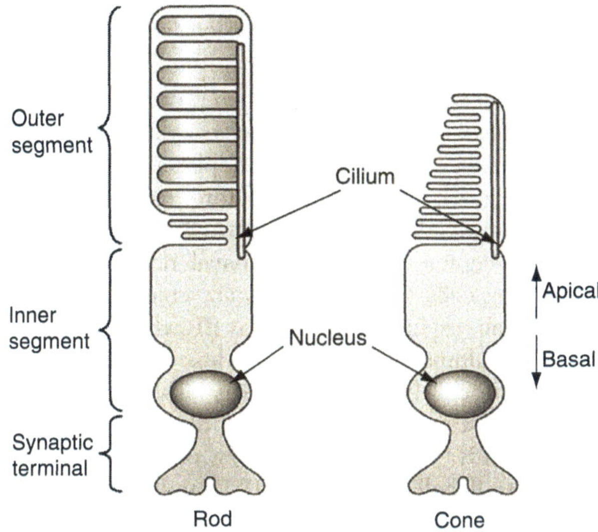

FIG. 5. Structure of rod and cone photoreceptors. Photoreceptor cells are composed of an outer segment that contains visual pigment-bound membrane disks. The inner segment is the polarized cell body of the photoreceptor, where the nucleus resides basally and the cilium originates apically and extends into the OS. The synaptic terminal forms synaptic connections to the interneurons of the retina, relaying visual input.[161,188]

with retinal photoreceptor dystrophies such as RP 12[197] and Leber's congenital amaurosis (LCA1; Ref. 198,199). Other zebrafish mutants, such as *nagie oko* (*nok*) and *heart and soul* (*has*) do not only phenocopy the *moe* mutant phenotype, but the proteins that they encode (Pals1 and PKCi, respectively) also interact biochemically with Moe.[20] Moreover, morpholino knockdown of *crb2b*, a zebrafish Crumbs paralog, results in the reduction of IS size.[14] These studies have therefore begun to uncover the role of the Crumbs complex in vertebrate photoreceptor development and disease.

The formation of correct apical–basal polarity depends not only on the proper function of cell polarity determinants, but also on their transport and localization. Photoreceptors in the *ale oko* (*ako*) mutant retina accumulate the apical determinants Pals1 and PKCλ in their cell bodies and exhibit extensive photoreceptor death late in development.[19] The *ako* locus encodes the p50 subunit of the dynactin complex, which serves an important role in the transport of cargo along microtubules as part of the dynein motor complex.[200] In another mutant, *mikre oko* (*mok*), which possesses a mutation in *dynactin-1*, retinal degeneration is at least partially due to mislocalization of photoreceptor nuclei. Surprisingly, unlike *ako* mutants, apical determinants such as Crumbs and aPKCλ are not mislocalized, suggesting that degeneration in these mutants is not due to loss of cell polarity. Experimental mislocalization of the nucleus by the overexpression of the dynein motor resulted in photoreceptor cell death, supporting the role of nuclear position in photoreceptor survival.[201] While nuclear positioning defects in *mok* photoreceptors seems to have a cell-autonomous role, other noncell autonomous components seem to play a role in *mok* degeneration. Mosaic analyses have revealed that mutant photoreceptors display over a 2.5-fold increase in survival when placed in a wild-type environment.[17] This finding suggests that photoreceptor health and survival depends on environmental cues from surrounding cells. These findings are consistent with defects associated with some human degenerations, such as RP, where an initial defect in the rod specific rhodopsin ultimately results in the loss of both rods and cones.[202]

Transport from the cell body to the OS is also important to replace OS components that are lost due to the continual phagocytosis of photoreceptor OS's by the overlying RPE.[203] In photoreceptors, as in cilia, transport occurs by a process known as Intraflagellar Transport (IFT) and mutations in IFT component genes such *ift57*, *ift80*, *ift88*, and *ift172* result in OS atrophy and/or complete loss of OS's in zebrafish.[24–26] Mutations in the *elipsa* gene, which encodes a protein that is thought to facilitate IFT,[35] result in early photoreceptor loss,[22] while morpholino knockdown of multiple subunits of the retrograde IFT motor dynein-2 has highlighted the importance of IFT in proper OS elongation and visual function.[204] Photoreceptor degenerations relating to defective IFT are some of the many cilia-related phenotypes associated with Bardet–Biedl Syndrome (BBS),[205] and the further study of the function of IFT components in zebrafish might prove to be instrumental in gaining a better understanding of the molecular causes underlying BBS-related ocular pathologies.

While the mutants described thus far in this section were isolated in genetic screens utilizing mainly histological methods to detect photoreceptor defects, others were isolated in behavioral screens.[15,82,206,207] One such screen took advantage of the ability of zebrafish to exhibit vision-dependent behavior as early as 3 dpf (days post fertilization). Brockerhoff *et al.*[15] first used the optokinetic response assay (OKR) on mutagenized embryos to detect defects in visual function. A second assay involved the use of electroretinogram (ERG) recordings to identify whether isolated mutations affect outer retinal function. From this screen, 18 mutants were isolated that were determined to have reduced visual function. One, *no optokinetic response a* (*noa*), possessed no gross photoreceptor abnormalities at 5 dpf (days post fertilization), but was found to be blind, as well as lethargic, and died prematurely. Analysis of the *noa* mutation revealed a deficiency in a subunit of the pyruvate dehydrogenase (PDH) complex, which regulates energy production in cells. The *noa* mutant line has been used as a model for the study of PDH deficiency (OMIM 245348), a human disorder that, like the zebrafish model, results in blindness, neurological defects, and early death.[16] Since current treatments for PDH deficiency in humans have resulted in limited success, Taylor *et al.* utilized *noa* mutants to test a ketogenic diet which has shown some success in alleviating PDH deficiency symptoms in a limited number of human patients. Administration of this special diet restored normal behavior in *noa* mutants, highlighting the potential of studying PDH deficiency therapies in the *noa* mutant line.

Another mutant, *no optokinetic response f* (*nof*) was found to possess a mutation in the α subunit of cone transducin (T$c\alpha$), a G-protein required for phototransduction. In human patients suffering from a condition known as achromatopsia (OMIM 139340), mutations in T$c\alpha$ underlie loss of color vision. In *nof* mutants, cone development occurs normally; they are, however, up to 1000$\times$ less sensitive to light, as detected by recordings of single photoreceptors. Extensive analysis of cone light responses revealed that residual phototransduction is light dependent, but transducin independent. Ca^{+2} influx, which is important for photoreceptor light adaptation and was previously thought to be controlled by transducin, was still detectable in *nof* mutants. This study therefore revealed that some Ca^{+2} influx in cone photoreceptors might be transducin independent.

In a separate screen, dominant mutations that cause photoreceptor degeneration in adult zebrafish was performed in order to isolate genetic mutants that could be later used to study human inherited night blindness such as RP.[37,38] In the case of RP, many genetic loci underlie this group of disorders. However, only about half of the cases of dominant RP have been linked to specific mutations at the time the screen was performed.[38] Li *et al.* therefore screened mutagenized adult zebrafish by utilizing a known escape response exhibited by the fish. A lack of such response to a threatening cue was interpreted as a loss of vision, which was later confirmed using ERG recordings. In all, seven heterozygous mutants were isolated (*nightblindness a*, *b*, *c*, *d*, *e*, and *f*), out of which six exhibited photoreceptor degeneration.[38,39,208,209] Four of the six were found to be embryonic lethal as homozygotes, suggesting that the mutated genes underlying the photoreceptor phenotypes have other critical functions during embryonic development. This finding highlights the importance of such screens for the identification of dominant mutations that might not be easily isolated in traditional screening due to early embryonic lethality.

C. RPE

The RPE is a monolayer of pigmented cells that serves in the protection and maintenance of photoreceptors, and is therefore essential for visual function. In addition to its role in absorbing excess light entering the eye, the RPE transports essential nutrients from the blood to photoreceptors, while removing ions, water, and metabolic end products from the retina to the blood. Vertebrate phototransduction depends on the conversion of all-*trans*-retinal to 11-*cis*-retinal through a series of biochemical reactions termed the retinoid cycle, and most of these essential reactions are carried out by the RPE rather than photoreceptors.[210] Finally, the RPE supports photoreceptor renewal by the daily phagocytosis of about 10% of OS volume, while the photoreceptor itself regenerates roughly the same volume each day to maintain proper OS length.[211] Failure of any of these functions often results in retinal degeneration and loss of vision ([212]).

An important question in the study of many types of retinal degenerations is whether photoreceptor degeneration occurs due to a primary defect in photoreceptors themselves or as a secondary consequence of RPE degeneration. Since in many cases defects in both photoreceptor and RPE are present, a cause-and-effect relationship cannot be easily established between these cell types. Krock *et al.*[42] studied this relationship in the zebrafish *rep1/chm* mutant line, which exhibits retinal phenotypes consistent with those found in humans suffering from choroideremia, a form of hereditary retinal degeneration associated with mutations in the human REP1 gene. Rep proteins are involved in the posttranslational modification of Rab protein, and are therefore critical for vesicle trafficking. Since both photoreceptors and the RPE depend on vesicle trafficking for transport of material (such as opsin) to the OS and degradation of phagocytosed OS material, respectively, the initial defect due to the *rep1* mutation could potentially lay in either tissue. Through the use of mosaic analysis, Krock *et al.* showed a loss of rep1 in the RPE is sufficient to cause photoreceptor degeneration and results in the localized accumulation of OS material. These findings, together with the observation that opsin was not mislocalized in *rep1* photoreceptors suggest that defective cellular processes within the RPE are the primary cause of photoreceptor degeneration in choroideremia patients. In a separate study, however, Moosajee *et al.*[43] showed that zebrafish do not possess an orthologue of a second human REP gene, *REP2*. *REP2* plays an essential role in the pathogenesis of choroideremia as its function is thought to compensate for the loss of *REP1*. The authors of the study therefore concluded that the *rep1* mutant line may not be a suitable model for the study of choroideremia in its present form.[43]

Defects in vesicle trafficking, formation, and/or fusion are known to be the underlying cause of many human ocular diseases such Chediak–Higashi syndrome (CHS, OMIM 214500), Hermansky–Pudlak syndrome (HPS, OMIM 203300), and Griscelli syndromes (GS1-3, OMIM 214450, 607624, and 609227). The pathologies of these human syndromes include hypopigmentation of the RPE, as well as loss of visual function. The RPE contains lysosome-related organelles known as melanosomes that synthesize and store melanin, the main pigment present in the melanosomes of the RPE.[213] Proper fusion of lysosome-related organelles depends on the homotypic fusion and vacuolar protein sorting (HOPS) complex, as it is required for SNARE complex assembly. A model for the study of human conditions such as CHS and HPS has emerged recently in the zebrafish *leberknödel* (*lbk*) mutant line. *lbk* mutants possess a mutation in the *vam6/vps39* gene, which encodes a component of the HOPS complex. Hypopigmentation of the *lbk* RPE, a common characteristic of vesicle traffic defects, results from a defect in melanosome maturation, and the RPE contains vesicles containing undigested photoreceptor OS's. As a result, OS's are shortened and *lbk* mutants display reduced visual function.[46] A mutant in another HOPS complex component, *vps18*, also possesses immature melanocytes and hypopigmentation, and can serve as an additional model for the study of the underlying cellular and molecular basis of ocular defects in hypopigmentation-related disorders.[45]

In addition to their role in melanin synthesis, melanosomes are also thought to play a role in the degradation and detoxification of phagocytosed photoreceptor OSs by the RPE.[214] In human patients suffering from ARMD, phagocytosed OS's are not properly digested by the RPE. This defect results in the accumulation of the lipofuscin component A2E, which is thought to further inhibit OS component degradation, contributing to the progression of ARMD.[215] It is thought that A2E might inhibit the activity of vacuolar ATPases (v-ATPase), whose known role is the acidification of lysosomes and lysosome-related organelles, such as melanosomes.[216] Acidification of these organelles is, in turn, important for the activity of proteases and therefore for proper degradation of accumulated materials. Consistent with this model, a recent analysis of five separate zebrafish mutant lines defective in different subunits of the v-ATPase complex revealed accumulations of undigested OS material in the RPE.[47] While no mutations in v-ATPase have been linked to blindness in humans, these results suggest that defects in v-ATPase might directly or indirectly cause disorders that affect RPE function.

Some of the other mutants that have been shown to possess RPE defects coupled to retinal degeneration are *gantenbein* (*gnn*), *bleached* (*blc*), *fading vision* (*fdv*), and *fade out* (*fad*).[48–50] Of those, only the mutation underlying the *fdv* phenotype has been cloned and identified. The *fdv* mutant was originally isolated during a mutagenesis screen due to reduced pigmentation in the eye as well as the rest its body.[217] Like the *lbk*, *vps18*, and v-ATPase mutants, the *fdv* RPE contains vesicular inclusions and their OS's are reduced. However, the visual defects identified in *fdv* mutants are thought to be due to disrupted recycling of visual pigment, a defect that can be attributed to primary defects in melanosome biogenesis.[44] Positional cloning identified a mutation in the *silver a/pmel17* gene, whose orthologue has been originally shown to be important for pigmentation in mice.[218] While *silver* orthologues in various species have been described and its overall function in melanosome biogenesis established, the molecular mechanisms underlying its roles in pigmentation in general and RPE function specifically are still controversial.[219] Further study of existing mutants, such as *fdv*, as well as the identification and cloning of additional pigmentation mutants, are therefore critical for our understanding of role of the RPE in ocular development and disease.

D. Müller Glia in Regeneration and Disease

The study of stem cells has been of major interest due to the growing potential of their use in therapeutic and regenerative medicine. The ability of zebrafish to regenerate various tissues after injury has attracted major attention from researchers seeking to identify the cellular and molecular mechanisms that enable this regenerative capacity, which is often absent in mammals.[220–222] In the vertebrate retina, two populations of stem/progenitor cells have been identified: the Müller glia of the central retina, and the ciliary marginal zone (CMZ) at the most peripheral edge of the retina. Both of these stem cell populations add new neurons and glia to the zebrafish retina throughout the lifetime of the animal. Müller glia give rise to rod photoreceptor precursors that migrate along Müller glial processes to the photoreceptor layer. All other retinal cell types are continually produced in the post-embryonic CMZ. While the study of the stem/progenitor populations of the CMZ in zebrafish has seen much progress in the past few years, including the identification of mutants with CMZ-specific phenotypes,[24,40,87,223–225] the remainder of this section will focus on recent findings regarding Müller glia.

Neuroprotection, maintenance of retinal homeostasis, and the establishment of proper retinal lamination are some of the many functions attributed to Müller glia in the healthy retina ([226]). In response to retinal disease or damage, Müller glia can become "reactive," as characterized by changes in gene expression that are often followed by dedifferentiation and proliferation.[227,228] Virtually every human retinal disease is associated to some degree with Müller glial reactivity.[226] In some cases, such as diabetic and proliferative retinopathies, and retinal detachment, Müller glia are thought to become reactive in response to primary detects arising in another cell type. In others, Müller glia are thought to be the primary cell type affected. Still, in many human eye diseases, the role of Müller glia in the onset of the disease is unclear. In humans suffering from Basal Cell Naevus Syndrome (BCNS), epiretinal membranes (ERMs)—proliferative and structural abnormalities at the boundary between the retina and the vitreous, have been shown to contain a major glial component. Studies in *Ptch* mutant mice have shown that Müller glial reactivity is locally associated with BCNS-related ocular pathologies.[229] However, it is yet unclear whether ocular defects in BCNS patients are a result of a primary defects at the level of Müller glia, or conversely, whether Müller glia are reactive in response to another ocular defect. Recently, a zebrafish model for the study of BCNS-related ocular pathologies has emerged in the *leprechaun* (*lep*) mutant line, which possesses a loss-of-function mutation in the *patched2* gene, the zebrafish orthologue of *PTCH*. *ptc2* mutants display retinal abnormalities that are similar to those observed in human BCNS patients, including disruptions at the vitreoretinal interface.[40] Further study of *ptc2* mutants is required to determine the role of Müller glia in BCNS-related ocular pathologies.

Müller glia have long been known to possess neurogenic potential, having an intrinsic ability to give rise to newborn neurons. In the adult retina of lower vertebrates, differentiated Müller glia continue to express molecular markers that are characteristic of early retinal progenitors, such as *pax6* and *rx1*.[230] Clusters of proliferating rod progenitors are closely associated with Müller glia and appear to migrate along their processes, which span the apical–basal width of the adult retina, to their final location in the photoreceptor layer.[231,232] A recent study utilizing transgenic zebrafish that express GFP under the control of a Müller glia-specific promoter (Tg(*gfap*:GFP)) as a lineage tracer has shown that rod progenitors retain low levels of GFP, suggesting that Müller glia indeed give rise to rod photoreceptor precursors in the adult retina.[233] These findings support the neurogenic potential of Müller glia and raise the intriguing possibility of using Müller glia as a source for cell replacement therapies for humans suffering from age-related photoreceptor loss.

While in the uninjured adult retina Müller glia give rise strictly to progenitors destined to the rod lineage, Müller glia are also able to dedifferentiate, proliferate, and replenish all neuronal cell types in response to retinal damage.[228,233] While conclusive lineage tracing to show Müller glia as the sole source of regenerated neurons have not yet been published, the use of zebrafish transgenic lines together with extensive immunohistochemical analysis has strongly supported the role of Müller glia as the stem cell population responsible for the regenerative response in the central retina.[233,234] Since mutant analysis in adults is labor intensive and time consuming, multiple labs have also performed microarray analyses after retinal damage in order identify genes that play a role in these processes. Consistent with the idea that regenerative processes often recapitulate embryonic development, these studies identified known developmental pathways such as BMP, Notch, Wnt, and Hedgehog as being transcriptionally regulated in reactive Müller glia.[235,236] It remains to be seen what exact roles of each of these pathways in retinal regeneration, and whether the genetic and/or chemical manipulation of these pathways might have therapeutic value in humans.

E. Intraocular Vasculature

The vertebrate eye is highly metabolically active, and the intraocular vasculature provides oxygen and other nutrients.[237,238] Many human diseases present with defects in intraocular vascularization, including diabetic retinopathy, retinopathy of prematurity, and ARMD.[238] Similarities between zebrafish and human intraocular vasculature make the zebrafish an exciting new model organism in this field. For instance, Alvarez *et al.*[239] recently showed that a zebrafish model of hyperglycemia[240] recapitulates some aspects of nonproliferative diabetic retinopathy; zebrafish may therefore be a useful model for this very common human disease.

In humans and in zebrafish, the hyaloid vessel network surrounds the posterior lens during development.[52,241] In humans, the hyaloid vasculature regresses completely by birth, during which time the retinal vasculature (which lines the inner limiting membrane) forms.[241] In contrast, hyaloid regression does not take place in zebrafish; the hyaloid vasculature instead remodels into the retinal vasculature.[52] In humans and in zebrafish, the choroid vasculature covers the outer surface of the retina.[134,238]

One study utilized a transgenic zebrafish line in which cells of the vasculature express GFP ($Tg(fli1a{:}EGFP)^{y1}$) to screen various mutant and morphant zebrafish for altered formation of the intraocular vasculature.[52] Utilizing this strategy, the authors demonstrated that several genes are required for hyaloid and retinal vasculature development and maintenance, including *syndecan 2* (morphants had no vasculature on the lens), and *mab21l2* (morphants displayed disrupted hyaloid vessel patterning). Another study which investigated the effects of *foxc1* depletion (by morpholino) on hyaloid vessel ultrastructure found that the integrity of the basement membrane of hyaloid vascular cells was disrupted.[51]

Also utilizing transgenic zebrafish lines which express GFP in the vasculature ($Tg(fli1a{:}EGFP)^{y1}$ and $Tg(kdrl{:}GFP)^{la116}$), Kitambi and colleagues[134] performed a small molecule screen to search for compounds which altered retinal vasculature without significant alteration in trunk vasculature. The authors reported two classes of compounds: one which induced degeneration of vessels, and one which increased vessel diameter.[134] This study provides an example of the powerful role that zebrafish small molecule screens can play in therapeutics research.

III. Anterior Segment

A. Lens

The lens is a specialized transparent tissue which, in conjunction with the cornea, focuses incoming light onto the retina.[242] Lens opacity, or cataract, is the leading cause of human visual impairment worldwide.[1] Most cases of cataracts are age-related, and likely have both genetic and environmental causes.[243] In addition, mutations in multiple genes are known to underlie congenital cataracts (which arise in children during the first year of life).[244] However, there are many cases of cataract and other lens disorders for which no causative gene is yet known.[245]

Development and adult morphology of the lens is similar in humans and in zebrafish, but there are important differences as well. During the early stages of lens development in humans and most other vertebrates, cells of the surface ectoderm invaginate to form the lens vesicle.[158,246] In zebrafish lens development, surface ectoderm cells within the lens placode delaminate as a solid cluster of cells rather than a vesicle (Fig. 3).[145,147,155] Despite this difference, cell fate analysis demonstrates that the relationship between zebrafish lens placode cell position and cell fate within the mature lens is consistent with the mammalian model,[146] which indicates that a common genetic program may exist for lens development in vertebrates.

In zebrafish and humans, the mature lens is made up of tightly packed lens fibers which are surrounded at the anterior periphery by a proliferative monolayer of lens epithelial cells (Fig. 4C). Throughout the life of the organism, epithelial cells near the transition zone exit the cell cycle, migrate, elongate, and degrade their light-scattering organelles to become new secondary fibers.[145,147,148,247–249] In zebrafish, this mature lens state is reached by 72 hpf,[146] by which time the lens appropriately focuses an image onto the plane of the retinal photoreceptors.[155] As lens opacity results from many defective cellular processes,[244] the early maturity of the zebrafish lens provides a convenient readout for gene requirement by the lens in mutants and knockdown experiments.

There are subtle differences in the mature structure of the lens between zebrafish and humans. The epithelial cell layer on the surface of the mature zebrafish lens extends further toward the posterior suture; in mammals, epithelial cells end near the lens equator.[147] The shape of the mature zebrafish lens is also more spherical than the human lens, and it is responsible for nearly all refraction to focus light onto the retina, unlike terrestrial vertebrates in which much of the light refraction is performed by the cornea.[144,250,251]

When genes whose mutation causes human congenital cataracts are knocked down in zebrafish embryos, cataracts or other lens abnormalities often result (Table I). Zebrafish therefore provide a useful model to characterize the molecular mechanism of cataract formation. Human mutations in transcription factors *PITX3* and *FOXE3* cause various anterior segment disorders, including cataracts.[252,253] Expression of both of these genes has been knocked down in zebrafish via antisense morpholino,[63,64,78,79] and for both genes this results in lens dysmorphogenesis. Further experiments utilizing these morpholino-injected embryos demonstrated that *foxe3* is genetically downstream of *pitx3* in the zebrafish lens.[63]

Zebrafish with lens defects identified in large-scale mutagenesis screens can also provide valuable information. Proteins known as crystallins contribute to the transparency and refractive power of the lens and cornea,[242] and mutations in many lens crystallins are associated with cataracts in human patients.[254] The zebrafish mutant *cloche* (the genetic basis of which is unknown) presents with cataracts as well as severe vascular defects.[85,255] In these mutants, both *αA-crystallin* mRNA and protein expression is decreased, and the γ-crystallin present within the lens is predominantly insoluble. Both the cataract phenotype and γ-crystallin insolubility were rescued in *cloche* mutants by overexpression of *αA-crystallin* by mRNA injection, likely due to the protein chaperone activity of αA-crystallin.[85] This study demonstrated the requirement for αA-crystallin for proper lens development, as well as providing an exciting example of cataract prevention.

The lens capsule is a transparent basement membrane which completely encloses the lens. Laminin is a component of the lens capsule, and the expression of specific Laminin subunits is conserved between zebrafish and humans.[256] One component of zebrafish and human lens capsules is Laminin-111, a heterotrimeric protein made up of Lama1, Lamb1, and Lamc1 subunits.[5,53,69,257–259] None of these subunit genes have so far been associated with an ocular disease in humans, although humans with mutations in *LAMB2* have lens defects including abnormal shape and cataract.[260,261] However, zebrafish mutants or morphants in *lama1*, *lamb1*, and *lamc1* all present with defective lens phenotypes.[5,24,53,58,69–72,82,262]

In *lama1* morpholino-injected embryos, the lens degenerates completely by 48 hpf,[69] and in various homozygous alleles of *lama1* mutants, the lens is dysmorphic and eventually degenerates completely.[53,72] Homozygous mutants in *lamb1* or *lamc1* present with very similar lens phenotypes characterized by lens dysplasia at 5 dpf,[5] and the lens degenerates completely by 6–9 dpf.[5,72] Zebrafish deficient in subunits of Laminin-111 provide a useful model system to elucidate the roles for laminin-containing ECM in the lens. For instance, one study found that reduced Focal Adhesion Kinase signaling likely underlies the lens phenotype in *lama1* mutants.[53]

B. Cornea

As the most anterior structure within the eye, the transparent cornea provides protection and, in terrestrial vertebrates, refraction of incoming light.[144,250,251] Major causes of corneal blindness include the infectious disease Trachoma, vitamin A deficiency, and injury.[1] Human genetic corneal dystrophies, though less common, are another cause of blindness due to corneal opacity.[263]

In humans and in zebrafish, the mature cornea contains five major layers: the stratified corneal epithelium at the anterior, Bowman's Layer, the stroma, Descemet's membrane, and the corneal endothelium.[148,150,151,154,264] At 72 hpf in zebrafish, a rudimentary cornea consisting of a two-cell-layer epithelium, an acellular stroma, and the endothelium is in place.[148,150] Bowman's layer is formed by 5 dpf, and the zebrafish cornea (which is much thinner than the human cornea) is fully mature at 2 months postfertilization.[150]

Many shared proteins are expressed in both the human and zebrafish cornea,[150] although many of the highly expressed corneal crystallins are taxon specific.[265] In most mammals, including humans, the most abundant corneal crystallin is Aldehyde Dehydrogenase 3a1.[266] In contrast, 50% of soluble protein in zebrafish adult cornea is made up of Scinderin-Like Protein A (also known as C/L-Gelsolin and Gelsolin-Like 1).[267]

There are several human corneal dystrophies which lead to opacity of the cornea. For many, the responsible gene is known, but few knockout or transgenic mouse mutants in the particular defective gene have been reported.[263] Some of these genes, including *pip5k3* (Fleck Corneal Dystrophy[268,269]) and *keratocan* (Cornea Plana[270,271]) are expressed in the zebrafish cornea. Using blastula-stage transplants (see Section I), it is possible to study the function of genes, the mutation of which would otherwise be embryonic lethal in the mature zebrafish cornea.[53] Semina and colleagues[53] utilized this technique to show that loss of functional *lama1* in eye tissues leads to focal corneal dysplasia in adult zebrafish. Utilizing early-stage lens ablation, the authors further showed that corneal and other eye defects in *lama1* mutants are not a secondary consequence of lens degeneration.

C. Glaucoma

Second in prevalence only to cataracts in cases of worldwide human blindness,[1] glaucoma encompasses a group of related progressive diseases characterized by death of retinal ganglion cells and subsequent degeneration of the optic nerve which leads to irreversible blindness.[272] Although many genes are associated with human cases of glaucoma, incomplete penetrance within families indicates that the various causes of glaucoma are multigenic.[273]

Elevated intraocular pressure (IOP) is a major risk factor associated with glaucoma.[274,275] IOP results from the balance between aqueous humor secretion (from the ciliary epithelium) and outflow, with defective outflow being the most common cause of elevated IOP.[276,277] In humans, aqueous humor provides nutrients and removes waste products, and its exit from the anterior segment is through the trabecular meshwork outflow pathways.[278]

Although one group has advised caution in using zebrafish as a model for glaucoma research due to the ultrastructural dissimilarity between the mammalian trabecular meshwork and the zebrafish annular ligament,[153] a second group has more recently utilized aqueous humor tracing experiments to meticulously detail the outflow pathway in the zebrafish eye.[149] Gray and colleagues[149] found that the appropriate zebrafish analog to the mammalian outflow pathway is in fact the ventral canalicular network, and not the annular ligament. Further, the authors found that the ultrastructure of this region was overall quite similar in zebrafish and humans, although the aqueous humor flow pathway in zebrafish is vectorial (flowing from dorsal to ventral) rather than circumferential as in humans. Despite this difference, the overall similarities in aqueous humor outflow tissue structure make the zebrafish a potentially valuable model organism for human glaucomas.

The zebrafish is an ideal model system in which to study multigenic traits.[279] In addition, methodology for measuring zebrafish IOP has been developed, and average intraocular pressure ranges are similar in mammals and zebrafish.[152] Utilizing this technique, the zebrafish mutant *brass* (genetic basis unknown) was shown to have elevated IOP along with anterior segment defects.[152] Although *brass* zebrafish do not develop glaucoma, their genetic background could be used to uncover glaucoma modifier loci.[152]

In humans, mutation of the gene *FOXC1* is associated with Axenfeld–Reiger Syndrome, which includes glaucoma in some individuals.[280–282] Zebrafish *foxc1* is expressed in the anterior segment and POM.[279] By utilizing morpholinos to knock down *foxc1* expression in zebrafish embryos along with other techniques, one study has provided evidence that FoxC1 directly activates expression of *fgf19* within the POM.[61] Fgf19 goes on to activate Fgfr4 receptors, and this interaction is required for appropriate anterior segment development.[61] Reduced *foxc1* expression during zebrafish eye development also results in reduced expression of the cellular homeostasis and apoptosis regulator *foxo1*, and in increased apoptosis within the zebrafish eye.[283]

Mutation of the gene *lmx1b* in humans is associated with Nail–Patella syndrome and an increased susceptibility to glaucoma.[284] Similar to the expression pattern in mice,[285] zebrafish orthologues of this gene are expressed in cells of the POM.[187] Reduction of *lmx1b* expression in zebrafish with morpholinos resulted in early eye morphological defects, enhanced ocular FGF activity, and in altered expression of two genes also implicated in glaucoma: *foxc1* and *pitx2*.[187,282,286,287] Results from the study further suggest that *lmx1b*-expressing cells in the zebrafish eye have migration defects in the absence of *lmx1b* expression.[187]

IV. Concluding Remarks

The zebrafish has emerged as a powerful model system for the study of the cellular and molecular underpinnings of human ocular pathologies. Large-scale genetic screens have identified genes whose study is relevant for understanding both eye development and disease, while the continual development of tools for the manipulation of gene function *in vivo* has enabled researchers to study their functions during eye development and maintenance. Current and future research efforts utilizing the zebrafish system promise to continue to provide important insights into human ocular conditions and contribute to the discovery and development of relevant therapeutics.

Acknowledgements

We apologize for the omission of other relevant zebrafish eye mutants and morphants due to space constraints, and the responsibility for this is borne entirely by the senior author. Work in the Gross lab has been supported by grants from the American Health Assistance Foundation Macular Degeneration Research Program, the Knights Templar Eye Foundation, the Retina Research Foundation, the Karl Kirchgessner Foundation, the E. Matilda Ziegler Foundation for the Blind, a CAREER Award from the National Science Foundation (IOS-0745782), and RO1-EY18005 from the National Institutes of Health. We are grateful to Marianna Grenadier for generating artwork, and to Robin Williamson, Edward Spaghetti and Simon Posford for editorial assistance.

References

1. Resnikoff S, Pascolini D, Etya'ale D, Kocur I, Pararajasegaram R, Pokharel GP, et al. Global data on visual impairment in the year 2002. *Bull World Health Organ* 2004;**82**:844–51.
2. Nadauld LD, Chidester S, Shelton DN, Rai K, Broadbent T, Sandoval IT, et al. Dual roles for adenomatous polyposis coli in regulating retinoic acid biosynthesis and Wnt during ocular development. *Proc Natl Acad Sci USA* 2006;**103**:13409–14.
3. Ng D, Thakker N, Corcoran CM, Donnai D, Perveen R, Schneider A, et al. Oculofaciocardiodental and Lenz microphthalmia syndromes result from distinct classes of mutations in BCOR. *Nat Genet* 2004;**36**:411–6.
4. Masai I, Lele Z, Yamaguchi M, Komori A, Nakata A, Nishiwaki Y, et al. N-cadherin mediates retinal lamination, maintenance of forebrain compartments and patterning of retinal neurites. *Development* 2003;**130**:2479–94.
5. Lee J, Gross JM. Laminin beta1 and gamma1 containing laminins are essential for basement membrane integrity in the zebrafish eye. *Invest Ophthalmol Vis Sci* 2007;**48**:2483–90.
6. Macdonald R, Scholes J, Strahle U, Brennan C, Holder N, Brand M, et al. The Pax protein Noi is required for commissural axon pathway formation in the rostral forebrain. *Development* 1997;**124**:2397–408.
7. Lee J, Willer JR, Willer GB, Smith K, Gregg RG, Gross JM. Zebrafish blowout provides genetic evidence for Patched1-mediated negative regulation of Hedgehog signaling within the proximal optic vesicle of the vertebrate eye. *Dev Biol* 2008;**319**:10–22.
8. Koudijs MJ, den Broeder MJ, Groot E, van Eeden FJ. Genetic analysis of the two zebrafish patched homologues identifies novel roles for the hedgehog signaling pathway. *BMC Dev Biol* 2008;**8**:15.
9. Chao R, Nevin L, Agarwal P, Riemer J, Bai X, Delaney A, et al. A male with unilateral microphthalmia reveals a role for TMX3 in eye development. *PLoS ONE* 2010;**5**:e10565.
10. Gestri G, Osborne RJ, Wyatt AW, Gerrelli D, Gribble S, Stewart H, et al. Reduced TFAP2A function causes variable optic fissure closure and retinal defects and sensitizes eye development to mutations in other morphogenetic regulators. *Hum Genet* 2009;**126**:791–803.
11. Sanek NA, Taylor AA, Nyholm MK, Grinblat Y. Zebrafish zic2a patterns the forebrain through modulation of Hedgehog-activated gene expression. *Development* 2009;**136**:3791–800.
12. Brown JD, Dutta S, Bharti K, Bonner RF, Munson PJ, Dawid IB, et al. Expression profiling during ocular development identifies 2 Nlz genes with a critical role in optic fissure closure. *Proc Natl Acad Sci USA* 2009;**106**:1462–7.

13. Brockerhoff SE, Rieke F, Matthews HR, Taylor MR, Kennedy B, Ankoudinova I, et al. Light stimulates a transducin-independent increase of cytoplasmic Ca2+ and suppression of current in cones from the zebrafish mutant nof. *J Neurosci* 2003;**23**:470–80.
14. Omori Y, Malicki J. oko meduzy and related crumbs genes are determinants of apical cell features in the vertebrate embryo. *Curr Biol* 2006;**16**:945–57.
15. Brockerhoff SE, Hurley JB, Janssen-Bienhold U, Neuhauss SC, Driever W, Dowling JE. A behavioral screen for isolating zebrafish mutants with visual system defects. *Proc Natl Acad Sci USA* 1995;**92**:10545–9.
16. Taylor MR, Hurley JB, Van Epps HA, Brockerhoff SE. A zebrafish model for pyruvate dehydrogenase deficiency: rescue of neurological dysfunction and embryonic lethality using a ketogenic diet. *Proc Natl Acad Sci USA* 2004;**101**:4584–9.
17. Doerre G, Malicki J. A mutation of early photoreceptor development, mikre oko, reveals cell-cell interactions involved in the survival and differentiation of zebrafish photoreceptors. *J Neurosci* 2001;**21**:6745–57.
18. Del Bene F, Wehman AM, Link BA, Baier H. Regulation of neurogenesis by interkinetic nuclear migration through an apical–basal notch gradient. *Cell* 2008;**134**:1055–65.
19. Jing X, Malicki J. Zebrafish ale oko, an essential determinant of sensory neuron survival and the polarity of retinal radial glia, encodes the p50 subunit of dynactin. *Development* 2009;**136**:2955–64.
20. Hsu YC, Willoughby JJ, Christensen AK, Jensen AM. Mosaic Eyes is a novel component of the Crumbs complex and negatively regulates photoreceptor apical size. *Development* 2006;**133**:4849–59.
21. Jensen AM, Walker C, Westerfield M. mosaic eyes: a zebrafish gene required in pigmented epithelium for apical localization of retinal cell division and lamination. *Development* 2001;**128**:95–105.
22. Doerre G, Malicki J. Genetic analysis of photoreceptor cell development in the zebrafish retina. *Mech Dev* 2002;**110**:125–38.
23. Pathak N, Obara T, Mangos S, Liu Y, Drummond IA. The zebrafish fleer gene encodes an essential regulator of cilia tubulin polyglutamylation. *Mol Biol Cell* 2007;**18**:4353–64.
24. Gross JM, Perkins BD, Amsterdam A, Egana A, Darland T, Matsui JI, et al. Identification of zebrafish insertional mutants with defects in visual system development and function. *Genetics* 2005;**170**:245–61.
25. Sukumaran S, Perkins BD. Early defects in photoreceptor outer segment morphogenesis in zebrafish ift57, ift88 and ift172 Intraflagellar Transport mutants. *Vision Res* 2009;**49**:479–89.
26. Hudak LM, Lunt S, Chang CH, Winkler E, Flammer H, Lindsey M, et al. The intraflagellar transport protein ift80 is essential for photoreceptor survival in a zebrafish model of jeune asphyxiating thoracic dystrophy. *Invest Ophthalmol Vis Sci* 2010;**7**:3792–9.
27. Tsujikawa M, Malicki J. Intraflagellar transport genes are essential for differentiation and survival of vertebrate sensory neurons. *Neuron* 2004;**42**:703–16.
28. Wei X, Malicki J. nagie oko, encoding a MAGUK-family protein, is essential for cellular patterning of the retina. *Nat Genet* 2002;**31**:150–7.
29. Zou J, Lathrop KL, Sun M, Wei X. Intact retinal pigment epithelium maintained by Nok is essential for retinal epithelial polarity and cellular patterning in zebrafish. *J Neurosci* 2008;**28**:13684–95.
30. Brockerhoff SE, Hurley JB, Niemi GA, Dowling JE. A new form of inherited red-blindness identified in zebrafish. *J Neurosci* 1997;**17**:4236–42.
31. Taylor MR, Kikkawa S, Diez-Juan A, Ramamurthy V, Kawakami K, Carmeliet P, et al. The zebrafish pob gene encodes a novel protein required for survival of red cone photoreceptor cells. *Genetics* 2005;**170**:263–73.

32. Nishiwaki Y, Komori A, Sagara H, Suzuki E, Manabe T, Hosoya T, et al. Mutation of cGMP phosphodiesterase 6alpha'-subunit gene causes progressive degeneration of cone photoreceptors in zebrafish. *Mech Dev* 2008;**125**:932–46.
33. Cui S, Otten C, Rohr S, Abdelilah-Seyfried S, Link BA. Analysis of aPKClambda and aPKCzeta reveals multiple and redundant functions during vertebrate retinogenesis. *Mol Cell Neurosci* 2007;**34**:431–44.
34. Seiler C, Finger-Baier KC, Rinner O, Makhankov YV, Schwarz H, Neuhauss SC, et al. Duplicated genes with split functions: independent roles of protocadherin15 orthologues in zebrafish hearing and vision. *Development* 2005;**132**:615–23.
35. Omori Y, Zhao C, Saras A, Mukhopadhyay S, Kim W, Furukawa T, et al. Elipsa is an early determinant of ciliogenesis that links the IFT particle to membrane-associated small GTPase Rab8. *Nat Cell Biol* 2008;**10**:437–44.
36. Otteson DC, Tsujikawa M, Gunatilaka T, Malicki J, Zack DJ. Genomic organization of zebrafish cone-rod homeobox gene and exclusion as a candidate gene for retinal degeneration in niezerka and mikre oko. *Mol Vis* 2005;**11**:986–95.
37. Maaswinkel H, Riesbeck LE, Riley ME, Carr AL, Mullin JP, Nakamoto AT, et al. Behavioral screening for nightblindness mutants in zebrafish reveals three new loci that cause dominant photoreceptor cell degeneration. *Mech Ageing Dev* 2005;**126**:1079–89.
38. Li L, Dowling JE. A dominant form of inherited retinal degeneration caused by a non-photoreceptor cell-specific mutation. *Proc Natl Acad Sci USA* 1997;**94**:11645–50.
39. Li L, Dowling JE. Disruption of the olfactoretinal centrifugal pathway may relate to the visual system defect in night blindness b mutant zebrafish. *J Neurosci* 2000;**20**:1883–92.
40. Bibliowicz J, Gross JM. Expanded progenitor populations, vitreo-retinal abnormalities, and Muller glial reactivity in the zebrafish leprechaun/patched2 retina. *BMC Dev Biol* 2009;**9**:52.
41. Faucherre A, Taylor GS, Overvoorde J, Dixon JE, Hertog J. Zebrafish pten genes have overlapping and non-redundant functions in tumorigenesis and embryonic development. *Oncogene* 2008;**27**:1079–86.
42. Krock BL, Bilotta J, Perkins BD. Noncell-autonomous photoreceptor degeneration in a zebrafish model of choroideremia. *Proc Natl Acad Sci USA* 2007;**104**:4600–5.
43. Moosajee M, Tulloch M, Baron RA, Gregory-Evans CY, Pereira-Leal JB, Seabra MC. Single choroideremia gene in nonmammalian vertebrates explains early embryonic lethality of the zebrafish model of choroideremia. *Invest Ophthalmol Vis Sci* 2009;**50**:3009–16.
44. Schonthaler HB, Lampert JM, von Lintig J, Schwarz H, Geisler R, Neuhauss SC. A mutation in the silver gene leads to defects in melanosome biogenesis and alterations in the visual system in the zebrafish mutant fading vision. *Dev Biol* 2005;**284**:421–36.
45. Maldonado E, Hernandez F, Lozano C, Castro ME, Navarro RE. The zebrafish mutant vps18 as a model for vesicle-traffic related hypopigmentation diseases. *Pigment Cell Res* 2006;**19**:315–26.
46. Schonthaler HB, Fleisch VC, Biehlmaier O, Makhankov Y, Rinner O, Bahadori R, et al. The zebrafish mutant lbk/vam6 resembles human multisystemic disorders caused by aberrant trafficking of endosomal vesicles. *Development* 2008;**135**:387–99.
47. Nuckels RJ, Ng A, Darland T, Gross JM. The vacuolar-ATPase complex regulates retinoblast proliferation and survival, photoreceptor morphogenesis, and pigmentation in the zebrafish eye. *Invest Ophthalmol Vis Sci* 2009;**50**:893–905.
48. Neuhauss SC, Seeliger MW, Schepp CP, Biehlmaier O. Retinal defects in the zebrafish bleached mutant. *Doc Ophthalmol* 2003;**107**:71–8.
49. Bahadori R, Rinner O, Schonthaler HB, Biehlmaier O, Makhankov YV, Rao P, et al. The Zebrafish fade out mutant: a novel genetic model for Hermansky–Pudlak syndrome. *Invest Ophthalmol Vis Sci* 2006;**47**:4523–31.

50. Biehlmaier O, Neuhauss SC, Kohler K. Double cone dystrophy and RPE degeneration in the retina of the zebrafish gnn mutant. *Invest Ophthalmol Vis Sci* 2003;**44**:1287–98.
51. Skarie JM, Link BA. FoxC1 is essential for vascular basement membrane integrity and hyaloid vessel morphogenesis. *Invest Ophthalmol Vis Sci* 2009;**50**:5026–34.
52. Alvarez Y, Cederlund ML, Cottell DC, Bill BR, Ekker SC, Torres-Vazquez J, et al. Genetic determinants of hyaloid and retinal vasculature in zebrafish. *BMC Dev Biol* 2007;**7**:114.
53. Semina EV, Bosenko DV, Zinkevich NC, Soules KA, Hyde DR, Vihtelic TS, et al. Mutations in laminin alpha 1 result in complex, lens-independent ocular phenotypes in zebrafish. *Dev Biol* 2006;**299**:63–77.
54. Babb SG, Kotradi SM, Shah B, Chiappini-Williamson C, Bell LN, Schmeiser G, et al. Zebrafish R-cadherin (Cdh4) controls visual system development and differentiation. *Dev Dyn* 2005;**233**:930–45.
55. Vihtelic TS, Hyde DR. Zebrafish mutagenesis yields eye morphological mutants with retinal and lens defects. *Vision Res* 2002;**42**:535–40.
56. Vihtelic TS, Murphy TR, Watson CT, Willer GB, Gregg RG, Hyde DR. Zebrafish lens opaque (lop) mutation mapping and gene identification. *Invest Ophthalmol Vis Sci* 2007;**48**:S2447.
57. Vihtelic TS, Yamamoto Y, Springer SS, Jeffery WR, Hyde DR. Lens opacity and photoreceptor degeneration in the zebrafish lens opaque mutant. *Dev Dyn* 2005;**233**:52–65.
58. Moosajee M, Gregory-Evans K, Ellis CD, Seabra MC, Gregory-Evans CY. Translational bypass of nonsense mutations in zebrafish rep1, pax2.1 and lamb1 highlights a viable therapeutic option for untreatable genetic eye disease. *Hum Mol Genet* 2008;**17**:3987–4000.
59. Cheng S, Shakespeare T, Mui R, White TW, Valdimarsson G. Connexin 48.5 is required for normal cardiovascular function and lens development in zebrafish embryos. *J Biol Chem* 2004;**279**:36993–7003.
60. Ye M, Berry-Wynne KM, Asai-Coakwell M, Sundaresan P, Footz T, French CR, et al. Mutation of the bone morphogenetic protein GDF3 causes ocular and skeletal anomalies. *Hum Mol Genet* 2010;**19**:287–98.
61. Tamimi Y, Skarie JM, Footz T, Berry FB, Link BA, Walter MA. FGF19 is a target for FOXC1 regulation in ciliary body-derived cells. *Hum Mol Genet* 2006;**15**:3229–40.
62. Nakayama Y, Miyake A, Nakagawa Y, Mido T, Yoshikawa M, Konishi M, et al. Fgf19 is required for zebrafish lens and retina development. *Dev Biol* 2008;**313**:752–66.
63. Shi X, Luo Y, Howley S, Dzialo A, Foley S, Hyde DR, et al. Zebrafish foxe3: roles in ocular lens morphogenesis through interaction with pitx3. *Mech Dev* 2006;**123**:761–82.
64. Swindell EC, Zilinski CA, Hashimoto R, Shah R, Lane ME, Jamrich M. Regulation and function of foxe3 during early zebrafish development. *Genesis* 2008;**46**:177–83.
65. Asai-Coakwell M, French CR, Berry KM, Ye M, Koss R, Somerville M, et al. GDF6, a novel locus for a spectrum of ocular developmental anomalies. *Am J Hum Genet* 2007;**80**:306–15.
66. French CR, Erickson T, French DV, Pilgrim DB, Waskiewicz AJ. Gdf6a is required for the initiation of dorsal-ventral retinal patterning and lens development. *Dev Biol* 2009;**333**:37–47.
67. Evans TG, Yamamoto Y, Jeffery WR, Krone PH. Zebrafish Hsp70 is required for embryonic lens formation. *Cell Stress Chaperones* 2005;**10**:66–78.
68. Evans TG, Belak Z, Ovsenek N, Krone PH. Heat shock factor 1 is required for constitutive Hsp70 expression and normal lens development in embryonic zebrafish. *Comp Biochem Physiol A Mol Integr Physiol* 2007;**146**:131–40.
69. Zinkevich NS, Bosenko DV, Link BA, Semina EV. laminin alpha 1 gene is essential for normal lens development in zebrafish. *BMC Dev Biol* 2006;**6**:13.
70. Vihtelic TS, Yamamoto Y, Sweeney MT, Jeffery WR, Hyde DR. Arrested differentiation and epithelial cell degeneration in zebrafish lens mutants. *Dev Dyn* 2001;**222**:625–36.

71. Stemple DL, Solnica-Krezel L, Zwartkruis F, Neuhauss SC, Schier AF, Malicki J, et al. Mutations affecting development of the notochord in zebrafish. *Development* 1996;**123**:117–28.
72. Biehlmaier O, Makhankov Y, Neuhauss SC. Impaired retinal differentiation and maintenance in zebrafish laminin mutants. *Invest Ophthalmol Vis Sci* 2007;**48**:2887–94.
73. Harding RL, Howley S, Baker LJ, Murphy TR, Archer WE, Wistow G, et al. Lengsin expression and function during zebrafish lens formation. *Exp Eye Res* 2008;**86**:807–18.
74. Kennedy BN, Stearns GW, Smyth VA, Ramamurthy V, van Eeden F, Ankoudinova I, et al. Zebrafish rx3 and mab21l2 are required during eye morphogenesis. *Dev Biol* 2004;**270**:336–49.
75. Kleinjan DA, Bancewicz RM, Gautier P, Dahm R, Schonthaler HB, Damante G, et al. Subfunctionalization of duplicated zebrafish pax6 genes by cis-regulatory divergence. *PLoS Genet* 2008;**4**:e29.
76. Yeyati PL, Bancewicz RM, Maule J, van Heyningen V. Hsp90 selectively modulates phenotype in vertebrate development. *PLoS Genet* 2007;**3**:e43.
77. Heisenberg CP, Brand M, Jiang YJ, Warga RM, Beuchle D, van Eeden FJ, et al. Genes involved in forebrain development in the zebrafish, *Danio rerio*. *Development* 1996;**123**:191–203.
78. Dutta S, Dietrich JE, Aspock G, Burdine RD, Schier A, Westerfield M, et al. pitx3 defines an equivalence domain for lens and anterior pituitary placode. *Development* 2005;**132**:1579–90.
79. Shi X, Bosenko DV, Zinkevich NS, Foley S, Hyde DR, Semina EV, et al. Zebrafish pitx3 is necessary for normal lens and retinal development. *Mech Dev* 2005;**122**:513–27.
80. Stigloher C, Ninkovic J, Laplante M, Geling A, Tannhauser B, Topp S, et al. Segregation of telencephalic and eye-field identities inside the zebrafish forebrain territory is controlled by Rx3. *Development* 2006;**133**:2925–35.
81. Skarie JM, Link BA. The primary open-angle glaucoma gene WDR36 functions in ribosomal RNA processing and interacts with the p53 stress-response pathway. *Hum Mol Genet* 2008;**17**:2474–85.
82. Neuhauss SC, Biehlmaier O, Seeliger MW, Das T, Kohler K, Harris WA, et al. Genetic disorders of vision revealed by a behavioral screen of 400 essential loci in zebrafish. *J Neurosci* 1999;**19**:8603–15.
83. Schonthaler HB, Franz-Odendaal TA, Hodel C, Gehring I, Geisler R, Schwarz H, et al. The zebrafish mutant bumper shows a hyperproliferation of lens epithelial cells and fibre cell degeneration leading to functional blindness. *Mech Dev* 2010;**127**:203–19.
84. Avanesov A, Dahm R, Sewell WF, Malicki JJ. Mutations that affect the survival of selected amacrine cell subpopulations define a new class of genetic defects in the vertebrate retina. *Dev Biol* 2005;**285**:138–55.
85. Goishi K, Shimizu A, Najarro G, Watanabe S, Rogers R, Zon LI, et al. AlphaA-crystallin expression prevents gamma-crystallin insolubility and cataract formation in the zebrafish cloche mutant lens. *Development* 2006;**133**:2585–93.
86. Catalano AE, Raymond PA, Goldman D, Wei X. Zebrafish dou yan mutation causes patterning defects and extensive cell death in the retina. *Dev Dyn* 2007;**236**:1295–306.
87. Wehman AM, Staub W, Meyers JR, Raymond PA, Baier H. Genetic dissection of the zebrafish retinal stem-cell compartment. *Dev Biol* 2005;**281**:53–65.
88. Spence R, Gerlach G, Lawrence C, Smith C. The behaviour and ecology of the zebrafish, *Danio rerio*. *Biol Rev Camb Philos Soc* 2008;**83**:13–34.
89. Haffter P, Nusslein-Volhard C. Large scale genetics in a small vertebrate, the zebrafish. *Int J Dev Biol* 1996;**40**:221–7.
90. Kimmel CB, Warga RM, Schilling TF. Origin and organization of the zebrafish fate map. *Development* 1990;**108**:581–94.

91. Woo K, Shih J, Fraser SE. Fate maps of the zebrafish embryo. *Curr Opin Genet Dev* 1995;**5**:439–43.
92. Carmany-Rampey A, Moens CB. Modern mosaic analysis in the zebrafish. *Methods* 2006;**39**:228–38.
93. Yamamoto Y, Jeffery WR. Probing teleost eye development by lens transplantation. *Methods* 2002;**28**:420–6.
94. Meng X, Noyes MB, Zhu LJ, Lawson ND, Wolfe SA. Targeted gene inactivation in zebrafish using engineered zinc-finger nucleases. *Nat Biotechnol* 2008;**26**:695–701.
95. Doyon Y, McCammon JM, Miller JC, Faraji F, Ngo C, Katibah GE, et al. Heritable targeted gene disruption in zebrafish using designed zinc-finger nucleases. *Nat Biotechnol* 2008;**26**:702–8.
96. Nasevicius A, Ekker SC. Effective targeted gene 'knockdown' in zebrafish. *Nat Genet* 2000;**26**:216–20.
97. Draper BW, Morcos PA, Kimmel CB. Inhibition of zebrafish fgf8 pre-mRNA splicing with morpholino oligos: a quantifiable method for gene knockdown. *Genesis* 2001;**30**:154–6.
98. Thummel R, Bai S, Sarras Jr. MP, Song P, McDermott J, Brewer J, et al. Inhibition of zebrafish fin regeneration using in vivo electroporation of morpholinos against fgfr1 and msxb. *Dev Dyn* 2006;**235**:336–46.
99. Shestopalov IA, Sinha S, Chen JK. Light-controlled gene silencing in zebrafish embryos. *Nat Chem Biol* 2007;**3**:650–1.
100. Stuart GW, McMurray JV, Westerfield M. Replication, integration and stable germ-line transmission of foreign sequences injected into early zebrafish embryos. *Development* 1988;**103**:403–12.
101. Stuart GW, Vielkind JR, McMurray JV, Westerfield M. Stable lines of transgenic zebrafish exhibit reproducible patterns of transgene expression. *Development* 1990;**109**:577–84.
102. Culp P, Nusslein-Volhard C, Hopkins N. High-frequency germ-line transmission of plasmid DNA sequences injected into fertilized zebrafish eggs. *Proc Natl Acad Sci USA* 1991;**88**:7953–7.
103. Kawakami K. Transgenesis and gene trap methods in zebrafish by using the Tol2 transposable element. *Methods Cell Biol* 2004;**77**:201–22.
104. Amsterdam A, Lin S, Hopkins N. The Aequorea victoria green fluorescent protein can be used as a reporter in live zebrafish embryos. *Dev Biol* 1995;**171**:123–9.
105. Halloran MC, Sato-Maeda M, Warren JT, Su F, Lele Z, Krone PH, et al. Laser-induced gene expression in specific cells of transgenic zebrafish. *Development* 2000;**127**:1953–60.
106. Scheer N, Campos-Ortega JA. Use of the Gal4–UAS technique for targeted gene expression in the zebrafish. *Mech Dev* 1999;**80**:153–8.
107. Hua JY, Smear MC, Baier H, Smith SJ. Regulation of axon growth in vivo by activity-based competition. *Nature* 2005;**434**:1022–6.
108. Hatta K, Tsujii H, Omura T. Cell tracking using a photoconvertible fluorescent protein. *Nat Protoc* 2006;**1**:960–7.
109. Kurita R, Sagara H, Aoki Y, Link BA, Arai K, Watanabe S. Suppression of lens growth by alphaA-crystallin promoter-driven expression of diphtheria toxin results in disruption of retinal cell organization in zebrafish. *Dev Biol* 2003;**255**:113–27.
110. Pisharath H, Rhee JM, Swanson MA, Leach SD, Parsons MJ. Targeted ablation of beta cells in the embryonic zebrafish pancreas using *E. coli* nitroreductase. *Mech Dev* 2007;**124**:218–29.
111. Brand AH, Perrimon N. Targeted gene expression as a means of altering cell fates and generating dominant phenotypes. *Development* 1993;**118**:401–15.
112. Halpern ME, Rhee J, Goll MG, Akitake CM, Parsons M, Leach SD. Gal4/UAS transgenic tools and their application to zebrafish. *Zebrafish* 2008;**5**:97–110.

113. Sauer B. Functional expression of the cre-lox site-specific recombination system in the yeast *Saccharomyces cerevisiae*. *Mol Cell Biol* 1987;**7**:2087–96.
114. Langenau DM, Feng H, Berghmans S, Kanki JP, Kutok JL, Look AT. Cre/lox-regulated transgenic zebrafish model with conditional myc-induced T cell acute lymphoblastic leukemia. *Proc Natl Acad Sci USA* 2005;**102**:6068–73.
115. Sinha DK, Neveu P, Gagey N, Aujard I, Le Saux T, Rampon C, et al. Photoactivation of the CreER T2 recombinase for conditional site-specific recombination with high spatiotemporal resolution. *Zebrafish* 2010;**7**:199–204.
116. Brenner S. The genetics of *Caenorhabditis elegans*. *Genetics* 1974;**77**:71–94.
117. Nusslein-Volhard C, Wieschaus E. Mutations affecting segment number and polarity in Drosophila. *Nature* 1980;**287**:795–801.
118. Walker C, Streisinger G. Induction of mutations by gamma-rays in pregonial germ cells of zebrafish embryos. *Genetics* 1983;**103**:125–36.
119. Streisinger G, Walker C, Dower N, Knauber D, Singer F. Production of clones of homozygous diploid zebra fish (*Brachydanio rerio*). *Nature* 1981;**291**:293–6.
120. Haffter P, Granato M, Brand M, Mullins MC, Hammerschmidt M, Kane DA, et al. The identification of genes with unique and essential functions in the development of the zebrafish, *Danio rerio*. *Development* 1996;**123**:1–36.
121. Driever W, Solnica-Krezel L, Schier AF, Neuhauss SC, Malicki J, Stemple DL, et al. A genetic screen for mutations affecting embryogenesis in zebrafish. *Development* 1996;**123**:37–46.
122. Mullins MC, Hammerschmidt M, Haffter P, Nusslein-Volhard C. Large-scale mutagenesis in the zebrafish: in search of genes controlling development in a vertebrate. *Curr Biol* 1994;**4**:189–202.
123. Patton EE, Zon LI. The art and design of genetic screens: zebrafish. *Nat Rev Genet* 2001;**2**:956–66.
124. Amsterdam A, Burgess S, Golling G, Chen W, Sun Z, Townsend K, et al. A large-scale insertional mutagenesis screen in zebrafish. *Genes Dev* 1999;**13**:2713–24.
125. Grunwald DJ, Streisinger G. Induction of recessive lethal and specific locus mutations in the zebrafish with ethyl nitrosourea. *Genet Res* 1992;**59**:103–16.
126. Solnica-Krezel L, Schier AF, Driever W. Efficient recovery of ENU-induced mutations from the zebrafish germline. *Genetics* 1994;**136**:1401–20.
127. Amsterdam A, Hopkins N. Mutagenesis strategies in zebrafish for identifying genes involved in development and disease. *Trends Genet* 2006;**22**:473–8.
128. Golling G, Amsterdam A, Sun Z, Antonelli M, Maldonado E, Chen W, et al. Insertional mutagenesis in zebrafish rapidly identifies genes essential for early vertebrate development. *Nat Genet* 2002;**31**:135–40.
129. Wienholds E, van Eeden F, Kosters M, Mudde J, Plasterk RH, Cuppen E. Efficient target-selected mutagenesis in zebrafish. *Genome Res* 2003;**13**:2700–7.
130. Moens CB, Donn TM, Wolf-Saxon ER, Ma TP. Reverse genetics in zebrafish by TILLING. *Brief Funct Genomic Proteomic* 2008;**7**:454–9.
131. Taylor KL, Grant NJ, Temperley ND, Patton EE. Small molecule screening in zebrafish: an in vivo approach to identifying new chemical tools and drug leads. *Cell Commun Signal* 2010;**8**:11.
132. Owens KN, Santos F, Roberts B, Linbo T, Coffin AB, Knisely AJ, et al. Identification of genetic and chemical modulators of zebrafish mechanosensory hair cell death. *PLoS Genet* 2008;**4**:e1000020.
133. Trompouki E, Zon LI. Small molecule screen in zebrafish and HSC expansion. *Methods Mol Biol* 2010;**636**:301–16.
134. Kitambi SS, McCulloch KJ, Peterson RT, Malicki JJ. Small molecule screen for compounds that affect vascular development in the zebrafish retina. *Mech Dev* 2009;**126**:464–77.

135. Kokel D, Bryan J, Laggner C, White R, Cheung CY, Mateus R, et al. Rapid behavior-based identification of neuroactive small molecules in the zebrafish. *Nat Chem Biol* 2010;**6**:231–7.
136. Peterson RT, Link BA, Dowling JE, Schreiber SL. Small molecule developmental screens reveal the logic and timing of vertebrate development. *Proc Natl Acad Sci USA* 2000;**97**:12965–9.
137. Murphey RD, Stern HM, Straub CT, Zon LI. A chemical genetic screen for cell cycle inhibitors in zebrafish embryos. *Chem Biol Drug Des* 2006;**68**:213–9.
138. Gross JM, Perkins BD. Zebrafish mutants as models for congenital ocular disorders in humans. *Mol Reprod Dev* 2008;**75**:547–55.
139. Bilotta J, Saszik S. The zebrafish as a model visual system. *Int J Dev Neurosci* 2001;**19**:621–9.
140. Fadool JM, Dowling JE. Zebrafish: a model system for the study of eye genetics. *Prog Retin Eye Res* 2008;**27**:89–110.
141. Goldsmith P, Harris WA. The zebrafish as a tool for understanding the biology of visual disorders. *Semin Cell Dev Biol* 2003;**14**:11–8.
142. Schmitt EA, Dowling JE. Early retinal development in the zebrafish, *Danio rerio*: light and electron microscopic analyses. *J Comp Neurol* 1999;**404**:515–36.
143. Lister JA. Development of pigment cells in the zebrafish embryo. *Microsc Res Tech* 2002;**58**:435–41.
144. Greiling TM, Clark JI. The transparent lens and cornea in the mouse and zebra fish eye. *Semin Cell Dev Biol* 2008;**19**:94–9.
145. Greiling TM, Clark JI. Early lens development in the zebrafish: a three-dimensional time-lapse analysis. *Dev Dyn* 2009;**238**:2254–65.
146. Greiling TM, Aose M, Clark JI. Cell fate and differentiation of the developing ocular lens. *Invest Ophthalmol Vis Sci* 2010;**51**:1540–6.
147. Dahm R, Schonthaler HB, Soehn AS, van Marle J, Vrensen GF. Development and adult morphology of the eye lens in the zebrafish. *Exp Eye Res* 2007;**85**:74–89.
148. Soules KA, Link BA. Morphogenesis of the anterior segment in the zebrafish eye. *BMC Dev Biol* 2005;**5**:12.
149. Gray MP, Smith RS, Soules KA, John SW, Link BA. The aqueous humor outflow pathway of zebrafish. *Invest Ophthalmol Vis Sci* 2009;**50**:1515–21.
150. Zhao XC, Yee RW, Norcom E, Burgess H, Avanesov AS, Barrish JP, et al. The zebrafish cornea: structure and development. *Invest Ophthalmol Vis Sci* 2006;**47**:4341–8.
151. Akhtar S, Schonthaler HB, Bron AJ, Dahm R. Formation of stromal collagen fibrils and proteoglycans in the developing zebrafish cornea. *Acta Ophthalmol* 2008;**86**:655–65.
152. Link BA, Gray MP, Smith RS, John SW. Intraocular pressure in zebrafish: comparison of inbred strains and identification of a reduced melanin mutant with raised IOP. *Invest Ophthalmol Vis Sci* 2004;**45**:4415–22.
153. Chen CC, Yeh LK, Liu CY, Kao WW, Samples JR, Lin SJ, et al. Morphological differences between the trabecular meshworks of zebrafish and mammals. *Curr Eye Res* 2008;**33**:59–72.
154. Swamynathan SK, Crawford MA, Robison Jr. WG, Kanungo J, Piatigorsky J. Adaptive differences in the structure and macromolecular compositions of the air and water corneas of the "four-eyed" fish (Anableps anableps). *FASEB J* 2003;**17**:1996–2005.
155. Easter Jr. SS, Nicola GN. The development of vision in the zebrafish (*Danio rerio*). *Dev Biol* 1996;**180**:646–63.
156. Schmitt EA, Dowling JE. Early eye morphogenesis in the zebrafish, *Brachydanio rerio*. *J Comp Neurol* 1994;**344**:532–42.
157. Li Z, Joseph NM, Easter Jr. SS. The morphogenesis of the zebrafish eye, including a fate map of the optic vesicle. *Dev Dyn* 2000;**218**:175–88.

158. Kuszak JR, Costello J. The structure of the vertebrate lens. In: Lovicu FJ, Robinson ML, editors. *Development of the ocular lens*. Cambridge: Cambridge University Press; 2004. p. 71–118.
159. Bilotta J, Saszik S, Sutherland SE. Rod contributions to the electroretinogram of the dark-adapted developing zebrafish. *Dev Dyn* 2001;**222**:564–70.
160. Robinson J, Schmitt EA, Harosi FI, Reece RJ, Dowling JE. Zebrafish ultraviolet visual pigment: absorption spectrum, sequence, and localization. *Proc Natl Acad Sci USA* 1993;**90**:6009–12.
161. Dowling JE. *The retina*. Cambridge, MA: Harvard University Press; 1987.
162. Fitzpatrick DR, van Heyningen V. Developmental eye disorders. *Curr Opin Genet Dev* 2005;**15**:348–53.
163. Gregory-Evans CY, Williams MJ, Halford S, Gregory-Evans K. Ocular coloboma: a reassessment in the age of molecular neuroscience. *J Med Genet* 2004;**41**:881–91.
164. Porges Y, Gershoni-Baruch R, Leibu R, Goldscher D, Zonis S, Shapira I, et al. Hereditary microphthalmia with colobomatous cyst. *Am J Ophthalmol* 1992;**114**:30–4.
165. Bermejo E, Martinez-Frias ML. Congenital eye malformations: clinical-epidemiological analysis of 1, 124, 654 consecutive births in Spain. *Am J Med Genet* 1998;**75**:497–504.
166. Stoll C, Alembik Y, Dott B, Roth MP. Congenital eye malformations in 212, 479 consecutive births. *Ann Génét* 1997;**40**:122–8.
167. Maumenee IH, Mitchell TN. Colobomatous malformations of the eye. *Trans Am Ophthalmol Soc* 1990;**88**:123–32, [Discussion: 133–125].
168. Chang L, Blain D, Bertuzzi S, Brooks BP. Uveal coloboma: clinical and basic science update. *Curr Opin Ophthalmol* 2006;**17**:447–70.
169. Sanyanusin P, Schimmenti LA, McNoe LA, Ward TA, Pierpont ME, Sullivan MJ, et al. Mutation of the PAX2 gene in a family with optic nerve colobomas, renal anomalies and vesicoureteral reflux. *Nat Genet* 1995;**9**:358–64.
170. Terzic J, Muller C, Gajovic S, Saraga-Babic M. Expression of PAX2 gene during human development. *Int J Dev Biol* 1998;**42**:701–7.
171. Macdonald R, Barth KA, Xu Q, Holder N, Mikkola I, Wilson SW. Midline signalling is required for Pax gene regulation and patterning of the eyes. *Development* 1995;**121**:3267–78.
172. Goodrich LV, Milenkovic L, Higgins KM, Scott MP. Altered neural cell fates and medulloblastoma in mouse patched mutants. *Science* 1997;**277**:1109–13.
173. Schimmenti LA, de la Cruz J, Lewis RA, Karkera JD, Manligas GS, Roessler E, et al. Novel mutation in sonic hedgehog in non-syndromic colobomatous microphthalmia. *Am J Med Genet A* 2003;**116A**:215–21.
174. Brown SA, Warburton D, Brown LY, Yu CY, Roeder ER, Stengel-Rutkowski S, et al. Holoprosencephaly due to mutations in ZIC2, a homologue of Drosophila odd-paired. *Nat Genet* 1998;**20**:180–3.
175. Ohsaki K, Morimitsu T, Ishida Y, Kominami R, Takahashi N. Expression of the Vax family homeobox genes suggests multiple roles in eye development. *Genes Cells* 1999;**4**:267–76.
176. Take-uchi M, Clarke JD, Wilson SW. Hedgehog signalling maintains the optic stalk-retinal interface through the regulation of Vax gene activity. *Development* 2003;**130**:955–68.
177. Hornby SJ, Ward SJ, Gilbert CE. Eye birth defects in humans may be caused by a recessively-inherited genetic predisposition to the effects of maternal vitamin A deficiency during pregnancy. *Med Sci Monit* 2003;**9**:23–6.
178. Hyatt GA, Schmitt EA, Marsh-Armstrong NR, Dowling JE. Retinoic acid-induced duplication of the zebrafish retina. *Proc Natl Acad Sci USA* 1992;**89**:8293–7.
179. Hyatt GA, Schmitt EA, Marsh-Armstrong N, McCaffery P, Drager UC, Dowling JE. Retinoic acid establishes ventral retinal characteristics. *Development* 1996;**122**:195–204.

180. Marsh-Armstrong N, McCaffery P, Gilbert W, Dowling JE, Drager UC. Retinoic acid is necessary for development of the ventral retina in zebrafish. *Proc Natl Acad Sci USA* 1994;**91**:7286–90.
181. Lampert JM, Holzschuh J, Hessel S, Driever W, Vogt K, von Lintig J. Provitamin A conversion to retinal via the beta, beta-carotene-15, 15'-oxygenase (bcox) is essential for pattern formation and differentiation during zebrafish embryogenesis. *Development* 2003;**130**:2173–86.
182. Fodde R, Smits R, Clevers H. APC, signal transduction and genetic instability in colorectal cancer. *Nat Rev Cancer* 2001;**1**:55–67.
183. Pittlik S, Domingues S, Meyer A, Begemann G. Expression of zebrafish aldh1a3 (raldh3) and absence of aldh1a1 in teleosts. *Gene Expr Patterns* 2008;**8**:141–7.
184. McCaffery P, Wagner E, O'Neil J, Petkovich M, Drager UC. Dorsal and ventral retinal territories defined by retinoic acid synthesis, break-down and nuclear receptor expression. *Mech Dev* 1999;**82**:119–30.
185. Fischer AJ, Wallman J, Mertz JR, Stell WK. Localization of retinoid binding proteins, retinoid receptors, and retinaldehyde dehydrogenase in the chick eye. *J Neurocytol* 1999;**28**:597–609.
186. Matt N, Ghyselinck NB, Pellerin I, Dupe V. Impairing retinoic acid signalling in the neural crest cells is sufficient to alter entire eye morphogenesis. *Dev Biol* 2008;**320**:140–8.
187. McMahon C, Gestri G, Wilson SW, Link BA. Lmx1b is essential for survival of periocular mesenchymal cells and influences Fgf-mediated retinal patterning in zebrafish. *Dev Biol* 2009;**332**:287–98.
188. Insinna C, Besharse JC. Intraflagellar transport and the sensory outer segment of vertebrate photoreceptors. *Dev Dyn* 2008;**237**:1982–92.
189. Rohlich P. The sensory cilium of retinal rods is analogous to the transitional zone of motile cilia. *Cell Tissue Res* 1975;**161**:421–30.
190. Badano JL, Teslovich TM, Katsanis N. The centrosome in human genetic disease. *Nat Rev Genet* 2005;**6**:194–205.
191. Hartong DT, Berson EL, Dryja TP. Retinitis pigmentosa. *Lancet* 2006;**368**:1795–809.
192. Fadool JM, Brockerhoff SE, Hyatt GA, Dowling JE. Mutations affecting eye morphology in the developing zebrafish (*Danio rerio*). *Dev Genet* 1997;**20**:288–95.
193. Malicki J, Neuhauss SC, Schier AF, Solnica-Krezel L, Stemple DL, Stainier DY, et al. Mutations affecting development of the zebrafish retina. *Development* 1996;**123**:263–73.
194. Tepass U, Theres C, Knust E. crumbs encodes an EGF-like protein expressed on apical membranes of Drosophila epithelial cells and required for organization of epithelia. *Cell* 1990;**61**:787–99.
195. Wodarz A, Hinz U, Engelbert M, Knust E. Expression of crumbs confers apical character on plasma membrane domains of ectodermal epithelia of Drosophila. *Cell* 1995;**82**:67–76.
196. Roh MH, Fan S, Liu CJ, Margolis B. The Crumbs3-Pals1 complex participates in the establishment of polarity in mammalian epithelial cells. *J Cell Sci* 2003;**116**:2895–906.
197. den Hollander AI, ten Brink JB, de Kok YJ, van Soest S, van den Born LI, van Driel MA, et al. Mutations in a human homologue of Drosophila crumbs cause retinitis pigmentosa (RP12). *Nat Genet* 1999;**23**:217–21.
198. Lotery AJ, Jacobson SG, Fishman GA, Weleber RG, Fulton AB, Namperumalsamy P, et al. Mutations in the CRB1 gene cause Leber congenital amaurosis. *Arch Ophthalmol* 2001;**119**:415–20.
199. Gerber S, Perrault I, Hanein S, Shalev S, Zlotogora J, Barbet F, et al. A novel mutation disrupting the cytoplasmic domain of CRB1 in a large consanguineous family of Palestinian origin affected with Leber congenital amaurosis. *Ophthalmic Genet* 2002;**23**:225–35.
200. Schroer TA. Dynactin. *Annu Rev Cell Dev Biol* 2004;**20**:759–79.
201. Tsujikawa M, Omori Y, Biyanwila J, Malicki J. Mechanism of positioning the cell nucleus in vertebrate photoreceptors. *Proc Natl Acad Sci USA* 2007;**104**:14819–24.

202. Dryja TP, Li T. Molecular genetics of retinitis pigmentosa. *Hum Mol Genet* 1995;**4 Spec No**:1739–43.
203. Young RW. The renewal of photoreceptor cell outer segments. *J Cell Biol* 1967;**33**:61–72.
204. Krock BL, Mills-Henry I, Perkins BD. Retrograde intraflagellar transport by cytoplasmic dynein-2 is required for outer segment extension in vertebrate photoreceptors but not arrestin translocation. *Invest Ophthalmol Vis Sci* 2009;**50**:5463–71.
205. Blacque OE, Leroux MR. Bardet–Biedl syndrome: an emerging pathomechanism of intracellular transport. *Cell Mol Life Sci* 2006;**63**:2145–61.
206. Muto A, Orger MB, Wehman AM, Smear MC, Kay JN, Page-McCaw PS, et al. Forward genetic analysis of visual behavior in zebrafish. *PLoS Genet* 2005;**1**:e66.
207. Li L. Zebrafish mutants: behavioral genetic studies of visual system defects. *Dev Dyn* 2001;**221**:365–72.
208. Maaswinkel H, Ren JQ, Li L. Slow-progressing photoreceptor cell degeneration in night blindness c mutant zebrafish. *J Neurocytol* 2003;**32**:1107–16.
209. Maaswinkel H, Mason B, Li L. ENU-induced late-onset night blindness associated with rod photoreceptor cell degeneration in zebrafish. *Mech Ageing Dev* 2003;**124**:1065–71.
210. Baehr W, Wu SM, Bird AC, Palczewski K. The retinoid cycle and retina disease. *Vision Res* 2003;**43**:2957–8.
211. Kevany BM, Palczewski K. Phagocytosis of retinal rod and cone photoreceptors. *Physiology (Bethesda)* 2010;**25**:8–15.
212. Strauss O. The retinal pigment epithelium in visual function. *Physiol Rev* 2005;**85**:845–81.
213. Raposo G, Marks MS. The dark side of lysosome-related organelles: specialization of the endocytic pathway for melanosome biogenesis. *Traffic* 2002;**3**:237–48.
214. Marmorstein AD, Finnemann SC, Bonilha VL, Rodriguez-Boulan E. Morphogenesis of the retinal pigment epithelium: toward understanding retinal degenerative diseases. *Ann NY Acad Sci* 1998;**857**:1–12.
215. Finnemann SC, Leung LW, Rodriguez-Boulan E. The lipofuscin component A2E selectively inhibits phagolysosomal degradation of photoreceptor phospholipid by the retinal pigment epithelium. *Proc Natl Acad Sci USA* 2002;**99**:3842–7.
216. Bergmann M, Schutt F, Holz FG, Kopitz J. Inhibition of the ATP-driven proton pump in RPE lysosomes by the major lipofuscin fluorophore A2-E may contribute to the pathogenesis of age-related macular degeneration. *FASEB J* 2004;**18**:562–4.
217. Kelsh RN, Brand M, Jiang YJ, Heisenberg CP, Lin S, Haffter P, et al. Zebrafish pigmentation mutations and the processes of neural crest development. *Development* 1996;**123**:369–89.
218. Spanakis E, Lamina P, Bennett DC. Effects of the developmental colour mutations silver and recessive spotting on proliferation of diploid and immortal mouse melanocytes in culture. *Development* 1992;**114**:675–80.
219. Theos AC, Truschel ST, Raposo G, Marks MS. The Silver locus product Pmel17/gp100/Silv/ME20: controversial in name and in function. *Pigment Cell Res* 2005;**18**:322–36.
220. Johnson SL, Weston JA. Temperature-sensitive mutations that cause stage-specific defects in Zebrafish fin regeneration. *Genetics* 1995;**141**:1583–95.
221. Poss KD, Wilson LG, Keating MT. Heart regeneration in zebrafish. *Science* 2002;**298**: 2188–90.
222. Vihtelic TS, Hyde DR. Light-induced rod and cone cell death and regeneration in the adult albino zebrafish (*Danio rerio*) retina. *J Neurobiol* 2000;**44**:289–307.
223. Gonzalez-Nunez V, Nocco V, Budd A. Characterization of DrCol15a1b, a novel component of the stem-cell niche in the zebrafish retina. *Stem Cells* 2010;**28**:1399–411.
224. Cerveny KL, Cavodeassi F, Turner KJ, de Jong-Curtain TA, Heath JK, Wilson SW. The zebrafish flotte lotte mutant reveals that the local retinal environment promotes the

differentiation of proliferating precursors emerging from their stem cell niche. *Development* 2010;**137**:2107–15.

225. Stephens WZ, Senecal M, Nguyen M, Piotrowski T. Loss of adenomatous polyposis coli (apc) results in an expanded ciliary marginal zone in the zebrafish eye. *Dev Dyn* 2010;**239**:2066–77.
226. Bringmann A, Pannicke T, Grosche J, Francke M, Wiedemann P, Skatchkov SN, et al. Muller cells in the healthy and diseased retina. *Prog Retin Eye Res* 2006;**25**:397–424.
227. Fischer AJ, Reh TA. Muller glia are a potential source of neural regeneration in the postnatal chicken retina. *Nat Neurosci* 2001;**4**:247–52.
228. Cameron DA. Cellular proliferation and neurogenesis in the injured retina of adult zebrafish. *Vis Neurosci* 2000;**17**:789–97.
229. Black GC, Mazerolle CJ, Wang Y, Campsall KD, Petrin D, Leonard BC, et al. Abnormalities of the vitreoretinal interface caused by dysregulated Hedgehog signaling during retinal development. *Hum Mol Genet* 2003;**12**:3269–76.
230. Raymond PA, Barthel LK, Bernardos RL, Perkowski JJ. Molecular characterization of retinal stem cells and their niches in adult zebrafish. *BMC Dev Biol* 2006;**6**:36.
231. Raymond PA, Rivlin PK. Germinal cells in the goldfish retina that produce rod photoreceptors. *Dev Biol* 1987;**122**:120–38.
232. Julian D, Ennis K, Korenbrot JI. Birth and fate of proliferative cells in the inner nuclear layer of the mature fish retina. *J Comp Neurol* 1998;**394**:271–82.
233. Bernardos RL, Barthel LK, Meyers JR, Raymond PA. Late-stage neuronal progenitors in the retina are radial Muller glia that function as retinal stem cells. *J Neurosci* 2007;**27**:7028–40.
234. Fausett BV, Goldman D. A role for alpha1 tubulin-expressing Muller glia in regeneration of the injured zebrafish retina. *J Neurosci* 2006;**26**:6303–13.
235. Kassen SC, Ramanan V, Montgomery JE, Christopher TB, Liu CG, Vihtelic TS, Hyde DR. Time course analysis of gene expression during light-induced photoreceptor cell death and regeneration in albino zebrafish. *Dev Neurobiol* 2007;**67**:1009–31.
236. Qin Z, Barthel LK, Raymond PA. Genetic evidence for shared mechanisms of epimorphic regeneration in zebrafish. *Proc Natl Acad Sci USA* 2009;**106**:9310–5.
237. Wangsa-Wirawan ND, Linsenmeier RA. Retinal oxygen: fundamental and clinical aspects. *Arch Ophthalmol* 2003;**121**:547–57.
238. Saint-Geniez M, D'Amore PA. Development and pathology of the hyaloid, choroidal and retinal vasculature. *Int J Dev Biol* 2004;**48**:1045–58.
239. Alvarez Y, Chen K, Reynolds AL, Waghorne N, O'Connor JJ, Kennedy BN. Predominant cone photoreceptor dysfunction in a hyperglycaemic model of non-proliferative diabetic retinopathy. *Dis Model Mech* 2010;**3**:236–45.
240. Gleeson M, Connaughton V, Arneson LS. Induction of hyperglycaemia in zebrafish (*Danio rerio*) leads to morphological changes in the retina. *Acta Diabetol* 2007;**44**:157–63.
241. Zhu M, Provis JM, Penfold PL. The human hyaloid system: cellular phenotypes and interrelationships. *Exp Eye Res* 1999;**68**:553–63.
242. Jonasova K, Kozmik Z. Eye evolution: lens and cornea as an upgrade of animal visual system. *Semin Cell Dev Biol* 2008;**19**:71–81.
243. Congdon N, West SK, Buhrmann RR, Kouzis A, Munoz B, Mkocha H. Prevalence of the different types of age-related cataract in an African population. *Invest Ophthalmol Vis Sci* 2001;**42**:2478–82.
244. Hejtmancik JF. Congenital cataracts and their molecular genetics. *Semin Cell Dev Biol* 2008;**19**:134–49.
245. Graw J. Congenital hereditary cataracts. *Int J Dev Biol* 2004;**48**:1031–44.
246. O'Rahilly R. The prenatal development of the human eye. *Exp Eye Res* 1975;**21**:93–112.
247. Rafferty NS, Rafferty Jr. KA. Cell population kinetics of the mouse lens epithelium. *J Cell Physiol* 1981;**107**:309–15.

248. Bassnett S, Beebe D. Lens fiber differentiation. In: Lovicu FJ, Robinson ML, editors. *Development of the ocular lens*. Cambridge: Cambridge University Press; 2004. p. 214–44.
249. Beebe DC, Compart PJ, Johnson MC, Feagans DE, Feinberg RN. The mechanism of cell elongation during lens fiber cell differentiation. *Dev Biol* 1982;**92**:54–9.
250. Land MF, Fernald RD. The evolution of eyes. *Annu Rev Neurosci* 1992;**15**:1–29.
251. Leonard DW, Meek KM. Refractive indices of the collagen fibrils and extrafibrillar material of the corneal stroma. *Biophys J* 1997;**72**:1382–7.
252. Semina EV, Brownell I, Mintz-Hittner HA, Murray JC, Jamrich M. Mutations in the human forkhead transcription factor FOXE3 associated with anterior segment ocular dysgenesis and cataracts. *Hum Mol Genet* 2001;**10**:231–6.
253. Semina EV, Ferrell RE, Mintz-Hittner HA, Bitoun P, Alward WL, Reiter RS, et al. A novel homeobox gene PITX3 is mutated in families with autosomal-dominant cataracts and ASMD. *Nat Genet* 1998;**19**:167–70.
254. Graw J. Genetics of crystallins: cataract and beyond. *Exp Eye Res* 2009;**88**:173–89.
255. Stainier DY, Weinstein BM, Detrich 3rd HW, Zon LI, Fishman MC. Cloche, an early acting zebrafish gene, is required by both the endothelial and hematopoietic lineages. *Development* 1995;**121**:3141–50.
256. Danysh BP, Duncan MK. The lens capsule. *Exp Eye Res* 2009;**88**:151–64.
257. Timpl R, Brown JC. The laminins. *Matrix Biol* 1994;**14**:275–81.
258. Bystrom B, Virtanen I, Rousselle P, Gullberg D, Pedrosa-Domellof F. Distribution of laminins in the developing human eye. *Invest Ophthalmol Vis Sci* 2006;**47**:777–85.
259. Kohno T, Sorgente N, Ishibashi T, Goodnight R, Ryan SJ. Immunofluorescent studies of fibronectin and laminin in the human eye. *Invest Ophthalmol Vis Sci* 1987;**28**:506–14.
260. Zenker M, Tralau T, Lennert T, Pitz S, Mark K, Madlon H, et al. Congenital nephrosis, mesangial sclerosis, and distinct eye abnormalities with microcoria: an autosomal recessive syndrome. *Am J Med Genet A* 2004;**130A**:138–45.
261. Zenker M, Aigner T, Wendler O, Tralau T, Muntefering H, Fenski R, et al. Human laminin beta2 deficiency causes congenital nephrosis with mesangial sclerosis and distinct eye abnormalities. *Hum Mol Genet* 2004;**13**:2625–32.
262. Paulus JD, Halloran MC. Zebrafish bashful/laminin-alpha 1 mutants exhibit multiple axon guidance defects. *Dev Dyn* 2006;**235**:213–24.
263. Klintworth GK. Corneal dystrophies. *Orphanet J Rare Dis* 2009;**4**:7.
264. Klyce SD, Beuerman RW. Structure and function of the cornea. In: Kaufman HE, Barron BA, McDonald MB, editors. *The cornea*. 2nd edition. Boston: Butterworth-Heinemann; 1998. p. 3–50.
265. Piatigorsky J. Enigma of the abundant water-soluble cytoplasmic proteins of the cornea: the "refracton" hypothesis. *Cornea* 2001;**20**:853–8.
266. Cuthbertson RA, Tomarev SI, Piatigorsky J. Taxon-specific recruitment of enzymes as major soluble proteins in the corneal epithelium of three mammals, chicken, and squid. *Proc Natl Acad Sci USA* 1992;**89**:4004–8.
267. Xu YS, Kantorow M, Davis J, Piatigorsky J. Evidence for gelsolin as a corneal crystallin in zebrafish. *J Biol Chem* 2000;**275**:24645–52.
268. Li S, Tiab L, Jiao X, Munier FL, Zografos L, Frueh BE, et al. Mutations in PIP5K3 are associated with Francois–Neetens mouchetee fleck corneal dystrophy. *Am J Hum Genet* 2005;**77**:54–63.
269. Boisset G, Polok BK, Schorderet DF. Characterization of pip5k3 fleck corneal dystrophy-linked gene in zebrafish. *Gene Expr Patterns* 2008;**8**:404–10.
270. Yeh LK, Liu CY, Chien CL, Converse RL, Kao WW, Chen MS, et al. Molecular analysis and characterization of zebrafish keratocan (zKera) gene. *J Biol Chem* 2008;**283**:506–17.

271. Pellegata NS, Dieguez-Lucena JL, Joensuu T, Lau S, Montgomery KT, Krahe R, et al. Mutations in KERA, encoding keratocan, cause cornea plana. *Nat Genet* 2000;**25**:91–5.
272. Ritch R, Shields M, Krupin T. *The glaucomas Vol. 2, clinical sciences*. 2nd ed. St. Louis, MO: Mosby; 1996.
273. Gould DB, John SW. Anterior segment dysgenesis and the developmental glaucomas are complex traits. *Hum Mol Genet* 2002;**11**:1185–93.
274. Gordon MO, Beiser JA, Brandt JD, Heuer DK, Higginbotham EJ, Johnson CA, et al. The Ocular Hypertension Treatment Study: baseline factors that predict the onset of primary open-angle glaucoma. *Arch Ophthalmol* 2002;**120**:714–20, discussion 829–730.
275. Le A, Mukesh BN, McCarty CA, Taylor HR. Risk factors associated with the incidence of open-angle glaucoma: the visual impairment project. *Invest Ophthalmol Vis Sci* 2003;**44**:3783–9.
276. Civan MM, Macknight AD. The ins and outs of aqueous humour secretion. *Exp Eye Res* 2004;**78**:625–31.
277. Gabelt BT, Kaufman PL. Pharmacologic enhancement of aqueous humor outflow. In: Buskirk M, Shields M, editors. *100 Years of progress in glaucoma*. Philadelphia: Lippincott-Raven; 1997.
278. Tamm ER. The trabecular meshwork outflow pathways: structural and functional aspects. *Exp Eye Res* 2009;**88**:648–55.
279. McMahon C, Semina EV, Link BA. Using zebrafish to study the complex genetics of glaucoma. *Comp Biochem Physiol C Toxicol Pharmacol* 2004;**138**:343–50.
280. Mears AJ, Jordan T, Mirzayans F, Dubois S, Kume T, Parlee M, et al. Mutations of the forkhead/winged-helix gene, FKHL7, in patients with Axenfeld-Rieger anomaly. *Am J Hum Genet* 1998;**63**:1316–28.
281. Mirzayans F, Gould DB, Heon E, Billingsley GD, Cheung JC, Mears AJ, et al. Axenfeld-Rieger syndrome resulting from mutation of the FKHL7 gene on chromosome 6p25. *Eur J Hum Genet* 2000;**8**:71–4.
282. Nishimura DY, Swiderski RE, Alward WL, Searby CC, Patil SR, Bennet SR, et al. The forkhead transcription factor gene FKHL7 is responsible for glaucoma phenotypes which map to 6p25. *Nat Genet* 1998;**19**:140–7.
283. Berry FB, Skarie JM, Mirzayans F, Fortin Y, Hudson TJ, Raymond V, et al. FOXC1 is required for cell viability and resistance to oxidative stress in the eye through the transcriptional regulation of FOXO1A. *Hum Mol Genet* 2008;**17**:490–505.
284. Vollrath D, Jaramillo-Babb VL, Clough MV, McIntosh I, Scott KM, Lichter PR, et al. Loss-of-function mutations in the LIM-homeodomain gene, LMX1B, in Nail–Patella syndrome. *Hum Mol Genet* 1998;**7**:1091–8.
285. Dunston JA, Reimschisel T, Ding YQ, Sweeney E, Johnson RL, Chen ZF, et al. A neurological phenotype in nail patella syndrome (NPS) patients illuminated by studies of murine Lmx1b expression. *Eur J Hum Genet* 2005;**13**:330–5.
286. Semina EV, Reiter R, Leysens NJ, Alward WL, Small KW, Datson NA, et al. Cloning and characterization of a novel bicoid-related homeobox transcription factor gene, RIEG, involved in Rieger syndrome. *Nat Genet* 1996;**14**:392–9.
287. Hjalt TA, Semina EV. Current molecular understanding of Axenfeld–Rieger syndrome. *Expert Rev Mol Med* 2005;**7**:1–17.

800 Facets of Retinal Degeneration

T. Cook,* A. Zelhof,† M. Mishra,† and J. Nie†

*Department of Pediatric Ophthalmology, Division of Developmental Biology Cincinnati Children's Hospital Medical Center, Cincinnati, Ohio, USA

†Department of Biology, Indiana University, Bloomington, Indiana, USA

In today's world of genomics and large computational analyses, rapid progress has been made in identifying genes associated with human retinal diseases. Nevertheless, before significant advances toward effective therapeutic intervention is made, a clearer understanding of the molecular and cellular role of these gene products in normal and diseased photoreceptor cell biology is required. Given the complexity of the vertebrate retina, these advancements are unlikely to be revealed in isolated human cell lines, but instead, will require

Progress in Molecular Biology and Translational Science, Vol. 100
DOI: 10.1016/B978-0-12-384878-9.00008-X

Copyright 2011, Elsevier Inc.
All rights reserved.
1877-1173/11 $35.00

the use of numerous model systems. Here, we describe several parallels between vertebrate and invertebrate photoreceptor cell biology that are beginning to emerge and advocate the use of *Drosophila melanogaster* as a powerful genetic model system for uncovering molecular mechanisms of human retinal pathologies, in particular photoreceptor neurodegeneration.

I. Retinal Diseases

> The health of the eye seems to demand a horizon. We are never tired, so long as we can see far enough.
>
> Ralph Waldo Emerson

Vision is one of our most dominant sensory systems and its impairment, whether it be light sensitivity or night blindness, loss of peripheral or central vision, or complete loss of sight, can significantly affect the quality of life. Vision can be affected in numerous ways, including damage of the optic nerve (e.g., glaucoma), retinopathies associated with vascular disease (e.g., diabetes), and retinal degenerative diseases targeting the two main photosensitive cells in the retina—rod and cone photoreceptors. To date, of the 100 genetic disorders connected to vision loss, the majority fall under this last category—the development and/or function of rod and cone cells (Retnet—www.sph.uth.tmc.edu/retnet/).

Rod and cone photoreceptors are specialized for different visual tasks. Rods are responsible for vision under dim light conditions and peripheral vision, whereas cones are responsible for central, high acuity image formation and color discrimination. Based on the type of photoreceptor cell that is affected, the age of onset, and whether the disease is progressive or stationary, most hereditary retinal disorders can be grouped into three overlapping categories: Leber's congenital amaurosis (LCA), cone–rod dystrophies (CORD), and rod–cone dystrophies, including retinitis pigmentosa (RP). LCA is the most severe of these diseases, resulting in the failure of both rod and cone photoreceptors differentiation and severe loss of sight or complete blindness within the first few years after birth. In contrast, CORD and RP are generally characterized by a later onset of neurodegeneration (e.g., second or third decade of life), and as the names imply, primarily affect cones or rods, respectively, followed by secondary loss of the other photoreceptor cell type. Together, these diseases can be X-linked, autosomal dominant, or autosomal recessive and affect up to 1:3000 individuals. Photoreceptor defects have also been associated with more complex developmental or degenerative syndromes, including Usher and Bardet-Biedl syndromes.[1] Thus, elucidating the molecular and genetic pathways underlying normal and diseased photoreceptor cell biology will have a broad impact on biomedical research. In this review, we highlight many critical questions related to photoreceptor biology that have been and will continue to be answered using the powerful genetics available in the common fruit fly, *Drosophila melanogaster*.

II. Understanding the Developmental Process of Retinal Cell Types

A. Introduction to Eye Types

During the course of evolution, various eye types arose to aid in visualizing an organism's environment.[2,3] The simplest of eyes only detect the presence or absence of light, and require just individual or small clusters of photoreceptors together with pigment cells. To form an image, however, requires a much more complex ocular structure.[4] To date, the best characterized of these complex eyes are the single-chambered, or camera eye, and the compound eye. A camera eye is present in most vertebrates, and has a single dioptic system comprised a lens and/or a cornea focusing light on a retina comprised of multiple photoreceptors and interneurons. Compound eyes, however, are found in most invertebrates and are built of numerous individual "eye" units, or ommatidia, each of which focuses light through a corneal lens onto a retina composed of a limited number of photoreceptors.

Two fundamental forms of photoreceptor cells also exist across the animal kingdom—ciliary and rhabdomeric. These cell types differ in the way in which the light-gathering apical membranes form, the light-sensitive opsin proteins they express, and the downstream phototransduction pathways used to convert light into a chemical signal.[2] Until fairly recently, these two forms were thought to largely segregate with vertebrate and invertebrate eyes, respectively, but now it is clear that many animals, including humans, have retinas with photosensitive cells of both types. Moreover, as will be discussed in this review, unforeseen similarities between these two seemingly distinct photoreceptor cell types are beginning to emerge that have significant implications for understanding retinal degenerative processes. To appreciate these similarities, however, we first briefly review what is known regarding how photoreceptors (PRs) develop in vertebrate and invertebrate retinas.

B. Vertebrate Retinogenesis and Photoreceptor Development

All vertebrate eyes contain a cornea, iris, lens, and retina. Light passes through the cornea, the primary focusing element, then through the iris, which controls how much light will enter the eye, and lastly, through the lens, which

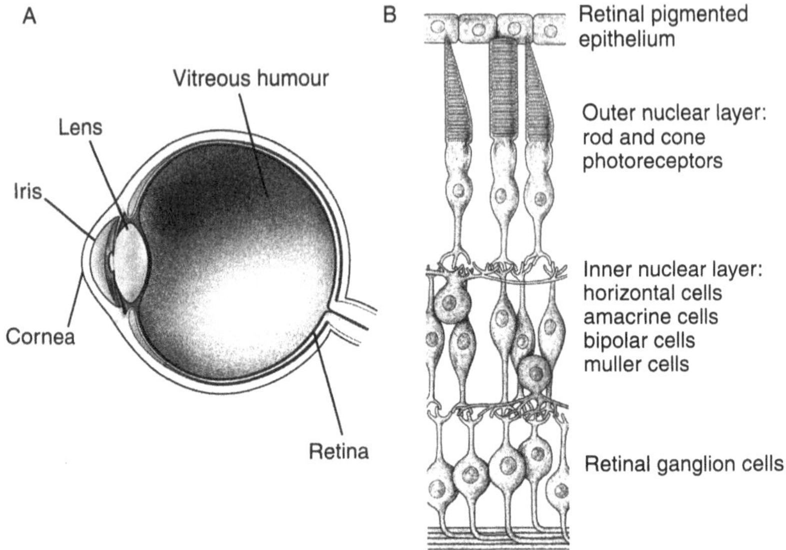

FIG. 1. The vertebrate eye. Schematic representation of a vertebrate camera eye (A) and cells of the retina (B).

provides an adjustable focusing element for the image before reaching the retina (Fig. 1). The adult vertebrate neuroretina is comprised of seven main cell types: retinal ganglion cells (RGCs), horizontal cells (HCs), amacrine cells (ACs), bipolar cells (BiPs), cone photoreceptors (cones), rod photoreceptors (rods), and Muller glia (MG). These cell types are layered into three distinct regions of the retina, with RGCs and a subset of ACs confined to the ganglion cell layer at the vitreous surface, facing the lens, the remaining ACs, BiPs, and HCs localized within the inner nuclear layer, and the rod and cone photoreceptors packed at the back of the retina, forming the outer nuclear layer (Fig. 1). Different cone photoreceptors are specified to maximally absorb distinct wavelengths. In humans, for instance, S-, M-, and L-cones capture short (blue), medium (green), and long (red) wavelengths, respectively, under bright light conditions, while rods absorb a broad range of wavelengths under dim light conditions. Once the photoreceptors absorb light, they convert this information to a chemical signal by phototransduction and transmit this information to the BiP, AC, and HC interneurons to begin integrating some of these signals. These cells then send their information to the RGCs, whose axons project through the optic nerve to central targets, sending visual input from the retina to the brain for final processing.

In general, the various retinal cell types that develop are specified in a similar temporal pattern in all chordate retinas: first, RGCs, followed shortly by other "early born" cell types including the cones, ACs, and HCs, and ending with "later born" cell types including rods, BiPs, and MG. Birthdating and lineage tracing studies helped develop the well-established "competence" model to explain how these diverse cell fates are determined. This model suggests that all initial retinal progenitors are pluripotent during early retinogenesis, but show a gradual restriction in their competence to form different cell types over time, due to both intrinsic and extrinsic factors.[5,6]

Based on the competence model, cones and rods arise from distinctly different retinal progenitors ("early born" vs. "later born"). However, relatively recent studies have revealed that, instead, all photoreceptors emerge from a common precursor. This change in paradigm largely arose from the studies of the transcription factor NRL (neural retina leucine zipper protein).[7–9] NRL is specifically expressed in rod photoreceptor nuclei, and in NRL knockout mice, rods are transformed into S-cones. This suggests that NRL blocks S-cone fate and promotes rod fate from a common "generic" photoreceptor precursor. Importantly, mutations in NRL and its direct downstream target and coregulatory protein, NR2E3, have been identified in human with RP and enhanced S-cone syndrome,[9–12] consistent with this pathway being conserved between mouse and human. Since the discovery of NRL, several additional factors have been defined that further delineate the progression of a photoreceptor precursor into distinct cone subtypes. Interestingly, many of these are members of the thyroid and retinoic acid receptor families of nuclear receptors.[7,13–15] Upstream of all of these factors, however, lies the Orthodenticle (Otd)-related homeodomain transcription factors Otx2 and -Crx (Cone-Rod homeobox).[7,16–21] As its name implies, Crx is largely restricted to photoreceptors, whereas Otx2 is more broadly expressed in several parts of the developing nervous system and in retinal BiPs. In photoreceptors, Otx2 activates Crx, and together they regulate the expression of many photoreceptor-specific gene products, including NRL.

Otd-related factors lie upstream of photoreceptor development in many systems, including *Drosophila*. In addition, many of the nuclear receptors are conserved for their roles in photoreceptor subtype specification in several vertebrate systems.[22–25] Currently, however, it remains unclear whether NRL is uniquely involved in mammalian rod genesis or not, since NRL loss-of-function studies have not been performed in other experimental models. Nevertheless, these studies of Otx2 and Crx do suggest that rods and cones are genetically related, a critical notion for understanding how different photoreceptor cell populations are differentially affected in individual retinal degenerative diseases.

C. *Drosophila* Retinogenesis and Photoreceptor Development

Groundbreaking work by Thomas H. Morgan in the early 1900s led to establishing the fruit fly, *D. melanogaster*, as a preeminent genetic model system for understanding the molecular basis of compound eye development. The structural and functional unit of the *Drosophila* fly eye is an ommatidium, which is repeated approximately 750–800 times in the adult compound eye (Fig. 2A). Each ommatidium consists of approximately 20 cells: 6 rod-like photoreceptor cells (R1-R6) and 2 cone-like photoreceptors (R7 and R8) comprise the retina, 6 epithelial cells produce the corneal lens, and approximately 6 highly pigmented cells are shared between ommatidia to limit light scatter and participate in chromophore formation (Fig. 2B). An interommatidial mechanosensory bristle is also present at alternating ommatidial apexes, but this structure is thought to arise from an independent lineage from the others in the eye[26,184] (Fig. 2).

Fly retinal tissue is initially specified as a small cluster of epithelial cells set aside in the embryo. This group of cells eventually forms the eye imaginal disc, and this proliferates throughout the first two stages of larval development. Prior to pupation in third instar larvae, Hedgehog signaling is activated in a dorso-ventral stripe of cells at the posterior margin of the eye disc. This signal instructs the proliferating retinal precursors to exit the cell cycle and initiate

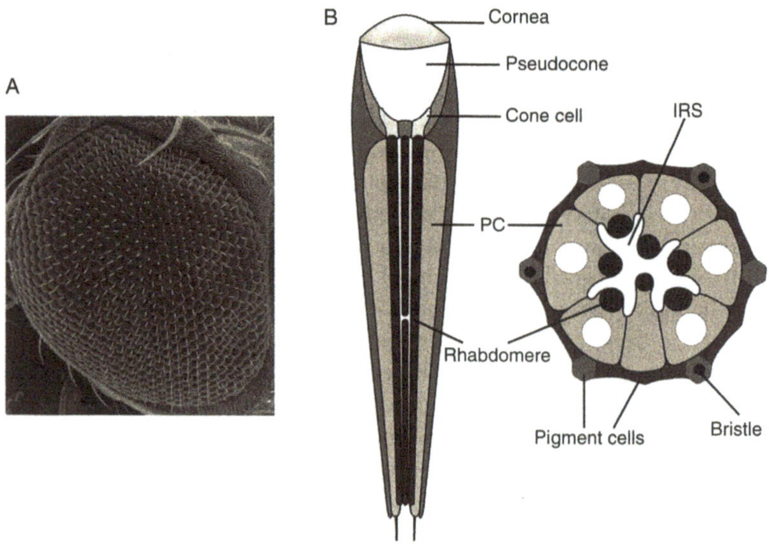

FIG. 2. The compound eye. (A) Scanning electron picture of a Drosophila compound eye. (B) Schematic representation of the cellular organization of a single ommatidium. PC = photoreceptor cell; IRS = interrhabdomeric space. Primary pigment cells and cone cells give rise to the corneal lens and psuedocone, whereas the secondary and tertiary pigment cells prevent light scattering between ommatidia.

neurogenesis. This signaling center, known as the morphogenetic furrow, progresses anteriorly across the disc, leaving new ommatidial clusters of photoreceptors in its wake. In each cluster, the first photoreceptor to be specified is the R8 cell, followed by pairwise recruitment of the R2/R5, R3/R4, and R1/R6 photoreceptors. Finally, the R8 together with the R1/R6 cells recruits the last photoreceptor, the R7 cell. Once photoreceptor recruitment is complete, the cells that generate the other parts of the eye are recruited (for recent reviews, see Refs. 27 and 28).

In the adult eye, the R1–R6 cells develop into what are called outer photoreceptors (OPRs). OPRs function much like rod photoreceptors in the vertebrate retina: they form large light-gathering apical surfaces (called rhabdomeres) that house the phototransduction machinery, they all express the same broad wavelength-sensitive opsin protein, Rhodopsin 1, and they are important for dim light and motion detection. The R7 and R8, called inner photoreceptors (IPRs) form cone-like photoreceptors.[29,30] The R7s express UV-sensitive opsins, Rh3 and/or Rh4, while R8 cells express either blue (Rh5)- or green (Rh6)-sensitive opsins.[31–36] Therefore, *Drosophila* has UV, blue, and green-sensitive photoreceptors for S-, M-, and L-wavelength-sensitive cells, much like humans have blue, green, and red sensitive cone photoreceptors. In contrast to the vertebrate retina, photoreceptors are the only neuronal cell types in the fly retina, but emerging evidence suggests that the interneurons present in the underlying optic lobes, where the photoreceptors project, are developmentally related to the other cell types in the vertebrate retina.[30,37,38]

Until recently, it was believed that OPRs and IPRs were specified using independent mechanisms, much like earlier models for rod and cone specification. In addition, because the R8 cell is the first cell to be specified and the R7 cell is the last, the "cone-like" IPR cells were also thought to be uniquely specified. However, in 2001, Mollereau and coworkers identified a complex of Spalt transcription factors that are restricted to both IPRs soon after their specification and are essential to prevent their differentiation into OPRs.[39] This showed that, like in vertebrates, cone- and rod-like photoreceptors differentiate from a common precursor. Moreover, it suggested that adult R7s and R8s are genetically related. Subsequent studies showed that R7s and R8s are later refined from a generic "IPR" fate by the reciprocal functions of the transcription factors Prospero (Pros) and Senseless (Sens).[40–42] Pros is expressed in R7 cells and is required to suppress R8 characteristics, including opsin expression, nuclear position, and axonal projections. Conversely, Sens is expressed in R8 cells to both block R7 characteristics and promote R8-specific characteristics. Following these cell fate decisions, the transcription factors Spineless and Otd in R7 cells and the signaling proteins Melted and Warts/Lats in R8 cells further restrict these cells to express the appropriate IPR opsins.[43–45]

Interestingly, although the factors involved in generating photoreceptor subtypes differ between vertebrate and invertebrate eyes, surprising parallels are beginning to emerge from the genetic relationships found within their different photoreceptor cell populations. For instance, temporally, S-cones form first in the mouse retina, followed by rods, and ending with M-cones, much like the cone-like R8 is specified first, followed by rod-like R1–R6 cells, and ending with the cone-like R7. Moreover, decisions of rod versus cone fates are made by a single factor—NRL in the case of mammals and Spalt in the case of flies. Thus, studies directly testing whether similar factors participate across species is an exciting future area of research.

III. Conserved Factors Involved in Eye Development

> To suppose that the eye with all its inimitable contrivances for adjusting the focus to different distances, for admitting different amounts of light and for the correction of spherical and chromatic aberration, could have been formed by natural selection, seems, I freely confess, absurd in the highest degree". [But] "Reason tells me, that if numerous gradations from a simple and imperfect eye to one complex and perfect can be shown to exist... then the difficulty of believing that a perfect and complex eye could be formed by natural selection, though insuperable by our imagination, should not be considered as subversive of the theory.
>
> –Charles Darwin, Chapter on "Organs of Extreme Perfection and Complication" in "The Origin of Species"

As late as the 1970s, it was widely believed that eyes were polyphyletic, that is, that they had independently arisen many times during evolution.[46] However, two seminal discoveries in the 1980s and 1990s inspired an entire research avenue that has permitted studying direct parallels between the development of the visual systems of vertebrates and invertebrates. The first was the finding that light detection in *Drosophila* is relegated to the same family of seven transmembrane G-protein coupled receptors as in vertebrates—Rhodopsin.[47,48] The second was the demonstration that eye specification is genetically conserved between vertebrates and invertebrates through the use of the same transcription factor, Pax6.[49–52] From this point, questions no longer centere around whether *Drosophila can* be utilized as a model system to understand vertebrate retina development and function, but rather, could be employed to conquer more complex questions such as: What can we learn about the specification of the vertebrate retina from the fly retina? What common mechanisms exist to create the light-gathering organelles? And how does the regulation of phototransduction influence the viability of the photoreceptor cell? These questions are currently very active areas of research and here, we will discuss the contribution and value of *Drosophila* in answering these questions.

A. Pax6 and the Retinal Determination Network

In the early 1990s, human geneticists studying patients with Aniridia and Peters' Anomaly, mouse geneticists studying a spontaneous "small eye" mutant, and fly geneticists studying the "eyeless" mutant almost simultaneously isolated a homeodomain-containing transcription factor that is evolutionarily conserved in eye formation, called Pax6.[49–52] Taking advantage of genetic tools uniquely available for *Drosophila* at the time, Halder et al.[53] showed a startling result-that misexpressing *Drosophila* or mouse Pax6 in other regions of the fly body, including legs, wings, and antennae, is sufficient to form a functioning ectopic compound eye.[53] Later, work in *Xenopus* showed that Pax6 is also capable of ectopically inducing eye structures in vertebrate systems.[54] Together, these studies revealed that Pax6 is both necessary and sufficient for eye formation, and suggested that Pax6 lies at the top of genetic hierarchy responsible for activating all the genes required to form an eye: camera or compound.

In the following years, fly geneticists performed screens to isolate other genes that were both necessary and sufficient for eye formation. This led to the identification of a "retinal determination network" that includes the transcription factors Sine oculis (vertebrate Six1/3), Eyes absent (vertebrate Eya 1–3), Dachsund (Dach1/2), and Optix (vertebrate Six4/6). With the exception of Optix, these factors lie downstream of Pax6, and exhibit extensive cross-regulation among themselves. Moreover, direct and conserved biochemical interactions have been shown between Pax6-Eya-So/Six and Eya-So-Dac. Importantly, all of these factors are critical for gene regulation. Pax6, Dac, and So bind DNA and thus are important for target gene selection, whereas Eya functions as a transcriptional coactivator and phosphatase to likely modify the ability of its DNA-binding partners to regulate gene expression (reviewed in Ref. 28).

Despite this knowledge, surprisingly little is known regarding downstream targets of the retinal determination network. Genetically, however, vertebrate and invertebrate Pax6 are expressed in proliferating progenitors and are important for maintaining their mitotic potential.[55–57] Similar functions have been attributed to Six3 and Sine oculis, suggesting that the retinal determination pathway is together involved in this process.[58,59] Unfortunately, because animals lacking Pax6 early in development fail to form an eye, dissecting possible later roles for it and the other retinal determination factors during ocular development have remained a challenge. In flies, this is further complicated by the fact that at least four genes encode Pax6 proteins: Eyeless and Twin of Eyeless, homologs of Pax6, as well as Eyegone and Twin-of-Eyegone, homologs of an alternatively spliced, homeodomain-minus Pax6 isoform, Pax6(5a). In contrast to flies, mice encode a single Pax6 gene. Thus, taking advantage of a dorsal retina-restricted Pax6 enhancer, Marquardt et al. showed that removing Pax6 from retinal precursor cells is critical for the specification of all cell types (except a subpopulation of ACs), likely through activation of proneural bHLH proteins.[56]

In the adult eye, vertebrate Pax6 expression persists in ACs and RGCs, suggesting that this factor plays recurring roles in the retina, and may play participate in maintaining these mature cell types.[56,60,61] Likewise, molecular studies in *Drosophila* have suggested that a Pax6-like molecule controls terminal differentiation events in photoreceptors by regulating Rhodopsin expression.[62,63] However, conflicting data have been reported for this function, at least for the Pax6 molecule, Eyeless.[64] Instead, recent studies suggest that another homeodomain-encoding factor, Pph13, may in fact be responsible for this function.[65] Clearly, many questions remain open regarding all of Pax6-dependent functions, but its central evolutionarily conserved function in defining ocular tissue across almost all animals tested thus far permits and encourages the use of *Drosophila* and other model organisms to define its function and role in eye ontogeny and phylogeny.

B. Atonal/Ath5: Insight into Conserved Pathways for Generating Neuronal Diversity

As mentioned, fly and human retinas contain rather distinct cell types. However, much like the common specification of retinal tissue by Pax6 in these two eye types, shared transcriptional mechanisms for generating neuronal diversity are also likely to exist. This is most clearly illustrated by a related familiy of basic helix-loop-helix (bHLH) transcription factors: Atonal in flies, and Ath5/Atoh7 in vertebrates. Atonal (Ato), is upregulated in the morphogenetic furrow by Hedgehog and is necessary and sufficient to specify the first cell to arise in the fly retina - the R8 cell.[66–69] Since subsequent eye development relies on R8-mediated signaling, *atonal* mutants lack any retinal tissue. Interestingly, like flies, retinogenesis follows Hedgehog-based signaling in the vertebrate retina, and the Atonal homolog 7 (Atoh7, or in mouse, Math5), is required for the formation of the first cell type to arise in this retina - RGCs.[70,71] Since RGCs are required for transferring visual information to the brain, Math5 knockout mice lack an optic nerve and are therefore blind. Importantly, mutations in human Ath5 (Atoh7) have recently been associated with optic nerve hypoplasia, one of the leading causes of blindness in children,[72] indicating that Atoh7-dependent RGC development is also conserved in humans.

Besides the conserved placement of ato/Ath5 in the genesis of the first cell type to be generated in their respective retinas, additional similarities have been found between R8 photoreceptors and RGCs.[71,73,74] One of the most striking of these findings is the recent discovery of a subset of intrinsically photoreceptive

RGCs (ipRGCs). These cells share many characteristics with *Drosophila* photoreceptors: they express a rhabdomeric-related opsin protein, melanopsin, and they use the same phototransduction pathway used by rhabdomeric photoreceptors.[2,75–79] Thus, R8s and RGCs may represent "sister" cell types, and thus, understanding R8 differentiation should provide insight into the events underlying the development of rhabdomeric ipRGCs.

C. Otx/Crx/Otd and Photoreceptor Differentiation

Another striking correlation observed between vertebrate and invertebrate retinal neurogenesis is the conserved role of Otd-related transcription factors in regulating photoreceptor development.[80,81] *Drosophila* Otd was the first member of the Otx/Crx family of homeodomain transcription factors to be identified, and since then, these factors have been shown to be evolutionarily conserved for the formation of anterior central nervous system structures in animals ranging from Cnidaria to humans.[81–85] During fly eye development, Otd is expressed in all photoreceptors soon after their recruitment in the larval imaginal disc, and is maintained in these cells throughout the lifetime of the animal.[86,87] Despite its early onset of expression, no known role for Otd is known until mid-pupation, where it is necessary for the formation of the light-gathering apical surface of all photoreceptors and for photoreceptor subtype-specific regulation of opsin gene expression.[43,87] In the vertebrate eye, two Otd-related factors are expressed in photoreceptors: Otx2 and Crx. Otx2 expression initiates very early in photoreceptor precursors and is required for Crx activation in these cells.[17] Like *Drosophila* Otd, Otx2 and Crx regulate PR-specific opsin gene expression and are necessary for the formation of all PRs. Thus, it is perhaps not surprising that mutations in both factors are associated with LCA, in which all photoreceptors fail to properly terminally differentiate.

Currently, it is unclear whether Otx2 and Crx play redundant or overlapping functions in different PRs, particularly since the majority of their known targets are the same.[185,186] However, differences in loss-of-function phenotypes in knockout mice for Otx2 and Crx suggest that they may regulate unique subsets of target genes.[16,17,187] Moreover, although Otx2 and Crx are both associated with LCA, Otx2 mutations can also lead microphthalmia, a more severe ocular defect, consistent with its broader expression pattern compared to Crx. Interestingly, though identical mutations in either Otx2 or Crx can associate with different ocular diseases, suggesting that additional genetic modifiers influence these factors during eye development. For instance, the same mutation in Otx2 that causes LCA in one patient causes microphthalmia in a sibling.[185] Similarly, a mutation in Crx associated with RP can also be associated with CORD.[186] This highlights that retinal degenerative diseases are likely to be multifactorial and that much remains unanswered regarding the genetic basis of these diseases.

Moreover, because of the genetic complexity of these diseases, it is becoming imperative that a flexible, high-throughput system such as *Drosophila* be developed for beginning to dissect the molecular mechanisms responsible for the formation and maintenance of distinct photoreceptor populations.

IV. Photoreceptor Structure Comparison

The most prominent feature of photoreceptor design is the presence of a specialized expansion of the apical membrane where millions of opsin receptor molecules are housed to detect and transduce light information. These subcellular compartments, called outer segments (OSs) in ciliary photoreceptors, and rhabdomeres in rhabdomeric photoreceptors (Fig. 3) are not only essential for phototransduction but are also necessary for the functional integrity and survival of the photoreceptor. In spite of their biological importance, surprisingly little is known about the molecular events choreographing the biogenesis and maintenance of these elaborate organelles. Nevertheless, recent studies have revealed unexpected conservation among the molecules required for the morphogenesis of the membrane discs of OSs in mammals and for the shape, size, and maintenance of rhabdomeres in *Drosophila*. Below, we discuss these different structures and the factors that commonly regulate their formation.

A. Cone and Rod Outer Segments

In mammals, the phototransduction machinery occupies the OS, a vastly expanded specialized cilia-based structure. Electron microscopy studies in the late 1940s and early 1950s provided the first images of these specialized structures and revealed the beautiful array of the membrane discs and tightly packed membrane evaginations of rod and cone cells, respectively.[88] These findings, combined with subsequent morphological studies and experimental data, have led to a model for their genesis.[89–91] The first signs of OS biogenesis in both cone and rod photoreceptor cells are the association of the centrioles with the apical membrane, the establishment of a basal body with a connecting cilium, and the induction of an overlying membrane to expand outward. EM studies suggest that in rod photoreceptor cells, the formation of individual membrane discs then proceeds in two sequential growth phases: membrane evagination and rim formation. First, the ciliary/plasma membrane extends outward as microvilli-like protrusions. A single disc is then formed by the fusion of membranes from two adjacent evaginations, leaving an extracellular space around each developing OS, which becomes the intradisc cavity. In the second phase, the disc is "closed" and separated from the plasma membrane, as the rim of ciliary membrane between each evagination grows around the circumference of each adjacent evagination to seal the disc. In cone photoreceptors, rim formation is incomplete, leaving the membrane evaginations still connected to the plasma membrane. Importantly, rod and cone OSs are very dynamic structures, constantly generating new and shedding old membrane discs. The distal/older discs are phagocytosed by the overlying retinal pigment epithelium (RPE) and replaced by new ones generated at the base of the OS. In rods, this can be up to 40% of the OS each day.[92] Thus, the machinery required to build OSs anew and digest shedded OSs is essential for maintaining healthy photoreceptors.

B. Rhabdomeres

The specialized rhabdomeric apical surface of *Drosophila* OPRs is composed of approximately 60,000 tightly packed microvilli, each approximately 50 nm in diameter, 1–2 μm in length, and 100 μm in depth. IPR rhabdomeres are narrower and shorter, but require many of the same factors for their generation as OPRs. These organelles develop as a result of finely choreographed membrane changes on the apical surface of the photoreceptor during pupation,[93] which lasts approximately 100 h. At the end of larval development, the eye imaginal disc is a pseudostratified epithelial monolayer, with the apical surfaces of all cells at the same plane. At approximately 15 h after puparium formation (APF), the epithelial cells that will generate the lens switch positions above the photoreceptors, which causes the apical–basal axis of the photoreceptors to turn inward 90°. Thus, the photoreceptor apical surfaces now face each other in the center of each ommatidium. Expansive apical surface growth is first detected at approximately 48 h APF and continues for the next 36–48 h. Two significant changes occur at the time growth begins. First, the apical membrane, bounded and defined by the zonula adherens, shows ruffling and accumulates F-actin. This apical membrane eventually subdivides into two regions: the rhabdomeric region that contains microvillli and the stalk membrane that lacks microvillli (Figs. 2 and 3). Next, both the apical membrane and the adherens junctions elongate down the entire length of the photoreceptor cell, producing the characteristic rod/bar morphology of the mature rhabdomere. Shortly thereafter, primordial microvilli with an actin cytoskeleton core extend into the center of the ommatidium, orienting their actin filament's plus (i.e., growing) ends distally, and extending their minus ends well into the cytoplasm of the photoreceptor cell.[94,95] These cytoplasmic extensions make up an actin-rich terminal web that is critical for rhabdomere formation and support, and is thought to serve as a docking area for delivery of material to the rhabdomere. By 72 h APF, the microvillar projections have elongated into the ommatidial cavity, and the developing rhabdomeres become clearly separated from each other due to the deposition of an extracellular matrix known as the inter-rhabdomeral space (IRS). Over the next 24 h, rhabdomeres achieve their adult shape and most of the phototransduction components are synthesized, assembled, and targeted to the microvilli. In contrast to vertebrate photoreceptors, it is not clear whether the rhabdomere is replenished during the life of the fly

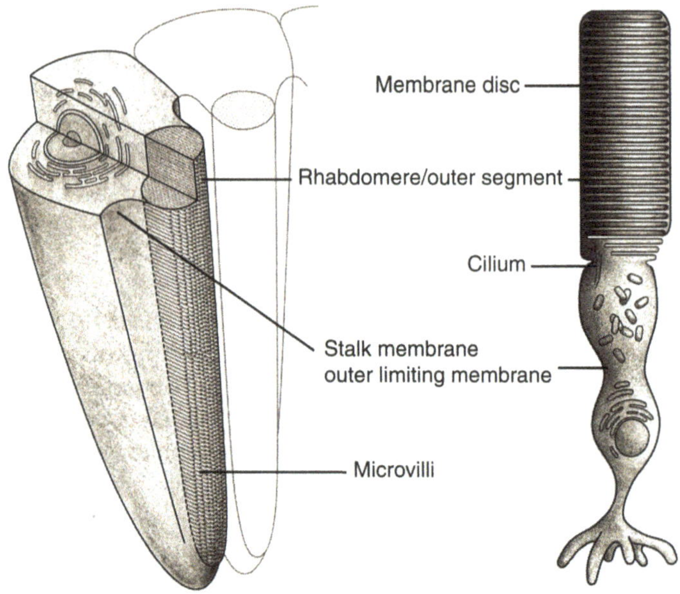

FIG. 3. Photoreceptor cell types. Comparison of a Drosophila rhabdomeric photoreceptor cell with a vertebrate rod photoreceptor cell.

or to what extent membrane turnover contributes to fly photoreceptor integrity. However, excessive light exposure does lead to retinal degeneration, suggesting that membrane regeneration occurs during the night.

Together, even though the fundamental difference between vertebrate and *Drosophila* photoreceptors is the presence or absence of a cilium, both cell types are highly polarized and they increase their apical membrane surfaces by the creation of specialized membrane protrusions. It is here within these two processes that we begin to elucidate the parallels between the two systems and the ability to utilize *Drosophila* as a model organism for understanding photoreceptor morphogenesis.

V. Factors Regulating Photoreceptor Morphogenesis and Maintenance

A. NinaE/Rhodopsin

Outer segments and rhabdomeres must house millions of opsin molecules and parts of the phototransduction machinery to mediate a visual response. Importantly, a wide range of mutations in rod opsin (Rhodopsin or Rho) contribute to it being one of the most often mutated genes in patients with

RP. In some instance, these mutations are degenerative due to their specific role as phototransduction molecules and this will be discussed later. Surprisingly, however, these molecules also play an active role in establishing and/or maintaining the structural integrity of both *Drosophila* rhabdomeres and OSs. This result was first formally demonstrated in *Drosophila* when Kumar et al. noted that a null allele of *ninaE* (the Rhodopsin gene expressed in the rod-like OPRs in *Drosophila*) displays a form of retinal degeneration that is different not only from other retinal degeneration mutants but also from other hypomorphic alleles of *ninaE.*[96] Specifically, the rhabdomere terminal web fails to form correctly, resulting in the rhabdomeric membrane intruding, unraveling, and collapsing into the cytoplasm. Later rescue experiments revealed that a surprisingly small developmental window is required for Rhodopsin expression to maintain the structural integrity of the rhabdomeres, and that supplying ample amounts of Rhodopsin after the initiation of degeneration cannot restore rhabdomere structure.[97] Similar findings are observed in knockout mice for Rhodopsin: photoreceptor specification is normal and components of the inner segment such as the connecting cilium are visible, but the cells are not capable of transforming the apical membrane into the membrane discs that would normally populate the OS.[98]

How does Rhodopsin influence the structural integrity of the rhabdomere/OS? Again, potential mechanisms come from the studies in *Drosophila*. Both fly and human Rhodopsin contain an Asn-Pro-X-X-Tyr motif located in the seventh transmembrane domain that, in other seven transmembrane G-protein coupled receptors, binds to the members of the Rho family of small guanosine triphosphatases (Rho GTPases).[99] Using this information, Chang and Ready demonstrated that the Rho GTPase Rac1 localizes to the actin-rich base of *Drosophila* rhabdomeres, and that a constitutively active Rac1 is sufficient to rescue the rhabdomeric defect seen in *ninaE* null mutants. Conversely, overexpressing a dominant-negative form of Rac1 phenocopies *ninaE* null mutants.[100] In mammals, Rac1 is also expressed in photoreceptor cells and has been associated with the fusion of Rhodopsin-bearing transport carriers to rod OS. In addition, Rac1 uses many of the same factors as *Drosophila* Rac1 to form OSs, including phosphoinositides and ezrin/moesin.[101] Unexpectedly, however, no structural defects are observed in a conditional knockout of Rac1 in mouse photoreceptors. Instead, mice with reduced Rac1 are protected against light-induced degeneration. Unfortunately, in these studies, it was not possible to eliminate the possibility of redundancy with other Rho GTPases or to verify that the conditional knockout occurred at an appropriate developmental time point for a phenotype to be seen. This latter point is important, given the timely requirement for fly Rhodopsin in mediating rhabdomere formation. Thus, whether Rac1 is conserved in regulating photoreceptor morphogenesis remains unclear. Nevertheless, further investigation in

Drosophila will likely reveal the molecular mechanisms of how Rhodopsins activates Rac1, and in turn, how Rac1 directs and promotes the structural integrity of the rhabdomeres. This may also provide mechanistic insight into how Rhodopsin-bearing transport carriers are targeted to OSs in vertebrate photoreceptors.

B. Crumbs/CRB1/arRp12 and LCA

During the biogenesis of rhabdomeres and ciliary OSs, compartmentalization of the photoreceptor cell membrane is necessary. In *Drosophila*, the apical membrane is defined by adherens junctions, which is later partitioned into stalk and rhabdomeric membranes (Figs. 1 and 3). In vertebrates, the homologous regions correspond to the outer limiting membrane of the inner and OSs (Fig. 2). In both systems, these two distinct domains are important for structural support and phototransduction, respectively. Characterizing how these regions are specified, therefore, is critical for understanding how the proper form and function of photoreceptors are established. In this regard, the identification of the transmembrane protein Crumbs and its associated complex is a prime example of how research in *Drosophila* has generated key insights into understanding photoreceptor cell biology and human disease.[102]

Crumbs was first identified in *Drosophila* based on the mottled appearance of embryos carrying a mutation in this gene.[103] Crumbs localizes to apical membranes and is required for the organization of epithelial structures. In particular, Crumbs is central in establishing the zonula adherens, a belt-like structure that demarcates the boundary between apical and basal-lateral surfaces. Crumbs is a transmembrane protein characterized by an extracellular domain consisting of 30 EGF-like repeats and 4 Laminin G domains. Its intracellular domain is small but contains two highly conserved motifs: a PDZ-binding motif required for the recruitment of the PDZ-containing protein Stardust, and a 4.1/ezrin/radixin/moesin (FERM)-binding domain that is thought to aid in linking Crumbs to the actin cytoskeleton.

Crumbs' association with retinal disease was first discovered when mutations in the human homolog CRB1 were shown in patients with autosomal recessive RP (RP12) and later, LCA.[104,105] Nevertheless, one could only speculate on the roles of the human homolog in photoreceptor biology until two key studies in *Drosophila* provided a framework for which vertebrate studies could be undertaken.[106,107] These studies showed that, in newly specified *Drosophila* photoreceptors, Crumbs localizes to the apical surface of the cell, and upon the initiation of rhabdomere morphogenesis, it is excluded from adherens junctions and becomes a component of the stalk membrane. *crumbs* mutants exhibit three key phenotypes. First, rhabdomeres remain confined to the distal region of the ommatidium, being 30–60% the length of wild-type retinas. This phenotype is directly related to the inability of the zonula adherens junctions to extend and remain intact during rhabdomere morphogenesis.[106,107] Second, loss of *crumbs* results in shorter stalk membranes, while overexpressing Crumbs results in expanded stalk membranes. This indicates that the length of the stalk membrane is directly proportional to the amount of Crumbs protein present.[107] Third, in addition to developmental defects in photoreceptor morphogenesis, *crumbs* mutants have late-onset defects, being particularly sensitive to light-induced photoreceptor degeneration.[108]

Molecular mapping of *Drosophila* Crumbs linked its various functions to specific regions of the protein. For instance, stalk length is regulated by the extracellular domain, while zonula adherens elongation requires the presence of the transmembrane and intracellular domains.[106,107,109] Although the mechanisms by which the extracellular domain controls stalk length are not understood, the intracellular domain has the potential to interact with many molecules that are also important in ZA formation, growth, and elongation.[102] In particular, the PDZ-binding protein Stardust (Sdt) directly interacts with Crumbs' four conserved C-terminal amino acids (ERLI) and recruits a third evolutionary conserved member of the Crumbs complex, *Drosophila* PATJ (D-PATJ).[110] As would be expected if Crumbs, Sdt, and Dpatj form a functional complex, mutations in either *sdt* or *dpatj* result in ZA and rhabdomeric defects and exhibit light-induced degeneration similar to *crumbs*.[111–113]

If *Drosophila* can serve as a model for human disease, correlations between Crumbs function should be observed between the two systems. Indeed, several lines of evidence indicate that this is the case. First and foremost, vertebrate models of CRB1 mutants have similar phenotypes as seen in *Drosophila*. For instance, both a mouse knockout model and a naturally occurring mutation in mouse CRB1 (the rd8 mouse) exhibit fragmentation of the outer limiting membrane, shorter inner segments, and retinal degeneration.[114,115] Second, the localization pattern of vertebrate CRB1 is homologous to that of *Drosophila* Crumbs: CRB1 is expressed in photoreceptor cells and is concentrated within the rod inner segment near the zonula adherens. This region, known as the outer limiting membrane, is the functional equivalent of the stalk membrane of *Drosophila* photoreceptor cells.[116,117] Third, human CRB1 can rescue fly *crumbs* mutant photoreceptors and produces the same overexpression phenotype as fly Crumbs, indicating that these proteins are functionally equivalent.[107,117,118] Fourth, just as seen in *Drosophila*, identical localization patterns and phenotypes can be extended to its associated proteins, Lin-7/Veli, MPP/Sdt, and PATJ.[102,116] And finally, mutational analyses in humans have mapped the most deleterious CRB1 mutations to the extracellular domain, highlighting another potential commonality between fly and human photoreceptor cells.[104,105]

Overall, the studies with Crumbs/CRB1 highlight several general key factors and advantages that *Drosophila* offers to study human retinal diseases. The first regards the question of redundancy. In humans, there are three Crumbs proteins (CRB1, 2, and 3), with CRB1 and CRB2 being coexpressed in the retina and CRB3 being expressed in a wide range of nonretinal epithelia cells. Thus, the characterization of a truly null phenotype for CRB function must involve the generation of mice mutant for all three proteins. Similarly, although the Crumb/CRB1-associated proteins are conserved (e.g., Stardust/MPP, Dpatj/PATJ, and Lin-7/Veli), these factors are found in multiple copies and are expressed outside of the retina, which may hinder analysis in the retina (for a review see Ref. 102). *Drosophila*, however, has only one gene for Crumbs and many of its associated proteins. The second advantage centers on the ease and speed in which genetic analysis can be performed in *Drosophila*. For instance, even though both fly and vertebrate models have systems to generate tissue-specific mutants, such methods in mice are far more cumbersome, timely, and expensive than in flies. Therefore, the ability to rapidly create transgenic flies and test these in various mutant backgrounds has proven to be critical in structure/function studies and pinpointing domains of possible therapeutic value. Finally, the ability to test epistatic relationships among several molecules via genetic interactions is invaluable in ordering these intricate signaling complexes and is significantly limited in vertebrates. Thus, to understand the pleiotropic phenotypes associated with Crumb mutations, it will be useful to investigate the entire network of Crumbs protein interactions in the genetically tractable system of *Drosophila*.

C. Prominin, arRP, and Macular Degeneration

Even though there is no cilium in *Drosophila* photoreceptor cells, it is clear that both the membrane discs of their OS and the microvilli of the rhabdomere are merely specialized versions of membrane protrusions. As such, a key question in photoreceptor cell biology is how are these membrane structures formed? Insight to this question has come from a family of proteins known as Prominins, evolutionary conserved proteins often associated with membrane protrusions and microvilli.[119–121] Prominin-1 was first identified in mouse as a protein concentrated at the apical domain of neuroepithelial progenitor cells, and in humans, as the AC133 surface antigen in hematopoietic stem cells.[122–124] Subsequent studies, however, showed that Prominin-1 is a very useful marker for various stem cells and is found on many types of differentiated epithelial and nonepithelial cell types, including photoreceptor cells. This latter expression seems to be particularly relevant, because several Prominin-1 mutations have been identified and these are all associated with retinal degenerative diseases.

In humans, mice, and flies, two members of the prominin family are expressed. Even though invertebrate Prominins show relatively low amino acid sequence homology with their vertebrate counterparts (20%), all family members share a very similar topology—five transmembrane domains with approximately 70% of the amino acids contained in two large extracellular loops that have numerous sites for N-glycosylation. The first mutation in Prominin-1 was isolated from a family with autosomal recessive RP.[125] These studies revealed that wild-type Prominin-1 localizes to the base of photoreceptor OSs and that the homozygous single nucleotide deletion identified in the Prominin-1 gene resulted in a truncated protein incapable of localizing to the cell surface. Given the topology of Prominin-1, and its localization to the protrusions that generate the membrane discs, according to the model for disc formation, the large extracellular loops of Prominin-1 would first contact the extracellular environment adjacent to the developing OS, and eventually become sequestered within the intradisc space. Since this initial finding, additional mutations in Prominin-1 have been identified as the causative agent in a number of retinopathies, including arRP, arCORD, and autosomal dominant macular degeneration.[126–128] Furthermore, a definitive role and importance of Prominin-1 in normal photoreceptor development and retinal degeneration has recently been demonstrated in mice.[129] Specifically, the absence of Prominin-1 results in correctly specified photoreceptors with a normal connecting cilium, but both the rod and cone OSs are morphologically abnormal and the photoreceptors progressively degenerate.

Drosophila Prominin was independently identified and isolated in a genetic screen for loci necessary for the transformation of the *Drosophila* photoreceptor apical membrane into a rhabdomere.[130] Like in humans, *Drosophila* Prominin localizes to photoreceptor apical membrane regions that are in direct contact with the extracellular matrix. However, although Prominin can be found on the entire apical stalk membrane, it is mainly concentrated on the circumference of the rhabdomere apical membrane. In addition, the loss of Prominin does not result in progressive degeneration as in humans, but rather, it varies the morphology and position of individual rhabdomeres. For example, instead of individual rhabdomeres being separated from each other, as in wild-type eyes, *prominin* mutant rhabdomeres make multiple inappropriate contacts with each other. This defect is a direct result of the failure in the extracellular matrix to be properly distributed (see arRP25/EYS/Spacemaker below). In addition, there is a second minor defect that would be consistent with the general role of Prominin molecules in shaping membrane protrusions—the microvilli of *prominin* mutant rhabdomeres no longer show their characteristic highly ordered parallel structure, but rather, the microvilli are often perpendicular to each other, disturbing the entire organization of the rhabdomere. These data are particularly exciting, as it demonstrates that the same molecule in flies and mammals are important for properly organizing the apical membrane structures in either rhabdomeres or OSs.

Whether we consider the results from humans, mouse, or *Drosophila*, the exact mechanism and function of Prominin in photoreceptor cell biology remains elusive. However, the similar localizations and phenotypic analyses of all three suggest an evolutionary conserved purpose. Thus, can we define the physiological role of Prominin using the fly? The first and critical experiment to answer this question is to establish whether Prominins are functionally conserved—for example, can the human protein rescue a *Drosophila prominin* mutant? Next, since these proteins are N-glycosylated and encode splice isoforms that affect the C-terminal intracellular domain, it will be important to ascertain whether these modifications/particular domains are required for its function. These questions can be rapidly tested in *Drosophila* and will undoubtedly provide much needed insights to understanding the fundamental role of Prominins in photoreceptor biology and disease. More importantly, *Drosophila* is a perfect model to perform genetic screens, including enhancer and suppression screens, to reveal partners of Prominin and potential downstream pathways. Indeed, as described below, the same screen that identified *Drosophila* Prominin not only uncovered a binding partner for Prominin but also helped to identify another key molecule involved in human retinal degeneration: EYS/Spacemaker.

D. arRP25/EYS/Spacemaker

As genetic screens for rhabdomere biogenesis continue, ever-increasing parallels between fly and human photoreceptor development are being uncovered. For example, in the same screen that identified Prominin, as well as in a behavioral screen for factors involved in the OPR-mediated optomotor response, a gene called *spacemaker/eyes shut* was isolated.[130,131] Importantly, the human ortholog for this gene product was later found to be responsible for arRP25.[132–134] While arRP25 is characterized by the onset of night blindness and progressive degeneration of rods like other RPs, a uniquely identifying characteristic of arRP25 is that its prevalence represents up to 11% of all arRP cases, compared to the 1–5% seen with most arRPs.[135–138] Given that 21 of the known 25 locus that result in arRP, excluding arRP25, account for only 30% of the arRP cases, these data indicate that arRP25 represents a significant portion of known arRP loci.[139] Notably, the arRP25 locus, located at position of 6q12, consists of 43 exons dispersed throughout 2 Mb of DNA, making it the fifth largest gene in the human genome, and as such, a large target for potential mutagenesis.

arRP25/*spacemaker/eyes shut-encoding* proteins are all characterized by an N-terminal signal peptide, a series of EGF-like domains, a spacer region that lacks sequence conservation between homologs, and a C-terminus that contains both EGF-like repeats and Laminin G domains. Similar mutations have been isolated in both vertebrate and invertebrate proteins, and mutations that truncate the C-terminus result in arRP in humans and null phenotypes in *Drosophila*, suggesting that this region is essential for function. To date, the possible functions of EYS/Spacemaker have been generated by studies in *Drosophila*. Spacemaker localizes to and is essential for the formation of the IRS, the extracellular matrix that separates each of the rhabdomeres in a single ommatidium (Fig. 1B). Examination of a null allele of Spacemaker reveals the total absence of the IRS, and juxtaposed rhabdomeres with irregular shapes and sizes. This result suggests that an interplay between the developing rhabdomeric membrane and the extracellular matrix is required to obtain the stereotypic structure and correct position of the rhabdomeres. Perhaps more importantly, is that *prominin* and *spacemaker* genetically interact and Prominin is required for the proper distribution of Spacemaker around the developing rhabdomeres.[130]

These data allow us to make some predictions about possible functions of EYS in human photoreceptor cell development. For instance, since EYS expression appears to be limited to photoreceptors and the protein localizes to OSs, then, based on the work in *Drosophila*, an intriguing model would propose that EYS binds to Prominin-1 and that Prominin-1 is required for the proper distribution of EYS. If this were true, this would place EYS in the intradisc space. As a result, EYS with Prominin-1 would provide the correct spacing and structural integrity to the membrane discs and folds of rod and cone photoreceptor cells. Opportunely, as with Prominin, we have begun to test whether human EYS can functionally substitute for Spacemaker in *Drosophila* (A. Zelhof, unpublished data) and therefore are in a position to define the domains necessary for the various aspects of EYS/Spacemaker biology. In addition, considering the differences between fly and human photoreceptor cells (non-ciliated *vs.* ciliated), since Spacemaker is also expressed and functions in *Drosophila* ciliated neurons[140] so any *in vivo* analysis of this factor in *Drosophila* will be able to include both types of cells, revealing nuances specific to ciliated cells. Overall, the identification of the locus for arRP25 and the parallel role of fly Spacemaker provide an invaluable tool for understanding new disease mechanisms, from identification of possible partners to defining molecular pathways that this family of molecules confers.

E. Otd/Crx and LCA

Not only are structural components common to invertebrate and vertebrate photoreceptors, but the spatial and temporal control of gene expression required for morphogenesis are also conserved. For example, as mentioned earlier, Otd in *Drosophila* and Otx2 and Crx in vertebrates are necessary for the morphogenesis of all photoreceptors, and mutations in human OTX2 and CRX are associated with LCA. Characterization of Crx knockout mice further demonstrated that although the connecting cilium is generated, the photoreceptor cells appear to stall and degenerate.[141] Otd mutant photoreceptors display a similar phenotype: they are properly specified but are unable to create a complete rhabdomere.[87] However, in contrast to vertebrates, Otd is not the sole homeodomain-containing transcription factor required for rhabdomere formation or photoreceptor cell function. Instead, recent data indicate that the function of Crx in vertebrate photoreceptor terminal differentiation is partitioned to two proteins in *Drosophila*, Otd and Pph13.[65] Thus, only when both Otd and Pph13 are mutant, is complete elimination of rhabdomere formation observed.

How Otd/Pph13- and Crx-dependent defects manifest is unclear, but microarray analysis of these factors has revealed that they are responsible for a number of factors required for building the rhabdomere/OS.[80,142–144] Notably, these targets are distinct between vertebrates and invertebrates, consistent with different requirements for building ciliary vs rhabdomeric structures. For example, Crx is required for Peripherin expression, a key molecule for generating the membrane discs of the OS, whereas in Otd mutants, there is a precipitous drop in expression of Chaoptin and Calphotin, proteins that are important in creating the rhabdomere and the IRS.[80] Interestingly, although Pph13 is required for rhabdomere morphogenesis, no previously identified morphogenetic proteins have been identified as Pph13 targets.[65,145] Instead, Pph13 regulates numerous genes necessary for phototransduction, which we will discuss next, can lead to degeneration, thus providing one possible mechanism underlying the rhabdomeric defects observed in Pph13 mutants. Likewise, Otd and Crx regulate the expression of phototransduction pathway factors, so lack of these targets may also contribute to morphogenetic defects observed in Otd/Crx mutants. Importantly, we have recently found that both OTX2 and CRX can rescue Otd-dependent rhabdomere morphogenesis in the fly eye (T. Cook, unpublished data), suggesting that Otd-related proteins sit high in the hierarchy of photoreceptor morphogenesis, prior to the divergence of ciliary and rhabdomeric subtypes. Thus, once again, the fly eye should serve as a useful model for further exploring the role of a wide range of factors thought to participate in photoreceptor morphogenesis, whether ciliary or rhabdomeric.

VI. Phototransduction and Disease

As mentioned earlier, mutations in Rhodopsin are the leading cause of autosomal dominant RP (ADRP), and its more than 100 identified mutations cause up to 25% of ADRP cases.[139] While some of these mutants contribute to morphogenesis specifically, it is clear that some of these mutants also cause retinal degeneration because of aberrant phototransduction. Interestingly, even though the downstream effectors of opsins are distinct between *Drosophila* and vertebrates, improper regulation of opsins and their underlying phototransduction cascade leads to retinal degeneration in both systems. Here, we examine the use of *Drosophila* to understand how mutations in Rhodopsin-mediated signaling lead to retinal degeneration.

A. Rhodopsin: Biosynthesis and Folding

The path that Rhodopsin takes in order to reach its light-gathering organelle begins in the endoplasmic reticulum (ER). Upon proper folding and N-glycosylation, Rhodopsin enters the Golgi, where more posttranslational modifications occur before it is targeted exclusively to the OS/rhabdomere. While it is clear that many RP-associated missense mutations affect residues that are conserved between human and *Drosophila* rhodopsins, and lead to improper folding and targeting of Rhodopsin,[146–149] the molecular mechanisms of how these mutations act in a dominant fashion or how they trigger retinal degeneration are less obvious.

One of the first identified and most common ADRP-associated missense mutations in Rhodopsin affects position 23 and converts a proline to a histidine, alanine, or leucine residue.[150] Numerous studies have demonstrated that such mutants do not associate with chromophore, form aggregates, and accumulate in the ER.[146,147,151,152] Even though models exist for how this occurs, many questions remain about how this mutation leads to the disease state.[153] For example, *in vivo*, does the mutant protein reach the OS and if so, is it active? Does it interfere with wild-type function? And, most importantly, how does it trigger degeneration? Fortunately, many of these questions can be and have been modeled in the fly. For instance, using a homologous mutation in *Drosophila* Rhodopsin (Rh1) (Pro37His or Rho^{P37H}), Galy *et al.* found that expressing this mutation under the control of its native promoter reproduces the same dominant phenotypes in *Drosophila* as it does in human patients; that is, it leads to obvious age- and light-dependent rhabdomeres degeneration and the majority of mutant protein localizes to the ER. Importantly, however, the protein that does reach the rhabdomere does not interfere with the localization of wild-type protein. Moreover, in the absence of wild-type protein, the mutant protein is able to provide the structural integrity necessary to form the rhabdomere and can even function in phototransduction, albeit at reduced levels.[154]

If mutant Rhodopsin proteins are able to provide structure and elicit a visual response, why then do they dominantly cause the cell to degenerate? Recent studies in both vertebrates and invertebrates suggest that at least one mechanism involves the unfolded protein response (UPR) pathway.[155,156] This pathway is often activated by misfolded proteins being retained in the ER, and indeed, it is well established that many degenerative Rhodopsin mutations lead to an unfolded protein.[157–161] However, is the activation of the UPR pathway what leads to degeneration? Hallmarks of UPR are the activation of both p38 and JNK, and subsequent induction of apoptosis. Similarly, in the studies with Rho^{P37H}, the authors noted an increase in activated forms of both p38 and JNK proteins, suggesting that ER stress was responsible for the observed photoreceptor cell death in this mutant. However, definitive proof for a role in UPR in Rhodopsin-dependent photoreceptor cell death was conclusively demonstrated by Ryoo et al.[162] This work first showed that ER stress leads to the alternative splicing of the *Drosophila* homologs of xbp1 by Ire-1, two factors critical for mediating a UPR response in both mammals and yeast. Then, by generating a GFP reporter assay based on xbp1 splicing, the authors tested whether specific Rhodopsin mutants that result in unfolded proteins activated the UPR response. Importantly, proteins that create cytoplasm aggregates did not activate the pathway, whereas heterozygous mutants for a Rhodopsin mutant that does accumulate in the ER, $Rh1^{G69D}$, did.[157] Moreover, the authors showed that the alternative splicing of xbp1 is required for protection from retinal degeneration, while the loss of xbp1 dominantly enhances $Rh1^{G69D}$-mediated retinal degeneration. Similarly, Yang *et al.* recently showed that in degenerating photoreceptors of the rd1 mouse, ER stress plays a critical role in photoreceptor cell apoptosis.[163] These results again exhibit the utility of *Drosophila* research to provide insight and a framework for elucidating the molecular mechanism of retinal degeneration, and more importantly, present a solid link between induction of apoptosis as a result of mutations in rhodopsin.

B. Chromophore Generation and Regeneration

The light-sensitive visual pigments in the eyes of both vertebrates and invertebrates consist of an opsin (the G-protein-coupled receptor) and a chromophore. In *Drosophila*, the chromophore is the derivative 11-*cis* 3-hydroxy-retinal, whereas in vertebrates, it is derived from 11-*cis*-retinal (Vitamin A). In either case, the first step in the phototransduction cascade is the absorption of a photon by the visual pigments, which isomerizes 11-*cis*- to all-*trans*-retinal to create an activated metarhodopsin molecule. This active Rhodopsin proceeds to interact and activate heterotrimeric G-proteins to initiate signal transduction. In vertebrates, light absorption leads to the dissociation of the chromophore from the opsin and another active Rhodopsin molecule is regenerated through an enzymatic cycle known as the visual cycle. In contrast, the absorption of light by Rhodopsin in *Drosophila* does not release the chromophore, but instead, the metarhodopsin state is stable and converted back to an active state by the absorption of another photon of light.

The vertebrate visual cycle begins after the chromophore is released from Rhodopsin upon light stimulation. The first step of the regeneration occurs in the OS. At least two retinol dehydrogenases (RDHs) convert the all-*trans* retinal to all-*trans* retinol, and this molecule is transferred to the RPE via IRBP. After several enzymatic steps, the RPE restores all-*trans* retinol to 11-*cis*-retinal, and this is transported back to the photoreceptor where it is then linked to a new Rhodopsin molecule (see Refs. 164 and 165 for recent reviews). Since chromophore is required for Rhodopsin function, many mutations that affect chromophore regeneration and the RPE are commonly associated with photoreceptor neurodegeneration (see Refs. 166 and 167).

In *Drosophila*, *de novo* synthesis of chromophore is also required for the proper expression and stability of Rhodopsin.[168,169] Nevertheless, because metarhodopsin is stable and can be regenerated back to an active Rhodopsin molecule without chromophore dissociation, and because all of the known chromophore synthesis pathway components that had been identified were not expressed in the retina, it was long believed that *Drosophila* did not have a visual cycle comparable to that performed in the vertebrate RPE. However, since chromophore is necessary for Rhodopsin function, ongoing genetic screens for retinal degeneration continue to uncover factors involved in chromophore synthesis. From such studies, Montell an suggest that a visual cycle is also present in *Drosophila*. The first of these factors was PINTA, named for its mutant phenotype: *prolonged depolarization afterpotential is not apparent*. PINTA is a CRAL-TRIO-related retinol binding protein important for the *de novo* synthesis of functional Rhodopsin protein, and perhaps more importantly, it is expressed in the pigmented interommatidial cells.[170] This study provided the first evidence that chromophore production does rely, in part, on functions in the retina and suggested that the interommatidial cells are equivalent to the vertebrate RPE. Even more excitingly, however, the same group recently characterized the function of a pigment cell-enriched dehydrogenase, PDH.[171,172] This work revealed that PDH prevents light-dependent retinal degeneration and progressive loss of Rhodopsin by regenerating chromophore in the fly "RPE," much like retinal dehydrogenases in the vertebrate visual cycle.[172] In addition, these studies demonstrated that human retinol dehydrogenase RDH12 partially rescues PDH mutants, supporting the idea that these processes are evolutionarily conserved. Overall, these results revealed that *Drosophila*, like vertebrates, do have a visual cycle, but in the fly, this cycle is primarily employed when nutrient deprivation would limit *de novo* synthesis (the dominant pathway for creating the chromophore), thus preventing retinal degeneration under these conditions. These results not only create another platform to explore mechanisms for progressive retinal dystrophies such as macular degeneration but also highlight that continued research in the *Drosophila* eye will continue to uncover parallels between vertebrate and invertebrate eye formation and function.

C. Rhodopsin Inactivation and Arrestin

Arrestins are key molecules in regulating the activity of G-protein coupled receptors.[173] As such, their regulation of Rhodopsin activity is a key component in preventing retinal degeneration. In *Drosophila*, there are two photoreceptor-specific arrestins, Arr1 and Arr2. Arr2, the major isoform, represents 85% of the total amount of arrestin, and its binding to Rhodopsin and subsequent displacement of Gα terminates Rhodopsin signaling. Upon the absorption of a second photon, metarhodopsin is converted back to Rhodopsin, Arr2 is released, and Rhodopsin is ready for the next round of excitation.[174]

Like Rhodopsin activity, the binding of Arr2 to Rhodopsin is tightly regulated by the phototransduction cascade. In particular, the phototransduction cascade leads to activation of calcium/calmodulin-dependent protein kinase II (CamKII), CamKII phosphorylates Arr2, and Arr2 subsequently dissociates from Rhodopsin.[175,176] This dissociation allows dephosphorylation of Rhodopsin and it becomes ready to signal again. Interestingly, mutations that affect the increase in calcium, like *norpA* (phospholipase C) and *calmodulin*, result in the formation of stable Arrestin2–Rhodopsin complexes, and this triggers a massive endocytosis of Rhodopsin, apoptosis, and photoreceptor degeneration.[177] Conversely, in Arr2 mutants, Rhodopsin fails to be inactivated and photoreceptors undergo light-dependent degeneration due to overstimulation of the phototransduction pathway. This nicely illustrates one of the confounding difficulties in understanding retinal degeneration: both overactivation of phototransduction (such as in Arr2 mutants) and inhibition of phototransduction (such as in norpA or Rhodopsin mutants) result in ultimate demise of functioning photoreceptors.

Arr1, even though the minor Arrestin, also plays a critical role in survival of photoreceptor cells.[178] Unlike the loss of Arr2 mutants, Arr1 mutants have little effect on deactivation rates. However, the loss of Arr1 causes light-independent photoreceptor cell degeneration. Arr1 is required for light-dependent endocytosis of Rhodopsin and interacts with Rhodopsin's phosphorylated C-terminus. Therefore, Arr1 is speculated to bind to phosphorylated Rhodopsin to target it for endocytosis, hence preventing the buildup of toxic Arr2–Rhodopsin complexes. In support of this model, the loss of *arr2* inhibits the photoreceptor cell degeneration in *arr1* mutants. Thus, in *Drosophila*, it appears that two visual arrestins provide distinct methods to prevent retinal degeneration: Arr2 quenches Rhodopsin signaling, whereas Arr1 promotes the endocytosis of activated Rhodopsin, thus limiting the number of Rhodopsin/Arr2 complexes that can form.

Unlike most of the examples already discussed, the interplay between Rhodopsin and Arrestins is not identical between fly and humans. While the role of Arrestin in vertebrate photoreceptor cells also uncouples the interactions between Rhodopsin and heterotrimeric G-protein complexes, Arrestin does not appear to be involved in Rhodopsin sequestration. However, two different mutations in Rhodopsin that lead to ADRP indicate potential for deleterious effects of stable Arrestin–Rhodopsin complexes. The first example was provided by the studies of several missense mutations at Arg135.[179] These mutants are hyperphosphorylated and have high affinity for Arrestin. When these mutants were expressed in photoreceptor cells, stable Arrestin–Rhodopsin complexes were internalized by endocytosis and Arrestin was redistributed from the OS cytoplasm to the photoreceptor inner segment. This redistribution of Rhodopsin/arrestin complex was thus postulated to trigger apoptosis and degeneration, rather similar to that observed in *Drosophila*. A second example was provided by the Rhodopsin mutant K296E (Rho^{K296E}). Originally, this mutant was thought to cause degeneration due to its constitutive activity,[180] but a transgenic mouse model did not support this *in vitro* data.[181] Instead, Chen et al. demonstrated that Rho^{K296E} associates with Arrestin, the complex mislocalizes to the inner segment of the photoreceptor cell, and this is toxic to the photoreceptor cell. This again suggests a direct parallel with *Drosophila* photoreceptor cells.[182] Thus, it will now be important to determine if endocytosis and other downstream components identified in the fly[183] are also at play in these forms of Rhodopsin-associated ADRP.

D. Phototransduction Gene Expression—Cone–Rod Homobox (CRX)/Otd

The downstream effectors of opsin signaling are distinct between *Drosophila* and vertebrates. Rhabdomeric photoreceptors, for instance, use Gαq and activate DAG and PLC, whereas ciliary photoreceptors use Gαt (transducin) to activate the PKC pathway. Regardless, it is obvious that mutants that affect either of these phototransduction cascades lead to retinal degeneration. Not surprisingly, transcription factors that control the expression of genes within the phototransduction pathway are also required to maintain a health retina. As mentioned earlier, Otd-related transcription factors are exactly such proteins. Crx, for instance, not only regulates peripherin, a key structural component of the OS, but it is also important for the expression of cone and rod opsins, arrestin, rod transducin, and phosphodiesterase gamma.[143] Similarly,

Otd regulates cone- and rod-like opsins, arrestin, and TRP channels.[80] Given the ability of Crx to regulate some genes in specific photoreceptor subpopulations, such as rod transducin or S-opsin, and its involvement in retinopathies that differentially affect one photoreceptor subtype over another, one could envision that particular disease-associated alleles would disrupt regulatory domains important for cell-specific target genes. Unfortunately, no genotype-phenotype correlation with Crx mutations have been possible, indicating that it will not be that simple. Indeed, even within a single photoreceptor type, Otd-related molecules must oppositely regulate distinct targets. For instance, in rods, Crx must activate some rod-specific genes and prevent cone-specific genes, yet in cones, this activity must have reversed. Thus, another possible model for how different Crx mutants affect distinct photoreceptor populations is due to de-repression of subtype-specific genes into the wrong photoreceptor class, much like what is observed in NRL mutants. Currently, however, it remains largely unclear how Crx regulates cone-specific target genes, and the majority of Crx cofactors are broadly expressed factors (reviewed in Ref. 21). How then can we begin to understand how Crx differentially controls cone and/or rod photoreceptor form and function? Fortunately, *Drosophila* Otd also exhibits complex regulatory functions in different photoreceptor populations which may allow us to begin to answer such questions. For instance, in rod-like OPRs, Otd represses the green-sensitive opsin, Rh6, and in cone-like photoreceptors, it is important for both the activation and repression of the same gene in different R8 subtypes.[40,86] Likewise, it activates the blue-sensitive opsin in some R8s but represses it in others. Yet, like Crx, Otd also regulates the morphology of all photoreceptors. Clearly, Otd (and Crx) must integrate several coregulatory complexes to control its distinct functions. Using a mapping strategy, since different domain combinations are used by Otd to mediate its different functions.[86] For instance, almost all of the protein must be removed to fully eliminate its function in morphogenesis, whereas repression of Rh6 in OPRs is restricted to its C-terminus. Importantly, since Otx2 and Crx rescue overlapping functions of Otd when expressed in the fly eye (T. Cook, unpublished data), this gives us a valuable system to determine what coregulatory complexes are used by all three Otd factors (Otd, Otx2, and Crx) to regulate distinct functions. Such studies should provide additional candidate factors that may be disrupted in patients suffering from retinopathies affecting different photoreceptor populations. It will also be interesting to explore the possibility that a Pph13-related ortholog is expressed in the vertebrate retina. As mentioned, this factor functions in parallel with Otd to regulate opsin expression and rhabdomeric photoreceptor cell terminal differentiation. Pph13 regulates a number of factors that are critical for photoreceptor form and function, including Arr2, Gβ, trp, and ninaC, and at least one of these targets, Arr2, is an overlapping target with Otd. Thus, one possibility is that a Pph13 ortholog is present in vertebrates, and in cases in which Crx is linked with LCA, where photoreceptor morphology is entirely disrupted, Pph13 is also disrupted. While this example is not meant to be all-inclusive, it helps to illustrate that further research in photoreceptor gene expression in *Drosophila* may provide insight into molecular mechanisms underlying retinal degenerative processes.

VII. Summary and Concluding Remarks

Retinal development is a complicated process. However, the recent finding that rod and cone photoreceptors share a genetically-related precursor provides a useful framework for better understanding how certain inherited retinal disorders arise. However, much still remains unclear about these diseases arise. For instance, how can the same gene mutation lead to complete loss of all photoreceptors in infants vs. loss of one photoreceptor cell type 20 years later? How can both too much and not enough phototransduction activity lead to degeneration? How can the loss of a factor expressed in both cones and rods differentially affect one cell type and not the other? What are the unknown genes that interact with the known culprits that can be used to begin to develop cell-specific therapies and be able to diagnose these late onset diseases at an early time?

Such questions are very difficult to ask in currently available mammalian models. However, here, we demonstrate that homologs not only exist in *Drosophila* but also share the same phenotypic consequences when mutant as compared to their vertebrate counterparts. As a result, it is now possible to do a more thorough of each of these molecules and determine their role in all important aspects of photoreceptor cell biology: behavior, physiology, and structure. Thus, *Drosophila* will be certain to serve as an important surrogate to provide additional insight into such questions. For instance, we can begin to understand the basic mechanisms of action for many of these factors, using rapid transgenic assays. We can begin to perform large-scale *in vivo* genotype–phenotype analysis of mutant genes associated with retinal degeneration. And we can begin to perform enhancer and suppressor genetic screens to identify full pathways of how photoreceptor gene expression and differentiation are coordinated. Together, these studies have a tremendous capacity to revolutionize ocular diagnostic and therapeutic approaches.

> None of God's creatures absolutely considered are in their own nature contemptible; the meanest fly, the poorest insect has its use and virtue.
>
> Mary Astell

REFERENCES

1. Adams NA, Awadein A, Toma HS. The retinal ciliopathies. *Ophthalmic Genet* 2007;**28**:113–25.
2. Arendt D. Evolution of eyes and photoreceptor cell types. *Int J Dev Biol* 2003;**47**:563–71.
3. Land MF, Nilsson DE. *Animal eyes.* Oxford University Press, Oxford; 2002.
4. Land MF. The optical structures of animal eyes. *Curr Biol* 2005;**15**:R319–23.
5. Livesey FJ, Cepko CL. Vertebrate neural cell-fate determination: lessons from the retina. *Nat Rev Neurosci* 2001;**2**:109–18.
6. Marquardt T, Gruss P. Generating neuronal diversity in the retina: one for nearly all. *Trends Neurosci* 2002;**25**:32–8.
7. Swaroop A, Kim D, Forrest D. Transcriptional regulation of photoreceptor development and homeostasis in the mammalian retina. *Nat Rev Neurosci* 2010;**11**:563–76.
8. Mears AJ, Kondo M, Swain PK, Takada Y, Bush RA, Saunders TL, et al. Nrl is required for rod photoreceptor development. *Nat Genet* 2001;**29**:447–52.
9. Bessant DA, Ali RR, Bhattacharya SS. Molecular genetics and prospects for therapy of the inherited retinal dystrophies. *Curr Opin Genet Dev* 2001;**11**:307–16.
10. DeAngelis MM, Grimsby JL, Sandberg MA, Berson EL, Dryja TP. Novel mutations in the NRL gene and associated clinical findings in patients with dominant retinitis pigmentosa. *Arch Ophthalmol* 2002;**120**:369–75.
11. Nishiguchi KM, Friedman JS, Sandberg MA, Swaroop A, Berson EL, Dryja TP. Recessive NRL mutations in patients with clumped pigmentary retinal degeneration and relative preservation of blue cone function. *Proc Natl Acad Sci USA* 2004;**101**:17819–24.
12. Wright AF, Reddick AC, Schwartz SB, Ferguson JS, Aleman TS, Kellner U, et al. Mutation analysis of NR2E3 and NRL genes in enhanced S cone syndrome. *Hum Mutat* 2004;**24**:439.
13. Jia L, Oh EC, Ng L, Srinivas M, Brooks M, Swaroop A, et al. Retinoid-related orphan nuclear receptor RORbeta is an early-acting factor in rod photoreceptor development. *Proc Natl Acad Sci USA* 2009;**106**:17534–9.
14. Srinivas M, Ng L, Liu H, Jia L, Forrest D. Activation of the blue opsin gene in cone photoreceptor development by retinoid-related orphan receptor beta. *Mol Endocrinol* 2006;**20**:1728–41.
15. Ng L, Hurley JB, Dierks B, Srinivas M, Salto C, Vennstrom B, et al. A thyroid hormone receptor that is required for the development of green cone photoreceptors. *Nat Genet* 2001;**27**:94–8.
16. Koike C, Nishida A, Ueno S, Saito H, Sanuki R, Sato S, et al. Functional roles of Otx2 transcription factor in postnatal mouse retinal development. *Mol Cell Biol* 2007;**27**:8318–29.
17. Nishida A, Furukawa A, Koike C, Tano Y, Aizawa S, Matsuo I, et al. Otx2 homeobox gene controls retinal photoreceptor cell fate and pineal gland development. *Nat Neurosci* 2003;**6**:1255–63.
18. Furukawa T, Morrow EM, Cepko CL. Crx, a novel otx-like homeobox gene, shows photoreceptor-specific expression and regulates photoreceptor differentiation. *Cell* 1997;**91**:531–41.
19. Morrow EM, Furukawa T, Cepko CL. Vertebrate photoreceptor cell development and disease. *Trends Cell Biol* 1998;**8**:353–8.
20. Hennig AK, Peng GH, Chen S. Regulation of photoreceptor gene expression by Crx-associated transcription factor network. *Brain Res* 2008;**1192**:114–33.
21. Hsiau TH, Diaconu C, Myers CA, Lee J, Cepko CL, Corbo JC. The cis-regulatory logic of the mammalian photoreceptor transcriptional network. *PLoS ONE* 2007;**2**:e643.
22. Prabhudesai SN, Cameron DA, Stenkamp DL. Targeted effects of retinoic acid signaling upon photoreceptor development in zebrafish. *Dev Biol* 2005;**287**:157–67.

23. Flamarique IN. Gradual and partial loss of corner cone-occupied area in the retina of rainbow trout. *Vis Res* 2001;**41**:3073–82.
24. Hyatt GA, Schmitt EA, Fadool JM, Dowling JE. Retinoic acid alters photoreceptor development in vivo. *Proc Natl Acad Sci USA* 1996;**93**:13298–303.
25. Harpavat S, Cepko CL. Thyroid hormone and retinal development: an emerging field. *Thyroid* 2003;**13**:1013–9.
26. Wolff T, Ready D. Pattern formation in the Drosophila retina. In: Bate M, Arias AM, editors. *The development of Drosophila melanogaster.* Cold Spring Harbor: Cold Spring Harbor Press; 1993. p. 1277–325.
27. Roignant JY, Treisman JE. Pattern formation in the Drosophila eye disc. *Int J Dev Biol* 2009;**53**:795–804.
28. Kumar JP. The molecular circuitry governing retinal determination. *Biochim Biophys Acta* 2009;**1789**:306–14.
29. Hardie, R. C. (1985). Functional organization of the fly retina. In Prog Sens Physiol (Ottoson, D., ed.), Vol. 5, pp. 1-79. Springer, Berlin, Heidelberg, New York, Toronto.
30. Sanes JR, Zipursky SL. Design principles of insect and vertebrate visual systems. *Neuron* 2010;**66**:15–36.
31. Montell C, Jones K, Zuker C, Rubin G. A second opsin gene expressed in the ultraviolet-sensitive R7 photoreceptor cells of *Drosophila melanogaster. J Neurosci* 1987;**7**:1558–66.
32. Papatsenko D, Sheng G, Desplan C. A new rhodopsin in R8 photoreceptors of Drosophila: evidence for coordinate expression with Rh3 in R7 cells. *Development* 1997;**124**:1665–73.
33. Fryxell KJ, Meyerowitz EM. An opsin gene that is expressed only in the R7 photoreceptor cell of Drosophila. *EMBO J* 1987;**6**:443–51.
34. Zuker CS, Montell C, Jones K, Laverty T, Rubin GM. A rhodopsin gene expressed in photoreceptor cell R7 of the Drosophila eye: homologies with other signal-transducing molecules. *J Neurosci* 1987;**7**:1550–7.
35. Chou WH, Hall KJ, Wilson DB, Wideman CL, Townson SM, Chadwell LV, et al. Identification of a novel Drosophila opsin reveals specific patterning of the R7 and R8 photoreceptor cells. *Neuron* 1996;**17**:1101–15.
36. Huber A, Schulz S, Bentrop J, Groell C, Wolfrum U, Paulsen R. Molecular cloning of Drosophila Rh6 rhodopsin: the visual pigment of a subset of R8 photoreceptor cells. *FEBS Lett* 1997;**406**:6–10.
37. Erclik T, Hartenstein V, McInnes RR, Lipshitz HD. Eye evolution at high resolution: the neuron as a unit of homology. *Dev Biol* 2009;**332**:70–9.
38. Morante J, Desplan C, Celik A. Generating patterned arrays of photoreceptors. *Curr Opin Genet Dev* 2007;**17**:314–9.
39. Mollereau B, Dominguez M, Webel R, Colley NJ, Keung B, de Celis JF, et al. Two-step process for photoreceptor formation in Drosophila. *Nature* 2001;**412**:911–3.
40. Xie B, Charlton-Perkins M, McDonald E, Gebelein B, Cook T. Senseless functions as a molecular switch for color photoreceptor differentiation in Drosophila. *Development* 2007;**134**:4243–53.
41. Cook T, Pichaud F, Sonneville R, Papatsenko D, Desplan C. Distinction between color photoreceptor cell fates is controlled by Prospero in Drosophila. *Dev Cell* 2003;**4**:853–64.
42. Morey M, Yee SK, Herman T, Nern A, Blanco E, Zipursky SL. Coordinate control of synaptic-layer specificity and rhodopsins in photoreceptor neurons. *Nature* 2008;**456**:795–9.
43. Tahayato A, Sonneville R, Pichaud F, Wernet MF, Papatsenko D, Beaufils P, et al. Otd/Crx, a dual regulator for the specification of ommatidia subtypes in the Drosophila retina. *Dev Cell* 2003;**5**:391–402.
44. Wernet MF, Mazzoni EO, Celik A, Duncan DM, Duncan I, Desplan C. Stochastic spineless expression creates the retinal mosaic for colour vision. *Nature* 2006;**440**:174–80.

45. Mikeladze-Dvali T, Wernet MF, Pistillo D, Mazzoni EO, Teleman AA, Chen YW, et al. The growth regulators warts/lats and melted interact in a bistable loop to specify opposite fates in Drosophila R8 photoreceptors. *Cell* 2005;**122**:775–87.
46. Salvini-Plawen L, Mayr E. On the evolution of photoreceptors and eyes. Vol. 10. In: Hecht MK, Steere WC, Wallace B, editors. *Evolutionary biology*. New York: Plenum Press; 1977.
47. O'Tousa JE, Baehr W, Martin RL, Hirsh J, Pak WL, Applebury ML. The Drosophila ninaE gene encodes an opsin. *Cell* 1985;**40**:839–50.
48. Zuker CS, Cowman AF, Rubin GM. Isolation and structure of a rhodopsin gene from *D. melanogaster*. *Cell* 1985;**40**:851–8.
49. Glaser T, Walton DS, Maas RL. Genomic structure, evolutionary conservation and aniridia mutations in the human PAX6 gene. *Nat Genet* 1992;**2**:232–9.
50. Hill RE, Favor J, Hogan BL, Ton CC, Saunders GF, Hanson IM, et al. Mouse small eye results from mutations in a paired-like homeobox-containing gene. *Nature* 1991;**354**:522–5.
51. Quiring R, Walldorf U, Kloter U, Gehring WJ. Homology of the eyeless gene of Drosophila to the Small eye gene in mice and Aniridia in humans. *Science* 1994;**265**:785–9.
52. Ton CC, Hirvonen H, Miwa H, Weil MM, Monaghan P, Jordan T, et al. Positional cloning and characterization of a paired box- and homeobox-containing gene from the aniridia region. *Cell* 1991;**67**:1059–74.
53. Halder G, Callaerts P, Gehring WJ. Induction of ectopic eyes by targeted expression of the eyeless gene in Drosophila. *Science* 1995;**267**:1788–92.
54. Chow RL, Altmann CR, Lang RA, Hemmati-Brivanlou A. Pax6 induces ectopic eyes in a vertebrate. *Development* 1999;**126**:4213–22.
55. Jang CC, Chao JL, Jones N, Yao LC, Bessarab DA, Kuo YM, et al. Two Pax genes, eye gone and eyeless, act cooperatively in promoting Drosophila eye development. *Development* 2003;**130**:2939–51.
56. Marquardt T, Ashery-Padan R, Andrejewski N, Scardigli R, Guillemot F, Gruss P. Pax6 is required for the multipotent state of retinal progenitor cells. *Cell* 2001;**105**:43–55.
57. Ashery-Padan R, Gruss P. Pax6 lights-up the way for eye development. *Curr Opin Cell Biol* 2001;**13**:706–14.
58. Dyer MA. Regulation of proliferation, cell fate specification and differentiation by the homeodomain proteins Prox1, Six3, and Chx10 in the developing retina. *Cell Cycle* 2003;**2**:350–7.
59. Pignoni F, Hu B, Zavitz KH, Xiao J, Garrity PA, Zipursky SL. The eye-specification proteins So and Eya form a complex and regulate multiple steps in Drosophila eye development. *Cell* 1997;**91**:881–91.
60. de Melo J, Qiu X, Du G, Cristante L, Eisenstat DD. Dlx1, Dlx2, Pax6, Brn3b, and Chx10 homeobox gene expression defines the retinal ganglion and inner nuclear layers of the developing and adult mouse retina. *J Comp Neurol* 2003;**461**:187–204.
61. Hitchcock PF, Macdonald RE, VanDeRyt JT, Wilson SW. Antibodies against Pax6 immunostain amacrine and ganglion cells and neuronal progenitors, but not rod precursors, in the normal and regenerating retina of the goldfish. *J Neurobiol* 1996;**29**:399–413.
62. Papatsenko D, Nazina A, Desplan C. A conserved regulatory element present in all Drosophila rhodopsin genes mediates Pax6 functions and participates in the fine-tuning of cell-specific expression. *Mech Dev* 2001;**101**:143–53.
63. Sheng G, Thouvenot E, Schmucker D, Wilson DS, Desplan C. Direct regulation of rhodopsin 1 by Pax-6/eyeless in Drosophila: evidence for a conserved function in photoreceptors. *Genes Dev* 1997;**11**:1122–31.
64. Punzo C, Kurata S, Gehring WJ. The eyeless homeodomain is dispensable for eye development in Drosophila. *Genes Dev* 2001;**15**:1716–23.

65. Mishra M, Oke A, Lebel C, McDonald EC, Plummer Z, Cook TA, et al. Pph13 and Orthodenticle define a dual regulatory pathway for photoreceptor cell morphogenesis and function. *Development* 2010;**137**:2895–904.
66. Jarman AP, Grell EH, Ackerman L, Jan LY, Jan YN. Atonal is the proneural gene for Drosophila photoreceptors. *Nature* 1994;**369**:398–400.
67. Jarman AP, Sun Y, Jan LY, Jan YN. Role of the proneural gene, atonal, in formation of Drosophila chordotonal organs and photoreceptors. *Development* 1995;**121**:2019–30.
68. Borod ER, Heberlein U. Mutual regulation of decapentaplegic and hedgehog during the initiation of differentiation in the Drosophila retina. *Dev Biol* 1998;**197**:187–97.
69. Baker NE, Yu SY. Proneural function of neurogenic genes in the developing Drosophila eye. *Curr Biol* 1997;**7**:122–32.
70. Wang SW, Kim BS, Ding K, Wang H, Sun D, Johnson RL, et al. Requirement for math5 in the development of retinal ganglion cells. *Genes Dev* 2001;**15**:24–9.
71. Brown NL, Patel S, Brzezinski J, Glaser T. Math5 is required for retinal ganglion cell and optic nerve formation. *Development* 2001;**128**:2497–508.
72. Macgregor S, Hewitt AW, Hysi PG, Ruddle JB, Medland SE, Henders AK, et al. Genome-wide association identifies ATOH7 as a major gene determining human optic disc size. *Hum Mol Genet* 2010;**19**:2716–24.
73. Hsiung F, Moses K. Retinal development in Drosophila: specifying the first neuron. *Hum Mol Genet* 2002;**11**:1207–14.
74. Sun Y, Kanekar SL, Vetter ML, Gorski S, Jan YN, Glaser T, et al. Conserved and divergent functions of Drosophila atonal, amphibian, and mammalian Ath5 genes. *Evol Dev* 2003;**5**:532–41.
75. Berson DM. Strange vision: ganglion cells as circadian photoreceptors. *Trends Neurosci* 2003;**26**:314–20.
76. Ecker JL, Dumitrescu ON, Wong KY, Alam NM, Chen SK, Legates T, et al. Melanopsin-expressing retinal ganglion-cell photoreceptors: cellular diversity and role in pattern vision. *Neuron* 2010;**67**:49–60.
77. Fu Y, Zhong H, Wang MH, Luo DG, Liao HW, Maeda H, et al. Intrinsically photosensitive retinal ganglion cells detect light with a vitamin A-based photopigment, melanopsin. *Proc Natl Acad Sci USA* 2005;**102**:10339–44.
78. Hattar S, Liao HW, Takao M, Berson DM, Yau KW. Melanopsin-containing retinal ganglion cells: architecture, projections, and intrinsic photosensitivity. *Science* 2002;**295**:1065–70.
79. Berson DM, Dunn FA, Takao M. Phototransduction by retinal ganglion cells that set the circadian clock. *Science* 2002;**295**:1070–3.
80. Ranade SS, Yang-Zhou D, Kong SW, McDonald EC, Cook TA, Pignoni F. Analysis of the Otd-dependent transcriptome supports the evolutionary conservation of CRX/OTX/OTD functions in flies and vertebrates. *Dev Biol* 2008;**315**:521–34.
81. Plouhinec JL, Sauka-Spengler T, Germot A, Le Mentec C, Cabana T, Harrison G, et al. The mammalian Crx genes are highly divergent representatives of the Otx5 gene family, a gnathostome orthology class of orthodenticle-related homeogenes involved in the differentiation of retinal photoreceptors and circadian entrainment. *Mol Biol Evol* 2003;**20**:513–21.
82. Acampora D, Annino A, Tuorto F, Puelles E, Lucchesi W, Papalia A, et al. Otx genes in the evolution of the vertebrate brain. *Brain Res Bull* 2005;**66**:410–20.
83. Hirth F, Reichert H. Conserved genetic programs in insect and mammalian brain development. *Bioessays* 1999;**21**:677–84.
84. Klein WH, Li X. Function and evolution of Otx proteins. *Biochem Biophys Res Commun* 1999;**258**:229–33.
85. Browne WE, Schmid BG, Wimmer EA, Martindale MQ. Expression of otd orthologs in the amphipod crustacean, *Parhyale hawaiensis*. *Dev Genes Evol* 2006;**216**:581–95.

86. McDonald EC, Xie B, Workman M, Charlton-Perkins M, Terrell DA, Reischl J, et al. Separable transcriptional regulatory domains within Otd control photoreceptor terminal differentiation events. *Dev Biol* 2010;**347**(1):122–32.
87. Vandendries ER, Johnson D, Reinke R. Orthodenticle is required for photoreceptor cell development in the Drosophila eye. *Dev Biol* 1996;**173**:243–55.
88. Sjostrand FS. The ultrastructure of the outer segments of rods and cones of the eye as revealed by the electron microscope. *J Cell Physiol* 1953;**42**:15–44.
89. Arikawa K, Molday LL, Molday RS, Williams DS. Localization of peripherin/rds in the disk membranes of cone and rod photoreceptors: relationship to disk membrane morphogenesis and retinal degeneration. *J Cell Biol* 1992;**116**:659–67.
90. Nilsson SE. Receptor cell outer segment development and ultrastructure of the disk membranes in the retina of the tadpole (*Rana pipiens*). *J Ultrastruct Res* 1964;**11**:581–602.
91. Steinberg RH, Fisher SK, Anderson DH. Disc morphogenesis in vertebrate photoreceptors. *J Comp Neurol* 1980;**190**:501–8.
92. Hollyfield JG, Besharse JC, Rayborn ME. Turnover of rod photoreceptor outer segments. I. Membrane addition and loss in relationship to temperature. *J Cell Biol* 1977;**75**:490–506.
93. Longley Jr. RL, Ready DF. Integrins and the development of three-dimensional structure in the Drosophila compound eye. *Dev Biol* 1995;**171**:415–33.
94. Arikawa K, Hicks JL, Williams DS. Identification of actin filaments in the rhabdomeral microvilli of Drosophila photoreceptors. *J Cell Biol* 1990;**110**:1993–8.
95. Williams DS. Actin filaments and photoreceptor membrane turnover. *Bioessays* 1991;**13**:171–8.
96. Kumar JP, Ready DF. Rhodopsin plays an essential structural role in Drosophila photoreceptor development. *Development* 1995;**121**:4359–70.
97. Kumar JP, Bowman J, O'Tousa JE, Ready DF. Rhodopsin replacement rescues photoreceptor structure during a critical developmental window. *Dev Biol* 1997;**188**:43–7.
98. Humphries MM, Rancourt D, Farrar GJ, Kenna P, Hazel M, Bush RA, et al. Retinopathy induced in mice by targeted disruption of the rhodopsin gene. *Nat Genet* 1997;**15**:216–9.
99. Mitchell R, McCulloch D, Lutz E, Johnson M, MacKenzie C, Fennell M, et al. Rhodopsin-family receptors associate with small G proteins to activate phospholipase D. *Nature* 1998;**392**:411–4.
100. Chang HY, Ready DF. Rescue of photoreceptor degeneration in rhodopsin-null Drosophila mutants by activated Rac1. *Science* 2000;**290**:1978–80.
101. Deretic D, Traverso V, Parkins N, Jackson F, Rodriguez de Turco EB, Ransom N. Phosphoinositides, ezrin/moesin, and rac1 regulate fusion of rhodopsin transport carriers in retinal photoreceptors. *Mol Biol Cell* 2004;**15**:359–70.
102. Bulgakova NA, Knust E. The Crumbs complex: from epithelial-cell polarity to retinal degeneration. *J Cell Sci* 2009;**122**:2587–96.
103. Tepass U, Theres C, Knust E. crumbs encodes an EGF-like protein expressed on apical membranes of Drosophila epithelial cells and required for organization of epithelia. *Cell* 1990;**61**:787–99.
104. den Hollander AI, Heckenlively JR, van den Born LI, de Kok YJ, van der Velde-Visser SD, Kellner U, et al. Leber congenital amaurosis and retinitis pigmentosa with Coats-like exudative vasculopathy are associated with mutations in the crumbs homologue 1 (CRB1) gene. *Am J Hum Genet* 2001;**69**:198–203.
105. den Hollander AI, ten Brink JB, de Kok YJ, van Soest S, van den Born LI, van Driel MA, et al. Mutations in a human homologue of Drosophila crumbs cause retinitis pigmentosa (RP12). *Nat Genet* 1999;**23**:217–21.
106. Izaddoost S, Nam SC, Bhat MA, Bellen HJ, Choi KW. Drosophila Crumbs is a positional cue in photoreceptor adherens junctions and rhabdomeres. *Nature* 2002;**416**:178–83.

107. Pellikka M, Tanentzapf G, Pinto M, Smith C, McGlade CJ, Ready DF, et al. Crumbs, the Drosophila homologue of human CRB1/RP12, is essential for photoreceptor morphogenesis. *Nature* 2002;**416**:143–9.
108. Johnson K, Grawe F, Grzeschik N, Knust E. Drosophila crumbs is required to inhibit light-induced photoreceptor degeneration. *Curr Biol* 2002;**12**:1675–80.
109. Richard M, Muschalik N, Grawe F, Ozuyaman S, Knust E. A role for the extracellular domain of Crumbs in morphogenesis of Drosophila photoreceptor cells. *Eur J Cell Biol* 2009;**88**:765–77.
110. Bulgakova NA, Kempkens O, Knust E. Multiple domains of Stardust differentially mediate localisation of the Crumbs-Stardust complex during photoreceptor development in Drosophila. *J Cell Sci* 2008;**121**:2018–26.
111. Berger S, Bulgakova NA, Grawe F, Johnson K, Knust E. Unraveling the genetic complexity of Drosophila stardust during photoreceptor morphogenesis and prevention of light-induced degeneration. *Genetics* 2007;**176**:2189–200.
112. Nam SC, Choi KW. Domain-specific early and late function of Dpatj in Drosophila photoreceptor cells. *Dev Dyn* 2006;**235**:1501–7.
113. Richard M, Grawe F, Knust E. DPATJ plays a role in retinal morphogenesis and protects against light-dependent degeneration of photoreceptor cells in the Drosophila eye. *Dev Dyn* 2006;**235**:895–907.
114. Mehalow AK, Kameya S, Smith RS, Hawes NL, Denegre JM, Young JA, et al. CRB1 is essential for external limiting membrane integrity and photoreceptor morphogenesis in the mammalian retina. *Hum Mol Genet* 2003;**12**:2179–89.
115. van de Pavert SA, Kantardzhieva A, Malysheva A, Meuleman J, Versteeg I, Levelt C, et al. Crumbs homologue 1 is required for maintenance of photoreceptor cell polarization and adhesion during light exposure. *J Cell Sci* 2004;**117**:4169–77.
116. Gosens I, den Hollander AI, Cremers FP, Roepman R. Composition and function of the Crumbs protein complex in the mammalian retina. *Exp Eye Res* 2008;**86**:713–26.
117. Richard M, Roepman R, Aartsen WM, van Rossum AG, den Hollander AI, Knust E, et al. Towards understanding CRUMBS function in retinal dystrophies. *Hum Mol Genet* 2006;**15** (Spec No 2):R235–43.
118. den Hollander AI, Johnson K, de Kok YJ, Klebes A, Brunner HG, Knust E, et al. CRB1 has a cytoplasmic domain that is functionally conserved between human and Drosophila. *Hum Mol Genet* 2001;**10**:2767–73.
119. Corbeil D, Roper K, Fargeas CA, Joester A, Huttner WB. Prominin: a story of cholesterol, plasma membrane protrusions and human pathology. *Traffic* 2001;**2**:82–91.
120. Jaszai J, Fargeas CA, Florek M, Huttner WB, Corbeil D. Focus on molecules: prominin-1 (CD133). *Exp Eye Res* 2007;**85**:585–6.
121. Kleinman ME, Ambati J. Fifty years later: the disk goes to the prom. *J Clin Invest* 2008;**118**:2681–4.
122. Miraglia S, Godfrey W, Yin AH, Atkins K, Warnke R, Holden JT, et al. A novel five-transmembrane hematopoietic stem cell antigen: isolation, characterization, and molecular cloning. *Blood* 1997;**90**:5013–21.
123. Weigmann A, Corbeil D, Hellwig A, Huttner WB. Prominin, a novel microvilli-specific polytopic membrane protein of the apical surface of epithelial cells, is targeted to plasmalemmal protrusions of non-epithelial cells. *Proc Natl Acad Sci USA* 1997;**94**:12425–30.
124. Yin AH, Miraglia S, Zanjani ED, Almeida-Porada G, Ogawa M, Leary AG, et al. AC133, a novel marker for human hematopoietic stem and progenitor cells. *Blood* 1997;**90**:5002–12.
125. Maw MA, Corbeil D, Koch J, Hellwig A, Wilson-Wheeler JC, Bridges RJ, et al. A frameshift mutation in prominin (mouse)-like 1 causes human retinal degeneration. *Hum Mol Genet* 2000;**9**:27–34.

126. Permanyer J, Navarro R, Friedman J, Pomares E, Castro-Navarro J, Marfany G, et al. Autosomal recessive retinitis pigmentosa with early macular affectation caused by premature truncation in PROM1. *Invest Ophthalmol Vis Sci* 2010;**51**:2656–63.
127. Pras E, Abu A, Rotenstreich Y, Avni I, Reish O, Morad Y, et al. Cone-rod dystrophy and a frameshift mutation in the PROM1 gene. *Mol Vis* 2009;**15**:1709–16.
128. Yang Z, Chen Y, Lillo C, Chien J, Yu Z, Michaelides M, et al. Mutant prominin 1 found in patients with macular degeneration disrupts photoreceptor disk morphogenesis in mice. *J Clin Invest* 2008;**118**:2908–16.
129. Zacchigna S, Oh H, Wilsch-Brauninger M, Missol-Kolka E, Jaszai J, Jansen S, et al. Loss of the cholesterol-binding protein prominin-1/CD133 causes disk dysmorphogenesis and photoreceptor degeneration. *J Neurosci* 2009;**29**:2297–308.
130. Zelhof AC, Hardy RW, Becker A, Zuker CS. Transforming the architecture of compound eyes. *Nature* 2006;**443**:696–9.
131. Husain N, Pellikka M, Hong H, Klimentova T, Choe KM, Clandinin TR, et al. The agrin/perlecan-related protein eyes shut is essential for epithelial lumen formation in the Drosophila retina. *Dev Cell* 2006;**11**:483–93.
132. Abd El-Aziz MM, Barragan I, O'Driscoll CA, Goodstadt L, Prigmore E, Borrego S, et al. EYS, encoding an ortholog of Drosophila spacemaker, is mutated in autosomal recessive retinitis pigmentosa. *Nat Genet* 2008;**40**:1285–7.
133. Collin RW, Littink KW, Klevering BJ, van den Born LI, Koenekoop RK, Zonneveld MN, et al. Identification of a 2 Mb human ortholog of Drosophila eyes shut/spacemaker that is mutated in patients with retinitis pigmentosa. *Am J Hum Genet* 2008;**83**:594–603.
134. Cook B, Zelhof AC. Photoreceptors in evolution and disease. *Nat Genet* 2008;**40**:1275–6.
135. Abd El-Aziz MM, El-Ashry MF, Chan WM, Chong KL, Barragan I, Antinolo G, et al. A novel genetic study of Chinese families with autosomal recessive retinitis pigmentosa. *Ann Hum Genet* 2007;**71**:281–94.
136. Barragan I, Marcos I, Borrego S, Antinolo G. Mutation screening of three candidate genes, ELOVL5, SMAP1 and GLULD1 in autosomal recessive retinitis pigmentosa. *Int J Mol Med* 2005;**16**:1163–7.
137. Khaliq S, Hameed A, Ismail M, Mehdi SQ, Bessant DA, Payne AM, et al. Refinement of the locus for autosomal recessive retinitis pigmentosa (RP25) linked to chromosome 6q in a family of Pakistani origin. *Am J Hum Genet* 1999;**65**:571–4.
138. Ruiz A, Borrego S, Marcos I, Antinolo G. A major locus for autosomal recessive retinitis pigmentosa on 6q, determined by homozygosity mapping of chromosomal regions that contain gamma-aminobutyric acid-receptor clusters. *Am J Hum Genet* 1998;**62**:1452–9.
139. Hartong DT, Berson EL, Dryja TP. Retinitis pigmentosa. *Lancet* 2006;**368**:1795–809.
140. Cook B, Hardy RW, McConnaughey WB, Zuker CS. Preserving cell shape under environmental stress. *Nature* 2008;**452**:361–4.
141. Morrow EM, Furukawa T, Raviola E, Cepko CL. Synaptogenesis and outer segment formation are perturbed in the neural retina of Crx mutant mice. *BMC Neurosci* 2005;**6**:5.
142. Blackshaw S, Fraioli RE, Furukawa T, Cepko CL. Comprehensive analysis of photoreceptor gene expression and the identification of candidate retinal disease genes. *Cell* 2001;**107**:579–89.
143. Livesey FJ, Furukawa T, Steffen MA, Church GM, Cepko CL. Microarray analysis of the transcriptional network controlled by the photoreceptor homeobox gene Crx. *Curr Biol* 2000;**10**:301–10.
144. Peng GH, Chen S. Chromatin immunoprecipitation identifies photoreceptor transcription factor targets in mouse models of retinal degeneration: new findings and challenges. *Vis Neurosci* 2005;**22**:575–86.

145. Wang T, Montell C. Phototransduction and retinal degeneration in Drosophila. *Pflugers Arch* 2007;**454**:821–47.
146. Kaushal S, Khorana HG. Structure and function in rhodopsin. 7. Point mutations associated with autosomal dominant retinitis pigmentosa. *Biochemistry* 1994;**33**:6121–8.
147. Kaushal S, Ridge KD, Khorana HG. Structure and function in rhodopsin: the role of asparagine-linked glycosylation. *Proc Natl Acad Sci USA* 1994;**91**:4024–8.
148. Sung CH, Davenport CM, Hennessey JC, Maumenee IH, Jacobson SG, Heckenlively JR, et al. Rhodopsin mutations in autosomal dominant retinitis pigmentosa. *Proc Natl Acad Sci USA* 1991;**88**:6481–5.
149. Sung CH, Davenport CM, Nathans J. Rhodopsin mutations responsible for autosomal dominant retinitis pigmentosa. Clustering of functional classes along the polypeptide chain. *J Biol Chem* 1993;**268**:26645–9.
150. Dryja TP, McGee TL, Reichel E, Hahn LB, Cowley GS, Yandell DW, et al. A point mutation of the rhodopsin gene in one form of retinitis pigmentosa. *Nature* 1990;**343**:364–6.
151. Liu X, Garriga P, Khorana HG. Structure and function in rhodopsin: correct folding and misfolding in two point mutants in the intradiscal domain of rhodopsin identified in retinitis pigmentosa. *Proc Natl Acad Sci USA* 1996;**93**:4554–9.
152. Sung CH, Schneider BG, Agarwal N, Papermaster DS, Nathans J. Functional heterogeneity of mutant rhodopsins responsible for autosomal dominant retinitis pigmentosa. *Proc Natl Acad Sci USA* 1991;**88**:8840–4.
153. Olsson JE, Gordon JW, Pawlyk BS, Roof D, Hayes A, Molday RS, et al. Transgenic mice with a rhodopsin mutation (Pro23His): a mouse model of autosomal dominant retinitis pigmentosa. *Neuron* 1992;**9**:815–30.
154. Galy A, Roux MJ, Sahel JA, Leveillard T, Giangrande A. Rhodopsin maturation defects induce photoreceptor death by apoptosis: a fly model for RhodopsinPro23His human retinitis pigmentosa. *Hum Mol Genet* 2005;**14**:2547–57.
155. Patil C, Walter P. Intracellular signaling from the endoplasmic reticulum to the nucleus: the unfolded protein response in yeast and mammals. *Curr Opin Cell Biol* 2001;**13**:349–55.
156. Schroder M, Kaufman RJ. The mammalian unfolded protein response. *Annu Rev Biochem* 2005;**74**:739–89.
157. Colley NJ, Cassill JA, Baker EK, Zuker CS. Defective intracellular transport is the molecular basis of rhodopsin-dependent dominant retinal degeneration. *Proc Natl Acad Sci USA* 1995;**92**:3070–4.
158. Davidson FF, Steller H. Blocking apoptosis prevents blindness in Drosophila retinal degeneration mutants. *Nature* 1998;**391**:587–91.
159. Kiselev A, Socolich M, Vinos J, Hardy RW, Zuker CS, Ranganathan R. A molecular pathway for light-dependent photoreceptor apoptosis in Drosophila. *Neuron* 2000;**28**:139–52.
160. Kurada P, O'Tousa JE. Retinal degeneration caused by dominant rhodopsin mutations in Drosophila. *Neuron* 1995;**14**:571–9.
161. Lin JH, Lavail MM. Misfolded proteins and retinal dystrophies. *Adv Exp Med Biol* 2010;**664**:115–21.
162. Ryoo HD, Domingos PM, Kang MJ, Steller H. Unfolded protein response in a Drosophila model for retinal degeneration. *EMBO J* 2007;**26**:242–52.
163. Yang LP, Wu LM, Guo XJ, Tso MO. Activation of endoplasmic reticulum stress in degenerating photoreceptors of the rd1 mouse. *Invest Ophthalmol Vis Sci* 2007;**48**:5191–8.
164. Ripps H. Light to sight: milestones in phototransduction. *FASEB J* 2010;**24**:970–5.
165. Sung CH, Chuang JZ. The cell biology of vision. *J Cell Biol* 2010;**190**:953–63.
166. Strauss O. The retinal pigment epithelium in visual function. *Physiol Rev* 2005;**85**:845–81.
167. Thompson DA, Gal A. Vitamin A metabolism in the retinal pigment epithelium: genes, mutations, and diseases. *Prog Retin Eye Res* 2003;**22**:683–703.

168. Harris WA, Ready DF, Lipson ED, Hudspeth AJ, Stark WS. Vitamin A deprivation and Drosophila photopigments. *Nature* 1977;**266**:648–50.
169. Ozaki K, Nagatani H, Ozaki M, Tokunaga F. Maturation of major Drosophila rhodopsin, ninaE, requires chromophore 3-hydroxyretinal. *Neuron* 1993;**10**:1113–9.
170. Wang T, Montell C. Rhodopsin formation in Drosophila is dependent on the PINTA retinoid-binding protein. *J Neurosci* 2005;**25**:5187–94.
171. Brunel CA, Madigan SJ, Cassill JA, Edeen PT, McKeown M. pcdr, a novel gene with sexually dimorphic expression in the pigment cells of the Drosophila eye. *Dev Genes Evol* 1998;**208**:327–35.
172. Wang X, Wang T, Jiao Y, von Lintig J, Montell C. Requirement for an enzymatic visual cycle in Drosophila. *Curr Biol* 2010;**20**:93–102.
173. Lefkowitz RJ, Shenoy SK. Transduction of receptor signals by beta-arrestins. *Science* 2005;**308**:512–7.
174. Byk T, Bar-Yaacov M, Doza YN, Minke B, Selinger Z. Regulatory arrestin cycle secures the fidelity and maintenance of the fly photoreceptor cell. *Proc Natl Acad Sci USA* 1993;**90**:1907–11.
175. Kahn ES, Matsumoto H. Calcium/calmodulin-dependent kinase II phosphorylates Drosophila visual arrestin. *J Neurochem* 1997;**68**:169–75.
176. Matsumoto H, Kurien BT, Takagi Y, Kahn ES, Kinumi T, Komori N, et al. Phosrestin I undergoes the earliest light-induced phosphorylation by a calcium/calmodulin-dependent protein kinase in Drosophila photoreceptors. *Neuron* 1994;**12**:997–1010.
177. Alloway PG, Howard L, Dolph PJ. The formation of stable rhodopsin-arrestin complexes induces apoptosis and photoreceptor cell degeneration. *Neuron* 2000;**28**:129–38.
178. Satoh AK, Ready DF. Arrestin1 mediates light-dependent rhodopsin endocytosis and cell survival. *Curr Biol* 2005;**15**:1722–33.
179. Chuang JZ, Vega C, Jun W, Sung CH. Structural and functional impairment of endocytic pathways by retinitis pigmentosa mutant rhodopsin-arrestin complexes. *J Clin Invest* 2004;**114**:131–40.
180. Robinson PR, Cohen GB, Zhukovsky EA, Oprian DD. Constitutively active mutants of rhodopsin. *Neuron* 1992;**9**:719–25.
181. Li T, Franson WK, Gordon JW, Berson EL, Dryja TP. Constitutive activation of phototransduction by K296E opsin is not a cause of photoreceptor degeneration. *Proc Natl Acad Sci USA* 1995;**92**:3551–5.
182. Chen J, Shi G, Concepcion FA, Xie G, Oprian D. Stable rhodopsin/arrestin complex leads to retinal degeneration in a transgenic mouse model of autosomal dominant retinitis pigmentosa. *J Neurosci* 2006;**26**:11929–37.
183. Chinchore Y, Mitra A, Dolph PJ. Accumulation of rhodopsin in late endosomes triggers photoreceptor cell degeneration. *PLoS Genet* 2009;**5**:e1000377.
184. Charlton-Perkins M and Cook TA. Building a Fly Eye: Terminal Differentiation Events of the Retina, Corneal Lens, and Pigmented Epithelia. *Curr Topics Dev Biol* 2010;**93**:129–73.
185. Ragge NK, Brown AG, Poloschek CM, Lorenz B, Henderson RA, Clarke MP, et al. Heterozygous mutations of OTX2 cause severe ocular malformations. *Am J Hum Genet* 2005 **76**:1008–22. Erratum in: *Am J Hum Genet* 2005;**77**:334.
186. Rivolta C, Berson EL, Dryja TP. Dominant Leber congenital amaurosis, cone-rod degeneration, and retinitis pigmentosa caused by mutant versions of the transcription factor CRX. *Hum Mutat* 2001;**18**:488–98.
187. Furukawa T, Morrow EM, Li T, Davis FC, Cepko CL. Retinopathy and attenuated circadian entrainment in Crx-deficient mice. *Nat Genet* 1999;**23**:466–70.

Animal Models of Glycogen Storage Disorders

H. Orhan Akman,* Adithya Raghavan,* and William J. Craigen*,†

*Department of Molecular and Human Genetics, Baylor College of Medicine, Houston, Texas, USA

†Department of Pediatrics, Baylor College of Medicine, Houston, Texas, USA

Glycogen is a polymer of glucose needed to provide for a continuous source of glucose during fasting. Glycogen synthesis and degradation are tightly controlled by complex regulatory mechanisms and any disturbance in this regulation can lead to an inadequate reservoir of glycogen or an accumulation of excess or abnormal glycogen stored either in the cytosol or in the lysosomes. Problems in the degradation or synthesis of glycogen are referred to as glycogen storage disorders (GSDs), which individually are rare diseases, yet collectively are a major category of inborn errors of metabolism in humans. To date,

Progress in Molecular Biology
and Translational Science, Vol. 100
DOI: 10.1016/B978-0-12-384878-9.00009-1

Copyright 2011, Elsevier Inc.
All rights reserved.
1877-1173/11 $35.00

11 distinct forms of GSDs are represented in animal models. These models provide a means to understand the mechanisms that regulate and execute the synthesis and degradation of glycogen. In this review, we summarize animal models that have arisen spontaneously in nature or have been engineered in the laboratory by recombinant DNA techniques, and categorize the disorders of glycogen metabolism as disorders of either synthesis or degradation.

I. Introduction

Glycogen is a complex branched polymer of glucose and is synthesized in the cell to store excess glucose during feeding in order to provide for a continuous source of glucose during fasting. Glycogen synthesis and degradation are tightly controlled by complex regulatory mechanisms and any disturbance in this regulation can lead to an inadequate reservoir of glycogen or an accumulation of excess or abnormal glycogen stored either in the cytosol or in the lysosomes. Problems in the degradation or synthesis of glycogen are referred to as glycogen storage disorders (GSDs), which individually are rare diseases, yet collectively are a major category of inborn errors of metabolism in humans. The numbering pattern for GSDs is not uniformly applied, and thus the literature for these disorders can be confusing. To date, there are 15 glycogenoses that have been described, with 11 being represented in animal models. These models provide a means to understand the mechanisms that regulate and execute the synthesis and degradation of glycogen. In this review, we summarize animal models that have arisen spontaneously in nature or have been engineered in the laboratory by recombinant DNA techniques, and categorize the disorders of glycogen metabolism as disorders of either synthesis or degradation, including defects in glycolysis and disorders where the mechanisms leading to glycogen accumulation are currently poorly understood at a molecular level.

A. Importance of Glycogen Synthesis and Degradation

The synthesis of glycogen is the primary means for storing carbohydrates in the cell, and carbohydrate is the primary source of cellular energy that is generated through glycolysis and via oxidative metabolism through the mitochondrial respiratory chain. The controlled synthesis of large molecules of glucose polymers is important not only for storing glucose in a compact state but also in order to properly regulate the osmolarity within the cell, since a comparable amount of individual glucose molecules would create far greater osmotic pressure within the cell. Animals rely on this stored carbohydrate for maintenance of glucose homeostasis, reflected in the fact that hypoglycemia is a common although not universal feature in disorders of glycogen metabolism. The primary sites of glycogen storage in animals are in the liver and muscle, although small amounts of glycogen can be found in other tissues such as the kidney and brain. The primary function of glycogen varies for each tissue. In skeletal muscle, glycogen is a source of energy for the cells in which it is stored, where it provides for short-term, high-energy muscle activity. In the brain, it provides an emergency supply of energy for use if hypoglycemia or hypoxia occurs. In contrast, the liver functions primarily as a reservoir for other tissues glucose needs, taking up glucose from the bloodstream during a meal and releasing glucose into the blood when blood glucose falls during fasting or high-energy consumption. Some deficiencies of glycogen metabolism affect only one tissue, for example, hepatic glycogen synthase deficiency, whereas others can affect several tissues, leading to more complex clinical consequences, an example being glucose 6 phosphatase deficiency (Fig. 1).

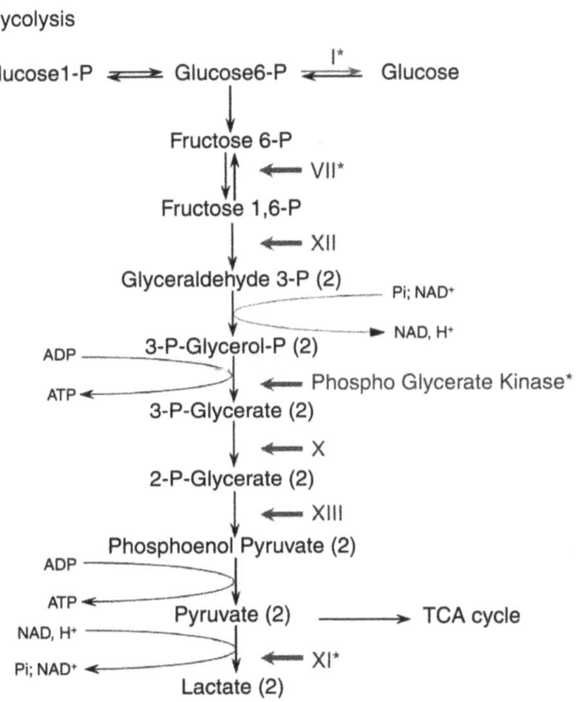

FIG. 1. Scheme of glycolysis described in humans and animals. Roman numerals denote glycogenoses due to defects in the following enzymes: I, glucose phosphatase; VII, phosphofructokinase; IX, phosphoglycerate kinase; X, phosphoglycerate mutase; XI, lactate dehydrogenase; XII, aldolase; XIII, b-enolase. Asterisk indicates there are available animal models for the enzyme defect.

B. Mechanism of Synthesis and Degradation

Different forms of glycogen are found in liver and muscle, with muscle glycogen taking on a structure referred to as β particles, which are visible in the electron microscope as spherical structures that contain up to 60,000 glucose residues per particle. However, in addition to isolated β particles, the liver contains aggregated β particles that appear as rosettes of glycogen and are referred to as α particles. In addition to glucose polymers, liver glycogen also contains glucosamine, which is not found in muscle glycogen. The presence of glucosamine may account for the larger molecular size of liver glycogen particles. The glycogen particles have been referred to as "glycosomes" to indicate that they are distinct organelles, and are composed of both glycogen and the proteins that make up the enzymatic machinery of glycogen synthesis and degradation. The initiation of glycogen synthesis requires a protein "primer" that is termed *glycogenin*, a uridine diphosphoglucose-requiring glucosyltransferase. The first step in creating glycogen is the autoglucosylation of glycogenin via the two-step attachment of eight glucose molecules to a tyrosine residue of the protein. Glycosylated glycogenin is the primer for the synthesis of glycogen by the sequential action of two enzymes: glycogen synthase and glycogen branching enzyme (GBE). Glycogen synthase and glycogenin exist as a complex in a 1:1 ratio, with one molecule of each per β particle of glycogen. Glycogen synthase adds glucose molecules via ester bonds between the first and fourth carbons of glucose, leading to α-1,4-glucosyl-linked chains. After the addition of 7–11 glucose units, GBE first removes one to four glucose units from the tip of the growing chain and attaches new glucose residues to the adjacent chain via covalent linkage of the sixth to first carbons of glucose, leading to α-1,6-glucosyl-linked branch points. This new branch is now a new primer site for glycogen synthase for the addition of more glucose to the growing chain. This cooperation between the two enzymes allows the glycogen molecule to be a highly branched, soluble molecule.

Similarly, degradation is mediated by the action of two key enzymes: glycogen phosphorylase and glycogen debranching enzyme. Glycogen phosphorylase phosphorylates the terminal glucose molecules of glycogen and catalyzes the sequential cleavage of glucosyl units from α-1,4-glucosyl-linked chains of glycogen, thereby liberating glucose 1-phosphate. Phosphorylase, which exists as a tissue-specific family of isoforms, degrades the glycogen chains until only four glucose units remain before an α-1-6 branch point. Debranching enzyme (amylo-1,6-glucosidase, 4-α-glucanotransferase, AGL), which has two distinct enzymatic activities, first transfers the three glucose residues from the short branch to the end of another branch using its glucosyltransferase activity, where phosphorylase can again degrade the glucosyl chain. Debranching enzyme then hydrolyzes the remaining α-1-6 branch point glucose residue using its amylo-1,6-glucosidase activity, thereby exposing additional glucosyl units to phosphorylase activity. These enzymes are tightly regulated by phosphorylation/dephosphorylation.

Glycogen breakdown is triggered either by a reduction in cellular energy, where hydrolyzed AMP activates cyclic AMP synthesis and, through a signaling cascade, phosphorylase activity is stimulated via phosphorylation, or via hormonal signaling by epinephrine or glucagon. Conversely, phosphorylase is inhibited via dephosphorylation by protein phosphatase 1 in conjunction with a phosphatase termed protein-targeting of glycogen (PTG). Glycogen synthesis is similarly activated by a signal transduction mechanism. Insulin initiates a kinase phosphorylation cascade that activates the glycogen synthase via glycogen synthase kinase (GSK). Glycogen synthase can also be allosterically activated by glucose 6 phosphate (G6P), whereas phosphorylase is inhibited, linking glycogen synthesis or degradation directly to glycolytic metabolism of glucose via the hexokinase step. A role for PTG in glycogen synthase activation via dephosphorylation has also been identified. The overall scheme for the regulation of glycogen metabolism is summarized (Fig. 2).

Allosteric regulation of glycogen synthase by G6P is stronger than its inhibition by phosphorylation. In the presence of persistent G6P, such as in the case of a defect in glycolysis, glycogen synthase remains active despite reduced energy levels (Fig. 3). In rare circumstances, the mechanism that senses the energy state of the cell can also be faulty and lead to glycogen accumulation. Activating

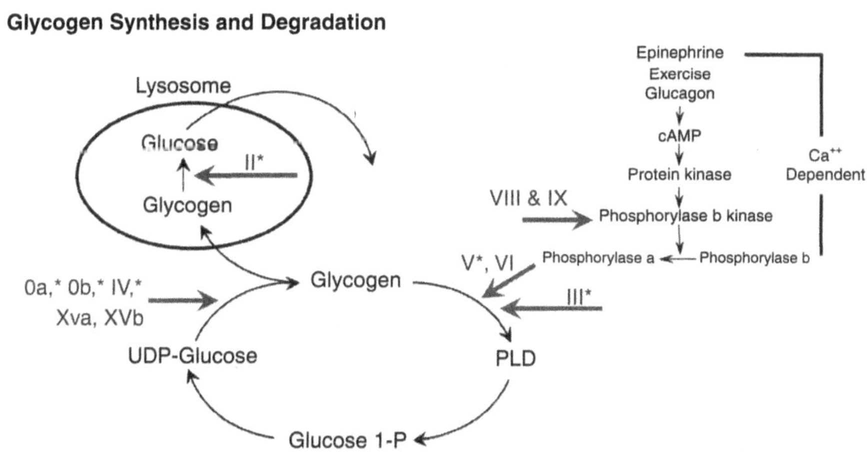

FIG. 2. Scheme of degradation and synthesis of glycogen; Roman numerals denote glycogenoses due to defects in the following enzymes: 0, glycogen synthase a, liver b, muscle; II, acid maltase; III, debrancher; IV, brancher; V, myophosphorylase; VI, liver phosphorylase; VIII, phosphorylase b kinase; XV, glycogenin a, liver b, muscle. Asterisk indicates there are available animal models for the enzyme defect.

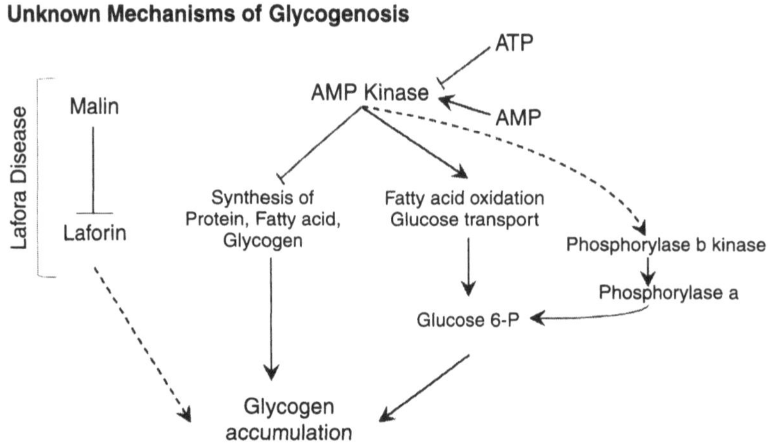

FIG. 3. Defects in malin or laforin cause accumulation of polyglucosan due to hyper phosphorylation of glycogen and its protection from lysosomal targetting. Defects of AMP-activated protein kinase cause familial hypertrophic cardiomyopathy with Wolff–Parkinson–White syndrome.

mutations that cause the inappropriate activation of AMP kinases lead to enhanced glucose uptake, activation of glycogen synthase, and excessive glycogen synthesis. Conversely, deficiency of AMP kinase activity inhibits the degradation of glycogen since it cannot sense the need for energy, again leading to glycogen accumulation due to a failure to degrade it appropriately.

Glycogen degradation by phosphorylase is not the only mechanism available to meet the demand for glucose. In the first few hours after birth, the neonate depends heavily on the energy from stored glycogen, leading to the degradation of glycogen via lysosomes. This mechanism, when perturbed, leads to a disorder termed glycogen storage disease type II or Pompe disease, in which glycogen accumulates in lysosomes, with the onset of disease varying from the newborn period to elderly adults.

With these mechanisms in mind, we will consider animal models of GSDs in four general categories.

II. Defects in Glycolysis

When there is a need for energy, glucose is metabolized via glycolysis to pyruvate. Pyruvate is then further oxidized to carbon dioxide in the Krebs or tricarboxylic acid cycle. Inadequate oxygen can shunt the aerobic degradation of pyruvate to the synthesis of lactate via lactate dehydrogenase (LDH). Glycolysis involves 13 interconnected enzymes, and, when one is lacking, the allosteric regulation of other glycolytic enzymes leads to a general inhibition of glycolysis that may result in the accumulation of glucose in the cell, leading to inappropriate synthesis of glycogen.

A. Phosphofructokinase Deficiency (GSDVII)

Deficiency of phosphofructokinase (PFK) leads to GSDVII. PFK is a key enzyme in glycolysis since it is the first irreversible reaction that channels G6P for degradation. Prior to this step, G6P can be utilized in the pentose phosphate pathway, or converted to glucose 1-phosphate and stored as glycogen. PFK has three isoforms encoded by three individual genes: the muscle (PFKM), liver (PFKL), and platelet (PFKP) isoforms, and based on the specific tissue PFK functions as homo- or heterotetramers. Adult muscle exclusively uses a homotetramer of the M isoform, while the liver and kidneys contain predominantly the L form. Erythrocytes use both the L and M isoforms, and thus the combinations of hetero- ($PFKML_3$, $PFKM_2L_2$, and $PFKM_3L$) and homo- ($PFKM_4$ and $PFKL_4$) tetramers are observed. Since muscle expresses only one PFK isoform, a loss of this isoform activity leads to a muscle disease that may have one of two clinical forms. One form of the disorder is a rare and severe disease with onset in infancy and characterized by hypotonia, limb weakness, progressive myopathy, and respiratory failure, leading to death early in the childhood.[1] The more common adult-onset form of PFK deficiency (Tarui disease) exhibits muscle weakness manifested mainly as exercise intolerance and myoglobinuria.[2]

Two animal models have been described for this disease. One is a canine model that highlights the compensatory expression of different isoforms seen in dogs. Unlike humans, canine muscle compensates for PFKM deficiency by expressing PFKL isoform; therefore, they appear not to suffer from muscle weakness or myoglobinuria on exertion. However, in canine erythrocytes, only the M and P isoforms exist, and because the P isoform cannot compensate for the loss of the muscle isoform, the dogs exhibit a chronic hemolytic anemia punctuated by acute hemolytic crises and hemoglobinuria.[3]

A mouse model of Tarui disease that was created by deleting exon 3 of the M isoform gene is more similar to the human disease, with PFKM null mice developing myopathic and hemolytic features similar to those noted in type VII glycogenosis patients.[4] The early lethality observed in PFK null mice also resembles the most severe variant of the disease, with a progressive myopathy, cardiac involvement, and rapid death in infancy.[1]

B. Phosphoglycerate Kinase Deficiency

Phosphoglycerate kinase (PGK1) (EC 2.7.2.3) is ubiquitously expressed and catalyzes one of the two ATP producing reactions in the glycolytic pathway via the reversible conversion of 1,3-diphosphoglycerate to 3-phosphoglycerate.

As an X-linked recessive disorder, PGK deficiency has very heterogeneous clinical characteristics; it typically leads to a nonspherocytic hemolytic anemia, but can also be associated with brain dysfunction and developmental delay, or with a pure myopathy. Finally, a few patients have a myopathy plus brain involvement.[5]

A *Drosophila* temperature-sensitive mutant of PGK has been reported by Wang *et al.*[6] They identified a PGK mutant line while screening for temperature-sensitive seizure mutants. Affected flies show a rapid loss of motor coordination and seizure-like activity referred to as "bang sensitivity," followed by paralysis within minutes of exposure to 38 °C. The authors concentrated on the neuronal phenotype of this mutation, emphasizing the importance of mitochondria-independent ATP production in the synapse and for vesicle trafficking. There are currently no other animal models available to study the underlying mechanisms of this disease.

C. Phosphoglycerate Mutase Deficiency (GSDX)

Phosphoglycerate mutase (PGAM) catalyzes the conversion of 3-phosphoglycerate to 2-phosphoglycerate. Deficiency of the muscle isoform of the enzyme (PGAM2) results in muscle disease that typically becomes apparent in childhood or adolescence and mimics the manifestations of McArdle disease (GSDV). Blood creatine kinase (CK) levels may be elevated and muscle biopsy shows mild glycogen storage. To date, no animal models have been reported.

D. Lactate Dehydrogenase Deficiency (GSDXI)

LDH (EC 1.1.1.27) catalyzes the interconversion of lactate and pyruvate. There are three isoforms expressed in skeletal muscle (LDHA, the M isoform), in cardiac muscle (LDHB, the H isoform), and in the testes (LDHC). LDH provides NAD+ for continued glycolytic activity in active muscle. The muscle isoform has a higher affinity for pyruvate than that of liver; therefore, LDHB in liver is more efficient in converting lactate produced by muscle to pyruvate for gluconeogenesis in order to sustain blood glucose levels during strenuous exercise.[7] Unlike other glycolytic enzyme defects, LDH deficiency has no effect on erythrocyte function. Mutations in the LDHA gene result in childhood-onset myopathy, manifested by exercise intolerance, muscle stiffness following strenuous exercise, intermittently elevated CK, and myoglobinuria.[8]

There are two animal models of LDH deficiency that were created by chemical mutagenesis of mice. Unlike humans, lack of LDHA causes embryonic lethality in the mouse. This severe phenotype is explained by very strict differential expression of LDH isoforms during mouse embryonic development that do not allow for compensation.[9]

E. Aldolase Deficiency (GSDXII)

Fructose bisphosphate aldolase (ALDOA) converts fructose 1,6 bisphosphate to dihydroxyacetone phosphate and glyceraldehyde 3 phosphate. Infantile hemolytic anemia, muscle weakness, and rhabdomyolysis have been observed in affected children.[10] No animal models have been reported.

F. β-Enolase Deficiency (GSDXIII)

β-Enolase (ENO3) interconverts 2-phosphoglycerate and phosphoenolpyruvate and its deficiency leads to exertional muscle pain and episodic elevations in CK, much like LDHA and PGAM deficiency.[11] No animal models have been reported.

III. Defects in Glycogen Synthesis

A. Glycogen Synthase Deficiency (GSD0a, Liver or GSD0b, Muscle)

Glycogenosis type 0 (GSD0) is a deficiency of glycogen synthase. Glycogen synthase (EC 2.4.1.11) catalyzes the addition of glucose monomers to the growing glycogen molecule through the formation of α-1,4-glycoside linkages.[12] There are two isoforms encoded by two genes: a muscle-specific GYS1 and a liver-specific GYS2. Depending on which isoform is deficient, the disease affects either the liver (GSD0a) or the muscle (GSD0b). Both of the liver and muscle glycogen synthase isoforms are regulated by a protein kinase and protein phosphatase signaling cascade that involves GSK3β or protein kinase A (PKA). GSK3β and PKA have been shown to phosphorylate GYS on amino- and carboxy-terminal serine residues, leading to inhibition of its activity. Both proteins are downstream of glucagon and adrenergic receptors that promote hepatic glucose release.[13] However, allosteric regulation of GYS by high concentrations of glucose-6 phosphate can overcome its inhibition by phosphorylation; therefore, glycogen accumulation is observed when there is a defect in glycolysis distal to hexokinase where, despite cellular ATP depletion, glycogen synthesis continues.

There are four animal models of GSD0, three of which are mouse models. Two mouse strains are knockouts of GYS1 or GYS2. For GYS1, a knockout was created by "gene trap" technology involving retroviral insertion downstream of the GYS1 promoter, leading to disruption of gene transcription. In the absence of GYS1, 90% of the pups die at birth, while the remaining 10% grow to adulthood and develop a cardiomyopathy.[12] For GYS2, a "conditional knockout" was created by use of the Loxp–Cre recombinase system. Homozygous deficient mice have almost complete absence of liver glycogen, but surprisingly develop normally. However, as would be expected, the mice cannot sustain blood glucose levels while fasting, and have high levels of liver triglycerides.[14] The third mouse model involving GYS is a mouse strain where the overexpression of constitutively active GYS is driven by a muscle-specific promoter. In this model, mice have increased glycogen content in the muscle, but, due to the high activity of GYS, glycogen accumulates fewer branches and has the microscopic characteristics of polyglucosan, an abnormal form of glycogen typically associated with glycogen brancher deficiency.[15] This emphasizes the importance of the stoichiometric balance needed between glycogen synthase activity and brancher activity in defining normal glycogen structure. The fourth model is the polysaccharide storage myopathy horse. Affected horses are heavily muscled and may exhibit rhabdomyolysis associated with exercise intolerance due to the accumulation of muscle glycogen and abnormal polysaccharide in skeletal muscles. It is seen in numerous breeds, including Quarter horses, American Paint Horses, Draft horses, and various crosses. Some, but not all, affected horses have a missense substitution in GYS1 leading to a form of glycogen synthase that is active in the absence of glucose-6 phosphate, similar to the GYS overexpressing mouse model. To date, 17 breeds have been found to harbor the mutation.[16]

B. GBE Deficiency (GSDIV)

Glycogenosis type IV (Andersen disease) is characterized by the accumulation of polyglucosan in tissues and 0-20% activity of GBE (EC 2.4.1.18). GBE is expressed in almost all cell types and works in concert with either GYS1 or GYS2 in the synthesis of glycogen. As indicated previously, it is crucial that GBE activity is matched to GYS activity. This comparable activity is important for the correct branching architecture of the glycogen molecule. When there is a reduction in the branching activity, branch points occur up to 16–20 glucose units further apart. This poorly branched new polymer develops the characteristics of starch, with low solubility and ends that are difficult to degrade in response to a demand for increased energy production. The accumulation of poorly branched glycogen can be identified by glycogen-specific periodic acid-schiff base (PAS) staining followed by amylase digestion. Normal glycogen disappears after amylase digestion, whereas polyglucosan is resistant to amylase and exhibits a persistent red stain in microscopic examination of tissues. To date, 25 mutations have been reported in the Human Gene Mutation Database (http://www.hgmd.cf.ac.uk/ac/index.php). Enzyme activity correlates with the phenotype; when there is no enzyme activity, the clinical features of Andersen disease[17] are observed, which is typically an embryonic or early infantile lethal phenotype and caused primarily by deletions that affects the enzyme structure and stability.[18] Low but detectable enzyme activity causes a juvenile form of GSDIV with cardiomyopathy and/or liver cirrhosis. Alternatively, an adult-onset form of GSDIV that affects muscle and peripheral nerve is associated with a partial loss of enzyme activity and is referred to as Adult Polyglucsan Body Disease, based upon its histological appearance. Several different point mutations have been described, including a common Y329S mutation found in the Ashkenazi Jewish population. In rare instances, APBD may cause dementia, but in addition there are cases with only muscle involvement (Dr. Salvatore Dimauro, personal communication). Progression of the adult-onset GBE deficiency with peripheral neuropathy and dementia can lead to a misdiagnosis of amyotrophic lateral sclerosis, multiple sclerosis, or Alzheimer disease.

There are two naturally occurring animal models for GBE1 deficiency. The first animal model was reported by Fyfe *et al.*[19] in Norwegian forest cats. This uncommon breed of domestic cats develops the signs of disease after 5 months of age, with skeletal muscle, heart, and central nervous system degeneration apparent to clinical examination, and histological changes evident in tissues. The enzyme deficiency is caused by a complex rearrangement of genomic DNA at the GBE1 locus, with a 334-bp insertion at the site of a 6.2-kb deletion that removes exon 12. A second animal model of the disease is found in the American Quarterhorse,[20] with fatal fetal and neonatal glycogen storage disease. Genetic analysis has revealed a homozygous missense mutation in the first exon of the GBE1 in the effected foals.

There is no reported mouse model of this disease; however in our laboratory, we have generated a conditional knockout of exon 7. Deletion of exon 7 removes 210 bases, which causes an in-frame deletion of a conserved region of GBE. Since it is a conditional mutation, either global ablation or tissue-specific deletion can be created. Global deletion causes the mouse equivalent of Andersen disease, with neonatal fatality. We also introduced the Y329S mutation, which is the most common mutation among adult onset GSDIV patients. This mutation decreases the GBE1 activity to 6% of normal activity in muscle and causes a severe accumulation of polyglucosan in all organs examined except the kidney. Affected animals are able to survive up to 6 months of age with this mutation. It is anticipated that this model will accelerate the development of therapeutic strategies for the disease.

C. Glycogenin Deficiency (GSDXV)

Glycogenin (GYG1 or GYG2) (EC 2.4.1.1.86) functions as the primer for glycogen synthesis and is a self-glycosylating enzyme that uses uridine diphosphoglucose to synthesize a short glucose polymer attached to a surface tyrosine residue. Once a 7- to 11-unit primer is synthesized, glycogen synthase continues the polymerization reaction.[21] There are two isoforms: GYG1, a muscle isoform, and GYG2, a liver isoform. There are currently no animal models available to study the function or pathophysiology of these genes. However, mutations affecting glycogenin function in muscle are described in patients. Symptoms are very similar to muscle glycogen synthase deficiency, with one patient reported to suffer from exercise intolerance, muscle weakness and loss of muscle mass, and cardiac failure at age 27.[22]

IV. Defects in Glycogen Degradation

A. Glycogen Phosphorylase Deficiency (GSDV and VI).

Glycogen phosphorylase (PYGM or PYGL) (EC 2.4.1.1) catalyzes the breakdown of glycogen to glucose-1-phosphate, providing the main source of glucose to the cell during fasting. There are three isoforms: muscle (PYGM), liver (PYGL), and brain (PYGB). During embryonic development, the brain isoform is expressed in all tissues. After tissue differentiation, the muscle and liver isoforms are expressed. Deficiency of PYGM causes McArdle disease, an autosomal recessive disorder that typically becomes clinically apparently in teens and young adults with symptoms of exercise intolerance and myoglobinuria, in particular during brief high-intensity exercise when glycogen breakdown provides the majority of energy in skeletal muscle. Recurrent episodes of myalgia and myoglobinuria associated with elevated CK are the norm and can progress to fixed muscle weakness in adulthood. Affected individuals often exhibit a "second wind" phenomenon, where muscle fatigue can improve with continued exercise as muscle energy switches from glycogenolysis to fatty acid oxidation.

A naturally occurring ovine model of McArdle disease was identified in a flock of sheep in Western Australia. The mutation occurs in the 3′ splice acceptor site of intron 19 of the PYGM gene, resulting in the activation of a cryptic splice-site site in exon 20 and the premature termination of the transcript.[23] Affected animals display exercise intolerance, and histological examination of muscle revealed excess subsarcolemmal storage of glycogen and absent phosphorylase activity, as observed in McArdle patients.

The second model has been described in Charolais cattle. Affected cattle are asymptomatic at rest, but when forced to exercise they collapse repeatedly. Much like McArdle patients, after a brief rest, the cattle are able to continue exercise (the "second wind"). The genetic defect is C-to-T substitution, changing arginine to trptophan at amino acid position 489. The mutant residue is adjacent to the pyridoxal phosphate binding site and an active site residue, and the sequence around this position is highly conserved across different species.

Due to their similarities in muscle mass to humans and the relative ease and low cost of maintenance, the ovine model of McArdle disease is preferable to the bovine model and important for testing therapeutic strategies.

PYGL deficiency causes Hers disease (GSDVI), which has the features of mild hypoglycemia, ketosis, growth retardation, and prominent hepatomegaly.[24] Given the liver-restricted expression of PYGL, heart and skeletal muscle are not affected. While targeted mouse embryonic stem cells are available for this locus from ES cell repositories, no reports of a mutant mouse strain exist.

B. Phosphorylase b Kinase Deficiency (GSDVIII and GSDIX)

There are several variants of phosphorylase kinase deficiency. The extensive clinical and genetic heterogeneity of phosphorylase kinase deficiency is explained by the complexity of the enzyme.[25] It is composed of four subunits: γ is the catalytic subunit, while the other three (α, β, and δ) are regulatory subunits. The δ-subunit is the calcium-binding protein calmodulin, providing calcium sensitivity. The multiple subunits and existence of various isoforms of each subunit explain why autosomal recessive and X-linked recessive inheritance are observed. The X-linked form (GSDIXa) is characterized by hepatomegaly and a mild hypoglycemia with fasting, growth retardation, and elevation of liver transaminases, hypercholesterolemia, and hypertriglyceridemia, all of which gradually improve with age. Muscle is not affected, but the enzyme deficiency is detectable in erythrocytes. It is caused by a defect in $\alpha 2$ subunit of hepatic phosphorylase kinase (PHKA2).[26] In GSDIXb, both liver and muscle are affected, there is hepatomegaly, stunted growth, and, in some patients, hypotonia and mild muscle weakness. It is inherited in an autosomal recessive manner and is due to mutations in the β subunit of phosphorylase kinase (PHKB).[27] No animal models have yet been reported.

C. Glycogen Debrancher Deficiency (GSDIII)

Glycogenosis type III (GSDIII or Cori disease) is an autosomal recessive disorder caused by the deficiency of glycogen debrancher enzyme (AGL). AGL catalyzes the hydrolysis of 1-6-glycosidic bonds in glycogen. This step is particularly important since glycogen phosphorylase can only hydrolyze α-1,4 glycosidic bonds in glycogen and pauses when it is four glucose residues away from the 1-6 branch point. AGL hydrolyzes and transfers the glucose at the branch to the tip of the chain, allowing degradation by phosphorylase to continue.[28] In the Human Gene Mutation Database, over 70 mutations have been reported (http://www.hgmd.cf.ac.uk/ac/index.php). Affected individuals have fasting hypoglycemia, growth retardation, and hepatomegaly. Up to 90% of affected patients have a progressive myopathy and may develop cardiomyopathy. Clinical heterogeneity is explained by the complex differential splicing of the AGL gene in different tissues, leading to at least six alternatively spliced transcripts. Therefore, GSDIII is divided into four subgroups: IIIa liver and muscle, IIIb liver only, and IIIc loss of glucosidase activity or IIId loss of transferase activity.

An animal model of GSDIII has been reported in curly coated retrievers. The animals are deficient of AGL activity in muscle and liver and they have intermittent mild hypoglycemia, increased serum liver enzymes, and high serum CK activity, as is observed in most affected humans. Molecular analysis of the AGL gene showed a single nucleotide deletion in exon 32, which introduces a frame shift and is predicted to lead to a premature truncation of the protein.[29] Metabolic disorders of muscle function are of considerable commercial interest. Since glycogen accumulation affects muscle quality and mass, the porcine AGL gene has been reported as a quantitative trait locus effecting growth and muscle quality. A polymorphic short interspersed repeat element within the AGL gene has been linked to effects on weight gain and meat quality, although the specific effect of this polymorphism on gene expression has not yet been reported.[30]

To date, there is no spontaneous or engineered mouse strain lacking AGL function or other convenient animal models available to study the effect of AGL on the accumulation of soluble glycogen in tissues.

D. Glucose 6 Phosphatase (GSDIa) or Transporter (GSDIb) Deficiency

GSD type I has two subtypes: type Ia (von Gierke Disease) and type Ib. Each is an autosomal recessive disorder caused by deficiency in the glucose-6-phosphatase catalytic subunit 1 (G6PC1; GSDIa) or the glucose-6-phosphate translocase (G6PT; GSDIb), respectively. Deficiency of either component results in excessive accumulation of glycogen and lipids in the liver, kidney, and intestinal mucosa. Affected infants exhibit hypoglycemia, hepatomegaly, hypertriglyceridemia, uricemia, growth retardation, and severe lactic acidosis (reviewed in Ref. 31). In addition, G6PT appears important for neutrophil function since the deficiency results in an abnormal myeloid phenotype characterized by neutropenia, neutrophil dysfunctions, and an abnormal increase in serum levels of granulocyte colony stimulating factor (G-CSF),[32] resulting in inflammatory bowel disease and frequent infections.

There is a naturally occurring canine model for GSDIa that is caused by a missense mutation (M121I) in the third transmembrane domain of the G6PC gene.[33] Two mouse models have been developed for GSDIa by: a global knockout of exon 3[34] or a conditional knockout of exon 3.[35] Chen *et al.* also produced a G6PT knockout by replacing exon 1 and the flanking intron with a neomycin resistance cassette.[36] These models have successfully replicated the human GSDIa and b and are used as models for dietary or gene therapy.

E. Acid α-Glucosidase Deficiency (GSDII)

Acid α-glucosidase (GAA) Deficiency (GSDII), or Pompe Disease, is the only glycogen storage disease directly involving abnormal lysosomal metabolism. GAA (also known as acid maltase) is an α-amylase that is localized exclusively to the lysosome. In the absence of this enzyme, glycogen accumulates in lysosomes, and although nearly every cell type accumulates some lysosomal glycogen, skeletal muscle and heart are the primary organs that exhibit features of this disease. There is a spectrum of disease in humans ranging from infantile-onset hypotonia and lethal cardiomyopathy to adult-onset progressive weakness that leads to respiratory failure. While there is no naturally occurring animal model for this disease, different mouse models have been developed by knocking out the GAA gene via homologous recombination in embryonic stem cells. In the first model, mice were created by inserting a neomycin drug resistance cassette into exon 13. These animals developed cardiomegaly as adults, but remained relatively asymptomatic.[37] A second strain was generated by deleting exon 6 via Cre recombinase-mediated recombination that exhibited a more severe phenotype, including adult-onset progressive weakness and cardiomegaly. However, if exon 6 was disrupted by the neomycin drug resistance cassette, an even more severe model was generated, with symptoms developing within weeks of birth corresponding to some features of the more severe infantile form observed in humans.[15b] Given the similar nature of these latter two strains, the observed difference in the mouse strains has been ascribed to strain-specific genetic background effects. Additional mouse models have been generated that maintain low-level expression of the GAA enzyme or a human GAA transgene in order to facilitate studies of enzyme replacement therapy (ERT). An immunologic reaction to ERT occurs rapidly in the absence of any endogenous GAA, while the expression of small amounts of the enzyme does not rescue the mouse disease but avoids the immune reaction to ERT, and the ERT itself does ameliorate the disease phenotype. These mouse models have also been used for the treatment of Pompe disease via gene replacement strategies,[38] both with adenoviral and with adenovirus-associated virus-based gene delivery systems. Recently, it has been shown that combination therapy with both rapamycin, an inhibiter of cellular autophagy, and ERT eliminates glycogen accumulation,[39] potentially providing yet another avenue for therapy.

V. Unknown Mechanisms of Glycogen Accumulation

A. Lafora Disease

Lafora disease is characterized by epilepsy and the accumulation of polyglucosan bodies, known as Lafora bodies, in the brain (specifically neurons), and other tissues such as liver, heart, and skin. Typically, the disease symptoms begin in the teenage years with myoclonic epilepsy and progress to ataxia, severe, refractory epilepsy, and dementia, leading to death within 10 years. It is an autosomal recessive disorder caused by the loss of either one of two genes, EPM2A and EPM2B (NHLRC1). EPM2A encodes a dual specificity phosphatase termed Laforin, while EPM2B encodes the protein Malin, an E3 ubiquitin ligase. While there was previously reported evidence that Malin interacts with the debrancher enzyme AGL to mediate its proteosomal degradation, neither the mechanisms that cause the accumulation of these polyglucosan bodies nor how this leads to epilepsy are known.

There are two groups of animal models for this disease: one is a naturally occurring canine model with a mutation in the canine EPM2B orthologue and the other is a mouse model generated by genetic engineering of mouse ES cells, including a knockout of Laforin,[40] overexpression of an enzymatically inactive laforin,[40] and a knockout of malin.[41] Mouse models have revealed functional properties of the Laforin protein, including polyglucosan accumulation not only in its absence of laforin but also when a phosphatase-deficient Laforin is overexpressed. While Malin–laforin complexes appear to promote the degradation of muscle glycogen synthase and protein-targeting to glycogen (PTG) that are required for glycogen synthesis, recently, a mouse lacking Malin was reported to demonstrate no changes in glycogen synthase, debranching enzyme, or PTG in the absence of malin, but an accumulation of laforin in the insoluble fraction of glycogen, indicating that malin does appear to regulate laforin levels.[41] Thus, it has been suggested that laforin acts to maintain the level of glycogen phosphorylation within tolerated limits so as to allow its normal cytosolic metabolism while preventing glycogen from becoming insoluble and metabolically inert. Hence, excess or inadequate amounts of laforin lead to an imbalance in the forms of glycogen, leading to homeostatic disruption of glycogen metabolism and disease.

Although Lafora disease shares the common feature of polyglucosan body formation with GSDIV, the localization of polyglucosan bodies is quite different between the disorders; even in late stage, Lafora disease polyglucosan bodies are found in the neuronal body while in GSDIV polyglucosan bodies are found both in the cell body and in axons. Likewise, the prognosis for the two diseases is quite different in spite of their histological similarities.

B. AMP-Activated Protein Kinases

AMP-activated protein kinases (AMPKs) are the sensors of the energy status of the cell, thus orchestrating the pathways that store or provide energy for the cell. Therefore, they are targets for therapeutics directed at diabetes and obesity. AMP kinase is a heterotetrameric protein. The kinase is composed of combinations of three subunits: α1 and 2 kinase domain, β1 and 2 carbohydrate binding domain, and γ1, 2, and 3 AMP and ATP-binding domain.[42]

The γ domain is the part of the protein that senses the AMP:ATP ratio in the cell, which is a measure of the energy state of the cell. Surprisingly, it has been shown that either inappropriate activation or inhibition of the protein can cause glycogen storage. Constitutively active mutations prevent the kinase from sensing the intracellular concentration of AMP (the gauge for energy) and thus it fails to activate the degradation of glycogen. However, excess AMPK activation increases the intracellular glucose concentration, which in return activates glycogen synthase, leading to the high glycogen accumulation that has been seen in heart and skeletal muscle of affected individuals.[43]

There is a spontaneously occurring porcine model where arginine has been replaced by glutamine (R225Q) in γ3 subunit of AMPK (PRKAG3) that creates a muscle-specific constitutively active AMP kinase complex that is similar to the human (R225W) constitutively active mutation. In both the human and porcine cases, muscle has excessive glycogen accumulation and increased oxidative capacity and glucose uptake in myotubes.[44] The PRKAG3 mutation does not directly affect glycogen synthesis or breakdown, only an increased glucose intake into the muscle that activates glycogen synthase and leads to glycogen accumulation.

There are mice models of AMP kinases that were generated by introducing Cre sites flanking the catalytic domain of AMPK α1 and α2 subunits. Whole body knockouts have been produced, are viable, and show that α2 catalytic subunit controls whole body insulin sensitivity.[42,44a] The same groups have also shown that the AMP analog 5-aminoimidazole-4-carboxamide-1-β-4-ribofuranoside (AICAR)-induced glucose uptake is lost in α2 subunit knockouts but not in α1 subunit knockouts. This is because α2 delivers most of the kinase activity in the resting state or after stimulation by AICAR; however, knockout was able to neither prevent nor reduce contraction-induced glucose uptake by the muscle. Glycogen storage was not observed in either model, but neither model has any change in the γ2 or 3 catalytic subunits that cause excessive glycogen accumulation in humans.

VI. Conclusion

Glycogen is a complex macromolecule that is an essential means of maintaining glucose levels within a tight physiologic range. The regulation of its synthesis and degradation is highly ordered and similarly complex. Disorders in virtually all steps of synthesis and degradation exist, as do appropriate animal models. These model organisms have been essential in providing mechanistic insights that undoubtedly will direct eventual therapies.

References

1. Servidei S, Bonilla E, Diedrich RG, Kornfeld M, Oates JD, Davidson M, et al. Fatal infantile form of muscle phosphofructokinase deficiency. *Neurology* 1986;**36**(11):1465–70.
2. Haller RG, Vissing J. No spontaneous second wind in muscle phosphofructokinase deficiency. *Neurology* 2004;**62**(1):82–6.
3. Vora S, Giger U, Turchen S, Harvey JW. Characterization of the enzymatic lesion in inherited phosphofructokinase deficiency in the dog: an animal analogue of human glycogen storage disease type VII. *Proc Natl Acad Sci USA* 1985;**82**(23):8109–13.
4. Garcia M, Pujol A, Ruzo A, Riu E, Ruberte J, Arbos A, et al. Phosphofructo-1-kinase deficiency leads to a severe cardiac and hematological disorder in addition to skeletal muscle glycogenosis. *PLoS Genet* 2009;**5**(8):e1000615.
5. (a) Sotiriou E, Greene P, Krishna S, Hirano M, DiMauro S. Myopathy and parkinsonism in phosphoglycerate kinase deficiency. *Muscle Nerve* 2010;**41**(5):707–10. (b) Spiegel R, Gomez EA, Akman HO, Krishna S, Horovitz Y, DiMauro S. Myopathic form of phosphoglycerate kinase (PGK) deficiency: a new case and pathogenic considerations. *Neuromuscul Disord* 2009;**19**(3):207–11.
6. Wang P, Saraswati S, Guan Z, Watkins CJ, Wurtman RJ, Littleton JT. A Drosophila temperature-sensitive seizure mutant in phosphoglycerate kinase disrupts ATP generation and alters synaptic function. *J Neurosci* 2004;**24**(19):4518–29.
7. Karpatkin S, Helmreich E, Cori CF. Regulation of Glycolysis in Muscle. II. Effect of Stimulation and Epinephrine in Isolated Frog Sartorius Muscle. *J Biol Chem* 1964;**239**:3139–45.
8. (a) Kanno T, Sudo K, Maekawa M, Nishimura Y, Ukita M, Fukutake K. Lactate dehydrogenase M-subunit deficiency: a new type of hereditary exertional myopathy. *Clin Chim Acta* 1988;**173**(1):89–98. (b) Takayasu S, Fujiwara S, Waki T. Hereditary lactate dehydrogenase M-subunit deficiency: lactate dehydrogenase activity in skin lesions and in hair follicles. *J Am Acad Dermatol* 1991;**24**(2 Pt 2):339–42.
9. (a) Merkle S, Favor J, Graw J, Hornhardt S, Pretsch W. Hereditary lactate dehydrogenase A-subunit deficiency as cause of early postimplantation death of homozygotes in *Mus musculus*. *Genetics* 1992;**131**(2):413–21. (b) Pretsch W, Merkle S, Favor J, Werner T. A mutation affecting the lactate dehydrogenase locus Ldh-1 in the mouse. II. Mechanism of the LDH-A deficiency associated with hemolytic anemia. *Genetics* 1993;**135**(1):161–70.
10. Yao DC, Tolan DR, Murray MF, Harris DJ, Darras BT, Geva A, et al. Hemolytic anemia and severe rhabdomyolysis caused by compound heterozygous mutations of the gene for erythrocyte/muscle isozyme of aldolase, ALDOA(Arg303X/Cys338Tyr). *Blood* 2004;**103**(6):2401–3.
11. Comi GP, Fortunato F, Lucchiari S, Bordoni A, Prelle A, Jann S, et al. Beta-enolase deficiency, a new metabolic myopathy of distal glycolysis. *Ann Neurol* 2001;**50**(2):202–7.
12. Pederson BA, Schroeder JM, Parker GE, Smith MW, DePaoli-Roach AA, Roach PJ. Glucose metabolism in mice lacking muscle glycogen synthase. *Diabetes* 2005;**54**(12):3466–73.
13. Niswender CM, Willis BS, Wallen A, Sweet IR, Jetton TL, Thompson BR, et al. Cre recombinase-dependent expression of a constitutively active mutant allele of the catalytic subunit of protein kinase A. *Genesis* 2005;**43**(3):109–19.
14. Irimia JM, Meyer CM, Peper CL, Zhai L, Bock CB, Previs SF, et al. Impaired glucose tolerance and predisposition to the fasted state in liver glycogen synthase knock-out mice. *J Biol Chem* 2010;**285**(17):12851–61.
15. (a) Manchester J, Skurat AV, Roach P, Hauschka SD, Lawrence Jr. JC. Increased glycogen accumulation in transgenic mice overexpressing glycogen synthase in skeletal muscle. *Proc Natl Acad Sci USA* 1996;**93**(20):10707–11. (b) Raben N, Danon M, Lu N, Lee E, Shliselfeld L, Skurat AV, et al. Surprises of genetic engineering: a possible model of polyglucosan body disease. *Neurology* 2001;**56**(12):1739–45.

16. McCue ME, Valberg SJ, Miller MB, Wade C, DiMauro S, Akman HO, et al. Glycogen synthase (GYS1) mutation causes a novel skeletal muscle glycogenosis. *Genomics* 2008;**91**(5):458–66.
17. Andersen DH. Familial cirrhosis of the liver with storage of abnormal glycogen. *Lab Invest* 1956;**5**(1):11–20.
18. Akman HO, Karadimas C, Gyftodimou Y, Grigoriadou M, Kokotas H, Konstantinidou A, et al. Prenatal diagnosis of glycogen storage disease type IV. *Prenat Diagn* 2006;**26**(10):951–5. (b)Tay SK, Akman HO, Chung WK, Pike MG, Muntoni F, Hays AP, et al. Fatal infantile neuromuscular presentation of glycogen storage disease type IV. *Neuromuscul Disord* 2004;**14**(4):253–60.
19. Fyfe JC, Giger U, Van Winkle TJ, Haskins ME, Steinberg SA, Wang P, et al. Glycogen storage disease type IV: inherited deficiency of branching enzyme activity in cats. *Pediatr Res* 1992;**32**(6):719–25.
20. Ward TL, Valberg SJ, Adelson DL, Abbey CA, Binns MM, Mickelson JR. Glycogen branching enzyme (GBE1) mutation causing equine glycogen storage disease IV. *Mamm Genome* 2004;**15**(7):570–7.
21. Smythe C, Watt P, Cohen P. Further studies on the role of glycogenin in glycogen biosynthesis. *Eur J Biochem* 1990;**189**(1):199–204.
22. Moslemi AR, Lindberg C, Nilsson J, Tajsharghi H, Andersson B, Oldfors A. Glycogenin-1 deficiency and inactivated priming of glycogen synthesis. *N Engl J Med* 2010;**362**(13):1203–10.
23. Tan P, Allen JG, Wilton SD, Akkari PA, Huxtable CR, Laing NG. A splice-site mutation causing ovine McArdle's disease. *Neuromuscul Disord* 1997;**7**(5):336–42.
24. Burwinkel B, Bakker HD, Herschkovitz E, Moses SW, Shin YS, Kilimann MW. Mutations in the liver glycogen phosphorylase gene (PYGL) underlying glycogenosis type VI. *Am J Hum Genet* 1998;**62**(4):785–91.
25. Beauchamp NJ, Dalton A, Ramaswami U, Niinikoski H, Mention K, Kenny P, et al. Glycogen storage disease type IX: high variability in clinical phenotype. *Mol Genet Metab* 2007;**92**(1–2):88–99.
26. Hendrickx J, Coucke P, Dams E, Lee P, Odievre M, Corbeel L, et al. Mutations in the phosphorylase kinase gene PHKA2 are responsible for X-linked liver glycogen storage disease. *Hum Mol Genet* 1995;**4**(1):77–83.
27. Burwinkel B, Maichele AJ, Aagenaes O, Bakker HD, Lerner A, Shin YS, et al. Autosomal glycogenosis of liver and muscle due to phosphorylase kinase deficiency is caused by mutations in the phosphorylase kinase beta subunit (PHKB). *Hum Mol Genet* 1997;**6**(7):1109–15.
28. Nakayama A, Yamamoto K, Tabata S. Identification of the catalytic residues of bifunctional glycogen debranching enzyme. *J Biol Chem* 2001;**276**(31):28824–8.
29. Gregory BL, Shelton GD, Bali DS, Chen YT, Fyfe JC. Glycogen storage disease type IIIa in curly-coated retrievers. *J Vet Intern Med* 2007;**21**(1):40–6.
30. Han SH, Shin KY, Lee SS, Ko MS, Jeong DK, Oh HS, et al. SINE indel polymorphism of AGL gene and association with growth and carcass traits in Landrace x Jeju Black pig F(2) population. *Mol Biol Rep* 2010;**37**(1):467–71.
31. Koeberl DD, Kishnani PS, Bali D, Chen YT. Emerging therapies for glycogen storage disease type I. *Trends Endocrinol Metab* 2009;**20**(5):252–8.
32. Kim SY, Jun HS, Mead PA, Mansfield BC, Chou JY. Neutrophil stress and apoptosis underlie myeloid dysfunction in glycogen storage disease type Ib. *Blood* 2008;**111**(12):5704–11.
33. Kishnani PS, Bao Y, Wu JY, Brix AE, Lin JL, Chen YT. Isolation and nucleotide sequence of canine glucose-6-phosphatase mRNA: identification of mutation in puppies with glycogen storage disease type Ia. *Biochem Mol Med* 1997;**61**(2):168–77.
34. Lei KJ, Chen H, Pan CJ, Ward JM, Mosinger Jr. B, Lee EJ, et al. Glucose-6-phosphatase dependent substrate transport in the glycogen storage disease type-1a mouse. *Nat Genet* 1996;**13**(2):203–9.

35. Peng WT, Pan CJ, Lee EJ, Westphal H, Chou JY. Generation of mice with a conditional allele for G6pc. *Genesis* 2009;**47**(9):590–4.
36. Chen LY, Shieh JJ, Lin B, Pan CJ, Gao JL, Murphy PM, et al. Impaired glucose homeostasis, neutrophil trafficking and function in mice lacking the glucose-6-phosphate transporter. *Hum Mol Genet* 2003;**12**(19):2547–58.
37. Bijvoet AG, van de Kamp EH, Kroos MA, Ding JH, Yang BZ, Visser P, et al. Generalized glycogen storage and cardiomegaly in a knockout mouse model of Pompe disease. *Hum Mol Genet* 1998;**7**(1):53–62.
38. Geel TM, McLaughlin PM, de Leij LF, Ruiters MH, Niezen-Koning KE. Pompe disease: current state of treatment modalities and animal models. *Mol Genet Metab* 2007;**92**(4):299–307.
39. Raben N, Schreiner C, Baum R, Takikita S, Xu S, Xie T, et al. Suppression of autophagy permits successful enzyme replacement therapy in a lysosomal storage disorder-murine Pompe disease. *Autophagy* 2010;**6**(8):1078–89.
40. Ganesh S, Delgado-Escueta AV, Sakamoto T, Avila MR, Machado-Salas J, Hoshii Y, et al. Targeted disruption of the Epm2a gene causes formation of Lafora inclusion bodies, neurodegeneration, ataxia, myoclonus epilepsy and impaired behavioral response in mice. *Hum Mol Genet* 2002;**11**(11):1251–62.
41. DePaoli-Roach AA, Tagliabracci VS, Segvich DM, Meyer CM, Irimia JM, Roach PJ. Genetic depletion of the malin E3 ubiquitin ligase in mice leads to lafora bodies and the accumulation of insoluble laforin. *J Biol Chem* 2010;**285**(33):25372–81.
42. Jorgensen SB, Viollet B, Andreelli F, Frosig C, Birk JB, Schjerling P, et al. Knockout of the alpha2 but not alpha1 5′-AMP-activated protein kinase isoform abolishes 5-aminoimidazole-4-carboxamide-1-beta-4-ribofuranosidebut not contraction-induced glucose uptake in skeletal muscle. *J Biol Chem* 2004;**279**(2):1070–9.
43. Akman HO, Sampayo JN, Ross FA, Scott JW, Wilson G, Benson L, et al. Fatal infantile cardiac glycogenosis with phosphorylase kinase deficiency and a mutation in the gamma2-subunit of AMP-activated protein kinase. *Pediatr Res* 2007;**62**(4):499–504.
44. (a) Viollet B, Andreelli F, Jorgensen SB, Perrin C, Geloen A, Flamez D, et al. The AMP-activated protein kinase alpha2 catalytic subunit controls whole-body insulin sensitivity. *J Clin Invest* 2003;**111**(1):91–8. (b) Ianzano L, Young EJ, Zhao XC, Chan EM, Rodriguez MT, Torrado MV, et al. Loss of function of the cytoplasmic isoform of the protein laforin (EPM2A) causes Lafora progressive myoclonus epilepsy. *Hum Mutat* 2004;**23**(2):170–6.

Modeling Disorders of Fatty Acid Metabolism in the Mouse

Eric S. Goetzman

Department of Pediatrics, University of Pittsburgh School of Medicine, Pittsburgh, Pennsylvania, USA

There are at least 17 enzymes involved in mitochondrial fatty acid oxidation encoded by at least 21 genes. For most of these genes, humans with genetic deficiencies have been identified. The mouse possesses a very similar fatty acid oxidation system and has served well as an organism for modeling genetic loss of function. Knockout mice have been created for 12 fatty acid oxidation genes, including all three carnitine palmitoyltransferase-1 genes, four of the acyl-CoA dehydrogenases, both subunits of trifunctional protein, short/medium-chain hydroxyacyl-CoA dehydrogenase, and two enzymes required for oxidation of polyunsaturated fatty acids (enoyl-CoA isomerase and 2,4 dienoyl-CoA reductase). This review covers the knowledge that has been gained from these mouse models in terms of understanding both single-gene fatty acid oxidation disorders and the contribution of the fatty acid oxidation pathway to polygenic diseases such as obesity and type 2 diabetes. Also reviewed are other mouse models displaying phenotypic aspects of a fatty acid oxidation disorder such as knockout mice lacking the carnitine transporter and knockouts of key regulators of the pathway such as

Progress in Molecular Biology
and Translational Science, Vol. 100
DOI: 10.1016/B978-0-12-384878-9.00010-8

Copyright 2011, Elsevier Inc.
All rights reserved.
1877-1173/11 $35.00

peroxisome proliferator-activated receptor-α and sirtuin-3. Finally, nongenetic means of manipulating fatty acid oxidation in the mouse are discussed, in particular the various chemical inhibitors that have been used successfully *in vivo*.

I. Introduction to Fatty Acid Oxidation

In humans, the catabolism of fatty acids serves two main purposes. First, fatty acids are an important source of energy, particularly for tissues with heavy bioenergetic demands such as heart, brown adipose tissue, skeletal muscle, and liver. It has been estimated that the heart derives 85–90% of its energy from fatty acids. Other tissues prefer glucose but turn to fatty acids or fatty acid-derived ketone bodies during times of starvation, illness, or other metabolic stress. A second purpose for fatty acid catabolism is to prevent toxic accumulation. Along with the saturated fatty acids 14–18 carbons in length that are so abundant in our diet come smaller amounts of polyunsaturated fatty acids, branched-chain fatty acids, and odd-chain fatty acids. Short branched-chain fatty acids are created as intermediates in the degradation of other molecules such as branched-chain amino acids. The cell must be equipped to degrade this wide array of fatty acid species to prevent toxicity. The task is shared by mitochondria and peroxisomes with peroxisomes generally handling the unusual species and mitochondria everything else.[1] The two organelles functionally interact in ways that are poorly understood. Many fatty acids are partially shortened in peroxisomes and finished in the mitochondria. The peroxisomal pathway produces acetyl-CoA which is funneled into synthetic pathways such as isoprenoid synthesis[2] but otherwise no energy is produced. The mitochondrial pathway produces acetyl-CoA and reduced NAD/FAD, with the acetyl-CoA progressing into the TCA cycle to yield additional reducing equivalents (and ultimately CO_2). Peroxisomal fatty acid oxidation disorders are typically diseases of toxicity while mitochondrial fatty acid oxidation disorders are largely diseases of energy metabolism. This review will focus on the mitochondrial fatty acid oxidation disorders, how they have been genetically and chemically modeled in the mouse over the past 35 years, and what the models have taught us.

II. Overview of the Mitochondrial Pathway

Degrading fatty acids in mitochondria is not a simple process. The three major phases are (1) transport, (2) β-oxidation, and (3) conveyance of reducing equivalents to the electron transport chain. The enzymology of the pathway has previously been reviewed.[3–5] All three phases are represented by human genetic diseases.

A. Transport into Mitochondria

Inside the cell, free fatty acids are esterified with coenzyme-A to produce acyl-CoAs which are less toxic. It is believed that short- and medium-chain acyl-CoAs (C4–C10) can diffuse across the mitochondrial membrane while longer chain acyl-CoAs must be carried across by the carnitine system. This system consists of three enzyme-mediated steps (Fig. 1). Carnitine palmitoyltransferase-1 (CPT-1), which resides in the outer mitochondrial membrane, converts long-chain acyl-CoAs to acyl-carnitines and translocates them to the intermitochondrial membrane space. Carnitine acyl-carnitine translocase (CACT) carries acyl-carnitines across the inner membrane in exchange for free carnitine, while carnitine palmitoyltransferase-2 (CPT-2) converts them back to acyl-CoAs. The carnitine cycle is absolutely essential and the oxidation of long-chain fatty acids cannot proceed without it. The cycle also works in reverse to prevent toxic intramitochondrial overload of acyl-CoAs in the face of defects in the β-oxidation pathway. Excess acyl-CoAs are exported back to the cytosol as acyl-carnitines and eventually find their way into the blood where they are a very useful diagnostic indicator of fatty acid oxidation disorders. Blood spot acyl-carnitines are the basis for newborn screening programs for these disorders.

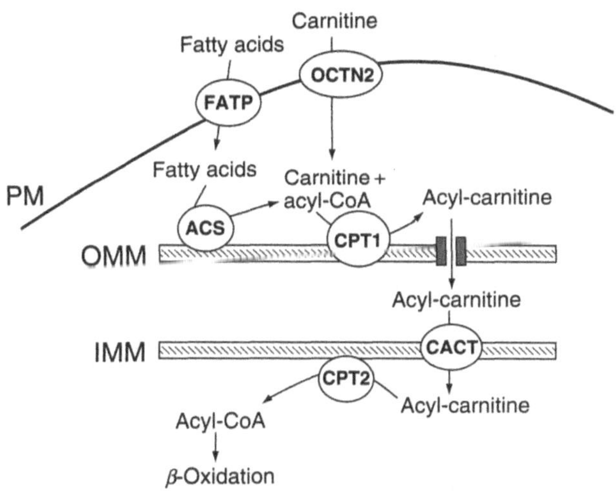

FIG. 1. The carnitine cycle for fatty acid import into mitochondria. Fatty acids and carnitine are transported across the plasma membrane (PM). Fatty acids are activated to acyl-CoAs by acyl-CoA synthetases (ACS) and then conjugated to carnitine at the outer mitochondrial membrane (OMM) by carnitine palmitoyltransferase-1 (CPT-1). Acyl-carnitines pass through the OMM into the intermembrane space and are subsequently transported across the inner mitochondrial membrane (IMM) by carnitine acyl-carnitine translocase (CACT) and converted back to acyl-CoA by carnitine palmitoyltransferase-2 (CPT-2).

B. Acyl-CoA Chain-Shortening by β-Oxidation

Fatty acid oxidation is called "β-oxidation" because the bond between the α (C_2) and β carbon (C_3) of the fatty acid is broken during each round of the cycle, which involves four enzymatic steps as illustrated in Fig. 2 and reviewed elsewhere.[6,7] First, a double bond is introduced between C_2 and C_3 by the action of the acyl-CoA dehydrogenases (ACADs). ACADs are flavoenzymes that have a noncovalently bound FAD moiety on each subunit to accept electrons during the

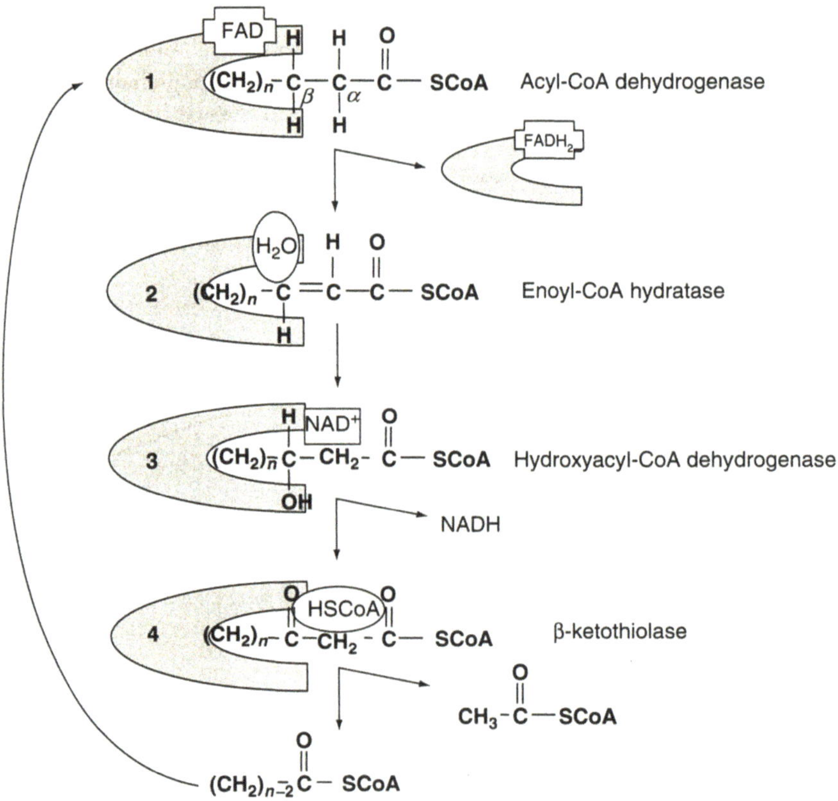

FIG. 2. The four enzymatic steps of mitochondrial fatty acid β-oxidation. (1) Dehydrogenation occurs between the α and β carbons, catalyzed by the acyl-CoA dehydrogenases with electrons going to an enzyme-bound FAD. The products are an enoyl-CoA and reduced acyl-CoA dehydrogenase enzyme. (2) Enoyl-CoA hydratases hydrate the newly formed double bond between the α and β carbons. (3) With NAD as the electron acceptor, hydroxyacyl-CoA dehydrogenases convert the hydroxyl group on the β carbon to a keto group. (4) In the final reaction β-ketothiolases complete the cleavage between the α and β carbons producing acetyl-CoA and an acyl-CoA that is now two carbons shorter.

dehydrogenation reaction. The next step is hydration of the newly introduced double bond. This reaction is stereospecific, forming only the L isomer, and is catalyzed by enoyl-CoA hydratases (EHs). The product of the reaction, L-3-hydroxyacyl-CoA, is oxidized in step three by hydroxyacyl-CoA dehydrogenases (HADs) using NAD+ as an electron acceptor and yielding 3-ketoacyl-CoA. Finally, β-ketothiolases complete the cycle using free CoA to generate a new acyl-CoA molecule which is two carbons shorter, and an acetyl-CoA. The shortened acyl-CoA can repeat this cycle until completely reduced to acetyl-CoA units. Fatty acids are thus a rich source of energy. A single C_{18} fatty acid is broken into 9 acetyl-CoA which by way of the TCA cycle and electron transport chain produces 90 ATP. The same number of carbons from glucose (three glucose molecules) would also produce 90 ATP. The advantage to fatty acid oxidation are the electrons captured at steps one and three of the β-oxidation cycle, which yield an additional 30 ATP for a single C_{18} fatty acid, bringing the total to 120 ATP per 18 carbons versus 90 ATP for the same number of glucose carbons.

C. Electron Transfer from β-Oxidation Cycle to the Electron Transport Chain

NADH produced by the HADs shares the same fate as NADH produced by the TCA cycle which is reoxidation to NAD via complex I of the electron transport chain (Fig. 3). The reduced FAD produced in step one of β-oxidation is handled by a specialized redox mechanism involving an intermediate electron carrier known as electron transferring flavoprotein (ETF), which like the ACADs is a flavoprotein with an enzyme-bound FAD. ETF interacts with and removes the electrons from ACADs through a mechanism that we are just coming to understand.[8–10] ETF in turn is reoxidized by ETF dehydrogenase which is located on the inner mitochondrial membrane in conjunction with the electron transport chain. ETF dehydrogenase passes the electrons to coenzyme Q and from there they go into complex III.

Thus far there are nine characterized ACADs that use ETF as an electron acceptor. Five of the nine ACADs function in the fatty acid oxidation pathway and the remaining four are involved in the degradation of branched-chain amino acids. During the degradation of lysine, leucine, isoleucine, and valine, short branched-chain acyl-CoAs are produced that are then β-oxidized in a manner that utilizes some components of the fatty acid oxidation machinery including ETF and ETF dehydrogenase. There are human disorders associated with these enzymes but they do not share the energy deficiency phenotype of fatty acid oxidation disorders and thus are outside the scope of this review.[11,12] Rather, the symptoms appear to be related to the toxic accumulation of branched-chain amino acids and their metabolites. A knockout mouse has been developed for one of these enzymes, glutaryl-CoA dehydrogenase.[13]

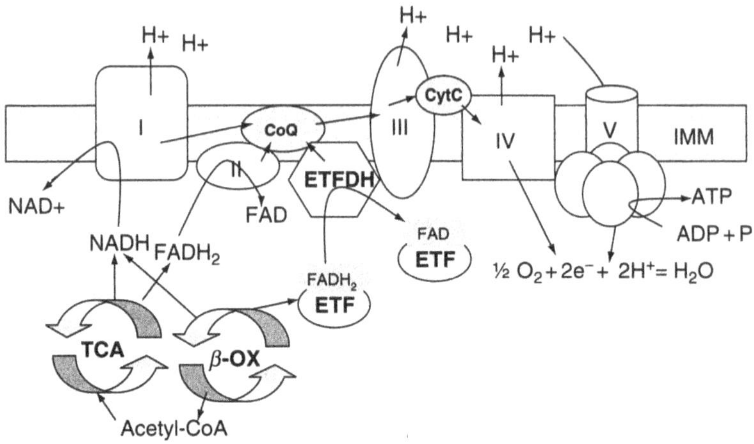

FIG. 3. Fate of the products of mitochondrial β-oxidation. The β-oxidation pathway (β-ox) produces NADH, enzyme-bound $FADH_2$, and acetyl-CoA. The latter is oxidized to completion by the TCA cycle which produces NADH and $FADH_2$. NADH from both the TCA and β-ox pathways is reoxidized by complex I of the electron transport chain. $FADH_2$ from the TCA cycle is reoxidized by complex II while a separate system exists for reoxidizing the enzyme-bound $FADH_2$ produced by β-ox. The electrons are passed to another FAD, bound to electron transferring flavoprotein (ETF), which then shuttles them to CoQ and into complex III by way of ETF dehydrogenase (ETFDH) which is bound to the inner mitochondrial membrane (IMM).

D. Evidence for a Fatty Acid Oxidation Enzyme Complex

Putting together the transport, β-oxidation, and electron disposal mechanisms gives the intact functional pathway of mitochondrial fatty acid oxidation. It has been proposed that the involved enzymes are organized into a large multienzyme complex on the inner mitochondrial membrane rather than floating individually about the mitochondrial matrix.[14,15] A fatty acid oxidation complex capable of channeling the substrate from one enzyme to the next would explain the observed difficulty in detecting any intermediates between the initial substrate acyl-CoA and the final product acetyl-CoA.[16] Of the fatty acid oxidation enzymes, five are known to be anchored to the inner mitochondrial membrane: CPT-2, mitochondrial trifunctional protein (MTP), very long-chain acyl-CoA dehydrogenase (VLCAD), ACAD9, and ETF dehydrogenase. The rest of the fatty acid oxidation enzymes are considered to be localized to the matrix but there is considerable evidence that they associate with other proteins on the inner membrane. Many years ago it was shown that antibodies against CPT-2 (believed at the time to be an isoform of CPT-1 localizing to the inner mitochondrial membrane) coimmunoprecipitate several other fatty acid oxidation enzymes including the short-chain specific EH and short/medium-chain

HAD.[17] Binding of these short-chain enzymes to CPT-2 may have accounted for the earlier observation by Sumegi and Srere that purified preparations of these enzymes bind isolated inner mitochondrial membranes but not membranes from other sources.[18] Additionally, the fatty acid oxidation complex may interact closely with the electron transport chain such that the electrons extracted in steps one and three (see Figs. 2 and 3) are efficiently passed into complexes III and I, respectively.[19,20] MTP has been shown to interact with complex I and ETF dehydrogenase interacts with complex III.[21,22] The electron transport chain is also subject to a higher order structure consisting of complexes I, III, and IV assembled into "supercomplexes".[23] Recently, Wang *et al.*[24] demonstrated that electron transport chain supercomplexes isolated by sucrose gradient centrifugation also contain the fatty acid oxidation machinery. These supercomplexes can completely oxidize acyl-CoAs of various chain lengths. For maximal activity exogenous ETF needed to be added suggesting that ETF is only very weakly associated with the larger complexes.

Figure 4 illustrates a hypothetical configuration for the fatty acid oxidation complex based on the available experimental evidence. In this model, an incoming typical C_{18} acyl-CoA would traverse the mitochondrial membrane with the help of CPT-1, CACT, and CPT-2, and then be channeled directly to VLCAD/ACAD9 which catalyze the first step in β-oxidation. The resulting enoyl-CoA product is channeled through MTP which possesses the catalytic activities of steps 2–4 and produces an acyl-CoA two carbons shorter (C_{16}) and an acetyl-CoA. From there, the C_{16} acyl-CoA product is passed either to long-chain acyl-CoA dehydrogenase (LCAD; highly expressed in rodents) or back to VLCAD to begin another round of chain-shortening. Once shortened to C_{12} the acyl-CoA channels to medium-chain acyl-CoA dehydrogenase (MCAD) and the medium/short-chain β-oxidation enzymes which continue the sequential chain-shortening. The final step is the cleavage of acetoacetyl-CoA into two acetyl-CoA by the multifunctional enzyme acetyl-CoA acetyltransferase (ACAT1).

Shown in Fig. 4 are 17 fatty acid oxidation enzymes. There are two genes encoding the α and β subunits of ETF, two for the MTP α and β subunits, and three separate CPT-1 genes (designated a, b, and c) that are expressed in different tissue types bringing the total number of genes to 21. In Fig. 4 the genes in bold font have had patients identified with disease-causing mutations. There are five genes with no described patients: CPT-1b, CPT-1c, LCAD, the short-chain EH, and enoyl-CoA isomerase (ECI) which is an auxiliary enzyme that catalyzes the shift of double bonds from the C_3 position to C_2. In Fig. 4, each gene rendered in gray color has a corresponding genetic mouse model. For two of the 12 models (HAD and MTP β) full characterizations have yet to be published.[25] The next section will summarize the literature on the remaining 10 models as well as some other mouse models that share aspects of the fatty acid oxidation disorder phenotype.

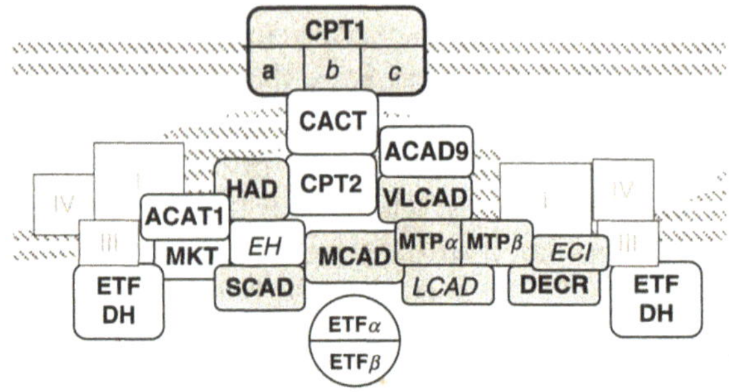

FIG. 4. A hypothetical model of the fatty acid oxidation enzyme complex interacting with electron transport chain supercomplexes (complexes I, III, and IV). Shown are 17 fatty acid oxidation enzymes encoded by 21 genes. Fatty acid oxidation genes rendered in italics have no associated human deficiencies to date (LCAD, ECI, CPT-1b, CPT-1c, EH). Genes filled with gray have an associated knockout mouse model. Long-chain substrates enter mitochondria through carnitine palmitoyltransferase-1 (CPT-1 genes a, b, c), carnitine acyl-carnitine translocase (CACT) and carnitine palmitoyltransferase-2 (CPT-2). Initial rounds of chain shortening are conducted by very long-chain acyl-CoA dehydrogenase (VLCAD), ACAD9, and long-chain acyl-CoA dehydrogenase (LCAD) working in conjunction with mitochondrial trifunctional protein (MTP). Unsaturated substrates are shortened with the help of enoyl-CoA isomerase (ECI) and 2,4 dienoyl-CoA reductase (DECR). Substrates 12 carbons and shorter are oxidized by medium-chain acyl-CoA dehydrogenase (MCAD) and short-chain acyl-CoA dehydrogenase (SCAD) working with the medium/short-chain enoyl-CoA hydratase (EH), hydroxyacyl-CoA dehydrogenase (HAD), and the medium and short-chain ketothiolases (MKT and ACAT1, respectively). Electron transferring flavoprotein (ETF) reoxidizes the acyl-CoA dehydrogenases and passes electrons into complex III via ETF dehydrogenase (ETFDH).

III. Genetic Models of Fatty Acid Oxidation Disorders

A. Short-Chain Acyl-CoA Dehydrogenase (SCAD)

SCAD is one of five ACAD enzymes involved in mitochondrial fatty acid oxidation. Like all fatty acid oxidation enzymes, it is a nuclear encoded protein that is imported into mitochondria. Following cleavage of the mitochondrial localization signal it folds, acquires FAD cofactors, and assembles into the active homotetrameric form. The optimal substrate for SCAD is butyryl-CoA, a four-carbon fatty acid. Human SCAD deficiency is a recessive disorder with an estimated frequency of at least 1:50,000 births.[26] The disease shows much clinical heterogeneity, with some patients displaying hallmark symptoms of fatty acid oxidation disorders (hypoglycemia, metabolic acidosis) and others presenting with failure to thrive, seizures, and neuromuscular symptoms. The one unifying feature of the disorder is excretion of ethylmalonic acid in the urine. With the advent of newborn screening programs it became clear that many children with SCAD deficiency are completely normal and asymptomatic. This has caused some controversy about how much of the clinical spectrum reported previously is actually caused by SCAD deficiency.

The SCAD-deficient mouse was the first genetic mouse model of a fatty acid oxidation disorder. It was identified in the late 1980s by two separate groups while screening inbred mouse strains for abnormal urine organic acids.[27,28] Like human patients, the SCAD-deficient mouse excretes ethylmalonic acid. And like many human fatty acid oxidation disorders, the mice are asymptomatic until fasted or otherwise metabolically stressed. After fasting the animals develop fatty liver and hypoglycemia.[29] Mouse SCAD deficiency is due to a naturally occurring deletion in the SCAD gene in the BALB/cByJ strain of mice. The mutation occurred between 1981 and 1982 in the BALB/cByJ colony at Jackson Laboratories.[30] As far back as 1984 there was a report of fatty liver and "Reye-like" syndrome in the BALB/cByJ strain.[31] These mice are commercially available through Jackson Laboratories although they are not explicitly described in their catalog as SCAD deficient. Thus, over the past 20 years many studies have been conducted on the BALB/cByJ strain by investigators unaware of the SCAD mutation and the metabolic phenotype, such as the experiments that led to the accidental discovery of cold sensitivity in the SCAD −/− mice. While studying brown adipose tissue nonshivering thermogenesis and uncoupling protein 1 (UCP1), investigators at the Jackson Laboratories created a UCP1 knockout mouse on the BALB/cByJ strain background. Because UCP1 is required to mediate heat production in brown fat, a cold-sensitive phenotype was expected. Surprisingly the wild-type BALB/cByJ controls were also cold sensitive.[32] The cold sensitivity in the BALB/cByJ strain was then mapped to a region of chromosome 5 containing the SCAD gene, at which time the investigators became aware of the existing literature describing the naturally occurring deletion in the SCAD gene. At that time, another ACAD mouse model had just been developed by gene targeting.[33] These mice, deficient in LCAD, were tested for cold sensitivity and found to be extremely sensitive.[32] Thus, cold sensitivity became established as an easy to follow indicator of fatty acid oxidation function in mouse models.

B. LCAD

With the advent of gene targeting and embryonic stem cell technologies in the late 1980s came the ability to "knock out" any gene of interest in the mouse genome. The first fatty acid oxidation gene to be successfully targeted with this approach was LCAD.[33] Since the 1950s when ACAD enzymes and ETF were first described in pig mitochondria it had been believed that there were three ACADs—SCAD for short-chain substrates, MCAD for medium-chain length substrates, and LCAD for palmitoyl-CoA (C_{16}-CoA) and other long-chain substrates.[34] Thus, when human patients were first described with impairments in the ability to dehydrogenate long-chain substrates LCAD was assumed to be the culprit.[35,36] The molecular basis for human LCAD deficiency was thought to be a substitution at residue 303 (Q303K), now recognized as a polymorphism. In 1992, Izai *et al.*[37] reported the discovery of a fourth ACAD that also had high activity toward palmitoyl-CoA. This enzyme was dubbed VLCAD and it was subsequently shown that the LCAD-deficient patients were really VLCAD deficient.[38,39] To the present day no true LCAD-deficient patients have been identified and it is one of only five known fatty acid oxidation genes without an associated genetic disease (Fig. 4). The lack of human LCAD patients coupled with studies demonstrating limited tissue distribution and low expression of LCAD in humans has resulted in controversy over its physiological role.[40–42]

In mice LCAD is highly expressed and plays a dominant role in dehydrogenation of long, straight-chain acyl-CoAs. The LCAD−/− mouse has a phenotype which closely models human VLCAD deficiency in terms of a severe energy deficiency, fasting-induced hypoglycemia, reduced ketone production, and cardiac involvement.[33] Due to the presence of VLCAD, LCAD−/− tissue extracts retain about 30–40% of normal ACAD activity with palmitoyl-CoA as substrate.[43] They accumulate free fatty acids in the liver and show elevated blood acyl-carnitine profiles similar to those seen in human VLCAD-deficient patients including the hallmark elevations in C14:1.[43] LCAD−/− mice show some early gestational loss with about 60% of mutants surviving to birth, indicating an important role for fatty acid oxidation in embryogenesis.[43]

As mentioned above, LCAD−/− mice are extremely cold sensitive, particularly if fasted prior to the acute cold exposure, and will succumb within a few hours after being transferred to a cold environment.[32,44] The cold sensitivity is a complex trait that likely reflects reduced energy metabolism in several organs, not just brown adipose tissue. Acute cold exposure results in a rapid release of fatty acids into circulation from adipose stores to be used by muscles for fueling shivering, by the liver for producing ketones, and by brown adipose tissue for nonshivering thermogenesis. These processes are interdependent and all rely upon an intact fatty acid oxidation pathway. For LCAD−/− mice, fasting alone is sufficient to induce a significant loss in core body temperature,[44] a phenomenon known as torpor which has also been noted in leptin-deficient ob/ob mice.[45,46] Torpor is a mechanism employed by rodents for surviving severe energy deprivation and is essentially a hibernation-like state.

Wood and colleagues used the LCAD−/− mice as a model to test the efficacy of treating a long-chain fatty acid oxidation disorder with a diet enriched in medium-chain length triglycerides (MCTs). Treatment of VLCAD-deficient patients and other patients with long-chain fatty acid oxidation disorders typically involves restricting dietary long-chain fats. Medium-chain fatty acids do not need the carnitine cycle to enter mitochondria and the rationale behind the diet is that MCT should provide a ready supply of energy to mitochondria of patients with long-chain fatty acid oxidation disorders. In the experiment, one group of mice received MCT diet beginning at weaning, a second group received gestational exposure to the MCT diet (mothers were maintained on the MCT diet), and the third group received a diet with long-chain triglycerides (LCT).[47] As young adults the mice were challenged with acute cold exposure. As expected, survival was lowest in LCAD−/− mice on the LCT diet (50%). The MCT diet began at weaning had little effect (57% survival). Surprisingly, exposure to the MCT diet *in utero* led to a metabolic buffering against the cold challenge with 93% of the LCAD−/− animals surviving. This was correlated with a fourfold increase in serum octanoic acid which was not present in mice receiving the diet beginning at weaning. The implications are that MCT has a limited potential for buffering against metabolic crises in patients with long-chain fatty acid oxidation disorders unless they are on the diet during development of adipose tissue such that medium-chain fatty acids can be incorporated into fat stores.

Finally, LCAD−/− mice may also be a useful model for studying the links between disrupted energy metabolism, ectopic lipid storage, and the development of insulin resistance and the metabolic syndrome. LCAD−/− mice accumulate triglycerides in liver and muscle and the effects of this accumulation on integrated metabolism were recently studied.[48] It was shown that LCAD−/− mice have increased body fat despite lower food intake. Circulating leptin levels were increased. The animals develop hepatic insulin resistance by a mechanism involving reduced insulin signal transduction and aberrant diacylglycerol synthesis. Interestingly, while blood free fatty acids and triglycerides were not different from wild-type mice, cholesterol levels were almost double in the LCAD−/− mice. This may be due to increased cholesterol synthesis in the liver as increased mRNA expression of several cholesterol synthesis genes was previously noted.[44] It remains to be seen whether such changes in insulin signaling or cholesterol synthesis are induced in human fatty acid oxidation disorders.

C. VLCAD

VLCAD deficiency has an estimated prevalence of 1:30,000 births.[49] The disease can manifest as a severe, early-onset form with symptoms appearing shortly after birth or as a later-onset, milder disease. The severe form involves cardiac hypertrophy which can be life-threatening as well as liver disease in the form of fasting intolerance (hypoketotic hypoglycemia). Patients with the milder forms can be asymptomatic until adulthood at which time they experience exercise-induced muscle or cardiac disease.[50] The severity of disease in general appears to correlate to residual enzyme activity. Two laboratories independently generated mice deficient in VLCAD.[43,51] As expected given the dominant role of LCAD in mice, VLCAD−/− have a milder disease phenotype. This phenotype can be greatly enhanced with metabolic stress such as fasting, exercise, or cold exposure. VLCAD−/− mice to not suffer from gestational loss or sudden death as LCAD−/− mice do.[43] They do not show increased hepatic free fatty acids like LCAD−/− mice but do accumulate long-chain acyl-carnitines in the blood, particularly C16 and C18 species. A 1-h bout of exercise or 8 h of fasting in the cold can greatly increase the blood acyl-carnitines and also results in hypoglycemia.[52]

Spiekerkoetter and her colleagues have used the VLCAD−/−mice to test some assumptions related to the management and treatment of fatty acid oxidation disorders. First they investigated the issue of "secondary carnitine deficiency." Humans with fatty acid oxidation disorders often have lower blood concentrations of free carnitine. It has been assumed that free carnitine is being depleted as a mechanism for conjugating and removing excess acyl-CoAs from mitochondria (secondary carnitine deficiency). This led to the practice of treating patients with carnitine as a long-term supplement, which has been controversial.[53–55] Spiekerkoetter and colleagues demonstrated using VLCAD−/− mice that tissue and blood carnitine levels do not necessarily correlate.[56–58] VLCAD−/− mice have reduced free carnitine in blood but not within tissues. On the contrary, in liver it is actually increased. This finding was substantiated in the same year by van Vlies *et al.*[59] using LCAD−/− mice which showed a similar pattern of reduced blood carnitine but either increased or normal carnitine content in tissues. Exercise stress, however, does induce a free carnitine deficiency in muscle and heart of VLCAD−/− mice. Interestingly, the drop in muscle/heart carnitine could not be prevented by carnitine supplementation in the drinking water. Recovery occurs naturally through carnitine production in the liver with levels restored to normal in heart and muscle within 24 h following the bout of exercise. The authors did demonstrate that carnitine supplementation increases the production and accumulation of tissue acyl-carnitines, even in wild-type mice. However, they propose that rather than being therapeutic, this may actually exacerbate toxicity issues. In cell culture studies they demonstrated that palmitoyl-carnitine reduces cell viability more than palmitoyl-CoA.[58]

Spiekerkoetter and colleagues have recently extended the work of Schuler *et al.*[47] regarding the use of MCT enriched diets to treat long-chain fatty acid oxidation disorders. Long-term use of the MCT diet in VLCAD−/− mice resulted in severe hepatic steatosis and higher levels of circulating blood free fatty acids.[60,61] The ability of the liver to mobilize fat stores during exercise was impaired. Finally, it was shown that mRNA levels of several genes involved in hepatic lipogenesis were increased, suggesting deranged regulation of hepatic lipid metabolism. Based on these results the authors recommend caution in the use of MCT for treating patients with long-chain fatty acid oxidation disorders. It did appear, however, that single bolus treatments of MCT are well tolerated and may be of use as a quick source of energy when patients are facing a bout of exercise or other metabolic stressor.

Like the LCAD knockout model, VLCAD−/− mice have also been useful for studying the contribution of impaired fat oxidation to polygenic diseases such as obesity, insulin resistance, and the metabolic syndrome. A recent study on VLCAD−/− mice by the same investigators who reported hepatic insulin resistance in LCAD−/− mice revealed some surprising findings. VLCAD−/− mice were shown to be resistant to high-fat diet-induced obesity (DIO).[62] They gained less fat mass and were more insulin sensitive than wild-type mice when maintained on a high-fat diet. Paradoxically, the mechanism behind the resistance to obesity is increased fatty acid oxidation. Since the impairment in long-chain ACAD activity is modest in VLCAD−/− mice—they retain about 70% residual activity due to the dominant role of LCAD—they are able to upregulate fatty acid oxidation. On a high-fat diet VLCAD−/− mice show constitutively increased expression of peroxisome proliferator-activated receptor alpha (PPARα) target genes in the fatty acid oxidation pathway. PPARα is a well known regulator of many genes in the pathway and is activated by long-chain fatty acids and fatty acid derivatives.[63–65] It appears that the C18 and particularly C18 unsaturated fatty acids that accumulate in VLCAD−/− mice activate PPARα. Additionally, AMP kinase (AMPK) activity is increased in VLCAD−/− mice resulting in significantly increased phosphorylation of acetyl-CoA carboxylase (ACC) which inactivates the enzyme and reduces synthesis of malonyl-CoA. Malonyl-CoA is a potent inhibitor of CPT-1 and is the primary regulatory mechanism governing entry of long-chain acyl-CoAs into mitochondria. Thus, VLCAD mice are geared for fatty acid oxidation through both an increase in the fatty acid oxidation machinery (PPARα effect) and an increase in substrate delivery across the mitochondrial membrane (reduced malonyl-CoA effect). The difference between VLCAD−/− and LCAD−/− in terms of insulin signaling highlights the fact that we know virtually nothing about how the accumulation of fatty acids of differing chain lengths might induce secondary effects in patients with these disorders. On a similar note, it was recently reported that VLCAD−/− mice have elevated serum cytokines, particularly TNFα, IL-4, and IL-5, while LCAD−/− mice do not show these changes.[66] Again, this may be due to the properties of the fatty acids accumulating in VLCAD−/− mice, particularly the C_{18} unsaturated species which may potentially be elongated to arachidonic acid and alter many different signaling and inflammatory pathways.

D. MCAD

MCAD is an ACAD enzyme with C_8 as its optimum substrate. MCAD deficiency is one of the most common inborn errors of metabolism, with an estimated prevalence of 1:17,000 in the United States and as high as 1:10,000 in parts of northern Europe.[67,68] The disease is largely hepatic in origin with symptoms arising whenever the liver is called upon to oxidize fats for ketone generation. Children with MCAD deficiency are often asymptomatic until a viral infection or other common illness forces an extended period of fasting. Lethargy and hypoglycemia occur, and if not corrected, can lead to death.

The MCAD-deficient mouse was developed and characterized in 2005.[69] The mouse model has no detectable MCAD antigen but retains about 25% residual liver enzyme activity toward octanoyl-CoA, indicating that other ACAD family members, most likely LCAD, have low level activity with medium-chain substrates. When fasted and then exposed to cold, death rapidly ensues (within 3 h of cold exposure) indicating a cold sensitivity almost as severe as that seen in LCAD−/− mice. Overall, the phenotypic severity among the four ACAD-deficient strains is thus LCAD > MCAD > VLCAD > SCAD.

MCAD−/− mice resemble human MCAD patients in many ways. The urine organic aciduria is similar with increased excretion of adipic, suberic, and sebacic acids. However, the serum acyl-carnitine profile showed some differences from that of human patients. Human patients have a large increase in C_8 acyl-carnitines while MCAD−/− mice have a predominant increase in $C_{10:1}$.[69] The mouse model also showed more cardiac involvement than seen in human patients. Some patients experience arrhythmias and mild cardiac dysfunction but not the sporadic cardiomyopathy, myocyte degeneration, and necrosis seen in the MCAD−/− mouse colony.

While the LCAD−/− model displays embryonic loss and thus smaller litter sizes, the MCAD−/− model has normal litter sizes but neonatal "failure to thrive" with about 60% of mutant pups dying before weaning age.[69] When fasted, MCAD−/− mice develop mild hypoglycemia and elevated serum free fatty acids but these changes do not reach statistical significance. Herrema *et al.*[70] performed a detailed study of gluconeogenesis and carbohydrate management in MCAD−/− mice under conditions of fasting and injections of lipopolysaccharide to induce the acute phase response. MCAD−/− mice did not become significantly more hypoglycemic than wild-type mice despite a 20% lower rate of glucose-6-phosphate synthesis during the acute phase response. However, the MCAD−/− mice showed a tendency to direct newly synthesized glucose into glycogen rather than releasing it. Some of these abnormalities in glucose handling by the liver may help explain the pronounced hypoglycemia seen in MCAD patients.

E. Carnitine Transporter (OCTN2)

Carnitine is a hydrophilic molecule that plays an essential role in the transport of fatty acids into mitochondria as well as into other organelles (microsomes and peroxisomes).[71] Intramitochondrial carnitine helps keep an appropriate balance between intramitochondrial acyl-CoA and free CoA which is needed to drive metabolism. In pathological situations carnitine aids in the elimination of accumulated fatty acids or organic acids and carnitine conjugates in the blood are an important diagnostic tool that forms the basis of newborn screening programs for fatty acid oxidation disorders. Carnitine is ingested in meat and dairy products and hepatic synthesis provides the remaining carnitine needed to facilitate metabolism.

Heart, muscle, liver, and kidney use a lot of carnitine and possess a high-affinity transporter known as OCTN2 to import it across the plasma membrane. Recessive genetic deficiency of OCTN2 occurs at a rate of approximately 1:40,000 births.[71] Children with these disorders exhibit severe carnitine wasting with carnitine being lost to the urine instead of being transported into the necessary organs. Overall the phenotype resembles a severe fatty acid oxidation disorder and is triggered by metabolic stressors such as fasting or viral illness. Metabolic decompensation quickly occurs and includes hypoketotic hypoglycemia, hyperammonemia, and lethargy which can progress to coma and death if the patient is not treated with intravenous glucose. Like long-chain fatty acid oxidation disorders, cardiomyopathy often occurs.

A model of OCTN2 deficiency arose through a spontaneous mutation in a mouse colony in Japan.[72] Since the gene responsible was unknown, the mouse model was referred to by the name "jvs" (juvenile visceral steatosis) referring to the major aspect of the phenotype which is profound hepatomegaly and fatty liver in young mice. Later a mutation was found in the OCTN2 gene.[73] The jvs mouse model has been extremely useful for understanding the role of carnitine homeostasis for health and disease with some 65 published papers using the model over the past 20 years.

F. CPT-1a, CPT-1b, and CPT-1c

Figures 1 and 4 simplistically depict CPT-1 as a single enzyme. In reality, there are three distinct CPT-1 genes producing proteins with about 70% similarity and these have been given the names CPT-1a, CPT-1b, and CPT-1c.[74] Based on differential patterns of expression CPT-1a is known as the liver isoform, -1b the muscle/heart isoform, and -1c the brain isoform. About 30 cases of human CPT-1a deficiency have been reported. CPT-1a patients present with a severe hypoketotic hypoglycemia phenotype of early onset.[75] There is little or no heart involvement because CPT-1b is the predominant isoform in heart and skeletal muscle. Deficiency of CPT-1b may be lethal. There is confusion in the literature regarding reports from the 1970s of patients with reduced muscle CPT activity.[76,77] At that time it was not known that CPT-1 and CPT-2 are separate proteins and those patients were most likely deficient in CPT-2.[75] No CPT-1c patients have yet been reported.

Because of the rate-limiting role of CPT-1 in mitochondrial fatty acid metabolism there has been much interest in studying the enzyme and developing means to pharmacologically control it for the treatment of heart disease, obesity, and the metabolic syndrome. Knockout mouse models have been developed for all three genes. Deficiency of either CPT-1a or -1b is lethal in the mouse.[78,79] Heterozygotes have been phenotyped and do display some features of fatty acid oxidation disorders. Interestingly, in CPT-1a+/− mice, males show about 50% reduction in CPT-1 activity in the liver but females have normal activity. Females apparently have a means of compensating for the loss of one CPT-1a allele. Another finding of note was that cold intolerance manifests in CPT-1b heterozygotes but not CPT-1a heterozygotes. Thus, perturbing hepatic metabolism alone is not sufficient to produce a cold-sensitive phenotype.

CPT-1c "brain isoform" null mice are viable and have an interesting phenotype.[74,80] CPT-1c has been shown to bind malonyl-CoA tightly like the other isoforms but has no detectable palmitoyltransferase activity. It is not yet known what this enzyme does. However, lack of CPT-1c influences whole-body metabolism. The mice are more prone to becoming obese on a high-fat diet and display reduced whole-body fatty acid oxidation. But paradoxically, when maintained on normal low-fat diet, they are hypophagic and have low body weights. This model has promise for understanding the elusive mechanism by which fat metabolism contributes to the regulation of feeding in the mouse.

G. Trifunctional Protein

MTP is a complex enzyme that is poorly understood. The active enzyme consists of four α subunits (HADHA gene) and four β units (HADHB gene) and resides on the inner mitochondrial membrane. The enzyme is capable of carrying out reactions 2–4 of the β-oxidation cycle depicted in Fig. 2 and is specific for long-chain substrates. For medium and short-chain substrates separate enzymes exist (Fig. 4). The α subunit of MTP has long-chain 3-enoyl-CoA hydratase activity and long-chain 3-hydroxyacyl-CoA dehydrogenase activity (LCHAD). The β subunit has long-chain 3-ketoacyl-CoA thiolase activity. Diseases associated with mutations in these genes can be categorized into two major groups, those with mutations that primarily affect LCHAD activity (presenting with Reye-like syndrome) and those with mutations that affect all three enzymatic activities (presenting with cardiomyopathy or neuromyopathy).[81,82] Sudden infant death is common for both groups of patients. The prevalence of disease-causing mutations in MTP genes is about 1:38,000 pregnancies.

A mouse model of complete MTP deficiency was developed by Strauss and colleagues by targeting the α subunit.[83] The resulting mouse had no detectable MTP complex. The phenotype was severe and consisted of intrauterine growth retardation, hypoglycemia, and neonatal death with all pups dying with the first 36 h after birth. Later, more detailed phenotyping was conducted on heterozygous mice.[84] Older mice showed accumulation of fat in the liver which, like the LCAD model, was associated with insulin resistance.

H. SCHAD

There is no MTP equivalent for conducting steps 2–4 of the β-oxidation cycle for medium and short-chain substrates. Rather, separate enzymes conduct each step. The enzyme best known as SCHAD is a 3-hydroxyacyl-CoA dehydrogenase with activity toward both short- and medium-chain hydroxyacyl-CoAs. Due to rampant confusion in the literature and multiple names, the enzyme and gene were recently redesignated HAD (gene name HADH, hydroxyacyl-CoA dehydrogenase).[85] Making matters more complicated, there are two isoforms of HAD.[85] The clinical picture of HAD deficiency consists of elevated medium-chain dicarboxylic and 3-hydroxydicarboxylic metabolites in the urine. Some attention has been focused on a small number of HAD patients exhibiting hyperinsulinemia.[86,87] It has been shown that HAD is expressed highly in the islets of langerhans and that it interacts with the mitochondrial enzyme glutamate dehydrogenase which has also been linked to a syndrome of hyperinsulinemia and hyperammonemia.[85] A knockout mouse has been created for HAD but no detailed phenotype has been published, thus it is not known whether the mouse has alterations in insulin secretion or pancreatic islet function.[88]

I. DECR/ECI

2,4-Dienoyl-CoA reductase (DECR) and ECI are two auxiliary fatty acid oxidation enzymes required for the full degradation of polyunsaturated fatty acyl-CoAs. Despite the popular notion that polyunsaturated fatty acids are important for human health very little is known about the mitochondrial degradation of these species, and the DECR/ECI enzymes have not been well characterized. There has been one patient with DECR deficiency and no reported ECI deficiencies. In 1990 Roe *et al.*[89] encountered a hypotonic infant with hyperlysinemia, hypocarnitinemia, and an unusual acyl-carnitine species in both urine and blood which they identified as 2-*trans*,4-*cis*-decadienoylcarnitine. This acyl-carnitine species accumulated from incomplete oxidation of linoleic acid. Efforts to treat the patient failed and she died at age 4 months. Subsequent postmortem analyses showed 60–80% reduction in DECR activity.

The recent development of knockout mouse models for ECI and DECR provides important insight into the oxidation of polyunsaturated fatty acids.[90,91] The ECI knockout has many similarities to the ACAD knockout strains. Fasting induces fatty liver, dicarboxylic aciduria, and activation of PPARα. The fatty acids that accumulate in liver are primarily $C_{16:0}$, $C_{18:1}$, and $C_{18:2}$, which is a similar pattern to that seen in VLCAD−/− mice. The DECR mutant mice also specifically accumulate $C_{18:1}$ and $C_{18:2}$ in liver. Serum acyl-carnitines are more generally elevated, including chain lengths C_{14} to C_{18}, both saturated and unsaturated. Like the ACAD-deficient strains, DECR mice become hypoglycemic and are cold intolerant. This indicates that polyunsaturated fatty acids contribute more than previously thought to energy production under conditions of metabolic stress.

J. Regulators of Fatty Acid Oxidation

Just as mutations in OCTN2 indirectly cause a fatty acid oxidation disorder phenotype by limiting the availability of a key cofactor (carnitine), mutations in regulatory proteins can induce a fatty acid oxidation disorder by limiting the abundance or activity of the β-oxidation machinery. To date, two such regulators have emerged: PPARα and sirtuin-3 (SIRT3).

The VLCAD, ECI, and DECR mutant mice all show signs of activated PPARα when fasted or stressed.[62,90,91] The important role of PPARα in regulating fatty acid metabolism has been elucidated over the past 15 years since the publication of the original phenotypic characterization of the PPARα knockout mouse.[92] This model is one of the most widely used knockout strains and has been described in several hundred publications. While PPARα does many things, one of the primary features of the mouse model is that of a fasting-induced fatty acid oxidation disorder.[64] PPARα transcriptionally regulates many genes in the mitochondrial, peroxisomal, and microsomal pathways of fatty acid oxidation.[63,93] Mice lacking PPARα cannot upregulate transcription of these genes under conditions that stimulate fatty acid oxidation, particularly fasting and fibrate therapy.

SIRT3 is one of seven known mammalian sirtuin enzymes. Sirtuins are NAD-dependent lysine deacetylases. Three of them, SIRT3, SIRT4, and SIRT5, are known to localize to mitochondria.[94] Lysine acetylation is a reversible posttranslational modification akin to phosphorylation. Several recent proteomics studies have demonstrated that lysine acetylation is particularly prevalent among the enzymes of metabolic pathways both in mitochondria and in the cytosol.[95] Virtually every fatty acid oxidation enzyme is acetylated to some degree. Recently, it was shown that both mouse and human LCAD are subject to acetylation on multiple lysines and that acetylation of the lysine at residue K42 of mouse LCAD can regulate enzymatic activity.[96] SIRT3 has been knocked out in the mouse and the model displays many signs of a fatty acid oxidation disorder including fasting-induced hypoketotic hypoglycemia, fatty liver, and cold intolerance.[96] It remains possible that mutations in SIRT3 could cause a fatty acid oxidation disorder-like disease in humans.

IV. Nongenetic Models of Fatty Acid Oxidation Disorders

Before the era of stem cell and gene targeting technology it was recognized that diet or drug manipulations could be used to disrupt metabolism in mice or other animal models. In the1970s it was demonstrated that maintaining rodents on a riboflavin (vitamin B2)-deficient diet severely compromised liver oxidative metabolism.[97] Numerous papers were published during the late 1970s and into the 1980s characterizing the effects of riboflavin deficiency on fatty acid oxidation and enzyme function.[98–101] Riboflavin-deficient diets greatly reduce the function of enzymes that rely upon FAD prosthetic groups to function, which includes all of the acyl-CoA dehydrogenases and ETF.[102] Riboflavin restriction may be considered a useful model of multiple acyl-CoA dehydrogenase deficiency (MADD) which in humans is caused by mutations in either subunit of ETF or in ETF dehydrogenase.[103]

While riboflavin-deficient diets have fallen out of favor as a method to induce fatty acid oxidation defects in rodent models, other drugs that target the carnitine cycle to inhibit entry of fatty acids into mitochondria are still in use. These include 2-tetradecylglycidic acid (TDGA) and etomoxir, which are both potent irreversible inhibitors of CPT-I.[104–108] These chemicals cause hallmark features of fatty acid oxidation disorders in mice including fatty liver and cardiac hypertrophy. There are also carnitine analogs such as L-aminocarnitine and acetyl-DL-aminocarnitine that competitively inhibit carnitine acetyltransferase and result in inhibition of fatty acid oxidation.[109] Several inhibitors of the ACADs have also been developed and a few of these used in animal models.[110,111] Methylenecyclopropaneacetic acid (MCPA) appears to inhibit SCAD, MCAD, and some or all of the ACADs involved in branched-chain amino acid degradation. Spiropentaneacetic acid (SPA) specifically inhibits MCAD and does not inhibit ACADs in the branched-chain amino acid degradation pathways. Mercaptoacetate has also been used *in vivo* and results in attenuated activity of several ACAD enzymes.[112–115]

A final chemical-induced model of interest is that of hopantenate. Hopantenate is an analog of panthothenate that potently inhibits the enzyme panthothenate kinase which catalyzes the rate-limiting step in CoA biosynthesis. Treating mice with hopantenate results in a liver/kidney-specific metabolic disorder caused by a 75% reduction in free CoA levels.[116] Symptoms include swollen mitochondria, large elevations in liver acyl-carnitines of all chain lengths, and severe hypoglycemia. Hopantenate was introduced clinically in 1978 in Japan to treat dementia. Later, several deaths from Reye-like encephalopathy and metabolic decompensation were reported.[117,118]

Chemical inhibition of fatty acid oxidation has been a useful method for modeling disease in the mouse over the past 35+ years. These methods may appeal to investigators who wish to reduce fatty acid oxidation in another genetic model they are studying, who do not wish to get involved in obtaining/breeding/maintaining genetic models of FAO, or who wish to inhibit FAO transiently in animals rather than chronically.

V. Legacy of Mouse Models and Future Directions

The models of fatty acid oxidation disorders described above have been very useful for understanding the corresponding human genetic diseases. They have allowed for objective evaluation of therapies such as carnitine supplementation or MCT diet in a disease state with age, sex, background genetics, and compliance controlled for.[57,60] They have been used in several studies to demonstrate the efficacy of gene therapy approaches for treating these diseases.[119–123] Additionally, these models have facilitated scientific discovery in many ways that could not have been anticipated at the time of their development. For example, the serendipitous discovery by Kozak and colleagues that BALB/cByJ mice are cold sensitive provided an easily followed metabolic phenotype that greatly facilitated the study of other phenomena such as synergistic heterozygosity. Synergistic heterozygosity was a concept promoted by Vockley *et al.*[124] 10 years ago as a possible explanation for patients that present with the symptoms of a fatty acid oxidation disorder but are not homozygous for disease-causing alleles in any known fatty acid oxidation gene. Carrying a single mutation in one gene and a single mutation in another gene of the same metabolic pathway could result in synergistic effects that ultimately produce disease. Mouse models of ACAD deficiency provided the ability to test the concept of synergistic heterozygosity. Mice bearing combinations of single mutations in SCAD, LCAD, and VLCAD were created and subjected to metabolic stress with cold sensitivity as the marker symptom. Indeed, combined heterozygosity was proven to be sufficient to produce a metabolic phenotype.[25,125]

The models have already been used to study the effect of disturbed fatty acid oxidation on insulin signaling and obesity.[48,62] The future will likely see an increase in investigations into the role of these enzymes in the etiology of complex, polygenic diseases such as obesity, diabetes, or cancer. Fatty acid oxidation enzymes have turned up in surprising places and in lists of disease-associated genes generated by microarray studies or proteomics. There are clear but unexplained links between short-chain fatty acid metabolism and cancer.[126–128] HAD (SCHAD) is known to be important in regulating insulin secretion.[86] HAD, SCAD, MCAD, and LCAD have all been shown to physically associate with glucose transporter-4 (GLUT4) or with components of vesicles that traffic GLUT4 to the plasma membrane.[129,130] Dogma would lead to the dismissal of these findings as artifacts since the fatty acid oxidation enzymes are mitochondrial and GLUT4 vesicles are cytosolic; however, the phenomena may yet find an explanation and the knockout mouse models may make important contributions. The list of potential uses for these mouse models outside the realm of inborn errors of metabolism is long. LCAD has been associated with type I diabetes, and has been found to be upregulated in white fat by PPARγ agonists.[131,132] SCAD mice show a distinct food preference for carbohydrates rather than fat, and have disturbed EEG patterns during sleep.[133,134]

These examples of novel applications for fatty acid oxidation mouse models are provided not only to highlight future uses of the existing models but also to advocate for the development of new models. Creating a knockout strain through gene targeting is a time-consuming and expensive endeavor. There is the risk of no phenotype on one end of the spectrum and lethality on the other end, leading to either no resulting publications or a single low-impact publication. In a pathway such as fatty acid oxidation, where multiple models already exist, there is a perception of diminishing returns and new models may not be considered to be a valuable contribution to the field. However, new mouse models could provide significant insight into the function of "orphan" enzymes such as ACAD9, ACAD10, and ACAD11.[135] ACAD9 deficiency has been reported in two patients and may have an important role in the brain.[136,137] ACAD10 was found to be linked to type 2 diabetes in Pima Indians.[138] Nothing is known about ACAD11. Mouse models have already contributed greatly to our understanding of mitochondrial fatty acid metabolism and hopefully will continue to do so for this new generation of enzymes and their associated diseases.

References

1. Rao MS, Reddy JK. Peroxisomal beta-oxidation and steatohepatitis. *Semin Liver Dis* 2001;**21**:43–55.
2. Kovacs WJ, Tape KN, Shackelford JE, Duan X, Kasumov T, Kelleher JK, et al. Localization of the pre-squalene segment of the isoprenoid biosynthetic pathway in mammalian peroxisomes. *Histochem Cell Biol* 2007;**127**:273–90.
3. Ghisla S, Thorpe C. Acyl-CoA dehydrogenases. A mechanistic overview. *Eur J Biochem* 2004;**271**:494–508.
4. Kunau WH, Dommes V, Schulz H. Beta-oxidation of fatty acids in mitochondria, peroxisomes, and bacteria: a century of continued progress. *Prog Lipid Res* 1995;**34**:267–342.
5. Kerner J, Hoppel C. Fatty acid import into mitochondria. *Biochim Biophys Acta* 2000;**1486**:1–17.

6. Eaton S, Bartlett K, Pourfarzam M. Mammalian mitochondrial beta-oxidation. *Biochem J* 1996;**320**(Pt 2):345–57.
7. Eaton S. Control of mitochondrial beta-oxidation flux. *Prog Lipid Res* 2002;**41**:197–239.
8. Chohan KK, Jones M, Grossmann JG, Frerman FE, Scrutton NS, Sutcliffe MJ. Protein dynamics enhance electronic coupling in electron transfer complexes. *J Biol Chem* 2001;**276**:34142–7.
9. Leys D, Basran J, Talfournier F, Sutcliffe MJ, Scrutton NS. Extensive conformational sampling in a ternary electron transfer complex. *Nat Struct Biol* 2003;**10**:219–25.
10. Toogood HS, van Thiel A, Basran J, Sutcliffe MJ, Scrutton NS, Leys D. Extensive domain motion and electron transfer in the human electron transferring flavoprotein medium chain Acyl-CoA dehydrogenase complex. *J Biol Chem* 2004;**279**:32904–12.
11. Vockley J. The changing face of disorders of fatty acid oxidation. *Mayo Clin Proc* 1994;**69**:249–57.
12. Vockley J, Mohsen AW, Binzak B, Willard J, Fauq A. Mammalian branched-chain acyl-CoA dehydrogenases: molecular cloning and characterization of recombinant enzymes. *Methods Enzymol* 2000;**324**:241–58.
13. Sauer SW, Okun JG, Schwab MA, Crnic LR, Hoffmann GF, Goodman SI, et al. Bioenergetics in glutaryl-coenzyme A dehydrogenase deficiency: a role for glutaryl-coenzyme A. *J Biol Chem* 2005;**280**:21830–6.
14. Liang X, Le W, Zhang D, Schulz H. Impact of the intramitochondrial enzyme organization on fatty acid oxidation. *Biochem Soc Trans* 2001;**29**:279–82.
15. Gregersen N, Olsen RK. Disease mechanisms and protein structures in fatty acid oxidation defects. *J Inherit Metab Dis* 2010;**33**:547–53.
16. Nada MA, Rhead WJ, Sprecher H, Schulz H, Roe CR. Evidence for intermediate channeling in mitochondrial beta-oxidation. *J Biol Chem* 1995;**270**:530–5.
17. Kerner J, Bieber L. Isolation of a malonyl-CoA-sensitive CPT/beta-oxidation enzyme complex from heart mitochondria. *Biochemistry* 1990;**29**:4326–34.
18. Sumegi B, Srere PA. Binding of the enzymes of fatty acid beta-oxidation and some related enzymes to pig heart inner mitochondrial membrane. *J Biol Chem* 1984;**259**:8748–52.
19. Sumegi B, Porpaczy Z, Alkonyi I. Kinetic advantage of the interaction between the fatty acid beta-oxidation enzymes and the complexes of the respiratory chain. *Biochim Biophys Acta* 1991;**1081**:121–8.
20. Parker A, Engel P. Preliminary evidence for the existence of specific functional assemblies between enzymes of the beta-oxidation pathway and the respiratory chain. *Biochem J* 2000;**345**:429–35.
21. Sumegi B, Srere PA. Complex I binds several mitochondrial NAD-coupled dehydrogenases. *J Biol Chem* 1984;**259**:15040–5.
22. Frerman FE. Reaction of electron-transfer flavoprotein ubiquinone oxidoreductase with the mitochondrial respiratory chain. *Biochim Biophys Acta* 1987;**893**:161–9.
23. Schafer E, Seelert H, Reifschneider NH, Krause F, Dencher NA, Vonck J. Architecture of active mammalian respiratory chain supercomplexes. *J Biol Chem* 2006;**281**:15370–5.
24. Wang Y, Mohsen AW, Mihalik SJ, Goetzman ES, Vockley J. Evidence for the physical association of mitochondrial fatty acid oxidation and oxidative phosphorylation complexes. *J Biol Chem* 2010;**285**:29834–41.
25. Spiekerkoetter U, Wood PA. Mitochondrial fatty acid oxidation disorders: pathophysiological studies in mouse models. *J Inherit Metab Dis* 2010;**33**:539–46.
26. Jethva R, Bennett MJ, Vockley J. Short-chain acyl-coenzyme A dehydrogenase deficiency. *Mol Genet Metab* 2008;**95**:195–200.
27. Wood P, Amendt B, Rhead W, Millington D, Inoue F, Armstrong D. Short chain acyl-CoA dehydrogenase deficiency in mice. *Pediatr Res* 1989;**25**:38–43.

28. Schiffer SP, Prochazka M, Jezyk PF, Roderick TH, Yudkoff M, Patterson DF. Organic aciduria and butyryl CoA dehydrogenase deficiency in BALB/cByJ mice. *Biochem Genet* 1989;**27**:47–58.

29. Armstrong DL, Masiowski ML, Wood PA. Pathologic characterization of short-chain acyl-CoA dehydrogenase deficiency in BALB/cByJ mice. *Am J Med Genet* 1993;**47**:884–92.

30. Reue K, Cohen RD. Acads gene deletion in BALB/cByJ mouse strain occurred after 1981 and is not present in BALB/cByJ-fld mutant mice. *Mamm Genome* 1996;**7**:694–5.

31. Brownstein DG, Johnson EA, Smith AL. Spontaneous Reye's-like syndrome in BALB/cByJ mice. *Lab Invest* 1984;**51**:386–95.

32. Guerra C, Koza RA, Walsh K, Kurtz DM, Wood PA, Kozak LP. Abnormal nonshivering thermogenesis in mice with inherited defects of fatty acid oxidation. *J Clin Investig* 1998;**102**:1724–31.

33. Kurtz DM, Rinaldo P, Rhead WJ, Tian L, Millington DS, Vockley J, et al. Targeted disruption of mouse long-chain acyl-CoA dehydrogenase gene reveals crucial roles for fatty acid oxidation. *Proc Natl Acad Sci USA* 1998;**95**:15592–7.

34. Crane FL, Mii S, Hauge JG, Green DE, Beinert H. On the mechanism of dehydrogenation of fatty acyl derivatives of coenzyme A. I. The general fatty acyl coenzyme A dehydrogenase. *J Biol Chem* 1956;**218**:701–6.

35. Hale DE, Batshaw ML, Coates PM, Frerman FE, Goodman SI, Singh I, et al. Long-chain acyl coenzyme A dehydrogenase deficiency: an inherited cause of nonketotic hypoglycemia. *Pediatr Res* 1985;**19**:666–71.

36. Treem WR, Stanley CA, Hale DE, Leopold HB, Hyams JS. Hypoglycemia, hypotonia, and cardiomyopathy: the evolving clinical picture of long chain acyl-CoA dehydrogenase deficiency. *Pediatric* 1991;**87**:328–33.

37. Izai K, Uchida Y, Orii T, Yamamoto S, Hashimoto T. Novel fatty acid b-oxidation enzymes in rat liver mitochondria. 1. Purification and properties of very-long-chain acyl-coenzyme A dehydrogenase. *J Biol Chem* 1992;**267**:1027–33.

38. Aoyama T, Uchida Y, Kelley RI, Marble M, Hofman K, Tonsgard JH, et al. A novel disease with deficiency of mitochondrial very-long-chain acyl-CoA dehydrogenase. *Biochem Biophys Res Commun* 1993;**191**:1369–72.

39. Yamaguchi S, Indo Y, Coates PM, Hashimoto T, Tanaka K. Identification of very-long-chain acyl-CoA dehydrogenase deficiency in 3 patients previously diagnosed with long-chain acyl-CoA dehydrogenase deficiency. *Pediatr Res* 1993;**34**:111–3.

40. Chegary M, Brinke H, Ruiter JP, Wijburg FA, Stoll MS, Minkler PE, et al. Mitochondrial long chain fatty acid beta-oxidation in man and mouse. *Biochim Biophys Acta* 2009;**1791**:806–15.

41. Maher AC, Mohsen AW, Vockley J, Tarnopolsky MA. Low expression of long-chain acyl-CoA dehydrogenase in human skeletal muscle. *Mol Genet Metab* 2010;**100**:163–7.

42. Lea WP, Abbas AS, Sprecher H, Vockley J, Schulz H. Long-chain acyl-CoA dehydrogenase is a key enzyme in the mitochondrial beta-oxidation of unsaturated fatty acids. *Biochim Biophys Acta: Mol Cell Biol Lipids* 2000;**31**:2–3.

43. Cox KB, Hamm DA, Millington DS, Matern D, Vockley J, Rinaldo P, et al. Gestational, pathologic and biochemical differences between very long-chain acyl-CoA dehydrogenase deficiency and long-chain acyl-CoA dehydrogenase deficiency in the mouse. *Hum Mol Genet* 2001;**10**:2069–77.

44. Goetzman ES, Tian L, Wood PA. Differential induction of genes in liver and brown adipose tissue regulated by peroxisome proliferator-activated receptor-alpha during fasting and cold exposure in acyl-CoA dehydrogenase-deficient mice. *Mol Genet Metab* 2005;**84**:39–47.

45. Nelson RJ. Leptin: the "skinny" on torpor. *Am J Physiol Regul Integr Comp Physiol* 2004;**287**: R6–7.

46. Swoap SJ, Weinshenker D. Norepinephrine controls both torpor initiation and emergence via distinct mechanisms in the mouse. *PLoS ONE* 2008;**3**:e4038.
47. Schuler AM, Gower BA, Matern D, Rinaldo P, Wood PA. Influence of dietary fatty acid chain-length on metabolic tolerance in mouse models of inherited defects in mitochondrial fatty acid beta-oxidation. *Mol Genet Metab* 2004;**83**:322–9.
48. Zhang D, Liu ZX, Choi CS, Tian L, Kibbey R, Dong J, et al. Mitochondrial dysfunction due to long-chain Acyl-CoA dehydrogenase deficiency causes hepatic steatosis and hepatic insulin resistance. *Proc Natl Acad Sci USA* 2007;**104**:17075–80.
49. Spiekerkoetter U. Mitochondrial fatty acid oxidation disorders: clinical presentation of long-chain fatty acid oxidation defects before and after newborn screening. *J Inherit Metab Dis* 2010.
50. Andresen BS, Olpin S, Poorthuis B, Scholte HR, Vianey-Saban C, Wanders R, et al. Clear correlation of genotype with disease phenotype in very-long-chain acyl-CoA dehydrogenase deficiency. *Am J Hum Genet* 1999;**64**:479–94.
51. Exil VJ, Roberts RL, Sims H, McLaughlin JE, Malkin RA, Gardner CD, et al. Very-long-chain acyl-coenzyme A dehydrogenase deficiency in mice. *Circ Res* 2003;**93**:448–55.
52. Spiekerkoetter U, Tokunaga C, Wendel U, Mayatepek E, Exil V, Duran M, et al. Changes in blood carnitine and acylcarnitine profiles of very long-chain acyl-CoA dehydrogenase-deficient mice subjected to stress. *Eur J Clin Investig* 2004;**34**:191–6.
53. Gillingham MB, Connor WE, Matern D, Rinaldo P, Burlingame T, Meeuws K, et al. Optimal dietary therapy of long-chain 3-hydroxyacyl-CoA dehydrogenase deficiency. *Mol Genet Metab* 2003;**79**:114–23.
54. Spiekerkoetter U, Lindner M, Santer R, Grotzke M, Baumgartner MR, Boehles H, et al. Treatment recommendations in long-chain fatty acid oxidation defects: consensus from a workshop. *J Inherit Metab Dis* 2009;**32**:498–505.
55. Spiekerkoetter U, Lindner M, Santer R, Grotzke M, Baumgartner MR, Boehles H, et al. Management and outcome in 75 individuals with long-chain fatty acid oxidation defects: results from a workshop. *J Inherit Metab Dis* 2009;**32**:488–97.
56. Spiekerkoetter U, Tokunaga C, Wendel U, Mayatepek E, Ijlst L, Vaz FM, et al. Tissue carnitine homeostasis in very-long-chain acyl-CoA dehydrogenase-deficient mice. *Pediatr Res* 2005;**57**:760–4.
57. Liebig M, Gyenes M, Brauers G, Ruiter JP, Wendel U, Mayatepek E, et al. Carnitine supplementation induces long-chain acylcarnitine production–studies in the VLCAD-deficient mouse. *J Inherit Metab Dis* 2006;**29**:343–4.
58. Primassin S, Ter Veld F, Mayatepek E, Spiekerkoetter U. Carnitine supplementation induces acylcarnitine production in tissues of very long-chain acyl-CoA dehydrogenase-deficient mice, without replenishing low free carnitine. *Pediatr Res* 2008;**63**:632–7.
59. van Vlies N, Tian L, Overmars H, Bootsma AH, Kulik W, Wanders RJ, et al. Characterization of carnitine and fatty acid metabolism in the long-chain acyl-CoA dehydrogenase-deficient mouse. *Biochem J* 2005;**387**:185–93.
60. Tucci S, Primassin S, Ter Veld F, Spiekerkoetter U. Medium-chain triglycerides impair lipid metabolism and induce hepatic steatosis in very long-chain acyl-CoA dehydrogenase (VLCAD)-deficient mice. *Mol Genet Metab* 2010;**101**:40–7.
61. Primassin S, Tucci S, Herebian D, Seibt A, Hoffmann L, ter Veld F, et al. Pre-exercise medium-chain triglyceride application prevents acylcarnitine accumulation in skeletal muscle from very-long-chain acyl-CoA-dehydrogenase-deficient mice. *J Inherit Metab Dis* 2010;**33**:237–46.
62. Zhang D, Christianson J, Liu ZX, Tian L, Choi CS, Neschen S, et al. Resistance to high-fat diet-induced obesity and insulin resistance in mice with very long-chain acyl-CoA dehydrogenase deficiency. *Cell Metab* 2010;**11**:402–11.

63. Gulick T, Cresci S, Caira T, Moore DD, Kelly DP. The peroxisome proliferator-activated receptor regulates mitochondrial fatty acid oxidative enzyme gene expression. *Proc Natl Acad Sci USA* 1994;**91**:11012–6.

64. Kersten S, Seydoux J, Peters JM, Gonzalez FJ, Desvergne B, Wahli W. Peroxisome proliferator-activated receptor alpha mediates the adaptive response to fasting. *J Clin Invest* 1999;**103**:1489–98.

65. Cherkaoui-Malki M, Meyer K, Cao WQ, Latruffe N, Yeldandi AV, Rao MS, et al. Identification of novel peroxisome proliferator-activated receptor alpha (PPARalpha) target genes in mouse liver using cDNA microarray analysis. *Gene Expr* 2001;**9**:291–304.

66. Mihalik SJ, Swigonova Z, Tian L, Wood PA, Vockley J. Mice null for very long chain acyl-CoA dehydrogenase (VLCAD) have upregulated immune functions, while mice null for long chain acyl-CoA dehydrogenase (LCAD) have upregulated expression of oxysterol associated genes including bile acid responsive LRH1. *Mol Genet Metabol* 2008;**93**:258.

67. Pourfarzam M, Morris A, Appleton M, Craft A, Bartlett K. Neonatal screening for medium-chain acyl-CoA dehydrogenase deficiency. *Lancet* 2001;**358**:1063–4.

68. Horvath GA, Davidson AG, Stockler-Ipsiroglu SG, Lillquist YP, Waters PJ, Olpin S, et al. Newborn screening for MCAD deficiency: experience of the first three years in British Columbia, Canada. *Can J Public Health* 2008;**99**:276–80.

69. Tolwani RJ, Hamm DA, Tian L, Sharer JD, Vockley J, Rinaldo P, et al. Medium-chain acyl-CoA dehydrogenase deficiency in gene-targeted mice. *PLoS Genet* 2005;**1**:e23.

70. Herrema H, Derks TG, van Dijk TH, Bloks VW, Gerding A, Havinga R, et al. Disturbed hepatic carbohydrate management during high metabolic demand in medium-chain acyl-CoA dehydrogenase (MCAD)-deficient mice. *Hepatology* 2008;**47**:1894–904.

71. Longo N, di San Ama, Filippo C, Pasquali M. Disorders of carnitine transport and the carnitine cycle. *Am J Med Genet C Semin Med Genet* 2006;**142C**:77–85.

72. Koizumi T, Nikaido H, Hayakawa J, Nonomura A, Yoneda T. Infantile disease with microvesicular fatty infiltration of viscera spontaneously occurring in the C3H-H-2(0) strain of mouse with similarities to Reye's syndrome. *Lab Anim* 1988;**22**:83–7.

73. Lu K, Nishimori H, Nakamura Y, Shima K, Kuwajima M. A missense mutation of mouse OCTN2, a sodium-dependent carnitine cotransporter, in the juvenile visceral steatosis mouse. *Biochem Biophys Res Commun* 1998;**252**:590–4.

74. Wolfgang MJ, Kurama T, Dai Y, Suwa A, Asaumi M, Matsumoto S, et al. The brain-specific carnitine palmitoyltransferase-1c regulates energy homeostasis. *Proc Natl Acad Sci USA* 2006;**103**:7282–7.

75. Bonnefont JP, Djouadi F, Prip-Buus C, Gobin S, Munnich A, Bastin J. Carnitine palmitoyltransferases 1 and 2: biochemical, molecular and medical aspects. *Mol Aspects Med* 2004;**25**:495–520.

76. DiMauro S, DiMauro PMM. Muscle carnitine palmitoyl transferase deficiency and myoglobinuria. *Science* 1973;**182**:929–31.

77. Hostetler KY, Hoppel CL, Romine JS, Sipe JC, Gross SR, Higginbottom PA. Partial deficiency of muscle carnitine palmitoyltransferase with normal ketone production. *N Engl J Med* 1978;**298**:553–7.

78. Nyman LR, Cox KB, Hoppel CL, Kerner J, Barnoski BL, Hamm DA, et al. Homozygous carnitine palmitoyltransferase 1a (liver isoform) deficiency is lethal in the mouse. *Mol Genet Metab* 2005;**86**:179–87.

79. Ji S, You Y, Kerner J, Hoppel CL, Schoeb TR, Chick WS, et al. Homozygous carnitine palmitoyltransferase 1b (muscle isoform) deficiency is lethal in the mouse. *Mol Genet Metab* 2008;**93**:314–22.

80. Wolfgang MJ, Cha SH, Millington DS, Cline G, Shulman GI, Suwa A, et al. Brain-specific carnitine palmitoyl-transferase-1c: role in CNS fatty acid metabolism, food intake, and body weight. *J Neurochem* 2008;**105**:1550–9.

81. den Boer MEJ, Dionisi-Vici C, Chakrapani A, van Thuijl AOJ, Wanders RJA, Wijburg FA. Mitochondrial trifunctional protein deficiency: a severe fatty acid oxidation disorder with cardiac and neurologic involvement. *J Pediatr* 2003;**142**:684–9.

82. Angdisen J, Moore VD, Cline JM, Payne RM, Ibdah JA. Mitochondrial trifunctional protein defects: molecular basis and novel therapeutic approaches. *Curr Drug Targets Immune Endocr Metab Disord* 2005;**5**:27–40.

83. Ibdah JA, Paul H, Zhao Y, Binford S, Salleng K, Cline M, et al. Lack of mitochondrial trifunctional protein in mice causes neonatal hypoglycemia and sudden death. *J Clin Invest* 2001;**107**:1403–9.

84. Ibdah JA, Perlegas P, Zhao Y, Angdisen J, Borgerink H, Shadoan MK, et al. Mice heterozygous for a defect in mitochondrial trifunctional protein develop hepatic steatosis and insulin resistance. *Gastroenterology* 2005;**128**:1381–90.

85. Filling C, Keller B, Hirschberg D, Marschall HU, Jornvall H, Bennett MJ, et al. Role of short-chain hydroxyacyl CoA dehydrogenases in SCHAD deficiency. *Biochem Biophys Res Commun* 2008;**368**:6–11.

86. Hardy OT, Hohmeier HE, Becker TC, Manduchi E, Doliba NM, Gupta RK, et al. Functional genomics of the beta-cell: short-chain 3-hydroxyacyl-coenzyme A dehydrogenase regulates insulin secretion independent of K+ currents. *Mol Endocrinol* 2007;**21**:765–73.

87. Kapoor RR, James C, Flanagan SE, Ellard S, Eaton S, Hussain K. 3-Hydroxyacyl-coenzyme A dehydrogenase deficiency and hyperinsulinemic hypoglycemia: characterization of a novel mutation and severe dietary protein sensitivity. *J Clin Endocrinol Metab* 2009;**94**:2221–5.

88. O'Brien L, Sims H, Bennett M, Strauss A. A mouse model for medium and short chain chain L-3-hydroxyacyl-CoA dehydrogenase deficiency. *J Inherit Metab Dis* 2000;**23**:127.

89. Roe CR, Millington DS, Kodo NN, Sprecher H, Mohammed BS, Nada M, et al. 2, 4-Dienoyl-coenzyme A reductase deficiency: a possible new disorder of fatty acid oxidation. *J Clin Invest* 1990;**85**:1703–7.

90. Janssen U, Stoffel W. Disruption of mitochondrial beta -oxidation of unsaturated fatty acids in the 3, 2-*trans*-enoyl-CoA isomerase-deficient mouse. *J Biol Chem* 2002;**277**:19579–84.

91. Miinalainen IJ, Schmitz W, Huotari A, Autio KJ, Soininen R, Loren Ve, et al. Mitochondrial 2, 4-dienoyl-CoA reductase deficiency in mice results in severe hypoglycemia with stress intolerance and unimpaired ketogenesis. *PLoS Genet* 2009;**5**:e1000543.

92. Lee SS, Gonzalez FJ. Targeted disruption of the peroxisome proliferator-activated receptor alpha gene, PPAR alpha. *Ann N Y Acad Sci* 1996;**804**:524–9.

93. Pyper SR, Viswakarma N, Yu S, Reddy JK. PPARalpha: energy combustion, hypolipidemia, inflammation and cancer. *Nucl Recept Signal* 2010;**8**:e002.

94. Schwer B, Verdin E. Conserved metabolic regulatory functions of sirtuins. *Cell Metab* 2008;**7**:104–12.

95. Zhao S, Xu W, Jiang W, Yu W, Lin Y, Zhang T, et al. Regulation of cellular metabolism by protein lysine acetylation. *Science* 2010;**327**:1000–4.

96. Hirschey MD, Shimazu T, Goetzman E, Jing E, Schwer B, Lombard DB, et al. SIRT3 regulates mitochondrial fatty-acid oxidation by reversible enzyme deacetylation. *Nature* 2010;**464**:121–5.

97. Hoppel CL, Tandler B. Riboflavin and mouse hepatic cell structure and function. Mitochondrial oxidative metabolism in severe deficiency states. *J Nutr* 1975;**105**:562–70.

98. Hoppel CL, Tandler B. Relationship between hepatic mitochondrial oxidative metabolism and morphology during riboflavin deficiency and recovery in mice. *J Nutr* 1976;**106**:73–6.

99. Hoppel C, DiMauco J, Tandler B. Riboflavin and rat hepatic cell structure and function. *J Biol Chem* 1979;**254**:4164–70.
100. Hoppel C, DiMarco JP, Tandler B. Riboflavin and rat hepatic cell structure and function. Mitochondrial oxidative metabolism in deficiency states. *J Biol Chem* 1979;**254**:4164–70.
101. Tandler B, Hoppel CL. Ultrastructural effects of riboflavin deficiency on rat hepatic mitochondria. *Anat Rec* 1980;**196**:183–90.
102. Ross NS, Hoppel CL. Acyl-CoA dehydrogenase activity in the riboflavin-deficient rat. Effects of starvation. *Biochem J* 1987;**244**:387–91.
103. Olsen RK, Andresen BS, Christensen E, Bross P, Skovby F, Gregersen N. Clear relationship between ETF/ETFDH genotype and phenotype in patients with multiple acyl-CoA dehydrogenation deficiency. *Hum Mutat* 2003;**22**:12–23.
104. Bressler R, Gay R, Copeland JG, Bahl JJ, Bedotto J, Goldman S. Chronic inhibition of fatty acid oxidation: new model of diastolic dysfunction. *Life Sci* 1989;**44**:1897–906.
105. van der Leij FR, Bloks VW, Grefhorst A, Hoekstra J, Gerding A, Kooi K, et al. Gene expression profiling in livers of mice after acute inhibition of beta-oxidation. *Genomics* 2007;**90**:680–9.
106. Yotsumoto T, Naitoh T, Kitahara M, Tsuruzoe N. Effects of carnitine palmitoyltransferase I inhibitors on hepatic hypertrophy. *Eur J Pharmacol* 2000;**398**:297–302.
107. Koteish A, Diehl AM. Animal models of steatosis. *Semin Liver Dis* 2001;**21**:89–104.
108. Djouadi F, Weinheimer CJ, Saffitz JE, Pitchford C, Bastin J, Gonzalez FJ, et al. A gender-related defect in lipid metabolism and glucose homeostasis in peroxisome proliferator-activated receptor alpha- deficient mice. *J Clin Invest* 1998;**102**:1083–91.
109. Jenkins DL, Griffith OW. Antiketogenic and hypoglycemic effects of aminocarnitine and acylaminocarnitines. *Proc Natl Acad Sci USA* 1986;**83**:290–4.
110. Lieu YK, Hsu BYL, Price WA, Corkey BE, Stanley CA. Carnitine effects on coenzyme a profiles in rat liver with hypoglycin inhibition of multiple dehydrogenases. *Am J Physiol* 1997;**35**:E359–66.
111. Tserng KY, Jin SJ, Hoppel CL. Spiropentaneacetic acid as a specific inhibitor of medium-chain acyl-CoA dehydrogenase. *Biochemistry* 1991;**30**:10755–60.
112. Bauche F, Sabourault D, Giudicelli Y, Nordmann J, Nordmann R. 2-Mercaptoacetate administration depresses the beta-oxidation pathway through an inhibition of long-chain acyl-CoA dehydrogenase activity. *Biochem J* 1981;**196**:803–9.
113. Bauche F, Sabourault D, Giudicelli Y, Nordmann J, Nordmann R. Inhibition in vitro of acyl-CoA dehydrogenases by 2-mercaptoacetate in rat liver mitochondria. *Biochem J* 1983;**215**:457–64.
114. Matsumura S, Saitou K, Miyaki T, Yoneda T, Mizushige T, Eguchi A, et al. Mercaptoacetate inhibition of fatty acid beta-oxidation attenuates the oral acceptance of fat in BALB/c mice. *Am J Physiol Regul Integr Comp Physiol* 2008;**295**:R82–91.
115. Fujikawa T, Fujita R, Iwaki Y, Matsumura S, Fushiki T, Inoue K. Inhibition of fatty acid oxidation activates transforming growth factor-beta in cerebrospinal fluid and decreases spontaneous motor activity. *Physiol Behav* 2010;**101**:370–5.
116. Zhang YM, Chohnan S, Virga KG, Stevens RD, Ilkayeva OR, Wenner BR, et al. Chemical knockout of pantothenate kinase reveals the metabolic and genetic program responsible for hepatic coenzyme A homeostasis. *Chem Biol* 2007;**14**:291–302.
117. Kimura A, Yoshida I, Ono E, Matsuishi T, Yoshino M, Yamashita F, et al. Acute encephalopathy with hyperammonemia and dicarboxylic aciduria during calcium hopantenate therapy: a patient report. *Brain Dev* 1986;**8**:601–5.
118. Matsumoto M, Kuhara T, Inoue Y, Shinka T, Matsumoto I. Abnormal fatty acid metabolism in patients in hopantenate therapy during clinical episodes. *J Chromatogr* 1991;**562**:139–45.

119. Conlon TJ, Walter G, Owen R, Cossette T, Erger K, Gutierrez G, et al. Systemic correction of a fatty acid oxidation defect by intramuscular injection of a recombinant adeno-associated virus vector. *Hum Gene Ther* 2006;**17**:71–80.
120. Beattie SG, Goetzman E, Tang Q, Conlon T, Campbell-Thompson M, Matern D, et al. Recombinant adeno-associated virus-mediated gene delivery of long chain acyl coenzyme A dehydrogenase (LCAD) into LCAD-deficient mice. *J Gene Med* 2008;**10**:1113–23.
121. Beattie SG, Goetzman E, Conlon T, Germain S, Walter G, Campbell-Thompson M, et al. Biochemical correction of short-chain acyl-coenzyme A dehydrogenase deficiency after portal vein injection of rAAV8-SCAD. *Hum Gene Ther* 2008;**19**:579–88.
122. Holm DA, Dagnaes-Hansen F, Simonsen H, Gregersen N, Bolund L, Jensen TG, et al. Expression of short-chain acyl-CoA dehydrogenase (SCAD) proteins in the liver of SCAD deficient mice after hydrodynamic gene transfer. *Mol Genet Metab* 2003;**78**:250–8.
123. Kelly CL, Rhead WJ, Kutschke DA, Hamm DA, Pinkert CA, Lindsey JR, et al. Functional correction of short-chain acyl-CoA dehydrogenase deficiency in transgenic mice. *Am J Hum Genet* 1995;**57**(Suppl.):A52.
124. Vockley J, Rinaldo P, Bennett MJ, Matern D, Vladutiu GD. Synergistic heterozygosity: disease resulting from multiple partial defects in one or more metabolic pathways. *Mol Genet Metab* 2000;**71**:10–8.
125. Schuler AM, Gower BA, Matern D, Rinaldo P, Vockley J, Wood PA. Synergistic heterozygosity in mice with inherited enzyme deficiencies of mitochondrial fatty acid beta-oxidation. *Mol Genet Metab* 2005;**85**:7–11.
126. Yeh CS, Wang JY, Cheng TL, Juan CH, Wu CH, Lin SR. Fatty acid metabolism pathway play an important role in carcinogenesis of human colorectal cancers by Microarray-Bioinformatics analysis. *Cancer Lett* 2006;**233**:297–308.
127. Liu X, Feng R, Du L. The role of enoyl-CoA hydratase short chain 1 and peroxiredoxin 3 in PP2-induced apoptosis in human breast cancer MCF-7 cells. *FEBS Lett* 2010;**584**:3185–92.
128. Roy MJ, Dionne S, Marx G, Qureshi I, Sarma D, Levy E, et al. In vitro studies on the inhibition of colon cancer by butyrate and carnitine. *Nutrition* 2009;**25**:1193–201.
129. Katagiri H, Asano T, Yamada T, Aoyama T, Fukushima Y, Kikuchi M, et al. Acyl-coenzyme A dehydrogenases are localized on GLUT4-containing vesicles via association with insulin-regulated aminopeptidase in a manner dependent on its dileucine motif. *Mol Endocrinol* 2002;**16**:1049–59.
130. Shi Y, Samuel SJ, Lee W, Yu C, Zhang W, Lachaal M, et al. Cloning of an L-3-hydroxyacyl-CoA dehydrogenase that interacts with the GLUT4 C-terminus. *Arch Biochem Biophys* 1999;**363**:323–32.
131. Goetzman ES. The regulation of acyl-CoA dehydrogenases in adipose tissue by rosiglitazone. *Obesity (Silver Spring)* 2009;**17**:196–8.
132. Irie J, Reck B, Wu Y, Wicker LS, Howlett S, Rainbow D, et al. Genome-wide microarray expression analysis of CD4+ T Cells from nonobese diabetic congenic mice identifies Cd55 (Dafl) and Acadl as candidate genes for type 1 diabetes. *J Immunol* 2008;**180**:1071–9.
133. Tafti M, Petit B, Chollet D, Neidhart E, de Bilbao F, Kiss JZ, et al. Deficiency in short-chain fatty acid beta-oxidation affects theta oscillations during sleep. *Nat Genet* 2003;**34**:320–5.
134. Smith Richards BK, Belton BN, York B, Volaufova J. Mice bearing Acads mutation display altered postingestive but not 5-s orosensory response to dietary fat. *Am J Physiol Regul Integr Comp Physiol* 2004;**286**:R311–9.
135. Swigonova Z, Mohsen AW, Vockley J. Acyl-CoA dehydrogenases: dynamic history of protein family evolution. *J Mol Evol* 2009;**69**:176–93.
136. He M, Rutledge SL, Kelly DR, Palmer CA, Murdoch G, Majumder N, et al. A new genetic disorder in mitochondrial fatty acid beta-oxidation: ACAD9 deficiency. *Am J Hum Genet* 2007;**81**:87–103.

137. Roomets E, Kivela T, Tyni T. Carnitine palmitoyltransferase I and Acyl-CoA dehydrogenase 9 in retina: insights of retinopathy in mitochondrial trifunctional protein defects. *Invest Ophthalmol Vis Sci* 2008;**49**:1660–4.
138. Bian L, Hanson RL, Muller YL, Ma L, Kobes S, Knowler WC, et al. Variants in ACAD10 are associated with type 2 diabetes, insulin resistance and lipid oxidation in Pima Indians. *Diabetologia* 2010;**53**:1349–53.

Genetic Mouse Models of Neurodegenerative Diseases

Alzbeta Trancikova,
David Ramonet, and
Darren J. Moore

Laboratory of Molecular Neurodegenerative Research, Ecole Polytechnique Fédérale de Lausanne (EPFL), School of Life Sciences, Brain Mind Institute, 1015, Lausanne, Switzerland

Neurodegenerative diseases are generally characterized by the selective degeneration of particular neuronal populations and the accumulation of abnormal or aggregated proteins within, but occasionally external to, neurons in affected brain regions. These diseases can be broadly classified as disorders of cognition and memory or movement, and both features can often coexist in a single disease. In recent years, the identification of genetic mutations that cause rare monogenic familial disease has revolutionized our understanding of the molecular basis of neurodegenerative disease and has provided new targets for the development of disease-modifying therapies. An essential part of this process has been the development of genetic animal models that accurately recapitulate the essential features of each disease, with particular emphasis on the use of mouse models. Such mouse models have provided unique insight into the molecular mechanism(s) through which genetic mutations precipitate neurodegeneration and produce associated clinical and pathological

Progress in Molecular Biology
and Translational Science, Vol. 100
DOI: 10.1016/B978-0-12-384878-9.00012-1

Copyright 2011, Elsevier Inc.
All rights reserved.
1877-1173/11 $35.00

phenotypes. In this review, we provide an overview of the current status, uses and limitations of genetic mouse models for understanding major neurodegenerative diseases, including Alzheimer's, Parkinson's, and Huntington's disease and amyotrophic lateral sclerosis.

I. Introduction

Neurodegenerative diseases are a collection of related disorders that are each characterized by the selective degeneration of one or more specific neuronal populations together with the accumulation of abnormal or aggregated proteins in affected neurons or brain regions. Such selective neurodegeneration and associated neuropathology give rise to distinct yet often overlapping clinical symptoms which most often affect cognition, memory, and movement. Alzheimer's disease (AD) is the most common progressive neurodegenerative disorder that is characterized by a profound dementia. The underlying neuropathology consists of cortical and subcortical neuronal degeneration with the appearance of extracellular plaques consisting of amyloid β (Aβ) peptide and intraneuronal neurofibrillary tangles (NFTs) composed of hyperphosphorylated tau. Frontotemporal dementia (FTD) and dementia with Lewy bodies (DLB) comprise other major causes of dementia characterized by selective neuronal loss together with tau, ubiquitin, or TDP-43-positive inclusions in FTD or α-synuclein-positive Lewy bodies in DLB. Parkinson's disease (PD) is the most common neurodegenerative movement disorder whereas Huntington's disease (HD) is comparatively less common. PD is characterized by the relatively selective loss of dopaminergic neurons of the substantia nigra pars compacta leading to a loss of dopaminergic input into the striatum. PD similar to DLB is characterized by the appearance of Lewy bodies in surviving neurons of the brainstem. PD pathology often extends to areas of the cerebral cortex and resulting dementia is a common symptom. HD is primarily due to the selective degeneration of medium-sized spiny output neurons in the caudate putamen and caudate nucleus leading to the abnormal control of movement, but subjects can also develop cognitive decline and dementia owing to cortical dysfunction and degeneration. HD is neuropathologically characterized by the appearance of neuronal nuclear inclusions composed of aggregated polyglutamine (polyQ)-expanded huntingtin (Htt) protein. Other less common neurodegenerative movement disorders include various forms of dystonia, progressive supranuclear palsy, corticobasal degeneration, and spinocerebellar ataxias. A final group of neurodegenerative disorders affect spinal cord motor neurons or the peripheral nervous system, including amyotrophic lateral sclerosis (ALS), spinal muscular atrophies, and peripheral neuropathies that impair motor and autonomic function. In the case of ALS, motor neurons of the spinal cord progressively degenerate with the appearance of cytoplasmic inclusions that contain ubiquitin and/or TDP-43 proteins. In each neurodegenerative disease, the inclusion-forming protein provides clues to the underlying mechanism of the sporadic form of disease and genetic mutations in the corresponding gene often cause a rare, earlier-onset, monogenic familial form of the disease. For example, mutations in the amyloid precursor protein (APP) which normally gives rise to Aβ peptide as well as Aβ plaques in sporadic AD cause autosomal-dominant familial AD, whereas mutations in α-synuclein, the major component of Lewy bodies in sporadic PD, cause autosomal-dominant familial PD. This suggests that the inclusion-forming protein may underlie the molecular basis of sporadic disease.

The identification of genetic mutations that cause rare familial monogenic forms of neurodegenerative disease has revolutionized our understanding of the underlying molecular basis of disease and has provided a number of potential targets for the development of disease-modifying therapies. Here, the development and utilization of genetic animal models that faithfully mimic disease, or at least important aspects of disease, have proved of immense value and have provided the basis for preclinical validation of new therapeutic compounds. A number of genes have been identified that when mutated cause familial forms of AD, PD, and ALS. In the case of HD, which is generally considered to be an inherited disease with some exceptions, polyQ expansions in the Htt protein above a critical threshold underlie the vast majority of cases. In most of these diseases, the mutated gene products tend to fall within common molecular pathway(s) or are in some way related to the underlying neuropathology, with the exception being PD where a common disease pathway has so far remained elusive. The development of genetic animal models serves three major roles: (i) the elucidation of the normal physiological function of the gene product (i.e., in knockout (KO) mice or transgenic mice overexpressing the wild-type (WT) protein); (ii) understanding how genetic mutations precipitate neuronal degeneration (i.e., knockout and knockin mice, or transgenic mice overexpressing the mutated protein); and (iii) the development of models with disease-relevant phenotypes that are critical models for preclinical testing of therapeutics. Mice have been widely used to model the effects of familial mutations because the introduction of genetic mutations is relatively straightforward and they are behaviorally and neuroanatomically well characterized. However, the use of rats is becoming ever more popular due to the advent of new methods for their genetic manipulation. Thus, genetic mouse models play an important role in providing formal proof that a particular genetic mutation is disease-causing *in vivo*, and in understanding how protein dysfunction owing to gain- or loss-of-function mutations induces neuronal degeneration. In this review, we provide a current overview of the various genetic mouse models that have been developed to model familial mutations associated with major neurodegenerative diseases.

II. Parkinson's Disease

A. Mouse Models for Autosomal-Dominant Parkinson's Disease (α-Synuclein and LRRK2)

Mutations in the *α-synuclein* and *leucine-rich repeat kinase 2* (*LRRK2*) genes unambiguously cause autosomal-dominant, familial forms of PD.[1] *α-Synuclein* (PARK1 and 4; OMIM 163890) was the first gene identified to cause familial PD.[2] Point mutations (A53T, A30P, and E46K) or gene multiplications (i.e., duplications and triplications) cause rare forms of familial PD.[2–5] In general, mutations in the *α-synuclein* gene are an exceedingly rare cause of familial PD. Recent genome-wide association studies (GWAS) have identified common variation in the *α-synuclein* gene as a major factor predisposing to the development of sporadic PD implying that variants which most likely lead to increased expression of the WT protein may increase disease risk.[6,7] Genetic multiplications and GWAS data suggest that enhanced expression of WT nonmutated α-synuclein plays an important role in the development of PD. Further linking α-synuclein to sporadic disease is the observation that fibrillar forms of α-synuclein protein comprise the major component of Lewy bodies and Lewy neurites, the main neuropathological hallmarks of sporadic and certain familial forms of PD.[8] Mutations in the *LRRK2* gene are a common cause of late-onset, autosomal-dominant familial PD with a clinical, neurochemical, and, for the most part, neuropathological disease spectrum that is indistinguishable from sporadic disease.[9–11] *LRRK2* mutations are also found in apparently sporadic PD subjects at varying frequency depending on ethnicity,[12,13] and to date *LRRK2* mutations represent the most common cause of familial and sporadic PD.[1] Unlike α-synuclein, the LRRK2 protein is not a major component of Lewy pathology in PD brains, whereas Lewy bodies are the predominant pathology associated with *LRRK2*-linked PD.[14–16] These observations suggest that mutations in LRRK2 are positioned upstream of α-synuclein aggregation and Lewy body formation in PD. To model autosomal-dominant familial PD, several knockout and transgenic mouse models have been developed to confirm and explore how mutations in α-synuclein and LRRK2 cause neurodegeneration and PD.[17]

1. α-Synuclein Transgenic Mice

α-Synuclein is a small, 140 amino acid neuronal protein that is typically associated with lipid membranes and is enriched in presynaptic nerve terminals in close association with synaptic vesicles.[18,19] The normal physiological function (s) of α-synuclein remain unclear. Numerous studies suggest a role for α-synuclein in intracellular vesicular trafficking especially at the synapse and deletion of α-synuclein in mice leads to impaired dopamine neurotransmission and alterations in the release and recycling of presynaptic vesicles.[19–23] α-Synuclein is a natively unstructured protein but can adopt an α-helical conformation when associated with cellular membranes.[24] Familial PD mutations tend to promote the misfolding and aggregation of α-synuclein.[25–27] Endogenous stimuli may also promote α-synuclein aggregation including oxidative and nitrosative stress and mitochondrial impairment which tend to increase with age.[27–31] α-Synuclein can assume various higher order structures (i.e., oligomers, protofibrils, fibrils, and filaments) and the stabilization of these higher order structures may play a crucial role in neuronal toxicity, Lewy body formation and the development of PD.[27,32] Current evidence supports the concept that soluble oligomers or protofibrils of α-synuclein are the major neurotoxic species in PD, although fibrillar α-synuclein may also contribute to neuronal dysfunction.[27,33–35]

Due to the important role of α-synuclein in familial and sporadic PD, a multitude of mouse models have been developed, including KO mice and transgenic mice overexpressing human α-synuclein harboring the familial A53T or A30P mutations (summarized in Table I).[17] α-Synuclein KO mice are viable, fertile, and generally unremarkable but exhibit subtle deficits including mild impairment of locomotor activity, reduced striatal dopamine levels and a reduction in the number of hippocampal presynaptic vesicles.[20] Double α- and β-synuclein KO mice also display a similar reduction in striatal dopamine levels.[36] Studies in KO mice support a role for α-synuclein in regulating dopaminergic neurotransmission, synaptic plasticity, and presynaptic vesicular release and recycling.[20,36] Importantly, the deletion of α-synuclein in mice fails to induce nigrostriatal dopaminergic neurodegeneration that is the major hallmark of PD supporting a gain-of-function rather than a loss-of-function mechanism for the actions of dominant familial mutations in PD.

A number of transgenic mice have been developed that express WT or familial mutant (A30P or A53T) human α-synuclein in neurons of the central nervous system (CNS) driven by different heterologous promoters (summarized in Table I). Such promoter elements include the broadly expressing prion protein (PrP), platelet-derived growth factor β-chain (PDGFβ) and Thy-1 promoters,[22,37–40,47] the catecholamine-specific tyrosine hydroxylase (TH) promoter which drives expression in dopaminergic neurons,[42–44,48] or inducible (tetracycline-regulatable cytomegalovirus [CMV] promoter)[46,49] and conditional (Cre-*loxP*-based endogenous ROSA26 promoter)[45] promoters (summarized in Table I). The resulting phenotypes of α-synuclein transgenic mice are largely dependent on the choice of promoter, the level and pattern of transgene expression, and the particular familial mutation with A53T producing stronger phenotypes than A30P, whereas WT α-synuclein rarely ever induces neurotoxicity.[17,50] α-Synuclein transgenic mice develop both nonfibrillar and fibrillar α-synuclein inclusions that may resemble, in part, early precursors of Lewy bodies. Transgenic mice can also develop progressive and selective

TABLE I

MODELS OF AUTOSOMAL DOMINANT PD: α-SYNUCLEIN TRANSGENIC AND KNOCKOUT MICE

Transgene/mutation	Promoter	LB-like inclusions	DAergic neuronal loss	Additional phenotypes	Reference
KO		–	–/+	↓ rearing activity, ↓ in striatal DA, ↓ number hippocampal synaptic vesicles	20
KO (α + β)		–	+	↓ in striatal DA	36
WT	PDGF-β	+	–	Nonfibrillar α-syn inclusions, motor impairment, modest degeneration of TH positive nerve terminals	37
WT	Thy-1	+	–	Age-related ↓ motor activity, ↓ in DAT, abnormal behavioral response to DA agonists,	38
A53T	PrP	+	–	Progressive motor impairment, paralysis, fibrillar α-syn inclusions, reduced survival	39
A53T	PrP	+	–	Progressive neurodegeneration, axonal degeneration, motor neuron loss, α-syn aggregates, mitochondrial damage	40,41
A53T/A30P	TH	–	–	↓ striatal DA, abnormal DAergic axons/terminals, motor impairment	42
WT 1–120	TH	+ (SN, OB)	–	↓ striatal DA content, ↓ motor activity, fibrillar/nonfibrillar α-syn inclusions	43
A53T 1–130	TH	–	+	Selective, nonprogressive loss of DAergic neurons, motor impairment	44
WT 1–119	ROSA26	–	–	No obvious pathological changes, ↓ striatal DA, and its metabolites	45
WT	TetO/CamKIIα	–	–	Modest degeneration of DAergic fibers, motor impairment	46

LB, Lewy bodies; DA, dopamine; DAergic, dopaminergic; DAT, dopamine transporter; SN, substantia nigra; OB, olfactory bulb.

neurodegeneration, often leading to premature survival, reduced levels of striatal dopamine, and associated deficits in locomotor activity and other behavioral parameters.[37–40,42,47,48,51] However, many of these phenotypes do not occur in the same transgenic model. In general, α-synuclein transgenic mice do not exhibit robust degeneration of substantia nigra dopaminergic neurons, and in the few models where this occurs, degeneration is either extremely subtle, is not progressive but instead developmental, or is caused by α-synuclein harboring multiple familial mutations which may not be relevant to disease.[42,51] In transgenic mice, premature survival is most likely due to a profound degeneration of spinal cord motor neurons leading to paralysis and death.[39,40] Such phenotypes are confined to mice overexpressing α-synuclein from the PrP promoter and, to a lesser extent, the Thy-1 promoter, both of which are known to drive high levels of transgene expression in neurons of the brainstem and spinal cord.[50] α-Synuclein transgenic mice have provided conclusive proof that familial mutations are pathogenic compared to the WT protein, act through a toxic gain-of-function mechanism, and that mutated α-synuclein can induce robust and progressive neurodegeneration albeit of neuronal populations that are not primarily affected in PD. At present, a α-synuclein transgenic model with degeneration of the nigrostriatal dopaminergic pathway and associated dopamine-dependent motoric deficits that recapitulates the key features of PD is still lacking.

α-Synuclein transgenic mice have proved useful for understanding the mechanism(s) underlying α-synuclein-induced neurodegeneration. In symptomatic mice expressing A53T α-synuclein, neuronal degeneration is associated with the formation of intraneuronal inclusions, mitochondrial impairment, DNA damage and degeneration, and apoptotic-like cell death.[41] At the presynapse, α-synuclein expression in transgenic mice was shown to reduce neurotransmitter release by inhibiting synaptic vesicle reclustering after endocytosis perhaps representing an initiating event of axonal degeneration.[52] Additional studies demonstrate that α-synuclein expression in mice leads to synaptic pathology and synaptic deficits through a critical loss of certain endogenous vesicular trafficking proteins at presynaptic boutons.[53,54] Similar to Lewy body pathology in PD brains, most transgenic models exhibit α-synuclein aggregates that are associated with α-synuclein phosphorylation predominantly at serine-129, ubiquitination, nitration and C-terminal truncation.[27,39,40,47,50,55] These posttranslational modifications correlate with regions of α-synuclein pathology and/or neuronal damage in transgenic mice and PD brains.[38–41,47,56,57] The role of α-synuclein phosphorylation in neuronal toxicity is unclear with support for both neurotoxic and protective effects, as well as suggestions that phosphorylation is a late event that occurs after Lewy body formation.[58–61] C-terminal truncation of α-synuclein is a normal physiological process but insoluble truncated species accumulate in PD brains and in A53T α-synuclein transgenic mice.[55] These physiological α-synuclein truncated species include 1–119, 1–122, and 1-123.[55,62] Mechanistically, prior *in vitro* studies with artificial α-synuclein truncated species suggest that truncated species aggregate more readily into filaments[62–64] and that familial PD mutations promote C-terminal truncation compared to WT α-synuclein.[55,62] α-Synuclein truncation species can promote aggregation of the full-length WT protein at substoichiometric ratios and may function to seed or nucleate α-synuclein aggregation thus providing a clear mechanistic basis for their pathogenic properties.[55,62,64]

These initial observations provided the rationale for the development of transgenic mice expressing physiological (1–119) or artificial (1–120 and 1–130) truncated forms of human α-synuclein to further explore the consequences of C-terminal truncation *in vivo*.[43–45] Mice expressing truncated human α-Syn120 driven by the TH promoter serendipitously on a genetic background lacking mouse α-synuclein exhibit both filamentous and nonfilamentous α-synuclein aggregates together with microglial activation, progressive locomotor impairment, and reduced striatal dopamine levels but in the absence of nigrostriatal dopaminergic neuronal loss.[43] Similarly, mice expressing truncated human α-Syn130 containing the A53T mutation from the TH promoter exhibit an early yet nonprogressive loss of nigrostriatal dopaminergic neurons, reduced striatal dopamine levels and mild locomotor impairment.[44] More recently, conditional Cre-*loxP*-based transgenic mice expressing truncated human α-Syn119 specifically in nigrostriatal dopaminergic neurons were developed that exhibit reduced levels of striatal dopamine and its metabolites but in the absence of dopaminergic neuronal loss perhaps suggesting presynaptic dopaminergic dysfunction.[45] Collectively, α-synuclein truncation models provide support for a pathogenic role for truncation variants in the nigrostriatal dopaminergic pathway and the development of PD.

α-Synuclein transgenic mice have also been employed to explore potential interactions with other PD-associated gene products, including parkin, DJ-1, synphilin-1, and LRRK2, in addition to other disease-related or aggregation-prone proteins. Deletion of the *parkin* gene largely fails to influence the lethal and pathological neurodegenerative phenotype of mutant α-synuclein transgenic mice although deletion of parkin may modestly delay sensorimotor impairment and the initial manifestation of neurodegeneration.[65,66] The lack of effect of parkin deletion in α-synuclein transgenic mice may relate to a less prominent role or reduced expression of parkin in motor neurons of the spinal cord. Likewise, deletion of DJ-1 fails to influence α-synuclein-induced neurodegeneration in transgenic mice.[67] In a viral-mediated gene transfer rat model of α-synuclein-induced dopaminergic neurodegeneration, the overexpression of parkin is neuroprotective.[68] This suggests that parkin and α-synuclein may act in a common pathway only in dopaminergic neurons with parkin located downstream of α-synuclein. LRRK2 has recently been shown to regulate the neuropathology induced by mutant α-synuclein expression in inducible transgenic mice.[49] Synphilin-1 interacts with α-synuclein to form Lewy body-like inclusions in cultured cells, is localized to Lewy bodies in PD, and mutations in synphilin-1 have been identified in two siblings with PD.[69–72] The overexpression of human synphilin-1 can attenuate mutant α-synuclein-induced neurodegeneration in a bigenic mouse model suggesting that synphilin-1 normally plays a neuroprotective role *in vivo*.[73] Finally, α-synuclein transgenic mice have been employed to reveal functional or pathological interactions with tau,[74] apolipoprotein E,[75] Aβ,[76] β-synuclein,[77] and cysteine-string protein α[56] *in vivo*.

As mentioned previously, the toxic properties of α-synuclein are associated with its posttranslational modification, aggregation, and fibrillization. These events could result from a number of factors including oxidative and nitrosative stress, DNA damage, and mitochondrial, proteasomal, autophagy or lysosomal dysfunction.[27,41,78–81] Impairment of proteasomal or autophagy–lysosomal function can lead to α-synuclein accumulation and nonselective neurodegeneration,[82–84] whereas conversely α-synuclein has been shown to impair chaperone-mediated autophagy and proteasomal activity.[79,85] An interest in connecting α-synuclein-induced toxicity with mitochondrial dysfunction resulted from the observations that sporadic PD brains exhibit mitochondrial complex I impairment,[86] and MPTP, a complex I inhibitor and neurotoxin, induces selective nigrostriatal dopaminergic neurodegeneration leading to an acute parkinsonian syndrome in humans, nonhuman primates, and mice with similarities to sporadic PD, including α-synuclein aggregation.[87] α-Synuclein KO mice are resistant to the neurotoxic actions of MPTP,[88,89] whereas transgenic mice overexpressing mutant α-synuclein are more sensitive to MPTP-induced toxicity and toxicity due to complex I inhibitors.[90,91] α-Synuclein accumulation also correlates with reduced mitochondrial activity in brains of PD subjects[92] and subjects with rare familial PD due to mutations in mitochondrial DNA develop Lewy pathology.[93] Therefore, α-synuclein aggregation may induce mitochondrial dysfunction, whereas oppositely mitochondrial impairment may promote α-synuclein aggregation. Mitochondria play an important role in the pathogenesis of sporadic PD with α-synuclein most likely located downstream of mitochondrial impairment where it may mediate neuronal toxicity and neurodegeneration.

Taken together, α-synuclein transgenic mice have so far failed to faithfully model the key pathology and clinical deficits of PD. These mice, however, represent powerful models for understanding the molecular basis of α-synuclein aggregation and α-synuclein-induced neuronal dysfunction and degeneration.

2. LRRK2 Transgenic Mice

Mutations in the *LRRK2* gene (PARK8, OMIM 609007) were first identified in 2004 and today represent the most frequent genetic cause of PD accounting on average for 5–6% of familial and 2–3% of sporadic PD cases.[10,11,94] *LRRK2* mutations cause late-onset, autosomal-dominant familial PD with mutations also found in sporadic cases at varying frequency.[12,13,94–97] *LRRK2*-linked PD is indistinguishable from sporadic PD both clinically and neurochemically.[9,94] *LRRK2*-linked disease is characterized by nigrostriatal dopaminergic degeneration and striatal dopamine depletion. Neuropathologically, *LRRK2*-linked PD is predominantly associated with Lewy pathology but in some cases can precipitate neurofibrillary tau pathology, ubiquitin-positive inclusions, or the distinct absence of protein inclusions depending on the mutation and/or individual.[11,14,16,98–100] The *LRRK2* gene encodes a large, 2527 amino acid protein containing multiple domains. LRRK2 is a member of the ROCO family, a subgroup of the *Ras* GTPase superfamily, characterized by a *Ras* of complex (ROC) GTPase domain and a C-terminal of ROC (COR) domain.[101] LRRK2 also contains a protein kinase domain with similarity to mixed-lineage kinase (MLK) and RIP kinase family members, in addition to protein–protein interaction domains including LRRK2-specific, ankyrin, leucine-rich (LRR), and WD40 repeat domains.[102,103]

In the mammalian brain, LRRK2 is expressed at highest levels in neurons of the cerebral cortex, striatum, and hippocampus and at lower levels in other regions including the substantia nigra.[104–106] Within neurons LRRK2 is predominantly localized to membranous and vesicular subcellular compartments including lysosomes, endosomes, multivesicular bodies, lipid rafts, mitochondria, microtubule-associated transport vesicles, endoplasmic reticulum (ER), and the Golgi complex.[104,107] Accordingly, LRRK2 may play a role in the regulation of vesicular trafficking including synaptic vesicle endocytosis and exocytosis, and autophagy.[108–110] LRRK2 contains two enzymatic domains and displays modest GTPase and kinase activity *in vitro*.[111,112] The kinase domain exhibits serine/threonine-directed activity and putative substrates including moesin, 4E-BP1, β-tubulin, and FoxO1 have been identified *in vitro*.[113–117] However, the physiological significance of these substrates requires confirmation *in vivo*. The GTPase domain of LRRK2 can induce kinase activity through binding of GTP and intact GTP-binding is a critical requirement for kinase activity.[117–119] LRRK2 appears to function as a GTP-dependent protein kinase but autophosphorylation of LRRK2 targets sites within or near the GTPase domain suggesting a complex crosstalk between both enzymatic domains.[120,121] Familial mutations associated with PD tend to cluster within the two enzymatic domains, with the most common G2019S mutation (kinase domain) consistently leading to enhanced kinase activity and the R1441C/G/H (GTPase) and

Y1699C (COR) mutations resulting in impaired GTP hydrolysis and increased GTP-binding.[110,112,115,117,118,122,123] Familial mutations also promote LRRK2-induced neurite shortening and neuronal toxicity in primary cultures in a GTPase- and kinase-dependent manner consistent with a toxic gain-of-function mechanism for mutations.[110,119,122,124,125] A recent study demonstrates that viral-mediated expression of G2019S mutant LRRK2 induces nigrostriatal dopaminergic neurodegeneration in mice which can be attenuated through genetic or pharmacological inhibition of LRRK2 kinase activity.[126] Therefore, GTPase and kinase activity are important for the development of PD due to familial *LRRK2* mutations.

To explore the role of LRRK2 *in vivo*, several KO and transgenic mouse models have recently been developed (summarized in Table II). Similar to α-synuclein KO mice, LRRK2 KO mice fail to exhibit motoric or neurochemical deficits nor dopaminergic neurodegeneration suggesting that LRRK2 does not play a role in the maintenance or survival of dopaminergic neurons.[127,128] The absence of neurodegeneration in KO mice supports a gain-of-function mechanism for dominant familial mutations. Deletion of LRRK2 promotes α-synuclein accumulation in the kidney but not in the brain which was suggested to result from a lack of functional compensation by the closely related LRRK1 protein.[128] The relevance of this observation for neurodegeneration and PD is unclear. LRRK2 is localized in part to the outer membrane of mitochondria in the mammalian brain,[104] but unlike α-synuclein KO mice, LRRK2 KO mice exhibit a normal sensitivity to MPTP-induced neurotoxicity.[127] This observation might suggest that LRRK2 activity does not play a role in mitochondrial function or alternatively that LRRK2 is located upstream of mitochondria where it would not be expected to influence MPTP toxicity. LRRK2 may also regulate protein homeostasis and the ubiquitin–proteasomal system.[128] LRRK2 KO mice show an age-dependent increase in protein ubiquitination, α-synuclein aggregation, impairment of the autophagy–lysosomal pathway, and apoptotic cell death confined to the kidney.[128] Primary cortical neurons derived from KO mice display increased neuritic length and branching perhaps due to altered phosphorylation and activity of the moesin/ezrin/radixin family which are involved in the remodeling of neuronal filipodia and neurite outgrowth.[133,134]

A number of LRRK2 transgenic mice have been developed to model the gain-of-function effects of dominant familial mutations. Bacterial artificial chromosome (BAC) transgenic mice expressing human R1441G mutant LRRK2 from its endogenous promoter develop subtle nigrostriatal dopaminergic axonal degeneration, L-DOPA-responsive motor deficits, modestly impaired striatal dopamine release, and hyperphosphorylated tau-positive inclusions up to 12 months of age.[132] It is not clear whether the motoric deficits are solely due to reduced striatal dopamine release or result from dysfunction

TABLE II

AUTOSOMAL DOMINANT PD MODELS: LRRK2 TRANSGENIC AND KNOCKOUT MICE

Transgene/mutation	Promoter	Neuropathology	DAergic neuronal loss	Additional phenotypes	Reference
KO		–	–	No major abnormalities, normal sensitivity to MPTP neurotoxicity	127
KO		+ (kidney)	–	No major abnormalities, α-syn aggregates in kidney, ↑ oxidative stress, impaired UPS, apoptosis	128
WT/G2019S	BAC	–	–	Abnormal DA neurotransmission, ↓ in striatal DA, ↑ tau phosphorylation, altered tau processing/modification	129
WT/G2019S	BAC	–	–	Enhanced DA neurotransmission and motor performance in WT not G2019S	130
R1441C	Knockin	–	–	↓ dopamine D2 receptor function, ↓ striatal DA neurotransmission	131
R1441G	BAC	+ (DAergic axonal pathology)	–	Age-dependent locomotor deficit with reversal with L-DOPA; ↓ DA striatal DA release, DAergic axonal degeneration, tau hyperphosphorylation/inclusions	132
G2019S	TetO/CamKIIα	–	–	Normal motor function, neuronal Golgi fragmentation, microtubule abnormalities, ↓ neuritic complexity	49,133

DA, dopamine; DAergic, dopaminergic; L-DOPA, levodopa.

or degeneration of other dopaminergic neuronal populations. BAC transgenic mice expressing G2019S mutant LRRK2 collectively exhibit impaired nigrostriatal dopaminergic neurotransmission, mild locomotor abnormalities, and alterations in tau phosphorylation and processing with advanced age, but without neurodegeneration or protein aggregation.[129,130] R1441C knockin mice also display impaired nigrostriatal dopaminergic neurotransmission.[131] G2019S mutant LRRK2 expressed in an inducible manner from a tetracycline-regulated promoter in neurons of the forebrain do not develop motor dysfunction, protein aggregation, or neurodegeneration but do display neuronal Golgi fragmentation and perturbed microtubule assembly in the striatum occuring at 1 month of age, and modest accumulation of ubiquitinated proteins in the brain at 18 months.[49] Our laboratory has also developed G2019S mutant LRRK2 transgenic mice with expression from a hybrid PDGFβ promoter. These mice exhibit a modest age-dependent degeneration of nigrostriatal dopaminergic neurons, the accumulation of autophagic vacuoles in the aged brain and reduced neuritic complexity of cultured midbrain dopaminergic neurons (D. Ramonet and D. J. Moore, personal communication). Despite the use of different transgene cassettes with various heterologous promoters that produce varied LRRK2 expression patterns and levels, current LRRK2 transgenic models fail to robustly replicate key features of PD but do demonstrate that mutant LRRK2 expression can impair the nigrostriatal dopaminergic pathway albeit subtely, disrupt intracellular vesicular trafficking pathways, and reduce neuritic complexity. What is lacking in these models is robust neurodegeneration and the formation of α-synuclein aggregates resembling Lewy pathology, which is the most common neuropathology associated with *LRRK2*-linked PD.[16] These mouse models neither confirm nor refute the hypothesis that LRRK2 is located upstream of α-synuclein aggregation in PD as suggested from *LRRK2*-linked PD brains.[14,16] Viral-mediated gene transfer models may offer promising alternatives for targeting mutant LRRK2 expression to the nigrostriatal dopaminergic pathway in rodents to induce neurodegeneration.[126]

While it is clear that dominant mutations in *LRRK2* and *α-synuclein* independently cause PD, little is known about the potential interaction of these proteins and whether they form part of a common pathway leading to neurodegeneration. The best evidence for the existence of such a pathway comes from human PD brains where *LRRK2* mutations precipitate α-synuclein-positive Lewy pathology thus potentially placing LRRK2 upstream of α-synuclein.[14,16] In this scenario, we might place mitochondria centrally between both genes as LRRK2 is not required for MPTP toxicity, whereas deletion of α-synuclein confers resistance to this neurotoxin.[88,127] To explore a pathological pathway connecting LRRK2 and α-synuclein, inducible bigenic mice have been developed overexpressing human A53T mutant α-synuclein together with human LRRK2 variants restricted to neurons of the forebrain.[49]

LRRK2 promotes the progression of neuropathology induced by A53T α-synuclein expression in the frontal cortex and striatum beginning at 1 month of age. In particular, LRRK2 strongly promoted A53T α-synuclein-induced neurodegeneration, neuronal Golgi fragmentation, mitochondrial abnormalities, and the somatic accumulation of α-synuclein in striatal neurons.[49] There was no clear distinction between WT and G2019S mutant LRRK2 in the development of these α-synuclein-induced phenotypes, making it unclear whether this relates to a normal physiological or disease-specific effect of LRRK2. As the A53T α-synuclein mice do not develop a lethal neurodegenerative phenotype due to forebrain-restricted transgene expression, it is not known whether the pathology-promoting effects of LRRK2 in this model are of functional relevance and are sufficient to promote striatal dopamine and motor deficits or reduced survival. Deletion of LRRK2 could attenuate mutant α-synuclein-induced striatal neurodegeneration, Golgi fragmentation and the somatic accumulation of α-synuclein.[49] These observations suggest that LRRK2 and α-synuclein are potentially linked together in a common pathological pathway. Conceptually, LRRK2 could act upstream to prime or promote α-synuclein-induced toxicity but may also act downstream of α-synuclein neurotoxicity suggesting that it may even be required to mediate the pathological effects of mutant α-synuclein. Perhaps, a more simple interpretation is that LRRK2 regulates α-synuclein protein turnover *in vivo*, with LRRK2 overexpression impairing clearance and causing an accumulation of α-synuclein and LRRK2 deletion, promoting increased α-synuclein clearance perhaps via the autophagy–lysosomal pathway. Indeed, LRRK2 expression in cell-based models leads to the accumulation of autophagic vacuoles which might indicate an impairment of autophagy.[108,110,135] A similar argument might also apply to the formation of soluble toxic oligomeric species of α-synuclein regulated by LRRK2 activity. The underlying mechanism involved in the LRRK2-mediated regulation of α-synuclein-induced neurotoxicity remains to be clarified and confirmed by independent studies.

Taken together, LRRK2 transgenic mice do not reliably reproduce key features of PD but have been informative for understanding more subtle phenotypes that perhaps precede overt neurodegeneration. In the coming years, LRRK2 mouse models will prove useful for identifying genetic, pathological or compensatory interactions, or pathways and may hold promise for the identification of physiological or pathological substrates of LRRK2-mediated phosphorylation. LRRK2 mouse models are still in their early development after the initial discovery of *LRRK2* mutations and more sophisicated strategies for targeting expression of more neurotoxic LRRK2 variants to the nigrostriatal pathway are likely to be developed.

B. Mouse Models for Autosomal Recessive Parkinson's Disease (Parkin, DJ-1, and PINK1)

Following the initial identification of autosomal-dominant mutations in the *α-synuclein* gene, mutations in three genes, *parkin*, *DJ-1*, and *PTEN-induced putative kinase 1* (*PINK1*), were shown to cause early-onset, autosomal recessive forms of familial PD.[1] Mutations in a fourth gene, *ATP13A2* (*PARK9*; OMIM 610513), cause juvenile-onset, autosomal recessive Kufor-Rakeb syndrome that is characterized by severe L-DOPA-responsive parkinsonism.[136] The genes encode proteins with different physiological functions and subcellular localization that at first glance do not obviously fit into a common pathogenic pathway. To explore the normal physiological function of these proteins, and to develop recessive loss-of-function animal models of PD, a number of KO mice have been created (summarized in Table III). At present, animal models have not yet been reported for ATP13A2.

1. Parkin Knockout Mice

Parkin (PARK2; OMIM 602544) was the first gene identified in which mutations cause autosomal recessive juvenile-onset parkinsonism (AR-JP).[147] AR-JP is characterized by pure nigral dopaminergic neuronal loss with the general absence of Lewy pathology.[148] This might suggest that parkin is either required for Lewy body formation or is located downstream of α-synuclein aggregation, the latter of which is supported by studies in rat α-synuclein viral models.[68] Presently, greater than 100 *parkin* mutations have been identified in PD subjects that are normally homozygous or compound heterozygous.[148] *Parkin* mutations account for up to 50% of recessive familial PD cases and 20% of all early-onset PD.[149] Familial mutations range from discrete missense and splice site mutations to truncations and large exonic deletions that are predicted to result in a loss-of-function.[148] Parkin can function as an E3 ubiquitin protein ligase and participates in the covalent attachment of ubiquitin to protein substrates through a highly coordinated process involving E1 ubiquitin-activating and E2 ubiquitin-conjugating enzymes.[150–152] There is evidence that parkin-mediated ubiquitination can promote protein degradation via the 26S proteasome and various nonproteasomal actions depending on the mode of ubiquitin attachment.[153] Parkin mutations act in a loss-of-function manner through impairing ligase activity, reducing protein stability or solubility, altering subcellular localization or relocalization, or through alterations in protein interactions.[154–156] Mutations in *parkin* have been suggested to impair substrate ubiquitination leading to the potentially neurotoxic accumulation of these substrates. However, there is currently little evidence for such a mechanism with only few substrates accumulating in parkin-deficient KO mice or AR-JP brains.[157,158] Alternatively, mutations may lead to a loss of

TABLE III
AUTOSOMAL RECESSIVE PD MODELS: PARKIN, PINK-1, AND DJ-1 KNOCKOUT MICE

Transgene/mutation	Exon deleted	PI/DAergic neuronal loss	Additional phenotypes	Reference
Parkin				
KO	3	−/−	↑ extracellular DA release, mild ↓ motor activity, ↓ mitochondrial respiration, ↓ antioxidant capacity	137,138
KO	3	−/−	↓ in DAT, ↑ extracellular striatal DA, mild ↓ in locomotor activity, and working memory	139
KO	2	−/−	Normal behavior, normal catecholamine levels	140
KO	2	−/−	Age-related ↓ striatal DA; ↑ in D1/D2 receptor binding	141
KO	7	−/−	Developmental loss of LC neurons, ↓ of NE in the SC and OB, ↓ acoustic startle	142
PINK-1				
KO	4–7	−/−	↓ striatal DA release and synaptic plasticity, age-dependent ↓ in mitochondrial respiration, ↓ in stress response	143
DJ-1				
KO	1–5	−/−	Motor deficits, ↓ striatal DA neurotransmission	144
KO	2	−/−	↓ locomotor activity,↓ in striatal DA neurotransmission, and synaptic plasticity	145
KO	2–5	−/−	↓ locomotor activity in response to amphetamine, hypersensitivity to MPTP-induced neurotoxicity, and oxidative stress	146

PI, protein inclusions; DA, dopamine; DAergic, dopaminergic; DAT, dopamine transporter; NE, norepinephrine; LC, locus coeruleus; SC, spinal cord; OB, olfactory bulb.

neuroprotective function involving ubiquitination activity but which may not involve proteasomal degradation. For example, parkin has been shown to translocate to damaged mitochondria and mediate their removal by autophagy perhaps initiated via the atypical nondegradative ubiquitination of voltage-dependent anion channel 1 (VDAC1) and/or mitofusins in cooperation with

PINK1.[159–162] Of relevance to sporadic PD, parkin activity can be impaired through various pathological stimuli including oxidative stress,[163,164] nitric oxide-mediated S-nitrosylation,[165,166] and covalent modification by dopamine,[167] perhaps suggesting that parkin inactivation may also contribute to sporadic disease.

Several KO mice have been generated to study the effects of *parkin* loss-of-function mutations *in vivo* (summarized in Table III). Surprisingly, none of these KO models recapitulate key PD-related phenotypes such as neuronal loss, obvious motor impairment, or protein inclusion pathology.[137,139,140,142] Deletion of exon 3 in one KO model causes a mild decrease in nigrostriatal dopaminergic transmission and reduced synaptic excitability of striatal neurons but without dopaminergic neuronal loss up to 2 years of age.[137] A second KO model generated through deletion of exon 7 exhibits a normal number of nigrostriatal dopaminergic neurons but displays a developmental loss of locus coeruleus (LC) noradrenergic neurons accompanied by a reduction of norepinephrine levels in the spinal cord and olfactory bulb which receive LC neuronal projections, and an impaired acoustic startle reflex.[142] In contrast to other parkin KO mice, an exon 2-deleted model initially developed on a mixed genetic background failed to reproduce any behavioral alterations following backcrossing onto a coisogenic 129/S4 genetic background.[140] The mice did not show altered catecholamine levels in the striatum, olfactory bulb, and spinal cord, with no evidence for nigrostriatal, cognitive or noradrenergic dysfunction.[140] A fourth KO mouse exhibited reduced striatal dopamine release and upregulation of D1 and D2 dopamine receptor binding in the striatum and increased dopamine levels in the midbrain.[141] Collectively, parkin-deficient mice do not provide a robust behavioral, neurochemical or neuropathological model of PD, but they have proved useful for understanding the normal function of parkin. Proteomic analysis of brains from parkin KO mice revealed a decreased abundance of proteins involved in mitochondrial function, including subunits of complexes I and IV, and impaired mitochondrial respiration and reduced antioxidant capacity.[138] Such observations are similar to parkin-deficient *Drosophila* models which show impaired mitochondrial function, increased oxidative stress, reduced life span, and locomotor defects owing to apoptotic cell death of muscle subsets.[168] Many studies now support a role for parkin in the regulation of mitochondrial activity, morphology and biogenesis, and protection against mitochondrial-dependent cell death.[169] However, parkin null mice are not more sensitive to MPTP-induced neurotoxicity suggesting that parkin may function independent of complex I to regulate mitochondrial function.[170] Biochemically, parkin KO mice were employed to identify parkin substrates with only nonubiquitinated forms of AIMP2 and FUSE-binding protein 1 accumulating in the KO brain and in parkin-linked AR-JP brains suggesting that both proteins might normally be targeted for degradation by parkin.[157,158] Viral-mediated overexpression of AIMP2 in the substantia nigra induces a loss of dopaminergic neurons supporting the idea that AIMP2 accumulation in the absence of parkin is neurotoxic.[158] Suppression of parkin levels in mice was shown to promote nigrostriatal and motor neuron loss induced by the overexpression of human tau harboring three FTDP-17-causing mutations.[171] The mechanism for this action is unclear since tau is not a substrate for parkin-mediated ubiquitination in cells but tau does accumulate in the brains of parkin KO mice.[171,172] Transgenic mice overexpressing the parkin substrate, Pael-R, on a parkin null background display progressive and selective catecholaminergic neuronal loss without protein inclusion formation and also exhibit age-dependent mitochondrial complex I inhibition.[173] This model replicates many of the key features of AR-JP and suggests that parkin KO mice may require a secondary insult to precipitate neurodegeneration.

2. PINK1 Knockout Mice

Mutations in the *PINK1* gene (PARK6; OMIM 608309) represent the second most common cause of autosomal recessive early-onset PD.[1,174] The *PINK1* gene encodes a 581 amino acid protein with a catalytic serine/threonine-directed kinase domain and a N-terminal mitochondrial signal peptide which targets PINK1 to the mitochondrial intermembrane space and outer membrane.[175,176] The kinase domain of PINK1 is orientated toward the cytosol suggesting that it may phosphorylate cytosolic or mitochondrial-associated proteins.[177] Two PINK1-deficient mouse models have been developed. One KO model contains a targeted deletion of exons 4–7, whereas a second model employed conditional RNA interference to silence PINK1 expression.[143,178] Both PINK1 models do not reveal obvious locomotor abnormalities, protein inclusions or aggregates, or loss of nigrostriatal dopaminergic neurons with age.[143,178–180] Instead, subtle abnormalities in nigrostriatal dopaminergic neurotransmission were apparent with impaired striatal dopamine release.[143] Striatal dopamine release deficits in PINK1 KO mice lead to an impairment of corticostriatal long-term potentiation (LTP) and long-term depression (LTD) which could be restored by administration of dopamine receptor agonists or through enhancing dopamine release.[143] PINK1 together with parkin has been shown to regulate mitochondrial morphology in *Drosophila* and mice although there are species-specific differences.[181–184] PINK1 KO mice reveal an increased number of enlarged mitochondria suggesting that PINK1 normally promotes mitochondrial fission or inhibits fusion.[179] Moreover, striatal-derived mitochondria from KO mice exhibit an age- and stress-dependent reduction in mitochondrial respiratory capacity due to defects in complexes I and IV.[179] Defects in mitochondrial integrity and resistance to oxidative stress, and a loss of flight muscle and dopaminergic neurons have been observed in PINK1-deficient *Drosophila.*[181,182,185] PINK1 null-dependent phenotypes can be rescued by overexpression of parkin or the mitochondrial fission factor Drp1.[181,182,185,186] Similar phenotypes in parkin-deficient flies could not be rescued by PINK1 overexpression. These and other observations have led to the identification of a PINK1-parkin pathway that regulates mitochondrial morphology and damage-induced mitochondrial autophagy (mitophagy).[187] Several mitochondrial proteins can be phosphorylated in a PINK1-dependent manner, including the mitochondrial chaperones HtrA2/Omi,[188] TNF receptor-associated protein 1 (TRAP1),[189] and Hsp90.[190] HrtA2 is a mitochondrial serine protease that is phosphorylated in an indirect PINK1-dependent manner upon activation of the p38/MAPK stress pathway.[188] Mice lacking HrtA2 reveal mitochondrial dysfunction with several characteristic parkinsonian features suggesting a normal neuroprotective role *in vivo*.[188,191] Moreover, phosphorylated HtrA2 is decreased in brains of PD subjects with familial PINK1 mutations.[188] TRAP1 colocalizes with PINK1 in mitochondria and is phosphorylated by PINK1 *in vitro* and *in vivo*.[189] In cultured cells, PINK1 overexpression protects against oxidative stress-induced cell death and this protective action correlates with PINK1-dependent phosphorylation of TRAP1. PINK1 mutations (G309D, L347P, and W437X) impair TRAP1 phosphorylation and PINK1-mediated protection.[189] Interaction of PINK1 with Hsp90 regulates PINK1 protein stability since Hsp90 inhibitors markedly reduce PINK1 levels.[190] Finally, PINK1 deletion in neurons causes mitochondrial calcium overload leading to increased reactive oxygen species and impaired respiration.[192] A reduced mitochondrial calcium capacity lowers the threshold for opening of the mitochondrial permeability transition pore thus rendering PINK1-deficient neurons more vulnerable to cell death.[192] Collectively, PINK1 null mice largely fail to recreate the hallmark features of PD but have revealed a more subtle role for PINK1 in maintaining normal nigrostriatal dopaminergic neurotransmission. PINK1 KO mice have proved useful for demonstrating an important role for PINK1 in maintaining mitochondrial integrity and turnover.

3. DJ-1 Knockout Mice

Mutations in the *DJ-1* gene (PARK7; OMIM 602533) represent a rare cause of autosomal recessive, early-onset PD.[193] Mutations comprise both missense (L166P, M26I, E64D, and E163K) and exonic deletions.[1,193] The L166P mutation results in the destabilization and enhanced proteasomal degradation of DJ-1, whereas other mutations may impair dimerization.[194–196] The *DJ-1* gene encodes a 189 amino acid protein belonging to the ThiJ/PfpI superfamily. DJ-1 is widely expressed in the mammalian brain where it is localized within the cytosol and the mitochondrial intermembrane space and matrix.[197] DJ-1 may function as a redox-sensitive molecular chaperone,[198] a cysteine protease,[199] a peroxiredoxin-like peroxidase to regulate hydrogen peroxide,[200] or as an RNA-binding protein to regulate transcription[201] but its exact function *in vivo* remains elusive. DJ-1 has been shown to inhibit the aggregation of α-synuclein *in vitro* and in cultured neurons,[198] although deletion of DJ-1 in mice fails to influence the lethal neurodegenerative phenotype induced by expression of A53T mutant α-synuclein.[67] Suppression of DJ-1 increases sensitivity to cell death induced by oxidative and ER stress or proteasome inhibition suggesting that DJ-1 is a neuroprotective protein.[202,203] Three independent KO mouse models have so far been reported (summarized in Table III). Similar to parkin and PINK1 null mice, deletion of DJ-1 fails to recreate the key features of PD, including age-dependent degeneration of nigrostriatal dopaminergic neurons, striatal dopamine depletion, and protein inclusion pathology.[144–146] One KO model exhibits modest deficits in motoric function, elevated striatal dopamine content, and enhanced striatal dopamine reuptake.[144] A second KO model also reveals a reduction in evoked dopamine release in the striatum owing to increased striatal dopamine reuptake due to impaired D2 autoreceptor stimulation at dopaminergic presynaptic terminals.[145] The impaired dopamine release in KO mice led to deficits in corticostriatal LTD due to diminished activation of postsynaptic D2 dopamine receptors. DJ-1 KO mice displayed normal corticostriatal LTP.[145] A third KO mouse exhibits a modest decrease of amphetamine-induced locomotor activity, and increased sensitivity to MPTP-induced dopaminergic neurotoxicity.[146] Primary cortical neurons derived from these KO mice were more sensitive to cell death induced by oxidative stress, whereas primary mesencephalic dopaminergic neurons were more sensitive to rotenone exposure. These KO phenotypes could be rescued by viral-mediated overexpression of WT DJ-1.[146] These observations suggest that DJ-1 may function *in vivo* to protect dopaminergic neurons from oxidative insult. Mitochondria isolated from DJ-1 KO mice contain elevated levels of hydrogen peroxide and reduced mitochondrial aconitase activity, with *in vitro* studies suggesting that DJ-1 acts as an atypical peroxiredoxin-like peroxidase via oxidation of cysteine 106.[200] Cys-106 oxidation of DJ-1 by various oxidative insults mediates translocation of DJ-1 to mitochondria similar to parkin translocation upon mitochondrial depolarization.[204] The mechanism through which DJ-1 normally maintains mitochondrial function and protects against mitochondrial oxidative insult is unclear. It is also not clear whether preservation of mitochondrial function is sufficient for the neuroprotective effects of DJ-1 against cellular insult, or whether antioxidant, chaperone, or other activities of DJ-1 are more important for neuroprotection. DJ-1 null mouse models have been instrumental in revealing a number potential neuroprotective functions for DJ-1.

As single KO mice lacking parkin, DJ-1, or PINK1 fail to recreate key pathological features of PD, this might suggest that there is functional redundancy between these proteins. To further determine if these three recessive

TABLE IV
DOUBLE AND TRIPLE MUTANT MICE AS MODELS OF PD

Double/triple mutant mice	PI/DAergic neuronal loss	Additional features	Reference
PrP-A53T-α-Syn/DJ-1 KO	+/– (α-syn aggregates)	No effect on lethal neurodegenerative phenotype of A53T-α-syn parent line	67
PrP-A53T-α-syn/parkin KO	+/– (α-syn aggregates)	No effect on lethal neurodegenerative phenotype of A53T-α-syn parent line	66
Thy-1-A30P-α-syn/parkin KO	+/– (α-syn aggregates)	Similar neurodegenerative phenotype to A30P-α-syn parent line, delayed motor decline, and disease manifestation	65
Parkin/PINK1/DJ-1 KO	–/–	Normal number of DAergic neurons, age-dependent ↑ in striatal DA levels	205

PI, protein inclusions; DA, dopamine; DAergic, dopaminergic; LC, locus coeruleus.

genes act in a common pathway and to potentially unmask PD-relevant phenotypes, these KO models have been crossed with each other (summarized in Table IV). Triple KO mice for parkin, DJ-1, and PINK1 were developed to determine whether inactivation of all three recessive PD genes might lead to a progressive PD-like phenotype.[205] Surprisingly, this model displays a normal morphology and number of nigrostriatal dopaminergic and LC noradrenergic neurons up to 24 months of age, together with an age-dependent increase in striatal dopamine levels. These observations show that inactivation of all three recessive PD genes is not sufficient to cause significant nigral degeneration, suggesting that these genes may function in a neuroprotective capacity rather than being important for the survival of dopaminergic neurons during aging.[205]

C. Viral-Mediated Gene Transfer Models of PD

Transgenic mice for α-synuclein and LRRK2 have largely failed to provide faithful models of PD with only modest effects on motor function, nigrostriatal dopaminergic neurodegeneration and protein aggregation. Strategies using viral vectors for the targeted delivery of mutated transgenes have proved more effective at recreating disease-related phenotypes. Viral-mediated gene transfer offers several advantages including precise and local delivery of the transgene to discrete brain regions, high-level transgene expression, acute-onset of transgene expression in adult animals to potentially bypass developmental compensatory mechanisms, direct comparison of multiple transgene variants, and the shorter duration of experiments compared to transgenic mouse studies.[206] Two viral-mediated gene transfer systems, lentivirus and adeno-associated virus (AAV), have been widely used to model α-synuclein-induced neurodegeneration in adult rats. Lentiviral-based expression of human WT, A30P and A53T α-synuclein via direct delivery to the substantia nigra caused a progressive and selective loss of nigral dopaminergic neurons and a loss of dopaminergic nerve terminals in the striatum over a 6-week period.[207] Neuronal degeneration was associated with α-synuclein-positive inclusions and extensive neuritic pathology.[207] Expression of rat rather than human α-synuclein also produced protein inclusions but without neuronal loss, suggesting that α-synuclein inclusions are not a primary cause of neurodegeneration but rather the byproduct of a pathologic pathway.[207] This lentiviral-based model has been used to assess the neuroprotective role of parkin and Hsp104. Coexpression of rat parkin or yeast Hsp104 attenuated neurodegeneration induced by A30P α-synuclein.[68,208] A limitation of lentivirus is the relatively low tropism for nigral neurons which has led to the development of alternate models using serotype 6 adeno-associated vectors (AAV2/6). Recombinant AAV2/6 vectors demonstrate a high tropism for nigral dopaminergic neurons (> 80% of neurons).[206] AAV2/6-mediated overexpression of mutant α-synuclein following intranigral delivery caused progressive dopaminergic neuronal loss, decreased striatal dopamine levels, α-synuclein-positive inclusions, and motoric impairment.[209] Viral models have been used to explore the role of pathological α-synuclein phosphorylation at serine-129. The nonphosphorylated form (S129A) exacerbates α-synuclein-induced dopaminergic degeneration, whereas α-synuclein phosphorylation (S129D) lacked neurotoxic effects.[58,60] Viral-based rodent models represent a promising approach to model key features of PD, and for understanding mechanism(s) underlying α-synuclein-induced neurodegeneration. Moreover, these models will serve as useful tools to rapidly evaluate neuroprotective therapies.

III. Mouse Models of Alzheimer's Disease

AD is the most common form of dementia and neurodegenerative disease. AD has a chronic and progressive course with manifestation of clinical symptoms usually above the age of 65 years although there are less common early-onset forms of AD (refer to www.alzforum.org). AD is characterized neuropathologically by extracellular senile plaques consisting of aggregated Aβ peptide (of 39–43 amino acids), intraneuronal NFTs consisting of filamentous hyperphosphorylated tau protein, and severe cortical and subcortical neurodegeneration resulting in gross atrophy of the affected regions. Cerebral amyloid angiopathy is a common pathology of AD resulting from deposition of insoluble aggregates of Aβ peptide in the walls of blood vessels, which often leads to cerebral hemorrhage.[210] AD brains can often exhibit cortical Lewy bodies or TDP-43-positive inclusions. These neuropathological features normally appear well before the onset of clinical symptoms during a prolonged asymptomatic period.[211,212]

AD is a genetically complex and heterogeneous disease with at least 16 loci identified so far that are associated with familial disease (AD1–AD16). Recent GWAS approaches have further identified variation in a number of genes that may predispose to the more common late-onset form of AD, including *APOE* (AD2), *clusterin*, *picalm*, *BIN1*, *SORL1*, and *MAPT*.[213–216] Mutations in three loci are responsible for 30–50% of all familial (FAD) and early-onset AD cases that represent approximately 0.5% of AD in general. These three genes are the *amyloid precursor protein* (*APP*, AD1),[217] *presenilin 1* (*PSEN1*, *PS1*, AD3),[218] and *presenilin 2* (*PSEN2*, *PS2*, AD4).[219,220] A fourth gene, *APOEε4* (AD2), substantially increases the risk of developing late-onset familial and sporadic forms of AD and reduces the age at onset of AD.[221] Mutations in *APP* are also associated with cerebral amyloid angiopathy which can occur independent of, or in association with, AD.[210,222] Several mouse models have been developed to model the effects of FAD mutations, including single, double, and triple transgenic mice that express various disease-associated mutant forms of human APP, PS1, PS2, or tau.

Mutations in the *tau* (*MAPT*) gene cause the related disorder, autosomal-dominant frontotemporal dementia with parkinsonism linked to chromosome 17 (FTDP-17),[223] but are not known to cause familial AD. In AD brains, the microtubule-associated protein tau becomes hyperphosphorylated and dissociates from microtubules where it forms paired helical filaments (PHFs).[224] PHF-tau forms the characteristic argyrophilic intraneuronal inclusions known as NFT. The loss of tau binding to microtubules leads to microtubule instability which together with the formation of NFTs results in disruption of the neuronal cytoskeleton and impaired axonal transport.[224] AD belongs to a collection of related diseases known as "tauopathies", including FTDP-17, progressive supranuclear palsy, corticobasal degeneration, and Pick's disease, that are characterized by the formation of pathological inclusions containing hyperphosphorylated and aggregated tau protein.[224] Thus, tau plays an important role in a number of related neurodegenerative diseases. Tau is of particular interest for AD pathophysiology since much evidence suggests that hyperphosphorylation of tau is downstream of Aβ deposition and forms a critical step in the progression of disease toward the onset of neurodegeneration. These pathological events in AD together with genetic data form the basis of the "amyloid cascade hypothesis" which suggests that altered metabolism of APP is the initiating event in the pathogenesis of AD, leading to the production and aggregation of Aβ (specifically $A\beta_{1-42}$).[225] The formation of soluble Aβ oligomers and/or Aβ neuritic plaques leads to tau hyperphosphorylation and the formation of NFTs, neuropil threads, and synaptic pathology which would eventually precipitate a loss of neurotransmitters, neuronal death, and dementia. The basis of this hypothesis is that the majority of mutations in *APP*, *PS1*, and *PS2*, which cause early-onset familial AD, increase the production of the fibrillogenic $A\beta_{1-42}$ peptide thus promoting Aβ aggregation.[226] While Aβ itself can exert neurotoxicity through a variety of pathways or mechanisms which most likely contributes to AD pathology, it is NFT formation *per se* which is required to precipitate frank neuronal degeneration. Furthermore, Aβ-induced cognitive impairments in mutant APP transgenic mice that occur in the absence of NFT formation are prevented by removing endogenous tau protein suggesting that Aβ toxicity is tau-dependent[227] and may rely on the actions of tau at the postsynapse to couple Aβ to NMDA receptor-mediated excitotoxicity.[228] Tau is clearly an important protein in the pathogenesis of AD but *tau* mutations are not a cause of AD. Therefore, a number of transgenic models of FTDP-17 expressing human mutant tau have been developed and have been used in combination with mice harboring FAD mutations in *APP* and *PS1* to simultaneously model Aβ and tau pathology in AD (summarized in Table V).

A. Mouse Models for APP

At least 16 mutations have been described for *APP* (OMIM 104760) of which the K670N/M671L (Swedish), V717I (London), E693G (Arctic), and V717F (Indiana) mutations have been used most widely for the development of transgenic mice. APP is a 770 amino acid transmembrane protein and APP mutations tend to increase the production and/or ratio of toxic Aβ42 peptides due to increased proteolytic processing of APP by β- and γ-secretases, or reduced cleavage by α-secretase. A major component of the intermembrane γ-secretase complex is PS1. Autosomal dominant FAD mutations in *APP* can be modeled in mice by transgenic overexpression of human mutant APP from heterologous promoter elements. PDAPP transgenic mice (line 109) express V717F mutant APP from the human PDGFβ promoter in the cortex, hippocampus, hypothalamus, and cerebellum.[242] PDAPP mice exhibit Aβ deposition and neuritic plaques in the cortex and limbic system first evident at 3 months, extensive gliosis, dystrophic neurites, and synaptic loss.[242] This occurred in the absence of NFT formation or neurodegeneration, while sparse and late phosphorylated tau could be observed in dystrophic neurites.[243] Another mouse

TABLE V

Selected Examples of Mouse Models for AD: APP, ApoE, PS1, and PS2 Mutant Mice

Model	Promoter	Mutation	Phenotype	Reference
APP				
APP23	Thy-1	APP Swedish	AD-like neurological and pathological alterations, including amyloidosis but not NFTs	229
APPArcSwe	Thy-1	APP Swedish + Artic	AD-like neurological and pathological alterations, Aβ aggregates are mostly intraneuronal. No NFTs	230
R1.40 (APP)	Human APP (YAC)	APP Swedish	AD-like neurological and pathological alterations, Aβ pathology	231,232
PS1				
PS1 (flox)	Conditional KO	Neuronal-specific KO	Normal phenotype with alterations in APP processing (global KO is lethal)	233
PS1(A246E) or PS1(A246E)/APPSwe	Thy-1	PS1 A246E/APP Swedish	PS1 mice have altered APP processing and reduced threshold for excitotoxicity, crossing with APPSwe mice enhances PS1 phenotypes	234,235
PS1 H163R	Human PS1 (YAC)	PS1 H163R	Homozygotes are lethal but heterozygotes are normal. Crossing with APP R1.40 mice exacerbates R1.40 Aβ-related phenotypes	236
5XFAD (Tg6799)	Thy-1	APP Swedish/Florida/London + PS1 M146L + PS1 L286V	AD-like neurological and pathological alterations, including neuroinflamation and hippocampal CA1 neurodegeneration. No NFTs	237
PS2				
PS2 KO	KO	Global KO	No AD-like phenotype, but phenotype similar to Notch KO mice	238
PS2 N141I	NSE	PS2 N141I	Elevated Aβ levels but no Aβ pathology or AD-like phenotype	239
ApoE				
ApoE KO	KO	Global KO	Broad effects on lipid metabolism including dysfunction of membrane generation but no AD-related pathology	240
ApoE ε4	Thy-1	C-terminally truncated ApoE ε4	Reduced lifespan and motor deficits, hyperphosphorylated tau forming proto-NFTs, hippocampal neurodegeneration, cognitive impairment	241

KO, knockout; PS1, presinilin-1; PS2, presenilin-2; APP, amyloid precursor protein; ApoE, apolipoprotein E; NSE, neuron-specific enolase.

model, Tg2576 (APPSWE), expresses human APP (1–695) with the Swedish mutation under the control of the hamster PrP promoter. Tg2576 mice have widespread CNS transgene expression and exhibit elevated Aβ production at 3 months, Aβ plaques and memory deficits at 9–12 months.[244] In contrast to PDAPP mice, Tg2576 mice display an equivalent ratio of Aβ40/Aβ42 peptide, whereas PDAPP mice showed abnormally increased Aβ42 levels.[245] Tg2576 mice develop amyloid angiopathy and increased markers of ROS-related damage including reactive microgliosis[246] that is absent from PDAPP mice.[247] Tg2576 mice also develop age-related memory impairments[244] but without neuronal loss.[248]

Transgenic APP mice containing multiple FAD mutations also have been generated to enhance the pathogenic effects of APP. For example, APP22 mice express human APP751 with the Swedish and London mutations driven from the human Thy-1 promoter. APP22 mice express low levels of APP and exhibit diffuse Aβ deposits, neuritic changes, dystrophic cholinergic fibers, and hyperphosphorylated tau.[229] The choice of promoter element, transgene copy number, and chromosome position effects leading to different patterns and levels of transgene expression results in phenotypic differences between models carrying the same APP mutation. APP23 mice, similar to Tg2576 mice, overexpress APP751 with the Swedish mutation at sevenfold over endogenous APP levels in a more restricted manner from the Thy-1 promoter. APP23 mice develop profound inflammation, neuritic and synaptic degeneration, and tau hyperphosphorylation. Furthermore, APP23 mice develop cerebral amyloid angiopathy as in AD[229,249] with microhemorrhages,[250,251] stroke,[252] and gliosis.[253–255] Surrounding Aβ plaques are dystrophic neurites, with aberrant axonal sprouting and ectopic formation of nerve terminals[256] but as with all APP-based mouse models, NFT formation and frank neuronal degeneration are lacking. APP23 mice robustly model several key features of AD and have been extensively employed for neuroimaging,[257–264] neurochemical,[265–269] and cognitive and behavioral analysis.[267,270–273] In the past 12 months alone, 13 years after the first description of APP23 mice, at least 18 studies were published relating to APP23 mice, ranging from the use of ibuprofen to treat AD[274] to the dysregulation of neuronal microRNAs in response to Aβ.[275] Similar AD-related phenotypes have been observed in APP transgenic models containing FAD mutations expressed from the Thy-1 promoter, including Tg-SwDI mice expressing APP770 with triple Swedish, Dutch (E693Q), and Iowa (D694N) mutations. This model is of particular interest because it develops Aβ pathology even though it expresses human APP below endogenous levels.[276] The phenotype of Tg-SwDI mice includes cerebral amyloid angiopathy, Aβ deposition, gliosis, reactive microglia, and cognitive impairment, but without NFTs or neurodegeneration.[276] Another interesting combination of APP-related FAD mutations is the APPArcSwe mouse expressing human APP with combined

Arctic and Swedish mutations from the Thy-1 promoter.[230] APPArcSwe mice display intraneuronal soluble Aβ aggregates, robust Aβ plaques, and although cerebral amyloid angiopathy is present, Aβ plaques are mostly parenchymal.[230,277,278]

Transgenic APP models have also been developed using the endogenous promoter and regulatory elements of the *APP* gene in yeast artificial chromosome (YAC) expression constructs to produce endogenous patterns of transgene expression.[279,280] B6-R1.40 mice contain four to six copies of a 650-kb YAC containing 300-kb of the human *APP* gene harboring the Swedish mutation.[231] These mice express human APP at two- to threefold endogenous APP levels but accumulate 25-fold total Aβ and 10-fold Aβ42 peptide. B6-R1.40 mice exhibit a high density of Aβ deposition (60–70-fold control), dystrophic neurites, altered tau immunoreactivity, gliosis, and microglial activation occurring at more than 1 year. ApoE immunoreactivity also overlapped with Aβ deposits.[232] Several coisogenic strains of the R1.40 line (B6-R1.40, D2-R1.40, and 129S1-R1.40) have been generated and, as they differ in the levels of Aβ production, have been useful for characterizing genetic differences in APP processing and subsequent Aβ deposition despite having identical APP transgene copy number.[281] Three genetic loci have been identified in linkage studies that are responsible for modulating APP processing, two of which are shared in the corresponding human genome as displaying linkage to PD.[282]

Another type of APP transgenic model is based on tetracycline-regulatable (Tet-Off) transgene expression which permits the spatial, temporal, and repressible expression of APP. This allows one to model the potential effects of therapies that inhibit APP processing. TetO-APPSwe/Ind mice (line 885) express human APP (1–695) with FAD-linked Swedish and Indiana mutations under the control of a tetracycline-responsive promoter (tetO) that permits mutant APP expression following crossbreeding to promoter-specific tetracycline transactivator (tTA) driver mice. Resulting APP/tTA bigenic mice expressed APP at 10–30-fold over endogenous levels resulting in robust Aβ pathology, and doxycycline administration inhibits APP expression up to 95% and reduces Aβ production.[283] Although Aβ pathology is halted in these mice, Aβ plaques appear to require far longer to disperse than to initially assemble with little signs of active Aβ clearance suggesting that Aβ deposits are in disequilibrium and are highly stable structures. This observation has important implications for arresting Aβ production in AD patients through inhibition of APP processing.

Taken together, transgenic mouse models overexpressing APP-related FAD mutations consistently reproduce neuronal and vascular Aβ deposition, and Aβ-related neuritic abnormalities, synaptic pathology, astroglial and microglial reactivity, enhanced tau phosphorylation, and cognitive impairments. However, APP mice fail to reproduce the tau-positive NFT pathology and neuronal degeneration that is characteristic of advanced AD.

B. Mouse Models for Apolipoprotein E ε4 (ApoEε4)

The *ApoEε4* allele is associated in a gene dosage-dependent manner with an increased risk of developing the common late-onset form of AD and reducing the age at disease onset.[221] Individuals who are homozygous for the *ApoEε4* allele display an increased risk by one order of magnitude for developing AD, and in some ethnicities it is a virtual determinant of AD.[221] Recent GWA studies identified *ApoEε4* as the major risk factor for AD by several orders of magnitude over other susceptibility factors.[215,284] ApoE binds to cholesterol- and lipid-containing particles, and mediates their incorporation into cells. In the brain, ApoE is synthesized not only by astrocytes and microglia but also by neurons under certain conditions,[285] where it mediates cholesterol storage and redistribution, and might modulate neuronal plasticity.[286] It was previously hypothesized that ApoE may function as an Aβ chaperone[287] because it strongly binds to Aβ[288] but this role remains controversial. *ApoE* is polymorphic with three major isoforms (*ApoE2*, *ApoE3*, and *ApoE4*) which translate into three alleles of the gene (ε 2, 3, and 4).

Deletion of the *ApoE* gene[240] results in hypercholesterolemia and arterial lesions[289] and in the CNS altered synaptic morphology,[290] cochlear neurodegeneration,[291] abnormal astroglial physiology,[292] blood–brain barrier dysfunction,[293] age-dependent synaptic damage,[247,294,295] exacerbated susceptibility to several insults, and diminished regenerative capacity.[296,297] Several transgenic mouse models overexpressing ApoE alleles and knockin (KI) mice have been generated. ApoE4 KI mice contain human ApoE4 in place of mouse ApoE4 under the control of the mouse ApoE promoter. ApoE4 KI mice show elevated brain cholesterol levels but this modification neither influenced Aβ production[298] nor were neuropathological abnormalities apparent.[299] Nevertheless, strong overexpression of the *ApoEε4* allele from the Thy-1 or PDGFβ promoters causes motor abnormalities, muscle wasting, and severely reduced lifespan that manifests in the absence of Aβ deposits, NFTs, and neuronal loss but with gliosis, ubiquitin-positive inclusions, and tau hyperphosphorylation.[300] C-terminal truncated forms of ApoE4 accumulate in AD brains, are associated with NFTs, and overexpression of these truncated forms in mammalian cells produced NFT-like structures.[301] Transgenic mice overexpressing truncated Δ272–299 and Δ241–299 forms of ApoE4 from the Thy-1 promoter have been developed[241] that exhibit AD-like neurodegenerative alterations in the cortex and hippocampus, including proto-NFTs containing hyperphosphorylated tau, and memory impairment.[241] While the *ApoEε4* allele is strongly associated with AD risk, the absence of disease-causing mutations makes it difficult to determine the exact contribution it plays in the pathogenesis of AD.

C. Mouse Models for Presenilin 1

Mutations in the *PS1* gene (OMIM 104311) at the AD3 locus cause autosomal-dominant early-onset AD.[218] Subjects with *PS1* mutations often exhibit atypical characteristics such as early-onset disease in their 1940s, spastic paraparesis which precedes dementia, Aβ plaques with characteristic cotton-like appearance that lack a congophilic dense core, and the absence of marked neuritic dystrophy.[302–304] PS1 is a component of the intermembrane γ-secretase complex that processes APP to produce an APP intracellular domain (AICD) and either an extracellular nontoxic p3 domain or a toxic Aβ peptide. γ-Secretase is an atypical protease in that it cleaves the processed C99 fragment of APP within the membrane and this digestion is not completely specific, producing Aβ peptides of different size (39–43 residues) with diverse aggregating properties. More than 160 FAD-linked *PS1* mutations have been identified so far that all tend to increase the production of Aβ42 peptide from APP.[305] Homozygous *PS1* KO mice exhibit perinatal lethality and embryos exhibit severely impaired neurogenesis, massive neuronal loss, CNS hemorrhages, and skeletal defects.[306] Heterozygous *PS1* mice are viable and fertile similar to WT mice implying that a single *PS1* allele is sufficient for development and survival. Conditional homozygous deletion of *PS1* in floxed mice through crossing to Thy-1-driven Cre mice to produce neuronal-specific deletion results in viable and fertile mice with decreased levels of Aβ40 and Aβ42 peptides but an accumulation of other APP fragments, indicative of altered APP processing.[233] PS1 is not essential for maintaining neuronal architecture or function.[233] Multiple transgenic mice overexpressing human PS1 harboring FAD mutations have been developed. For example, mice expressing human PS1 with the FAD-linked A246E mutation from the mouse Thy-1 promoter exhibit a normal neuropathological profile but display increased sensitivity to excitotoxicity *in vivo* and *in vitro* together with alterations in calcium homeostasis.[234] PS1 transgenic mice containing the M146L or Δexon9 mutations from a PDGFβ promoter display no abnormal pathology up to 2.5 years, whereas similar mice with M146V PS1 exhibit elevated Aβ42, altered mitochondrial activity, and dysregulation of calcium homeostasis.[307] Thus, expression of FAD-linked mutant PS1 in mice fails to produce significant Aβ and NFT pathology or neurodegeneration. Combining mutant forms of PS1 and APP in 5XFAD mice (Tg6799), which overexpress APP with the Swedish (KM670/671NL), Florida (I716V), and London (V717I) mutations together with PS1 with the M146L and L286V mutations, results in neurological and pathological alterations including neuroinflammation and neurodegeneration within the hippocampal CA1 region but without NFT formation.[237] Similarly, APP/PS1 bigenic mice coexpressing APP with the KM670/671NL Swedish mutation and PS1 with an L166P mutation from the Thy-1 promoter develop early cerebral amyloidosis beginning at 6–8 weeks, dystrophic synaptic boutons, hyperphosphorylated tau-positive neuritic structures, gliosis, and subtle loss of dentate gyrus neurons but without NFT formation or overt cortical neurodegeneration.[308] Similarly, PS1 ΔE9 mice crossed with APPSwe mice produced early Aβ plaque deposition at 6 months, and at later ages, a mild reduction in cortical and hippocampal cholinergic markers, reduced somatostatin levels, and impaired cognition and memory.[309] Thus, combining PS1 and APP mutations has succeeded in producing mice with enhanced Aβ deposition, Aβ-related synaptic pathology, and cognitive deficits.[310]

YAC transgenic mice have been developed for expressing FAD-linked PS1 mutations. B6-G9 mice containing the entire human *PS1* gene with the H163R mutation express PS1 at half the levels of endogenous PS1 and produced a number of alternate mRNA and protein species.[236] Heterozygous B6-G9 PS1 mice are viable and fertile, whereas homozygous mice are not viable. Combining B6-G9 PS1 mice with FAD-linked APP mice (B6-R1.40) elevated the levels of Aβ42 peptide and produced earlier Aβ deposition compared to the parent lines.[236] *PS1* knockin mice have been developed with, for example, the FAD-linked P246L mutation where, unlike conditional KO mice, primary cortical neurons are not more sensitive to Aβ, staurosporine or excitotoxic insults. Crossing PS1 P246L knockin mice with APPSwe knockin mice led to increased Aβ42 production and accelerated Aβ deposition.[311] Another PS1 knockin model bearing the M146V mutation is generally normal without neuropathological or behavioral abnormalities but which are hypersensitive to excitotoxicity[312] and can exhibit mild behavioral impairments depending on the genetic strain background.[313]

Taken together, mice harboring FAD-linked PS1 mutations exhibit altered APP processing, elevated Aβ42 peptide but without robust Aβ deposition. Combining FAD-linked APP and PS1 mutations in transgenic mice can enhance Aβ pathology, synaptic abnormalities, and associated cognitive deficits but fails to elicit NFT formation and overt neurodegeneration. Collectively, these models have challenged the validity of the amyloid cascade hypothesis because it has not yet been possible to demonstrate that enhanced Aβ accumulation and aggregation due to FAD-linked APP, PS1, or both mutations precipitate NFT formation and/or significant neurodegeneration. The lack of NFT formation correlates with the general absence of neurodegeneration. This has prompted the incorporation of mutant tau transgenic mice to facilitate the development of NFTs in AD mouse models. For example, 3xTg-AD triple transgenic mice with knockin of PS1 (M146V) and transgenes expressing human APP (Swe), and tau (P301L) mutations develop Aβ plaques and tau-positive NFTs together with synaptic dysfunction that precedes pathology and cognitive deficits that both correlate with the early accumulation of intraneuronal Aβ.[314–316] Although 3xTg-AD mice reproduce many of the key pathological hallmarks of AD, the use of tau mutations associated with the neurodegenerative tauopathy, FTDP-17, raises questions of whether the progressive NFT pathology in these mice is fully representative of AD. Nevertheless, these mice provide an elegant model for understanding the relationship between Aβ and tau-positive NFT pathology.

D. Mouse Models for Presenilin 2

Mutations in the *PS2* gene (OMIM 600759) at the AD4 locus cause autosomal dominant, familial AD with a mean age at onset of 55 years that overlaps with sporadic AD.[219,220] *PS2* mutations are a much less common cause of AD than mutations in *APP* and *PS1*. The characteristic neuropathology of *PS2*-linked AD is the development of moderate to severe amyloid angiopathy that often progresses to a lethal acute intracerebral hemorrhage. Neuritic plaques and Lewy bodies are less frequent than in other forms of AD.[317,318] Homodimers of PS2 can potentially substitute for PS1 in the γ-secretase complex to process APP but might also play a role in Notch signaling.[238] PS2 KO mice are viable similar to Notch-deficient mice and do not exhibit neuropathological or behavioral abnormalities. PS2 null mice exhibit lung abnormalities at 3–6 months with hemorrhages and ongoing apoptosis in lung parenchyma and vascular endothelium, bronchial epithelium, and alveoli, which develops into pulmonary fibrosis.[319] Another PS2-deficient model shares a phenotype similar to Notch KO mice and also exhibits Aβ42 accumulation only upon crossing to PS2 transgenic mice with the FAD-linked D366A mutation.[238] Several PS2 transgenic mice have been developed containing the FAD-linked N141I mutation from different promoters including the chicken β-actin or neuron-specific enolase promoters that exhibit increased Aβ42 levels at advanced ages but lack obvious neuropathology.[239,320] Mice coexpressing mutant PS2 (N141I) and APP (Swedish) exhibit an age-dependent increase in Aβ42 levels, severe cerebral amyloidosis and gliosis, and cognitive impairment.[321] Collectively, PS2 deletion or overexpression, similar to PS1, fails to produce significant Aβ pathology and Aβ-related abnormalities unless combined with FAD-linked APP mutations.

Mouse models based upon APP present cerebral amyloidosis and Aβ-related deficits but lack tau-positive NFTs and neuronal loss. Certain ApoEε4 models develop early tau pathology but it has been difficult to establish a mechanistic relationship between ApoE and APP processing. Mouse models based upon presenilins reveal that while PS1 is essential for development, PS2 may only play a subsidiary role in APP processing. PS1 and PS2 mouse models do not reproduce AD-like pathology but are useful tools for dissecting the molecular basis of APP processing. Transgenic models with combinations of FAD-linked mutant forms of APP, PS1, and PS2 tend to faithfully reproduce Aβ pathology, cerebral amyloidosis, and cognitive deficits similar to AD but without the NFT pathology, significant neurodegeneration and severe cortical atrophy that is characteristic of disease. The introduction of FTDP-17-linked mutant tau to these APP and/or PS1 models may help to recreate NFT pathology and associated abnormalities. APP/PS1 models with robust cerebral amyloidosis may require a subsequent insult such as oxidative stress, prolonged aging, or a permissive genetic background, to initiate tau hyperphosphorylation in order for disease to progress. Tau hyperphosphorylation and NFT formation are critical steps in AD for producing neuronal degeneration and full cognitive decline. ApoEε4 can provide such a genetic insult but alone fails to reproduce the full disease without the appropriate genetic background. Differences in genetic background between mouse strains may determine the extent of secondary damage required to precipitate the full disease spectrum. At this stage, a genetic mouse model that faithfully recreates all key neuropathological features of AD is still lacking, aside from 3xTg-AD mice containing FTDP-17 mutant tau,[322] perhaps suggesting that mice are not able to appropriately model the pathophysiology of AD, or that significant aging could be required beyond the lifespan of mice to model the chronic nature of disease in humans. Current AD mouse models serve important roles in the validation of therapeutic approaches to alter APP processing, lower Aβ accumulation and aggregation, and target tau hyperphosphorylation and tau-positive NFT pathology.

IV. Mouse Models for HD

HD (OMIM 143100) is a progressive neurodegenerative disorder with age at onset between 30 and 40 years with juvenile-onset in some cases. The vast majority of HD cases are familial with autosomal-dominant inheritance, while sporadic cases are rare. There is a prodromal phase of disease lasting for up to 10 years with mild psychotic and behavioral symptoms that precede the characteristic motor symptoms. The clinical symptoms of HD are characterized by a progressive chorea, dystonia, impaired coordination often together with cognitive deficits, and behavioral difficulties, including personality changes. The primary mechanism of the disease is impaired cellular energy production, characterized by dramatic metabolic deficits caused by a negative energy balance, glucose hypometabolism, reduced mitochondrial activity, and severe muscle wasting. The hallmark pathology of HD consists of the selective loss of medium-sized spinal output neurons in the caudate putamen that results in a severe atrophy. The underlying cause of HD is usually genetic and is characterized by a triplet CAG repeat expansion encoding a polyQ expansion, occurring in exon 1 of the *huntingtin* (*Htt*) gene (OMIM 613004) in a heterozygous manner.[323] Homozygosity for the expanded *Htt* gene does not result in a more aggressive disease. The length of the polyQ expansion directly correlates with the age of disease onset and disease is fully penetrant with more than 41 repeats, partial with 36–40 repeats, whereas normal individuals have between 9 and 36 repeats.[324] A number of transgenic models have been generated to model autosomal-dominant HD which employ exon 1 or full-length Htt (summarized in Table VI).

Htt null mice are embryonic lethal in the homozygous state suggesting that Htt is essential for development and survival. Heterozygous KO mice are viable and otherwise normal.[325] Transgenic models have been developed expressing Htt with disease-associated polyQ expansions from heterologous or YAC-based promoters. Expression of a truncated N-terminal fragment of human Htt

TABLE VI
OVERVIEW OF MOUSE MODELS OF HD

Model	Promoter	PolyQ length	Phenotype	Reference
Knockout				
KO	–	KO	Lethal at E8.5, ectoderm fails to madurates	325
Transgenic (cDNA)				
HD R6/1	Native Htt	116	Progresive full HD phenotype, onset at 12 weeks (homozygous), 32 weeks (heterozygous)	326
HD R6/2	Native Htt	144	Progresive full HD phenotype at 8 weeks with rapid onset and progression	326
N171-82Q	PrP	82	Full HD phenotype, onset at 12 weeks, lethal at 25 weeks	327
HD100	NSE	100	Full HD phenotype	328
Transgenic (YAC)				
YAC72	Native Htt	72	Progresive full HD phenotype w/onset at 54 weeks	329
YAC128	Native Htt	128	Progressive full HD phenotype w/onset at 36 weeks	329
Knockin				
HdhQ20	Endogenous Htt	18	No HD phenotype	330
HdhQ50	Endogenous Htt	48	No HD phenotype	331
HdhQ92	Endogenous Htt	90	Histopathological and biochemical changes at 54 weeks, progressive, incomplete HD	332
HdhQ111	Endogenous Htt	111	Histopathological and biochemical changes at 40 weeks, progressive full HD phenotype	332
HdhQ150	Endogenous Htt	150	Progressive full HD with disease onset at 54–64 weeks	333

Htt, huntingtin; Q, glutamine; NSE, neuron-specific enolase; YAC, yeast artificial chromosome.

(N171) containing 18, 44, or 82 glutamines from the PrP promoter produced abnormal phenotypes for N171-82Q and N171-44Q mice but not N171-18Q mice. N171-82Q mice display a reduced lifespan of 5–6 months, decreased body size, abnormal gait, bradykinesia, impaired coordination, and tremor. The mice exhibit neurodegeneration in several hypothalamic areas, nuclear inclusions composed of Htt, and neuritic Htt aggregates in the cortex and striatum.[327] A similar transgenic model expresses the initial N-terminal third of the mutant human *Htt* gene (IT15) from a neuron-specific enolase promoter with an expansion of 100 CAG repeats. These HD100 mice exhibit a phenotype similar to that of N171-82Q mice with motor deficits evident from 3 to 10 months and similar neuropathology including striatal intranuclear Htt inclusions in striatum and cortical regions, with cytoplasmic accumulation of Htt in the cortex.[328] Dendrites in these regions were dysmorphic with plasma membrane blebs and vacuolation. Striatal neurons further exhibited altered synaptic transmission.[328] Transgenic mice (i.e., R6/1 and R6/2 lines) have also been developed using the 5′-end of the human Htt gene carrying 115–150 CAG repeat expansions[326] that develop a progressive neurological phenotype. Hemizygous R6/1 mice express a 116 CAG repeat expansion that produces neuronal intranuclear inclusions in striatum and cortex, and abnormal striatal morphology,[334] abnormal spatial learning, and impaired motor coordination[335] at 12 weeks of age. Hemizygous R6/2 mice express a 144 CAG repeat expansion producing a more robust phenotype similar to that of R6/1 mice with early functional and cognitive impairments at 3 weeks, a full HD-like phenotype by 9–11 weeks, and premature death at 10–13 weeks.[326]

Mouse models expressing the full *Htt* gene have been developed using YAC constructs, including normal Htt with 18 CAG repeats and mutant Htt with 46, 72, and 128 CAG repeat expansions.[329] YAC72 mice express human Htt in many tissues at levels similar to endogenous Htt. At 6 months, YAC72 mice display electrophysiological abnormalities (abnormal excitatory postsynaptic potential [EPSP] and LTP), behavioral abnormalities at 7 months, and at 12 months they develop selective neurodegeneration in the striatum with intranuclear Htt aggregates.[329] YAC128 mice develop behavioral and motor deficits as early as 2 months that progress to a full HD-like phenotype with an age-dependent reduction in striatal and cortical volume, significant neuronal loss at 12 months, and intranuclear Htt inclusions at 18 months.[336] Markers of excitotoxicity similar to human patients (3-hydroxykynurenine and quinolinate) are detected in YAC128 mice.[337] BAC transgenic mice contain the full *Htt* gene with a 97 polyQ expansion exhibit phenotypes similar to that of YAC128 mice at 12 months, with age-dependent motor impairment as early as 2 months and severe brain atrophy at 12 months.[338] Htt knockin mice with introduction of 20, 50, 92, or 111 polyQ expansions have been developed,[332] where a polyQ expansion of 20 or 50 produces a normal phenotype,[330,331] and polyQ of 92 produces intranuclear inclusions and neurological abnormalities at 12 months[339] and increased 3-hydroxykynurenine and quinolinate levels at 15 months.[337] Knockin mice with a 111 polyQ expansion show similar phenotypes but develop hypoactivity, tremors, irritability, and limb grasping beginning at 12 months.[330] Another Htt knockin model with a 150 polyQ expansion (Hdh (CAG)150) exhibits reduced body size and gliosis by 13 months of age but without evidence of brain atrophy or neurodegeneration.[333] A loss of striatal dopamine receptor density and neuronal number, and significant motor abnormalities are observed at advanced ages in Hdh(CAG)150 mice.[340] Collectively, mouse models of HD based upon polyQ expansions within the N-terminal region of the Htt protein have succeeded in robustly and faithfully reproducing many key pathological and behavioral hallmarks of HD, including age-dependent disease progression, intranuclear inclusions, overt neurodegeneration, motoric deficits, and premature death, with a strong correlation between polyQ expansion length and disease severity. These robust HD models have proved instrumental in exploring the molecular basis of disease and in the identification of potential disease-modifying pathways and therapies. For example, neurotoxicity elicited by polyQ-expanded Htt proteins in mice may act through mechanisms involving transcriptional regulation,[341] neurotrophic factors (i.e., brain-derived neurotrophic factor),[341,342] cholesterol metabolism,[343,344] PGC-1α,[345,346] sirtuins,[343,347] histone deacetylases/chromatin modification,[348] autophagy,[349] CREB signaling,[350] and many other pathways. The major challenge now is to identify the primary events that initiate neuronal dysfunction and degeneration in HD.

V. Mouse Models for ALS

ALS is a rare adult-onset motor neuron disease, which affects both upper and lower neurons of the spinal cord and is clinically characterized by progressive weakness, spasticity, paralysis, and death within 3–5 years of initial symptom onset.[351] Most ALS cases are sporadic (SALS) with unknown etiology, whereas 10% are familial (FALS).[351,352] The majority of familial cases are caused by missense mutations in the *Cu-Zn-superoxide dismutase* (*SOD1*, *ALS1*) gene (OMIM 147450) and occur in an autosomal dominant, adult-onset manner.[353,354] Four additional genes are clearly associated with FALS, including *angiogenin* (*ANG*, *ALS9*),[355–357] *TAR DNA-binding protein 43* (*TDP-43*, *ALS10*; OMIM 605078), *fused in sarcoma/translated in liposarcoma* (*FUS/TLS*, *ALS6*),[358,359] and more recently *optineurin* (*OPTN*, *ALS12*).[360] Collectively, mutations in these five genes account for only 20–30% of all FALS.

TDP-43 was initially discovered as a component of ubiquitin-positive inclusions in frontotemporal lobar degeneration (FTLD) and ALS.[361] Approximately, 20% of FALS subjects develop clinical features of FTLD.[362,363] Pathological alterations of SOD1 and TDP-43 proteins are observed in the majority of sporadic ALS cases suggesting that both proteins may also play an important role in sporadic disease.[364–367] In the past 2 years, 13 mutations in the *FUS/TLS* gene have been identified as a cause of familial ALS, with the characteristic accumulation of FUS/TLS protein in the cytoplasm of neurons, a pathology similar to that caused by TDP-43 mutations.[358,359] Recently, three mutations in *OPTN*, including a homozygous exon 5 deletion or Q398X nonsense mutation and a heterozygous E478G missense mutation, have been shown to cause familial forms of ALS. Interestingly, TDP-43- or SOD1-positive inclusions found in sporadic and familial SOD1-linked brains of ALS subjects are also noticeably immunoreactive for OPTN, thus supporting its role in ALS.[360]

A small proportion of familial ALS cases harbor mutations in other genes including loss-of function mutations in *Alsin* (ALS2), encoding a small GTPase,[368,369] *senataxin* (ALS4), encoding a DNA/RNA helicase,[370] *vesicle-associated membrane protein/synaptobrevin-associated membrane protein B* (*VAPB*, ALS8),[371] and several cytoskeletal proteins including tau and the p150 dynactin-1 subunit.[223,372] It is important to note that only SOD1-linked familial ALS cases are referred to as classical adult-onset Charcot-type ALS, while familial ALS due to mutations in other genes represent atypical ALS syndromes.

A. SOD1 Transgenic Mouse Models

SOD1 functions as an antioxidant protein and is ubiquitously expressed with predominant localization in the cytoplasm but it is also localized within mitochondria,[373,374] the nucleus,[375] and the ER.[376] Presently, over 150 mutations in the *SOD1* gene have been identified in association with FALS.[377–379] FALS mutations include 110 missense mutations, 3 in-frame coding deletions, and 7 mutations in noncoding regions which promote aberrant mRNA splicing.[379,380] Certain clinical features, disease onset and duration, are closely linked with particular pathogenic mutations.[381] For example, the G93A mutation causes rapid disease progression and severe neurological symptoms over 1–4 years, whereas the H46R mutation causes slower disease progression and milder symptoms over 1–24 years.[381] Red cells derived from subjects with heterozygous SOD1 mutations display less than 50% of normal SOD1 activity[382] which prompted the generation of SOD1 null mice.[383–385] KO mice are viable without obvious motor abnormalities, while adult mice show increased sensitivity to axotomy[385] and paraquat exposure[384] and display age-dependent peripheral axonopathy and muscle atrophy that correlates with progressive motor deficits.[386–388] Transgenic mice overexpressing WT human SOD1[389,390] exhibit mild age-dependent axonal loss, motor neuron degeneration,[389,391] and impaired regeneration after neuronal injury.[392]

A number of transgenic mice expressing FALS-linked mutant SOD1 have been generated to model disease (summarized in Table VII). Transgenic mice expressing human SOD1 harboring the G93A mutation, which causes rapid and robust phenotypes in humans are commonly used in ALS research.[389,401–403,407] As with most transgenic mouse studies, transgene copy number and expression levels strongly influence the severity of clinical and neuropathological phenotypes in these SOD1 mice. Models with high-level expression of G93A mutant SOD1 develop early-onset disease with rapid progression leading to premature death, together with neuronal loss, mitochondrial vacuolation, neurofilament-positive inclusions, and SOD1-positive aggregates.[401–403] Models with low expression of mutant SOD1 display a later onset and slower progression of disease,[389] highlighting the gene dosage-dependent pathological effects of FALS-linked SOD1 mutations. Mutations in the copper-binding domain cause impaired dismutase activity of SOD1. Accordingly, several transgenic mice expressing single, double, or quadruple FALS-linked SOD1 mutations have been developed.[395–397] These three models develop late-onset disease with characteristic ALS-like cytopathology, suggesting a normal neuroprotective role for SOD1 activity in the preventing neuronal degeneration. Mice expressing the SOD1 G37R mutation with normal enzymatic activity[394,408] still develop early-onset disease with mitochondrial-derived vacuoles in axons and dendrites similar to G93A SOD1 mice.[394] Importantly, mitochondrial pathology that is typically observed in G93A and G37R SOD1 mice represents an extremely rare feature of human ALS, suggesting that such phenotypes observed in transgenic mice might result from secondary effects due to nonphysiological levels of SOD1 expression. Mice expressing G85R or G86R mutant SOD1 that are enzymatically inactive develop a rapidly progressive disease.[398,399] Transgenic SOD1 mice with the A4V mutation were developed as this mutation produces the most aggressive phenotype in human ALS subjects.[379] Surprisingly, however, A4V mice do not develop clinical or pathological features of disease. Coexpression of A4V SOD1 with WT SOD1 leads to a dramatic conversion to an ALS-like phenotype comparable to G93A mice,[393] suggesting the formation of toxic A4V/WT heterodimers. These observations suggest that FALS-linked mutations may not primarily act through causing a loss of SOD1 activity but may instead acquire toxic properties unrelated to its antioxidant capacity.

Among FALS-linked mutations, a small number of mutations (in codon 126) result in a frameshift and premature truncation of the SOD1 protein. Truncations may lead to structural changes, disruption of critical disulphide bonds (between C57 and C146 residues), and impaired dimerization that may

TABLE VII

SOD1 Knockout and Transgenic Mice as Models of ALS

Mutation	Disease onset (months)	Disease duration (months)	Additional phenotypes	Reference
KO	–	–	No marked pathologic changes, motor neuron loss only after axonal injury	385
WT	–	–	Hypotonia, neuromuscular pathology	390
WT	–	–	Age-dependent vacuolar pathology, axonal loss, motor neuron degeneration	389
A4V + WT	8	3	Coexpression with WT converts unaffected A4V mice to ALS-like phenotype	393
G37R	4–6	ND	Mitochondrial-derived vacuoles in axons and dendrites	394
H46R	5	1	No superoxide scavenging activity	395
H46R/H48Q	4–6	ND	No superoxide scavenging activity, develop motor neuron disease, thioflavin S-positive inclusions	396
H46R/H48Q /H63G/H120G	8–12	ND	No superoxide scavenging activity, protein inclusions	397
G85R	8–14	0.5	No changes in SOD1 activity, SOD1-/ubiquitin-positive inclusions, and aggregates in neurons	398
G86R murine	3–4	1	Degeneration in spinal cord motor neurons, normal enzymatic activity	399
D90A	12	2	SOD1 aggregates in spinal cord and abundant hSOD1 inclusions and vacuoles	400
G93A (TC-18)	3–4	1–2	Mice become paralyzed due to spinal cord motor neuron loss with premature death at 5–6 months of age	401
G93A (TC-25)	3	1–2	Mitochondrial vacuoles accumulating in motor neurons, significant death of somatic motor neurons innervating limb muscles	402
G93A (TC-8)	6–8	2–3	Motor neurons sensitive to excitotoxic effects of ambient glutamate, progressive neurological disease, and paralysis	403
G93A homozygous (TC-10)	10	3–4	Severe vacuolar degeneration of motor neurons, Lewy-like pathology and swollen axons	403
G93A hemizygous (TC-2)	> 10	ND	Atrophy, astrogliosis, ubiquitinated Lewy body-like inclusions and axonal swellings, but minimal vacuolar pathology	389
L126Z	7–9	ND	Late-onset disease with rapid progression to death	404
L126delTT	15	1	Loss of motor neurons, reactive gliosis, inclusions in surviving motor neurons	405
G127X	8	0.25	Late-onset disease with rapid progression to death	406

TC, transgene copy number; ND, not detected.

alter SOD1 function. Several C-terminally truncated SOD1 transgenic mice have been developed to model these effects.[404–406] Overexpression of these truncated SOD1 variants (L126X, L126delTT and G127X) causes late-onset disease (9–15 months) but with rapid disease progression (1–4 weeks), even though such mutants are highly unstable and only accumulate in minute quantities. This might further suggest a role for SOD1 monomers in the pathogenesis of ALS.[404–406]

Taken together, transgenic expression of mutant human SOD1 consistently results in a fatal ALS-like motor neuron disease and supports a dominant gain-of function mechanism for familial SOD1 mutations. The underlying mechanism and specificity of motor neuron degeneration in ALS and transgenic models still remain unclear. Recently, a critical role of the ER, responsible for posttranslational modification and protein sorting, has been highlighted for the pathogenesis of ALS and other neurodegenerative diseases.[409–411] Moreover, markers of ER stress and induction of the unfolded protein response (UPR) are commonly observed in brains of ALS subjects and mutant SOD1 transgenic models.[412] ER stress precedes clinical symptoms in G93A mice, and certain UPR pathway proteins, including protein disulphide isomerase (PDI) and C/EBP homologous protein (CHOP), are upregulated in affected brain regions of G93A mice. PDI is also upregulated in sporadic ALS brains.[409,410] Inactivation of PDI by S-nitrosylation due to nitrosative stress results in protein misfolding and accumulation in PD and AD brains perhaps suggesting a similar mechanism might apply in ALS.[413] SOD1 transgenic mice have proved extremely useful for understanding the molecular basis of ALS and for identification of disease-modifying genes and compounds.

B. TDP-43 Transgenic Mouse Models

TDP-43 is a 43-kDa protein normally localized to the nucleus, which was identified as a major component of ubiquitin-positive, tau-, and α-synuclein-negative neuronal and glial intracytoplasmic inclusions in FTLD as well as in ALS.[361,414,415] Several mutations in the *TDP-43* gene were identified that cause rare familial forms of ALS, including A315T,[416] Q331K, M337V, G294A, and A90V.[417] To further understand the role of TDP-43 in FTLD and ALS, a number of transgenic mice have been developed.[418–421] TDP-43 deletion through gene-targeting in mice results in peri-implantation lethality between embryonic days 3.5 and 6.5 suggesting an essential role for TDP-43 in embryonic development.[421] Overexpression of WT human TDP-43 from the forebrain-specific CaMKIIα promoter results in impaired learning and memory, progressive motor deficits, hippocampal atrophy, and the characteristic formation of TDP-43-/ubiquitin-positive neuronal cytoplasmic inclusions (NCIs) and TDP-43-depleted nuclei.[419] These mice provide evidence that elevation of WT TDP-43 levels is sufficient to precipitate TDP-43-/ubiquitin-positive NCIs and neurodegeneration similar to FTLD and ALS. Moreover, additional WT TDP-43 transgenic models display transgene dosage-dependent degeneration of cortical, subcortical, and spinal cord motor neurons, accumulation of TDP-43-/ubiquitin-positive nuclear and cytoplasmic aggregates that also phosphorylated, and the formation of characteristic C-terminally truncated TDP-43 species in nuclear fractions similar to those observed in ALS and FTLD brains.[420] Thus, nonmutated WT TDP-43 is neurotoxic and most likely acts through a toxic gain-of-function mechanism suggesting an active role for TDP-43 in the development of familial and sporadic forms of FTLD and ALS. The appearance of nuclear C-terminal fragments of TDP-43 in transgenic mice correlates with disease development and progression suggesting that they are pathologically relevant species that may act through a gain of aberrant nuclear function. Transgenic mice expressing human TDP-43 bearing pathogenic mutations (A315T and M337V) develop a severe progressive myopathy due to high levels of transgene expression in skeletal muscle.[418] WT, A315T, and M337V TDP-43 mice develop progressive weakness with characteristic CNS pathology, including cytoplasmic ubiquitin inclusions and TDP-43 fragmentation.[418] Collectively, TDP-43 plays an important role in ALS and FLTD, and TDP-43 mice provide valuable models to explore the pathological role of TDP-43 in neurodegenerative diseases.

VI. Conclusions

Genetic mouse models of neurodegenerative disease provide important tools for understanding the molecular basis of familial disease with the expectation that similar mechanisms are conserved and account for sporadic disease. Mouse models have provided critical proof that disease-associated mutations identified in human disease are responsible for, and can often recapitulate, neurodegeneration and associated behavioral deficits and pathologies. Such models should also ideally reproduce key pathological features of neurodegenerative disease to permit a detailed analysis of disease onset and progression and the critical molecular and cellular changes involved at each step. Similar studies are not possible in humans as it is difficult to diagnose disease in its early stages before symptom onset, and because only noninvasive imaging approaches are available for assessing brain structure and function. Mouse models of disease provide a critical tool for translational research for the initial validation of disease mechanisms and subsequently for testing disease-modifying therapies. Models of HD and ALS are at a sufficiently advanced stage in that they faithfully model various key aspects of disease and have been used for preclinical studies. Models of AD are evolving rapidly but do not yet model the full spectrum of disease pathology and critically lack substantial neurodegeneration. Current AD models have proved valuable for understanding the mechanisms of APP processing, Aβ production and aggregation, cerebral amyloidosis, and Aβ-related cognitive impairments. Mouse models of PD have suffered from a lack of robust disease-relevant phenotypes but efforts are ongoing to identify critical endogenous, exogenous, or genetic factors that can unmask or exacerbate disease phenotypes. Continuing refinement of PD disease models, utilization of new technologies and/or species, and less emphasis on traditional dopaminergic-related phenotypes will hopefully lead to advances in developing appropriate disease models. Genetic mouse models will continue to provide invaluable tools for understanding disease mechanisms and for identifying and validating novel therapeutic approaches for treating neurodegenerative diseases.

Acknowledgments

We apologize to our colleagues whose important research we were unable to cite due to space limitations. The authors are grateful for funding support from the Ecole Polytechnique Fédérale de Lausanne, Swiss National Science Foundation (grant no. 310030_127478), Michael J. Fox Foundation for Parkinson's Research, and Parkinson Schweiz.

References

1. Gasser T. Mendelian forms of Parkinson's disease. *Biochim Biophys Acta* 2009;**1792**:587–96.
2. Polymeropoulos MH, Lavedan C, Leroy E, Ide SE, Dehejia A, et al. Mutation in the alpha-synuclein gene identified in families with Parkinson's disease. *Science* 1997;**276**:2045–7.
3. Kruger R, Kuhn W, Muller T, Woitalla D, Graeber M, et al. Ala30Pro mutation in the gene encoding alpha-synuclein in Parkinson's disease. *Nat Genet* 1998;**18**:106–8.
4. Singleton AB, Farrer M, Johnson J, Singleton A, Hague S, et al. alpha-Synuclein locus triplication causes Parkinson's disease. *Science* 2003;**302**:841.
5. Zarranz JJ, Alegre J, Gomez-Esteban JC, Lezcano E, Ros R, et al. The new mutation, E46K, of alpha-synuclein causes Parkinson and Lewy body dementia. *Ann Neurol* 2004;**55**:164–73.
6. Satake W, Nakabayashi Y, Mizuta I, Hirota Y, Ito C, et al. Genome-wide association study identifies common variants at four loci as genetic risk factors for Parkinson's disease. *Nat Genet* 2009;**41**:1303–7.
7. Simon-Sanchez J, Schulte C, Bras JM, Sharma M, Gibbs JR, et al. Genome-wide association study reveals genetic risk underlying Parkinson's disease. *Nat Genet* 2009;**41**:1308–12.
8. Spillantini MG, Crowther RA, Jakes R, Hasegawa M, Goedert M. alpha-Synuclein in filamentous inclusions of Lewy bodies from Parkinson's disease and dementia with lewy bodies. *Proc Natl Acad Sci USA* 1998;**95**:6469–73.
9. Haugarvoll K, Wszolek ZK. Clinical features of LRRK2 parkinsonism. *Parkinsonism Relat Disord* 2009;**15**:S205–8.
10. Paisan-Ruiz C, Jain S, Evans EW, Gilks WP, Simon J, et al. Cloning of the gene containing mutations that cause PARK8-linked Parkinson's disease. *Neuron* 2004;**44**:595–600.

11. Zimprich A, Biskup S, Leitner P, Lichtner P, Farrer M, et al. Mutations in LRRK2 cause autosomal-dominant parkinsonism with pleomorphic pathology. *Neuron* 2004;**44**:601–7.
12. Gilks WP, Abou-Sleiman PM, Gandhi S, Jain S, Singleton A, et al. A common LRRK2 mutation in idiopathic Parkinson's disease. *Lancet* 2005;**365**:415–6.
13. Nichols WC, Pankratz N, Hernandez D, Paisan-Ruiz C, Jain S, et al. Genetic screening for a single common LRRK2 mutation in familial Parkinson's disease. *Lancet* 2005;**365**:410–2.
14. Giasson BI, Covy JP, Bonini NM, Hurtig HI, Farrer MJ, et al. Biochemical and pathological characterization of Lrrk2. *Ann Neurol* 2006;**59**:315–22.
15. Higashi S, Biskup S, West AB, Trinkaus D, Dawson VL, et al. Localization of Parkinson's disease-associated LRRK2 in normal and pathological human brain. *Brain Res* 2007;**1155**:208–19.
16. Ross OA, Toft M, Whittle AJ, Johnson JL, Papapetropoulos S, et al. Lrrk2 and Lewy body disease. *Ann Neurol* 2006;**59**:388–93.
17. Moore DJ, Dawson TM. Value of genetic models in understanding the cause and mechanisms of Parkinson's disease. *Curr Neurol Neurosci Rep* 2008;**8**:288–96.
18. Clayton DF, George JM. The synucleins: a family of proteins involved in synaptic function, plasticity, neurodegeneration and disease. *Trends Neurosci* 1998;**21**:249–54.
19. Murphy DD, Rueter SM, Trojanowski JQ, Lee VM. Synucleins are developmentally expressed, and alpha-synuclein regulates the size of the presynaptic vesicular pool in primary hippocampal neurons. *J Neurosci* 2000;**20**:3214–20.
20. Abeliovich A, Schmitz Y, Farinas I, Choi-Lundberg D, Ho WH, et al. Mice lacking alpha-synuclein display functional deficits in the nigrostriatal dopamine system. *Neuron* 2000;**25**:239–52.
21. Irizarry MC, Kim TW, McNamara M, Tanzi RE, George JM, et al. Characterization of the precursor protein of the non-A beta component of senile plaques (NACP) in the human central nervous system. *J Neuropathol Exp Neurol* 1996;**55**:889–95.
22. Kahle PJ, Neumann M, Ozmen L, Muller V, Jacobsen H, et al. Subcellular localization of wild-type and Parkinson's disease-associated mutant alpha-synuclein in human and transgenic mouse brain. *J Neurosci* 2000;**20**:6365–73.
23. Sharon R, Goldberg MS, Bar-Josef I, Betensky RA, Shen J, et al. alpha-Synuclein occurs in lipid-rich high molecular weight complexes, binds fatty acids, and shows homology to the fatty acid-binding proteins. *Proc Natl Acad Sci USA* 2001;**98**:9110–5.
24. Uversky VN. A protein-chameleon: conformational plasticity of alpha-synuclein, a disordered protein involved in neurodegenerative disorders. *J Biomol Struct Dyn* 2003;**21**:211–34.
25. Conway KA, Harper JD, Lansbury PT. Accelerated in vitro fibril formation by a mutant alpha-synuclein linked to early-onset Parkinson disease. *Nat Med* 1998;**4**:1318–20.
26. Conway KA, Lee SJ, Rochet JC, Ding TT, Williamson RE, et al. Acceleration of oligomerization, not fibrillization, is a shared property of both alpha-synuclein mutations linked to early-onset Parkinson's disease: implications for pathogenesis and therapy. *Proc Natl Acad Sci USA* 2000;**97**:571–6.
27. Waxman EA, Giasson BI. Molecular mechanisms of alpha-synuclein neurodegeneration. *Biochim Biophys Acta* 2009;**1792**:616–24.
28. Betarbet R, Sherer TB, MacKenzie G, Garcia-Osuna M, Panov AV, et al. Chronic systemic pesticide exposure reproduces features of Parkinson's disease. *Nat Neurosci* 2000;**3**:1301–6.
29. Fornai F, Schluter OM, Lenzi P, Gesi M, Ruffoli R, et al. Parkinson-like syndrome induced by continuous MPTP infusion: convergent roles of the ubiquitin-proteasome system and alpha-synuclein. *Proc Natl Acad Sci USA* 2005;**102**:3413–8.
30. Giasson BI, Duda JE, Murray IV, Chen Q, Souza JM, et al. Oxidative damage linked to neurodegeneration by selective alpha-synuclein nitration in synucleinopathy lesions. *Science* 2000;**290**:985–9.

31. Manning-Bog AB, McCormack AL, Li J, Uversky VN, Fink AL, et al. The herbicide paraquat causes up-regulation and aggregation of alpha-synuclein in mice: paraquat and alpha-synuclein. *J Biol Chem* 2002;**277**:1641–4.
32. Lee VM, Trojanowski JQ. Mechanisms of Parkinson's disease linked to pathological alpha-synuclein: new targets for drug discovery. *Neuron* 2006;**52**:33–8.
33. Conway KA, Rochet JC, Bieganski RM, Lansbury Jr. PT. Kinetic stabilization of the alpha-synuclein protofibril by a dopamine-alpha-synuclein adduct. *Science* 2001;**294**:1346–9.
34. Lashuel HA, Hartley D, Petre BM, Walz T, Lansbury Jr. PT. Neurodegenerative disease: amyloid pores from pathogenic mutations. *Nature* 2002;**418**:291.
35. Sharon R, Bar-Joseph I, Frosch MP, Walsh DM, Hamilton JA, et al. The formation of highly soluble oligomers of alpha-synuclein is regulated by fatty acids and enhanced in Parkinson's disease. *Neuron* 2003;**37**:583–95.
36. Chandra S, Fornai F, Kwon HB, Yazdani U, Atasoy D, et al. Double-knockout mice for alpha- and beta-synucleins: effect on synaptic functions. *Proc Natl Acad Sci USA* 2004;**101**:14966–71.
37. Masliah E, Rockenstein E, Veinbergs I, Mallory M, Hashimoto M, et al. Dopaminergic loss and inclusion body formation in alpha-synuclein mice: implications for neurodegenerative disorders. *Science* 2000;**287**:1265–9.
38. van der Putten H, Wiederhold KH, Probst A, Barbieri S, Mistl C, et al. Neuropathology in mice expressing human alpha-synuclein. *J Neurosci* 2000;**20**:6021–9.
39. Giasson BI, Duda JE, Quinn SM, Zhang B, Trojanowski JQ, et al. Neuronal alpha-synucleinopathy with severe movement disorder in mice expressing A53T human alpha-synuclein. *Neuron* 2002;**34**:521–33.
40. Lee MK, Stirling W, Xu Y, Xu X, Qui D, et al. Human alpha-synuclein-harboring familial Parkinson's disease-linked Ala-53 –> Thr mutation causes neurodegenerative disease with alpha-synuclein aggregation in transgenic mice. *Proc Natl Acad Sci USA* 2002;**99**:8968–73.
41. Martin LJ, Pan Y, Price AC, Sterling W, Copeland NG, et al. Parkinson's disease alpha-synuclein transgenic mice develop neuronal mitochondrial degeneration and cell death. *J Neurosci* 2006;**26**:41–50.
42. Richfield EK, Thiruchelvam MJ, Cory-Slechta DA, Wuertzer C, Gainetdinov RR, et al. Behavioral and neurochemical effects of wild-type and mutated human alpha-synuclein in transgenic mice. *Exp Neurol* 2002;**175**:35–48.
43. Tofaris GK, Garcia Reitbock P, Humby T, Lambourne SL, O'Connell M, et al. Pathological changes in dopaminergic nerve cells of the substantia nigra and olfactory bulb in mice transgenic for truncated human alpha-synuclein(1-120): implications for Lewy body disorders. *J Neurosci* 2006;**26**:3942–50.
44. Wakamatsu M, Ishii A, Iwata S, Sakagami J, Ukai Y, et al. Selective loss of nigral dopamine neurons induced by overexpression of truncated human alpha-synuclein in mice. *Neurobiol Aging* 2008;**29**:574–85.
45. Daher JP, Ying M, Banerjee R, McDonald RS, Hahn MD, et al. Conditional transgenic mice expressing C-terminally truncated human alpha-synuclein (alphaSyn119) exhibit reduced striatal dopamine without loss of nigrostriatal pathway dopaminergic neurons. *Mol Neurodegener* 2009;**4**:34.
46. Nuber S, Petrasch-Parwez E, Winner B, Winkler J, von Horsten S, et al. Neurodegeneration and motor dysfunction in a conditional model of Parkinson's disease. *J Neurosci* 2008;**28**:2471–84.
47. Neumann M, Kahle PJ, Giasson BI, Ozmen L, Borroni E, et al. Misfolded proteinase K-resistant hyperphosphorylated alpha-synuclein in aged transgenic mice with locomotor deterioration and in human alpha-synucleinopathies. *J Clin Invest* 2002;**110**:1429–39.

48. Matsuoka Y, Vila M, Lincoln S, McCormack A, Picciano M, et al. Lack of nigral pathology in transgenic mice expressing human alpha-synuclein driven by the tyrosine hydroxylase promoter. *Neurobiol Dis* 2001;**8**:535–9.
49. Lin X, Parisiadou L, Gu XL, Wang L, Shim H, et al. Leucine-rich repeat kinase 2 regulates the progression of neuropathology induced by Parkinson's-disease-related mutant alpha-synuclein. *Neuron* 2009;**64**:807–27.
50. Chesselet MF. In vivo alpha-synuclein overexpression in rodents: a useful model of Parkinson's disease? *Exp Neurol* 2008;**209**:22–7.
51. Thiruchelvam MJ, Powers JM, Cory-Slechta DA, Richfield EK. Risk factors for dopaminergic neuron loss in human alpha-synuclein transgenic mice. *Eur J Neurosci* 2004;**19**:845–54.
52. Nemani VM, Lu W, Berge V, Nakamura K, Onoa B, et al. Increased expression of alpha-synuclein reduces neurotransmitter release by inhibiting synaptic vesicle reclustering after endocytosis. *Neuron* 2010;**65**:66–79.
53. Garcia-Reitbock P, Anichtchik O, Bellucci A, Iovino M, Ballini C, et al. SNARE protein redistribution and synaptic failure in a transgenic mouse model of Parkinson's disease. *Brain* 2010;**133**:2032–44.
54. Scott DA, Tabarean I, Tang Y, Cartier A, Masliah E, et al. A pathologic cascade leading to synaptic dysfunction in alpha-synuclein-induced neurodegeneration. *J Neurosci* 2010;**30**:8083–95.
55. Li W, West N, Colla E, Pletnikova O, Troncoso JC, et al. Aggregation promoting C-terminal truncation of alpha-synuclein is a normal cellular process and is enhanced by the familial Parkinson's disease-linked mutations. *Proc Natl Acad Sci USA* 2005;**102**:2162–7.
56. Chandra S, Gallardo G, Fernandez-Chacon R, Schluter OM, Sudhof TC. Alpha-synuclein cooperates with CSPalpha in preventing neurodegeneration. *Cell* 2005;**123**:383–96.
57. Freichel C, Neumann M, Ballard T, Muller V, Woolley M, et al. Age-dependent cognitive decline and amygdala pathology in alpha-synuclein transgenic mice. *Neurobiol Aging* 2007;**28**:1421–35.
58. Azeredo da Silveira S, Schneider BL, Cifuentes-Diaz C, Sage D, Abbas-Terki T, et al. Phosphorylation does not prompt, nor prevent, the formation of alpha-synuclein toxic species in a rat model of Parkinson's disease. *Hum Mol Genet* 2009;**18**:872–87.
59. Chen L, Feany MB. Alpha-synuclein phosphorylation controls neurotoxicity and inclusion formation in a Drosophila model of Parkinson disease. *Nat Neurosci* 2005;**8**:657–63.
60. Gorbatyuk OS, Li S, Sullivan LF, Chen W, Kondrikova G, et al. The phosphorylation state of Ser-129 in human alpha-synuclein determines neurodegeneration in a rat model of Parkinson disease. *Proc Natl Acad Sci USA* 2008;**105**:763–8.
61. Oueslati A, Fournier M, Lashuel HA. Role of post-translational modifications in modulating the structure, function and toxicity of alpha-synuclein: implications for Parkinson's disease pathogenesis and therapies. *Prog Brain Res* 2010;**183**:115–45.
62. Liu CW, Giasson BI, Lewis KA, Lee VM, Demartino GN, et al. A precipitating role for truncated alpha-synuclein and the proteasome in alpha-synuclein aggregation: implications for pathogenesis of Parkinson disease. *J Biol Chem* 2005;**280**:22670–8.
63. Crowther RA, Jakes R, Spillantini MG, Goedert M. Synthetic filaments assembled from C-terminally truncated alpha-synuclein. *FEBS Lett* 1998;**436**:309–12.
64. Murray IV, Giasson BI, Quinn SM, Koppaka V, Axelsen PH, et al. Role of alpha-synuclein carboxy-terminus on fibril formation in vitro. *Biochemistry* 2003;**42**:8530–40.
65. Fournier M, Vitte J, Garrigue J, Langui D, Dullin JP, et al. Parkin deficiency delays motor decline and disease manifestation in a mouse model of synucleinopathy. *PLoS ONE* 2009;**4**: e6629.
66. von Coelln R, Thomas B, Andrabi SA, Lim KL, Savitt JM, et al. Inclusion body formation and neurodegeneration are parkin independent in a mouse model of alpha-synucleinopathy. *J Neurosci* 2006;**26**:3685–96.

67. Ramsey CP, Tsika E, Ischiropoulos H, Giasson BI. DJ-1 deficient mice demonstrate similar vulnerability to pathogenic Ala53Thr human alpha-syn toxicity. *Hum Mol Genet* 2010;**19**:1425–37.
68. Lo Bianco C, Schneider BL, Bauer M, Sajadi A, Brice A, et al. Lentiviral vector delivery of parkin prevents dopaminergic degeneration in an alpha-synuclein rat model of Parkinson's disease. *Proc Natl Acad Sci USA* 2004;**101**:17510–5.
69. Chung KK, Zhang Y, Lim KL, Tanaka Y, Huang H, et al. Parkin ubiquitinates the alpha-synuclein-interacting protein, synphilin-1: implications for Lewy-body formation in Parkinson disease. *Nat Med* 2001;**7**:1144–50.
70. Engelender S, Kaminsky Z, Guo X, Sharp AH, Amaravi RK, et al. Synphilin-1 associates with alpha-synuclein and promotes the formation of cytosolic inclusions. *Nat Genet* 1999;**22**:110–4.
71. Marx FP, Holzmann C, Strauss KM, Li L, Eberhardt O, et al. Identification and functional characterization of a novel R621C mutation in the synphilin-1 gene in Parkinson's disease. *Hum Mol Genet* 2003;**12**:1223–31.
72. Wakabayashi K, Engelender S, Yoshimoto M, Tsuji S, Ross CA, et al. Synphilin-1 is present in Lewy bodies in Parkinson's disease. *Ann Neurol* 2000;**47**:521–3.
73. Smith WW, Liu Z, Liang Y, Masuda N, Swing DA, et al. Synphilin-1 attenuates neuronal degeneration in the A53T alpha-synuclein transgenic mouse model. *Hum Mol Genet* 2010;**19**:2087–98.
74. Giasson BI, Forman MS, Higuchi M, Golbe LI, Graves CL, et al. Initiation and synergistic fibrillization of tau and alpha-synuclein. *Science* 2003;**300**:636–40.
75. Gallardo G, Schluter OM, Sudhof TC. A molecular pathway of neurodegeneration linking alpha-synuclein to ApoE and Abeta peptides. *Nat Neurosci* 2008;**11**:301–8.
76. Masliah E, Rockenstein E, Veinbergs I, Sagara Y, Mallory M, et al. beta-amyloid peptides enhance alpha-synuclein accumulation and neuronal deficits in a transgenic mouse model linking Alzheimer's disease and Parkinson's disease. *Proc Natl Acad Sci USA* 2001;**98**:12245–50.
77. Hashimoto M, Rockenstein E, Mante M, Mallory M, Masliah E. beta-Synuclein inhibits alpha-synuclein aggregation: a possible role as an anti-parkinsonian factor. *Neuron* 2001;**32**:213–23.
78. Cooper AA, Gitler AD, Cashikar A, Haynes CM, Hill KJ, et al. Alpha-synuclein blocks ER-Golgi traffic and Rab1 rescues neuron loss in Parkinson's models. *Science* 2006;**313**:324–8.
79. Cuervo AM, Stefanis L, Fredenburg R, Lansbury PT, Sulzer D. Impaired degradation of mutant alpha-synuclein by chaperone-mediated autophagy. *Science* 2004;**305**:1292–5.
80. Smith WW, Jiang H, Pei Z, Tanaka Y, Morita H, et al. Endoplasmic reticulum stress and mitochondrial cell death pathways mediate A53T mutant alpha-synuclein-induced toxicity. *Hum Mol Genet* 2005;**14**:3801–11.
81. Tanaka Y, Engelender S, Igarashi S, Rao RK, Wanner T, et al. Inducible expression of mutant alpha-synuclein decreases proteasome activity and increases sensitivity to mitochondria-dependent apoptosis. *Hum Mol Genet* 2001;**10**:919–26.
82. Bedford L, Hay D, Devoy A, Paine S, Powe DG, et al. Depletion of 26S proteasomes in mouse brain neurons causes neurodegeneration and Lewy-like inclusions resembling human pale bodies. *J Neurosci* 2008;**28**:8189–98.
83. Crews L, Spencer B, Desplats P, Patrick C, Paulino A, et al. Selective molecular alterations in the autophagy pathway in patients with Lewy body disease and in models of alpha-synucleinopathy. *PLoS ONE* 2010;**5**:e9313.
84. Komatsu M, Kominami E, Tanaka K. Autophagy and neurodegeneration. *Autophagy* 2006;**2**:315–7.
85. Lindersson E, Beedholm R, Hojrup P, Moos T, Gai W, et al. Proteasomal inhibition by alpha-synuclein filaments and oligomers. *J Biol Chem* 2004;**279**:12924–34.

86. Schapira AH, Cooper JM, Dexter D, Clark JB, Jenner P, et al. Mitochondrial complex I deficiency in Parkinson's disease. *J Neurochem* 1990;**54**:823–7.
87. Przedborski S, Tieu K, Perier C, Vila M. MPTP as a mitochondrial neurotoxic model of Parkinson's disease. *J Bioenerg Biomembr* 2004;**36**:375–9.
88. Dauer W, Kholodilov N, Vila M, Trillat AC, Goodchild R, et al. Resistance of alpha-synuclein null mice to the parkinsonian neurotoxin MPTP. *Proc Natl Acad Sci USA* 2002;**99**:14524–9.
89. Klivenyi P, Siwek D, Gardian G, Yang L, Starkov A, et al. Mice lacking alpha-synuclein are resistant to mitochondrial toxins. *Neurobiol Dis* 2006;**21**:541–8.
90. Norris EH, Uryu K, Leight S, Giasson BI, Trojanowski JQ, et al. Pesticide exposure exacerbates alpha-synucleinopathy in an A53T transgenic mouse model. *Am J Pathol* 2007;**170**:658–66.
91. Song DD, Shults CW, Sisk A, Rockenstein E, Masliah E. Enhanced substantia nigra mitochondrial pathology in human alpha-synuclein transgenic mice after treatment with MPTP. *Exp Neurol* 2004;**186**:158–72.
92. Devi L, Raghavendran V, Prabhu BM, Avadhani NG, Anandatheerthavarada HK. Mitochondrial import and accumulation of alpha-synuclein impair complex I in human dopaminergic neuronal cultures and Parkinson disease brain. *J Biol Chem* 2008;**283**:9089–100.
93. Betts-Henderson J, Jaros E, Krishnan KJ, Perry RH, Reeve AK, et al. Alpha-synuclein pathology and Parkinsonism associated with POLG1 mutations and multiple mitochondrial DNA deletions. *Neuropathol Appl Neurobiol* 2009;**35**:120–4.
94. Healy DG, Falchi M, O'Sullivan SS, Bonifati V, Durr A, et al. Phenotype, genotype, and worldwide genetic penetrance of LRRK2-associated Parkinson's disease: a case-control study. *Lancet Neurol* 2008;**7**:583–90.
95. Kachergus J, Mata IF, Hulihan M, Taylor JP, Lincoln S, et al. Identification of a novel LRRK2 mutation linked to autosomal dominant parkinsonism: evidence of a common founder across European populations. *Am J Hum Genet* 2005;**76**:672–80.
96. Lesage S, Durr A, Tazir M, Lohmann E, Leutenegger AL, et al. LRRK2 G2019S as a cause of Parkinson's disease in North African Arabs. *N Engl J Med* 2006;**354**:422–3.
97. Ozelius LJ, Senthil G, Saunders-Pullman R, Ohmann E, Deligtisch A, et al. LRRK2 G2019S as a cause of Parkinson's disease in Ashkenazi Jews. *N Engl J Med* 2006;**354**:424–5.
98. Funayama M, Hasegawa K, Kowa H, Saito M, Tsuji S, et al. A new locus for Parkinson's disease (PARK8) maps to chromosome 12p11.2-q13.1. *Ann Neurol* 2002;**51**:296–301.
99. Khan NL, Jain S, Lynch JM, Pavese N, Abou-Sleiman P, et al. Mutations in the gene LRRK2 encoding dardarin (PARK8) cause familial Parkinson's disease: clinical, pathological, olfactory and functional imaging and genetic data. *Brain* 2005;**128**:2786–96.
100. Rajput A, Dickson DW, Robinson CA, Ross OA, Dachsel JC, et al. Parkinsonism, Lrrk2 G2019S, and tau neuropathology. *Neurology* 2006;**67**:1506–8.
101. Moore DJ. The biology and pathobiology of LRRK2: implications for Parkinson's disease. *Parkinsonism Relat Disord* 2008;**14**(Suppl. 2):S92–8.
102. Bosgraaf L, Van Haastert PJ. Roc, a Ras/GTPase domain in complex proteins. *Biochim Biophys Acta* 2003;**1643**:5–10.
103. Marin I. The Parkinson disease gene LRRK2: evolutionary and structural insights. *Mol Biol Evol* 2006;**23**:2423–33.
104. Biskup S, Moore DJ, Celsi F, Higashi S, West AB, et al. Localization of LRRK2 to membranous and vesicular structures in mammalian brain. *Ann Neurol* 2006;**60**:557–69.
105. Galter D, Westerlund M, Carmine A, Lindqvist E, Sydow O, et al. LRRK2 expression linked to dopamine-innervated areas. *Ann Neurol* 2006;**59**:714–9.
106. Higashi S, Moore DJ, Colebrooke RE, Biskup S, Dawson VL, et al. Expression and localization of Parkinson's disease-associated leucine-rich repeat kinase 2 in the mouse brain. *J Neurochem* 2007;**100**:368–81.

107. Hatano T, Kubo S, Imai S, Maeda M, Ishikawa K, et al. Leucine-rich repeat kinase 2 associates with lipid rafts. *Hum Mol Genet* 2007;**16**:678–90.
108. Plowey ED, Cherra 3rd SJ, Liu YJ, Chu CT. Role of autophagy in G2019S-LRRK2-associated neurite shortening in differentiated SH-SY5Y cells. *J Neurochem* 2008;**105**:1048–56.
109. Shin N, Jeong H, Kwon J, Heo HY, Kwon JJ, et al. LRRK2 regulates synaptic vesicle endocytosis. *Exp Cell Res* 2008;**314**:2055–65.
110. Xiong Y, Coombes CE, Kilaru A, Li X, Gitler AD, et al. GTPase activity plays a key role in the pathobiology of LRRK2. *PLoS Genet* 2010;**6**:e1000902.
111. Li X, Tan YC, Poulose S, Olanow CW, Huang XY, et al. Leucine-rich repeat kinase 2 (LRRK2)/PARK8 possesses GTPase activity that is altered in familial Parkinson's disease R1441C/G mutants. *J Neurochem* 2007;**103**:238–47.
112. West AB, Moore DJ, Biskup S, Bugayenko A, Smith WW, et al. Parkinson's disease-associated mutations in leucine-rich repeat kinase 2 augment kinase activity. *Proc Natl Acad Sci USA* 2005;**102**:16842–7.
113. Gillardon F. Leucine-rich repeat kinase 2 phosphorylates brain tubulin-beta isoforms and modulates microtubule stability—a point of convergence in Parkinsonian neurodegeneration? *J Neurochem* 2009;**110**:1514–22.
114. Imai Y, Gehrke S, Wang HQ, Takahashi R, Hasegawa K, et al. Phosphorylation of 4E-BP by LRRK2 affects the maintenance of dopaminergic neurons in Drosophila. *EMBO J* 2008;**27**:2432–43.
115. Jaleel M, Nichols RJ, Deak M, Campbell DG, Gillardon F, et al. LRRK2 phosphorylates moesin at threonine-558: characterization of how Parkinson's disease mutants affect kinase activity. *Biochem J* 2007;**405**:307–17.
116. Kanao T, Venderova K, Park DS, Unterman T, Lu B, et al. Activation of FoxO by LRRK2 induces expression of proapoptotic proteins and alters survival of postmitotic dopaminergic neuron in Drosophila. *Hum Mol Genet* 2010;**19**:3747–58.
117. West AB, Moore DJ, Choi C, Andrabi SA, Li X, et al. Parkinson's disease-associated mutations in LRRK2 link enhanced GTP-binding and kinase activities to neuronal toxicity. *Hum Mol Genet* 2007;**16**:223–32.
118. Ito G, Okai T, Fujino G, Takeda K, Ichijo H, et al. GTP binding is essential to the protein kinase activity of LRRK2, a causative gene product for familial Parkinson's disease. *Biochemistry* 2007;**46**:1380–8.
119. Smith WW, Pei Z, Jiang H, Dawson VL, Dawson TM, et al. Kinase activity of mutant LRRK2 mediates neuronal toxicity. *Nat Neurosci* 2006;**9**:1231–3.
120. Greggio E, Taymans JM, Zhen EY, Ryder J, Vancraenenbroeck R, et al. The Parkinson's disease kinase LRRK2 autophosphorylates its GTPase domain at multiple sites. *Biochem Biophys Res Commun* 2009;**389**:449–54.
121. Kamikawaji S, Ito G, Iwatsubo T. Identification of the autophosphorylation sites of LRRK2. *Biochemistry* 2009;**48**:10963–75.
122. Greggio E, Jain S, Kingsbury A, Bandopadhyay R, Lewis P, et al. Kinase activity is required for the toxic effects of mutant LRRK2/dardarin. *Neurobiol Dis* 2006;**23**:329–41.
123. Lewis PA, Greggio E, Beilina A, Jain S, Baker A, et al. The R1441C mutation of LRRK2 disrupts GTP hydrolysis. *Biochem Biophys Res Commun* 2007;**357**:668–71.
124. MacLeod D, Dowman J, Hammond R, Leete T, Inoue K, et al. The familial Parkinsonism gene LRRK2 regulates neurite process morphology. *Neuron* 2006;**52**:587–93.
125. Smith WW, Pei Z, Jiang H, Moore DJ, Liang Y, et al. Leucine-rich repeat kinase 2 (LRRK2) interacts with parkin, and mutant LRRK2 induces neuronal degeneration. *Proc Natl Acad Sci USA* 2005;**102**:18676–81.
126. Lee BD, Shin JH, Vankampen J, Petrucelli L, West AB, et al. Inhibitors of leucine-rich repeat kinase-2 protect against models of Parkinson's disease. *Nat Med* 2010;**16**:998–1000.

127. Andres-Mateos E, Mejias R, Sasaki M, Li X, Lin BM, et al. Unexpected lack of hypersensitivity in LRRK2 knock-out mice to MPTP (1-methyl-4-phenyl-1,2,3,6-tetrahydropyridine). *J Neurosci* 2009;**29**:15846–50.
128. Tong Y, Yamaguchi H, Giaime E, Boyle S, Kopan R, et al. Loss of leucine-rich repeat kinase 2 causes impairment of protein degradation pathways, accumulation of alpha-synuclein, and apoptotic cell death in aged mice. *Proc Natl Acad Sci USA* 2010;**107**:9879–84.
129. Melrose HL, Dachsel JC, Behrouz B, Lincoln SJ, Yue M, et al. Impaired dopaminergic neurotransmission and microtubule-associated protein tau alterations in human LRRK2 transgenic mice. *Neurobiol Dis* 2010;**40**:503–17.
130. Li X, Patel JC, Wang J, Avshalumov MV, Nicholson C, et al. Enhanced striatal dopamine transmission and motor performance with LRRK2 overexpression in mice is eliminated by familial Parkinson's disease mutation G2019S. *J Neurosci* 2010;**30**:1788–97.
131. Tong Y, Pisani A, Martella G, Karouani M, Yamaguchi H, et al. R1441C mutation in LRRK2 impairs dopaminergic neurotransmission in mice. *Proc Natl Acad Sci USA* 2009;**106**:14622–7.
132. Li Y, Liu W, Oo TF, Wang L, Tang Y, et al. Mutant LRRK2(R1441G) BAC transgenic mice recapitulate cardinal features of Parkinson's disease. *Nat Neurosci* 2009;**12**:826–8.
133. Parisiadou L, Xie C, Cho HJ, Lin X, Gu X-L, et al. Phosphorylation of ezrin/radixin/moesin proteins by LRRK2 promotes the rearrangement of actin cytoskeleton in neuronal morphogenesis. *J Neurosci* 2009;**29**:13971–80.
134. Wang L, Xie C, Greggio E, Parisiadou L, Shim H, et al. The chaperone activity of heat shock protein 90 is critical for maintaining the stability of leucine-rich repeat kinase 2. *J Neurosci* 2008;**28**:3384–91.
135. Alegre-Abarrategui J, Christian H, Lufino MM, Mutihac R, Venda LL, Ansorge O, et al. LRRK2 regulates autophagic activity and localizes to specific membrane microdomains in a novel human genomic reporter cellular model. *Hum Mol Genet* 2009;**18**:4022–34.
136. Ramirez A, Heimbach A, Grundemann J, Stiller B, Hampshire D, et al. Hereditary parkinsonism with dementia is caused by mutations in ATP13A2, encoding a lysosomal type 5 P-type ATPase. *Nat Genet* 2006;**38**:1184–91.
137. Goldberg MS, Fleming SM, Palacino JJ, Cepeda C, Lam HA, et al. Parkin-deficient mice exhibit nigrostriatal deficits but not loss of dopaminergic neurons. *J Biol Chem* 2003;**278**:43628–35.
138. Palacino JJ, Sagi D, Goldberg MS, Krauss S, Motz C, et al. Mitochondrial dysfunction and oxidative damage in parkin-deficient mice. *J Biol Chem* 2004;**279**:18614–22.
139. Itier JM, Ibanez P, Mena MA, Abbas N, Cohen-Salmon C, et al. Parkin gene inactivation alters behaviour and dopamine neurotransmission in the mouse. *Hum Mol Genet* 2003;**12**:2277–91.
140. Perez FA, Palmiter RD. Parkin-deficient mice are not a robust model of parkinsonism. *Proc Natl Acad Sci USA* 2005;**102**:2174–9.
141. Sato S, Chiba T, Nishiyama S, Kakiuchi T, Tsukada H, et al. Decline of striatal dopamine release in parkin-deficient mice shown by ex vivo autoradiography. *J Neurosci Res* 2006;**84**:1350–7.
142. Von Coelln R, Thomas B, Savitt JM, Lim KL, Sasaki M, et al. Loss of locus coeruleus neurons and reduced startle in parkin null mice. *Proc Natl Acad Sci USA* 2004;**101**:10744–9.
143. Kitada T, Pisani A, Porter DR, Yamaguchi H, Tscherter A, et al. Impaired dopamine release and synaptic plasticity in the striatum of PINK1-deficient mice. *Proc Natl Acad Sci USA* 2007;**104**:11441–6.
144. Chen L, Cagniard B, Mathews T, Jones S, Koh HC, et al. Age-dependent motor deficits and dopaminergic dysfunction in DJ-1 null mice. *J Biol Chem* 2005;**280**:21418–26.
145. Goldberg MS, Pisani A, Haburcak M, Vortherms TA, Kitada T, et al. Nigrostriatal dopaminergic deficits and hypokinesia caused by inactivation of the familial Parkinsonism-linked gene DJ-1. *Neuron* 2005;**45**:489–96.

146. Kim RH, Smith PD, Aleyasin H, Hayley S, Mount MP, et al. Hypersensitivity of DJ-1-deficient mice to 1-methyl-4-phenyl-1,2,3,6-tetrahydropyrindine (MPTP) and oxidative stress. *Proc Natl Acad Sci USA* 2005;**102**:5215–20.
147. Kitada T, Asakawa S, Hattori N, Matsumine H, Yamamura Y, et al. Mutations in the parkin gene cause autosomal recessive juvenile parkinsonism. *Nature* 1998;**392**:605–8.
148. West AB, Maidment NT. Genetics of parkin-linked disease. *Hum Genet* 2004;**114**:327–36.
149. Lucking CB, Durr A, Bonifati V, Vaughan J, De Michele G, et al. Association between early-onset Parkinson's disease and mutations in the parkin gene. *N Engl J Med* 2000;**342**:1560–7.
150. Imai Y, Soda M, Takahashi R. Parkin suppresses unfolded protein stress-induced cell death through its E3 ubiquitin-protein ligase activity. *J Biol Chem* 2000;**275**:35661–4.
151. Shimura H, Hattori N, Kubo S, Mizuno Y, Asakawa S, et al. Familial Parkinson disease gene product, parkin, is a ubiquitin-protein ligase. *Nat Genet* 2000;**25**:302–5.
152. Zhang Y, Gao J, Chung KK, Huang H, Dawson VL, et al. Parkin functions as an E2-dependent ubiquitin-protein ligase and promotes the degradation of the synaptic vesicle-associated protein, CDCrel-1. *Proc Natl Acad Sci USA* 2000;**97**:13354–9.
153. Moore DJ. Parkin: a multifaceted ubiquitin ligase. *Biochem Soc Trans* 2006;**34**:749–53.
154. Hampe C, Ardila-Osorio H, Fournier M, Brice A, Corti O. Biochemical analysis of Parkinson's disease-causing variants of Parkin, an E3 ubiquitin-protein ligase with monoubiquitylation capacity. *Hum Mol Genet* 2006;**15**:2059–75.
155. Sriram SR, Li X, Ko HS, Chung KK, Wong E, et al. Familial-associated mutations differentially disrupt the solubility, localization, binding and ubiquitination properties of parkin. *Hum Mol Genet* 2005;**14**:2571–86.
156. Wang C, Tan JM, Ho MW, Zaiden N, Wong SH, et al. Alterations in the solubility and intracellular localization of parkin by several familial Parkinson's disease-linked point mutations. *J Neurochem* 2005;**93**:422–31.
157. Ko HS, Kim SW, Sriram SR, Dawson VL, Dawson TM. Identification of far upstream element-binding protein-1 as an authentic Parkin substrate. *J Biol Chem* 2006;**281**:16193–6.
158. Ko HS, von Coelln R, Sriram SR, Kim SW, Chung KK, et al. Accumulation of the authentic parkin substrate aminoacyl-tRNA synthetase cofactor, p38/JTV-1, leads to catecholaminergic cell death. *J Neurosci* 2005;**25**:7968–78.
159. Geisler S, Holmstrom KM, Skujat D, Fiesel FC, Rothfuss OC, et al. PINK1/Parkin-mediated mitophagy is dependent on VDAC1 and p62/SQSTM1. *Nat Cell Biol* 2010;**12**:119–31.
160. Narendra D, Tanaka A, Suen DF, Youle RJ. Parkin-induced mitophagy in the pathogenesis of Parkinson disease. *Autophagy* 2009;**5**:706–8.
161. Vives-Bauza C, Zhou C, Huang Y, Cui M, de Vries RL, et al. PINK1-dependent recruitment of Parkin to mitochondria in mitophagy. *Proc Natl Acad Sci USA* 2010;**107**:378–83.
162. Ziviani E, Tao RN, Whitworth AJ. Drosophila parkin requires PINK1 for mitochondrial translocation and ubiquitinates mitofusin. *Proc Natl Acad Sci USA* 2010;**107**:5018–23.
163. Wang C, Ko HS, Thomas B, Tsang F, Chew KC, et al. Stress-induced alterations in parkin solubility promote parkin aggregation and compromise parkin's protective function. *Hum Mol Genet* 2005;**14**:3885–97.
164. Wong ES, Tan JM, Wang C, Zhang Z, Tay SP, et al. Relative sensitivity of parkin and other cysteine-containing enzymes to stress-induced solubility alterations. *J Biol Chem* 2007;**282**:12310–8.
165. Chung KK, Thomas B, Li X, Pletnikova O, Troncoso JC, et al. S-nitrosylation of parkin regulates ubiquitination and compromises parkin's protective function. *Science* 2004;**304**:1328–31.
166. Yao D, Gu Z, Nakamura T, Shi ZQ, Ma Y, et al. Nitrosative stress linked to sporadic Parkinson's disease: S-nitrosylation of parkin regulates its E3 ubiquitin ligase activity. *Proc Natl Acad Sci USA* 2004;**101**:10810–4.

167. LaVoie MJ, Ostaszewski BL, Weihofen A, Schlossmacher MG, Selkoe DJ. Dopamine covalently modifies and functionally inactivates parkin. *Nat Med* 2005;**11**:1214–21.
168. Greene JC, Whitworth AJ, Kuo I, Andrews LA, Feany MB, et al. Mitochondrial pathology and apoptotic muscle degeneration in Drosophila parkin mutants. *Proc Natl Acad Sci USA* 2003;**100**:4078–83.
169. Dawson TM, Dawson VL. The role of parkin in familial and sporadic Parkinson's disease. *Mov Disord* 2010;**25**(Suppl. 1):S32–39.
170. Thomas B, von Coelln R, Mandir AS, Trinkaus DB, Farah MH, et al. MPTP and DSP-4 susceptibility of substantia nigra and locus coeruleus catecholaminergic neurons in mice is independent of parkin activity. *Neurobiol Dis* 2007;**26**:312–22.
171. Menendez J, Rodriguez-Navarro JA, Solano RM, Casarejos MJ, Rodal I, et al. Suppression of Parkin enhances nigrostriatal and motor neuron lesion in mice over-expressing human-mutated tau protein. *Hum Mol Genet* 2006;**15**:2045–58.
172. Moore DJ, West AB, Dikeman DA, Dawson VL, Dawson TM. Parkin mediates the degradation-independent ubiquitination of Hsp70. *J Neurochem* 2008;**105**:1806–19.
173. Wang HQ, Imai Y, Inoue H, Kataoka A, Iita S, et al. Pael-R transgenic mice crossed with parkin deficient mice displayed progressive and selective catecholaminergic neuronal loss. *J Neurochem* 2008;**107**:171–85.
174. Valente EM, Abou-Sleiman PM, Caputo V, Muqit MM, Harvey K, et al. Hereditary early-onset Parkinson's disease caused by mutations in PINK1. *Science* 2004;**304**:1158–60.
175. Beilina A, Van Der Brug M, Ahmad R, Kesavapany S, Miller DW, et al. Mutations in PTEN-induced putative kinase 1 associated with recessive parkinsonism have differential effects on protein stability. *Proc Natl Acad Sci USA* 2005;**102**:5703–8.
176. Silvestri L, Caputo V, Bellacchio E, Atorino L, Dallapiccola B, et al. Mitochondrial import and enzymatic activity of PINK1 mutants associated to recessive parkinsonism. *Hum Mol Genet* 2005;**14**:3477–92.
177. Zhou C, Huang Y, Shao Y, May J, Prou D, et al. The kinase domain of mitochondrial PINK1 faces the cytoplasm. *Proc Natl Acad Sci USA* 2008;**105**:12022–7.
178. Zhou H, Falkenburger BH, Schulz JB, Tieu K, Xu Z, et al. Silencing of the Pink1 gene expression by conditional RNAi does not induce dopaminergic neuron death in mice. *Int J Biol Sci* 2007;**3**:242–50.
179. Gautier CA, Kitada T, Shen J. Loss of PINK1 causes mitochondrial functional defects and increased sensitivity to oxidative stress. *Proc Natl Acad Sci USA* 2008;**105**:11364–9.
180. Gispert S, Ricciardi F, Kurz A, Azizov M, Hoepken HH, et al. Parkinson phenotype in aged PINK1-deficient mice is accompanied by progressive mitochondrial dysfunction in absence of neurodegeneration. *PLoS ONE* 2009;**4**:e5777.
181. Clark IE, Dodson MW, Jiang C, Cao JH, Huh JR, et al. Drosophila pink1 is required for mitochondrial function and interacts genetically with parkin. *Nature* 2006;**441**:1162–6.
182. Park J, Lee SB, Lee S, Kim Y, Song S, et al. Mitochondrial dysfunction in Drosophila PINK1 mutants is complemented by parkin. *Nature* 2006;**441**:1157–61.
183. Poole AC, Thomas RE, Andrews LA, McBride HM, Whitworth AJ, et al. The PINK1/Parkin pathway regulates mitochondrial morphology. *Proc Natl Acad Sci USA* 2008;**105**:1638–43.
184. Yang Y, Ouyang Y, Yang L, Beal MF, McQuibban A, et al. Pink1 regulates mitochondrial dynamics through interaction with the fission/fusion machinery. *Proc Natl Acad Sci USA* 2008;**105**:7070–5.
185. Yang Y, Gehrke S, Imai Y, Huang Z, Ouyang Y, et al. Mitochondrial pathology and muscle and dopaminergic neuron degeneration caused by inactivation of Drosophila Pink1 is rescued by Parkin. *Proc Natl Acad Sci USA* 2006;**103**:10793–8.

186. Deng H, Dodson MW, Huang H, Guo M. The Parkinson's disease genes pink1 and parkin promote mitochondrial fission and/or inhibit fusion in Drosophila. *Proc Natl Acad Sci USA* 2008;**105**:14503–8.

187. Whitworth AJ, Pallanck LJ. The PINK1/Parkin pathway: a mitochondrial quality control system? *J Bioenerg Biomembr* 2009;**41**:499–503.

188. Plun-Favreau H, Klupsch K, Moisoi N, Gandhi S, Kjaer S, et al. The mitochondrial protease HtrA2 is regulated by Parkinson's disease-associated kinase PINK1. *Nat Cell Biol* 2007;**9**:1243–52.

189. Pridgeon JW, Olzmann JA, Chin LS, Li L. PINK1 protects against oxidative stress by phosphorylating mitochondrial chaperone TRAP1. *PLoS Biol* 2007;**5**:e172.

190. Weihofen A, Ostaszewski B, Minami Y, Selkoe DJ. Pink1 Parkinson mutations, the Cdc37/Hsp90 chaperones and Parkin all influence the maturation or subcellular distribution of Pink1. *Hum Mol Genet* 2008;**17**:602–16.

191. Martins LM, Morrison A, Klupsch K, Fedele V, Moisoi N, et al. Neuroprotective role of the Reaper-related serine protease HtrA2/Omi revealed by targeted deletion in mice. *Mol Cell Biol* 2004;**24**:9848–62.

192. Gandhi S, Wood-Kaczmar A, Yao Z, Plun-Favreau H, Deas E, et al. PINK1-associated Parkinson's disease is caused by neuronal vulnerability to calcium-induced cell death. *Mol Cell* 2009;**33**:627–38.

193. Bonifati V, Rizzu P, van Baren MJ, Schaap O, Breedveld GJ, et al. Mutations in the DJ-1 gene associated with autosomal recessive early-onset parkinsonism. *Science* 2003;**299**:256–9.

194. Miller DW, Ahmad R, Hague S, Baptista MJ, Canet-Aviles R, et al. L166P mutant DJ-1, causative for recessive Parkinson's disease, is degraded through the ubiquitin-proteasome system. *J Biol Chem* 2003;**278**:36588–95.

195. Moore DJ, Zhang L, Dawson TM, Dawson VL. A missense mutation (L166P) in DJ-1, linked to familial Parkinson's disease, confers reduced protein stability and impairs homo-oligomerization. *J Neurochem* 2003;**87**:1558–67.

196. Moore DJ, Zhang L, Troncoso J, Lee MK, Hattori N, et al. Association of DJ-1 and parkin mediated by pathogenic DJ-1 mutations and oxidative stress. *Hum Mol Genet* 2005;**14**:71–84.

197. Zhang L, Shimoji M, Thomas B, Moore DJ, Yu SW, et al. Mitochondrial localization of the Parkinson's disease related protein DJ-1: implications for pathogenesis. *Hum Mol Genet* 2005;**14**:2063–73.

198. Shendelman S, Jonason A, Martinat C, Leete T, Abeliovich A. DJ-1 is a redox-dependent molecular chaperone that inhibits alpha-synuclein aggregate formation. *PLoS Biol* 2004;**2**: e362.

199. Tao X, Tong L. Crystal structure of human DJ-1, a protein associated with early onset Parkinson's disease. *J Biol Chem* 2003;**278**:31372–9.

200. Andres-Mateos E, Perier C, Zhang L, Blanchard-Fillion B, Greco TM, et al. DJ-1 gene deletion reveals that DJ-1 is an atypical peroxiredoxin-like peroxidase. *Proc Natl Acad Sci USA* 2007;**104**:14807–12.

201. van der Brug MP, Blackinton J, Chandran J, Hao LY, Lal A, et al. RNA binding activity of the recessive parkinsonism protein DJ-1 supports involvement in multiple cellular pathways. *Proc Natl Acad Sci USA* 2008;**105**:10244–9.

202. Martinat C, Shendelman S, Jonason A, Leete T, Beal MF, et al. Sensitivity to oxidative stress in DJ-1-deficient dopamine neurons: an ES-derived cell model of primary Parkinsonism. *PLoS Biol* 2004;**2**:e327.

203. Yokota T, Sugawara K, Ito K, Takahashi R, Ariga H, et al. Down regulation of DJ-1 enhances cell death by oxidative stress, ER stress, and proteasome inhibition. *Biochem Biophys Res Commun* 2003;**312**:1342–8.

204. Canet-Aviles RM, Wilson MA, Miller DW, Ahmad R, McLendon C, et al. The Parkinson's disease protein DJ-1 is neuroprotective due to cysteine-sulfinic acid-driven mitochondrial localization. *Proc Natl Acad Sci USA* 2004;**101**:9103–8.

205. Kitada T, Tong Y, Gautier CA, Shen J. Absence of nigral degeneration in aged parkin/DJ-1/PINK1 triple knockout mice. *J Neurochem* 2009;**111**:696–702.

206. Schneider B, Zufferey R, Aebischer P. Viral vectors, animal models and new therapies for Parkinson's disease. *Parkinsonism Relat Disord* 2008;**14**(S2):S169–171.

207. Lo Bianco C, Ridet JL, Schneider BL, Deglon N, Aebischer P. alpha-Synucleinopathy and selective dopaminergic neuron loss in a rat lentiviral-based model of Parkinson's disease. *Proc Natl Acad Sci USA* 2002;**99**:10813–8.

208. Lo Bianco C, Shorter J, Regulier E, Lashuel H, Iwatsubo T, et al. Hsp104 antagonizes alpha-synuclein aggregation and reduces dopaminergic degeneration in a rat model of Parkinson disease. *J Clin Invest* 2008;**118**:3087–97.

209. Kirik D, Rosenblad C, Burger C, Lundberg C, Johansen TE, et al. Parkinson-like neurodegeneration induced by targeted overexpression of alpha-synuclein in the nigrostriatal system. *J Neurosci* 2002;**22**:2780–91.

210. Shepherd C, McCann H, Halliday GM. Variations in the neuropathology of familial Alzheimer's disease. *Acta Neuropathol* 2009;**118**:37–52.

211. Bennett DA, Schneider JA, Arvanitakis Z, Kelly JF, Aggarwal NT, et al. Neuropathology of older persons without cognitive impairment from two community-based studies. *Neurology* 2006;**66**:1837–44.

212. Price JL, McKeel Jr. DW, Buckles VD, Roe CM, Xiong C, et al. Neuropathology of non-demented aging: presumptive evidence for preclinical Alzheimer disease. *Neurobiol Aging* 2009;**30**:1026–36.

213. Beecham GW, Martin ER, Li YJ, Slifer MA, Gilbert JR, et al. Genome-wide association study implicates a chromosome 12 risk locus for late-onset Alzheimer disease. *Am J Hum Genet* 2009;**84**:35–43.

214. Bertram L, Tanzi RE. Genome-wide association studies in Alzheimer's disease. *Hum Mol Genet* 2009;**18**:R137–145.

215. Harold D, Abraham R, Hollingworth P, Sims R, Gerrish A, et al. Genome-wide association study identifies variants at CLU and PICALM associated with Alzheimer's disease. *Nat Genet* 2009;**41**:1088–93.

216. Seshadri S, Fitzpatrick AL, Ikram MA, DeStefano AL, Gudnason V, et al. Genome-wide analysis of genetic loci associated with Alzheimer disease. *JAMA* 2010;**303**:1832–40.

217. Mullan M, Houlden H, Windelspecht M, Fidani L, Lombardi C, et al. A locus for familial early-onset Alzheimer's disease on the long arm of chromosome 14, proximal to the alpha 1-antichymotrypsin gene. *Nat Genet* 1992;**2**:340–2.

218. Sherrington R, Rogaev EI, Liang Y, Rogaeva EA, Levesque G, et al. Cloning of a gene bearing missense mutations in early-onset familial Alzheimer's disease. *Nature* 1995;**375**:754–60.

219. Levy-Lahad E, Wasco W, Poorkaj P, Romano DM, Oshima J, et al. Candidate gene for the chromosome 1 familial Alzheimer's disease locus. *Science* 1995;**269**:973–7.

220. Rogaev EI, Sherrington R, Rogaeva EA, Levesque G, Ikeda M, et al. Familial Alzheimer's disease in kindreds with missense mutations in a gene on chromosome 1 related to the Alzheimer's disease type 3 gene. *Nature* 1995;**376**:775–8.

221. Corder EH, Saunders AM, Strittmatter WJ, Schmechel DE, Gaskell PC, et al. Gene dose of apolipoprotein E type 4 allele and the risk of Alzheimer's disease in late onset families. *Science* 1993;**261**:921–3.

222. Pantoni L. Cerebral small vessel disease: from pathogenesis and clinical characteristics to therapeutic challenges. *Lancet Neurol* 2010;**9**:689–701.

223. Hutton M, Lendon CL, Rizzu P, Baker M, Froelich S, et al. Association of missense and 5′-splice-site mutations in tau with the inherited dementia FTDP-17. *Nature* 1998;**393**:702–5.
224. Ballatore C, Lee VM, Trojanowski JQ. Tau-mediated neurodegeneration in Alzheimer's disease and related disorders. *Nat Rev Neurosci* 2007;**8**:663–72.
225. Hardy J, Selkoe DJ. The amyloid hypothesis of Alzheimer's disease: progress and problems on the road to therapeutics. *Science* 2002;**297**:353–6.
226. Scheuner D, Eckman C, Jensen M, Song X, Citron M, et al. Secreted amyloid beta-protein similar to that in the senile plaques of Alzheimer's disease is increased in vivo by the presenilin 1 and 2 and APP mutations linked to familial Alzheimer's disease. *Nat Med* 1996;**2**:864–70.
227. Roberson ED, Scearce-Levie K, Palop JJ, Yan F, Cheng IH, et al. Reducing endogenous tau ameliorates amyloid beta-induced deficits in an Alzheimer's disease mouse model. *Science* 2007;**316**:750–4.
228. Ittner LM, Ke YD, Delerue F, Bi M, Gladbach A, et al. Dendritic function of tau mediates amyloid-beta toxicity in Alzheimer's disease mouse models. *Cell* 2010;**142**:387–97.
229. Sturchler-Pierrat C, Abramowski D, Duke M, Wiederhold K-H, Mistl C, et al. Two amyloid precursor protein transgenic mouse models with Alzheimer disease-like pathology. *Proc Natl Acad Sci USA* 1997;**94**:13287–92.
230. Lord A, Kalimo H, Eckman C, Zhang X-Q, Lannfelt L, et al. The Arctic Alzheimer mutation facilitates early intraneuronal Abeta aggregation and senile plaque formation in transgenic mice. *Neurobiol Aging* 2006;**27**:67–77.
231. Lamb B, Call L, Slunt H, Bardel K, Lawler A, et al. Altered metabolism of familial Alzheimer's disease-linked amyloid precursor protein variants in yeast artificial chromosome transgenic mice. *Hum Mol Genet* 1997;**6**:1535–41.
232. Kulnane LS, Lamb BT. Neuropathological characterization of mutant amyloid precursor protein yeast artificial chromosome transgenic mice. *Neurobiol Dis* 2001;**8**:982–92.
233. Dewachter I, Reversé D, Caluwaerts N, Ris L, Kuipéri C, et al. Neuronal deficiency of presenilin 1 inhibits amyloid plaque formation and corrects hippocampal long-term potentiation but not a cognitive defect of amyloid precursor protein [V717I] transgenic mice. *J Neurosci* 2002;**22**:3445–53.
234. Schneider I, Reversé D, Dewachter I, Ris L, Caluwaerts N, et al. Mutant presenilins disturb neuronal calcium homeostasis in the brain of transgenic mice, decreasing the threshold for excitotoxicity and facilitating long-term potentiation. *J Biol Chem* 2001;**276**:11539–44.
235. Herms J, Schneider I, Dewachter I, Caluwaerts N, Kretzschmar H, et al. Capacitive calcium entry is directly attenuated by mutant presenilin-1, independent of the expression of the amyloid precursor protein. *J Biol Chem* 2003;**278**:2484–9.
236. Lamb BA, Bardel KA, Kulnane LS, Anderson JJ, Holtz G, et al. Amyloid production and deposition in mutant amyloid precursor protein and presenilin-1 yeast artificial chromosome transgenic mice. *Nat Neurosci* 1999;**2**:695–7.
237. Oakley H, Cole SL, Logan S, Maus E, Shao P, et al. Intraneuronal beta-amyloid aggregates, neurodegeneration, and neuron loss in transgenic mice with five familial Alzheimer's disease mutations: potential factors in amyloid plaque formation. *J Neurosci* 2006;**26**:10129–40.
238. Steiner H, Duff K, Capell A, Romig H, Grim MG, et al. A loss of function mutation of presenilin-2 interferes with amyloid β-peptide production and Notch signaling. *J Biol Chem* 1999;**274**:28669–73.
239. Hwang DY, CHAE KR, KANG TS, HWANG JH, LIM CH, et al. Alterations in behavior, amyloid {beta}-42, caspase-3, and Cox-2 in mutant PS2 transgenic mouse model of Alzheimer's disease. *FASEB J* 2002;**16**:805–13.
240. Piedrahita JA, Zhang SH, Hagaman JR, Oliver PM, Maeda N. Generation of mice carrying a mutant apolipoprotein E gene inactivated by gene targeting in embryonic stem cells. *Proc Natl Acad Sci USA* 1992;**89**:4471–5.

241. Harris FM, Brecht WJ, Xu Q, Tesseur I, Kekonius L, et al. Carboxyl-terminal-truncated apolipoprotein E4 causes Alzheimer's disease-like neurodegeneration and behavioral deficits in transgenic mice. *Proc Natl Acad Sci USA* 2003;**100**:10966–71.

242. Games D, Adams D, Alessandrini R, Barbour R, Borthelette P, et al. Alzheimer-type neuropathology in transgenic mice overexpressing V717F β-amyloid precursor protein. *Nature* 1995;**373**:523–7.

243. Masliah E, Sisk A, Mallory M, Games D. Neurofibrillary pathology in transgenic mice overexpressing V717F beta-amyloid precursor protein. *J Neuropathol Exp Neurol* 2001;**60**:357–68.

244. Hsiao K, Chapman P, Nilsen S, Eckman C, Harigaya Y, et al. Correlative memory deficits, Abeta elevation, and amyloid plaques in transgenic mice. *Science* 1996;**274**:99–103.

245. Kawarabayashi T, Younkin LH, Saido TC, Shoji M, Ashe KH, et al. Age-dependent changes in brain, CSF, and plasma amyloid beta protein in the Tg2576 transgenic mouse model of Alzheimer's disease. *J Neurosci* 2001;**21**:372–81.

246. Frautschy SA, Yang F, Irrizarry M, Hyman B, Saido TC, et al. Microglial response to amyloid plaques in APPsw transgenic mice. *Am J Pathol* 1998;**152**:307–17.

247. Masliah E, Sisk A, Mallory M, Mucke L, Schenk D, et al. Comparison of neurodegenerative pathology in transgenic mice overexpressing V717F beta-amyloid precursor protein and Alzheimer's disease. *J Neurosci* 1996;**16**:5795–811.

248. Irizarry MC, McNamara M, Fedorchak K, Hsiao K, Hyman BT. APPSw transgenic mice develop age-related A beta deposits and neuropil abnormalities, but no neuronal loss in CA1. *J Neuropathol Exp Neurol* 1997;**56**:965–73.

249. Calhoun ME, Burgermeister P, Phinney AL, Stalder M, Tolnay M, et al. Neuronal overexpression of mutant amyloid precursor protein results in prominent deposition of cerebrovascular amyloid. *Proc Natl Acad Sci USA* 1999;**96**:14088–93.

250. Burbach GJ, Vlachos A, Ghebremedhin E, Del Turco D, Coomaraswamy J, et al. Vessel ultrastructure in APP23 transgenic mice after passive anti-Abeta immunotherapy and subsequent intracerebral hemorrhage. *Neurobiol Aging* 2007;**28**:202–12.

251. Winkler DT, Biedermann L, Tolnay M, Allegrini PR, Staufenbiel M, et al. Thrombolysis induces cerebral hemorrhage in a mouse model of cerebral amyloid angiopathy. *Ann Neurol* 2002;**51**:790–3.

252. Winkler DT, Bondolfi L, Herzig MC, Jann L, Calhoun ME, et al. Spontaneous hemorrhagic stroke in a mouse model of cerebral amyloid angiopathy. *J Neurosci* 2001;**21**:1019–27.

253. Burbach GJ, Dehn D, Del Turco D, Staufenbiel M, Deller T. Laser microdissection reveals regional and cellular differences in GFAP mRNA upregulation following brain injury, axonal denervation, and amyloid plaque deposition. *Glia* 2004;**48**:76–84.

254. Burbach GJ, Hellweg R, Haas CA, Del Turco D, Deicke U, et al. Induction of brain-derived neurotrophic factor in plaque-associated glial cells of aged APP23 transgenic mice. *J Neurosci* 2004;**24**:2421–30.

255. Gärtner U, Brückner MK, Krug S, Schmetsdorf S, Staufenbiel M, et al. Amyloid deposition in APP23 mice is associated with the expression of cyclins in astrocytes but not in neurons. *Acta Neuropathol* 2003;**106**:535–44.

256. Phinney AL, Deller T, Stalder M, Calhoun ME, Frotscher M, et al. Cerebral amyloid induces aberrant axonal sprouting and ectopic terminal formation in amyloid precursor protein transgenic mice. *J Neurosci* 1999;**19**:8552–9.

257. Beckmann N, Schuler A, Mueggler T, Meyer EP, Wiederhold K-H, et al. Age-dependent cerebrovascular abnormalities and blood flow disturbances in APP23 mice modeling Alzheimer's disease. *J Neurosci* 2003;**23**:8453–9.

258. Krucker T, Schuler A, Meyer EP, Staufenbiel M, Beckmann N. Magnetic resonance angiography and vascular corrosion casting as tools in biomedical research: application to transgenic mice modeling Alzheimer's disease. *Neurol Res* 2004;**26**:507–16.
259. Mueggler T, Baumann D, Rausch M, Staufenbiel M, Rudin M. Age-dependent impairment of somatosensory response in the amyloid precursor protein 23 transgenic mouse model of Alzheimer's disease. *J Neurosci* 2003;**23**:8231–6.
260. Mueggler T, Meyer-Luehmann M, Rausch M, Staufenbiel M, Jucker M, et al. Restricted diffusion in the brain of transgenic mice with cerebral amyloidosis. *Eur J Neurosci* 2004;**20**:811–7.
261. Mueggler T, Sturchler-Pierrat C, Baumann D, Rausch M, Staufenbiel M, et al. Compromised hemodynamic response in amyloid precursor protein transgenic mice. *J Neurosci* 2002;**22**:7218–24.
262. Okamura N, Suemoto T, Shimadzu H, Suzuki M, Shiomitsu T, et al. Styrylbenzoxazole derivatives for in vivo imaging of amyloid plaques in the brain. *J Neurosci* 2004;**24**:2535–41.
263. Okamura N, Suemoto T, Shiomitsu T, Suzuki M, Shimadzu H, et al. A novel imaging probe for in vivo detection of neuritic and diffuse amyloid plaques in the brain. *J Mol Neurosci* 2004;**24**:247–55.
264. Suemoto T, Okamura N, Shiomitsu T, Suzuki M, Shimadzu H, et al. In vivo labeling of amyloid with BF-108. *Neurosci Res* 2004;**48**:65–74.
265. Boncristiano S, Calhoun ME, Kelly PH, Pfeifer M, Bondolfi L, et al. Cholinergic changes in the APP23 transgenic mouse model of cerebral amyloidosis. *J Neurosci* 2002;**22**:3234–43.
266. Diez M, Danner S, Frey P, Sommer B, Staufenbiel M, et al. Neuropeptide alterations in the hippocampal formation and cortex of transgenic mice overexpressing beta-amyloid precursor protein (APP) with the Swedish double mutation (APP23). *Neurobiol Dis* 2003;**14**:579–94.
267. Van Dam D, D'Hooge R, Staufenbiel M, Van Ginneken C, Van Meir F, et al. Age-dependent cognitive decline in the APP23 model precedes amyloid deposition. *Eur J Neurosci* 2003;**17**:388–96.
268. Van Dam D, Marescau B, Engelborghs S, Cremers T, Mulder J, et al. Analysis of cholinergic markers, biogenic amines, and amino acids in the CNS of two APP overexpression mouse models. *Neurochem Int* 2005;**46**:409–22.
269. Van Dam D, Vloeberghs E, Abramowski D, Staufenbiel M, De Deyn PP. APP23 mice as a model of Alzheimer's disease: an example of a transgenic approach to modeling a CNS disorder. *CNS Spectr* 2005;**10**:207–22.
270. Dumont M, Strazielle C, Staufenbiel M, Lalonde R. Spatial learning and exploration of environmental stimuli in 24-month-old female APP23 transgenic mice with the Swedish mutation. *Brain Res* 2004;**1024**:113–21.
271. Kelly PH, Bondolfi L, Hunziker D, Schlecht H-P, Carver K, et al. Progressive age-related impairment of cognitive behavior in APP23 transgenic mice. *Neurobiol Aging* 2003;**24**:365–78.
272. Lalonde R, Dumont M, Staufenbiel M, Sturchler-Pierrat C, Strazielle C. Spatial learning, exploration, anxiety, and motor coordination in female APP23 transgenic mice with the Swedish mutation. *Brain Res* 2002;**956**:36–44.
273. Parvathy S, Davies P, Haroutunian V, Purohit DP, Davis KL, et al. Correlation between Abetax-40-, Abetax-42-, and Abeta-43-containing amyloid plaques and cognitive decline. *Arch Neurol* 2001;**58**:2025–32.
274. Van Dam D, Coen K, De Deyn PP. Ibuprofen modifies cognitive disease progression in an Alzheimer's mouse model. *J Psychopharmacol* 2010;**24**:383–8.
275. Schonrock N, Ke YD, Humphreys D, Staufenbiel M, Ittner LM, et al. Neuronal microRNA deregulation in response to Alzheimer's disease amyloid-β. *PLoS ONE* 2010;**5**:e11070.

276. Davis J, Xu F, Deane R, Romanov G, Previti ML, et al. Early-onset and robust cerebral microvascular accumulation of amyloid β-protein in transgenic mice expressing low levels of a vasculotropic Dutch/Iowa mutant form of amyloid β-protein precursor. *J Biol Chem* 2004;**279**:20296–306.

277. Philipson O, Hammarström P, Nilsson KPR, Portelius E, Olofsson T, et al. A highly insoluble state of Abeta similar to that of Alzheimer's disease brain is found in Arctic APP transgenic mice. *Neurobiol Aging* 2009;**30**:1393–405.

278. Sahlin C, Lord A, Magnusson K, Englund H, Almeida CG, et al. The Arctic Alzheimer mutation favors intracellular amyloid-beta production by making amyloid precursor protein less available to alpha-secretase. *J Neurochem* 2007;**101**:854–62.

279. Peterson KR. transgenic mice carrying yeast artificial chromosomes. *Expert Rev Mol Med* 2003;**5**:1–25.

280. Peterson KR, Clegg CH, Li Q. Production of transgenic mice with yeast artificial chromosomes. *Trends Genet* 1997;**13**:61–6.

281. Lehman EJH, Kulnane LS, Gao Y, Petriello MC, Pimpis KM, et al. Genetic background regulates beta-amyloid precursor protein processing and beta-amyloid deposition in the mouse. *Hum Mol Genet* 2003;**12**:2949–56.

282. Ryman D, Gao Y, Lamb BT. Genetic loci modulating amyloid-beta levels in a mouse model of Alzheimer's disease. *Neurobiol Aging* 2008;**29**:1190–8.

283. Jankowsky JL, Slunt HH, Gonzales V, Savonenko AV, Wen JC, et al. Persistent amyloidosis following suppression of Abeta production in a transgenic model of Alzheimer disease. *PLoS Med* 2005;**2**:e355.

284. Sillen A, Forsell C, Lilius L, Axelman K, Bjork BF, et al. Genome scan on Swedish Alzheimer's disease families. *Mol Psychiatry* 2005;**11**:182–6.

285. Kim J, Basak JM, Holtzman DM. The role of apolipoprotein E in Alzheimer's disease. *Neuron* 2009;**63**:287–303.

286. Beffert U, Danik M, Krzywkowski P, Ramassamy C, Berrada F, et al. The neurobiology of apolipoproteins and their receptors in the CNS and Alzheimer's disease. *Brain Res Rev* 1998;**27**:119–42.

287. Wisniewski T, Frangione B. Apolipoprotein E: a pathological chaperone protein in patients with cerebral and systemic amyloid. *Neurosci Lett* 1992;**135**:235–8.

288. Näslund J, Thyberg J, Tjernberg LO, Wernstedt C, Karlström AR, et al. Characterization of stable complexes involving apolipoprotein E and the amyloid beta peptide in Alzheimer's disease brain. *Neuron* 1995;**15**:219–28.

289. Zhang SH, Reddick RL, Piedrahita JA, Maeda N. Spontaneous hypercholesterolemia and arterial lesions in mice lacking apolipoprotein E. *Science* 1992;**258**:468–71.

290. Igbavboa U, Avdulov NA, Chochina SV, Wood WG. Transbilayer distribution of cholesterol is modified in brain synaptic plasma membranes of knockout mice deficient in the low-density lipoprotein receptor, apolipoprotein E, or both proteins. *J Neurochem* 1997;**69**:1661–7.

291. Guo Y, Zhang C, Du X, Nair U, Yoo T-J. Morphological and functional alterations of the cochlea in apolipoprotein E gene deficient mice. *Hear Res* 2005;**208**:54–67.

292. Fagan AM, Holtzman DM, Munson G, Mathur T, Schneider D, et al. Unique lipoproteins secreted by primary astrocytes from wild type, apoE (−/−), and human apoE transgenic mice. *J Biol Chem* 1999;**274**:30001–7.

293. Fullerton SM, Shirman GA, Strittmatter WJ, Matthew WD. Impairment of the blood-nerve and blood-brain barriers in apolipoprotein e knockout mice. *Exp Neurol* 2001;**169**:13–22.

294. Masliah E, Mallory M, Ge N, Alford M, Veinbergs I, et al. Neurodegeneration in the central nervous system of apoE-deficient mice. *Exp Neurol* 1995;**136**:107–22.

295. Raber J, Wong D, Buttini M, Orth M, Bellosta S, et al. Isoform-specific effects of human apolipoprotein E on brain function revealed in ApoE knockout mice: increased susceptibility of females. *Proc Natl Acad Sci USA* 1998;**95**:10914–9.

296. Buttini M, Orth M, Bellosta S, Akeefe H, Pitas RE, et al. Expression of human apolipoprotein E3 or E4 in the brains of ApoE−/− mice: isoform-specific effects on neurodegeneration. *J Neurosci* 1999;**19**:4867–80.

297. Laskowitz DT, Sheng H, Bart RD, Joyner KA, Roses AD, et al. Apolipoprotein E-deficient mice have increased susceptibility to focal cerebral ischemia. *J Cereb Blood Flow Metab* 1997;**17**:753–8.

298. Mann KM, Thorngate FE, Katoh-Fukui Y, Hamanaka H, Williams DL, et al. Independent effects of APOE on cholesterol metabolism and brain A{beta} levels in an Alzheimer disease mouse model. *Hum Mol Genet* 2004;**13**:1959–68.

299. Hamanaka H, Katoh-Fukui Y, Suzuki K, Kobayashi M, Suzuki R, et al. Altered cholesterol metabolism in human apolipoprotein E4 knock-in mice. *Hum Mol Genet* 2000;**9**:353–61.

300. Tesseur I, Van Dorpe J, Spittaels K, Van Den Haute C, Moechars D, et al. Expression of human apolipoprotein E4 in neurons causes hyperphosphorylation of protein tau in the brains of transgenic mice. *Am J Pathol* 2000;**156**:951–64.

301. Huang Y, Liu XQ, Wyss-Coray T, Brecht WJ, Sanan DA, et al. Apolipoprotein E fragments present in Alzheimer's disease brains induce neurofibrillary tangle-like intracellular inclusions in neurons. *Proc Natl Acad Sci USA* 2001;**98**:8838–43.

302. Crook R, Verkkoniemi A, Perez-Tur J, Mehta N, Baker M, et al. A variant of Alzheimer's disease with spastic paraparesis and unusual plaques due to deletion of exon 9 of presenilin 1. *Nat Med* 1998;**4**:452–5.

303. O'Riordan S, McMonagle P, Janssen JC, Fox NC, Farrell M, et al. Presenilin-1 mutation (E280G), spastic paraparesis, and cranial MRI white-matter abnormalities. *Neurology* 2002;**59**:1108–10.

304. Rogaeva E, Bergeron C, Sato C, Moliaka I, Kawarai T, et al. PS1 Alzheimer's disease family with spastic paraplegia: the search for a gene modifier. *Neurology* 2003;**61**:1005–7.

305. Borchelt DR, Thinakaran G, Eckman CB, Lee MK, Davenport F, et al. Familial Alzheimer's disease-linked presenilin 1 variants elevate Abeta1-42/1-40 ratio in vitro and in vivo. *Neuron* 1996;**17**:1005–13.

306. Shen J, Bronson RT, Chen DF, Xia W, Selkoe DJ, et al. Skeletal and CNS defects in presenilin-1-deficient mice. *Cell* 1997;**89**:629–39.

307. Duff K, Eckman C, Zehr C, Yu X, Prada CM, et al. Increased amyloid-beta42(43) in brains of mice expressing mutant presenilin 1. *Nature* 1996;**383**:710–3.

308. Radde R, Bolmont T, Kaeser SA, Coomaraswamy J, Lindau D, et al. Abeta42-driven cerebral amyloidosis in transgenic mice reveals early and robust pathology. *EMBO Rep* 2006;**7**:940–6.

309. Savonenko A, Xu GM, Melnikova T, Morton JL, Gonzales V, et al. Episodic-like memory deficits in the APPswe/PS1dE9 mouse model of Alzheimer's disease: relationships to beta-amyloid deposition and neurotransmitter abnormalities. *Neurobiol Dis* 2005;**18**:602–17.

310. Borchelt DR, Ratovitski T, van Lare J, Lee MK, Gonzales V, et al. Accelerated amyloid deposition in the brains of transgenic mice coexpressing mutant presenilin 1 and amyloid precursor proteins. *Neuron* 1997;**19**:939–45.

311. Siman R, Reaume AG, Savage MJ, Trusko S, Lin YG, et al. Presenilin-1 P264L knock-in mutation: differential effects on abeta production, amyloid deposition, and neuronal vulnerability. *J Neurosci* 2000;**20**:8717–26.

312. Guo Q, Fu W, Sopher BL, Miller MW, Ware CB, et al. Increased vulnerability of hippocampal neurons to excitotoxic necrosis in presenilin-1 mutant knock-in mice. *Nat Med* 1999;**5**:101–6.

313. Wang R, Dineley KT, Sweatt JD, Zheng H. Presenilin 1 familial Alzheimer's disease mutation leads to defective associative learning and impaired adult neurogenesis. *Neuroscience* 2004;**126**:305–12.

314. Billings LM, Oddo S, Green KN, McGaugh JL, LaFerla FM. Intraneuronal Abeta causes the onset of early Alzheimer's disease-related cognitive deficits in transgenic mice. *Neuron* 2005;**45**:675–88.

315. Oddo S, Caccamo A, Shepherd JD, Murphy MP, Golde TE, et al. Triple-transgenic model of Alzheimer's disease with plaques and tangles: intracellular Abeta and synaptic dysfunction. *Neuron* 2003;**39**:409–21.

316. Oddo S, Caccamo A, Tran L, Lambert MP, Glabe CG, et al. Temporal profile of amyloid-beta (Abeta) oligomerization in an in vivo model of Alzheimer disease. A link between Abeta and tau pathology. *J Biol Chem* 2006;**281**:1599–604.

317. Leverenz JB, Fishel MA, Peskind ER, Montine TJ, Nochlin D, et al. Lewy body pathology in familial Alzheimer disease: evidence for disease- and mutation-specific pathologic phenotype. *Arch Neurol* 2006;**63**:370–6.

318. Nochlin D, Bird TD, Nemens EJ, Ball MJ, Sumi SM. Amyloid angiopathy in a Volga German family with Alzheimer's disease and a presenilin-2 mutation (N141I). *Ann Neurol* 1998;**43**:131–5.

319. Herreman A, Hartmann D, Annaert W, Saftig P, Craessaerts K, et al. Presenilin 2 deficiency causes a mild pulmonary phenotype and no changes in amyloid precursor protein processing but enhances the embryonic lethal phenotype of presenilin 1 deficiency. *Proc Natl Acad Sci USA* 1999;**96**:11872–7.

320. Oyama F, Sawamura N, Kobayashi K, Morishima-Kawashima M, Kuramochi T, et al. Mutant presenilin 2 transgenic mouse: effect on an age-dependent increase of amyloid β-protein 42 in the brain. *J Neurochem* 1998;**71**:313–22.

321. Richards JG, Higgins GA, Ouagazzal A-M, Ozmen L, Kew JNC, et al. PS2APP transgenic mice, coexpressing hPS2mut and hAPPswe, show age-related cognitive deficits associated with discrete brain amyloid deposition and inflammation. *J Neurosci* 2003;**23**:8989–9003.

322. Oddo S, Caccamo A, Kitazawa M, Tseng BP, LaFerla FM. Amyloid deposition precedes tangle formation in a triple transgenic model of Alzheimer's disease. *Neurobiol Aging* 2003;**24**:1063–70.

323. HDCRG. A novel gene containing a trinucleotide repeat that is expanded and unstable on Huntington's disease chromosomes. The Huntington's Disease Collaborative Research Group. *Cell* 1993;**72**:971–83.

324. Brinkman RR, Mezei MM, Theilmann J, Almqvist E, Hayden MR. The likelihood of being affected with Huntington disease by a particular age, for a specific CAG size. *Am J Hum Genet* 1997;**60**:1202–10.

325. Duyao MP, Auerbach AB, Ryan A, Persichetti F, Barnes GT, et al. Inactivation of the mouse Huntington's disease gene homolog Hdh. *Science* 1995;**269**:407–10.

326. Mangiarini L, Sathasivam K, Seller M, Cozens B, Harper A, et al. Exon 1 of the HD gene with an expanded CAG repeat is sufficient to cause a progressive neurological phenotype in transgenic mice. *Cell* 1996;**87**:493–506.

327. Schilling G, Becher MW, Sharp AH, Jinnah HA, Duan K, et al. Intranuclear inclusions and neuritic aggregates in transgenic mice expressing a mutant N-terminal fragment of huntingtin. *Hum Mol Genet* 1999;**8**:397–407.

328. Laforet GA, Sapp E, Chase K, McIntyre C, Boyce FM, et al. Changes in cortical and striatal neurons predict behavioral and electrophysiological abnormalities in a transgenic murine model of Huntington's disease. *J Neurosci* 2001;**21**:9112–23.

329. Hodgson JG, Agopyan N, Gutekunst C-A, Leavitt BR, LePiane F, et al. A YAC mouse model for Huntington's disease with full-length mutant Huntingtin, cytoplasmic toxicity, and selective striatal neurodegeneration. *Neuron* 1999;**23**:181–92.
330. Brustovetsky N, LaFrance R, Purl KJ, Brustovetsky T, Keene CD, et al. Age-dependent changes in the calcium sensitivity of striatal mitochondria in mouse models of Huntington's disease. *J Neurochem* 2005;**93**:1361–70.
331. White JK, Auerbach W, Duyao MP, Vonsattel J-P, Gusella JF, et al. Huntingtin is required for neurogenesis and is not impaired by the Huntington's disease CAG expansion. *Nat Genet* 1997;**17**:404–10.
332. Wheeler VC, Auerbach W, White JK, Srinidhi J, Auerbach A, et al. Length-dependent gametic CAG repeat instability in the Huntington's disease knock-in mouse. *Hum Mol Genet* 1999;**8**:115–22.
333. Lin C-H. Neurological abnormalities in a knock-in mouse model of Huntington's disease. *Hum Mol Genet* 2001;**10**:137–44.
334. Davies SW, Turmaine M, Cozens BA, DiFiglia M, Sharp AH, et al. Formation of neuronal intranuclear inclusions underlies the neurological dysfunction in mice transgenic for the HD mutation. *Cell* 1997;**90**:537–48.
335. Nithianantharajah J, Barkus C, Murphy M, Hannan AJ. Gene-environment interactions modulating cognitive function and molecular correlates of synaptic plasticity in Huntington's disease transgenic mice. *Neurobiol Dis* 2008;**29**:490–504.
336. Slow EJ, van Raamsdonk J, Rogers D, Coleman SH, Graham RK, et al. Selective striatal neuronal loss in a YAC128 mouse model of Huntington disease. *Hum Mol Genet* 2003;**12**:1555–67.
337. Guidetti P, Bates GP, Graham RK, Hayden MR, Leavitt BR, et al. Elevated brain 3-hydroxykynurenine and quinolinate levels in Huntington disease mice. *Neurobiol Dis* 2006;**23**:190–7.
338. Gray M, Shirasaki DI, Cepeda C, Andre VM, Wilburn B, et al. Full-length human mutant Huntingtin with a stable polyglutamine repeat can elicit progressive and selective neuropathogenesis in BACHD mice. *J Neurosci* 2008;**28**:6182–95.
339. Wheeler VC, White JK, Gutekunst CA, Vrbanac V, Weaver M, et al. Long glutamine tracts cause nuclear localization of a novel form of huntingtin in medium spiny striatal neurons in HdhQ92 and HdhQ111 knock-in mice. *Hum Mol Genet* 2000;**9**:503–13.
340. Heng MY, Tallaksen-Greene SJ, Detloff PJ, Albin RL. Longitudinal evaluation of the Hdh (CAG)150 knock-in murine model of Huntington's disease. *J Neurosci* 2007,**27**:8989–98.
341. Kuhn A, Goldstein DR, Hodges A, Strand AD, Sengstag T, et al. Mutant Huntingtin" effects on striatal gene expression in mice recapitulate changes observed in human Huntington's disease brain and do not differ with mutant huntingtin length or wild-type huntingtin dosage. *Hum Mol Genet* 2007;**16**:1845–61.
342. Strand AD, Baquet ZC, Aragaki AK, Holmans P, Yang L, et al. Expression profiling of Huntington's disease models suggests that brain-derived neurotrophic factor depletion plays a major role in striatal degeneration. *J Neurosci* 2007;**27**:11758–68.
343. Luthi-Carter R, Taylor DM, Pallos J, Lambert E, Amore A, et al. SIRT2 inhibition achieves neuroprotection by decreasing sterol biosynthesis. *Proc Natl Acad Sci USA* 2010;**107**:7927–32.
344. Valenza M, Leoni V, Karasinska JM, Petricca L, Fan J, et al. Cholesterol defect is marked across multiple rodent models of Huntington's disease and is manifest in astrocytes. *J Neurosci* 2010;**30**:10844–50.
345. Cui L, Jeong H, Borovecki F, Parkhurst CN, Tanese N, et al. Transcriptional repression of PGC-1alpha by mutant huntingtin leads to mitochondrial dysfunction and neurodegeneration. *Cell* 2006;**127**:59–69.

346. Weydt P, Pineda VV, Torrence AE, Libby RT, Satterfield TF, et al. Thermoregulatory and metabolic defects in Huntington's disease transgenic mice implicate PGC-1alpha in Huntington's disease neurodegeneration. *Cell Metab* 2006;**4**:349–62.
347. Chopra V, Fox JH, Lieberman G, Dorsey K, Matson W, et al. A small-molecule therapeutic lead for Huntington's disease: preclinical pharmacology and efficacy of C2-8 in the R6/2 transgenic mouse. *Proc Natl Acad Sci USA* 2007;**104**:16685–9.
348. Thomas EA, Coppola G, Desplats PA, Tang B, Soragni E, et al. The HDAC inhibitor 4b ameliorates the disease phenotype and transcriptional abnormalities in Huntington's disease transgenic mice. *Proc Natl Acad Sci USA* 2008;**105**:15564–9.
349. Bauer PO, Goswami A, Wong HK, Okuno M, Kurosawa M, et al. Harnessing chaperone-mediated autophagy for the selective degradation of mutant huntingtin protein. *Nat Biotechnol* 2010;**28**:256–63.
350. Steffan JS, Kazantsev A, Spasic-Boskovic O, Greenwald M, Zhu YZ, et al. The Huntington's disease protein interacts with p53 and CREB-binding protein and represses transcription. *Proc Natl Acad Sci USA* 2000;**97**:6763–8.
351. Rowland LP, Shneider NA. Amyotrophic lateral sclerosis. *N Engl J Med* 2001;**344**:1688–700.
352. Pasinelli P, Brown RH. Molecular biology of amyotrophic lateral sclerosis: insights from genetics. *Nat Rev Neurosci* 2006;**7**:710–23.
353. Rosen DR, Siddique T, Patterson D, Figlewicz DA, Sapp P, et al. Mutations in Cu/Zn superoxide dismutase gene are associated with familial amyotrophic lateral sclerosis. *Nature* 1993;**362**:59–62.
354. Siddique T, Figlewicz DA, Pericak-Vance MA, Haines JL, Rouleau G, et al. Linkage of a gene causing familial amyotrophic lateral sclerosis to chromosome 21 and evidence of genetic-locus heterogeneity. *N Engl J Med* 1991;**324**:1381–4.
355. Conforti FL, Sprovieri T, Mazzei R, Ungaro C, La Bella V, et al. A novel Angiogenin gene mutation in a sporadic patient with amyotrophic lateral sclerosis from southern Italy. *Neuromuscul Disord* 2008;**18**:68–70.
356. Greenway MJ, Andersen PM, Russ C, Ennis S, Cashman S, et al. ANG mutations segregate with familial and 'sporadic' amyotrophic lateral sclerosis. *Nat Genet* 2006;**38**:411–3.
357. Paubel A, Violette J, Amy M, Praline J, Meininger V, et al. Mutations of the ANG gene in French patients with sporadic amyotrophic lateral sclerosis. *Arch Neurol* 2008;**65**:1333–6.
358. Kwiatkowski Jr. TJ, Bosco DA, Leclerc AL, Tamrazian E, Vanderburg CR, et al. Mutations in the FUS/TLS gene on chromosome 16 cause familial amyotrophic lateral sclerosis. *Science* 2009;**323**:1205–8.
359. Vance C, Rogelj B, Hortobagyi T, De Vos KJ, Nishimura AL, et al. Mutations in FUS, an RNA processing protein, cause familial amyotrophic lateral sclerosis type 6. *Science* 2009;**323**:1208–11.
360. Maruyama H, Morino H, Ito H, Izumi Y, Kato H, et al. Mutations of optineurin in amyotrophic lateral sclerosis. *Nature* 2010;**465**:223–6.
361. Neumann M, Sampathu DM, Kwong LK, Truax AC, Micsenyi MC, et al. Ubiquitinated TDP-43 in frontotemporal lobar degeneration and amyotrophic lateral sclerosis. *Science* 2006;**314**:130–3.
362. Lomen-Hoerth C, Anderson T, Miller B. The overlap of amyotrophic lateral sclerosis and frontotemporal dementia. *Neurology* 2002;**59**:1077–9.
363. Lomen-Hoerth C, Murphy J, Langmore S, Kramer JH, Olney RK, et al. Are amyotrophic lateral sclerosis patients cognitively normal? *Neurology* 2003;**60**:1094–7.
364. Andersen PM, Nilsson P, Ala-Hurula V, Keranen ML, Tarvainen I, et al. Amyotrophic lateral sclerosis associated with homozygosity for an Asp90Ala mutation in CuZn-superoxide dismutase. *Nat Genet* 1995;**10**:61–6.

365. Jackson M, Al-Chalabi A, Enayat ZE, Chioza B, Leigh PN, et al. Copper/zinc superoxide dismutase 1 and sporadic amyotrophic lateral sclerosis: analysis of 155 cases and identification of a novel insertion mutation. *Ann Neurol* 1997;**42**:803–7.
366. Jones CT, Shaw PJ, Chari G, Brock DJ. Identification of a novel exon 4 SOD1 mutation in a sporadic amyotrophic lateral sclerosis patient. *Mol Cell Probes* 1994;**8**:329–30.
367. Jones CT, Swingler RJ, Brock DJ. Identification of a novel SOD1 mutation in an apparently sporadic amyotrophic lateral sclerosis patient and the detection of Ile113Thr in three others. *Hum Mol Genet* 1994;**3**:649–50.
368. Hadano S, Hand CK, Osuga H, Yanagisawa Y, Otomo A, et al. A gene encoding a putative GTPase regulator is mutated in familial amyotrophic lateral sclerosis 2. *Nat Genet* 2001;**29**:166–73.
369. Yang Y, Hentati A, Deng HX, Dabbagh O, Sasaki T, et al. The gene encoding alsin, a protein with three guanine-nucleotide exchange factor domains, is mutated in a form of recessive amyotrophic lateral sclerosis. *Nat Genet* 2001;**29**:160–5.
370. Chen YZ, Bennett CL, Huynh HM, Blair IP, Puls I, et al. DNA/RNA helicase gene mutations in a form of juvenile amyotrophic lateral sclerosis (ALS4). *Am J Hum Genet* 2004;**74**:1128–35.
371. Nishimura AL, Mitne-Neto M, Silva HC, Richieri-Costa A, Middleton S, et al. A mutation in the vesicle-trafficking protein VAPB causes late-onset spinal muscular atrophy and amyotrophic lateral sclerosis. *Am J Hum Genet* 2004;**75**:822–31.
372. Puls I, Jonnakuty C, LaMonte BH, Holzbaur EL, Tokito M, et al. Mutant dynactin in motor neuron disease. *Nat Genet* 2003;**33**:455–6.
373. Okado-Matsumoto A, Fridovich I. Subcellular distribution of superoxide dismutases (SOD) in rat liver: Cu, Zn-SOD in mitochondria. *J Biol Chem* 2001;**276**:38388–93.
374. Sturtz LA, Diekert K, Jensen LT, Lill R, Culotta VC. A fraction of yeast Cu, Zn-superoxide dismutase and its metallochaperone, CCS, localize to the intermembrane space of mitochondria. A physiological role for SOD1 in guarding against mitochondrial oxidative damage. *J Biol Chem* 2001;**276**:38084–9.
375. Crapo JD, Oury T, Rabouille C, Slot JW, Chang LY. Copper, zinc superoxide dismutase is primarily a cytosolic protein in human cells. *Proc Natl Acad Sci USA* 1992;**89**:10405–9.
376. Kikuchi H, Almer G, Yamashita S, Guegan C, Nagai M, et al. Spinal cord endoplasmic reticulum stress associated with a microsomal accumulation of mutant superoxide dismutase-1 in an ALS model. *Proc Natl Acad Sci USA* 2006;**103**:6025–30.
377. Andersen PM. Amyotrophic lateral sclerosis associated with mutations in the CuZn superoxide dismutase gene. *Curr Neurol Neurosci Rep* 2006;**6**:37–46.
378. Andersen PM, Sims KB, Xin WW, Kiely R, O'Neill G, et al. Sixteen novel mutations in the Cu/Zn superoxide dismutase gene in amyotrophic lateral sclerosis: a decade of discoveries, defects and disputes. *Amyotroph Lateral Scler Other Motor Neuron Disord* 2003;**4**:62–73.
379. Turner BJ, Talbot K. Transgenics, toxicity and therapeutics in rodent models of mutant SOD1-mediated familial ALS. *Prog Neurobiol* 2008;**85**:94–134.
380. Restagno G, Gomez A, Lombardo F, Cocco E, Calvo A, et al. The IVS1 + 319 t>a of SOD1 gene is not an ALS causing mutation. *Amyotroph Lateral Scler Other Mot Neuron Disord* 2005;**6**:45–9.
381. Kato S. Amyotrophic lateral sclerosis models and human neuropathology: similarities and differences. *Acta Neuropathol* 2008;**115**:97–114.
382. Deng HX, Hentati A, Tainer JA, Iqbal Z, Cayabyab A, et al. Amyotrophic lateral sclerosis and structural defects in Cu, Zn superoxide dismutase. *Science* 1993;**261**:1047–51.
383. Ho YS, Gargano M, Cao J, Bronson RT, Heimler I, et al. Reduced fertility in female mice lacking copper-zinc superoxide dismutase. *J Biol Chem* 1998;**273**:7765–9.
384. Ho YS, Magnenat JL, Gargano M, Cao J. The nature of antioxidant defense mechanisms: a lesson from transgenic studies. *Environ Health Perspect* 1998;**106**(Suppl. 5):1219–28.

385. Reaume AG, Elliott JL, Hoffman EK, Kowall NW, Ferrante RJ, et al. Motor neurons in Cu/Zn superoxide dismutase-deficient mice develop normally but exhibit enhanced cell death after axonal injury. *Nat Genet* 1996;**13**:43–7.

386. Flood DG, Reaume AG, Gruner JA, Hoffman EK, Hirsch JD, et al. Hindlimb motor neurons require Cu/Zn superoxide dismutase for maintenance of neuromuscular junctions. *Am J Pathol* 1999;**155**:663–72.

387. Muller FL, Song W, Liu Y, Chaudhuri A, Pieke-Dahl S, et al. Absence of CuZn superoxide dismutase leads to elevated oxidative stress and acceleration of age-dependent skeletal muscle atrophy. *Free Radic Biol Med* 2006;**40**:1993–2004.

388. Shefner JM, Reaume AG, Flood DG, Scott RW, Kowall NW, et al. Mice lacking cytosolic copper/zinc superoxide dismutase display a distinctive motor axonopathy. *Neurology* 1999;**53**:1239–46.

389. Dal Canto MC, Gurney ME. Neuropathological changes in two lines of mice carrying a transgene for mutant human Cu, Zn SOD, and in mice overexpressing wild type human SOD: a model of familial amyotrophic lateral sclerosis (FALS). *Brain Res* 1995;**676**:25–40.

390. Epstein CJ, Avraham KB, Lovett M, Smith S, Elroy-Stein O, et al. Transgenic mice with increased Cu/Zn-superoxide dismutase activity: animal model of dosage effects in Down syndrome. *Proc Natl Acad Sci USA* 1987;**84**:8044–8.

391. Jaarsma D, Haasdijk ED, Grashorn JA, Hawkins R, van Duijn W, et al. Human Cu/Zn superoxide dismutase (SOD1) overexpression in mice causes mitochondrial vacuolization, axonal degeneration, and premature motoneuron death and accelerates motoneuron disease in mice expressing a familial amyotrophic lateral sclerosis mutant SOD1. *Neurobiol Dis* 2000;**7**:623–43.

392. Kotulska K, LePecheur M, Marcol W, Lewin-Kowalik J, Larysz-Brysz M, et al. Overexpression of copper/zinc-superoxide dismutase in transgenic mice markedly impairs regeneration and increases development of neuropathic pain after sciatic nerve injury. *J Neurosci Res* 2006;**84**:1091–7.

393. Deng HX, Shi Y, Furukawa Y, Zhai H, Fu R, et al. Conversion to the amyotrophic lateral sclerosis phenotype is associated with intermolecular linked insoluble aggregates of SOD1 in mitochondria. *Proc Natl Acad Sci USA* 2006;**103**:7142–7.

394. Wong PC, Pardo CA, Borchelt DR, Lee MK, Copeland NG, et al. An adverse property of a familial ALS-linked SOD1 mutation causes motor neuron disease characterized by vacuolar degeneration of mitochondria. *Neuron* 1995;**14**:1105–16.

395. Chang-Hong R, Wada M, Koyama S, Kimura H, Arawaka S, et al. Neuroprotective effect of oxidized galectin-1 in a transgenic mouse model of amyotrophic lateral sclerosis. *Exp Neurol* 2005;**194**:203–11.

396. Wang J, Xu G, Gonzales V, Coonfield M, Fromholt D, et al. Fibrillar inclusions and motor neuron degeneration in transgenic mice expressing superoxide dismutase 1 with a disrupted copper-binding site. *Neurobiol Dis* 2002;**10**:128–38.

397. Wang J, Slunt H, Gonzales V, Fromholt D, Coonfield M, et al. Copper-binding-site-null SOD1 causes ALS in transgenic mice: aggregates of non-native SOD1 delineate a common feature. *Hum Mol Genet* 2003;**12**:2753–64.

398. Bruijn LI, Becher MW, Lee MK, Anderson KL, Jenkins NA, et al. ALS-linked SOD1 mutant G85R mediates damage to astrocytes and promotes rapidly progressive disease with SOD1-containing inclusions. *Neuron* 1997;**18**:327–38.

399. Ripps ME, Huntley GW, Hof PR, Morrison JH, Gordon JW. Transgenic mice expressing an altered murine superoxide dismutase gene provide an animal model of amyotrophic lateral sclerosis. *Proc Natl Acad Sci USA* 1995;**92**:689–93.

400. Jonsson PA, Graffmo KS, Brannstrom T, Nilsson P, Andersen PM, et al. Motor neuron disease in mice expressing the wild type-like D90A mutant superoxide dismutase-1. *J Neuropathol Exp Neurol* 2006;**65**:1126–36.
401. Gurney ME, Pu H, Chiu AY, Dal Canto MC, Polchow CY, et al. Motor neuron degeneration in mice that express a human Cu, Zn superoxide dismutase mutation. *Science* 1994;**264**:1772–5.
402. Chiu AY, Zhai P, Dal Canto MC, Peters TM, Kwon YW, et al. Age-dependent penetrance of disease in a transgenic mouse model of familial amyotrophic lateral sclerosis. *Mol Cell Neurosci* 1995;**6**:349–62.
403. Dal Canto MC, Gurney ME. A low expressor line of transgenic mice carrying a mutant human Cu, Zn superoxide dismutase (SOD1) gene develops pathological changes that most closely resemble those in human amyotrophic lateral sclerosis. *Acta Neuropathol* 1997;**93**:537–50.
404. Wang J, Xu G, Li H, Gonzales V, Fromholt D, et al. Somatodendritic accumulation of misfolded SOD1-L126Z in motor neurons mediates degeneration: alphaB-crystallin modulates aggregation. *Hum Mol Genet* 2005;**14**:2335–47.
405. Watanabe Y, Yasui K, Nakano T, Doi K, Fukada Y, et al. Mouse motor neuron disease caused by truncated SOD1 with or without C-terminal modification. *Brain Res Mol Brain Res* 2005;**135**:12–20.
406. Jonsson PA, Ernhill K, Andersen PM, Bergemalm D, Brannstrom T, et al. Minute quantities of misfolded mutant superoxide dismutase-1 cause amyotrophic lateral sclerosis. *Brain* 2004;**127**:73–88.
407. Gurney ME. The use of transgenic mouse models of amyotrophic lateral sclerosis in preclinical drug studies. *J Neurol Sci* 1997;**152**(Suppl. 1):S67–73.
408. Borchelt DR, Lee MK, Slunt HS, Guarnieri M, Xu ZS, et al. Superoxide dismutase 1 with mutations linked to familial amyotrophic lateral sclerosis possesses significant activity. *Proc Natl Acad Sci USA* 1994;**91**:8292–6.
409. Atkin JD, Farg MA, Turner BJ, Tomas D, Lysaght JA, et al. Induction of the unfolded protein response in familial amyotrophic lateral sclerosis and association of protein-disulfide isomerase with superoxide dismutase 1. *J Biol Chem* 2006;**281**:30152–65.
410. Atkin JD, Farg MA, Walker AK, McLean C, Tomas D, et al. Endoplasmic reticulum stress and induction of the unfolded protein response in human sporadic amyotrophic lateral sclerosis. *Neurobiol Dis* 2008;**30**:400–7.
411. Saxena S, Cabuy E, Caroni P. A role for motoneuron subtype-selective ER stress in disease manifestations of FALS mice. *Nat Neurosci* 2009;**12**:627–36.
412. Walker AK, Farg MA, Bye CR, McLean CA, Horne MK, et al. Protein disulphide isomerase protects against protein aggregation and is S-nitrosylated in amyotrophic lateral sclerosis. *Brain* 2010;**133**:105–16.
413. Uehara T, Nakamura T, Yao D, Shi ZQ, Gu Z, et al. S-nitrosylated protein-disulphide isomerase links protein misfolding to neurodegeneration. *Nature* 2006;**441**:513–7.
414. Arai T, Hasegawa M, Akiyama H, Ikeda K, Nonaka T, et al. TDP-43 is a component of ubiquitin-positive tau-negative inclusions in frontotemporal lobar degeneration and amyotrophic lateral sclerosis. *Biochem Biophys Res Commun* 2006;**351**:602–11.
415. Forman MS, Farmer J, Johnson JK, Clark CM, Arnold SE, et al. Frontotemporal dementia: clinicopathological correlations. *Ann Neurol* 2006;**59**:952–62.
416. Gitcho MA, Baloh RH, Chakraverty S, Mayo K, Norton JB, et al. TDP-43 A315T mutation in familial motor neuron disease. *Ann Neurol* 2008;**63**:535–8.
417. Sreedharan J, Blair IP, Tripathi VB, Hu X, Vance C, et al. TDP-43 mutations in familial and sporadic amyotrophic lateral sclerosis. *Science* 2008;**319**:1668–72.
418. Stallings NR, Puttaparthi K, Luther CM, Burns DK, Elliott JL. Progressive motor weakness in transgenic mice expressing human TDP-43. *Neurobiol Dis* 2010;**40**:404–14.

419. Tsai KJ, Yang CH, Fang YH, Cho KH, Chien WL, et al. Elevated expression of TDP-43 in the forebrain of mice is sufficient to cause neurological and pathological phenotypes mimicking FTLD-U. *J Exp Med.* 2010;**207**:1661–73.
420. Wils H, Kleinberger G, Janssens J, Pereson S, Joris G, et al. TDP-43 transgenic mice develop spastic paralysis and neuronal inclusions characteristic of ALS and frontotemporal lobar degeneration. *Proc Natl Acad Sci USA* 2010;**107**:3858–63.
421. Wu LS, Cheng WC, Hou SC, Yan YT, Jiang ST, et al. TDP-43, a neuro-pathosignature factor, is essential for early mouse embryogenesis. *Genesis* 2010;**48**:56–62.

Demise of the Flies: Why *Drosophila* Models Still Matter

Mathieu F. Bakhoum and
George R. Jackson

Mitchell Center for Neurodegenerative Diseases, Department of Neurology, The University of Texas Medical Branch, Galveston, Texas, USA

Over the past decade, *Drosophila melanogaster* has emerged as a widely used model for human disease via targeted misexpression of human disease-associated proteins. The chief advantage of creating such models is that once a suitable phenotype has been obtained, the genetic toolkit of fly genetics can be used to dissect underlying disease pathways. Although some critics of this approach have argued that it has not generated many novel insights, we would argue that fly models of human neurodegenerative disorders have provided valuable information when viewed within the context of other models and systems of analysis. Here, we will provide a brief overview of some *Drosophila* models of neurodegenerative disorders with a special focus on our own work.

I. The Making of a Fly Model

A. Ease of Genetic Manipulation Has Facilitated Model Development

Neurodegenerative diseases are characterized by loss of specific neuronal populations, with resulting signs and symptoms corresponding to these affected cells. Most of these diseases share a common pathological feature: abnormal protein accumulations variously referred to as aggregates or inclusions. Misfolding of disease-associated proteins may alter their function and disrupt

Progress in Molecular Biology
and Translational Science, Vol. 100
DOI: 10.1016/B978-0-12-384878-9.00011-X

Copyright 2011, Elsevier Inc.
All rights reserved.
1877-1173/11 $35.00

crucial cellular pathways, eventually leading to neurodegeneration. Familial forms of some neurodegenerative diseases have been linked to mutations in specific genes. These disease-associated mutations have facilitated the development of many animal models for neurodegenerative diseases, some of which have utilized the simple vinegar fly *Drosophila melanogaster.*

Drosophila is an attractive organism for establishing models of human disease and studying genetic interactions for several reasons. First is the functional similarity between the genomes of *Homo sapiens* and *Drosophila*; more than 75% of genes associated with human diseases have a homolog in *Drosophila.*[1] Given this high similarity between human genes and their *Drosophila* counterparts, fly models using misexpression of human genes or mutations of their fly homologs have been engineered for many classes of human diseases, including cancer and developmental, cardiac, immunological, and neurodegenerative disorders.[2] Second, many well-studied biological pathways are conserved between humans and *Drosophila*; these include endocrine, intracellular signaling, cell death, and pattern formation pathways, for example, the transforming growth factor-β (TGF-β)/*decapentaplegic* pathway, the *wingless*/Wnt pathway, insulin signaling, and the *Notch* pathway.[3] A third attractive feature of using *Drosophila* is the versatile genetic toolkit that has been developed over more than a century of use. Genetic manipulation of *Drosophila* can generate a wide variety of mutants and transgenes. Gene loss of function can be accomplished via RNAi, transposon insertion, or imprecise excision, as well as by chemical and X-ray mutagenesis.[4] Gain of function effects can be studied using various systems, including the binary GAL4/UAS system, which allows overexpression in a tissue-specific manner.[5] The relatively short life cycle of the fly also provides advantages for genetic modeling. It takes 9–10 days for an egg to reach the adult stage, and a single cross yields abundant progeny, facilitating generation of genetic variants and the efficient production of *Drosophila* models for human diseases; moreover, the relatively short lifespan of the fly (no more than 120 days, depending upon conditions) facilitates examination of gene effects in aged individuals.

Keeping these attractive features of modeling using the fly in mind, here we provide a brief overview of some fly models of neurodegenerative work, with attention to the contributions of our own laboratory.

B. The Eye as a Model System

Generation of a useful fly model for a disease requires the development of a morphological, behavioral, or physiological phenotype in response to a transgene or a mutation. Several factors determine the potential for success of a model. First, the similarity between the phenotype and the clinical signs observed in patients is considered by many to be a critical test of a model. An additional factor that can be used to assess the quality of a model is the degree of similarity between pathology formed in the fly and that observed in humans with regard to tissue specificity, formation of inclusions/aggregates, and cell dysfunction or death. The ease with which the phenotype can be screened and the specificity of scoring are important additional factors in designing screens for genetic modifiers.

The *Drosophila* eye has been widely used as a model for neurodegenerative diseases and is well suited to this purpose for several reasons. The retina is of neural origin and contains photoreceptor neurons similar to their mammalian counterparts. The final assembly of the adult compound eye requires precisely coordinated developmental events in larvae and pupae. These developmental stages are well characterized, and each stage can be analyzed at a single cell level. It has been estimated that at least two-third of *Drosophila* vital genes play a role in pattern formation and/or connectivity of the adult retina; thus, if a genetic manipulation (e.g., misexpression of a human gene) impacts even one such vital pathway, it may reasonably be expected to yield an abnormal eye phenotype.[6,7] Under laboratory conditions, the eye is dispensable for survival; thus, knockdown of vital genes does not affect the overall survival or fertility of the fly. Finally, many eye phenotypes can be readily scored under the dissecting microscope or by using techniques such as the deep pseudopupil or optical neutralization.

The compound eye of *Drosophila* is comprised of approximately 760 repeating subunits called ommatidia. Each ommatidium includes photoreceptor neurons, cone cells, and pigment cells, as well as interommatidial cells formed in early developmental stages and removed by cell death. Because patterning in the eye results from a wave of morphogenesis that starts in the larval stage, identical cells are not derived from a common precursor cell. Therefore, different genetic clones can be analyzed in the same eye, that is, clones homozygous for a lethal mutation can be identified adjacent to wild-type clones, facilitating phenotypic analysis.

A phenotype commonly used to score degeneration in the eye is the "rough" eye. Many transgenes induce a rough eye phenotype, which consists of ommatidial fusion and loss of interommatidial bristles. The rough phenotype varies in severity and (ideally) is dosage-sensitive, making it suitable for genetic screening.

II. Alzheimer Disease

Alzheimer disease (AD) is the most common cause of dementia among aged populations. AD is characterized by the deposition of amyloid plaques and neurofibrillary tangles (NFT). Histologically, disease progression follows an anticipated regional progression described by Braak and Braak.[8] NFT contain the microtubule-associated protein tau (MAPT). Tau hyperphosphorylation is thought to precede and perhaps cause tau aggregation. Tau phosphorylation negatively regulates its interactions with microtubules.[9] Other aberrant post-translational modifications, for example, nitrosylation and truncation, are thought to occur in AD and other tauopathies.

A. Tauopathies

Tauopathies are a group of neurodegenerative diseases characterized by pathologic accumulation of tau; AD is one such tauopathy. Dominant mutations in tau have been identified in frontotemporal dementia with parkinsonism linked to chromosome 17 (FTDP-17, now also referred to as frontotemporal lobar degeneration-tau (FTLD-T)). In other tauopathies, such as AD, Pick disease, corticobasal degeneration, and progressive supranuclear palsy, tau mutations have not been identified, although NFT and other forms of tau accumulations are common pathological hallmarks of the diseases.[10] Recent genome-wide association studies have also identified tau polymorphisms as important contributors to risk for sporadic Parkinson's disease (PD).[11–13]

In *Drosophila*, several models of tauopathy have been developed. One of the early models used the binary GAL4/UAS expression system[14] and was reported by Shepherd and colleagues.[15] In this model, targeted expression of human tau resulted in age-dependent loss of sensory axons and axonal pathology such as abnormal axonal bundling and swelling (beading). Feany and colleagues reported more robust toxicity of FTLD-associated R406W and V337M mutations as compared to wild-type tau; phenotypes included early mortality and vacuolization of brain but without formation of obvious NFT.[16]

We used a direct fusion construct driving expression of wild-type human tau in the *Drosophila* eye without the need for GAL4, which produced a dosage-sensitive rough eye phenotype.[17] In this model, overexpression of the *Drosophila* glycogen synthase kinase-3β (GSK-3β) homolog *shaggy* exacerbated the toxic effects of tau. The engineering of these models has provided useful reagents for genetic screens. Kinases and phosphatases have been identified as modifiers of tau toxicity,[18] lending credence to the assertion that tau phosphorylation plays an important role in mediating its toxicity. This assertion is further supported by the finding that a phosphorylation-incompetent form of tau, obtained by mutating all the kinase sites to alanine, has reduced neurotoxicity as compared to wild-type tau.[19] Pseudo-hyperphosphorylated tau produced by substituting all serines and threonines with glutamate has increased toxicity.[20] Our laboratory, on the other hand, found that a slightly different form of tau resistant to GSK-3β retained substantial toxicity, perhaps because of an increased affinity for microtubules.[21] Lecourtois and colleagues, however, showed that human tau, unlike *Drosophila* tau, has low affinity for *Drosophila* microtubules.[22] Mudher and coworkers recently demonstrated that soluble, non-filamentous hyperphosphorylated tau causes toxicity through disruption of axonal transport. [23]

Skoulakis and colleagues recently investigated the role of tau phosphorylation in the CNS. Wild-type tau overexpression causes developmental mushroom body ablation in the fly.[24] Tau phosphorylation at Ser^{238} and Thr^{245} was necessary for toxicity in this model. These flies exhibited severe impairment of associative learning and memory. FTLD-linked mutants were significantly less toxic to the mushroom body than wild-type tau. Another recent tau model which may be useful for modifier screening was developed by Yeh *et al.* using a GAL4 driver to express tau in the dorsal mesothorax (notum) of *Drosophila*. Overexpression of tau caused loss of mechanosensory bristles on the dorsal surface of the fly.[25] The phenotype was enhanced by tau phosphorylation and was suppressed by reduction of shaggy dosage.

B. Aβ Amyloid

Mutations in the amyloid precursor protein (APP) cause familial AD.[26] Amyloid plaques are mainly composed of Aβ40 or Aβ42 amino acid peptides derived from APP cleavage.

Amyloidogenic Aβ peptides arise from APP cleavage through β- and γ-secretase.[27] Presenilins, which have a single fly homolog (Psn), are a component of the γ-secretase complex. Mutations in PS1 and PS2 cause early onset familial AD.[26] *Drosophila* homologs of other components of the γ-secretase, including nicastrin, Pen-2, and Aph-1, have been identified.[28] The fly homolog of β-secretase, dBACE, has been identified recently[29]; this enzyme cleaves human APP at a different site than that at which human β-secretase acts. However, dBACE overexpression cleaves dAPPl, the APP fly homolog, and produces an amyloidogenic form that aggregates. It is noteworthy that dAppl does not contain the Aβ domain found in human APP. This suggests that although the sequence is not conserved between humans and *Drosophila*, amyloidogenic processing may still be conserved.[29] dAPPl may function similar to APP, since large deletions of dAppl result in reduced locomotion and phototaxis deficiency that are rescued by human APP.[30]

Aβ toxicity in flies also has been modeled by directed overexpression of Aβ fragments. $A\beta_{1-42}$ (but not $A\beta_{1-40}$) overexpression reduces lifespan in the fly. However, both Aβ fragments cause progressive loss of associative learning. Aβ overexpression in the eye causes retinal phenotypes.[31] Moreover, Aβ overexpression causes long-term depression. Chiang *et al.* showed through immunostaining of PI3K that Aβ induces PI3K hyperactivity. Knockdown or pharmacologic inhibition of PI3K function leads to rescue of the long-term depression phenotype.[32] Crowther *et al.* reported that Aβ forms nonamyloidogenic aggregates that resemble diffuse plaques. Immunohistochemistry using a conformation-dependent antibody indicates that oligomers may actually be the entities mediating neurotoxic effects.[33] The downstream effects of Aβ have not been identified with certainty, although recent findings suggest tau phosphorylation to be an important downstream effect of Aβ-induced neurotoxicity. A recent paper by Iijima *et al.* showed that coexpression of Aβ42 with tau increased tau phosphorylation at Ser262, an AD-related phosphoepitope.[34] Ser262 phosphorylation enhanced tau-induced neurotoxicity, whereas coexpression of Aβ42 and a nonphosphorylatable form of tau, Ser262Ala, did not cause any neurodegenerative phenotypes.

III. Parkinson's Disease

PD is characterized by bradykinesia, tremor, postural instability, and rigidity.[35] Pathological hallmarks of PD include degeneration of dopaminergic neurons in the substantia nigra pars compacta and the presence of Lewy bodies, eosinophilic cytoplasmic inclusions comprised in part of α-synuclein.[36] Although PD is usually sporadic, mutations have been linked to familial forms of PD. Autosomal dominant PD is caused by the A53T[37] and A30P[38] mutations in SNCA/α-synuclein; the E46K mutation appears to be associated with a more widespread synucleinopathy called dementia with Lewy bodies.[39] Synucleins (α-, β-, and γ-synuclein) are soluble proteins found primarily in neural tissue; α-synuclein is associated with PD. Genomic duplication[40] and triplication[41] have also been identified as causes for familial PD. Overexpression of either wild type or mutant forms of α-synuclein in the fly leads to loss of tyrosine hydroxylase-immunoreactive neurons.[42,43] Moreover, intracytoplasmic accumulation of α-synuclein suggestive of Lewy bodies is observed. These transgenic flies also exhibit age-dependent motor impairment and mild retinal abnormalities. Climbing defects are more severe in the A30P mutant as compared to wild-type α-synuclein or the A53T mutant form.[42]

Autosomal recessive juvenile parkinsonism, a familial form of PD that lacks Lewy body pathology, has been linked to loss-of-function mutations in the *parkin* gene, which encodes a ubiquitin ligase.[44] *Drosophila* parkin null mutants show reduced lifespan, death of indirect flight muscles, and male sterility.[45] The dying cells exhibit mitochondrial pathology. Despite the recessive nature of parkin-linked PD, overexpression of certain human parkin mutants can also have toxic effects. Our laboratory generated a fly PD model that manifested dominant toxicity of mutant human parkin. In this model, mutant forms of human parkin derived from familial PD, Q311X and T240R, cause selective degeneration of dopaminergic neurons, accompanied by progressive motor impairment.[46] The toxic effects of parkin are age-dependent. *Drosophila* vesicular monoamine transporter (DVMAT) knockdown exacerbates the toxic effects of mutant parkin, whereas its overexpression reduced the toxic effects of mutant parkin.[46] This observation suggests that the selective vulnerability of dopaminergic neurons to mutant parkin is associated with dopamine and/or its metabolites. Recent models have also demonstrated the dominant toxic effects of other parkin mutations. Wang *et al.* demonstrated that expression of missense mutant parkin R275W but not the G328E mutant caused dopaminergic neurodegeneration and mitochondrial abnormalities.[47]

Another gene implicated in recessive familial PD is PTEN-induced kinase 1, *PINK1*.[48] PINK1 encodes a serine/threonine kinase with a mitochondrial targeting sequence. In *Drosophila*, knockout of the PINK1 homolog also results in male sterility and apoptotic muscle degeneration with mitochondrial pathology.[49–51] Both PINK1 and human parkin were both able to rescue the phenotype, suggesting that there is a functional homology between *Drosophila* and human PINK1 and that parkin and PINK1 function in the same pathway. Transgenic expression of parkin rescued the PINK1 loss-of-function phenotypes, but not vice versa, suggesting that parkin functions downstream of PINK1. Chung and coworkers also reported that expression of *Drosophila* Bcl-2 was able to rescue most of the phenotypes resulting from PINK1 loss of function, suggesting that mitochondrial dysfunction accounts for the degenerative phenotypes caused by PINK1 mutant.[50]

A form of PD phenotypically similar to the sporadic form of the disease has been linked to mutations in Leucine-rich repeat kinase 2, LRRK2.[52,53] The pathological hallmarks are variable and do not always include Lewy bodies.[52–54] Several reports of fly models of LRRK2-related PD have appeared. Smith and coworkers found that misexpression of wild-type G2019S mutant LRRK2 in flies caused photoreceptor and dopaminergic degeneration.[55]

Ng *et al.* showed that expression of other mutant forms of LRRK, such as G2019S, Y1699C, or G2385R LRRK2, but not wild-type LRRK2, caused age-dependent loss of dopaminergic neurons in addition to reduced lifespan and increased sensitivity to rotenone, a mitochondrial complex I inhibitor.[56] Interestingly, these investigators showed that coexpression of human parkin with the G2019S mutant of LRRK2 conferred some protection, suggesting a link between LRRK2 and parkin and a possible role for mitochondria in the pathogenesis of LRRK2-related parkinsonism. Another possible mechanism for LRRK2 toxicity was presented by Imai *et al.*[57] These investigators showed that both human LRRK2 and its *Drosophila* ortholog phosphorylate eukaryotic initiation factor 4E (eIF4E)-binding protein (4E-BP), a negative regulator of eIF4E-mediated protein translation and a key mediator of various stress responses. The resultant dysregulation of protein translation may thus contribute to age-dependent loss of dopaminergic neurons.

A rare cause for PD is associated with mutations in the *DJ-1* gene.[58] *Drosophila* has two DJ-1 homologs: DJ-1A and DJ-1B. Double-knockout of both DJ-1 homologs results in flies that are hypersensitive to environmental toxins such as paraquat and rotenone.[59] Menzies *et al.* showed that DJ-1A is upregulated in the DJ-1B knockout flies. DJ-1A overexpression may explain the resistance of this model to paraquat.[60] This is supported by the findings from DJ-1A knockdown flies which showed stronger phenotypes than the double knockout flies. RNAi knockdown of DJ-1A results in dysfunction and degeneration of dopaminergic and photoreceptors neurons, hypersensitivity to oxidative stress, and accumulation of reactive oxygen species.[61]

IV. Polyglutamine Disease

Polyglutamine diseases are those neurodegenerative disorders resulting from a CAG expansion leading to a large polyglutamine tract. The polyglutamine expansion within the open reading frame of the disease protein causes a conformation change to a β-sheet structure.[62]

Drosophila models of polyglutamine diseases share several characteristics of human disease, including ubiquitinated aggregates and nuclear inclusions in neurons. Also, *Drosophila* models share length-dependent pathology.[63]

Huntington disease (HD) is caused by a polyglutamine expansion near the amino terminus of huntingtin[64] and is characterized by chorea, cognitive impairment, and affective changes. HD is the most prevalent autosomal dominant polyglutamine neurodegenerative disorder. Expressing the huntingtin protein with a polyglutamine tract expansion, but not wild-type repeat lengths, results in degeneration of photoreceptors. Polyglutamine-induced cell death can be suppressed in a mutant Dark background.[65] Dark is the fly homolog of human Apaf-1, a regulator of apoptosis. Chaperones have also been identified as modifiers of polyglutamine toxicity. J-domain-containing proteins such as *Drosophila* heat-shock protein 40/HDJ1 and dTPR2 have been identified in a forward genetic screen as suppressors of polyglutamine toxicity. Hsp70, another chaperone, suppresses polyglutamine-induced degeneration in the context of huntingtin and ataxin-3.[66,67]

Spinocerebellar ataxia type 3 (SCA-3) is caused by polyglutamine expansion in ataxin-3, leading to neuronal dysfunction. Expression of pathogenic ataxin-3 in *Drosophila* recapitulates the characteristics of the human disease. Transgenic flies expressing repeat expansions show late onset degeneration with nuclear inclusions. Wild-type full-length ataxin-3 rescues the phenotype of polyglutamine-expanded ataxin-3. This suppression requires an intact ubiquitin-binding domain of ataxin-3 as well as ubiquitin protease, both located in the amino terminus of ataxin-3.[68] A genetic modifier screen of ataxin-3 neurodegeneration in *Drosophila* has identified several genes that play a role in protein quality control. The majority of genes were part of the chaperone and ubiquitin–proteasome pathway, whereas others were involved in autophagy.[69]

Drosophila models have also been developed for spinocerebellar ataxia types 1, 2, and 7 (SCA-1, -2, and -7). SCA-2 is caused by a CAG repeat in the polyglutamine tract of ataxin-2. Satterfield *et al.* showed that mutations in the *Drosophila* homolog of ataxin-2, dAtx-2, result in female sterility, aberrant sensory bristle morphology, and lethality. These effects were associated with actin filament disruption, yet dAtx-2 does not assemble with actin filaments. This points to a potential indirect role of Atx-2 in mediating actin cytoskeleton dysregulation and eventually neurodegeneration, since loss of cytoskeleton-dependent dendritic structure marks an early event in SCA-2 pathogenesis.[70]

SCA-1 is caused by a polyglutamine expansion in ataxin-1. Overexpression of human polyglutamine-expanded ataxin-1 in the *Drosophila* eye causes retinal phenotypes with nuclear inclusions.[71] The nuclear inclusions were also observed when ataxin-1 was expressed in the CNS or salivary glands. In the same study, a P-element insertion screen identified modifiers of ataxin-1 toxicity which underscore the role of protein folding and clearance as well as RNA processing and transcriptional regulation in SCA-1.

Endogenous ataxin-2 also has been shown to affect expanded ataxin-1 toxicity in flies. Increased dAtx-2 levels enhance expanded ataxin-1 toxicity, while decreased dAtx-2 levels suppress it.[72] In postmortem neurons from SCA-1 brains, human Ataxin-2 accumulates in nuclei. This is paralleled by the findings that human expanded Atx-1 expression in *Drosophila* induces intranuclear accumulation of dAtx-2. These observations suggest that intranuclear accumulation of Ataxin-2, almost exclusively found in the cytoplasm, contributes to expanded Ataxin-1-induced toxicity.

SCA-7 is caused by a polyglutamine expansion in ataxin-7, which is involved in histone acetylation. Latouche *et al.* created a conditional *Drosophila* model where a truncated and expanded Ataxin-7 is expressed in adult flies neurons. Similar to the pathology seen in SCA-7 brains, mutant ataxin-7 accumulated in intranuclear inclusions containing ubiquitin, the 19S proteasome subunit, and HSP70.[73] The flies expressing mutant ataxin-7 show locomotion defects as well as reduced lifespan.

Spinobulbar muscular atrophy (SBMA) or Kennedy's disease is an X-linked disease characterized by progressive weakness caused by the degeneration of motor neurons in brainstem and spinal cord.[74] The disease is caused by an unstable CAG repeat in the androgen receptor (AR) gene.[75] Overexpression of full-length AR protein with a 52-glutamine stretch in the *Drosophila* eye does not result in any degeneration; however, when these flies were exposed to dihydrotestosterone (DHT), an AR ligand, the flies exhibit retinal degeneration in polyglutamine length-dependent manner. Chan *et al.* showed that a truncated form of polyglutamine-expanded AR results in retinal degeneration without the use of DHT as an AR ligand.[76] Taylor and colleagues reported that DNA binding to the AR is essential for toxicity.[77] A mutation in the DNA-binding domain (A574D), which disrupts DNA binding to AR without affecting ligand binding, does not result in degeneration even in the presence of DHT. Nuclear translocation, which ensues after the binding of DHT to AR, is necessary but not sufficient for toxicity—an AR construct that locates to the nucleus independently of DHT does not result in a toxic phenotype. Modifiers of polyglutamine-expanded AR toxicity, as well as mutation of the AF-2 domain, have indicated that an intact AF-2 domain is necessary for toxicity. The authors have proposed a mechanism[77] in which ligand binding to AR induces a conformational change in the ligand-binding domain, thus creating the AF-2 surface and translocation of the receptor to the nucleus.

V. Amyotrophic Lateral Sclerosis

Amyotrophic lateral sclerosis (ALS), also known as Lou Gehrig's disease in the USA or Charcot's disease in Europe, affects upper and lower motor neurons. Ten percent of cases are familial. Superoxide dismutase (SOD) mutations have been identified in familial cases, and SOD neuronal inclusions are found in all ALS cases, suggesting that oxidative stress plays an important role in the pathogenesis of the disease. *Drosophila* with a null SOD allele exhibit increased sensitivity to oxidative stress, as well as reduced longevity.[78] Mutant or wild-type human SOD expression in *Drosophila* resulted in neuronal toxicity and aggregation.[78,79]

Other genes associated with familial ALS include the vesicle-associated membrane protein/synaptobrevin-associated membrane protein B, VAPB. Expression of wild-type and mutant VAPB in *Drosophila* results in neurodegeneration, with the mutant form more susceptible to aggregate and form inclusions.[80–82] Our laboratory generated a model for ALS by increasing expression of wild-type VAPB in sensory neurons.[81] This leads to loss of mechanosensory bristles, providing an external phenotype which is a useful tool which can be used in screens for modifiers of VAPB-mediated neurodegeneration. Our data suggest that mutant VAPB exerts dominant-negative effects on the wild type.[81]

Recently, mutations in TARDBP have been linked to both familial and sporadic ALS. TARDBP encodes a heterogeneous nuclear ribonuclear protein (hnRNP) transactive response–DNA binding protein.[83,84] Knockdown or knockout of *Drosophila* TDP results in larval lethality which is rescued by expression of human TDP-43. This observation suggests that *Drosophila* and human TDP-43 are functionally homologous. However, overexpression of human TDP-43 in a wild-type background also results in larval lethality.[85,86] Both knockout of *Drosophila* TDP-43 and overexpression of human TDP-43 in a wild-type background reduce dendritic branching and synaptic boutons. Recently, additional *Drosophila* models of TDP-43 proteinopathy were developed by Li *et al.*[86] Overexpression of hTDP-43 in the *Drosophila* eye causes degeneration in an age-dependent manner. Mushroom body overexpression of hTDP-43 also resulted in age-dependent neurodegeneration manifested by axonal loss and neuronal loss. Moreover, motor neuron death, bouton reduction, and axon swelling were also caused by overexpression of hTDP-43 in motor neurons. Motor activity of these flies was impaired at both larval and adult stages. A recent study by Elden *et al.*[87] showed that ataxin-2 is a potential regulator of TDP-43 toxicity. Upregulation of Atx-2, the *Drosophila* homolog of ataxin-2, enhances the toxicity of wild-type and Q331K mutant human TDP-43. This effect is manifest as increasing retinal degeneration and reduced lifespan. A reduction of one Atx-2 null allele, however, suppresses TDP-43 toxicity in the eye and extended lifespan. The Atx-2-mediated enhanced toxicity of TDP-43 is not due to increased levels of TDP-43, since increasing protein levels of the latter did not result in the same level of toxicity.

VI. Concluding Remarks

The pathological hallmarks of many neurodegenerative diseases have been well described for decades, and the past two decades have seen major strides forward in understanding the molecular bases of these diseases; nonetheless, the mechanisms underlying neurodegenerative disorders largely remain unclear. As described in this brief chapter, *Drosophila* models of neurodegeneration have proved useful in recapitulating many aspects of human neurodegenerative disorders. The utility of such models has not proved to be in uncovering startling new mechanistic pathways, but rather in contributing along with other models to incremental advances in understanding of neurodegenerative disorders. Certainly not every modifier identified using fly models has been validated as a key component of disease pathways; nonetheless, critics are likely to find that rumors of the demise of fly models have been exaggerated. Exciting developments in the rapidly advancing genetics of FTLD and ALS, in particular, are likely to provide opportunities for the application of new invertebrate models to the study of disease pathogenesis.

References

1. Reiter LT, Potocki L, Chien S, Gribskov M, Bier E. A systematic analysis of human disease-associated gene sequences in *Drosophila melanogaster*. *Genome Res* 2001;**11**:1114–25.
2. Bier E. *Drosophila*, the golden bug, emerges as a tool for human genetics. *Nat Rev Genet* 2005;**6**:9–23.

3. Adams MD, Celniker SE, Holt RA, Evans CA, Gocayne JD, Amanatides PG, et al. The genome sequence of *Drosophila melanogaster*. *Science* 2000;**287**:2185–95.
4. Bellen HJ, Levis RW, Liao G, He Y, Carlson JW, Tsang G, et al. The BDGP gene disruption project: single transposon insertions associated with 40% of *Drosophila* genes. *Genetics* 2004;**167**:761–81.
5. Venken KJ, Bellen HJ. Transgenesis upgrades for *Drosophila melanogaster*. *Development* 2007;**134**:3571–84.
6. Thomas BJ, Wassarman DA. A fly's eye view of biology. *Trends Genet* 1999;**15**:184–90.
7. Thaker HM, Kankel DR. Mosaic analysis gives an estimate of the extent of genomic involvement in the development of the visual system in *Drosophila melanogaster*. *Genetics* 1992;**131**:883–94.
8. Braak H, Braak E. Morphological criteria for the recognition of Alzheimer's disease and the distribution pattern of cortical changes related to this disorder. *Neurobiol Aging* 1994;**15**:355–6, (discussion 379–80).
9. Buee L, Bussiere T, Buee-Scherrer V, Delacourte A, Hof PR. Tau protein isoforms, phosphorylation and role in neurodegenerative disorders. *Brain Res Brain Res Rev* 2000;**33**:95–130.
10. Ballatore C, Lee VM, Trojanowski JQ. Tau-mediated neurodegeneration in Alzheimer's disease and related disorders. *Nat Rev Neurosci* 2007;**8**:663–72.
11. Satake W, Nakabayashi Y, Mizuta I, Hirota Y, Ito C, Kubo M, et al. Genome-wide association study identifies common variants at four loci as genetic risk factors for Parkinson's disease. *Nat Genet* 2009;**41**:1303–7.
12. Simon-Sanchez J, Schulte C, Bras JM, Sharma M, Gibbs JR, Berg D, et al. Genome-wide association study reveals genetic risk underlying Parkinson's disease. *Nat Genet* 2009;**41**:1308–12.
13. Edwards TL, Scott WK, Almonte C, Burt A, Powell EH, Beecham GW, et al. Genome-wide association study confirms SNPs in SNCA and the MAPT region as common risk factors for Parkinson disease. *Ann Hum Genet* 2010;**74**:97–109.
14. Brand AH, Perrimon N. Targeted gene expression as a means of altering cell fates and generating dominant phenotypes. *Development* 1993;**118**:401–15.
15. Williams DW, Tyrer M, Shepherd D. Tau and tau reporters disrupt central projections of sensory neurons in *Drosophila*. *J Comp Neurol* 2000;**428**:630–40.
16. Wittmann CW, Wszolek MF, Shulman JM, Salvaterra PM, Lewis J, Hutton M, et al. Tauopathy in *Drosophila*: neurodegeneration without neurofibrillary tangles. *Science* 2001;**293**:711–4.
17. Jackson GR, Wiedau-Pazos M, Sang T-K, Wagle N, Brown CA, Massachi S, et al. Human wild-type tau interacts with wingless pathway components and produces neurofibrillary pathology in *Drosophila*. *Neuron* 2002;**34**:509–19.
18. Shulman JM, Feany MB. Genetic modifiers of tauopathy in *Drosophila*. *Genetics* 2003;**165**:1233–42.
19. Steinhilb ML, Dias-Santagata D, Mulkearns EE, Shulman JM, Biernat J, Mandelkow E-M, et al. S/P and T/P phosphorylation is critical for tau neurotoxicity in *Drosophila*. *J Neurosci Res* 2007;**85**:1271–8.
20. Fulga TA, Elson-Schwab I, Khurana V, Steinhilb ML, Spires TL, Hyman BT, et al. Abnormal bundling and accumulation of F-actin mediates tau-induced neuronal degeneration in vivo. *Nat Cell Biol* 2007;**9**:139–48.
21. Chatterjee S, Sang TK, Lawless GM, Jackson GR. Dissociation of tau toxicity and phosphorylation: role of GSK-3beta, MARK and Cdk5 in a *Drosophila* model. *Hum Mol Genet* 2009;**18**:164–77.
22. Feuillette S, Miguel L, Frébourg T, Campion D, Lecourtois M. *Drosophila* models of human tauopathies indicate that Tau protein toxicity in vivo is mediated by soluble cytosolic phosphorylated forms of the protein. *J Neurochem* 2010;**113**:895–903.

23. Cowan CM, Chee F, Shepherd D, Mudher A. Disruption of neuronal function by soluble hyperphosphorylated tau in a *Drosophila* model of tauopathy. *Biochem Soc Trans* 2010;**38**:564–70.
24. Kosmidis S, Grammenoudi S, Papanikolopoulou K, Skoulakis EMC. Differential effects of Tau on the integrity and function of neurons essential for learning in *Drosophila*. *J Neurosci* 2010;**30**:464–77.
25. Yeh P-A, Chien J-Y, Chou C-C, Huang Y-F, Tang C-Y, Wang H-Y, et al. *Drosophila* notal bristle as a novel assessment tool for pathogenic study of Tau toxicity and screening of therapeutic compounds. *Biochem Biophys Res Commun* 2010;**391**:510–6.
26. Bertram L, Lill CM, Tanzi RE. The genetics of Alzheimer disease: back to the future. *Neuron* 2010;**68**:270–81.
27. Hardy J. Alzheimer's disease: The amyloid cascade hypothesis: an update and reappraisal. *J Alzheimers Dis* 2006;**9**:151–3.
28. Hu Y, Fortini ME. Different cofactor activities in gamma-secretase assembly: evidence for a nicastrin-Aph-1 subcomplex. *J Cell Biol* 2003;**161**:685–90.
29. Carmine-Simmen K, Proctor T, Tschäpe J, Poeck B, Triphan T, Strauss R, et al. Neurotoxic effects induced by the *Drosophila* amyloid-beta peptide suggest a conserved toxic function. *Neurobiol Dis* 2009;**33**:274–81.
30. Luo L, Tully T, White K. Human amyloid precursor protein ameliorates behavioral deficit of flies deleted for Appl gene. *Neuron* 1992;**9**:595–605.
31. Iijima K, Liu HP, Chiang AS, Hearn SA, Konsolaki M, Zhong Y. Dissecting the pathological effects of human Abeta40 and Abeta42 in *Drosophila*: a potential model for Alzheimer's disease. *Proc Natl Acad Sci USA* 2004;**101**:6623–8.
32. Chiang H-C, Wang L, Xie Z, Yau A, Zhong Y. PI3 kinase signaling is involved in Abeta-induced memory loss in *Drosophila*. *Proc Natl Acad Sci USA* 2010;**107**:7060–5.
33. Crowther DC, Kinghorn KJ, Miranda E, Page R, Curry JA, Duthie FAI, et al. Intraneuronal Abeta, non-amyloid aggregates and neurodegeneration in a *Drosophila* model of Alzheimer's disease. *Neuroscience* 2005;**132**:123–35.
34. Iijima K, Gatt A, Iijima-Ando K. Tau Ser262 phosphorylation is critical for Abeta42-induced tau toxicity in a transgenic *Drosophila* model of Alzheimer's disease. *Hum Mol Genet* 2010;**19**:2947–57.
35. Jankovic J. Parkinson's disease: clinical features and diagnosis. *J Neurol Neurosurg Psychiatry* 2008;**79**:368–76.
36. Braak H, Del Tredici K, Rub U, de Vos RA, Jansen Steur EN, Braak E. Staging of brain pathology related to sporadic Parkinson's disease. *Neurobiol Aging* 2003;**24**:197–211.
37. Polymeropoulos MH, Lavedan C, Leroy E, Ide SE, Dehejia A, Dutra A, et al. Mutation in the alpha-synuclein gene identified in families with Parkinson's disease. *Science* 1997;**276**:2045–7.
38. Krüger R, Kuhn W, Müller T, Woitalla D, Graeber M, Kösel S, et al. Ala30Pro mutation in the gene encoding alpha-synuclein in Parkinson's disease. *Nat Genet* 1998;**18**:106–8.
39. Zarranz JJ, Alegre J, Gómez-Esteban JC, Lezcano E, Ros R, Ampuero I, et al. The new mutation, E46K, of alpha-synuclein causes Parkinson and Lewy body dementia. *Ann Neurol* 2004;**55**:164–73.
40. Ibáñez P, Bonnet A-M, Débarges B, Lohmann E, Tison F, Pollak P, et al. Causal relation between alpha-synuclein gene duplication and familial Parkinson's disease. *Lancet* 2004;**364**:1169–71.
41. Singleton AB, Farrer M, Johnson J, Singleton A, Hague S, Kachergus J, et al. alpha-Synuclein locus triplication causes Parkinson's disease. *Science* 2003;**302**:841.
42. Feany MB, Bender WW. A *Drosophila* model of Parkinson's disease. *Nature* 2000;**404**:394–8.
43. Auluck PK, Chan HY, Trojanowski JQ, Lee VM, Bonini NM. Chaperone suppression of alpha-synuclein toxicity in a *Drosophila* model for Parkinson's disease. *Science* 2002;**295**:865–8.

44. Kitada T, Asakawa S, Hattori N, Matsumine H, Yamamura Y, Minoshima S, et al. Mutations in the parkin gene cause autosomal recessive juvenile parkinsonism. *Nature* 1998;**392**:605–8.
45. Greene JC, Whitworth AJ, Kuo I, Andrews LA, Feany MB, Pallanck LJ. Mitochondrial pathology and apoptotic muscle degeneration in *Drosophila* parkin mutants. *Proc Natl Acad Sci USA* 2003;**100**:4078–83.
46. Sang T-K, Chang H-Y, Lawless GM, Ratnaparkhi A, Mee L, Ackerson LC, et al. A *Drosophila* model of mutant human parkin-induced toxicity demonstrates selective loss of dopaminergic neurons and dependence on cellular dopamine. *J Neurosci* 2007;**27**:981–92.
47. Wang C, Lu R, Ouyang X, Ho MWL, Chia W, Yu F, et al. *Drosophila* overexpressing parkin R275W mutant exhibits dopaminergic neuron degeneration and mitochondrial abnormalities. *J Neurosci* 2007;**27**:8563–70.
48. Valente EM, Abou-Sleiman PM, Caputo V, Muqit MM, Harvey K, Gispert S, et al. Hereditary early-onset Parkinson's disease caused by mutations in PINK1. *Science* 2004;**304**:1158–60.
49. Clark IE, Dodson MW, Jiang C, Cao JH, Huh JR, Seol JH, et al. *Drosophila* pink1 is required for mitochondrial function and interacts genetically with parkin. *Nature* 2006;**441**:1162–6.
50. Park J, Lee SB, Lee S, Kim Y, Song S, Kim S, et al. Mitochondrial dysfunction in *Drosophila* PINK1 mutants is complemented by parkin. *Nature* 2006;**441**:1157–61.
51. Yang Y, Gehrke S, Imai Y, Huang Z, Ouyang Y, Wang JW, et al. Mitochondrial pathology and muscle and dopaminergic neuron degeneration caused by inactivation of *Drosophila* Pink1 is rescued by Parkin. *Proc Natl Acad Sci USA* 2006;**103**:10793–8.
52. Paisán-Ruíz C, Jain S, Evans EW, Gilks WP, Simón J, van der Brug M, et al. Cloning of the gene containing mutations that cause PARK8-linked Parkinson's disease. *Neuron* 2004;**44**:595–600.
53. Zimprich A, Biskup S, Leitner P, Lichtner P, Farrer M, Lincoln S, et al. Mutations in LRRK2 cause autosomal-dominant parkinsonism with pleomorphic pathology. *Neuron* 2004;**44**:601–7.
54. Yue Z. LRRK2 in Parkinson's disease: in vivo models and approaches for understanding pathogenic roles. *FEBS J* 2009;**276**:6445–54.
55. Liu Z, Wang X, Yu Y, Li X, Wang T, Jiang H, et al. A *Drosophila* model for LRRK2-linked parkinsonism. *Proc Natl Acad Sci USA* 2008;**105**:2693–8.
56. Ng C-H, Mok SZS, Koh C, Ouyang X, Fivaz ML, Tan E-K, et al. Parkin protects against LRRK2 G2019S mutant-induced dopaminergic neurodegeneration in *Drosophila*. *J Neurosci* 2009;**29**:11257–62.
57. Imai Y, Gehrke S, Wang H-Q, Takahashi R, Hasegawa K, Oota E, et al. Phosphorylation of 4E-BP by LRRK2 affects the maintenance of dopaminergic neurons in *Drosophila*. *EMBO J* 2008;**27**:2432–43.
58. Bonifati V, Rizzu P, van Baren MJ, Schaap O, Breedveld GJ, Krieger E, et al. Mutations in the DJ-1 gene associated with autosomal recessive early-onset parkinsonism. *Science* 2003;**299**:256–9.
59. Meulener M, Whitworth AJ, Armstrong-Gold CE, Rizzu P, Heutink P, Wes PD, et al. *Drosophila* DJ-1 mutants are selectively sensitive to environmental toxins associated with Parkinson's disease. *Curr Biol* 2005;**15**:1572–7.
60. Menzies FM, Yenisetti SC, Min K-T. Roles of *Drosophila* DJ-1 in survival of dopaminergic neurons and oxidative stress. *Curr Biol* 2005;**15**:1578–82.
61. Yang Y, Gehrke S, Haque ME, Imai Y, Kosek J, Yang L, et al. Inactivation of *Drosophila* DJ-1 leads to impairments of oxidative stress response and phosphatidylinositol 3-kinase/Akt signaling. *Proc Natl Acad Sci USA* 2005;**102**:13670–5.
62. Perutz MF. Glutamine repeats and neurodegenerative diseases: molecular aspects. *Trends Biochem Sci* 1999;**24**:58–63.
63. Chan HY, Warrick JM, Gray-Board GL, Paulson HL, Bonini NM. Mechanisms of chaperone suppression of polyglutamine disease: selectivity, synergy and modulation of protein solubility in *Drosophila*. *Hum Mol Genet* 2000;**9**:2811–20.

64. Gusella JF, MacDonald ME, Ambrose CM, Duyao MP. Molecular genetics of Huntington's disease. *Arch Neurol* 1993;**50**:1157–63.
65. Sang T-K, Li C, Liu W, Rodriguez A, Abrams JM, Zipursky SL, et al. Inactivation of *Drosophila* Apaf-1 related killer suppresses formation of polyglutamine aggregates and blocks polyglutamine pathogenesis. *Hum Mol Genet* 2005;**14**:357–72.
66. Warrick JM, Chan HY, Gray-Board GL, Chai Y, Paulson HL, Bonini NM. Suppression of polyglutamine-mediated neurodegeneration in *Drosophila* by the molecular chaperone HSP70. *Nat Genet* 1999;**23**:425–8.
67. Bilen J, Bonini NM. *Drosophila* as a model for human neurodegenerative disease. *Annu Rev Genet* 2005;**39**:153–71.
68. Warrick JM, Morabito LM, Bilen J, Gordesky-Gold B, Faust LZ, Paulson HL, et al. Ataxin-3 suppresses polyglutamine neurodegeneration in *Drosophila* by a ubiquitin-associated mechanism. *Mol Cell* 2005;**18**:37–48.
69. Bilen J, Bonini NM. Genome-wide screen for modifiers of ataxin-3 neurodegeneration in *Drosophila*. *PLoS Genet* 2007;**3**:1950–64.
70. Satterfield TF, Jackson SM, Pallanck LJ. A *Drosophila* homolog of the polyglutamine disease gene SCA2 is a dosage-sensitive regulator of actin filament formation. *Genetics* 2002;**162**:1687–702.
71. Fernandez-Funez P, Nino-Rosales ML, de Gouyon B, She WC, Luchak JM, Martinez P, et al. Identification of genes that modify ataxin-1-induced neurodegeneration. *Nature* 2000;**408**:101–6.
72. Al-Ramahi I, Pérez AM, Lim J, Zhang M, Sorensen R, de Haro M, et al. dAtaxin-2 mediates expanded Ataxin-1-induced neurodegeneration in a *Drosophila* model of SCA1. *PLoS Genet* 2007;**3**:e234.
73. Latouche M, Lasbleiz C, Martin E, Monnier V, Debeir T, Mouatt-Prigent A, et al. A conditional pan-neuronal *Drosophila* model of spinocerebellar ataxia 7 with a reversible adult phenotype suitable for identifying modifier genes. *J Neurosci* 2007;**27**:2483–92.
74. Kennedy WR, Alter M, Sung JH. Progressive proximal spinal and bulbar muscular atrophy of late onset. A sex-linked recessive trait. *Neurology* 1968;**18**:671–80.
75. La Spada AR, Wilson EM, Lubahn DB, Harding AE, Fischbeck KH. Androgen receptor gene mutations in X-linked spinal and bulbar muscular atrophy. *Nature* 1991;**352**:77–9.
76. Chan HYE, Warrick JM, Andriola I, Merry D, Bonini NM. Genetic modulation of polyglutamine toxicity by protein conjugation pathways in *Drosophila*. *Hum Mol Genet* 2002;**11**:2895–904.
77. Nedelsky NB, Pennuto M, Smith RB, Palazzolo I, Moore J, Nie Z, et al. Native functions of the androgen receptor are essential to pathogenesis in a *Drosophila* model of spinobulbar muscular atrophy. *Neuron* 2010;**67**:936–52.
78. Phillips JP, Campbell SD, Michaud D, Charbonneau M, Hilliker AJ. Null mutation of copper/zinc superoxide dismutase in *Drosophila* confers hypersensitivity to paraquat and reduced longevity. *Proc Natl Acad Sci USA* 1989;**86**:2761–5.
79. Watson MR, Lagow RD, Xu K, Zhang B, Bonini NM. A *drosophila* model for amyotrophic lateral sclerosis reveals motor neuron damage by human SOD1. *J Biol Chem* 2008;**283**:24972–81.
80. Chai A, Withers J, Koh YH, Parry K, Bao H, Zhang B, et al. hVAPB, the causative gene of a heterogeneous group of motor neuron diseases in humans, is functionally interchangeable with its *Drosophila* homologue DVAP-33A at the neuromuscular junction. *Hum Mol Genet* 2008;**17**:266–80.
81. Ratnaparkhi A, Lawless GM, Schweizer FE, Golshani P, Jackson GR. A *Drosophila* model of ALS: human ALS-associated mutation in VAP33A suggests a dominant negative mechanism. *PLoS ONE* 2008;**3**:e2334.

82. Tsuda H, Han SM, Yang Y, Tong C, Lin YQ, Mohan K, et al. The amyotrophic lateral sclerosis 8 protein VAPB is cleaved, secreted, and acts as a ligand for Eph receptors. *Cell* 2008;**133**:963–77.
83. Kabashi E, Valdmanis PN, Dion P, Spiegelman D, McConkey BJ, Vande Velde C, et al. TARDBP mutations in individuals with sporadic and familial amyotrophic lateral sclerosis. *Nat Genet* 2008;**40**:572–4.
84. Sreedharan J, Blair IP, Tripathi VB, Hu X, Vance C, Rogelj B, et al. TDP-43 mutations in familial and sporadic amyotrophic lateral sclerosis. *Science* 2008;**319**:1668–72.
85. Feiguin F, Godena VK, Romano G, D'Ambrogio A, Klima R, Baralle FE. Depletion of TDP-43 affects *Drosophila* motoneurons terminal synapsis and locomotive behavior. *FEBS Lett* 2009;**583**:1586–92.
86. Li Y, Ray P, Rao EJ, Shi C, Guo W, Chen X, et al. A *Drosophila* model for TDP-43 proteinopathy. *Proc Natl Acad Sci USA* 2010;**107**:3169–74.
87. Elden AC, Kim H-J, Hart MP, Chen-Plotkin AS, Johnson BS, Fang X, et al. Ataxin-2 intermediate-length polyglutamine expansions are associated with increased risk for ALS. *Nature* 2010;**466**:1069–75.

Watching Worms Whither: Modeling Neurodegeneration in *C. elegans*

Benjamin Wolozin,[*,†]
Christopher Gabel,[*,‡]
Andrew Ferree,[*]
Maria Guillily,[*]
and Atsushi Ebata[*,§]

[*]*Department of Pharmacology, Boston University School of Medicine, Boston, Massachusetts, USA*

[†]*Department of Neurology, Boston University School of Medicine, Boston, Massachusetts, USA*

[‡]*Department of Physiology and Biophysics, Boston University School of Medicine, Boston, Massachusetts, USA*

[§]*Department of Pathology, Boston University School of Medicine, Boston, Massachusetts, USA*

Caenorhabditis elegans is increasingly being used to study neurodegenerative diseases. Nematodes are translucent, which facilitates study of particular neurons in the living animal, and easy to manipulate genetically. Despite vast evolutionary divergence, human proteins are functionally active when expressed in *C. elegans*, and disease-linked mutations in these proteins also cause phenotypic changes in the nematode. In this chapter, we review use of *C. elegans* to investigate the pathophysiology of Alzheimer's disease, Parkinson's disease, and axonal degeneration. Studies of presenilin, β-amyloid, tau, α-synuclein, and LRRK2 all produce strong phenotypic effects in *C. elegans*, and in many cases reproduce selective neuronal vulnerability observed in humans. Disease-linked mutations enhance degeneration in the *C. elegans* models. These studies are

Progress in Molecular Biology
and Translational Science, Vol. 100
DOI: 10.1016/B978-0-12-384878-9.00015-7

Copyright 2011, Elsevier Inc.
All rights reserved.
1877-1173/11 $35.00

increasingly leading to high-throughput screens that identify novel genes and novel pharmaceuticals that modify the disease course.

I. Introduction

Caenorhabditis elegans is a small roundworm with a precise, predetermined anatomy, a rapid reproductive cycle, a short lifespan, and a translucent body (http://www.wormbook.org/).[1] Each of these features contributes to the power of *C. elegans* as a valuable tool for understanding the functional genomics of disease. Gene expression can be readily modulated by promoter-driven expression, knockdown, or gene deletion. Transgenes can be directed to particular neuronal subtypes, such as dopaminergic neurons or touch neurons. The integrity of the neurons can be readily examined in live animals using fluorescent indicators, and over 3 weeks these neurons can be followed through the nematode life cycle from the larval stage until the animal dies from senescence (Fig. 1). These features combine together to make *C. elegans* a powerful system for studying diseases of aging where particular genes are typically expressed in a manner that gradually induces pathological changes as the animal senesces.

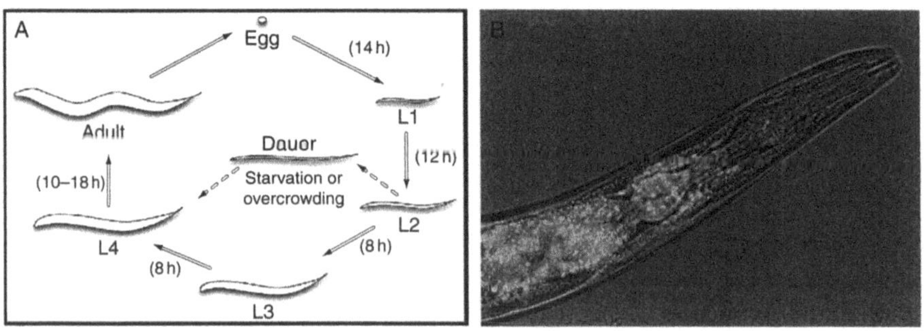

FIG. 1. (A). *C. elegans* life cycle. *C. elegans* progress through four larval stages until it reaches adulthood, whereupon it lays eggs. Larval development takes 2–3 days, depending on the temperature at which the nematode is grown. Adult nematodes can live for 2–3 weeks. Nematodes that encounter food deprivation during early development enter an alternate stage, termed dauer. The dauers can survive for extended periods of time on very little food, and do not continue through development until nutrient resources are restored. (B) Neurons in the nematode are mainly present in a region in the head termed the nerve ring. Panel (B) shows the head region in a line of *C. elegans* the human protein, Tar DNA-binding protein-43 (TDP-43, wild type) linked to GFP driven by the neuronal-selective synaptobrevin promoter. TDP-43 localizes exclusively to neuronal nuclei. The mouth and pharynx are observed in the background, with the pharynx surrounded by GFP-positive neurons.

Human transgenes show a remarkable ability to function appropriately within the context of *C. elegans*. At the simplest level, polyglutamine peptides aggregate in similar manners in human and nematode cells, and these aggregation properties have been elegantly used to elucidate mechanisms of proteostasis that smoothly translate from *C. elegans* to eukaryotic cells.[2] Functional equivalency across species extends beyond stretches of homomeric peptides, such as polyglutamines. Human proteins, such as presenilin, LRRK2 and TDP-43 exhibit functional activities that are shared between *C. elegans* and humans despite sequence homologies that are often 30–45%.[3–5] Genetic mutations that cause dysfunction and disease in humans, frequently exhibit similar types of dysfunction in *C. elegans*. The cross-species translation of function might derive from two factors: functional domains and conserved structure. Seemingly, low levels of total protein homology frequently hide much higher levels of cross-species homology present in functional domains that are shared between humans and nematodes. For instance, presenilins regulate notch in both human and *C. elegans.*[6] Mutations linked to Alzheimer's disease (AD) cause a loss of presenilin-mediated notch cleavage in human, and also cause a phenotype associated with the loss of notch function nematode.[6]

Applying results from *C. elegans* to understand diseases of the human central nervous system requires a nuanced view of anatomy. Nematodes models provide outstanding systems for understanding the behavior of individual cell types *in vivo* because key elements of the biology of neurons are conserved between nematode and man. These similarities mean that studies of neurodegeneration in the nematode provide strong insight into mechanisms of degeneration that apply across species. For instance, dopaminergic neurons show vulnerabilities to toxins and aggregating human proteins (e.g., α-synuclein) that are largely similar to that of dopaminergic neurons in the human brain.[7–9] However, the synaptic connectivity in nematodes is entirely different than that of the human brain. Nematodes have no substantia nigra, striatum, or cortex for modeling circuits present in the human brain. These differences preclude using the nematode to understand how disease processes affect nerve circuit characteristics of the human brain. For instance, in humans dopamine (DA) neurons are required to initiate movement, and loss of dopaminergic neurons leads to motor impairment. Loss of dopaminergic neurons in nematodes, though, increases motor activity, removing the ability of hungry nematodes to slow down and feed when encountering a new food supply.[10]

II. Modeling the Pathophysiology of AD: β-Amyloid, Tau, and Presenilin

Tau, β-amyloid, and presenilin are three major proteins that have been implicated in AD. The amyloid cascade hypothesis describes the pathophysiological relationship between each of these proteins. β-Amyloid is generated from its precursor protein, amyloid precursor protein (APP), by cleavage at two sites, termed the β and γ-cleavage sites, which occur just outside the membrane (the β-site) and within the membrane (the γ-site).[11] The enzymes that generate these cleavages are termed β-secretase and γ-secretase, respectively. Presenilin is the proteolytic subunit of γ-secretase. Cleavage of APP generates Aβ, which is secreted as a mixture of monomers and oligomers, and the oligomers are thought to toxic to neurons. Aβ toxicity leads to phosphorylation and aggregation of tau in neurons with subsequent formation of neurofibrillary tangles. Tau accumulates in neurons as an intracellular aggregate, and overexpressing tau in neurons in *C. elegans* produces neuronal dysfunction that is functionally apparent, and biochemical changes similar to those seen in the brains of subjects with AD or frontotemporal dementia.[12] β-Amyloid is produced in neurons, can accumulate in neurons, but predominantly accumulates as an extracellular deposit. β-Amyloid is generated from its precursor, APP, by cleavage at the β and γ sites. *C. elegans* naturally expresses the homologous protein that generates the γ cleavage, termed presenilin (sel-12), but lacks the homolog of β-secretase, which cleaves at the β-site.[13–15] Consequently, β-amyloid accumulation has been modeled in *C. elegans* by directly producing the β-amyloid peptide and the peptide remains intracellular. The effects of intracellular accumulation of β-amyloid, though, are dramatic, with marked growth stunting and movement deficits.

Analysis of the biology of the *C. elegans* homolog of presenilin 1, Sel-12, represents one of the most striking examples of the utility of *C. elegans* functional genomics. Presenilin is an aspartyl protease that makes up one of four proteins of the γ-secretase complex, which carries out intramembranous cleavages of transmembrane proteins (via presenilin-mediated cleavage). When presenilin was first discovered its function and substrates were unknown. Functional insights rapidly evolved based on homologies between human presenilin 1, and *C. elegans* Sel-12, which functions in a pathway that regulates notch activity.[13–15] Loss of Sel-12 leads to deficits in Notch function, which cause reduced egg-laying. This nematode assay provided the first link between presenilin and notch function, and set the stage for future papers defining the biochemical role of presenilin in γ-secretase in regulating Notch by intramembranous cleavage. Transgenic expression of human wild-type and mutant presenilin in *C. elegans* confirmed the relationship by showing that wild-type human presenilin rescued *C. elegans* lines defective for Sel-12 activity. These same assays indicated that disease-linked mutations in presenilins represent a loss of function. The hypothesis that disease-linked mutations represented a loss of function was difficult to reconcile with the dominant mode of action of these mutant genes and their tendency to increase production of Aβ42 in humans. However, advances in our understanding of presenilin biology have validated the *C. elegans* work. Presenilin 1 cleaves APP by a process of multiple cleavages that are progressively closer to the classic γ-secretase cleavage site generating Aβ. Reduced presenilin activity slows down the cleavage progression leading to production of longer Aβ peptides. Disease-linked mutations in presenilin 1 appear to reduce proteolytic activity, which increases production of Aβ42 and reduces production of Aβ40. This story exemplifies the value of *C. elegans* in analyzing structure and function.

C. elegans has also been used to analyze factors regulating aggregation of Aβ and tau proteins, *in vivo*. Work on β-amyloid toxicity has been elegantly pioneered by Christopher Link. Modeling β-amyloidosis in nematodes requires a different approach than in mammalian cells because *C. elegans* lacks β-secretase, which is required to generate β-amyloid from its precursor, APP. Link and colleagues circumvented this issue by expressing constructs that directly produce Aβ in muscle via the myo-3 promoter.[16] These nematode lines exhibit many functional deficits, including movement dysfunction, growth retardation, reduced movement, and premature death. Link's group used the *C. elegans* system to analyze the molecular changes associated with expressing Aβ. Transcriptome analysis indicated that expressing Aβ increased expression of the small heat shock protein, hsp-16, which is the *C. elegans* homolog of α/β-crystallin. They also observed increased expression of a number of pro-apoptotic proteins including a protein homologous to TNFAIP1, which is induced by tumor necrosis factor.[17] Proteomic analysis showed strong evidence of mitochondrial dysfunction induced by Aβ expression, with increased levels of oxidized mitochondrial proteins evident by 2D gel analysis.[18] In each case, similar changes are observed in brains from subjects who died with AD, which indicates that the results observed in *C. elegans* translate back to human pathophysiology.[17,18] Work in *C. elegans* also suggests potential avenues of therapy, because cellular responses that protect against oxidative stress, such as inhibiting the insulin receptor daf-2 signaling cascade, protect against Aβ toxicity in these models.[19] These studies demonstrate the utility of using *C. elegans* to study Aβ toxicity, and also convey the ability of results from *C. elegans* to translate to the human context.

Studies for investigating the effects of transgenic expression of human tau protein in *C. elegans* provide elegant examples of cross-platform translation. Brian Kraemer's group expressed M337V or P301L tau driven by the neuronal specific aex-3 promoter in *C. elegans* neurons.[20] The nematodes showed strong evidence of tau toxicity, with motor dysfunction, loss of GABAergic neurons, and premature death. The biochemical changes in tau protein were associated with the motor dysfunction paralleled biochemical changes observed in the brains of subjects with AD. Tau from the transgenic nematode lines accumulated in an insoluble fraction and showed increased phosphorylation at sites classically hyperphosphorylated in the Alzheimer brain, including phosphorylation of sites identified by the antibodies CP13 (phospho-S202) and PHF1 or

AT8 (phospho-394/404).[20] These data suggest that the signaling changes impinging on tau were similar in both nematode and human. Tau pathology in the nematodes did not fully reflect that of the human AD brain because there was no evidence of paired helical filament formation. The presence of a strong phenotype and clear biochemical changes provides a good platform for genetic studies, which is one of the strengths of the nematode system. Kraemer's group used the tau phenotype in a forward genetic screen to identify genetic modifiers of tau pathology. These screens highlighted putative roles of the enabled, a cytoskeletal regulatory protein, and HOOK1-3, a family of proteins with no known function, in regulating tau toxicity.[21,22] These data provide solid evidence that proteins associated with disease in humans can elicit pathological reactions in nematodes that share many similarities with the events occurring in humans, and that nematode models can identify novel putative modifiers of disease function.

III. Modeling Dopaminergic Dysfunction

C. elegans is a readily used model for evaluating the effect of toxins and/or genetic factors on the survival of DA neurons. The hermaphrodite has 8 DA neurons, while the male worm has 12.[1] The additional neurons in the male are utilized in the distinctive mating behavior of the male. Six neurons are found in the head of the animal, four CEP and two ADE neurons, which mediate locomotion behavior in response to the presence of food (Fig. 2). The survival of DA neurons can be evaluated by the expression of GFP driven by the DA transporter gene, dat-1. GFP fluorescence has been shown by Nass *et al.* to

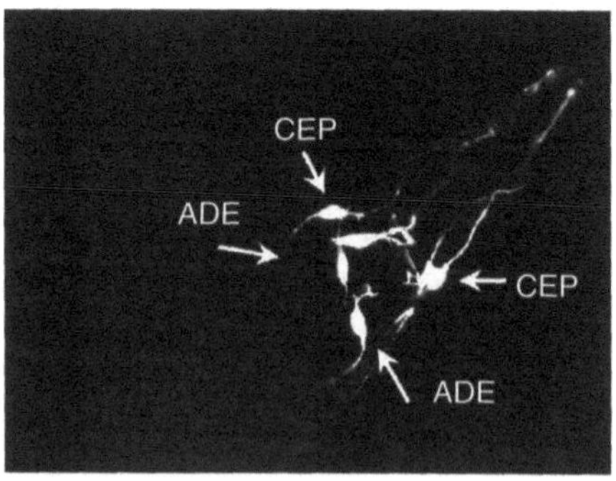

FIG. 2. Localization of dopaminergic neurons in C. *elegans* shown using GFP driven by the dopamine transporter promoter. *C. elegans* has six dopaminergic neurons in the head: four neurons are termed CEP and two are termed ADE (arrows).

TABLE I
PARKINSON'S DISEASE-RELATED *C. ELEGANS* MODELS: DA NEURON SURVIVAL ASSAYS

Reference	Neurons	Method of evaluation	Treatment
Saha et al.[24]	2 CEP	3 mM levamisole Fluorescence intensity	20 μM rotenone hLRRK2
Ruan et al.[60]	4 CEP, 2 ADE	3 mM levamisole Wild type versus degenerating	10 mM 6-OHDA
Pu and Le[61]	4 CEP, 2 ADE	% Worms with intact neurons	5 mM MPP+
Hamamichi et al.[25]	4 CEP, 2 ADE	2 mM levamisole Rescued if all neurons intact	α-Synuclein
Cao et al.[62]	4 CEP	3 mM levamisole	10 mM 6-OHDA
Lakso et al.[7]	2 CEP, 4 ADE, 2 PDE	Cell bodies and dendrites	α-Synuclein
Nass et al.[23]	4 CEP	2% Na-azide Wild type if dendrites were intact	10 or 50 mM 6-OHDA

The diverse methodologies used to evaluate DA neurons in Parkinson's disease-related *C. elegans* models. The CEP and ADE neurons in the head of the worm have been shown to be sensitive to both PD-related gene expression and toxins. These studies utilized the expression of GFP driven by the dat-1 promoter as a marker of DA neuron survival. A variety of methodologies have been used including counting cell bodies, evaluating degeneration of dendrites, and analyzing fluorescence intensity levels. Both levamisole and Na-azide have been utilized to paralyze the worms to facilitate the assay.

correlate with neuron survival as measured by EM.[23] A number of studies have shown that the DA neurons are specifically sensitive to toxins used to model aspects of Parkinson's disease (PD), including 6-OHDA, MPP+, and rotenone[8,23] (Table I). In addition, the modulation of DA survival by the expression of proteins involved in PD etiology such as α-synuclein has been evaluated[9,24,25] (Table I).

IV. Models of Parkinson's Disease: Neurotoxins, α-Synuclein, and LRRK2

Multiple laboratories have used *C. elegans* to query molecular and genetic mechanisms contributing to PD. The translucence of *C. elegans* provides an excellent system for monitoring the physiology and pathophysiology of dopaminergic neurons *in vivo*, by selective expression of green fluorescent protein driven by the dopamine transporter (dat-1) promoter.[8,9,23–29] The well-characterized dopaminergic neurons of *C. elegans* are highly sensitive to many classic neurotoxins used to model PD in higher organisms, such as 6-hydroxydopamine, MPP+, and rotenone.[24,27,28,30] In the case of rotenone, we observe that dopaminergic neurons are 50–100-fold more sensitive than the entire organism, which allows one to screen for effects of rotenone on dopaminergic neuronal death under conditions in which nematode survival is not impaired (Wolozin, personal communication).

Transgenic approaches have also been used to model PD in *C. elegans.* Orthologs of many genes linked to PD exist in the nematode including, parkin (*pdr-1*), PINK1 (*pink-1*), DJ-1 (*djr-1.1, 1.2*), LRRK2 (*lrk-1*), ATP13A2 (*capt-6*), and GBA (*F11E6.1, TO4A8.7*). Knockout models of these orthologs have produced some important insights into their normal function.[24,30–33] In each case, mutations that cause PD either cause direct degeneration of dopaminergic neurons or sensitize the nematodes to mitochondrial toxins.[24,30–32,34,43]. The phenotype observed can vary, depending on the model.

An example of the types of phenotypes observed is shown in Fig. 3, where age-synchronized lines of *C. elegans* expressing GFP driven by the DA transporter promoter were subjected knockdown by empty vector (control) or two different RNAi vectors. The combination of fluorescence intensity and morphology of the dopaminergic neurons have been shown by us and others to strongly correlate with survival of dopaminergic neurons.[24] L1 nematodes were placed on plates, and allowed to feed on bacteria containing RNAi for two different constructs or a vector control. The nematodes were aged at 20 °C for 3 days, until approximately day 1 of adulthood, and then exposed to 6-hydroxydopamine (1 h, 50 mM 6-OHDA, plus 10 mM ascorbic acid; followed by neutralization). After replating, the nematodes were aged for 4 days and survival of the dopaminergic neurons was assessed by microscopy. As shown in Fig. 3, one RNAi construct (4) greatly reduced survival of the dopaminergic neurons, while the other construct (16) was actually trophic for the neurons.

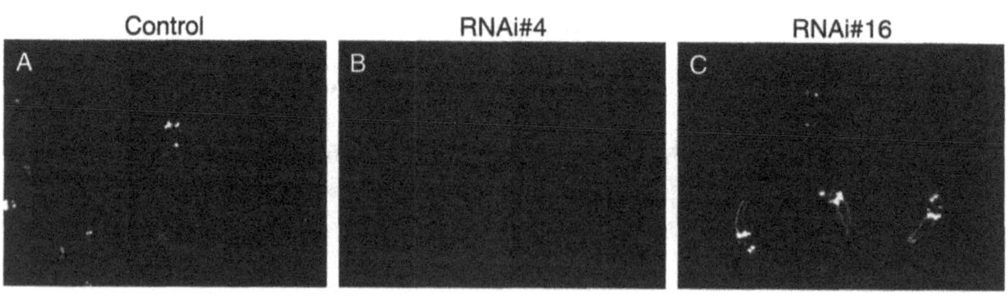

FIG. 3. Fluorescence of dopaminergic neurons following treatment with three different RNAi constructs (empty vector control (A), RNAi #4 (B), and RNAi #16 (C)), followed by exposure to 6-hydroxydopamine (50 mM) and subsequent aging for 4 more days with continued exposure to RNAi. RNAi #4 sensitized the dopaminergic neurons to 6-hydroxydopamine, while RNAi #16 protected the dopaminergic neurons against 6-hydroxydopamine.

Work from our laboratory showed increased sensitivity to rotenone, which is a pattern that we have observed with multiple genes linked to PD.[30] Multiple groups have also observed age-dependent degeneration of dopaminergic neurons in nematode lines expressing α-synuclein.[7,9,25,35] Degeneration of the dopaminergic neurons presents as a combination of three different outcomes, which are present to differing degrees in each nematode: loss of neuronal bodies, loss of neuronal processes, and reduced expression of dopaminergic markers. One issue that arises is marked variability in the time to degeneration among lines, with some groups observing loss of dopaminergic neurons in as little as 5 days and other groups observing survival up to 14 days.[9,25] Many of these studies would benefit from use of a comparison to the parent line expressing the reporter but not α-synuclein. Despite some experimental issues, the cumulative results consistently indicate that expressing α-synuclein renders dopaminergic neurons susceptible to degeneration.

Three findings are emerging from the models of PD in *C. elegans*. The first finding is the tendency of different PD-linked genes to produce a similar phenotype of dopaminergic vulnerability and degeneration. A second finding is the ability of different PD-linked genes to complement each other in genetic studies, and the third finding is linkage of the degeneration to dysfunction of membrane bound organelles, such as the mitochondria and autophagasomes. Work from our laboratory provided one of the first examples of different PD-linked genes producing similar phenotypes. We observed that three different models of PD in *C. elegans* all showed enhanced vulnerability to mitochondrial toxins, including lines expressing A53T α-synuclein, a line lacking pdr-1 (the *C. elegans* homolog of parkin), and a line subject to knockdown for DJ-1.[30] In more recent studies, we observed that lines of *C. elegans* expressing mutant forms of LRRK2 (G2019S or R1441C) are more sensitive than lines of *C. elegans* expressing WT LRRK2.[24] These data raise the possibility that enhanced vulnerability to mitochondrial dysfunction is a general feature of genes linked to PD. Complementation is observed in studies of parkin, pink-1, DJ-1, and lrk-1. LRRK2 has not been formally linked to pink-1 yet, but using an lrk-1 knockout line, Samann and colleagues demonstrated genetic complementation by *pink1* of sensitivity to the endoplasmic reticulum toxin, tunicamycin.[32] Two different studies noted that expressing α-synuclein alters the levels of parkin.[35,36] Genetic complementation between PINK1 and parkin has also been shown in *drosophila* and in mammalian cells, and interaction between LRRK2 and parkin/PINK1 is also evident in *drosophila*.[37–41] These data contribute to an emerging view that many of the PD-linked genes share mechanisms and biochemical pathways in common.

Groups have made particular use of α-synuclein expressing nematode lines for forward genetic screens[7,25,42] and to monitor changes gene expression associated with α-synuclein expression and/or toxicity.[33,35,43,44] A unifying theme that has arisen from these synuclein experiments is a strong connection with mitochondrial function, proteostatis, and vesicle/vacuolar biology. Mitochondrial genes, such as *ymel-1* (a mitochondrial protease), F59F4.1 (acyl-CoA oxidase), ubiquitin/proteasomal proteins, such as hrdl-1 and HIP (a ubiquitin ligase and a transcription factor, respectively), and vacuolar proteins, such as vps41 and yptlp (Rab1, a vesicle-linked GTPase) all suppress α-synuclein toxicity.[25,42,45,46] In addition, overexpressing the lysosomal protease cathepsin D also strongly protects against α-synuclein toxicity.[47] These forward screens also contribute to the growing consensus placing all PD genes in convergent pathways, because in most screens other PD-linked genes modify α-synuclein toxicity. PINK1, pdr-1 (parkin), djr-1.1 (dj-1), and capt-6 (ATP13A2, PARK9) consistently appear as protective factors in screens of α-synuclein toxicity.[25,33,42] Finally, chemical screen also suggest that chemicals that protect mitochondria or increase autophagy (Wolozin, personal communication) protect against α-synuclein toxicity.[30] Thus, genes linked to PD appear to act through common pathways that involve the mitochondria and proteostasis. An emerging concept is the hypothesis that deficits in autophagy might unify each of these disparate areas of biology. In this context, the genes mentioned above that are linked to vacuolar function might act by enhancing the autophagic cascade.

V. Models of Axonal Degeneration: Laser Ablation

C. elegans has recently emerged as a powerful model system for the study of neuronal damage and regeneration. Using advanced laser ablation techniques and time-lapse fluorescence imaging, researchers can selectively damage target neurons and observe subsequent regrowth *in vivo*. The advent of femtosecond laser surgery allows submicron ablation of biological tissue deep within a sample.[48] Applied to the *C. elegans* nervous system, this enables the precise cutting of individual nerve fibers within an adult animal, with minimal collateral damage to the surrounding tissue. Remarkably, severed *C. elegans*' neurons display robust regeneration within a matter of hours following laser damage (Fig. 4). The simple stereotyped pattern of the *C. elegans* nervous system and the precision of laser ablation allow for exquisite single cell experimental repeatability and quantitative cell-specific comparison across genetic backgrounds. Coupled with the unprecedented power of *C. elegans* genetics these techniques are opening the door for the study of nerve damage and recovery on a genetic and molecular basis, at a level unattainable in higher systems.

Regeneration of *C. elegans* neurons was first observed by Yanik *et al.*[49] Using a femtosecond laser, they snipped motor neurons within the body of *C. elegans* and observed the subsequent regrowth and apparent restoration of

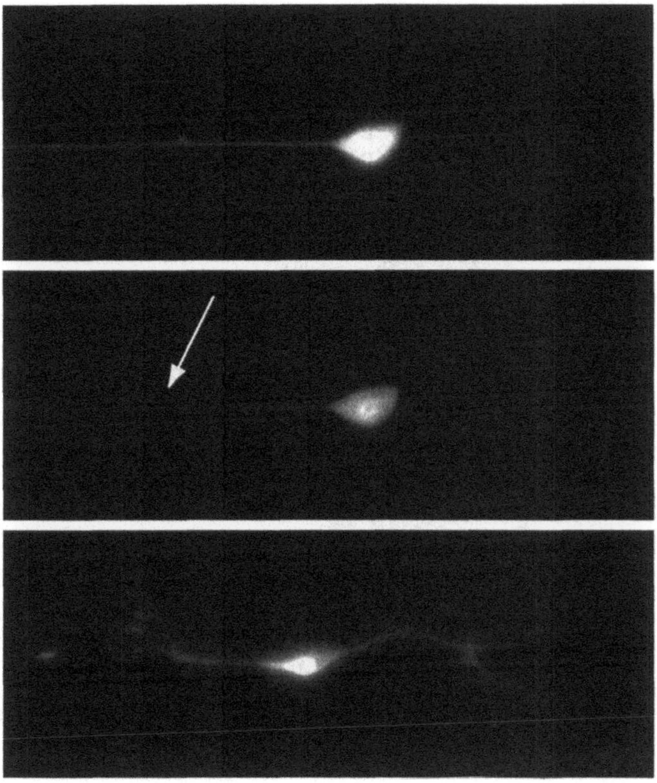

FIG. 4. *In vivo* neuronal regeneration in *C. elegans* after femtosecond laser surgery. *Top:* An intact mechanosensory neuron imaged *in vivo* using transgenically expressed fluorescent markers. *Middle:* Immediately following laser exposure, a clean break is seen at the target point (arrow). Femtosecond laser surgery allows *in vivo* ablation on the submicron scale and snipping of single nerve fibers within an intact adult *C. elegans*. *Bottom:* 24 h after laser surgery substantial regenerative growth is observed. In this example, growth is seen both at the original cut point and sprouting directly from the cell soma.

nerve function. Numerous studies have reported the remarkable ability of *C. elegan's* neurons to reconnect with the severed distal fragment of the damage process.[50,51] In such cases, the cell appears to effectively repair itself without regrowing the entire nerve process. Regenerative ability in *C. elegans* neurons also appears to vary with neuron type. Motor neurons in the body of the worm, controlling body movement and egg-laying, readily regrow as do mechanosensory and some chemosensory neurons. In contrast, many sensory neurons in the head and tail of the animal do not show significant regrowth.[49,52,53]

The simplicity and stereotype morphology of the *C. elegans* nervous system as well as the submicron precision of femtosecond laser surgery allow targeting not only of a single neuron but a particular location along a specific nerve fiber as well. The cellular response to damage depends in part on the location of the ablation point. For example, cut points close to the cell soma tend to result in new collateral branches extending from the cell body. In contrast, cuts farther along the same neuron result in regenerative outgrowth from the severed end of the original fiber.[53] Branch points within a single nerve process can also delineate regions of regenerative ability. For example, beyond a particular branch point of the PLM neuron (a posterior mechanosensory neuron in the worm's body) regeneration does not readily occur, while cuts proximal to this point show substantial regrowth.[50] These findings, as with cell typed characteristics, are striking as similar variability in regeneration is also seen in the vertebrate nervous system. The peripheral nervous system of higher animals displays substantial regrowth following traumatic damage, while the central nervous system tragically does not.[54]

Cell-specific comparison of neuronal recovery and regeneration after laser surgery across genetic backgrounds is fueling an array of new discoveries on the genetic and molecular level. For example, the *vab-1* gene, which encodes an enhrin protein and has been shown to be a regenerative deterrent in mammals, appears to detrimentally effect axon guidance during regeneration. *C. elegans* lacking the *vab-1* gene displays substantially more accurate neuronal regeneration following laser damage.[50] Likewise, researchers have found that many of the genes involved in developmental axon outgrowth and guidance play similar roles in adult stage axon regeneration.[53] Reports have also demonstrated the critical nature of cAMP signaling in damaged neurons for efficient regeneration.[51]

The *C. elegans* system greatly facilitates the direct discovery of novel factors specifically relevant to cell damage and regeneration. A prominent example is the demonstration of the DLK-1 MAP kinase pathway as an essential element of neural regeneration.[55] Components of this pathway, and specifically the DLK-1 protein, are required for growth cone formation after neuron damage in *C. elegans* but are dispensable during initial developmental axon outgrowth. Moreover, overexpression of DLK-1 resulted in improved regeneration in both speed of growth cone formation and accuracy of the regrowth. Further studies along these lines have identified additional factors in this signaling pathway,[56] showing that the DLK-1 pathway triggers local synthesis of key proteins within the severed axon. While large-scale screening for neural regenerative factors using single cell laser surgery presents numerous technical challenges, several systems are in development that will automate much of the process. In particular, microfluidic devices designed to automatically manipulate and restrain *C. elegans* for laser surgery and imaging show great potential in increasing experimental throughput.[57–59] These advances will bring the full power of *C. elegans* genetics to bear on the study of neural regeneration, allowing genome-wide screening and analysis.

VI. Conclusion

C. elegans is clearly a versatile organism that has strong utility for investigations of the neurobiology of disease. From a conceptual standpoint, study of *C. elegans* has the greatest strength for studies of human neurodegenerative disease aimed at understanding the biology of neuron *in vivo*, but is less applicable to studies whose goal is to understand neuronal connectivity. Despite a vast evolutionary distance between nematodes and humans, human proteins appear to function appropriately when expressed in nematodes. Surprisingly, mutations that cause disease in humans also cause phenotypic abnormalities in nematodes. This cross-species translation of proteins greatly facilitates study of human disease in such simple organisms as *C. elegans*.

Nematodes will continue to serve as an important experimental system as technology advances. Nematodes often represent a superb platform for preliminary screening of putative therapeutic medications before moving into transgenic rodents. New techniques such as optogenetics are readily applied to *C. elegans*, which is translucent and very amenable to photoactivation of compounds. Nematodes represent an outstanding platform for testing advanced physical techniques, such as femtosecond laser ablation (described above) under conditions that can be tightly controlled. Complex fields, such as systems biology, are increasingly applied to eukaryotic systems. The simplicity of the nematode and the ability to genetically manipulate the organisms makes it a valuable stepping-stone bridging the gap between the simplicity of *in vitro* studies and the complexity of the human brain.

Acknowledgments

This work was supported by awards to B. W. (Alzheimer Association, the Michael J. Fox Foundation, NIEHS ES15567, NINDS NS060872).

References

1. Wood WB. *The Nematode Caenorhabditis elegans*. Cold Spring Harbor Monograph Series 17. Cold Spring Harbor, NY: Cold Spring Harbor Press. 1988, p. 667.
2. Morimoto RI. Proteotoxic stress and inducible chaperone networks in neurodegenerative disease and aging. *Genes Dev* 2008;**22**:1427–38.
3. Kaether C, Haass C, Steiner H. Assembly, trafficking and function of gamma-secretase. *Neurodegener Dis* 2006;**3**:275–83.
4. Biskup S, West AB. Zeroing in on LRRK2-linked pathogenic mechanisms in Parkinson's disease. *Biochim Biophys Acta* 2009;**1792**:625–33.

5. Chen-Plotkin AS, Lee VM, Trojanowski JQ. TAR DNA-binding protein 43 in neurodegenerative disease. *Nat Rev Neurol* 2010;**6**:211–20.
6. Annaert W, De Strooper B. Presenilins: molecular switches between proteolysis and signal transduction. *Trends Neurosci* 1999;**22**:439–43.
7. Lakso M, Vartiainen S, Moilanen AM, Sirvio J, Thomas JH, Nass R, et al. Dopaminergic neuronal loss and motor deficits in *Caenorhabditis elegans* overexpressing human alpha-synuclein. *J Neurochem* 2003;**86**:165–72.
8. Nass R, Blakely RD. The *Caenorhabditis elegans* dopaminergic system: opportunities for insights into dopamine transport and neurodegeneration. *Annu Rev Pharmacol Toxicol* 2003;**43**:521–44.
9. Kuwahara T, Koyama A, Gengyo-Ando K, Masuda M, Kowa H, Tsunoda M, et al. Familial Parkinson mutant alpha-synuclein causes dopamine neuron dysfunction in transgenic *Caenorhabditis elegans*. *J Biol Chem* 2006;**281**:334–40.
10. Sawin ER, Ranganathan R, Horvitz HR. *C. elegans* locomotory rate is modulated by the environment through a dopaminergic pathway and by experience through a serotonergic pathway. *Neuron* 2000;**26**:619–31.
11. Selkoe DJ, Schenk D. Alzheimer's disease: molecular understanding predicts amyloid-based therapeutics. *Annu Rev Pharmacol Toxicol* 2003;**43**:545–84.
12. Stoothoff WH, Johnson GV. Tau phosphorylation: physiological and pathological consequences. *Biochim Biophys Acta* 2005;**1739**:280–97.
13. Levitan D, Doyle T, Brousseau D, Lee M, Thinakaran G, Slunt H, et al. Assessment of normal and mutant human presenilin function in *Caenorhabditis elegans*. *Proc Natl Acad Sci USA* 1996;**93**:14940–4.
14. Levitan D, Greenwald I. Facilitation of lin-12 mediated signalling by sel-12, a *Caenorhabditis elegans* S182 Alzheimer's disease gene. *Nature* 1995;**377**:351–5.
15. Li X, Greenwald I. Membrane topology of the *C. elegans* SEL-12 presenilin. *Neuron* 1996;**17**:1015–21.
16. Link CD. Expression of human beta-amyloid peptide in transgenic *Caenorhabditis elegans*. *Proc Natl Acad Sci USA* 1995;**92**:9368–72.
17. Link CD, Taft A, Kapulkin V, Duke K, Kim S, Fei Q, et al. Gene expression analysis in a transgenic *Caenorhabditis elegans* Alzheimer's disease model. *Neurobiol Aging* 2003;**24**:397–413.
18. Boyd-Kimball D, Poon HF, Lynn BC, Cai J, Pierce Jr. WM, Klein JB, et al. Proteomic identification of proteins specifically oxidized in *Caenorhabditis elegans* expressing human Abeta(1-42): implications for Alzheimer's disease. *Neurobiol Aging* 2000;**27**.1230–40.
19. Florez-McClure ML, Hohsfield LA, Fonte G, Bealor MT, Link CD. Decreased insulin-receptor signaling promotes the autophagic degradation of beta-amyloid peptide in *C. elegans*. *Autophagy* 2007;**3**:569–80.
20. Kraemer BC, Zhang B, Leverenz JB, Thomas JH, Trojanowski JQ, Schellenberg GD. Neurodegeneration and defective neurotransmission in a *Caenorhabditis elegans* model of tauopathy. *Proc Natl Acad Sci USA* 2003;**100**:9980–5.
21. Guthrie CR, Schellenberg GD, Kraemer BC. SUT-2 potentiates tau-induced neurotoxicity in *Caenorhabditis elegans*. *Hum Mol Genet* 2009;**18**:1825–38.
22. Kraemer BC, Schellenberg GD. SUT-1 enables tau-induced neurotoxicity in *C. elegans*. *Hum Mol Genet* 2007;**16**:1959–71.
23. Nass R, Hall DH, Miller 3rd DM, Blakely RD. Neurotoxin-induced degeneration of dopamine neurons in *Caenorhabditis elegans*. *Proc Natl Acad Sci USA* 2002;**99**:3264–9.
24. Saha S, Guillily MD, Ferree A, Lanceta J, Chan D, Ghosh J, et al. LRRK2 modulates vulnerability to mitochondrial dysfunction in *Caenorhabditis elegans*. *J Neurosci* 2009;**29**:9210–8.

25. Hamamichi S, Rivas RN, Knight AL, Cao S, Caldwell KA, Caldwell GA. Hypothesis-based RNAi screening identifies neuroprotective genes in a Parkinson's disease model. *Proc Natl Acad Sci USA* 2008;**105**:728–33.
26. Berkowitz LA, Hamamichi S, Knight AL, Harrington AJ, Caldwell GA, Caldwell KA. Application of a *C. elegans* dopamine neuron degeneration assay for the validation of potential Parkinson's disease genes. *J Vis Exp* 2008;**835**: doi: 10.3791/835.
27. Marvanova M, Nichols CD. Identification of neuroprotective compounds of *Caenorhabditis elegans* dopaminergic neurons against 6-OHDA. *J Mol Neurosci* 2007;**31**:127–37.
28. Braungart E, Gerlach M, Riederer P, Baumeister R, Hoener MC. *Caenorhabditis elegans* MPP+ model of Parkinson's disease for high-throughput drug screenings. *Neurodegener Dis* 2004;**1**:175–83.
29. Kritzer JA, Hamamichi S, McCaffery JM, Santagata S, Naumann TA, Caldwell KA, et al. Rapid selection of cyclic peptides that reduce alpha-synuclein toxicity in yeast and animal models. *Nat Chem Biol* 2009;**5**:655–63.
30. Ved R, Saha S, Westlund B, Perrier C, Burnam L, Sluder A, et al. Similar patterns of mitochondrial vulnerability and rescue induced by genetic modification of α-synuclein, Parkin and DJ-1 in *C.* elegans. *J Biol Chem* 2005;**280**:42655–68.
31. Sakaguchi-Nakashima A, Meir JY, Jin Y, Matsumoto K, Hisamoto N. LRK-1, a *C. elegans* PARK8-related kinase, regulates axonal-dendritic polarity of SV proteins. *Curr Biol* 2007;**17**:592–8.
32. Samann J, Hegermann J, von Gromoff E, Eimer S, Baumeister R, Schmidt E. *Caenorhabditis elegans* LRK-1 and PINK-1 act antagonistically in stress response and neurite outgrowth. *J Biol Chem* 2009;**284**:16482–91.
33. Gitler AD, Bevis BJ, Shorter J, Strathearn KE, Hamamichi S, Su LJ, et al. The Parkinson's disease protein alpha-synuclein disrupts cellular Rab homeostasis. *Proc Natl Acad Sci USA* 2008;**105**:145–50.
34. Karpinar DP, Balija MB, Kugler S, Opazo F, Rezaei-Ghaleh N, Wender N, et al. Pre-fibrillar alpha-synuclein variants with impaired beta-structure increase neurotoxicity in Parkinson's disease models. *EMBO J* 2009;**28**:3256–68.
35. Vartiainen S, Pehkonen P, Lakso M, Nass R, Wong G. Identification of gene expression changes in transgenic *C. elegans* overexpressing human alpha-synuclein. *Neurobiol Dis* 2006;**22**:477–86.
36. Asikainen S, Rudgalvyte M, Heikkinen L, Louhiranta K, Lakso M, Wong G, et al. Global microRNA expression profiling of *Caenorhabditis elegans* Parkinson's disease models. *J Mol Neurosci* 2010;**41**:210–8.
37. Yang Y, Gehrke S, Imai Y, Huang Z, Ouyang Y, Wang JW, et al. Mitochondrial pathology and muscle and dopaminergic neuron degeneration caused by inactivation of Drosophila Pink1 is rescued by Parkin. *Proc Natl Acad Sci USA* 2006;**103**:10793–8.
38. Ziviani E, Tao RN, Whitworth AJ. Drosophila Parkin requires PINK1 for mitochondrial translocation and ubiquitinates Mitofusin. *Proc Natl Acad Sci USA* 2010;**107**:5018–23.
39. Venderova K, Kabbach G, Abdel-Messih E, Zhang Y, Parks RJ, Imai Y, et al. Leucine-rich repeat kinase 2 interacts with Parkin, DJ-1 and PINK-1 in a *Drosophila melanogaster* model of Parkinson's disease. *Hum Mol Genet* 2009;**18**:4390–404.
40. Narendra D, Tanaka A, Suen DF, Youle RJ. Parkin is recruited selectively to impaired mitochondria and promotes their autophagy. *J Cell Biol* 2008;**183**:795–803.
41. Narendra DP, Jin SM, Tanaka A, Suen DF, Gautier CA, Shen J, et al. PINK1 is selectively stabilized on impaired mitochondria to activate Parkin. *PLoS Biol* 2010;**8**:e1000298.
42. van Ham TJ, Thijssen KL, Breitling R, Hofstra RM, Plasterk RH, Nollen EA. *C. elegans* model identifies genetic modifiers of alpha-synuclein inclusion formation during aging. *PLoS Genet* 2008;**4**:e1000027.

43. Gitler AD, Chesi A, Geddie ML, Strathearn KE, Hamamichi S, Hill KJ, et al. Alpha-synuclein is part of a diverse and highly conserved interaction network that includes PARK9 and manganese toxicity. *Nat Genet* 2009;**41**:308–15.
44. Ichibangase T, Saimaru H, Takamura N, Kuwahara T, Koyama A, Iwatsubo T, et al. Proteomics of *Caenorhabditis elegans* over-expressing human alpha-synuclein analyzed by fluorogenic derivatization-liquid chromatography/tandem mass spectrometry: identification of actin and several ribosomal proteins as negative markers at early Parkinson's disease stages. *Biomed Chromatogr* 2008;**22**:232–4.
45. Cooper AA, Gitler AD, Cashikar A, Haynes CM, Hill KJ, Bhullar B, et al. Alpha-synuclein blocks ER-Golgi traffic and Rab1 rescues neuron loss in Parkinson's models. *Science* 2006;**313**:324–8.
46. Roodveldt C, Bertoncini CW, Andersson A, van der Goot AT, Hsu ST, Fernandez-Montesinos R, et al. Chaperone proteostasis in Parkinson's disease: stabilization of the Hsp70/alpha-synuclein complex by Hip. *EMBO J* 2009;**28**:3758–70.
47. Qiao L, Hamamichi S, Caldwell KA, Caldwell GA, Yacoubian TA, Wilson S, et al. Lysosomal enzyme cathepsin D protects against alpha-synuclein aggregation and toxicity. *Mol Brain* 2008;**1**:17.
48. Gabel CV. Femtosecond lasers in biology: nanoscale surgery with ultrafast optics. *Contemp Phys* 2008;**49**:391–411.
49. Yanik MF, Cinar H, Cinar HN, Chisholm AD, Jin Y, Ben-Yakar A. Neurosurgery: functional regeneration after laser axotomy. *Nature* 2004;**432**:822.
50. Wu Z, Ghosh-Roy A, Yanik MF, Zhang JZ, Jin Y, Chisholm AD. *Caenorhabditis elegans* neuronal regeneration is influenced by life stage, ephrin signaling, and synaptic branching. *Proc Natl Acad Sci USA* 2007;**104**:15132–7.
51. Ghosh-Roy A, Wu Z, Goncharov A, Jin Y, Chisholm AD. Calcium and cyclic AMP promote axonal regeneration in *Caenorhabditis elegans* and require DLK-1 kinase. *J Neurosci* 2010;**30**:3175–83.
52. Chung SH, Clark DA, Gabel CV, Mazur E, Samuel AD. The role of the AFD neuron in *C. elegans* thermotaxis analyzed using femtosecond laser ablation. *BMC Neurosci* 2006;**7**:30.
53. Gabel CV, Antoine F, Chuang CF, Samuel AD, Chang C. Distinct cellular and molecular mechanisms mediate initial axon development and adult-stage axon regeneration in *C. elegans*. *Development* 2008;**135**:1129–36.
54. Case LC, Tessier-Lavigne M. Regeneration of the adult central nervous system. *Curr Biol* 2005;**15**:R749–53.
55. Hammarlund M, Nix P, Hauth L, Jorgensen EM, Bastiani M. Axon regeneration requires a conserved MAP kinase pathway. *Science* 2009;**323**:802–6.
56. Yan D, Wu Z, Chisholm AD, Jin Y. The DLK-1 kinase promotes mRNA stability and local translation in *C. elegans* synapses and axon regeneration. *Cell* 2009;**138**:1005–18.
57. Guo SX, Bourgeois F, Chokshi T, Durr NJ, Hilliard MA, Chronis N, et al. Femtosecond laser nanoaxotomy lab-on-a-chip for in vivo nerve regeneration studies. *Nat Methods* 2008;**5**:531–3.
58. Hulme SE, Shevkoplyas SS, Apfeld J, Fontana W, Whitesides GM. A microfabricated array of clamps for immobilizing and imaging *C. elegans*. *Lab Chip* 2007;**7**:1515–23.
59. Zeng F, Rohde CB, Yanik MF. Sub-cellular precision on-chip small-animal immobilization, multi-photon imaging and femtosecond-laser manipulation. *Lab Chip* 2008;**8**:653–6.
60. Ruan QL, et al. Evaluation of pesticide toxicities with differing mechanisms using *C. elegans*. *J Toxicol Environ Health A* 2009;**72**:746–51.
61. Pu P, Le W. Dopamine neuron degeneration induced by MPP+ is independent of CED-4 pathway in *C. elegans*. *Cell Res* 2008;**18**:978–81.
62. Cao S, et al. Torsin-mediated protection from cellular stress in the dopaminergic neurons of *C. elegans*. *J Neurosci* 2005;**25**:3801–12.

Index

B

C

G

H

I

K

L

M

N

R

S

T

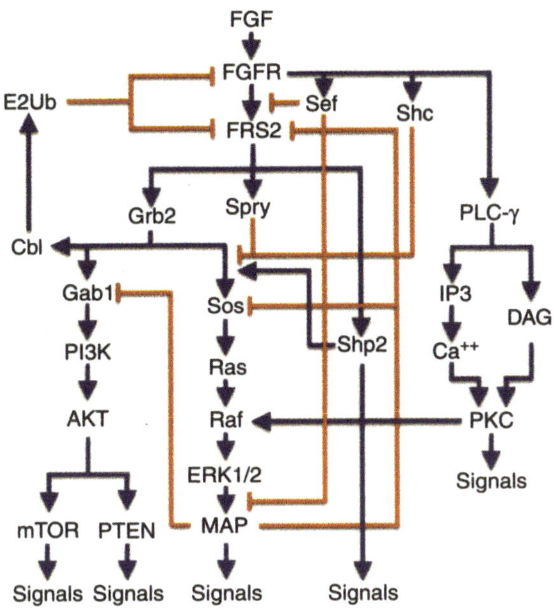

Fen Wang, Fig. 3. The cascade of FGFR signaling. Blue lines represent positive regulation, and red lines represent negative effects.

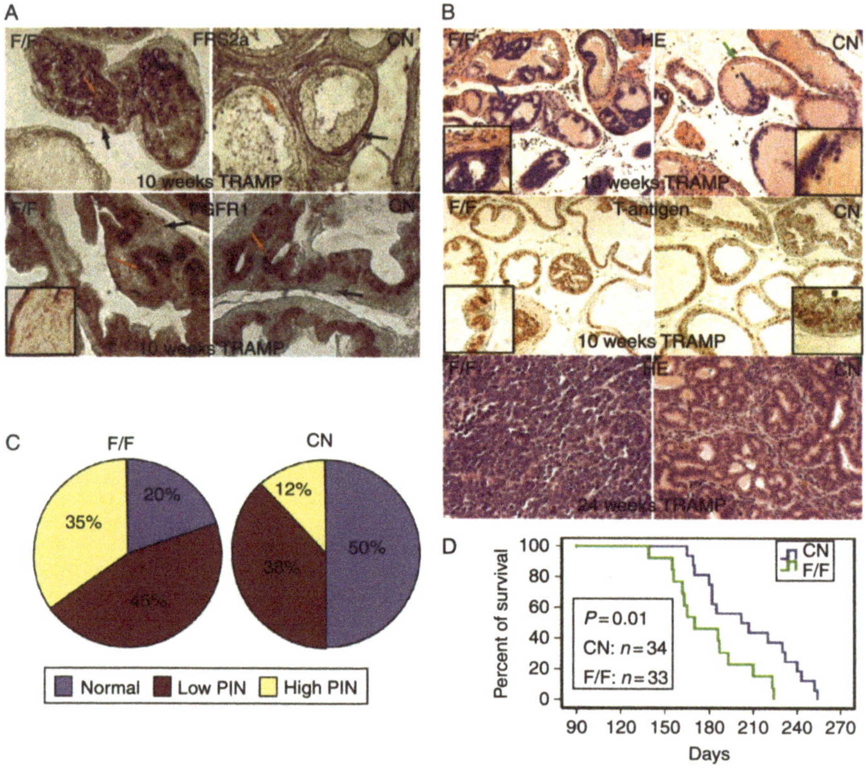

FEN WANG, FIG. 5. Ablation of *Frs2α* in the prostatic epithelium inhibited tumorigenesis in the TRAMP prostatic tumor model. (A) Expression of *Frs2α* and *Fgfr1* at the mRNA level was assessed in TRAMP prostates by *in situ* hybridization. Red arrows indicate epithelial cells, and black arrows indicate stromal cells. *Inset*, wild-type control showing no expression of *Fgfr1* in the epithelial compartment. (B) Prostate tissue sections were prepared from TRAMP mice with or without *Frs2α* null alleles at the indicated ages and stained with HE or analyzed by immunohistochemical staining with anti-T-antigen antibody. Blue arrows indicate focal lesions, and green arrows indicate normal epithelial cells. (C) The percentage of areas occupied by the PIN foci in prostates of 10-week-old TRAMP mice. The data are mean and standard deviation from five prostates. (D) Mortality of TRAMP mice with the indicated *Frs2α* alleles was determined from daily observation over 250 days. The percentage of mice that survived to the respective age is shown. F/F, homozygous TRAMP-$Frs2\alpha^{flox}$ mice; CN, TRAMP-$Frs2\alpha^{cn}$ mice. (Adopted from Zhang *et al.*, 2008, Development. 135 (4): 782).

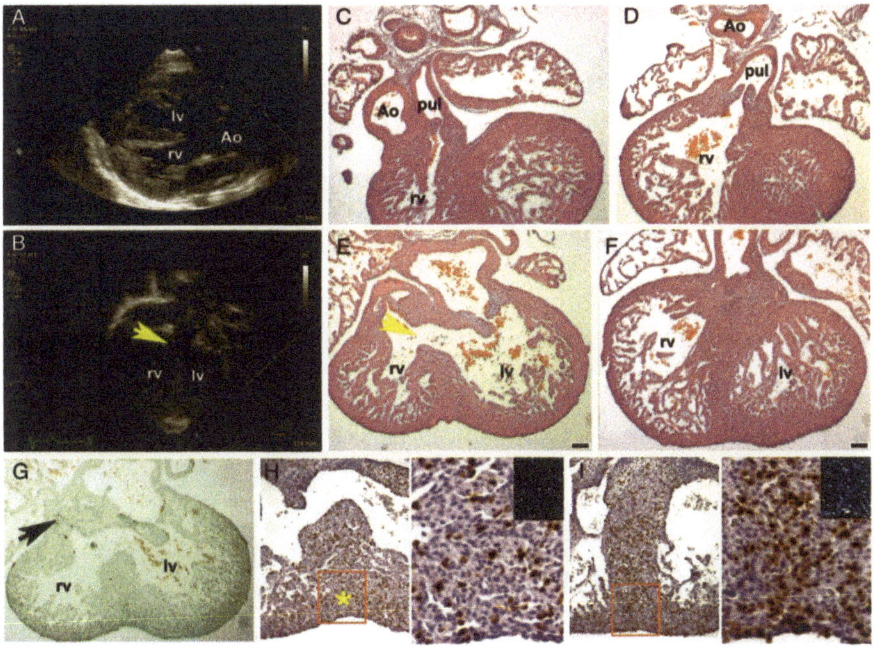

Paige Snider and Simon J. Conway, Fig. 1. (A) Echocardiography of 2-day-old newborn patient with DORV. Note aorta (Ao) is exiting from the right ventricle (rv). (B) Echocardiography of 5-month-old infant with perimembranous VSD. Note large hole connecting right and left ventricles (indicated via yellow arrow). (C–F) Histological hematoxylin/eosin stained sections of E14 *Hey2* knockout mutant and control littermate hearts. Note large VSD in (E) (indicated via yellow arrow) and DORV in (C) (as both the aorta and pulmonary (pul)) vessels exit the right ventricle. (G) Analysis of apoptosis via TUNEL immunohistochemistry in E14 Hety2 knockout heart. Note that there is no elevation of cell death associated with VSD, but there are a few apoptotic (brown) cells localized at the base of the DORV. (H and I) Characterization of cell proliferation in E14 *Hey2* cardiovascular system via BrdU immunohistochemistry. *Hey2* knockouts (H) exhibit VSDs, while the septum is intact in wild-type littermates (I). Heat map inserts (in H and I) indicated the relative differences between the mutant (~13% proliferating cells) when compared to normal control hearts (~27% proliferating cells). Note that BrdU staining is reduced by approximately 50% in the stunted mutant interventricular septum. Abbreviation: lv, left ventricle. Bar in (E) and (F) is 100 μM.

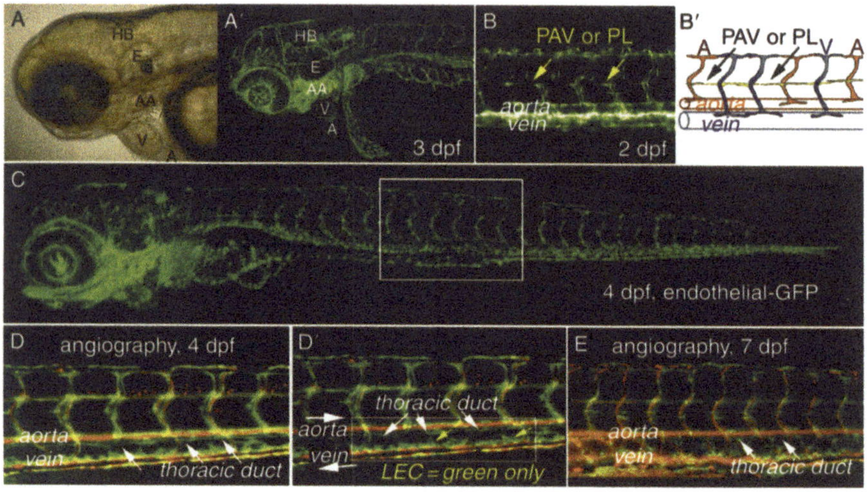

Joanne Chan and John D. Mably, Fig. 2. Blood and lymphatic vessel development. Lateral views shown with stages in days postfertilization (dpf). Head region in phase (A), or by endothelial-GFP-labeling of cardiovascular system (A′). Anatomical locations for hindbrain (HB), ear (E), aortic arches (AA), ventricle (V), and atrium (A) are indicated. (B) Fluorescent and schematic views of intersegmental arteries (red), veins (blue), parachordal vessel (PAV), or lymphangioblasts (PL). (C) Endothelially labeled-GFP indicates vascular development at 4 dpf (box indicates trunk region). (D–E) Microangiography using red-fluorescent dextran to indicate blood flow in vessels, green indicates both blood and lymphatic endothelial cells. The thoracic duct forms just ventral to the aorta by 4 dpf. (E) At 7 dpf, the thoracic duct is patent.

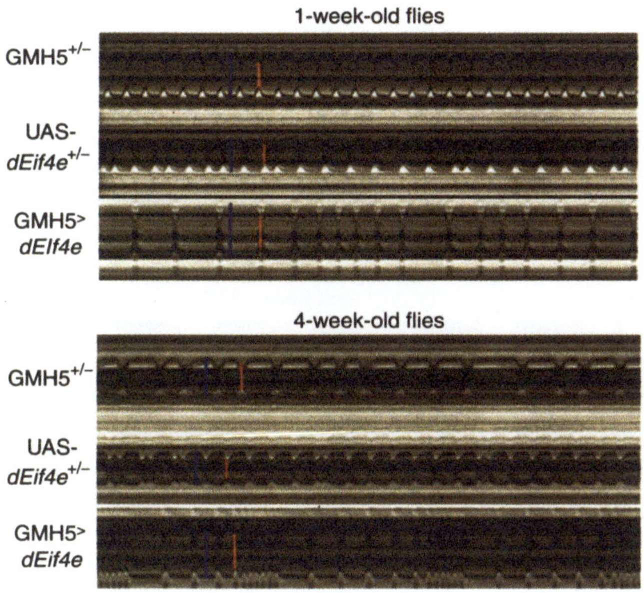

NICOLE PIAZZA AND R.J. WESSELLS, FIG. 6. Representative M-mode records showing the movement of the heart tube walls (y-axis) over time (x-axis). Blue bars indicate the diastolic diameter of each heart and the red bars indicate the systolic diameter. Records for 1-week-old flies (A) show predominately regular heart beat patterns as the GMH5 heterozygotes (as in Ref. 42), whereas most flies overexpressing UAS-*dEif4e* in the heart show arrhythmic heart beat patterns.[55]

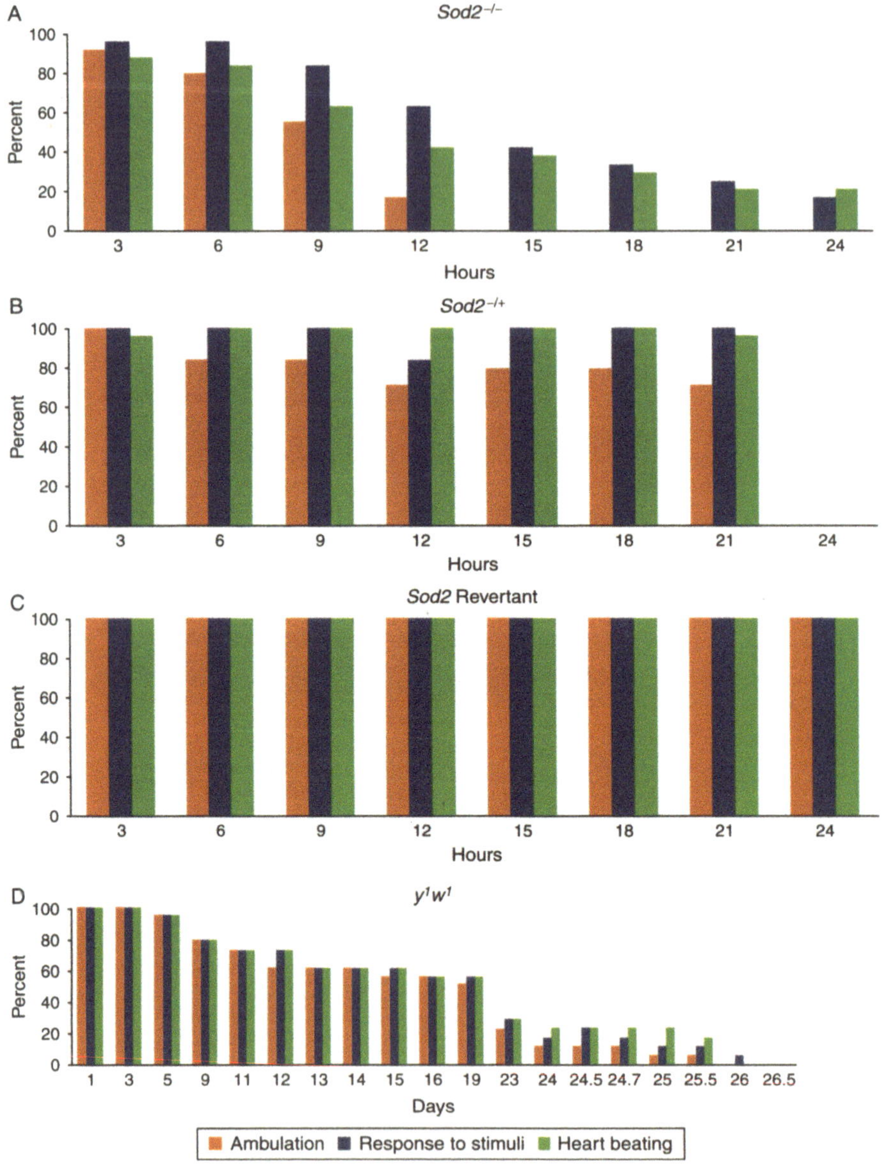

NICOLE PIAZZA AND R.J. WESSELLS, FIG. 7. Various measures of functionality decline at different rates in *Sod2*$^{-/-}$ flies. Percentage of flies observed to exhibit either spontaneous ambulation (red), response to gentle stimuli (blue), or spontaneous heartbeat (green) is charted for (A) *Sod2*$^{-/-}$, (B) *Sod2*$^{-/+}$, (C) *Sod2* revertants or (D) y^1w^1. (A), (B) and (C) are shown over a 24-hour time period, whereas (D) is shown throughout a lifetime. In (D), day 1 in the graph represents day 1 of the experiment, but day 64 of the age of flies.[156]

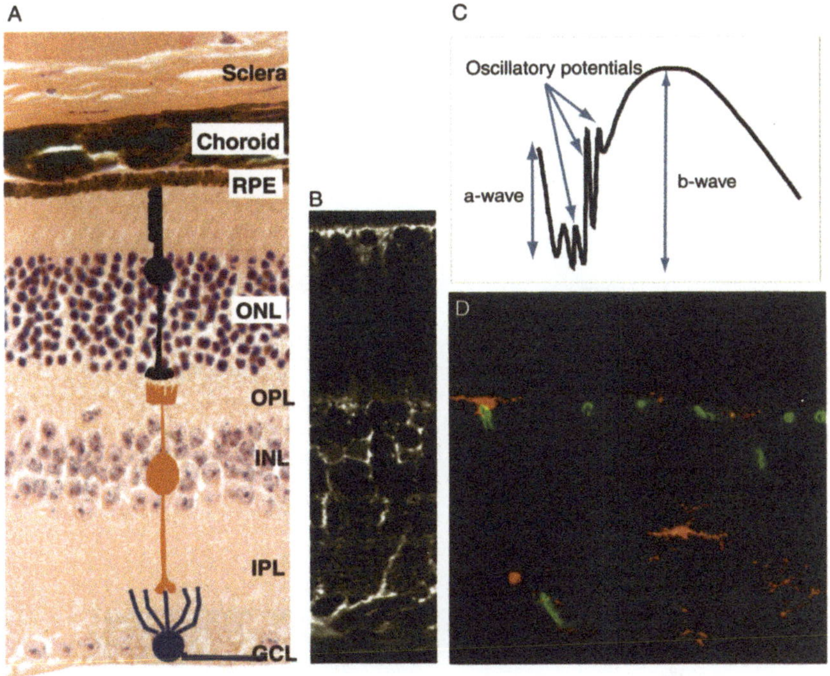

ERICA L. FLETCHER *ET AL.*, FIG. 1. Summary of the cellular structure and function of the retina. (A) A vertical section of the adult rat retina showing the retinal layers and major neuronal subclasses. The neuronal layers of the retina include the outer nuclear layer (ONL) corresponding to the location of photoreceptor nuclei (black cell), the inner nuclear layer (INL), where horizontal, bipolar and amacrine cells are located, and the ganglion cell layer (GCL), where nuclei of ganglion and displaced amacrine cells are located. The two synaptic layers include the outer plexiform layer (OPL) and the inner plexiform layer (IPL). Beneath the retina, are located the retinal pigment epithelium (RPE), choroid, and sclera. The major neuronal circuit that carrier visual information through the retina is the retinal through pathway which comprises photoreceptors (indicated as a black cell), bipolar cells (indicated as a red cell), and ganglion cells (indicated as a blue cell). (B) Vertical section of rat retina immunolabeled for the glial cell marker, glutamine synthetase. Müller cells span the entire thickness of the retina and have their somata located in the middle of the inner nuclear layer. (C) Schematic diagram of the major wavelet components of the flash ERG, including the a-wave (derived from the activity of photoreceptors), b-wave (attributed to postreceptoral neurons especially ON Bipolar cells) and oscillatory potentials (attributed to the function of amacrine cells). (D) Vertical section of the rat retina immunolabeled for the microglial marker, IBA-1 (red) and a marker of blood vessels (green). Microglia, the main resident immune cell of the retina, are stellate cells located within the two plexiform layers.

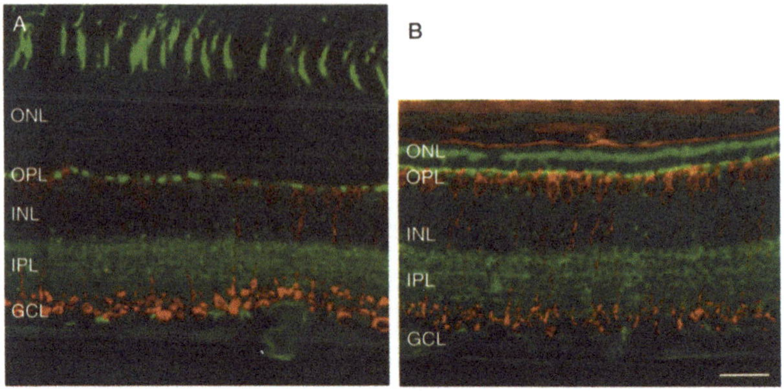

ERICA L. FLETCHER *ET AL.*, FIG. 4. Vertical sections of a (A) control C57Bl6 and (B) Postnatal day 25 rd1 mouse retina that have been labeled for the cone photoreceptor marker, Peanut Agglutin (green). Shown in red are rod bipolar cells immunolabeled for Protein Kinase C. By postnatal day 25, only a single layer of cones remain in the outer nuclear layer. All the inner layers of the retina remain intact. Abbreviations as per Fig. 3. Scale bar = 50 μm.

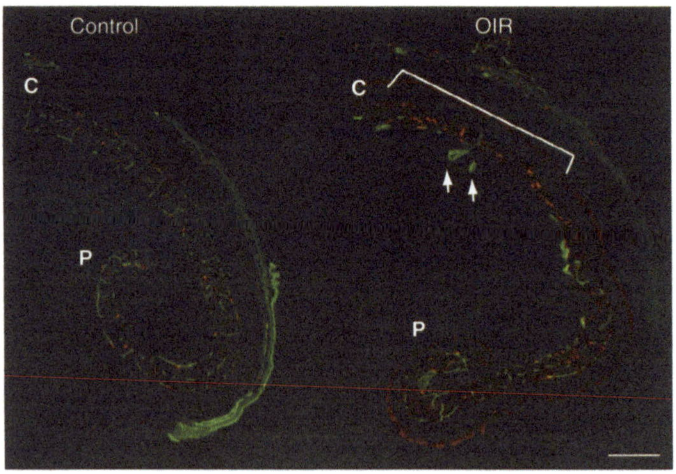

ERICA L. FLETCHER *ET AL.*, FIG. 11. Low power vertical sections of a control P18 mouse retina and a P18 mouse retina that has received oxygen induced retinopathy that have been immunolabeled to identify changes in blood vessels (green). In the normal mouse retina there are two main plexuses of blood vessels-a superficial layer located in the ganglion cell layer and a deep plexus in the Outer plexiform layer. In OIR mice, the central retina is rendered avascular (White line) and there is aberrant vessel growth in to the vitreous (white arrows). Scale bar = 200 μm.

FSC
www.fsc.org
MIX
Papier | Fördert
gute Waldnutzung
FSC® C083411

Kolibri 360 GmbH
Ferdinand-Jühlke-Straße 7
99095 Erfurt, Deutschland
produktsicherheit@kolibri360.de